Lecture Notes in Energy

Volume 24

For further volumes:
http://www.springer.com/series/8874

Lecture Notes in Energy (LNE) is a series that reports on new developments in the study of energy: from science and engineering to the analysis of energy policy. The series' scope includes but is not limited to, renewable and green energy, nuclear, fossil fuels and carbon capture, energy systems, energy storage and harvesting, batteries and fuel cells, power systems, energy efficiency, energy in buildings, energy policy, as well as energy-related topics in economics, management and transportation. Books published in LNE are original and timely and bridge between advanced textbooks and the forefront of research. Readers of LNE include postgraduate students and non-specialist researchers wishing to gain an accessible introduction to a field of research as well as professionals and researchers with a need for an up-to-date reference book on a well-defined topic. The series publishes single- and multi-authored volumes as well as advanced textbooks.

Ling Bing Kong · Tao Li
Huey Hoon Hng · Freddy Boey
Tianshu Zhang · Sean Li

Waste Energy Harvesting

Mechanical and Thermal Energies

Springer

Ling Bing Kong
Tao Li
Huey Hoon Hng
Freddy Boey
School of Materials Science
and Engineering
Nanyang Technological University
Singapore
Singapore

Tianshu Zhang
Sean Li
School of Materials Science and
Engineering
The University of New South Wales
Sydney, NSW
Australia

ISSN 2195-1284 ISSN 2195-1292 (electronic)
ISBN 978-3-662-52226-4 ISBN 978-3-642-54634-1 (eBook)
DOI 10.1007/978-3-642-54634-1
Springer Heidelberg New York Dordrecht London

Printed on acid-free paper

Springer is part of Springer Science+Business Media (www.springer.com)

Preface

Concern has grown in recent decades about energy consumption and environmental impact. This is the reason why governments all over the world are heavily investing in R&D in energy and environment. As a result, the labor market will require more and more manpower with extensive and in-depth knowledge in energy. This creates an unprecedented opportunity for graduates of Materials Science and Engineering. In schools or departments in universities around the world, the proportion of courses regarding energy harvesting and storage has been significantly increased. Common courses include solar cells, lithium-ion batteries, and supercapacitors. However, waste energy harvesting is rarely found as a course for students in departments of materials science and engineering, although all the related materials have been included in other courses in one way or another.

As a result, two years ago, we proposed to open a new course, Waste Energy Harvesting, for year-three and year-four undergraduate students, in the School of Materials Science and Engineering, Nanyang Technological University. The purpose was to instill the students with more application-oriented knowledge on materials science and engineering, focusing on waste energy harvesting. While we drafted the syllabus, we tried to find a textbook as reference, but we found that there were no books with similar names and contents covering all topics of interest. Therefore, we decided to compile a book based on the outline we proposed for the above-mentioned course.

This book consists of six chapters, with the first chapter serving as an *Introduction* and Chaps. 2–6 each for one type of energy harvesting and related materials. Chapters 2 and 3 deal with waste mechanical energy harvesting by using piezoelectric effect, while Chaps. 4–6 cater to waste thermal energy (heat) harvesting. The five chapters follow a similar structure that is aimed to demonstrate interrelationships among principle, materials, processing/fabrication, performance (enhancement strategies), and applications. Therefore, it can be used as textbook or reference book for senior undergraduate and postgraduate students in materials science and engineering, as well as students of other disciplines, such as applied physics, chemical engineering, electric and electronic engineering, mechanical engineering, and so on, who are involved in materials-related research or training. It also can be a reference book for engineers and scientists in R&D of energy materials and applications.

Ling Bing Kong
Tao Li
Huey Hoon Hng
Freddy Boey
Tianshu Zhang
Sean Li

Contents

Chapter 1
Introduction

1.1 Overview on Global Energy Demand

Energy is an integral part of our everyday lives. It is directly used to light and heat our homes, cook our meals, and travel to and from our workplaces. Energy is also indirectly in the goods we purchase and the services provided to us. Energy makes things happen and is one of the most fundamental parts of our universe. Almost everything and any activity is dependent on energy. How we produce and consume energy also has far-reaching implications, which affect our economy and demonstrate our responsibility to the environment and future generations and help to define us as a society.

In other words, we demand reliable and affordable energy all the time. To meet this demand, it is necessary to have a foresight and effective long-term planning, followed by huge investments and years of work to build reliable infrastructure that is required to produce and deliver the energy. Global demand for energy has risen unprecedentedly in the last 150 years, due to industrial development and the rapid growth of population. As shown in Table. 1.1, the demand for energy has been predicted to rise continuously.

1.2 Energy Sources

Energy has several forms, including chemical, electrical, thermal (heat), radiant (light), nuclear and mechanical (potential and kinetic), and so on. When something is able to produce energy, it is called potential energy, like water. When something is producing energy through motion, it is called kinetic energy, like water moving through a dam. Energy can be converted from one form to another. For example, a piece of magnesium has energy stored in chemical form. If the magnesium is ignited, that energy is converted into heat, light, and sound. A hydroelectric facility makes use of the potential energy in stored water. It converts the kinetic

L. B. Kong et al., *Waste Energy Harvesting*, Lecture Notes in Energy 24,
DOI: 10.1007/978-3-642-54634-1_1, © Springer-Verlag Berlin Heidelberg 2014

Table 1.1 World total primary energy consumption by region (Quadrillion Btu[a])

Region/Country	2015	2020	2025	2030	2035
OECD[b]					
OECD Americas	126.1	131.0	135.9	141.6	147.7
United States	102.0	104.9	108.0	111.0	114.2
Canada	14.6	15.7	16.4	17.6	18.8
Mexico/Chile	9.5	10.4	11.5	13.0	14.7
OECD Europe	83.6	86.9	89.7	91.8	93.8
OECD Asia	40.7	42.7	44.2	45.4	46.7
Japan	22.2	23.2	23.7	23.7	23.8
South Korea	11.1	11.6	12.4	13.1	13.9
Australia/New Zealand	7.4	7.8	8.1	8.5	8.9
Total OECD	250.4	260.6	269.8	278.7	288.2
Non-OECD					
Non-OECD Europe and Eurasia	51.4	52.3	54.0	56.0	58.4
Russia	31.1	31.3	32.3	33.7	35.5
Other	20.4	21.0	21.7	22.3	22.9
Non-OECD Asia	188.1	215.0	246.4	274.3	298.8
China	124.2	140.6	160.9	177.9	191.4
India	27.8	33.1	38.9	44.3	49.2
Other	36.2	41.3	46.7	52.1	58.2
Middle East	31.0	33.9	37.3	41.3	45.3
Africa	21.5	23.6	25.9	28.5	31.4
Central and South America	31.0	34.2	38.0	42.6	47.8
Brazil	15.5	17.3	19.9	23.2	26.9
Other	15.6	16.9	18.1	19.5	20.8
Total Non-OECD	323.1	358.9	401.7	442.8	481.6
Total world	573.5	619.5	671.5	721.5	769.8

US energy information administration or EIA website

[a] British thermal unit, a traditional unit of energy equal to about 1055 J

[b] Organization for economic cooperation and development

energy of falling water into the electrical energy to power our homes and offices. The energy we use every day comes from a wide array of sources and can be classified into two categories: nonrenewable and renewable [1–3].

1.2.1 Nonrenewable Energy Sources

The supply of nonrenewable sources is limited because they cannot be created over and over again. Once they are used up, the resources are gone forever. Nonrenewable energy sources include oil, natural gas, coal, and nuclear energy. Among these nonrenewable sources, oil, natural gas, and coal are called fossil fuels, which have been the primary commercial energy source in the world today. The geologic processes that created fossil fuels are no longer occurring at a rate useful to humans. Fossil fuels are so called because they were formed many millions of years ago, before the time of the dinosaurs. The age in which the fossil fuels were formed is called the Carboniferous Period. It was part of the Paleozoic Era. Carboniferous is named so because carbon is the basic element in coal and other fossil fuels.

During the Carboniferous Period, all trees and plants died and sank to the bottom of the oceans where they formed layers of a spongy material known as peat. Over many hundreds of years, the peat then were covered with sand, clay, and other minerals, which over time were transformed into a layer of sedimentary rock. Over the following millennia, surface materials continued to build up and increased pressure was applied onto the carboniferous layer, which eventually was evolved into the fossil fuels that we are using now. Oil and natural gas are found underground between folds of rock and in areas where the rocks are porous. Therefore, they can be found by drilling through the Earth to the deposits deep below the surface. The oil and natural gas are then extracted from below the ground by oil rigs and shipped to a holding terminal for redistribution or refining.

1.2.1.1 Oil

Oil is also called petroleum or crude oil. The word petroleum comes from the Greek word petra, meaning rock and the Latin word oleum, meaning oil. It is a naturally occurring flammable liquid that is found within layers of rock formations beneath the Earth's surface. It a complex mixture, consisting of hydrocarbons and other liquid organic compounds. After extraction, the crude oil is split into various types of products in refineries, so that the hydrocarbon molecules can be separated by weight and size, by a process called cracking.

The smallest and lightest molecules rise to the top of the giant distillation towers at the refineries, which is the source of propane, butane, and other gases. The medium weight fractions condense near the middle part of the tower, leading to gasoline and aviation fuel. The heavier petroleum products condense near the bottom of the towers, which are mainly used as diesel oil and home heating oil. The residue is made into greases, waxes, and asphalt. The main products of the processed crude oil are supplied as energy sources, such as gasoline, diesel fuel, aviation or jet fuels, home heating oil, oil for ships, and oil to burn in power plants to generate electricity. Crude oil is also used to synthesize most of the fertilizers, pesticides, and plastics, which are then used to produce daily use articles such as clothes, toothbrush, and plastic bottles.

Oil has been used to generate electricity using three technologies. The first technology is conventional steam, in which oil is burned to heat water to create steam to generate electricity. The second technology is combustion turbine, in which oil is burned under pressure to produce hot exhaust gases, which spin a turbine to generate electricity. The last one is combined-cycle technology. Oil is first burned in a combustion turbine, using the heated exhaust gases to generate electricity. After these exhaust gases are recovered, they heat water in a boiler to create steam to drive a second turbine, or they are used for heating purposes.

Whenever a fossil fuel is burned, carbon dioxide (CO_2) is produced. CO_2 is a greenhouse gas that is responsible for climate change. Burning oil to generate electricity leads to significant air pollution during the burning of oil in several forms, including nitrogen oxides (NO_x), sulfur dioxide (SO_2), particulates, (e.g., fine particles in the air that are linked to health hazards), toxic heavy metals, such as mercury, and volatile organic compound—which contribute to ground-level ozone, a major component of smog.

Oil-fired power plants also have a strong impact on water, land use, and solid waste disposal. Oil-fired conventional steam plants require large amounts of water for steam and cooling, which can harm local water resources and aquatic habitats. Oil residues and sludge that are not consumed during combustion contain toxic and hazardous wastes that are dangerous to the environment. Drilling and extraction of oil produces air pollutants, toxic and hazardous materials, and emissions of hydrogen sulfide (H_2S). Oil exploration and drilling also disturb wildlife habitats, water, and land. Finally, oil transportation accidents such as tanker spills can result in catastrophic damage to wildlife and wildlife habitat.

1.2.1.2 Natural Gas

Natural gas is a gaseous form of petroleum, containing primarily methane—the lightest compound of the hydrocarbons. Because it is a gas, natural gas is harder to handle than liquid oil or solid coal. The most efficient means of containing and moving natural gas is transportation through pipelines which are built to connect the gas fields directly to major urban areas. Gas can also be compressed in tanks and moved about as liquefied natural gas (LNG).

Liquefied petroleum gas (LPG) is a product of oil fields. Although it is a mix of several gases, it is commonly called propane. LPG is found dissolved in crude oil. It is separated from the liquid oil and stored under pressure in heavy metal cylinders. These containers can easily be transported to homes for use in heating and cooking. Like natural gas, propane is a relatively clean burning fuel and produces few contaminants. Similar to oil, natural gas is also found underground and under the ocean floors. Natural gas reservoirs are tapped by drilling wells. Once a pocket of natural gas is hit by drill, it can be pumped to the surface of the Earth like oil.

Natural gas is cleaner when burning. It is less expensive than oil, so that electricity generated by using natural gas is more cost-effective. More and more homes and buildings are heated by using natural gas. Like other fossil fuels, natural gas can be used as a fuel in electricity generators with conventional steam boilers. New technologies are being developed, such as combustion turbines, to further reduce the consumption of fossil fuels. For instance, in jet engines, superheated gas is used in the combustion turbine technology. The gas is pressurized in pipes and used to drive the turbine. Waste heat from this process is further used to heat steam so as to power a second generator. By doing this, efficiency can be increased by as much as 40 %. The efficiency can be even further increased by using a process called cogeneration. In this case, a fuel source such as natural gas is used to create electricity and the waste heat is used for residential or commercial heating applications.

The negative impact of natural gas on the environment has been found to be significantly smaller than that of oil and coal. The burning of natural gas generates about half carbon dioxide as the burning of coal. It also produces less particulate matters, sulfur dioxide (SO_2), and other toxic emissions. However, there may be nitrogen oxides (NO_x) and carbon monoxide (CO) produced during burning natural gas, whose quantities are comparable to those of burning coal.

The use of natural gas inevitably leads to methane (CH_4) emissions. Methane is a very strong greenhouse gas, similar to CO_2, contributing to global climate change. Similar to oil, the extraction of natural gas can also have a negative impact on wildlife and wilderness habitats. Because natural gas has no smell and is invisible, it is necessary to be mixed with a chemical that gives a strong odor before it is fed to pipelines and storage tanks. The odor is like rotten eggs, which makes it easy to discover a leak. Cars fueled by natural gas run up to 90 % cleaner than those fueled by gasoline. Natural gas can be used to heat homes and buildings. It can be used for heating water, drying clothes, and cooking. It can also be used for outdoor activities such as barbecuing, gas lighting, heating hot tub, and even swimming pool.

1.2.1.3 Coal

Coal appears as a hard solid black rock-like substance made up essentially of carbon (C), hydrogen (H), oxygen (O), nitrogen (N), and varying amounts of sulfur (S). There are three main types of coal: anthracite, bituminous, and lignite. Anthracite coal is the hardest, which contains the most carbon, thus possessing the highest energy content among the three types. Lignite is the softest and contains the lowest amount of carbon, but relatively high contents of hydrogen and oxygen. The properties of bituminous coal are between those of the other two coals.

Coal is mined from the ground using underground mines and strip mines. Strip mines use huge shovels to strip away the layers of Earth above the coal in order to remove it. Most coal is transported in its solid form either by trains or ships. On some occasions, pipelines are used to connect coalfields directly to electric power plants. In these cases, the coal is mixed with water to form slurries than can flow through the pipelines. The formation of the slurries consumes a large quantity of water and can generate considerable pollution to the surrounding water resources. Coal has been widely used to generate electric power all over the world. For example, coal-fired generators produce over half of the electricity used in the United States. The iron and steel manufacturing industry consumes large amounts of coal, mostly as a mixture called coke. A very small amount of coal is used directly in houses and commercial buildings as a fuel source for heating applications.

A number of serious environmental problems have been encountered with the use of coal, all the way from mining to the end-use. For instance, strip mining likely destroys wildlife habitat and damages the soil. Transporting and processing coal could affect the quality of water systems for both aquatic life and humans. Mining coal releases methane (CH_4) gas, which is a greenhouse gas and contributes to climate change. Other greenhouse gases, such as nitrous oxide (NO_x), are also released during the production and use of coal. There are also heavy metals (such as mercury) in coal, which are released as toxic compounds into the atmosphere after burning. They are toxic to both humans and animals, especially as contaminants of food or water systems.

Several technologies have been developed to minimize these negative environmental impacts. One example is to create a pressurized fluid and blast hot air into the pulverized coal dust to facilitate a floating burning environment. This process can significantly increase the burning efficiency at a lower burning temperature, thus leading to decreased generation of NO_x. Sulfur-absorbing limestone is added to the coal mixture to remove the pollutant by forming calcium sulfate ($CaSO_4$), which is the base material for dry wall. Treatment of flue gases by using catalysts and scrubbers can further reduce the degree of pollution.

1.2.1.4 Nuclear

Nuclear energy is also known as atomic energy, which is released when an atom is split. A tremendous amount of energy is released in the form of both heat and light when atoms are split. Release of such a huge amount of energy in an uncontrolled way is dangerous. However, when controlled, this energy can be harvested in a safe way to generate electricity at high efficiencies. Like fossil fuels, the uranium used to fuel nuclear power plants is a mineral of limited quantity and thus belongs to nonrenewable sources.

When nuclear energy is used to produce electricity, inside the reactor of an atomic power plant, uranium atoms are split apart through a controlled chain reaction. Neutrons bombard atoms of uranium-235 (U-235) to trigger their split, thus forming two new elements and additionally two neutrons. These newly formed neutrons then continue to bombard more U-235 atoms, so as to maintain the chain reaction, because the number of neutron increases geometrically. During this chain reaction, a huge amount of heat energy is released. The heat is used to boil water in the core of the reactor, by transferring it into a separate and radiation-free hot water system through a heat exchanger. The steam in this second set of pipes drives turbines to generate electricity. Sometimes, gaseous fluids are also used as the heat transfer medium.

It seems that nuclear power plants are a clean electricity source because there is no release of pollutants during the operation of nuclear plants. However, mining the uranium has environmental impacts, similar to that of fossil fuels. In addition, there is an important potential environmental impact in the form of radioactive waste. It is even serious especially when accidents happen. Radioactive wastes are present throughout the nuclear fuel cycle, such as in mining and milling of uranium ore, processing and fabrication of uranium ore into nuclear fuel, using nuclear fuel in the reactor, treating the spent fuel from the reactor after use, and disposing of wastes. Although the amount of solid wastes generated in nuclear plants is relatively small, radioactive wastes increase health risks that could be unimaginably more serious than any of the other sources of electricity. A typical 1,000 (MW) reactor produces 20–25 tons of spent fuel per year. If this spent fuel is not well reprocessed, it can have radioactive risks for as long as 250,000 years. The storage of that highly toxic waste still remains challenging.

Additionally, the amount of water usage, in a nuclear plant that uses water for once-through cooling system, is about 2.5 times of that required by a typical fossil fuel plant. Therefore, a nuclear power plant poses a more significant impact on water resources, aquatic habitats, and fish than any other power generation technology. Furthermore, in the processing of nuclear fuel, the enrichment process of uranium itself consumes great amounts of electricity, which must be provided by the fossil fuel plants that have air pollution emissions although they are not released by the nuclear reactor itself.

1.2.2 Renewable Energy Sources

Renewable sources are those that can be replaced or regenerated from a naturally occurring source. Renewable energy sources include hydro, geothermal, solar, wind, biomass and tidal, and sea waves. Different sources have different characteristics and need to be harvested by using different technologies, which will be described and discussed as follows.

1.2.2.1 Hydro

Streams, rivers, and lakes on Earth are usually formed due to rain water and melting snow flows by gravity. Hydropower facilities are designed and built in such a way that the water can be intercepted on its downward path, and its mechanical energy is then converted into electricity through turbines. Hydropower is often considered a renewable energy resource, simply because the cycle of water evaporating from the heat of the Sun and falling back to Earth is continuously renewed by the Sun's energy.

Because hydroelectric plants take advantage of falling water to generate electricity, it is easily understood that the higher the drop (head) and the greater the flow, the more the power that can be generated. Therefore, to generate hydroelectricity, a reservoir is created by using a dam. As the water behind the dam is released and flows through the intake into a pipe called a penstock, the water pushes against blades in a turbine, causing them to turn, and thus electricity is produced. The speed of the turbine must be constant, so that the amount of electricity generated is also constant. The speed and volume of the water flowing to the turbine are controlled by a speed governor, which opens and closes the gates surrounding the turbine. It is not possible to have a universal evaluation on the type and severity of impacts of hydroelectricity plants, because different plants in different places should be different in characteristics of rivers and dams. The environmental impact of a dam is also dependent on the facility design, sensitivity of the local environment, and modifications to the design or operation of the hydropower facility. Therefore, dams can affect the environment in different ways.

Dams can alter the water temperature and levels of dissolved gases within a river, which in turn contributes to a variety of related water quality problems and thus are harmful to the aquatic habitat. Because the dam cuts off the flow of the river, sediment drops and accumulates at the bottom of the river. The absence of sediment downstream can degrade the in-stream habitat and cause the disappearance of delta areas at the mouth of the river. With the deposition of the sediment, toxic chemical or industrial residues from upstream sources may also be accumulated, which could result in the build up of highly concentrated toxins. For example, the use of small hydroelectric generators can effectively reduce the environmental impacts. These are usually known as run-of-river designs, because they do not require a reservoir to impound water. The flow of the river is partly channeled through the turbine, so that the rest remains in the streambed. In this case, the normal river process continues and thus eliminates the relocation of the upstream population.

1.2.2.2 Geothermal

Geothermal energy is due to heat from the Earth's molten core. Geothermal energy is a renewable energy source because it is almost limitless. The temperature of the rock increases about 3 °C for every 100 m from the surface to the core of the Earth. Therefore, nearly everywhere on the planet, the temperature 3 m below ground level is the same, from 10 to 16 °C. At 3 km below ground, the temperature of the rock is sufficiently high to boil water. Due to the constant temperature, a geothermal source can be used as a heat pump to directly heat or cool a building. In other words, the Earth can be used as a heat sink in the summer and as a heat source in the winter. As a result, geothermal source is also called ground-source heat pump.

The working principle of a ground-source heat pump is similar to the refrigerators we use. However, it has extra valves that allow heat-exchange fluids to follow separate paths for heating and cooling, respectively. In the winter, a ground-source heat pump transfers heat from the warmer Earth to buildings, while in the summer, the heat of the buildings is taken away and discharged into the cooler ground. These processes are realized through a system of underground pipes. Because ground-source heat pumps do not use any fuel, they are cost-effective compared to most other methods.

Geothermal energy can also be used to generate electricity under certain geologic conditions. In some regions, the Earth's naturally occurring heat flow is very near to the Earth's surface, where water is heated and even steam is formed. Geothermal reservoirs are classified into two types: low temperature (<150 °C) and high temperature (>150 °C). High-temperature reservoirs are suitable for commercial production of electricity. If geothermal systems create steam, there could be gases that are released into the atmosphere. The chemical composition of the gases varies in geologic sites. Dissolved gases usually include carbon dioxide (CO_2), methane (CH_4), hydrogen sulfide (H_2S), ammonia (NH_3), nitrogen (N_2), and hydrogen (H_2). However, these air emissions represent only very tiny quantities and thus do not pose a serious threat to the environment. Additionally, in most cases, used geothermal fluids are collected and reinjected, for recycling and reuse.

1.2.2.3 Wind

Wind power is the world's fastest growing technology for the generation of electricity. Wind is a result of air convection caused by the Sun's heating effect and the rotation of the Earth. Daily and seasonal variations in temperature consistently generate wind, producing an energy source that is never depleted. Therefore, wind is a renewable source of energy.

Wind power plants use large spinning blades to capture the kinetic energy in moving air, which is then transferred to turbines to generate electricity. The best sites for wind power plants should have steady wind patterns and average wind speeds of higher than 16 km/h^{-1}. Wind power is one of the cheapest renewable energy technologies available on the market today. The cost of wind-generated electricity at prime sites is close to that of coal-fired electricity. Because wind power is clean and the technology is always improving, wind power generation is expected to grow continuously in the future.

Although wind plants do not produce air pollution, do not use water, and do not impact the land, there are environmental problems with some wind plants. Three types of environmental impacts can be expected when using wind power plants, although they might vary from site to site. Because wind plants use many turbines mounted at top tall towers, which can often be seen from afar, these man-made structures are viewed as an intrusion on the natural landscape. Furthermore, wind turbines emit noise that can be heard in the vicinity of the wind plants. Although the frequency and volume of this noise can be minimized, they cannot be eliminated completely.

It is worth mentioning that early wind turbines posed a detrimental impact on bird populations. However, this problem has been well addressed by new turbine designs. Currently, solid tubular towers are used so that birds cannot perch on them. The rotating speed of the current turbine blades is slower than those of earlier ones, and thus reduces the potential of collisions with birds. It is still important to locate wind plants away from migratory bird routes to avoid possible bird kills.

1.2.2.4 Tidal Power and Wave Energy

The oceans represent another type of continuous and renewable source of energy, which can be used to provide power for home and industrial applications. The energy of the oceans can be harvested from two basic aspects: waves and tides.

Waves form a free and sustainable energy resource that are created when strong winds blow over the ocean surface. The greater the distances involved, the higher and longer the waves will be. Energy is stored in this way until it reaches the shallow costal areas where it is released. Tidal power can be used to generate electricity through the power of tidal motion. The ultimate source of this energy is gravity. The Moon and Sun exert gravity on the nearest ocean mass, creating an imbalance on the water further away, thus resulting in a relative height difference known as tidal movement. The range and magnitude of these tides are dependent on the relative location of the Moon and the Sun to the Earth.

The kinetic energy of ocean waves is usually harvested through the use of a wave chamber. As wave rises into or goes away from a chamber, the rising water forces the air out of the chamber and the receding water brings the air into the chamber, which is repeated at the frequency of the tidal. The moving air spins a turbine that can turn a generator. Other wave-energy systems use the up-and-down motion of the wave to power a piston that moves up and down inside a cylinder. That piston, in turn, spins a generator to produce electricity.

During the earlier time, the wave-energy systems are usually very small, designed to create only enough energy to power a warning buoy or a small lighthouse. However, larger scale projects have been developed more recently, such as the 750 kW wave farm project off Islay, Scotland and the 2 MW project off the coast of Vancouver Island, Canada. Therefore, it is reasonably anticipated that such kinds of huge projects will form the basis of large-scale multimachined wave-energy plants in the near future.

Tidal power generation is often related to a bay. Usually, a barrier is placed across the entrance to the bay, so that the incoming flood tide rises to a level that is higher than that of the bay itself, leading to a difference in water level. When the tide retreats, the water behind the dam released. The difference in the water level is used to generate electricity, because the fast running water passes through the turbines located in the water passages within the barrier. In this respect, it is just like a regular hydroelectric power plant. The amount of power generated by a tidal system is directly related to the size of the tide. Therefore, only areas with large tidal ranges are considered economically viable tidal energy sources.

Tidal power is often difficult to match supply and demand of electricity. Electricity demands are peaked mostly in the certain times of day, depending on the climate and the application characteristics. The peak times of the power output are not always matching these peak times of power demand, because the tides are dictated by the lunar cycle. As a result, tidal power plants usually require a backup system so that it can be supplied according to the demand of electricity. These backup plants often operate on fuel oil or natural gas. When the costs of the tidal plant and the backup plant are considered together, tidal power will not be very cost-effective.

Tidal power plants could have significant impacts on the local ecosystem as they are usually built across large estuaries. Estuaries often are a host to migratory birds that feed on worms and other invertebrates found on the exposed mudflats. Building a large barrier across an estuary affects variations of water level, shrinking the areas of mudflats. The barriers also decrease the flow of sediments from the estuary, which changes the biological productivity of the ecosystem. Although wave-energy systems do not consume fuels for their operation and thus have no emissions of harmful pollutants, they can impact their immediate environment. Large-scale wave-energy systems can make wide areas of ocean to be unnavigable for ships, affect the feeding patterns of aquatic and marine bird life and reduce the energy of the waves, thus changing the patterns of erosion and sedimentation of nearby coastal environments.

1.2.2.5 Biomass

Biomass is organic matter, which is usually from waste materials, such as wood, wood waste from manufacturing activities, agricultural wastes and residues, and municipal wastes. In the future, much larger quantities of biomass power could come from crops that are planted and harvested specifically for use as an energy fuel source in short-rotation plantations, such as poplars and willows.

Biomass is considered to be a renewable source of energy because it is derived from materials that can be regenerated from naturally occurring sources. Biomass materials either can be used as solid fuels or can be converted into liquids or gases. Biomass has been used to generate electric power, heat or chemicals as vehicle fuels. Most biomass power plants burn lumber, agricultural waste or wood waste from construction and demolition. Some power plants burn the biomass fuel directly in boilers that supply steam for generators. This is called direct combustion. Other power plants convert biomass into gases, (e.g., methane) to fuel steam generators, combustion turbines, combined-cycle technologies or fuel cells. This is called biomass gasification.

Large municipal or industrial landfills produce gas that can be tapped to generate electricity. Microorganisms that live in decomposing organic materials such as food waste, paper, or yard clippings produce this gas. Landfill gas typically comprises approximately 60 % methane (CH_4) and 40 % carbon dioxide (CO_2). It is collected by drilling wells into the landfills and collecting the gases through pipes. Once landfill gas is processed, it can be combined with natural gas to fuel conventional combustion turbines. It can also be used to fuel small combustion or combined-cycle turbines. Methane generated from animal or human waste can also be captured for power generation or as heating fuels. Because biomass technologies use combustion processes to generate electricity, they can generate electricity at any time. In comparison, wind and solar technologies only produce energy when the wind is blowing or the Sun is shining.

Biofuels are used in direct combustion and are typically used as liquid fuels for transportation. The two most common types of biofuels are bioethanol made from carbohydrates and biodiesel made from fats or oils. Commercially available ethanol is already making a contribution to renewable domestic transportation fuels in some countries. It can be used at low concentrations (<10 %) in regular gasoline engines or at higher concentrations in specially adapted engines. Biofuels have several advantages. When produced on a small scale, they can be good for the environment because they add fewer emissions to the atmosphere than petroleum fuels. They also utilize waste materials that have no other use. However, their environmental benefit depends on how they are used. In some cases, a modest benefit is possible.

Biomass energy production can have a negative effect on the environment. For example, biomass power plants generate air emissions. These emissions are dependent on the fuel and the technology used. If wood is the primary biomass resource, there is very little emission of sulfur dioxide (SO_2), which is one of the sources that contribute to acid rain. Some biomass power plants that use combustion technologies have high nitrogen oxide (NO_x) emissions. This is because many biomass fuels have high nitrogen content. This high NO_x rate is one of the top air-quality concerns associated with biomass.

Biofuel power plants also emit carbon monoxide (CO), sometimes at levels higher than coal plants. They also produce carbon dioxide (CO_2), which is the primary greenhouse gas responsible for climate change. However, the cycle of growing, processing, and burning biomass recycles CO_2 from the atmosphere. If this cycle continues, the level of CO_2 in the atmosphere will not increase significantly. Although there is significant impact during the use, the collection and fabrication of biomass fuels can have serious environmental impacts. Harvesting timber and growing agricultural products for fuel requires large space and storehouse to collect, transport, process and store the products.

It is also highly possible that the biomass fuel source contains toxic contaminants. For example, agricultural waste handling pollutes local water resources. Burning biomass deprives local ecosystems of nutrients that forest or agricultural waste may otherwise provide. In the view that food is increasingly scarce world widely, the production of biofuels can divert agricultural crops away from food. Managed production of biomass fuels recycles CO_2 in one-third less time than natural processes. This is because short-rotation crops of woody plants can be matured and harvested in less time than natural growth forests. Biomass power plants also divert wood waste from landfills. This wood waste produces methane, another potent greenhouse gas, into the atmosphere.

Biomass fuels may be obtained from supplies of clean, uncontaminated wood that otherwise would be sent to the landfill. They may also come from sustainable harvests. In both of these fuel collection examples, the environmental benefits of biomass are significant. This is especially true when compared to fossil fuel collection alternatives. Biomass power is the largest source of non-hydro renewable electricity in the world. It generates an estimated 47,000 MW worldwide.

1.2.2.6 Solar

The Sun is the ultimate source of much of the world's energy. It provides the Earth with light and heat in the form of electromagnetic radiation. Solar energy is the Earth's primary source of renewable energy. The enormous flow of radiant energy supplied by the sun far exceeds what the world requires for electricity fuel. Solar energy can be used directly as a thermal energy source and indirectly as a source of electricity. Since generating electricity directly from sunlight does not deplete any of the Earth's natural resources, solar energy is a renewable source of energy.

Using sunlight to heat water has the potential to be the most cost-effective use of solar energy. There are numerous, increasingly sophisticated technologies that can capture solar energy to provide households and offices with a significant percentage of their hot water needs. Active solar hot water heaters use roof panels that contain a network of black heat absorbent pipes. The sun heats a liquid in the pipes which is then pumped through a heat exchanger to warm household water. Domestic hot water and pool heating systems are the two most common uses of solar thermal collectors. Solar hot water heaters are more promising than photovoltaic (PV) solar cells, which convert solar energy into electricity, because they are less expensive and capture a much higher proportion of the sun's energy. Passive solar provides both heating and cooling to keep building occupants comfortable without mechanical equipment. Passive solar buildings use windows, walls, and floors to absorb, store, release, and distribute the Sun's heat in the winter and reject solar heat in the summer. It also maximizes the use of sunlight for interior illumination.

Active solar heating systems use mechanical equipment to collect and distribute solar energy to heat the interior of buildings. Active systems use collectors, which look like flat boxes covered with glass. The collectors are usually placed high up on roofs where they can collect the greatest amount of sunlight. Dark-colored metal plates inside the boxes absorb heat from the sun. Air or water is then passed through the collectors, warmed, and distributed to the rest of the house. As in the ordinary furnace system, thermostats are used to control the delivery of heat. There are two different technologies to generate electricity from the sun—photovoltaic (PV) and solar-thermal.

PV solar technology relies on chemical reactions to generate electricity. PV cells are small, square-shaped semiconductors manufactured in thin film layers. They are made using silicon and other conductive materials. When sunlight strikes the PV cell, chemical reactions release electrons and generate an electric current. The small current from individual PV cells, which are installed in modules, can power individual homes and businesses. They can also be plugged into the bulk electricity grid. PV technology has limited efficiency and is expensive. However, as the technology improves, PV materials are being incorporated into construction materials like roofing, tiles, and cladding. This makes economic sense for new building construction and major renovation projects.

Solar-thermal technologies are a traditional electricity generating technology using the sun's heat to create steam or pressurized gas to drive a turbine generator. These new solar-thermal technologies are now in use in some parts of the world. Parabolic dish systems concentrate sunlight to heat gaseous hydrogen (H_2), helium (He) or liquid sodium (Na) to create pressurized gas or steam to drive a turbine to generate electricity. This is accomplished with a central receiver system which feature mirrors that reflect sunlight on to a large tower filled with fluid that when heated creates steam to drive a turbine.

PV and solar-thermal technologies generate zero emissions themselves but greenhouse gas emissions and other pollutants are created by their construction. PV cell manufacturing generates some hazardous materials that need to be handled appropriately to minimize risks of exposure to humans and the environment. Solar panel construction requires rare metals, which leads to mining impacts on wildlife and water quality. Water use for solar thermal plants is similar to amounts needed for a comparably sized coal or nuclear plant.

1.3 Potential Global Energy Crisis

According to thermodynamics, energy can be neither created nor destroyed, but can only be transformed from one form to another. About 80 % of global energy is supplied by coal, oil, and gas. As discussed above, they are all nonrenewable sources that will one day be exhausted. Pressure to replace fossil fuels has focused more attention on renewable sources – for example, solar and wind. Nonpolluting and enduring, they offer an attractive alternative to the traditional nonrenewable energy sources. However, there are tremendous challenges in making use of the renewable energies. Viable technologies are still under developing and start-up costs are still unaffordably high. Renewables are not expected to achieve a significant market share in the near future.

Currently, not everyone is fortunate enough to be able to enjoy the use of the fossil trio in the world. Almost a third of today's world population have no electricity or other modern energy supplies, while another third have only limited access. About 2.5 billion people have only wood or other biomass as their energy source—often bad for the environment, almost always bad for their health. It is reasonably understandable that they have the right to live a better life that cheap and accessible energy offers. However, if everyone in developing countries used the same amount of energy as the average consumer in developed countries does, the developing world's energy consumption would increase more than eightfold between 2000 and 2050.

Energy sources are often long distances away from the point of consumption. The energy for fuel, heat, and light can travel crossing continents. These long distances create a whole host of challenges, such as the environmental risks of long distance pipelines and the high cost of long-distance transportation. Furthermore, some analysts suggest that oil wells might be depleted within 70–80 years. Natural gas might run out a little later. The present reserves of uranium might be adequate for only 80–90 years. Of course, the fears of energy running out might be based on the present economic models. If higher costs are acceptable, oil could be extracted from oil sands, and lean uranium ores could be mined to recover the metal. There are also no imminent dangers of running out of coal, which remains a vital workhorse for energy generation. Moreover, one hour of solar radiation has energy equivalent to the world's annual primary energy consumption.

1.4 Climate Change and Sustainability

The current outlook for energy adds a crucial new dimension that was not present in the first oil shock—carbon dioxide (CO_2) emissions and climate change. CO_2 level is now 30 % higher than they were before the Industrial Revolution and is rising at an accelerating rate, driven by the increasing use of fossil fuels. The direct consequence of excessive CO_2 emission is global warming, thus leading to the rising of sea levels, shrinking snow cover in the northern hemisphere, and higher surface temperatures. One of the immediate consequences of sea level rising is the loss of habitable land in most coastal areas. Other potential implications of global warming and climate change include production of dislocations in agricultural, trade and demographic patterns that define global economic and social structures [4–6].

Therefore, the problem of greenhouse gas emission and the depletion of the nonrenewal energy sources, i.e., the dual challenges of energy and climate, point to in a single term—sustainability. Our current reliance on oil and other fossil fuels and our unfettered emission of CO_2 to the atmosphere are not sustainable activities. As mentioned earlier, we are using oil and fossil fuels at far greater rates than nature creates them. Our CO_2 emission is growing by 22 % per decade, which will be far beyond the capability of the ocean—atmosphere system to absorb. Therefore, these activities not only deplete the fossil resources required for our current energy system, but also undermine the environment and climate essential to our future prosperity.

Are there ways to control the greenhouse gas emissions without harming the environment? What are the energy technologies that emit no or minimal CO_2? Are there technologies and policies that help to minimize energy demand and consumption? These questions along with a few of corollaries have all been subjects of research on energy all around the world.

Many new energy-saving technologies are now emerging. Light-emitting diodes that can replace incandescent bulbs, electric cars and hybrids that substitute for petrol engines, and high-voltage direct current transmission of electric power instead of alternating-current transmission are some of the energy-saving options. There are also concerns about the availability of more efficient energy storage systems. Storage is going to become increasingly important because some of the renewable resources generate power only intermittently. Exploring new renewable energy sources has become another research topic.

1.5 Waste Energies and Their Harvesting

The definition of waste energy could include two aspects: waste to produce energy or energy from waste and energy that can be used but wasted instead, the latter of which is the focus in this book. Waste energy could be useful if it can be captured. Recovering even a fraction of this energy would have a significant economic and environmental impact. This is where energy harvesting comes in. Energy harvesting, also known as energy scavenging, is a process that captures small amounts of energy

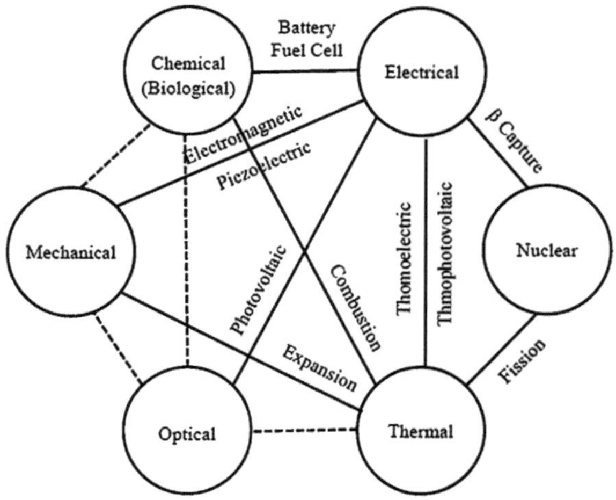

Fig. 1.1 Types of energy and the transformation mechanisms among them

that would otherwise be lost as heat, light, sound, vibration or movement. This captured energy can be used for various applications in several ways.

For instance, it can be used as power source for computer or wireless sensor devices. With increasing speed and shrinkage of components, the use of computers creates a huge amount of heat, which is dissipated into the environment as waste. If this waste heat is harvested and used to power the computer, it can help cut the cost of computer and reduce the energy consumption of use of computers. Waste energy harvesting also has the potential to replace batteries with other small and low power electronic devices. This provides several advantages. Free maintenance is one of them. It is also environmentally friendly and the problem of disposal of batteries is avoided. Batteries usually contain chemicals and metals that are harmful to the environment and hazardous to human health. It has potential applications to monitor remote or underwater locations.

Different types of waste energy should be captured using different technologies and materials. Figure 1.1 shows types of energy and transfer mechanisms from one of them to the other. Among them, the most promising energy harvesting technologies in development include vibration, movement, and sound that can be captured and transformed into electrical power using piezoelectric materials, heat that can be captured and transformed into electrical power using thermoelectric and pyroelectric materials or directly stored as thermal energy, which are the main scopes of this book.

1.6 Outline of this Book

This book consists of six chapters. The first chapter serves as an introduction to the book. Chapter 2 presents a summary on progress in harvesting mechanical waste energy with piezoelectric effect. A detailed description of piezoelectrics, including history of piezoelectric materials, principle of piezoelectric effect, types of piezoelectric materials, and materials processing and fabrication, is discussed first. Then, the principle of piezoelectric effect for analysis of mechanical waste energy harvesting is described. Various potential sources of mechanical waste energy, such as human motions, vehicle suspensions, civil structures, railway tracks, roads, streams, and ocean waves, which have been reported in the literature, are summarized.

Chapter 3 serves to provide an overview of the latest development in nanogenerators, a new research direction of small-scale mechanical energy harvesting, in which nanosized piezoelectric materials, especially zinc oxide (ZnO), are used. The devices convert various types of ambient mechanical energy into electricity based on the piezoelectric effect. A similar structure is used in this chapter to show the feasibility of nanogenerators in mechanical energy harvesting. For different applications, the requirements for morphology, microstructure, and geometry of the nanosized materials are different, which are synthesized and fabricated using different methods and techniques. Potential applications of nanogenerators that have been explored are summarized and discussed.

Chapter 4 is aimed to update the progress in thermoelectric effect and materials as a potential technique to harvest waster thermal energy. Various concepts related to thermoelectric effect are explained. Thermoelectric materials that have been developed in order to search for high figure of merit for practical applications are listed and summarized. Theoretical foundations leading to high performance thermoelectric materials and structures are elaborated in detail. Nanostructuring as a unique technology in developing thermoelectric materials is discussed with successful examples reported in the open literature.

Pyroelectric effect as an alternative technology to harvest waste thermal energy is presented in Chap. 5. Principles of pyroelectric effect, including definition, pyroelectric coefficients, coefficient measurement, phase transition, and so on, is presented first, followed by a brief description of the type and characteristics of pyroelectric materials. The latest progress in pyroelectric waste heat energy harvesters based on the Olsen cycle and other types of principles, based on pyroelectric polymers, perovskite ceramics, and single crystals, are summarized and discussed.

In the last chapter, thermal energy harvesting through storage by using phase change materials (PCM) is reviewed. PCMs that have been used for applications in various thermal energy storage systems are summarized and discussed according to their composition. Design criteria, heat transfer phenomena, configuration of various PCM thermal energy storage systems, which have been reported in the open literature, are presented, with specific attention to exergy analysis. The main applications of the PCM storage systems are discussed and analyzed. Strategies of enhancing heat transfer efficiency in PCM thermal energy storage systems are evaluated.

References

1. http://en.wikipedia.org/wiki/World_energy_consumption
2. http://en.wikipedia.org/wiki/Renewable_energy
3. K.J. Holmes, L. Papay, Prospects for electricity from renewable resources in the United States. J. Renew. Sustain. Energy **3**, 042701 (2011)
4. V.A. Arunachalam, E.L. Fleischer, Harnessing materials for energy. Mater. Res. Bull. **33**, 261 (2008)
5. V.A. Arunachalam, E.L. Fleischer, The global energy landscape and materials innovation. Mater. Res. Bull. **33**, 264–276 (2008)
6. A.M. Omer, Energy use and environmental impacts: a general review. J. Renew. Sustain. Energy **1**, 053101 (2009)

Chapter 2
Waste Mechanical Energy Harvesting (I): Piezoelectric Effect

2.1 Introduction

Mechanical energy is one of the most ubiquitous energies that can be reused in our surroundings. The sources of mechanical energy can be a vibrating structure, a moving object, and vibration induced by flowing air or water. The energies related to induced vibrations or movement by flow of air and water at large-scale are wind energy and hydroelectric energy, respectively, which are not within the scope of this book. Instead, the mechanical energies here can be classified as so-called "low-level" vibrations and movements. Such potential "low-level" vibrations and movements are summarized in Table 2.1 [1] and Table 2.2 [2].

Mechanical waste energies usually can be harvested by using vibration-to-electricity conversion [2–4]. The most distinguished characteristic of this kind of waste energy harvesting is initially identified for low power generations. Therefore, one of the targeted applications is to power small electronic devices. However, recent development indicates that it can also be used for large-scale applications [5].Vibration-to-electricity conversion can be realized through three basic mechanisms, including electromagnetic [6–8], electrostatic [9], and piezoelectric [10, 11] transductions.

Among the three mechanisms, piezoelectric transduction has received the greatest attention. This is because piezoelectric materials have larger power densities and higher feasibility for practical applications than the materials used in the other two mechanisms [3]. For example, voltage outputs in electromagnetic energy harvesting are typically very low and thus must be amplified to a level sufficiently high to charge storage devices. In contrast, however, piezoelectric energy harvesters output voltages that can be used directly. In electrostatic energy harvesting, the materials should be subject to an external applied voltage to trigger the relative vibratory motion of the capacitor elements, which outputs alternative electrical currents [9]. Such external applied voltages are required in piezoelectric energy harvesting. Another advantage over electromagnetic devices is that piezoelectric harvesting devices can be fabricated at both macro-scale and micro-scale, due to the well-established deposition techniques for thick-films and thin-films piezoelectric

L. B. Kong et al., *Waste Energy Harvesting*, Lecture Notes in Energy 24,
DOI: 10.1007/978-3-642-54634-1_2, © Springer-Verlag Berlin Heidelberg 2014

Table 2.1 Sources of mechanical energy around us each and every day that can be harvested for electricity [1]

Human body/motion	Transportation	Infrastructure	Industry	Environment
Breathing, blood flow/pressure, exhalation, walking, arm motion, finger motion, jogging, talking	Aircraft, automobile, train, tires, tracks, peddles, brakes, turbine engine, vibration, noises	Bridges, roads, tunnels, farm, house structure, control-switch, water/gas pipes, AC system	Motors, compressor, chillers, pumps, fans, vibrations, cutting and dicing, noise	Wing, ocean current/wave, acoustic wave

Table 2.2 Acceleration ($m\ s^{-2}$) magnitude and frequency of vibration mode of potential vibration sources

Vibration source	A ($m\ s^{-2}$)	f_{peak} (Hz)
Car engine compartment	12.0	200
Base of 3-axis machine tool	10.0	70
Blender casing	6.4	121
Clothes dryer	3.5	121
Person nervously tapping their heel	3.0	1
Car instrument panel	3.0	13
Door frame just after door closes	3.0	125
Small microwave oven	2.5	121
HVAC vents in office building	0.2–1.5	60
Windows next to a busy road	0.7	100
CD on notebook computer	0.6	75
Second story floor of busy office	0.2	100

Reproduced with permission from [2]. Copyright @ 2003, Elsevier

materials [10, 12]. Comparatively, it is hard to fabricate electromagnetic energy harvesters at micro-scale, due to the poor properties of planar magnets and the limited number of turns that can be realized when using planar coils [3].

In this chapter, harvesting mechanical waste energy with piezoelectric effect will be systematically presented. The main content will be started with a detailed description of piezoelectrics, including history of piezoelectric materials, principle of piezoelectric effect, and types and preparation of piezoelectric materials. After that, the progress in the development of mechanical waste energy harvesters based on piezoelectric effect will be summarized. Potential mechanical waste energy sources, including vehicle suspensions, civil structures, rail way tracks, roads, ocean waves, streams, and human motions, have been explored for respective potential applications. The chapter will wound up with some concluding remarks.

2.2 Piezoelectricity and Piezoelectric Materials

2.2.1 *Piezoelectricity*

The word "piezoelectricity" is derived from the Greek "*piezein*", which means to "squeeze" or "press" [13–16]. There are two piezoelectric effects: direct effect and converse effect. The direct effect (designated as a generator) is identified with the phenomenon whereby electrical charge (polarization) is generated from a mechanical stress, whereas the converse effect (designated as a motor) is associated with the mechanical movement generated by the application of an electrical field. Therefore, piezoelectric energy harvesting is to use the direct effect (generator). Properties of piezoelectric materials are generally characterized by k_p, k_{33}, d_{33}, d_{31}, and g_{33}.

The k factors (e.g., k_{33}, k_{31}, and k_p), which are also called piezoelectric coupling factors, are convenient and direct measurements of the overall strength of the electromechanical effects, i.e., the ability of the ceramic transducer to convert one form of energy to another. They are defined as the square root of the ratio of energy output in electrical form to the total mechanical energy input (direct effect), or the square root of the ratio of the energy available in mechanical form to the total electrical energy input (converse effect). Because the conversion of electrical to mechanical energy (or vice versa) is always incomplete, k is always less than unity. Commonly used as a figure-of-merit for piezoelectrics, the higher k values are most desirable and constantly sought after in new materials. For ceramics, k_p is a typical measure used to compare piezoelectric properties of ferroelectric materials-values ranging from 0.35 for $BaTiO_3$ to as high as 0.72 for PLZT [13].

The d coefficients are called piezoelectric coefficients, having magnitudes of $\times 10^{12}$ C N^{-1} (or pC N^{-1}) for the direct effect and $\times 10^{-12}$ m V^{-1} (or pm V^{-1}) for the converse effect, respectively. Subscript is used to describe the relative direction of inputs and outputs. For example, d_{31} means that this piezoelectric coefficient relates to the generation of polarization (direct effect) in the electrodes perpendicular to the vertical direction (3) and to the stress mechanically applied in the lateral direction (1), while d_{33} indicates the polarization generated in the vertical direction (3) when the stress is applied in the same direction. There are also other similar symbols [13].

g factors are called open-circuit coefficients, another parameters used to evaluate piezoelectric ceramics for their ability to generate large amounts of voltage per unit of input stress. The g constant is related to d constant: $g = d/K\varepsilon_0$ (K is relative dielectric constant and ε_0 is the dielectric constant of free space that equals to unit). High-g-constant piezoelectric ceramics are usually ferroelectrically hard materials whose polarizations are not readily switched and thus they possess lower K values.

Piezoelectricity is usually discussed together with ferroelectricity and pyroelectricity, because they have interesting inter-relationships in terms of crystal structures. All crystals can be categorized into 32 different classes. In the theory of

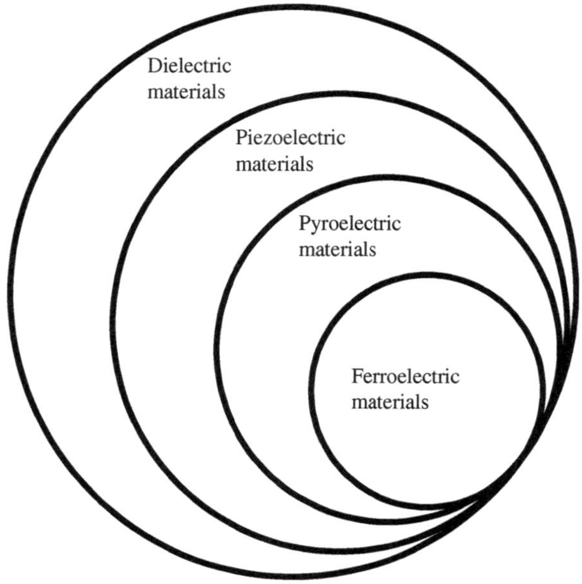

Fig. 2.1 Inter-relationship among piezoelectric, ferroelectric, pyroelectric and dielectric materials. Ferroelectric materials have superior pyroelectric, piezoelectric and dielectric properties than nonferroelectric materials

point groups, these classes are determined by using several symmetry elements: (i) center of symmetry, (ii) axis of rotation, (iii) mirror planes, and (iv) several combinations of them. The 32 point groups are subdivisions of seven basic crystal systems that are, in order of ascending symmetry, triclinic, monoclinic, orthorhombic, tetragonal, rhombohedral (trigonal), hexagonal, and cubic. 21 classes out of the 32 point groups are noncentrosymmetric, which is a necessary condition for piezoelectricity to exist. 20 of them are piezoelectric. Of the 20 piezoelectric crystal classes, 10 crystals are of pyroelectric properties. Within a given temperature range, this group of materials is permanently polarized. Compared to the general piezoelectric polarization produced under stress, the pyroelectric polarization is developed spontaneously and kept as permanent dipoles in the structure. Because this polarization varies with temperature, the response is termed as pyroelectricity. Within the pyroelectric group, there is a subgroup that has spontaneous polarization, which is called ferroelectric materials. On one hand, the polarization in a ferroelectric material is similar to the polarization in a pyroelectric one. On the other hand, there is difference between the two polarizations because the ferroelectric polarization is reversible by an external applied electric field, provided that the applied field is less than the dielectric breakdown of the materials. Therefore, materials that can be defined as ferroelectrics must have two characteristics: the presence of spontaneous polarization and reversibility of the polarization under electric field [13, 14]. Figure 2.1 shows their inter-relationship, together with general dielectrics. This inter-relationship is important because ferroelectric materials have best piezoelectric properties than nonferroelectric materials.

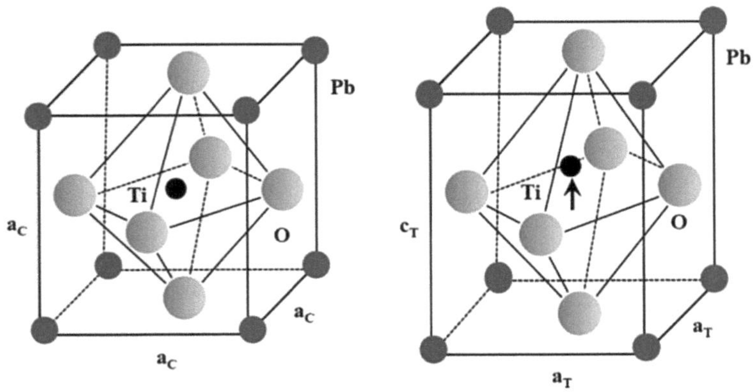

Fig. 2.2 Schematic perovskite structure of $PbTiO_3$, with cubic (*C*) structure in the paraelectric state ($P_s = 0$) and tetragonal (*T*) structure in the ferroelectric state ($P_s \neq 0$)

Ferroelectric materials can be grouped into four subcategories according to their crystal structures: perovskite group, pyrochlore group, tungsten-bronze group, and bismuth layer structure group, among which the perovskite group is the most important and thus the most widely studied. Perovskite is usually expressed as ABO_3. A typical ABO_3 unit-cell structure is shown in Fig. 2.2, taking $PbTiO_3$ as an example [13, 14, 16]. It consists of a corner-linked network of oxygen octahedra, creating an octahedral cage (B-site) and the interstices (A-sites). Ti^{4+} ions occupy the B-site while Pb^{2+} ions occupy the A-site. Figure 2.2 also shows the paraelectric and ferroelectric states of $PbTiO_3$. Most ferroelectric materials undergo a structural phase transition from a high temperature paraelectric phase into a low temperature ferroelectric phase. The paraelectric phase always has a higher symmetry than the ferroelectric phase. The temperature of the phase transition is called as the Curie temperature (T_C). Different ferroelectric materials have different values of T_C, which can be either lower than liquid nitrogen (LN) temperature or higher than 1,000 °C. For a given material (composition), the T_C is closely related to microstructure (grain size and distribution, density, porosity, and pore size and distribution, and impurity, and so on). Generally, the T_C of a given material decreases with decreasing grain size [54].

In the ferroelectric state, the ability of displacement of the central Ti^{4+} ion is the cause for the reversibility of polarization. The switch of many adjacent unit cells is referred to as domain reorientation or switching. The homogeneous areas of the material with the same polarization orientation are referred to as domains, with domain walls existing between areas of unlike polarization orientation [1, 2]. For as-prepared piezoelectric ceramics, the domains are randomly oriented and thus the net polarization of materials is zero because of their cancellation effect. Therefore, the as-prepared piezoelectric ceramics have no piezoelectric effect. To show piezoelectric properties, the newly obtained polycrystalline piezoelectric ceramics must be poled at strong external DC electric fields (10–100 kV cm^{-1}).

Poling is to electrically align the orientation of the randomly distributed domains within the grains and to make ferroelectric ceramics act like a single crystal possessing both ferroelectric and piezoelectric properties. Poling is usually conducted at elevated temperatures, because polarization would be more compliant at high temperatures [13, 14]. After poling during cooling, the applied voltages should be remained until the temperature is sufficiently low.

2.2.2 Brief History of Modern Piezoelectric Ceramics

The history of modern piezoelectricity and piezoelectric ceramics is close related to the history of ferroelectricity and ferroelectric ceramics. The history of ferroelectrics can be tracked back to Rochelle salt (sodium potassium tartrate tetrahydrate, $KNa(C_4H_4O_6)\cdot 4H_2O$), which was synthesized more than 400 years ago, initially for medicinal purposes [13, 14, 16, 17]. It is in this same crystalline material that pyroelectric (thermal-polar), piezoelectric (stress-polar), and ferroelectric were discovered subsequently. Before this discovery, ferroelectricity was only a hypothetical property of solid materials at the turn of the twentieth century. However, the practical application of this material is largely limited due to its water solubility. It was after the discovery of ferroelectric ceramics (barium titanate, $BaTiO_3$), this class of materials became extremely useful for a variety applications.

The first ferroelectric ceramic material is barium titanate ($BaTiO_3$ or BT), which was discovered in the mid-1940s [13, 14, 16, 17]. Before the discovery of $BaTiO_3$, the most widely used materials for capacitors were steatite, mica, TiO_2, $MgTiO_3$, and $CaTiO_3$, with dielectric constant of not higher than 100. During the World War II, there was pressing needs for high dielectric constant materials to fabricate high capacitance capacitors. Before publication was available in the literature, $BaTiO_3$ had already been studied as a high dielectric constant material concurrently. In the later open publications, it was concluded that the source of the high dielectric constant in $BaTiO_3$ is due to its ferroelectric properties [13].

The history of ferroelectric ceramics also includes the report of lead zirconatetitanate ($PbZr_{1-x}Ti_xO_3$, or PZT) piezoelectric ceramics, the development of transparent electro-optical lead lanthanum zirconate titanate ($Pb_{1-x}La_xZr_{1-y}Ti_yO_3$, or PLZT), the research on lead magnesium niobate ($PbMg_{1/3}Nb_{2/3}O_3$, or PMN) relaxor ferroelectric ceramics and the discovery of many other nonperovskite ferroelectric ceramics [13, 16]. Among these, PZT has been demonstrated to possess best performances as piezoelectric ceramics.

Recently, there has been a concern with PZT, due to the toxicity of Pb. Regulations and legislations have been established globally to restrict the use of lead-containing materials. For example, according to the directive for the Restriction of the use of certain Hazardous Substances in electrical and electronic equipment (RoHS) adopted by the European Parliament in the year 2006, the maximum allowed concentration of lead is established to be 0.1 wt% in homogeneous materials for electrical and electronic equipment used in households as well as industry. Therefore, lead-containing piezoeletric materials will be prohibited eventually. Similar regulations have been established worldwide. As a consequence, there is significantly increasing interest in developing lead-free piezoelectric ceramics all around the world. Although major progress has been made in materials research [18, 19], there are still obstacles blocking the successful industrial implementation of lead-free piezoelectric ceramics. One key problem is their poor piezoelectric properties and fatigue degradation problem. Therefore, PZT will still be dominant materials for piezoelectric applications at least in the near future.

2.2.3 Microstructures of Piezoelectric Ceramics

Fully dense piezoelectric ceramics (>95 % of theoretical density) are required by most applications for several reasons. Firstly, a full densification ensures the ceramics to have their maximum performance. For example, the piezoelectric constant of piezoceramics usually decreases with decreasing density. This is because the pores or vacuum have no piezoelectric effect. Secondly, the presence of pores is generally a cause of high loss tangent, since porosity could provide a conduction path for electricity and thus could be a main contribution to dielectric loss tangent, as conduction loss. In this case, external electric fields cannot be applied to the materials effectively. Moreover, electrical conduction caused by porosity is also responsible for degraded electrical breakdown strength. Finally, mechanical strength, sometimes a critical requirement for some specific applications, especially as actuators, is directly related to density of the materials used.

Generally, the densities of piezoelectric ceramics increase with increasing sintering temperature. However, very high temperature is not suitable for lead- and bismuth-containing piezoelectric materials, which is due to the volatility characteristic of lead and bismuth. High temperature sintering also leads to abnormal grain growth (secondary grain growth). The presence of exaggeratedly grown grains is harmful to the performance for most piezoelectric ceramics. To effectively enhance the densification of piezoelectric ceramics at relatively low temperatures, various attempts have been made and have been widely reported in the literature. The main strategies that have been employed to reduce the sintering temperature of piezoelectric ceramics are the use of fine/ultrafine powders and the addition of sintering aids. If sintering aids are used, they must have relatively low melting points and no reaction with the piezoelectric phases during sintering process. Grain size is another important factor that determines piezoelectric properties of piezoceramics [20, 21]. The variation in grain size with sintering temperature is similar to that of density, i.e., grain size increases with increasing sintering temperature. There is a critical grain size for most piezoelectric ceramics. These properties can be optimized by adjusting materials processing parameters.

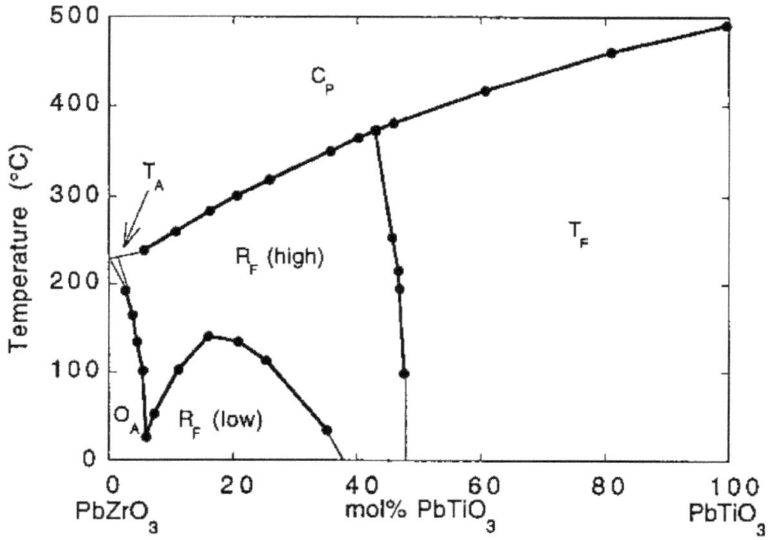

Fig. 2.3 Phase diagram of $PbZr_{1-x}Ti_xO_3$ [23]. There is a morphotropic phase boundary (MPB) at $x = 0.52$–0.53, with highest electrical properties

2.2.4 Typical Piezoceramics

As mentioned above, lead zirconate titanate ($Pb(Zr_xTi_{1-x})O_3$ or PZT, $x = 0$–1) is the most widely used to fabricate piezoelectric ceramics. PZT is a solid solution of $PbTiO_3$ (PT) and $PbZrO_3$ (PZ). PT and PZ can form solid solution over composition range. Figure 2.3 shows a phase diagram of PZT. There exists an almost temperature-independent phase boundary at $x = 0.52$–0.53, which separates a rhombohedral Zr-rich phase from a tetragonal Ti-rich phase. Dielectric constant, piezoelectric constant, and electromechanical coupling coefficient all exhibit a pronounced maximum value for the composition corresponding to this phase boundary, which is generally referred to as the morphotropic phase boundary (MPB) [22, 23]. This is mainly attributed to the existence of a mixture of phases at the boundary and the presence of a larger number of reorientable polarization directions existing in the MPB mixed-phase region. This is reason why PZT is usually meant to be $PbZr_{0.52}Ti_{0.48}O_3$ or $PbZr_{0.53}Ti_{0.47}O_3$ used as piezoceramics in the open literature.

2.2.5 PZT Ceramics

Conventionally, PZT powder is synthesized by using solid-state reaction method, with PbO (Pb_3O_4), ZrO_2, and TiO_2 oxide powders as starting materials. Mixtures of the three oxide powders with nominated compositions of PZT are ball milled and then calcined at a suitable temperature to form the desired perovskite phase. PZT piezoelectric ceramics are obtained by sintering the calcined powders at higher temperatures to achieve full densification.

Based on this basic composition, a large number of modifications have been adapted to cater for the requirements of different applications. According to international convention, piezoceramics are divided into two groups: soft and hard PZT ceramics. Soft and hard PZT ceramics refer to their mobility of the dipoles or domains and hence also to the polarization and depolarization behavior. Soft piezoceramics have high domain mobility and thus are easy to be poled. They usually have large piezoelectric charge coefficient, moderate values of permittivity, and high coupling factors. Hard PZT materials can be subjected to high electrical and mechanical stresses. Their properties change only little under these conditions, which makes them more suitable for high-power applications. They possess large piezoelectric coupling factors, high qualities, and very good stability at high mechanical loads and operating fields.

Commercially available oxide powders have relatively large grain/particle sizes. The synthesis of PZT with these powders requiring a high calcination temperature form the perovskite phase [24]. Figure 2.4 shows representative microstructural properties of a set of PZT ceramics synthesized by using the conventional solid-state reaction method [24]. PZT powders thus have rough particles that lead to high sintering temperature, which is not favorite owing to the volatility of Pb. Therefore, special carefulness should be taken when sintering these rough PZT powders to avoid the loss of Pb during the sintering process. One of the ways to address this problem is the use of fine PZT powders, which can be synthesized by using wet-chemical synthesis methods, such as chemical co-precipitation [25] and sol-gel [26]. Due to the high chemical homogeneity provided by these wet-chemical processing routes, PZT powders have ultrafine or nanosized grain/particles and thus can be sintered at significantly low temperatures. However, these chemical methods are relatively less cost-effective, because they require the use expensive starting chemicals and they also involve a multistep processing.

More recently, a novel technique, mechanochemical synthesis with high-energy ball milling, was employed to synthesize PZT powders [16]. In this method, the formation of designed compounds is due to the reactions of oxide precursors which are activated by mechanical energy, instead of the heat energy required in the conventional solid-state reaction process. The mechanical technique is superior to both the conventional solid-state reaction and the wet-chemistry-based processing routes for several reasons. Firstly, it uses cost-effective and widely available oxides as the starting materials. Secondly, it skips the intermediate temperature calcination step, leading to a simpler process. Thirdly, it takes place at room temperature in closely sealed containers, thus effectively alleviating the loss of Pb. Furthermore, due to their nanometer scale size and very high homogeneity, the mechanochemically derived PZT powders demonstrate much better sintering behavior than those synthesized by the conventional solid-state reaction and wet-chemical processes.

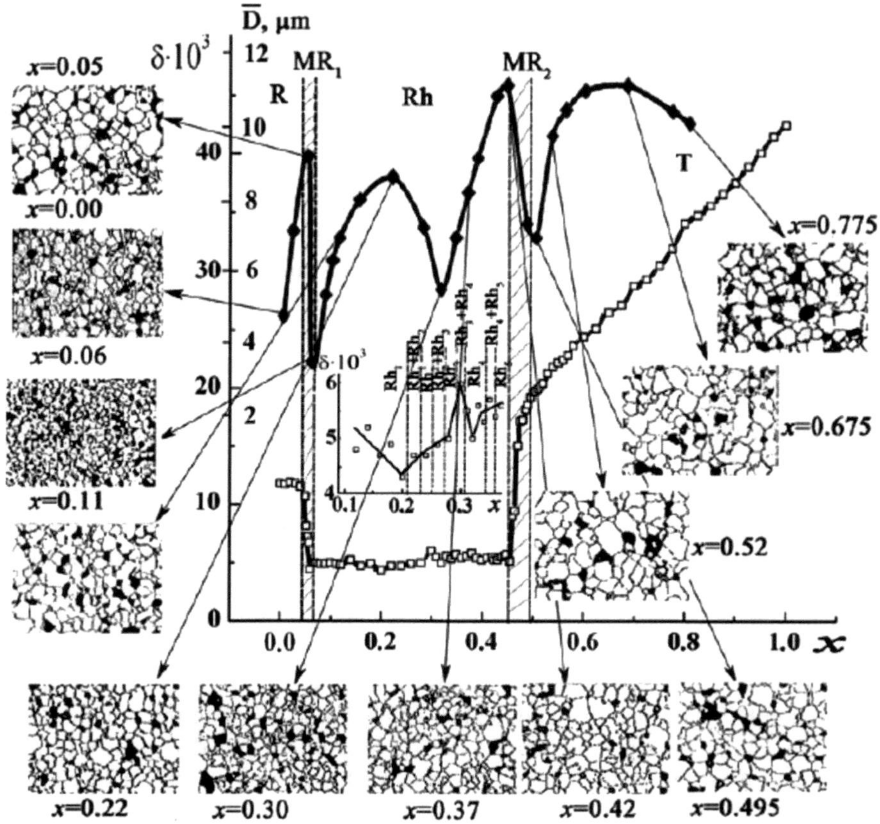

Fig. 2.4 Variation in mean grain size, D, and homogeneous deformation parameter, d, as a function of the content of $PbTiO_3$ in the PZT system. The symmetries are designated as *R* rhombic, *Rh* rhombohedral, *T* tetragonal, and *MR1* and *MR2* denote morphotropic regions. Reproduced with permission from [24]. Copyright @ 2013, Elsevier

As an example, Fig. 2.5 shows XRD patterns of the mixtures (PbO, ZrO_2, and TiO_2) for the composition of $PbZr_{0.52}Ti_{0.48}O_3$, milled for different time durations, using a Fritsch Pulverisette five planetary high-energy ball milling system, with tungsten carbide (WC) vials and balls as the milling media [27]. After milling for 4 h, no PZT is formed, but the diffraction peaks from PbO are greatly broadened and weakened, indicating that the starting oxides have been significantly refined as a result of the high-energy ball milling. The diffraction peaks of PbO are further widened and reduced in the 8-h milled sample. At the same time, a trace of PZT can be observed. PZT with perovskite structure comes to the predominant phase in the samples milled for 15 and 24 h. It is worth mentioning that all the fours samples can be used to make PZT ceramics, which means that complete reaction of the precursor oxides is not necessary. This is because the reaction can be completed during the sintering processing. In this case, the milling time can be significantly reduced to save energy and time.

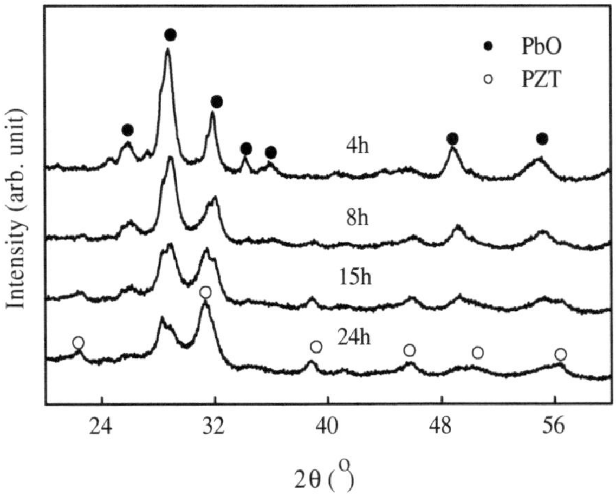

Fig. 2.5 XRD patterns of the mixtures for $PbZr_{0.52}Ti_{0.48}O_3$ milled for different times, using a planetary high-energy ball mill and tungsten carbide (WC) media. Reproduced with permission from [27]. Copyright @ 2001, Materials Research Society

2.2.6 PZT Films

For applications in microelectromechanical system (MEMS), PZT thin films or thick films should be used [28]. Although there is no restricted definition, thin films usually have thicknesses of <1 μm and thick films are those with thickness of >1 μm. In practice, the thickness of a thick film can be up to hundreds of μm. PZT films can be deposited by using physical or chemical methods. Physical deposition includes sputtering, physical vapor deposition (PVD), and pulsed laser deposition (PLD), while chemical method includes solution, sol-gel, hydrothermal, and chemical vapor deposition (CVD). Among these physical and chemical methods, sol-gel has been most widely used to deposit PZT films, due to its effectiveness and efficiency.

Figure 2.6 shows a flow chart for deposition of PZT films by using a sol-gel technique [29]. $Pb(CH_3COO)_2{\cdot}5H_2O$ is dissolved in CH_3COOH at 120 °C, while $Zr[CH_3(CH_2)_2CO]_4$ and $Ti[(CH_2)_2CHO]_4$ are stabilized with acetylaceton. They are then mixed to form 5 M solution with a composition of PZT, which is used to deposit PZT films by using spin-coating. After every coating, the samples are pyrolyzed at 350 °C for 10 min, followed by annealing at 650 °C for 30 min. A final annealing at 700 °C for 30 min is conducted. PZT films with desired thickness can be obtained by repeating the pyrolysis-annealing step.

Figure 2.7 shows XRD patterns of the PZT films deposited for 1–6 times [29]. Pervoskite phase has been well formed in all films, which means that the annealing temperature of 700 °C is sufficient. There is no pyrochlore phase in all samples. The sample coated for just one time has a random orientation. However, with increasing thickness, it becomes (100) orientated gradually. Cross-sectional SEM images of selected films are shown in Fig. 2.8 [29]. The thickness of the sample deposited for one time is ~0.5 μm. The total thickness of the films increases almost linearly with the number of deposition layers, which provides a very

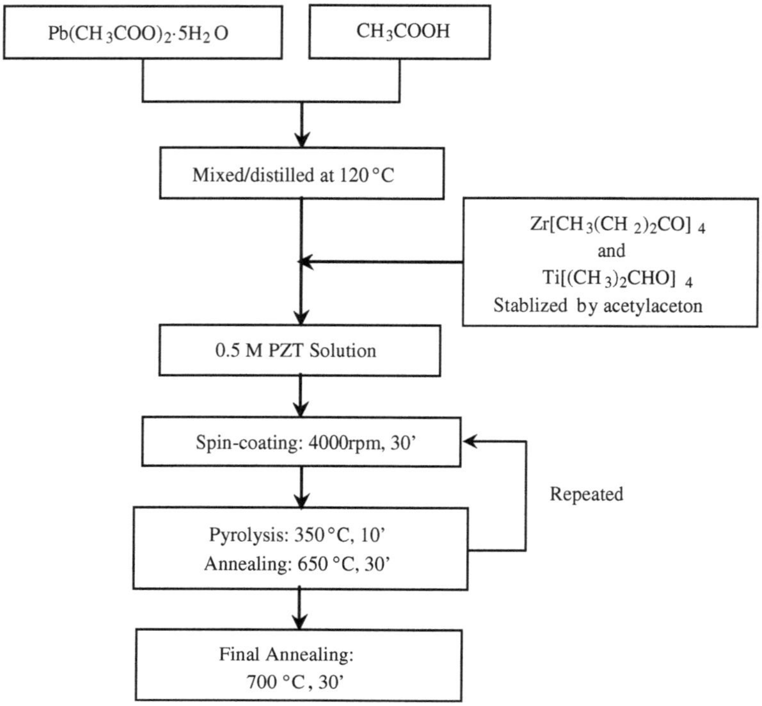

Fig. 2.6 Flow chart for deposition of PZT films with a modified sol-gel technique to produce single layer of PZT with sufficient thickness for thick film fabrication [29]

effective way to deposit thick PZT films. Comparatively, a single layer of PZT film deposited by using normal sol-gel process is much less than 0.1 μm. The use of acetic acid is the key to deposit thicker single layer of Pb containing films [30–32]. By the way, the films have a columnar microstructure, indicating their good crystallinity. Nevertheless, deposition of PZT thick films is still a challenge.

Figure 2.9 shows surface SEM images of selected PZT films [29]. All films have dense and crack-free microstructure. The average grain size slightly increases with increasing thickness, which can be readily attributed to increasing number of annealing times.

More recently, a new approach, called nanocomposite processing route, has been proposed to incorporated high-energy ball milling technique with sol-gel process to deposit thick films, which cannot be realized using the typical sol-gel with solution precursors [33–37]. This approach has combined the advantages of both high-energy milling and the sol-gel process. In this processing, commercial PZT powder is milled using a high-energy mill to produce nano-sized PZT powder. The nano-sized PZT powder is then milled for one more time with appropriate dispersants, which are used to modify its surficial characteristics. The modified PZT nano-sized powder can be readily mixed with sol-gel solutions

Fig. 2.7 XRD patterns of the PZT films deposited for 1–6 times (layers) with film thickness indicated [29]. There is a gradual increase in (100) orientation

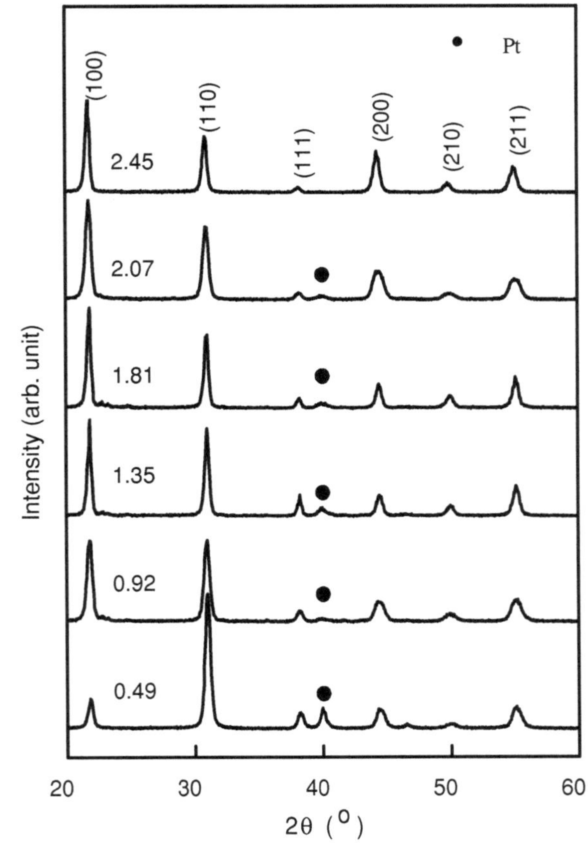

without the presence of agglomerations or precipitations. This kind of mixture can be used as same as the normal sol-gel solutions to deposit thick films on various substrates. The slurries may also directly be used to other processing, such as tape casing, screen printing, and molding [34] (Fig. 2.10).

Thick films with thickness of up to 25 μm have been prepared through multi-layer deposition with one layer being ~2 μm, which is nearly 50–100 times the normal sol-gel solution process. Figure 2.11 shows an example of such thick films [35]. The film is very uniform in thickness, with a dense microstructure and narrow distribution of grain sizes (~100 nm). Compared with the thick films prepared by using a precursor made of micro-sized PZT powder, the nanocomposite film has electrical properties, which has been attributed to the factor that the latter has a dense microstructure than the former. Such kind of composite thick films have been successfully combined with silicon technology to fabricate MEMS piezo-electric devices, so that they are suitable for applications in mechanical energy harvesting devices.

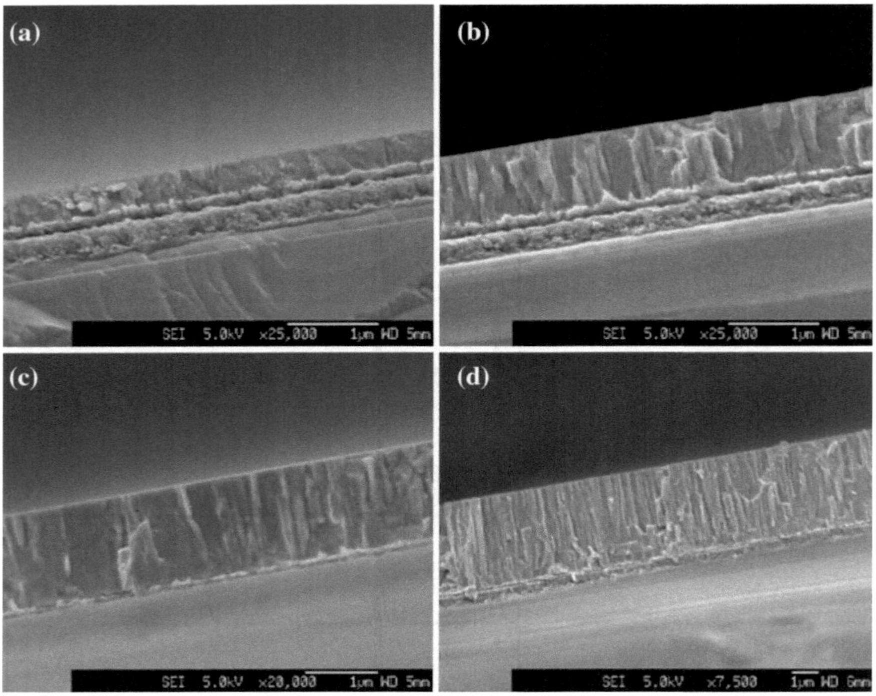

Fig. 2.8 Cross-sectional SEM images of the PZT films deposited for different times (layers): **a** 1, **b** 2, **c** 3, and **d** 6 [29]. All the films have a columnar structure

PZT ceramics have been prepared from the nanocomposite precursors at much lower temperature than that required by the conventional ceramic process [34]. Fully dense PZT ceramics can be achieved after sintering at a temperature as low as 800 °C. Interestingly, no obvious grain growth is observed in the samples sintered at temperatures from 850 to 1,000 °C, which means that the grain growth is stopped at ~850 °C. The absence of grain growth at high temperatures has been attributed to the uniform size distribution and nonagglomeration of the nanocomposite precursors. This feature is particularly of importance when this kind of precursor is used to fabricate multilayer structured devices in which small thickness of single layer is required.

2.2.7 Piezoelectric Polymers

Polyvinylidene fluoride or polyvinylidene difluoride (PVDF) has been found to show piezoelectric (ferroelectric) effect, due to the presence of a non-centrosymmetric unit cell and a net polarization in the material [38–40]. The common characteristics of piezoelectric polymeric materials are: a large dipole moment in

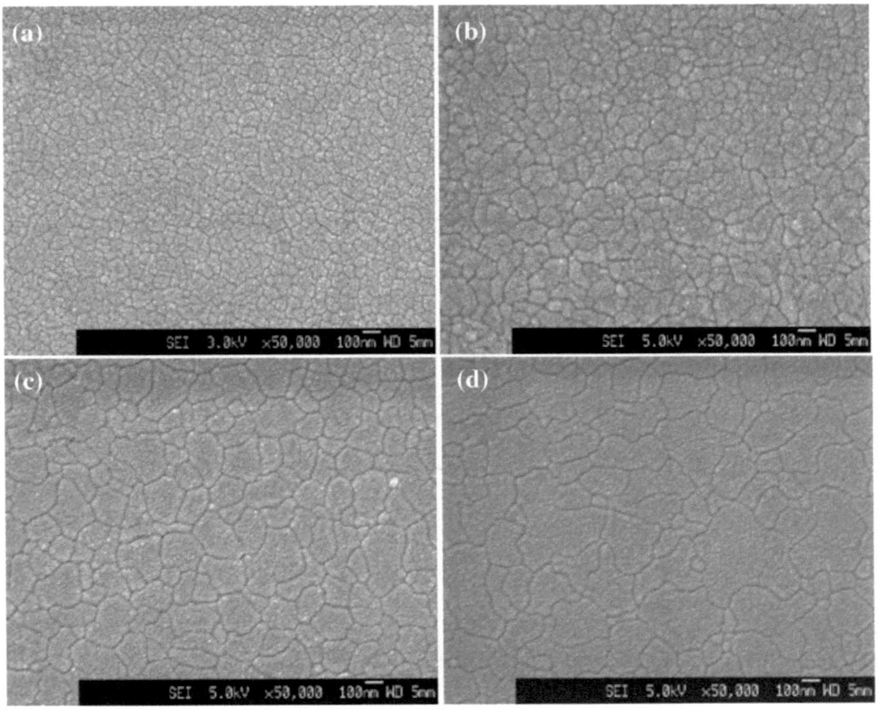

Fig. 2.9 Surface SEM images of the PZT films deposited for different times (layers): **a** 1, **b** 2, **c** 3 and **d** 5 [29]. Grain size of the films gradually increases with increasing thickness

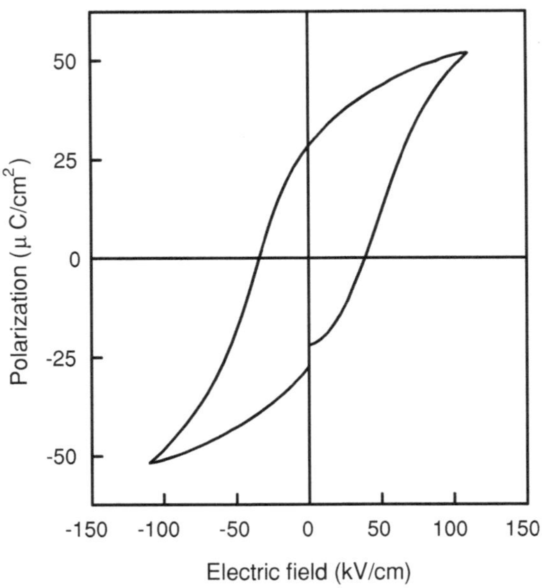

Fig. 2.10 Representative polarization-field (P-E) hysteresis curve of the PZT films showing their high quality [29]

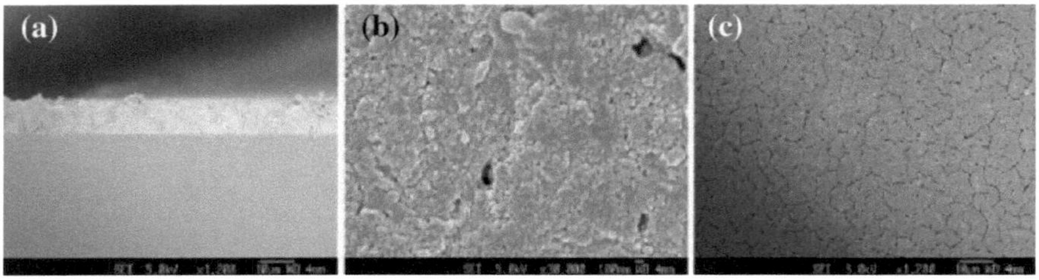

Fig. 2.11 SEM images of the thick film prepared with nanocomposite precursors: **a** and **b** cross-section and **c** surface. Reproduced with permission from [35]. Copyright @ 2003 Elsevier

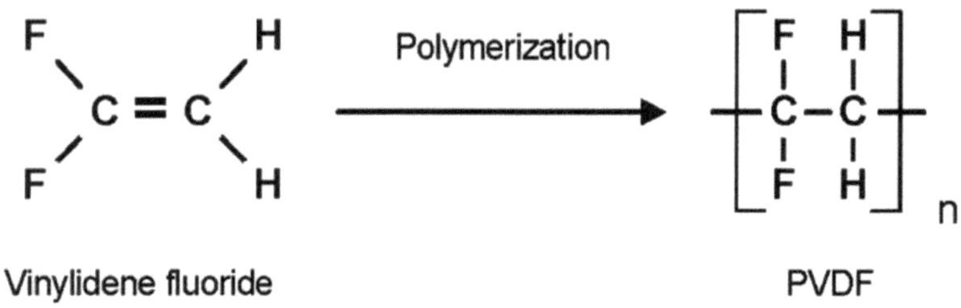

Fig. 2.12 Polymerization of vinylidene fluoride to form PVDF

the repeating unit, feasibility of crystallization in a noncentrosymmetric unit cell, and alignment of molecular dipoles. A dipole moment arises from a charge separation between adjacent atoms. In polymers, this can occur between covalently bonded atoms along the polymer chain or between adjacent chains.

PVDF has chains of CH_2CF_2 and is a semi crystalline polymer. Figure 2.12 shows the formation of PVDF polymer through the polymerization of monomer vinylidene fluoride. The hydrogen atoms having net positive charge and the fluorine atoms with net negative charge end up on opposite sides, as shown in Fig. 2.13. This develops a pole direction (indicated by the small ***p*** in Fig. 2.13). If the external applied electric field is in the opposite direction of the poled direction of the PVDF sheet, it will be stretched in the length direction. If the electric field is in the direction of the poled direction of the PVDF sheet, there will be a contraction in the length direction of the sheet. These two responses are shown in Figs. 2.14 and 2.15, respectively, which is the mechanism of piezoelectric effect of PVDF.

It was later found that copolymers of PVDF with trifluoroethylene (TrFE) are new polymer materials with stronger piezoelectric effect. The copolymers have advantages over pure PVDF. The P(VDF-co-TrFE) copolymers can be modified by adjusting the compositions of the two components. The random copolymers P(VDF-co-TrFE)s are usually synthesized by copolymerization of VDF and TrFE monomers.

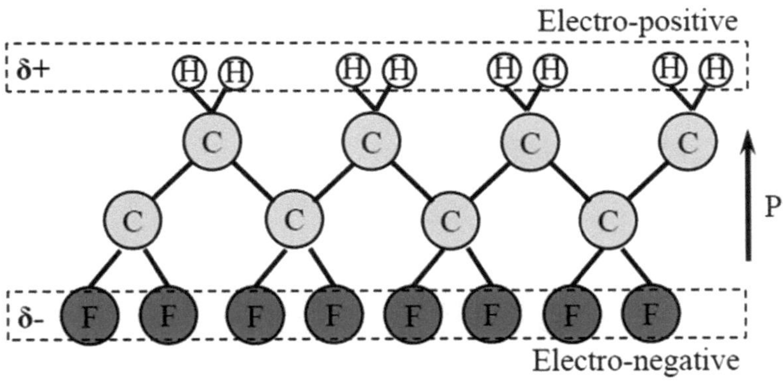

Fig. 2.13 Schematic diagram of atom arrangement in PVDF molecule

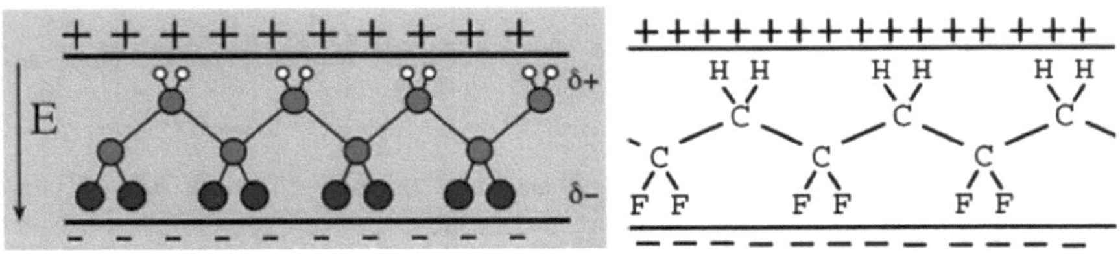

Fig. 2.14 Stretch in length direction occurred if the applied electric field is in the opposite direction of the poled direction of the PVDF sheet

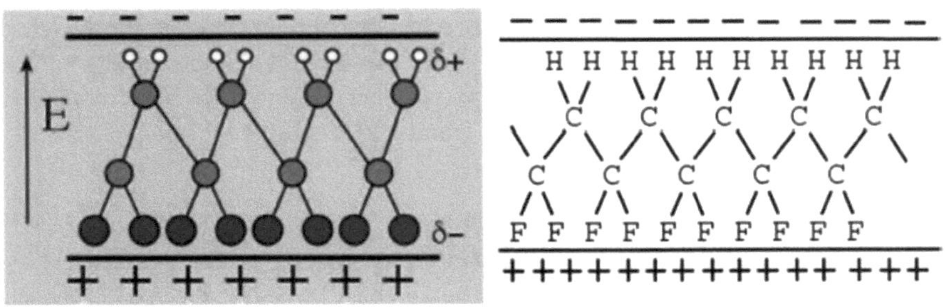

Fig. 2.15 Contraction in length direction occurred if the electric field is in the direction of the poled direction of the PVDF sheet

2.2.8 Composites

Piezoelectric ceramics have high piezoelectric performances, but are hard and rigid for some applications that require flexibility and complicated shapes. To address this problem, composite materials are proposed, which consist of

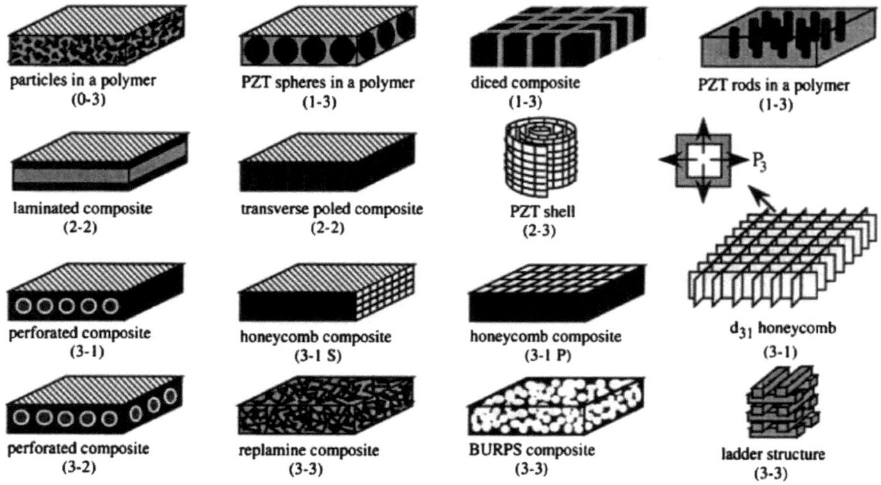

Fig. 2.16 Connectivity of constituent phases in piezoelectric ceramic-polymer composites. Reproduced with permission from [41]. Copyright @ 1999, Elsevier

piezoelectric ceramic particles and polymer matrix. The presence of polymer matrix makes the composites to have mechanical flexibility. Composites are usually described by using a notation-connectivity. Figure 2.16 shows schematics of possible connectivity of composites with PZT and a polymer [41]. 0, 1, 2, and 3 are used to represent spherical particles, 1D wires, rods or fibers, 2D sheets and 3D matrix, respectively. To date, eight types of two-phase piezoelectric composites (piezocomposites) have been studied: 0–3, 1–3, 2–2, 2–3, 3–0, 3–1, 3–2, and 3–3. In these piezocomposites, the first number in the notation denotes the physical connectivity of the active phase (PZT) and the second number refers to the physical connectivity of the passive phase (polymer). There have been increasing reports on piezocomposites based on PZT and PVDF [42].

2.3 Principle of Piezoelectric Effect for Energy Harvesting

2.3.1 General Theory of Mechanical Energy Conversion

Harvesting of mechanical energy is to convert it into electrical energy, which requires a mechanical system that couples motion or vibration to a transduction mechanism. The mechanical system should be designed to be able to maximize the coupling between the mechanical energy sources and the transduction mechanism, depending on the characteristics of the environmental motions. For example, energy due to vibration can be converted by using inertial generators, with the mechanical component attached to an inertial frame that acts as a fixed reference.

The inertial frame transmits the vibrations to a suspended inertial mass to produce a relative displacement between them. System like this usually has a resonant frequency, which can be designed to match the characteristic frequency of the environmental motions. Detailed analysis on mechanisms of mechanical energy harvesting can be found in Refs. [43, 44]. A brief description is presented as follows.

These inertial-based generators can be well described as second-order spring-mass systems. For a system with a seismic mass of m on a spring with a stiffness of k, its energy loss, consisting of parasitic loss, c_p and electric energy generated by the transduction mechanism, c_e, can be represented by damping coefficient, c_T. The system is excited by an external sinusoidal vibration, $y(t) = Y\sin(\omega t)$. At resonant frequency, there is a net displacement, $z(t)$, between the mass and the frame. If the mass of the vibration source is greatly larger than that of the seismic mass, the latter can be ignored. If the external excitation is harmonic, the differential equation of the motion is given by:

$$m\ddot{z}(t) + c\ddot{z}(t) + kz(t) = -m\ddot{y}(t). \tag{2.1}$$

Standard solution for the mass displacement will be:

$$z(t) = \frac{\omega^2}{\sqrt{\left(\frac{k}{m} - \omega^2\right)^2 + \left(\frac{c_T\omega}{m}\right)^2}} y \sin(\omega t - \phi), \tag{2.2}$$

where ϕ is phase angle, given by:

$$\phi = \tan^{-1}\left(\frac{c_T\omega}{k - \omega^2 m}\right). \tag{2.3}$$

Energy conversion can be maximized when the excitation frequency matches the natural frequency of the system, ω_n, given by:

$$\omega_n = \sqrt{k/m}. \tag{2.4}$$

The powder dissipated with the system is:

$$P_d = \frac{m\zeta_T Y^2 \left(\frac{\omega}{\omega_n}\right)^3 \omega^3}{\left[1 - \left(\frac{\omega}{\omega_n}\right)^2\right]^2 + \left[2\zeta_T\left(\frac{\omega}{\omega_n}\right)\right]^2}, \tag{2.5}$$

where ς_T is the total damping ration, which is $\varsigma_T = c_T/2m\omega_n$. Maximum powder is achieved when the system is operated at ω_n, while P_d can be expressed as:

$$P_d = \frac{mY^2\omega_n^3}{4\zeta_T}, \tag{2.6}$$

$$P_d = \frac{mA^2}{4\omega_n\zeta_T}, \tag{2.7}$$

where A is the excitation acceleration level, with $A = \omega_n^2 Y$. Noting that these are steady-state solutions, powder will not tend to be infinity when the damping ratio approaches zero. The maximum powder generation can be evaluated by considering the parasitic and system damping ratio, which is given by:

$$P_e = \frac{m\zeta_e A^2}{4\omega_n(\zeta_p + \zeta_e)^2}, \tag{2.8}$$

P_e is maximized at $\varsigma_p = \varsigma_e$. When there is sufficient acceleration, increased damping effects will lead to a response with broadened bandwidth, so that the generator will be less sensitive to frequency. An excessive device amplitude can lead to nonlinear behavior of the generator, which will make it difficult in keeping the generator working at resonance frequency. For specific applications, both the frequency of the generator and the level of damping should be specifically designed to maximize the power output. The power generation can also be maximized by maximizing the mass of the mechanical structure.

The piezoelectric damping coefficient can be estimated by using the following equation:

$$c_e = \frac{2m\omega_n^2 k^2}{2\sqrt{\omega_n^2 + (1/(R_{\text{load}}C_{\text{load}})^2)}}. \tag{2.9}$$

where k is the piezoelectric coupling factor of the materials, while R_{load} and C_{load} are load resistance and capacitance. At maximum powder generation, there is optimal load, given by:

$$R_{\text{opt}} = \frac{1}{\omega_n C}\frac{2m\omega_n^2 k^2}{\sqrt{4\zeta_p^2 + k^4}}. \tag{2.10}$$

Specifically, for piezoelectric materials, the mechanical and electrical behaviors can be described by using the following linear constitutive equations:

$$S_{ij} = s_{ijkl}^E T_{kl} + d_{kij}E_k, \tag{2.11}$$

$$D_i = d_{ikl}^E T_{kl} + \varepsilon_{ik}^T E_k. \tag{2.12}$$

The subscripts i, j, k and l can be values of 1, 2, and 3. S and T are strain and stress tensors, respectively. The stresses, represented by T with a unit of N m^{-2}, are induced by the mechanical and electrical effects. D and E are the electric displacement and electric field vectors, with a unit of C m^{-2} and V m^{-1}, respectively. Also, s^E is the elastic compliance matrix evaluated at a constant electric field with a unit of m^2 N^{-1}, d is a matrix of piezoelectric strain coefficients with a unit of m V^{-1} and ε^T is a matrix of permittivity values that are evaluated at a constant stress with a unit of N V^{-2}. In Eqs. (2.11) and (2.12), d represents the charge created by an external force in the absence of an electric field (short circuit electrical condition) or the displacement caused by an applied voltage in the absence of an applied force (stress free mechanical condition).

For piezoelectric composite materials, effective electromechanical coupling factor is used, which is given by:

$$k_{\mathrm{eff}} = \sqrt{1 - \left(\frac{F_r}{F_a}\right)^2} \tag{2.13}$$

where F_r is the resonance frequency (Hz) and F_a is the antiresonace frequency (Hz) of a piezoelectric cantilever beam. The voltage coefficient, g (V m N^{-1}) is given by [45]:

$$g = {}^{d}/_{\varepsilon_T}. \tag{2.14}$$

Mechanical quality factor, Q_M, is defined as [45]:

$$Q_M = 2\pi \frac{\text{energy stored/cycle}}{\text{energy dissipated/cycle}}. \tag{2.15}$$

The amount of energy stored in a piezoelectric device, E_C is given by:

$$E_C = \frac{1}{2} CV^2, \tag{2.16}$$

where C is capacitance of the piezoelectric element and V is the voltage produced.

The maximum efficiency of the piezoelectric devices can also be estimated by the following equation:

$$\eta = \frac{\frac{1}{2}\left(\frac{k^2}{1-k^2}\right)}{\frac{1}{Q_M} + \frac{1}{2}\left(\frac{k^2}{1-k^2}\right)}. \tag{2.17}$$

This equation indicates that the efficiency can be increased by increasing k and Q_M, which are properties of the piezoelectric materials. Therefore, selection of materials is a very important step toward energy harvesters with high efficiencies. This is also the reason why PZT has been the most promising piezoelectric materials for mechanical energy harvesting applications.

2.3.2 Piezoelectric Energy Harvesting Devices

Piezoelectric materials can produce electrical charges when they are subject to external mechanical loads. Figure 2.17 shows working principle of a piece of piezoelectric material. The magnitude and direction of the electrical current are determined by the magnitude and direction of the external mechanical stress/strain applied to the materials. There have been various modes of vibration that can be used to construct piezoelectric harvesting devices. The common modes of vibration are summarized in Fig. 2.18 [43]. With given modes of vibration, there are

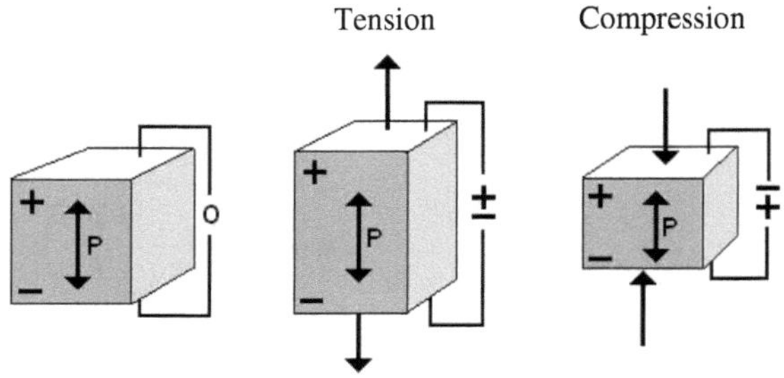

Fig. 2.17 Schematic showing the response of a piece of piezoelectric ceramics to external mechanical stimulation

different piezoelectric structures. Figure 2.19 shows typical piezoelectric structures that can be found in open literature [43].

Among the various piezoelectric structures for energy harvesters, the cantilevered beams with one or two piezoelectric ceramic thin sheets, which are named unimorph and bimorph (Fig. 2.19a), respectively, are the simplest ones. As discussed above, the harvester beam is positioned onto a vibrating host, where the dynamic strain induced in the piezoceramic layer(s) results in an alternating voltage output across their electrodes. Figure 2.20 shows a schematic of a cantilever tested under base excitation [46]. When a harmonic base motion is applied to the structure, an alternating voltage output is produced.

Cantilevered piezoelectric energy harvesters can work in two modes: d_{33} mode and d_{31} mode, as shown in Fig. 2.21 [47]. In d_{31} mode, a lateral force is applied in the direction perpendicular to the polarization direction. In this case, the bending beam has electrodes on its top and bottom surfaces, as in Fig. 2.21a. In d_{33} mode, forces are applied in the same direction as the polarization direction, where the bending beam has all electrodes on its top surface, as in Fig. 2.21b. Although piezoelectric materials working in d_{31} mode normally have lower coupling coefficients than in d_{33} mode, d_{31} mode is more commonly used. This is because when a single-layer cantilever or a double-clamped beam bends, more lateral stress is produced than vertical stress, which makes it easier to couple in d_{31} mode [47]. Similar principle can be applied to the harvesters with other structures.

Beam structures are usually used for low stress levels, whereas at high stress levels, another type of device, ceramic–metal composites, is preferred. Ceramic–metal composites generally have a simple design with a metal faceplate, called shell or cap, which couples to both the ceramic and the surrounding medium. The metal component transfers the incident stress to the ceramic or the displacement to the medium. Flextensional transducers are good examples of ceramic–metal composites. In flextensional transducers, the flexural vibration of the metal shell causes an extensional (or contractional) vibration of the piezoelectric element. The miniaturized versions of flextensionals, the moonie (Fig. 2.19c), and cymbal (Fig. 2.19e) transducers possess 2-(0)-2 connectivity. These transducers, shown in Fig. 2.20, consist of a poled piezoelectric disk (fully electroded on both faces) which is sandwiched between two metal endcaps, each containing a shallow

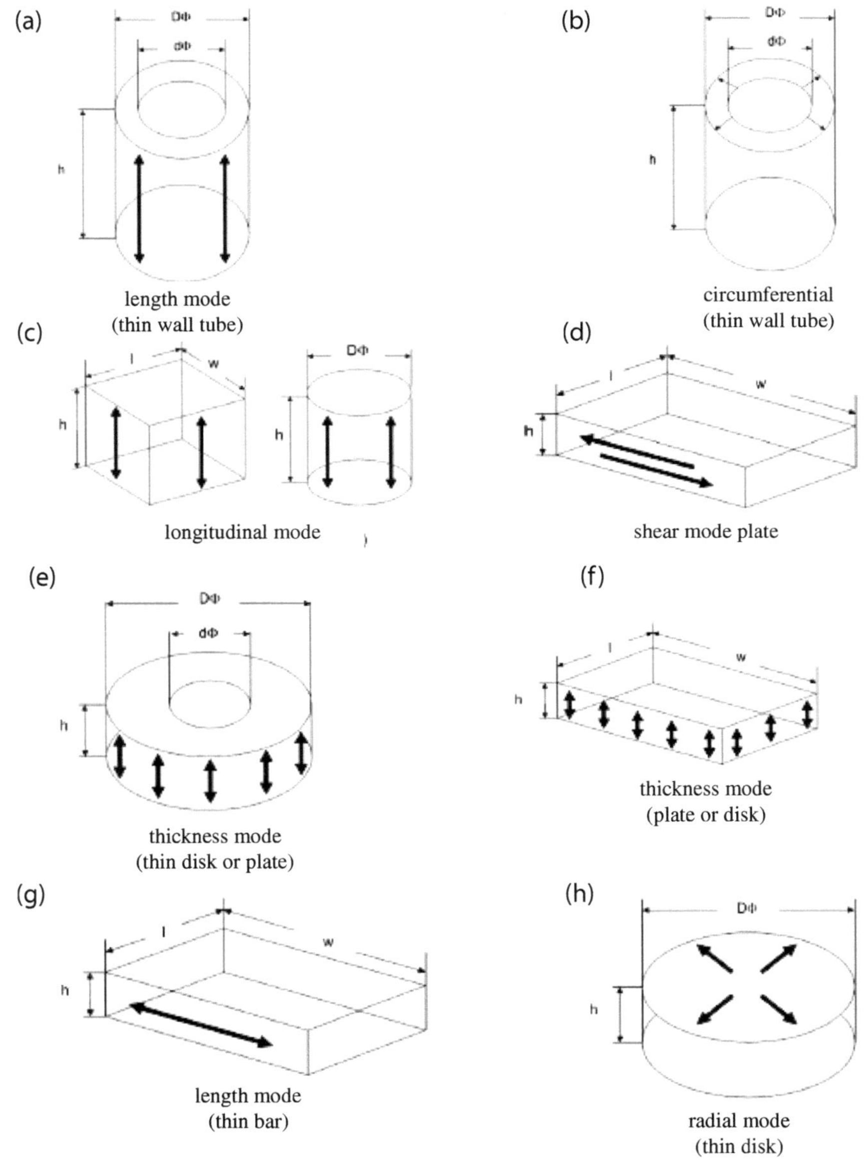

Fig. 2.18 Common modes of vibration. Reproduced with permission from [43]. Copyright @ 2008, IoP Publishing

air-filled cavity on their inner surface [41, 43]. In the case of the moonie, the cavities are in the shape of a half moon, whereas the cymbal has a truncated cone-shaped cavity. The presence of these cavities allows the metal caps to serve as mechanical transformers for transforming and amplifying a portion of the incident

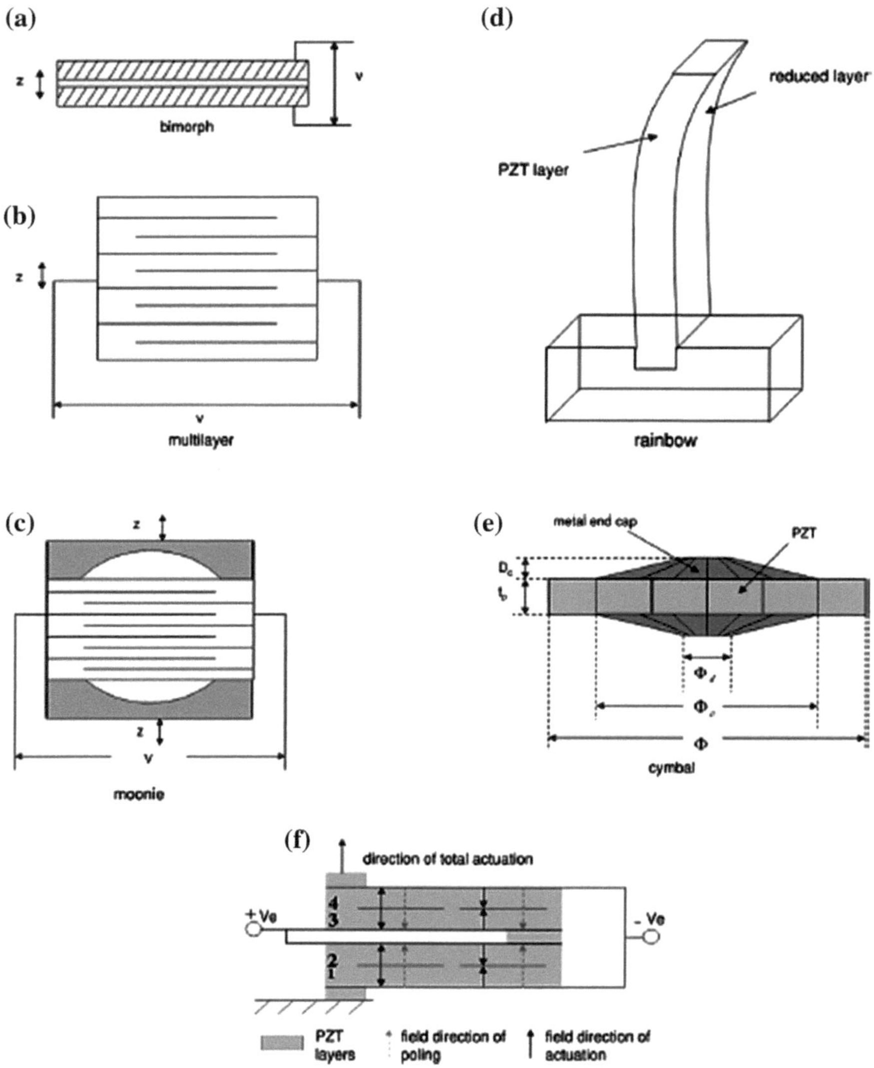

Fig. 2.19 Piezoelectric structures: **a** bimorph, **b** multilayer, **c** Moonie, **d** RAINBOW, **e** cymbal and **f** S-morph. Reproduced with permission from [43]. Copyright @ 2008, IoP Publishing

axial-direction stress into tangential and radial stresses of opposite sign. Thus, the d_{33} and d_{31} contributions of the PZT now add together (rather than subtracting) in the effective d_h of the device.

To theoretically study the mechanics of piezoelectric energy harvesting and experimentally evaluate the performances of alternative current (AC) power generation, the devices are usually considered to be subject to a resistive load in the electrical domain [48–50]. To use the electricity produced by a piezoelectric

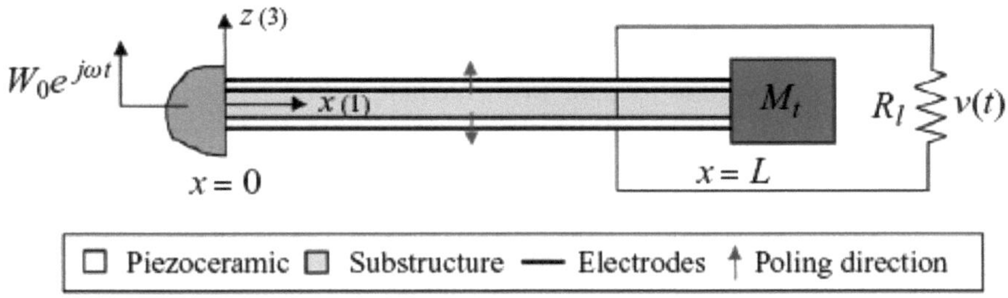

Fig. 2.20 Schematic cantilevered piezoelectric energy harvester tested under base excitation. Reproduced with permission from [46]. Copyright @ 2011, John Wiley & Sons

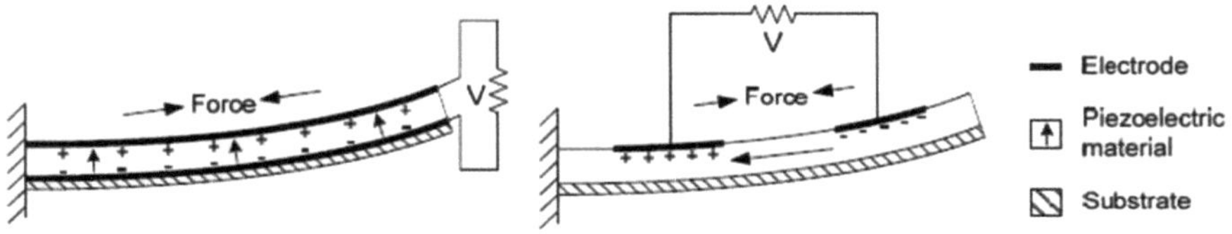

Fig. 2.21 Two types of piezoelectric energy harvesters: **a** d_{31} mode and **b** d_{33} mode. Reproduced with permission from [47], Courtesy of Dr. D. Zhu, University of Southampton

energy harvester, the alternating voltage output should be converted to a stable rectified voltage. This can be realized by using a rectifier bridge and a smoothing capacitor to form an AC–DC converter. The energy can be used to charge small batteries or stored in capacitors. To maximize the power transfer to the energy storage devices, it is also necessary to use a DC–DC converter to regulate the voltage outputs of the rectifier. These electrical circuit and power electronics are very important for practical applications of the energies harvested [51–53].

2.3.3 Modeling of Piezoelectric Energy Harvesting Devices

Modeling is an important approach to predict the performances of a piezoelectric energy harvesting device, which in turn can be used to optimize the design parameters to achieve further improvement. Therefore, tremendous efforts have been made to develop models of piezoelectric devices. In the early stage of mathematical modeling, [54, 55] lumped parameter solutions with a single mechanical degree of freedom were used to predict the coupled system dynamics of piezoelectric energy harvesters. Lumped parameter modeling is a convenient modeling approach, because the electrical domain already comprised of lumped parameters: a capacitor due to the internal (or inherent) capacitance of the piezoelectric materials and a resistor due to an external load resistance. Hence, the only prerequisite is to obtain the lumped parameters representing the mechanical domain so that the mechanical equilibrium and the electrical loop equations can be coupled through the piezoelectric constitutive equations [45] and thus a transformer relation can be built up. One of the advantages of the lumped-parameter modeling is its use of simple expressions. However, it is limited only to a single vibration mode due to its approximation characteristics. It also misses several important aspects of a coupled physical system, such as the dynamic mode shapes and accurate strain distributions, as well as their effects on the electrical response.

Cantilevered energy harvesters are basically excited under base motion. The lumped-parameter harmonic base excitation relation for elementary vibration can be used directly [56, 57]. It has been used widely for both theoretical modeling [55] and experimental studying to maximize power generation and optimize parameter [58]. In lumped-parameter models, the contribution of the distributed mass to the forcing amplitude in the base excitation is usually neglected [54, 55]. The inertial contribution of the distributed mass to the excitation amplitude should be considered, especially when a harvester has not a large proof mass [59].

As an improved modeling approach was later developed to model cantilevered piezoelectric energy harvesters [60, 61]. Also, the vibration mode shapes obtained from the Euler–Bernoulli beam theory, along with the piezoelectric constitutive equation that gives the electric displacement to relate the electrical outputs to the vibration mode shape, have been used to obtain analytical expressions [62–64]. However, these approaches have several shortcomings, including the missing of the resonance phenomenon, misrepresentation of the forcing due to the base excitation, oversimplified modeling of the piezoelectric coupling in the beam equation by taking it as viscous damping and use of static sensing/actuation equations to describe the fundamentally dynamic problems [65].

Further developments in modeling of piezoelectric energy harvesters include analytical solutions based on distributed-parameter electromechanical modeling, which have been experimentally validated [66, 67], and others [68–71]. The linear electromechanical models have later been modified to cater for the mechanical nonlinearities in vibration-based energy harvesting devices [72–74]. A more recent modeling problem of interest is the stochastic excitation of vibration-based energy harvesters because some types of waste energy due to vibration often appear in nondeterministic forms [75].

2.4 Energy from Human Activity

2.4.1 Shoe-Mounted Harvesters

2.4.1.1 Examples of Shoe-Mounted Piezoelectric Harvesters

Several examples have been reported that use piezoelectric harvesters mounted in shoe to harvest mechanical energy due to human walk or running [76–82]. A representative photograph of shoe-mounted piezoelectric energy harvesters is

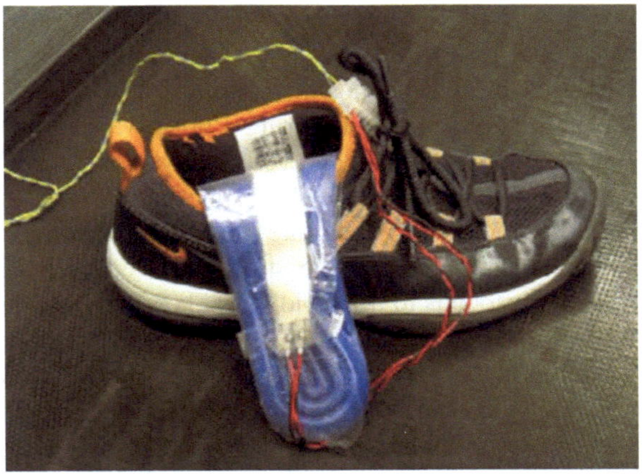

Fig. 2.22 Prototype of the piezoelectric film energy harvesting device inserted in a shoe. There are two DT4-028 K/L piezoelectric films from MSIUSA in parallel inside the shoe. Reproduced with permission from [82]. Copyright @ 2006, Elsevier

shown in Fig. 2.22 [82]. One of the earlier examples is described in. In Ref. [76], where two main methods are explored, with 31-mode bending operation. One method is to harness the energy dissipated in bending the ball of the foot, using a flexible, multilaminar polyvinylidene fluoride (PVDF) bimorph stave mounted under the insole. The second method is to harness foot strike energy by flattening curved, pre-stressed spring metal strips laminated with a semiflexible form of piezoelectric lead zirconate titanate (PZT) under the heel. This device is called dimorph, consisting of two back-to-back single-sided unimorphs.

The multilaminate piezoelectric foil stave has an elongated hexagonal shape. It consists of two eight-layer stacks of 28 μm PVDF sandwiching a 2 mm flexible plastic substrate bounded with epoxy. The hexagonal design is to maximize the integrated stress across the bending distribution of a typical insole with a shape of shoe. Bending the stave elongates its outside surface and compresses its inside surface with respect to its plastic core due to the difference in curvature radii. The PVDF sheets on both sides of the core are connected with silver electrodes in parallel. The charges in the 16 layers add up and a voltage drop builds up across the two electrodes. Bending and restoring the stave create energy due to the deformation of the piezoelectric PVDF layers.

The PZT device is to harvest the energy due to heel strikes. A simple unimorph is used first and nonbending compressive dimorph is developed. Two commercially available PZT transducers, a heel-shaped 0.025 in. beryllium-copper midplate and two aluminum rivets are used to assemble the device. The transducers consists of a 5 cm × 5 cm, 0.015 in. PZT strip bonded to a prestressed neutrally curved sheet of spring steel with a size of 5 cm × 8.5 cm. The whole structure is trimmed to fit the beryllium-copper midplate and then mounted under the heel-strike force center with two rivets. It is thus ready to be connected with electrodes with electric parallel. When the dimorph is compressed first and then released, charges are accumulated across the surfaces of the PZT strips.

The PVDF stave has been tested under the insole of a standard athletic sneaker and the PZT dimorph is in an orthopedic insole in work boot. The athletic shoe's insole is flexible to accommodate the flexible stave, while the rigid heel cup of work boot is suitable to the rigid dimorph. Energy harvesting efficiency can be measured by terminating the transducers with matched resistive load and measuring the voltage output during a brisk walk of a wearer.

The PVDF stave has an average power of 1.3 mW in a 250 kΩ load at 0.9 Hz walking pace, whereas the PZT dimorph offers an average power of 8.4 mW in a 500 kΩ load under the same mechanical excitation. The footfall has evident characteristics, with a large spike followed by a smaller one. The larger power spike corresponds to the rapid initial bending or compression of the devices, while the following smaller spike is caused by the restoration of the transducer when the wearer shifts weight from one foot to the other. Power efficiencies of the PVDF and PZT harvesters are 0.5 and 20 %, respectively.

The energy harvested by using the PVDF and PZT shoe-mounted harvesters has been used to power a radio frequency (RF) tag system to demonstrate the feasibility for potential self-powering applications. An active RF tag, which can transmit a short range 12-bit wireless identification (ID) code while the bearer walks, is used in the demonstration. Such a system can be used in a smart environment, in which multiple users transmit their identities to the local surrounding. Because the RF-based design does not require line of sight to the reader, it can be mounted in a shoe. The piezoelectric harvesters provide energy to power the system, without the use of batteries.

Low frequency mechanical energy sources, like walking of human, are purely capacitive, which usually produce high voltage, low energy density and low level of current at about one cycle per second. The excitation corresponds to an extremely high source impedance, with voltage outputs of hundreds of volts and currents on the order of 10^{-7} A. Therefore, linear regulation is not sufficient to make the energy for real applications during a brisk walk. To further make use of the energy harvested by using the shoe-mounted piezoelectric harvesters, various switching conversion strategies have been developed, including switched-capacitor converters, direct DC–DC down converters and forward-switching converters. The forward-switching converter is comprised of a small number of inexpensive widely available components and materials.

Another example of shoe-mounted piezoelectric energy harvester is made with PVDF unimorph [77]. The unimorph strips are constructed with one 0.5 in. tall, 52 μm thick silver laminated PVDF films, bonded with cyanoacrylate to the side of a slightly wider and longer 4 mil thick PET plastic film substrate. The strips vary in length with the changing space available in the cutout from 1 to 2.25 in. The substrate has suitable stiffness and spring-like qualities for the harvester applications. During a heel strike, the polycarbonate plates are compressed together, in which all the PET strips aligned between them are bent. A strain is thus applied to the bonded PVDF film, due to the bending of the plastic strips. After optimization, the shoe-mounted harvester can have a power of 4 mW and energy efficiency of 8 %.

The last example of shoe-mounted piezoelectric harvester is also made of PVDF [78]. PVDF with β phase is dissolved in N,N-dimethyl formamide or dimethyl acetamide at temperature at temperatures of <70 °C, to make thin films of 20–60 μm. PVDF thin films with oriented crystal structure can be obtained by stretching. The films are then poled to have improved piezoelectric performance. Metallic films (such as Au) are deposited on both sides of the PVDF thin films as electrodes by magnetron sputtering or thermal evaporation. Piezoelectric harvester made with the PVDF thin film is then mounted at bottom of a shoe. A shaker is used to apply mechanical excitation similar to that of human walking. A maximum output voltage of about 6 V is produced by a harvester made with 28 μm thick PVDF films. The signal can be rectified so that the harvested energy can be stored for real applications, which has been demonstrated by using a rechargeable lithium battery. The lithium battery is a thin film all-solid-state storage device, having a capacity of 35 μAh cm^{-2}, with current densities of above 200 μA cm^{-2} and less than 10 μm in thickness. The batter consists of $LiCoO_2$ as cathode, Li_3PO_4 as electrolyte and Li as anode. Although the energy generated by the shoe-mounted harvester is still not sufficient for practical applications, it has potential to be further improved.

2.4.1.2 Gait Analysis for Shoe-Mounted Harvester

To effectively make use the kinetic energy from human working by using shoe-mounted piezoelectric harvester, it is very useful to have a deeper understanding on the gait pattern. Previous studies have indicated that foot movement of human is characterized by large displacements and accelerations, which is thus very promising sources of vibrating mechanical energy that can be harvested by using shoe-mounted devices. The high velocities and accelerations are observed in foot movement, especially during the foot/ground contact phase [81]. As an example, Fig. 2.23 plots a typical acceleration time history measured in the calcaneous of one limb, in one gait during normal walking (the gravitation acceleration has been subtracted already). Several significant features are mentioned as follows. First, the acceleration peak at heel strike (A, G) is followed by certain oscillation after heel strike (B). Second, there area baseline during stance phase (C) and an acceleration during foot movement (D–E). Finally, there is a downward acceleration at heel strike (F) [81].

To effectively harvest the energy with shoe-mounted piezoelectric harvesters, it is import to understand the characteristics of gait cycles of human movement. Human foot movements are characterized by gait cycles as a well-defined sequence. Within a sequence of two consecutive heel strikes, i.e., "contact phase," of the same limb, the gait actually consists of a "stance phase" when the foot is at rest and a subsequent "swing phase" when the foot starts to move [83]. The contact phase is usually instantaneous, while the stance and swing phases take approximately 60 and 40 % of a single gait cycle [84]. The heel starts moving at approximately 40 % of the stride while the foot is still in contact with the ground and reaches its maximum displacement just after toe-off, when the velocity of the heel is approximately zero. During the upward and downward heel movements in

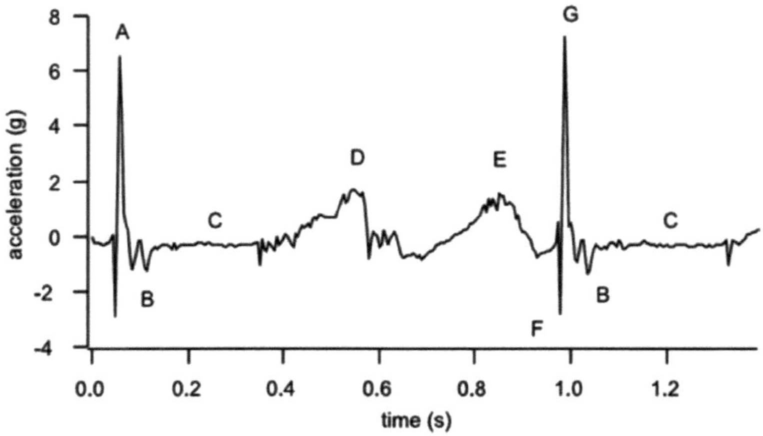

Fig. 2.23 Acceleration signal of the calcaneus for one complete gait (heel strike to heel strike of the same limb). The accelerations are normalized to that of gravity $g = 9.81\ \mathrm{ms}^{-2}$. Reproduced with permission from [81]. Copyright @ 2001, Elsevier

the swing phase, the velocity has a sinusoidal trend, while the acceleration shows some oscillation consequent to velocity change [85].

The vertical and horizontal accelerations in the foot are potential vibration sources of energy to be harvested. The acceleration signal shown in Fig. 2.23 that every heel strike has a large acceleration, followed by certain oscillation during the swing phase. The observed acceleration spikes have been confirmed by the measurement of ground reaction forces under the foot [86]. The heel acceleration has been shown to vary due to several factors, such as characteristics of the ground surface, conditions of gait, types of shoes, status of the walkers. Therefore, different walking conditions lead to acceleration signals with different characteristics. A common sense is that a variation in these walking conditions has only influences on the measured values, without significant change in the overall trend [81]. In this respect, an experimental measurement of gait acceleration time histories is essential to identify all the relevant features, so as to have a standardized acceleration signal that can be used as a reference excitation in numerical evaluation of the electromechanical response of a shoe-mounted vibrating device.

Such experiments have been conducted by using a custom-made clamp system to fix on the heel pad, while an accelerometer is aligned along the tibial axis. Gait accelerations, of a male subject (1.75 m in height and 70 kg in weight) walking on a level walkway at a natural pace (about 0.8 steps s^{-1} for the same limb), have been measured with five repeated tests [79, 80]. The signal is virtually periodic and contains all the relevant features observed in the acceleration time history of Fig. 2.23. At contact phase, a large acceleration spike is followed by a negative valley, which is most likely caused by the mid-foot and toe impact to the ground. During the stance phase, the acceleration is almost zero, while it exhibits a sinusoidal-shaped oscillation during the swing phase.

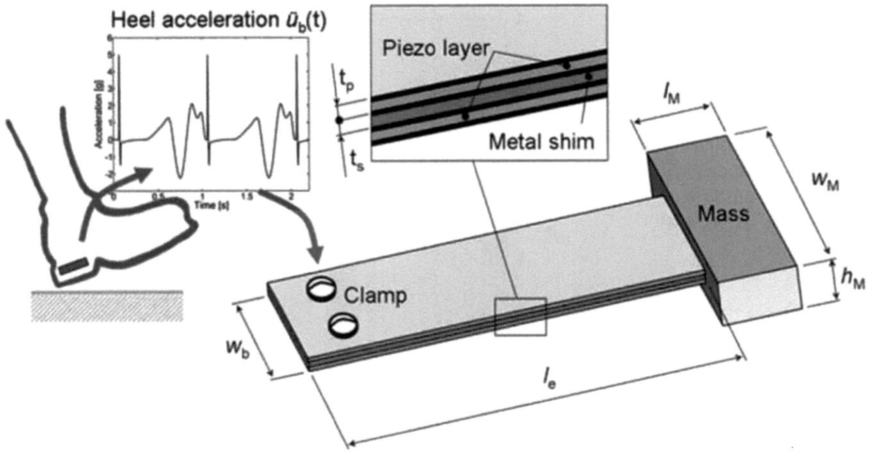

Fig. 2.24 Rectangular bimorph excited by heel acceleration at its base. Reproduced with permission from [79]. Copyright @ 2013, Springer Science+Business Media

The information has been used to synthesize all the relevant features of gait acceleration into a periodic acceleration signal, as a reference input in numerical simulations of an electromechanical bimorph. The standard acceleration signal can be established by combining the values of vertical and horizontal displacements, velocity and acceleration, as well as the typical time variation of the foot–ground angle during walking. For example, the average step distance is 65 cm [87] and the maximum vertical heel elevation is about 25 cm [85], where the foot–ground angle reaches its maximum of 46° [86], with other aspects being also taken into account [81]. All these have been validated by experimental measurements, for the definition of acceleration during the contact phase. Finally, the periodic standard acceleration signal can be obtained. There are certain typical features, including two sharp positive/negative peaks in the contact phase, an oscillation in the swing phase, followed by a small double peak, due to the vertical and horizontal accelerations [79, 80].

Having these data, the electromechanical response of a shoe-mounted vibrating scavenger excited by heel acceleration during human gait can be simulated, with bimorph as an example [79, 80]. As shown in Fig. 2.24, the bimorph has a rectangular shape, made of two piezoelectric layers bounded to a metallic shim. One end is clamped, a tip mass is mounted on the other end, which reduces the natural frequency of the bender. Material properties are typical PZT-5A piezoceramic and stainless steel.

Figure 2.25 shows the results of theoretical simulation and experimental measurement for the shoe-mounted bimorph excited by the nonharmonic heel acceleration [79]. The simulation and the experiment are in a very good agreement, which means the standardized acceleration signal, even if somewhat simplified, represents all the relevant features of gait acceleration to reproduce the real measured signal. Therefore, it can be conveniently used as a reference of the

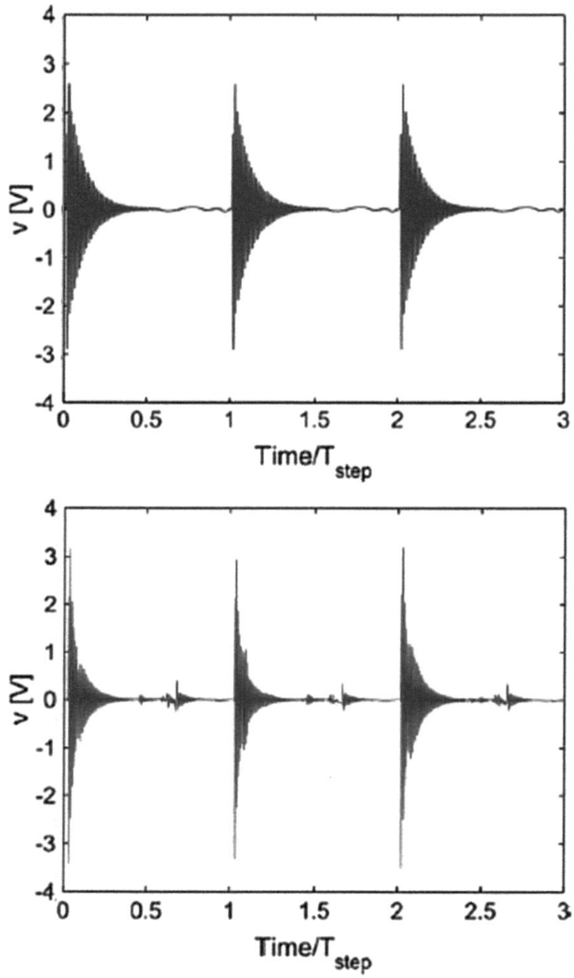

Fig. 2.25 Transient output voltage at heel acceleration for three heel strikes of the same foot: **a** numerical simulation and **b** experimental measurement results. Reproduced with permission from [79]. Copyright @ 2010, IoP Publishing

measured signals to simulate and predict the dynamic responses of shoe-mounted harvesters.

The performance of a shoe-mounted piezoelectric harvester can be represented by the average power per footstep dissipated across the resistance R_L, given by:

$$P_{\text{ave.}} = \frac{1}{n_s} \int_0^{T_{\text{step}}} \frac{v^2(t)}{R_L} \mathrm{d}t, \tag{2.18}$$

where n_s is the number of footsteps counted in the time period T_{step}. This definition of power is independent of pace frequency or period. It can be used to compare the performance of harvesters with different materials and configurations.

Experimental data indicate that the power harvested per step is about 13 μW, just about 6 % higher than the power calculated numerically with the standard acceleration input. Such a small difference can be attributed to the unpredictable low-level oscillations observed in the measured signal. Also, the consecutive steps may provide slightly different acceleration signals, even under same stable walking conditions. Although for practical applications the obtained power levels are still not sufficiently high, there are rooms for further improvement. For example, the performance of the harvester can be optimized through materials selection and structure design. Nevertheless, the simulation method can be used to estimate energy harvesting performance of a given device.

2.4.2 Energy from Human Body Motion

Human activities, with the exception of vocalization, occur at very low frequencies, only up to a few hertz in most cases. It has been well known that piezoelectric bimorphs operating for sensing or energy generation are most effective at quite high frequencies. Devices of acceptable dimensions and mass for human integration and offering power outputs in the order of milliwatts have resonance frequencies of hundreds of hertz. This frequency mismatch between excitation and harvester makes it challenging to efficiently harvest energy from human motion with piezoelectric devices. As a result, various strategies have been proposed to address this problem. One of them is called frequency up-conversion method, where the low frequency motions are used as an excitation to trigger the piezoelectric components to vibrate at their resonant frequencies, which is the focus of this subsection. The harvesters in the following discussion are grouped according to the types of target energy sources instead of the harvesting principles.

2.4.2.1 Impulse-Excited Harvesters

In this section, three examples will be discussed on energy harvested from human body motion by using piezoelectric harvesters [88–90]. The first example is a piezoelectric impulse-excited approach, in which a cylindrical proof mass actuates an array of piezoelectric bi-morph beams through magnetic attraction [88]. With an initial excitation, the transducers start to vibrate at natural frequency. By using this approach, the operational frequency range can be widened while the electromechanical coupling is significantly increased. The principle of the impulse excitation is demonstrated by using a commercially available integrated circuit for voltage regulation. At a frequency of 2 Hz and an acceleration of 2.7 m s^{-2}, the harvester has a maximal power output of 2.1 mW. This type of harvesters can be used for promising medical applications, for example, as power suppliers for wearable and implantable sensors for heart rate, blood glucose level, blood pressure and oxygen saturation.

There is a limit in the achievable power output for inertial devices, which is given by:

$$P_{\max} = \pi f a_0 Z_L m, \quad (2.19)$$

where $f = \omega/2\pi$ is the excitation frequency in Hz, a_0 is the external acceleration, Z_L is the internal displacement limit and m is mass of the seismic element. Here, a sinusoidal excitation, with acceleration $a(t) = a_0 \sin(\omega t)$, is assumed. More commonly, a_0 is expressed in terms of external motion amplitude Y_0 and angular excitation frequency ω as $a_0 = Y_0\omega^2$. This equation gives an upper limit regardless of actual transduction mechanism, because only the mechanical kinetic energy is calculated. Moreover, for resonant designs, Y_0 is smaller than Z_L, whereas for nonresonant devices, Y_0 can be much larger than Z_L.

There are several characteristics of human motion that must be taken into account. First, the excitation frequencies that determine the power output are usually very low. Second, different points of human body have different frequencies and accelerations. For instance, in normal gait, the highest excitations are about 2.5 g and 1.7 Hz at the ankle, while for chest, shoulder, and wrist, it is about 0.3–0.5 g and between 1 and 2 Hz [88]. At the same time, the displacement limit and mass are should be also considered in practical designs.

The theory only predicts the amount of the kinetic energy stored in the proof mass, while the performance of actual devices is related to the conversion mechanism. Therefore, to maximize the efficiency of a harvester, it is necessary to the effectiveness of the transducer used, i.e., the percentage of the available inertial energy that can be actually converted into electrical energy. In this respect, impulse excitation of the transducer has such characteristics. The approach involves two steps. A large proof mass moves under external acceleration at low frequency. This mass periodically transmits an impulse to the actual transducer, which then vibrates at a higher frequency [88]. However, such a device is still limited by the kinetic energy of the large proof mass. To achieve the maximum performance, the transducer needs to work under its ideal conditions.

The impulse-excited harvesters have another advantage, broad range of working frequency. Every time the transducer is actuated, a fixed amount of energy is converted. If the excitation frequency is increased, the number of actuations of transducer is increased, so that the power output is increased. At resonance frequency, the power output of a device with continuously optimized damping and spring constant increases as cube of frequency. However, the power output of the impulse-excited devices is linearly proportional to frequency. Theoretically, if the excitation is too slow, the proof mass fails to actuate the transducer at all, corresponding to a drop power output at a specific frequency, which is similar to a critical frequency.

Figure 2.26 shows a prototype of the piezoelectric impulse excited harvester [88]. This design has taken into account the challenges for human body energy harvesting. An external steel cylinder is used as a proof mass due to its high density. An electrostatic device with a rolling rod is specifically developed, with a

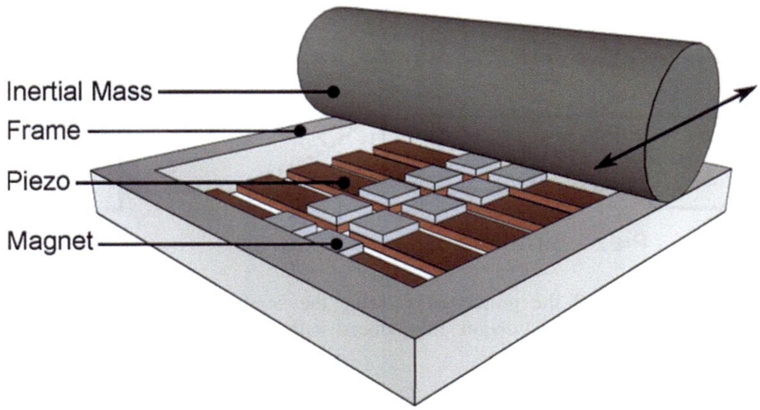

Fig. 2.26 Impulse-excited piezoelectric harvester prototype. Courtesy of P. Pillatsch, Imperical College London. Reproduced with permission from [88]. Copyright @ 2012, IoP Publishing

high coupling over the given travel range. In this piezoelectric approach for energy conversion, a series of bi-morph piezoelectric beams are used transducers. The tip of each beam is attached with a permanent magnet. When the proof mass rolls over one element, the magnet snaps onto the steel cylinder and the beam bends, as shown in Fig. 2.27. As the proof mass continues its travel, the beam is released and oscillates freely at its damped natural frequency.

In this design, the device is in perpendicular direction of proof mass motion and transducer actuation. As a result, the travel range of the proof mass is not limited by the actuation range of the transducer. Also, the actuation forces can be spread over a larger number of individual transducers and adjusted by changing their shapes. According to above theory, with an initial gap, for example, h = 1 mm, the required force on each single beam for optimal operation as a function of excitation frequency and total number of beams can be obtained. This segmented transduction is the reason that the bandwidth can be effectively widened.

Figure 2.28 shows a macroscale functional model as a proof of concept [88]. Two cylinders machined from mild steel have masses of $m_1 = 285$ g and $m_2 = 143$ g (half mass of m_1), which are used to check the influence of mass on power output of the device. Two rails, with adjustable heights, constrain the travel to linear motion with two blocks acting as variable end stops. At each side, a clamping mechanism holds eight piezoelectric beams. The beams having a dimension of $72 \times 5 \times 0.5$ mm^3 are cut out of a series of connected bi-morph plate. The top and bottom layers are made of PZT 507 material (Morgan electroceramics) with a thickness of 0.2 mm. The center shim consists of a thin layer FeNi alloy with thickness of 0.1 mm. These beams are clamped to a free beam with a length of 60 mm. The tip mass is N52 type neodymium magnets with a dimension of $5 \times 5 \times 1$ mm^3. Consequently, the device has a damped natural frequency of 46.3 Hz. Each single beam has a capacitance $C = 30$ nF. Therefore,

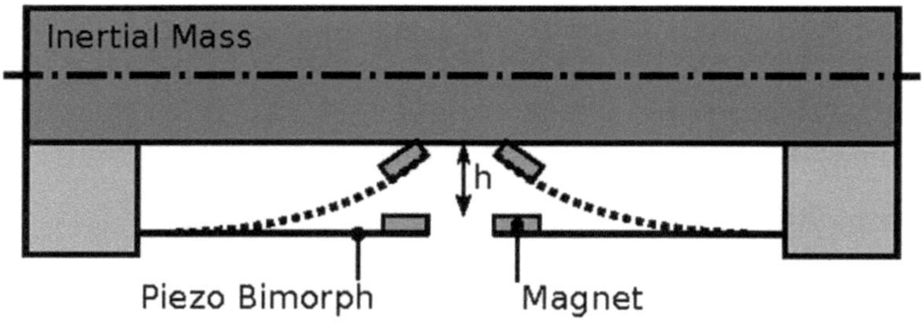

Fig. 2.27 Section view of the prototype, showing the bending of the beams. Courtesy of P. Pillatsch, Imperical College London. Reproduced with permission from [88]. Copyright @ 2012, IoP Publishing

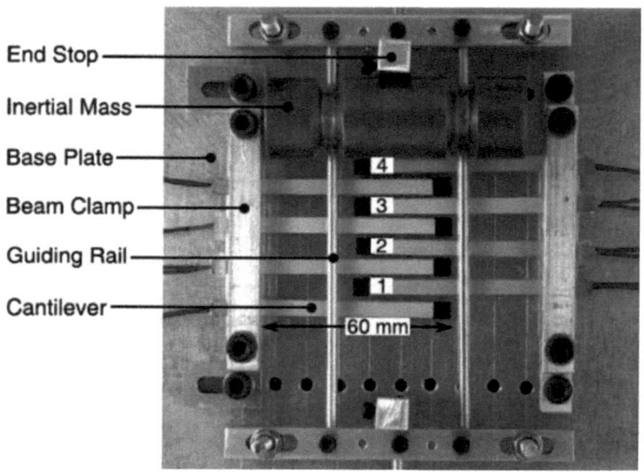

Fig. 2.28 Functional model of the impulse-excited harvester. Courtesy of P. Pillatsch, Imperical College London. Reproduced with permission from [88]. Copyright @ 2012, IoP Publishing

according to the output impedance of the beam, i.e., $Z = 1/\omega C$, the matched resistive load for maximal power output is $R = 120\ \mathrm{k}\Omega$. The corresponding values of the results are calculated as the measured root-mean-squared (rms) voltage squared divided by this load resistance.

A rocking table has been used to describe the functional model, which is used for reproducible excitations [88]. The angular velocity ω of the driving wheel is adjustable between 20 and 120 rpm, and 0.33 and 2 Hz, respectively. This motion is transferred to the platform through a connection rod giving rise to a sinusoidal excitation:

$$a(t) = g\frac{r}{L}\sin(\omega t) = a_o \sin(\omega t), \tag{2.20}$$

where L is the distance between the rotational axis of the platform and the mounting point of the connection rod and g is the gravitational acceleration. The radius r stands for the eccentric position of the connection rod on the driving wheel. By adjusting r, the tilt angle of the table can be changed. At angles of 16.1°

and 5.1°, a high acceleration $a_1 = 2.7\ \mathrm{m\ s^{-2}} \approx 0.28$ g and a low acceleration $a_2 = 0.87\ \mathrm{m\ s^{-2}} \approx 0.089$ g are obtained. This allows for four possible configurations between masses, m_1, m_2, and accelerations, a_1, a_2.

The measurement setup also has a piezoelectric energy harvesting power supply (Linear Technology LTC3588-1). It contains a full wave bridge rectifier and a DC–DC converter with selectable regulated output and an input voltage range between 2.7 and 20 V, with a dimension of $3 \times 3\ \mathrm{mm}^2$. The values are: $C1 = 1\ \mu\mathrm{F}$, $C2 = 4.7\ \mu\mathrm{F}$, $C3 = 47\ \mu\mathrm{F}$, $C4 = 10\ \mu\mathrm{F}$ and $L1 = 10\ \mu\mathrm{H}$.

The open circuit voltage on a single beam for one actuation has been measured and explained [88]. As the proof mass approaches the beam, the magnet latches onto it, pulling the tip of the beam up. Initially, the beam stays in this deflected position as the cylinder continues its traveling. In second phase, the corner of the magnet holds onto the proof mass and the beam is pulled up further due to the curvature of the cylinder, before is finally released and vibrate freely. This is an ideal scenario because the release occurs just after the equilibrium at which point the beam force matches to the magnetic attraction.

When the device is continuously operated at 0.33 Hz, beam 4 is actuated first, followed by beam 3, and then beam 2 and finally beam 1. The oscillation of the individual beams has completely stopped before the next actuation is started. If the continuous operation is at a higher frequency of 1.66 Hz, the above-mentioned advantage of the segmented transduction is demonstrated. Although beams 1 and 2 are not actuated at all, beams 3 and 4 are wording very well. It means that the device can also be used to harvest energy at this higher frequency, indicating its wide range of operating frequency.

The measured power output of the whole device and the corresponding rms voltage of a single beam have been characterized for four described measurement configurations with different proof masses and external accelerations [88]. It has been found that the power output curves, especially in the cases of large accelerations, the expected linear behavior is observed, i.e., the power output doubles at the frequency is doubled. The rms voltages on each beam are within a usable range of 2–6 V.

Ideally, the device should be operated at low acceleration with light proof mass, because this is the case in which the amount of the available kinetic energy matches that of the energy that can be extracted by the transducers in each cycle. This has been confirmed by experimental results. Up to about 0.5 Hz all four curves are very close to each other. At high accelerations, although the amount of the energy stored in the heavier proof mass m_1 is high, the transducers cannot extract it. Therefore, no energy is harvested, in contrast to the case of the lighter proof mass m_2 at frequencies above 1.7 Hz. At this frequency, m_2 does not travel the full range and thus the level of the output power drops. Similar behavior is observed for lower acceleration a_2, but with the breakdown point occurs at a lower frequency of about 0.5 Hz for m_2 and about 0.7 Hz for m_1. This can be readily understood that the kinetic energy of the proof masses is smaller at lower accelerations. The harvester has a maximum output power of 2.1 mW at 2 Hz.

The device has a maximum power efficiency of about 8 %, when the proof mass just about touches the end stops and energy lost in collisions is at the minimized level. This implies that a proper match between the transducers with optimized excitations and the intended proof mass is important for high power efficiencies. This can be taken into account during device design, for example, higher proof masses can be realized by simply using thicker beams. At certain combinations, for instance, a_1 plus m_2, pretty high level of power efficiency ($\sim$3 %) can be maintained over the whole frequency range, confirming again the wide bandwidth capability of the impulse excitation approach. The device exhibits a power density of 3.8–13 μW cm^{-3}, with proof mass m_2 at high acceleration $a_1 = 2.7$ m s^{-2} $\approx$ 0.28 g and excitation frequencies of 0.33–2 Hz. The low operation frequency, wide frequency range, and relatively high power density, make the harvester very favorable for human body applications.

The LTC3588-1 voltage regulator has been tested on a single beam at high acceleration a_1 and with the lighter proof mass m_1 between 0.33 and 2 Hz. The voltage from the piezoelectric actuator is first rectified to charge the input capacitor connected to V_{in}. Once a threshold of 5 V is reached, a certain amount of charge is transferred to the output capacitor. As a result, the output voltage rises in steps until it reaches the predefined level of about 3.6 V.

The power output after voltage regulation is determined by using a variable resistor with initial value of 1,050 kΩ. After the target level is reached, the resistance is decreased, so as to increase the load current, until the voltage at the output terminals starts to drop. The power can then be calculated as the chosen output voltage squared divided by the resistance ($P = V^2/R$). An efficiency of close to 40 % is observed. Furthermore, at 0.33 Hz, because the consumption of the load is too high, the level of output voltage is always below 3.6 V and an equilibrium instead of a drop at a lower level is attained after every transfer of charge coming from the input capacitor.

2.4.2.2 Impact-Driven Harvesters

The second type of piezoelectric devices is called impact-driven harvesters, specifically harvesting mechanical energy due to human body motion [89, 90]. A design based on the impact of a moving mass on piezoelectric bending structures will be discussed first [89]. With the aid of theoretical analysis, the parameters influencing the device performances in terms of energy harvesting have been optimized. Experimental results on a prototype of the impact harvester have shown that a generator with a volume of 25 cm^3 and weight of 60 g has an output power of 47 μW, which is measured across a resistive load when the device is rotated by 180°/s. At 10 Hz, 10-cm amplitude linear motion results in a power of 600 μW.

The piezoelectric element in this impact-driven harvester is a bimorph, consisting of two piezoelectric capacitors supported by an elastic cantilever. An additional mass can be added at the tip of the bimorph. The piezoelectric capacitors can be arranged in series or parallel configuration, which are connected to the powered device through a signal conditioner. The principal deformation of the piezoelectric layer occurs along the direction (axis-1) perpendicular to the direction (axis-3) of polarization in the piezoelectric layer, so that the piezoelectric material is excited in the transverse e_{31} mode. The parameters of the electrical load can be tuned to optimize the transfer efficiency of energy from mechanical to electrical domain.

For energy harvesting due to vibration, piezoelectric bimorphs should be designed according to resonant frequencies. In this case, the behavior of the bender can be approximately treated a mass-spring system damped by electrical dissipations. Therefore, the bimorph has to be excited at a frequency close to its fundamental frequency ω_0 in order to maximize the output power. If the displace is not limited, the power generated is dependent on ω_0^3. For this reason, resonant vibration harvesters are usually used to harvest energy from vibrations in an 'industrial' environment, for example, on the motor of a vehicle. This type of harvesters is not able to harvest the energy due to the motion of human body occurring at very low-frequencies (<25 Hz) and irregular motion of the limbs.

Although this problem has been addressed by applying mechanical excitation directly to piezoelectric transducers, such as shoe-mounted harvesters discussed above, this type of devices is relatively large and the energy harvested must be transmitted to devices to be powered in on another part of human body. Therefore, inertial principle has been used as an alternative method, where the harvesters can be located with the sensors to be powered. Also, if the slow body vibrating can be transferred to high-frequency cantilevers, the energy can then be harvested. The impact-driven harvester is developed by taking these into account. Figure 2.29 shows conceptual representation of the system. It consists of a frame with a channel inside which allows for free sliding of a mass and two piezoelectric bimorphs located at two sides of the frame. If the frame is shaken along the direction of the channel, the mass strikes the piezoelectric cantilevers alternatively. Kinetic energy of the moving object can be harvested by the piezoelectric harvester. The device to be powered is also included inside the frame.

The impact-driven harvester can be described by using an electrical network model [89]. There are several requirements for piezoelectric bimorph to be represented by an electrical network. The constitutive equations relating mechanical and electrical variables are linear. The magnitude of displacements is small. The mechanics of the bimorph can be described by using Euler–Bernoulli beam theory with assumption of plane stress or plane strain. There is an ideal interface between the elastic and piezoelectric layer. The mechanical input applied to the harvester can be accurately represented by a concentrated force and excites only the fundamental vibration mode of the bimorph.

Because all these requirements are met for the impact-driven harvester, the behavior of the piezoelectric bimorph can be described by a circuit shown in Fig. 2 in Ref. [89]. The mechanical part of the system is represented by the series association of an inductor m and a capacitor k. m and k correspond to the effective mass and the stiffness of the bender, respectively. k is a complex quantity, which can be expressed as $k = k'(1 + j/Q)$, so that parasitic dissipations are represented

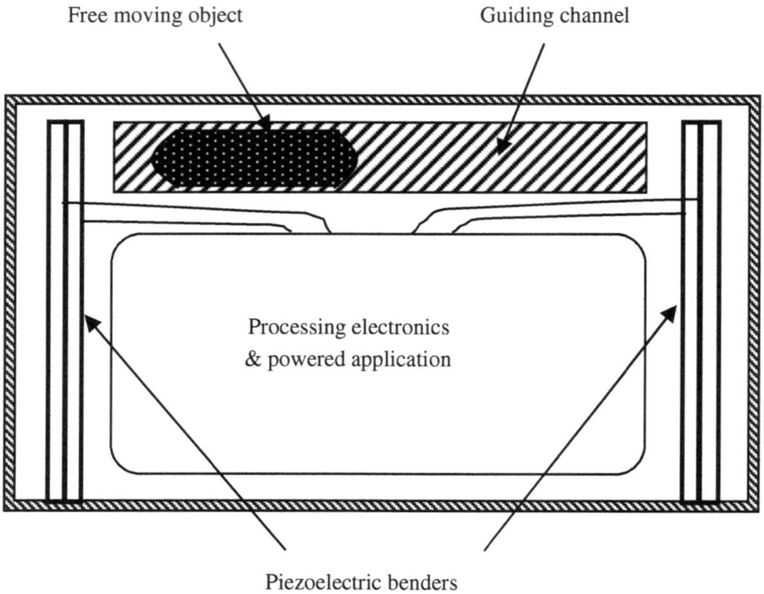

Fig. 2.29 Conceptual representation of the impact-based vibration energy harvester. Courtesy of M. Renaud, Interuniversity Microelectronics Center (IMEC). Reproduced with permission from [89]. Copyright @ 2009, IoP Publishing

in the model through the quality factor Q. F and δ stand for the concentrated force due to the mechanical excitation and the deflection at the position of the force applied. The mechanical elements are connected to the piezoelectric capacitor with a capacitance C_p through a perfect transformer with a ratio of Γ:1. C_p is the clamped capacitance, which is measured when the piezoelectric layer is not deformed. The coefficient Γ is directly proportional to the piezoelectric constant e_{31}. Finally, I and V are the electrical current injected into the capacitor and the voltage developed across the electrodes of the piezoelectric layers.

The analysis has been limited to one-dimensional collision on a flexible beam. The velocities of the objects and the mechanical excitation due to the collision are assumed to be in the direction perpendicular to the surface of the beam. The contact surfaces are supposed to be perfectly smooth, so that the friction mechanisms can be neglected. The impact has different phases. First, the moving object approaches the beam at a velocity which is dependent on the particular excitation applied to the frame of the device in Fig. 2.29. The impact process consists of two separate phases.

In the first phase, the two objects tend to interpenetrate each other and a local compression referred to as indentation δ_i is observed in the area surrounding the contact point. The contact area becomes larger as the magnitude of the indentation is increased. The compression phase ends when the amplitude of the restoring elastic force $F(t)$ is large enough to induce a local expansion of the two bodies which tend to repulse each other at this moment. During the contact time, radial waves due to the indentation start propagating away from the impact location. The contact area and the magnitude of the indentation are small, so the corresponding radial waves do not have an important influence on the macroscopic behavior of the two bodies and thus can be ignored. Another wave consisting in a vertical displacement is initiated in the beam during the impact process; it is referred to as a bending wave that results from the exchange of momentum between the two objects. After the expansion phase, the two objects separate with velocities in opposite directions. The object also oscillates along its own eigenmodes after the separation. Because the moving object is assumed to be very stiff, these oscillations do not have any effects on the general dynamics. In the beam, the bending wave propagates further away from the impacted vertical segment. The behavior of the bender prior to a second impact can be treated as free oscillations.

The analysis of the model has revealed several important aspects. The contact force $F(t)$ can be expressed by the Hertz model. The dynamics of two spherical contact surfaces during an elastic impact can be represented by a simple model consisting of two lumped masses connected by a nonlinear spring k_i, which creates a contact force of $k_i\delta_i^{3/2}$. The lumped mass associated with the impacted segment of the beam corresponds to its effective mass m. For the impacting process, this corresponds to its full mass M. The indentation stiffness k_i can be expressed in terms of the curvature radius of the impacting surfaces and certain compliance components.

The moving object can be treated to be perfectly rigid, so the equations of motion for the indentation δ_i and the displacements of the two bodies have been established through the Euler–Bernoulli beam model. The equations can be solved analytically by having certain assumptions. If the nonlinear contact force can be described by a sinusoidal pulse of angular frequency ω, there is $F(t) \approx F_c \sin \omega(t)$, with $t_i < t < t_i + t_c$, and otherwise $F(t) = 0$. F_c is the maximum amplitude of the impact force, t_i is the time at which the two objects collide, and $t_c = \pi\omega$ is the contact time. F_c and ω can be obtained by solving the equation of motion for the indentation.

Numerical values of F_c and t_c can be obtained with the dimensions of the bimorphs: $W = 10$ mm, $l = 25$ mm, $t_s + 2t_p = 800$ μm, without an attached mass. The relevant values of stainless steel moving object and NavyIII-PZT5A of the beam are: $\varepsilon_{33}^E s_{33}^E = 15.6$ pPa^{-1}, $\varepsilon_{13}^E = -4.8$ pPa^{-1}. The radius of curvature at the contact surface is 5 mm while the beam is flat. M is 4 g and the average density of the beam ρ is 7,600 kg m^{-3}. The relative approach velocity U_0 is set to 1 m s^{-1} ($U_0 = v_{b0} - v_{M0}$, with v_{b0} and v_{M0} being the velocity of the impact points of the beam and the sliding object just before impact, respectively). With these values, there are $F_c = 15$ N and $t_c = 7 \times 10^{-6}$ s. The numerical expression of $F(t)$ can be used to check the validity of the last assumption to develop the network model of the bimorph. Combined with finite element (FE) simulation, it is found that the single-mode approximation lead to negligible errors in the general dynamics of the system [89].

Because the contact time is very short as compared to the fundamental period of the bender, the velocity of the impacted segment in the beam and of the moving object, due to the impact process and the corresponding pulse, can be redefined quasi-instantaneously. In this case, the dynamic behavior of the system can be treated to be discontinuous at the moment of the collision. The velocities of the two objects before (v_{b0} and v_{M0}) and after (v_{b0}' and v_{M0}') the impact can be obtained with the fact that the variation in momentum during the compression-expansion phase is equal to the time integral of the pulse $F(t)$. However, the momentum principle is only valid when it is a perfect elastic impact without energy dissipations. In practice, there have complex dissipation mechanisms in the collision, such as heat due to viscoelastic or internal friction and local plastic deformations. All these factors result in a decrease in the time integral of the impulse.

To account for these loss mechanisms, an energetic coefficient of restitution e is defined for each body. It is the square root of the ratio of the work W_c done on the contact surfaces during the compression over the work W_e done by the force in the expansion phase. That is: $e_b = (W_{e,\,b}/W_{c,\,b})^{1/2}$ for the beam and $e_M = (W_{e,\,M}/W_{c,\,M})^{1/2}$ for the sliding object, with e_b and e_M being the fractions of energy dissipated in each body during the compression-expansion. Similarly, it is further to define an effective coefficient of restitution e_{eff}, which relates the energy in the system before $(W_{c,\,b} + W_{c,\,M})$ and after $(W_{e,\,b} + W_{e,\,M})$ impact. This is particularly useful to the impact-driven device due to the very short duration of the impact process. The magnitude of e_{eff} depends on the effective masses, the curvatures and areas of the contact surfaces, the material properties, and the relative collision velocity. Generally, e_{eff} decreases as U_0 is increased. The values of e_{eff} have been determined experimentally. According to the energy conservation principle, the velocities of the two bodies after the impact can be expressed in terms of their velocities before the impact as follows:

$$v_{b0}' = \frac{mv_{b0} + Mv_{M0} - e_{\text{eff}}M(v_{b0} - v_{M0})}{M + m}, \tag{2.21}$$

$$v_{b0}' = \frac{mv_{b0} + Mv_{M0} + e_{\text{eff}}M(v_{b0} - v_{M0})}{M + m}. \tag{2.22}$$

With all these discussed above, equations to describe the dynamics of the impact-driven harvester can be established [89]: $M\mathrm{d}_2\delta M/\mathrm{d}t^2 + D_v\mathrm{d}\delta_M/\mathrm{d}t = F_{\text{ext}}$, where D_v is a viscous damping factor, δ_M is the position of the mass with respect to that of the piezoelectric bender (δ_M is assumed to be one-dimensional), and F_{ext} is the apparent force due to the motion of the frame. The behavior of the whole system can be presented by coupling the two network models of the bimorph with the impact representation consisting of the velocities at the moment of collision.

By having specific types of load circuitry and input forces, the power delivered to the load can be analyzed and optimized [89]. Because the equations of motion are transcendental, it is not possible to develop an analytical model to represent the exact behavior of the human motion harvesters. However, to have a reference for device fabrication, it is still necessary to conduct theoretical analysis with simplified situations. To do so, a single-impact approximation has been used, in which the sliding object hits one of the benders and but does not interact with it thereafter. That is, the piezoelectric beam oscillates freely without perturbation after the impact. It is assumed that there is only a purely resistive load. With this impact, it seems that an initial quantity of mechanical energy has been applied to the cantilever, which is $E_o = 1/2mv_{b0}'^2$. If there is no energy loss, the average output power of the harvester per impulse can be given by $P = E_o/T$, with T being the time interval between two consecutive impacts. The piezoelectric cantilever should be designed in such a way that its output power is as close to this theoretical limit as possible.

One way to increase the electromechanical energy conversion is to maximize the value of the generalized electromechanical coupling factor (GEMC) K^2 [89]. The GEMC can be expressed in terms of the circuit parameters as $K^2 = \Gamma^2/(kC_p)$, which is the equivalent for a piezoelectric bending structure, with the effective electromechanical coupling (EMC) to be $k_{31}^2/\left(1 - k_{31}^2\right)$, for a suspended and axial piezoelectric transducer, with $k_{31}^2 = d_{31}^2 s_{11}^E/\varepsilon_{33}^T$.

To the first approximation, the optimum value of the GEMC is directly proportional to k_{31}^2, which can be obtained by implementing a specific ratio of the thickness of the elastic to that of the piezoelectric layers. This thickness ratio depends on the ratio of the compliances between the two layers. With these parameters, the optimum thickness ratio and the optimum value of K^2 normalized to k_{31}^2 can be expressed as a function of the compliance ratio. The maximum values of the GEMC that can be obtained with state-of-the-art piezoelectric and elastic materials are between 0.1 and 0.3. The value of the GEMC is not affected no matter whether the piezoelectric capacitors in the bimorph are connected in parallel or series.

In addition, the mechanical quality factor Q has to be maximized, so that the maximum amount of the vibration energy initially applied to the beam is harvested by the device and not dissipated otherwise. An approximate expression for the optimum resistive load R_{opt} is given as follows, where $\omega_0 = (k/m)^{1/2}$ is the fundamental frequency of the short-circuited piezoelectric bender.

$$R_{\text{opt}} = \frac{1}{\omega_0 C_P\left(1 + \frac{K^2}{2}\right)}. \tag{2.23}$$

This equation is similar to Eq. (2.10), but with somehow simplification. Although these results are derived from a simplified situation; the conclusions are applicable to most representative cases. Although the real behavior of the device is very complicated, it is assumed that the conditions of maximum GEMC and Q for optimum power derived in the case of single impact are also applicable to the cases of multiple impacts. The parameters, k, Γ, C_p and Q, are 1,000 N m^{-1}, 0.5 m N V^{-1}, 3 nF and 50, so there is $K^2 = 0.08$. Additionally, the dissipations of the sliding object (D_v) must be minimized due to the negative influence on the output power of the device.

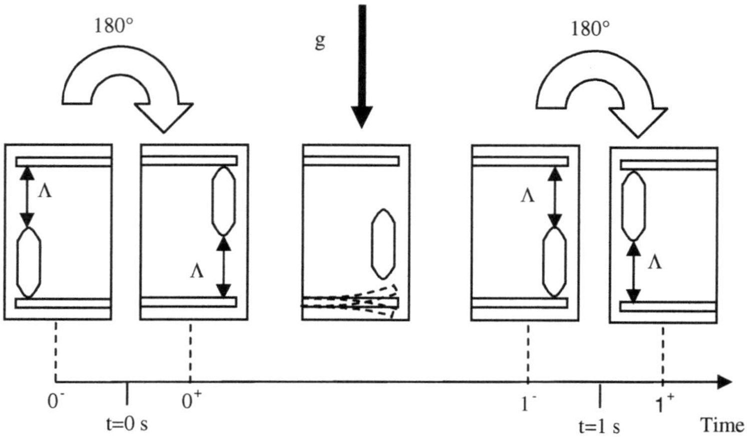

Fig. 2.30 Assumed motion of the harvester for the numerical simulations. Courtesy of M. Renaud, Interuniversity Microelectronics Center (IMEC). Reproduced with permission from [89]. Copyright @ 2009, IoP Publishing

Figure 2.30 shows the assumed motion of the frame for the numerical analysis, which is a quasi-instantaneous 180° rotation per second [89]. Between every two rotations, the frame is supposed to be oriented in such a way that the sliding channel is aligned with the gravity field g, so the missile experiences a gravitational force $F_{\mathrm{ext}} = -Mg$, which makes it move from one cantilever to the other, with the action of gravity on the beams being neglected. There are a lot of such body motions in human daily activity.

Simulation results indicate that the performance of the harvester has a slight decrease with increasing traveling distance Λ of the sliding object. This is because longer traveling distance means longer traveling time required. For example, if the object cannot reach its rest position on the second cantilever before the second rotation starts, the potential energy of the mass will not be totally transferred to the piezoelectric cantilever. As a result, the efficiency of the energy conversion is reduced. This does not necessarily mean a lower output power but in any case a lower output power per unit volume. A constant value $\Lambda = 15$ mm has been used for further analysis.

It is also found that when the natural frequency of the piezoelectric beam is much higher than that of the rotations of the frame, the performance of the harvester in dependent only on the mass ratio M/m, being irrelevant to their individual values. The electrical load used for analysis should be a simple circuit consisting of only a pure resistor. The use of two cantilevers makes the system complicated. For example, if the piezoelectric benders are connected to the same load circuit, they can be arranged in either parallel or series. In this case, when one of the cantilevers is excited, part of the electrical energy it harvests flows toward the second cantilever, rather than toward the load. This inevitably reduces the conversion efficiency. This problem can only be addressed by design, for example, the use of appropriate switching. For simplicity in analysis, each cantilever is connected to its own load. It is also usually assumed that the sliding object is at rest on one of the cantilevers before every rotation, the power generated by the harvester solely can be estimated by studying the behavior of the system between two successive rotations [89].

In practice, every real impact is followed by a certain number of collisions that are separated by short-time intervals, until the sliding object is ejected from the neighborhood of the cantilever. At this moment, the beam oscillates freely till a second real impact occurs. Depending on the values of the different parameters, the sliding object might apparently "stick" to the cantilever for a short duration. If it is a perfectly inelastic case ($e_{\mathrm{eff}} = 0$), the sliding object and the beam merge into a single oscillator just after the impact and they then separate when the acceleration of the beam becomes zero. For nonzero values of the coefficient of restitution, it is difficult to predict the details of the impact, because they are dependent on different characteristic parameters. However, when the effective mass of the cantilever is negligible as compared with that of the sliding object, the behavior is very similar to the inelastic one, which is independent of the value of the coefficient of restitution.

It is important to estimate the maximum value $\delta_{\max}$, of the deflection of the cantilevers, because too large displacements lead to failure of the device. The upper and lower limits of $\delta_{\max}$ can be estimated according to energy conservation principles. The upper limit of $\delta_{\max}$ is obtained when the impact is perfectly elastic ($e_{\mathrm{eff}} = 1$), while its lower limit can be derived from inelastic impact. By equating the kinetic energy of the sliding object and the beam just after an inelastic or elastic impact to the elastic energy of the cantilever when it has reached its maximum deflection $\delta_{\max}$, an inequality can be obtained in Eq. (2.24), noting that the electrical energy in the piezoelectric beam is neglected [89]. Therefore, the maximum deflection increases proportionally with the impact velocity and the inverse of the open circuited resonance frequency of the cantilevers. Also, the larger the mass of the sliding object, the larger the $\delta_{\max}$ is. Equation (2.24) also indicates that when M/m is very large, the upper and lower limits of $\delta_{\max}$ become the same value, i.e., there is no difference between elastic and inelastic impact.

$$\frac{U_0}{\omega_0}\sqrt{\frac{M}{m_{\mathrm{eff}}}\left(1-\frac{m_{\mathrm{eff}}}{m_{\mathrm{eff}}+M}\right)} \le \delta_{\max} \le \frac{U_0}{\omega_0}\sqrt{\frac{M}{m_{\mathrm{eff}}}}. \tag{2.24}$$

The open circuit voltage of the piezoelectric beam as a function of time has been developed on the electrical ports [89]. Every first real impact is characterized by a sharp peak of voltage followed by decaying sinusoidal oscillations. The open circuit voltage V can be related to the deflection δ by $V/\delta = \Gamma/C_p$, if an open circuit and mechanically unconstrained situation are present. In this case, the voltage is directly proportional to the magnitude of the deflection, so a maximum voltage $V_{\max}$ corresponds to a maximum deflection δ and thus the value of $V_{\max}$ can be can be estimated by using Eq. (2.24).

The effects of M/m, e_{eff} and R_L on the efficiency of the energy conversion have also been studied [89]. The initial energy E_0 in the system at the beginning of a cycle can be estimated by the potential energy of the sliding object $E_0 = Mg\Lambda$, with $g = 9.8$ m s^{-2} and $\Lambda = 15$ mm. The energy E_L dissipated in the load resistor during a period of the motion is obtained by integrating $V(t)^2/R_L$. The efficiency η is defined as E_L/E_0 and the average output power P is equal to $\eta E_0/T_i (T_i = 1$ s). With these parameters, waveform of the voltage and the energy dissipated in the resistive load can be well simulated.

The analytical expression of the optimum load resistor for the single-impact approximation has been evaluated for the validation in multiple-impact situations. For such a purpose, the power conversion efficiency η has been systematically studied, with various values of coefficient of restitution and mass ratio, at a constant load resistor. It is found that the effect of the coefficient of restitution on the value of the optimum load is not noticeable, whereas that of the mass ratio M/m is very significant. This important observation has been explained by considering the characteristics of the behavior of the cantilevers. The efficiency does not vary monotonically with M/m [89].

In summary, the efficiency of the energy harvester from human body motion can be optimized by maximizing the GEMC, minimizing the parasitic losses, selecting an appropriate load and using an appropriate mass ratio. In such a way, an optimum efficiency of 40–50 % has been predicted.

The value of the coefficient of restitution e_{eff} has been obtained by measuring the velocity of a steel sliding object before and after impact by using a laser detection method. The excited piezoelectric beam is clamped, so there is no need to measure the velocity of the cantilever. The sliding object is dropped from a height of 15 mm onto the surface of the clamped piezoelectric cantilever. A laser beam coupled to a light detector is used to detect the successive times t_1 and t_2 at which the moving object crosses and exits the beam. From the basic law of motion $L_M = g(t_2 - t_1)/2 + v_1(t_2 - t_1)$, and, as t_1 and t_2 are measured, v_1 can be derived, which is the velocity at time t_1. The velocity v_2 at time t_2 can be obtained from $v_2 = v_1 + g(t_2 - t_1)$ with $v_1 = L_M/(t_2 - t_1) - g(t_2 - t_1)/2$. The impact time t_3 is determined according to the surface wave created at the moment of the collision, which leads to a voltage across the open circuit piezoelectric laminate. The distance between the laser beam and the piezoelectric cantilever L_L is chosen just to be a bit larger than L_M. In this way, the dissipations due to air damping can be neglected in this portion of the motion of the sliding object, and its velocity just before impact v_3 can be approximated by $v_2 + g(t_3 - t_2)$. Similarly, the velocity just after impact, v_3', can be estimated from the time interval between the impact and t_4, so $v_3' = (L_L - L_M)/(t_4 - t_3) - g(t_4 - t_3)/2$. If the impact of a sliding object is on a clamped beam ($v_{b0} = v_{b0'} = 0$ and $m \gg M$), there is $e_{\mathrm{eff}} = -v_3'/v_3$, which mean that the energetic coefficient of restitution is equivalent to the kinematic one. In addition, the angle of drop of the sliding object also has effect on the measured values of e_{eff}, with a standard deviation of 5 %. The average coefficient of restitution is 0.55.

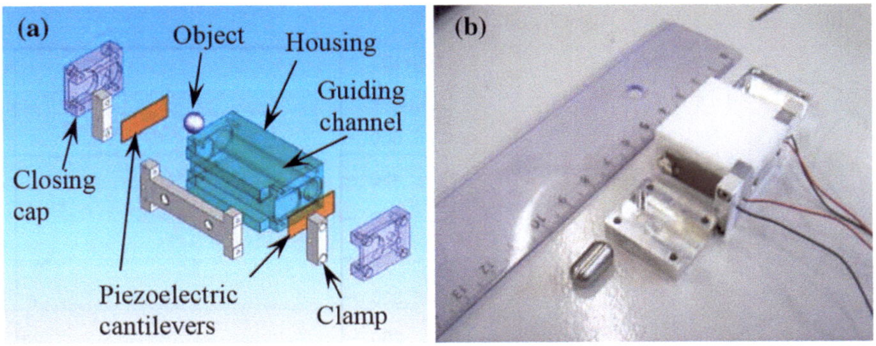

Fig. 2.31 **a** Conceptual representation of the harvester prototype. **b** Photograph of the actual prototype. Courtesy of M. Renaud, Interuniversity Microelectronics Center (IMEC). Reproduced with permission from [89]. Copyright @ 2009, IoP Publishing

The voltage predicted theoretically across the electrodes of the piezoelectric beam and the power dissipated in the load resistor has been compared with experimental measurements. The equivalent network parameters of the piezoelectric beam used in the experiments are: $k = 980$ N m^{-1}, $m = 460$ mg, $M = 4$ g, $Q = 50$, $\Gamma = 0.5$ mN V^{-1}, and $C_p = 3.1$ nF, which lead to $K^2 = 0.08$. They are determined from the analysis of the frequency dependence of the impedance. Experimental results indicate that, with a drop distance of 15 mm, $R_{\text{opt}}^{m} \approx 400$ kΩ $\approx 2/\omega_0 C_p$, corresponding to $e_{\text{eff}} = 0.5$ and $M/m = 10$, the predicted efficiency is about 50 % higher than the measured one, which has been attributed to the losses occurring during the motion of the sliding object. These losses have been neglected in the numerical simulation. To include these losses in the model, a viscous damping term D_v in the motion of the sliding object should be introduced.

A drawing and photograph of a prototype of the impact-driven energy harvester are shown in Fig. 2.31a and b [89]. The frame of the device is made of Teflon, while Al is used for the clamps and the closing caps. The sliding object is made of steel ($M = 4$ g) with an oblong shape, so that it occupies approximately half of the length of the guiding channel, 30 mm. The electronics to be powered are built inside the frame, so as to have a high output power per unit volume or mass. The characteristics of the equivalent network are: $k = 6{,}000$ N m^{-1}, $m = 600$ mg, $Q = 40$, $\Gamma = 1$ mN V^{-1}, and $C_p = 9.6$ nF, $e_{\text{eff}} = 0.55$, corresponding to $K^2 = 1.7 \times 10^{-2}$.

The kinetic energy of the mass when hitting one of the cantilevers is equal to $Mg = 593$ μJ. Due to the low value of the GEMC, the maximum theoretical value of the efficiency is reduced to 13 % instead of 40 %. The measured maximum efficiency is 8 %. The discrepancy between the predictions and the measurements is slightly lower, which can be attributed to the fact that the sliding object is fully guided so that the losses are reduced.

The device is tested when it is shaken in a hand, as shown in Fig. 2.32a. The output power of the device, as a function of the load resistance, is demonstrated in

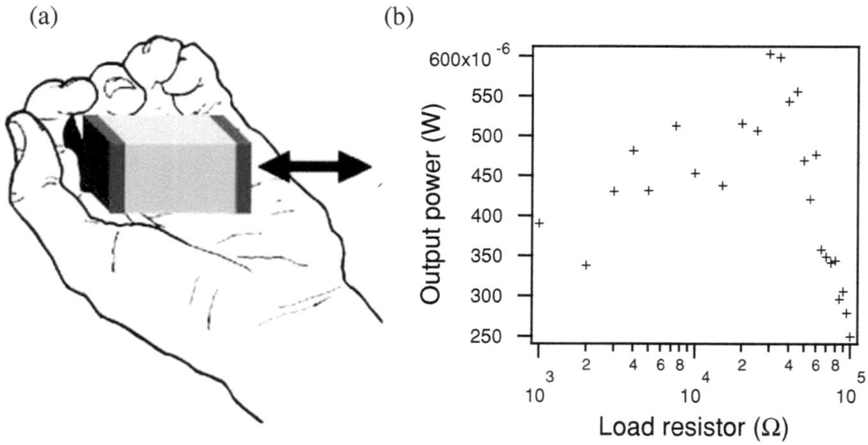

Fig. 2.32 **a** Illustration of the motion applied to the impact energy harvester. **b** Measured output power. Courtesy of M. Renaud, Interuniversity Microelectronics Center (IMEC). Reproduced with permission from [89]. Copyright @ 2009, IoP Publishing

Fig. 2.32b. The frequency of the motion is about 10 Hz, while the amplitude is about 10 cm. A maximum of 600 μW can be achieved. As the resistor is ranging from 3 to 60 kΩ, the average power is up to 400 μW. This amount of power is sufficient to supply energy to simple sensor nodes [89].

There are still rooms to improve the power density of the device, for example, reducing the volume of the device, decreasing its weight, optimizing the performance of the piezoelectric cantilevers, increasing the frequency of the motion and circuit designing [89]. A power density of 10 μW cm^{-3} or 4 μW g^{-1} is expectable for such an impact-driven harvester in human motion energy harvesting applications.

The second example of impact-driven piezoelectric energy harvester from human motion represents the latest development of the type of energy harvesting devices [90]. It uses high-frequency PZT-5A bimorph cantilever beam with attached proof mass at the free end. An aluminum prototype, with a dimension of $90 \times 40 \times 24$ mm^3, has been designed for attaching to human leg to harvest energy produced by waling. An open circuit voltage of 2.47 V and maximum average power of 51 μW can be achieved across a 20 kΩ external load resistance at walking speed of 5 km h^{-1}.

Figure 2.33 shows schematic drawing of the impact-driven energy harvester [90]. It consists of a ridged cylinder, which can slide freely on a shaft and a PZT bimorph, with a tip that can be fixed in parallel with the cylinder. When the cylinder is excited by external impacts, the ridge on it hits the tip and the PZT bimorph vibrates at its resonant frequency after they separate. As the PZT bimorph vibrates, the stored mechanical energy is converted into electrical energy, which is thus transferred to external circuits. The asymmetric design of the tip is to increase the vertical impulse force when the cylinder moves backwards. The parallel layout

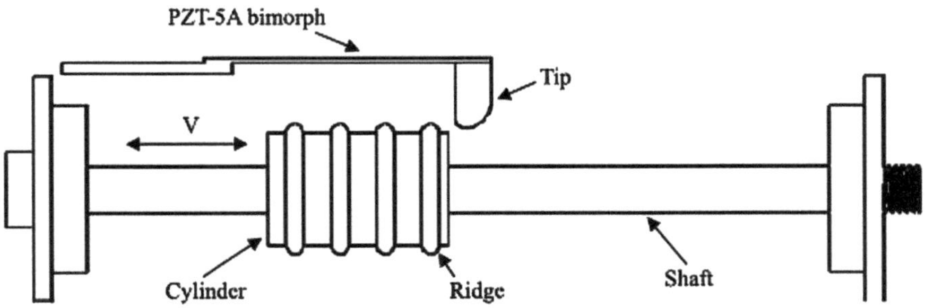

Fig. 2.33 Schematic drawing of the impact-driven piezoelectric energy harvester. Reproduced with permission from [90]. Copyright @ 2013, IoP Publishing

of the piezoelectric cantilever beam and the sliding mass take the unused space so as to achieve high power density. Numerous ridges on the cylinder ensure a repetition of the impact actions to achieve sustained output power. The suitable overlap between the ridge and the tip restricts the impulse force to prevent the PZT bimorphs from damaging.

Figure 2.34a shows a general piezoelectric cantilever beam structure of bimorph, where the attached proof mass on the free end of the beam is used to reduce the resonant frequency [90]. In this configuration, the piezoelectric layer is excited in a bending mode, e.g., d_{31} mode. When the cantilever bean is excited at its resonant frequency, maximum deflection is observed and an electric field is built up between the upper and the lower electrodes.

An analytical model has been established to predict the response of the piezoelectric beam. Because it is a vibration-based energy harvester, lumped parameter models can be used, as shown in Fig. 2.34b. In this approach, the electromechanical properties of the device can be simply modeled as a spring, mass, damper and a piezoelectric element, as discussed earlier. The energy in the piezoelectric layers is measured by using an external resistive load R.

A coupled piezoelectric-circuit finite element method has been used to predict optimized load resistance, as well as amplitude of displacement and resonant frequency of the piezoelectric beam. The PZT bimorph is a 0.13 mm thickness brass shim sandwiched by two piezoelectric layers, with each having a thickness of 0.19 mm. Its overall dimension is 31.8 × 6.4 × 0.51 mm^3. The open circuit resonant frequency is about 260 Hz in modal simulation, which is a reference for further simulation and experiments. Maximum output power is observed as the electrical load is 30 kΩ.

Components used to assemble the impact vibration harvester prototype are shown in Fig. 2.35a, including a PZT bimorph, a cylinder with four ridges, and the tip and the shaft. An aluminum rectangle frame is used encapsulate the PZT bimorph and the sliding cylinder. The structure has a through hole to support the smooth shaft, on which the cylinder can move freely. A piece of rubber is used to protect the harvester, as shown in Fig. 2.35b. The experimental setup to test the

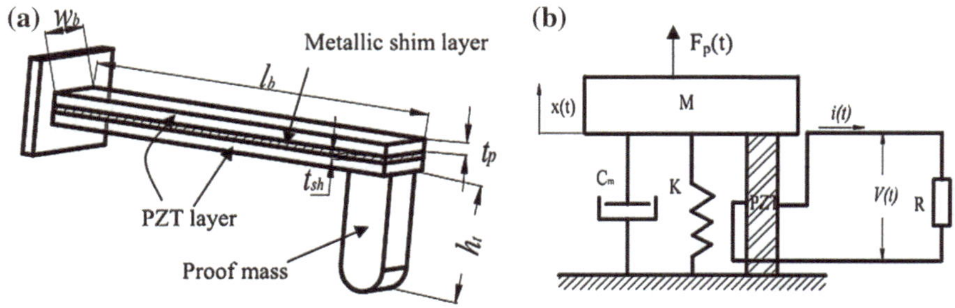

Fig. 2.34 **a** Schematic diagram of a piezoelectric bimorph. **b** General electromechanical model of the piezoelectric vibration energy harvester. Reproduced with permission from [90]. Copyright @ 2013, IoP Publishing

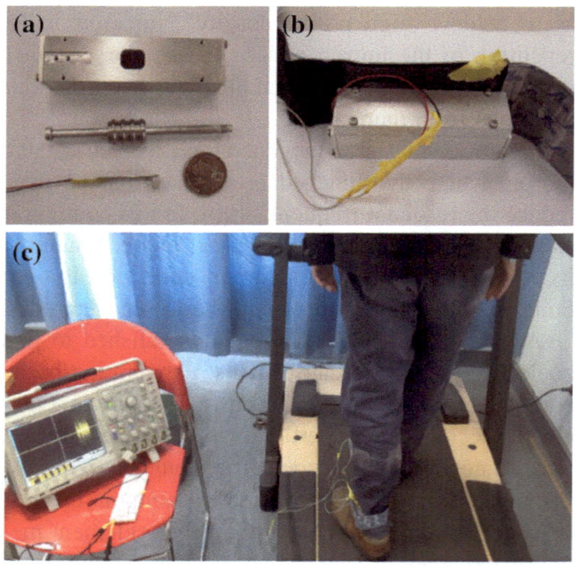

Fig. 2.35 **a** Photograph of components of the harvester, showing the PZT bimorph, shaft, cylinder, cover and frame. **b** An assembled prototype of the harvester. **c** Experimental layout for measuring voltage and power of the harvester, showing an oscilloscope, a circuit board and a treadmill. Reproduced with permission from [90]. Copyright @ 2013, IoP Publishing

impact-driven energy harvester is shown in Fig. 2.35c. The energy harvester is mounted on human leg and the cylinder moves repeatedly along the shaft when the human walks on the treadmill and the ridges in the cylinder impact the tip, so that the PZT bimorph beam vibrates. The output voltage is measured by using the oscilloscope.

The performance of the harvester prototype has been tested at four walking speeds, 3.5, 4.0, 4.5, and 5.0 km h^{-1}. The actual measured resonance frequency of the PZT bimorph with 30 kΩ electric load is 265.5 Hz, very close to the value predicted by the simulation. The output voltage of the harvester is proportional to the magnitude of the deflection of the piezoelectric bimorphs and the impulse force increases with increasing walking speed. Therefore, the peak amplitude of voltage increases with increasing walking speed. A maximum of 51 μW is observed at walking speed of 5 km h^{-1}. Further improvement in performance of the impact-driven piezoelectric harvester is expectable through materials and design optimization.

2.4.3 Energy from Joint Motion

Piezoelectric harvesters have been designed and fabricated for energy harvesting from joint motion of human body, including knee-joint [91–94] and elbow joint and finger [94]. PZT piezoelectric bimorph is used in the former case, while PVDF sheet is used in the latter case. The PZT bimorph is excited to vibrate with the frequency up-conversion technique. The PVDF device works due to the bending and releasing process of the polymer piezoelectric sheet. Both can be used for wearable applications.

2.4.3.1 Rotary Knee-Joint Harvester

Figure 2.36 shows schematic diagram of the mechanical plucking process, constituting of a sequence of phases [91]. In the approach phase, the distance between bimorph and plectrum is reduced until they come into contact. At the point they contact, the loading phase is immediately started, during which both elastic elements are deflected, according to their mechanical compliances. With their further deflection, the overlap between the two items is reduced, i.e., their contact area becomes smaller and smaller. A maximum deflection is reached when the contact is just lost and the two items are free to return to their un-deformed states, which is the releasing point. At this instant, the bimorph starts to vibrate at its resonance frequency as a cantilever beam. As the bimorph vibrates, the stored mechanical energy is converted through the direct piezoelectric effect into electrical energy, which is then transferred to the external circuit. Part of the mechanical energy is dissipated through various forms of damping, such as air damping, dielectric losses and material internal damping. The outcome of the plucking is frequency up-conversion, because a large number of vibrations are created at high frequency by one single slow movement of the plectrum.

Figure 2.37 shows a knee-joint harvester that can apple repeat plucking actions to achieve continuous power harvesting during the motion of walking or running [91]. During normal gait, the knee alternatively bends and extends nearly once per second, covering a rotation angle of approximately 70°. The outer ring that carries the plectra is fixed to the thigh, while the internal hub that is connected to the bimorphs rotates with the knee as it is fixed to the shank. As the person walks, the knee alternatively extends and bends, each of the multiple plectra on the outer ring plucks the bimorphs in the sequence outlined in Fig. 2.36. This is the principle why the harvester is able to harvest the energy due to the slow motion of walking.

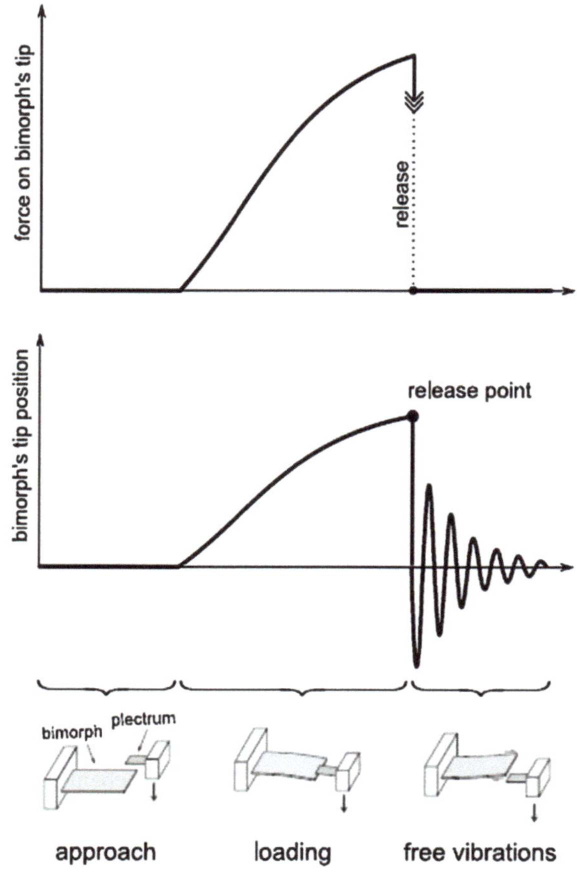

Fig. 2.36 Illustration of the plucking action: during approach, the force on the bimorph is zero, force increases during loading, where mechanical energy is input in the bimorph (and the plectrum), after release, the external force on the bimorph is limited to air damping and to its constraints (mounting) and the device can vibrate unhindered. Courtesy of Dr. M. Pozzi, Newcastle University. Reproduced with permission from [91]. Copyright @ 2011, IoP Publishing

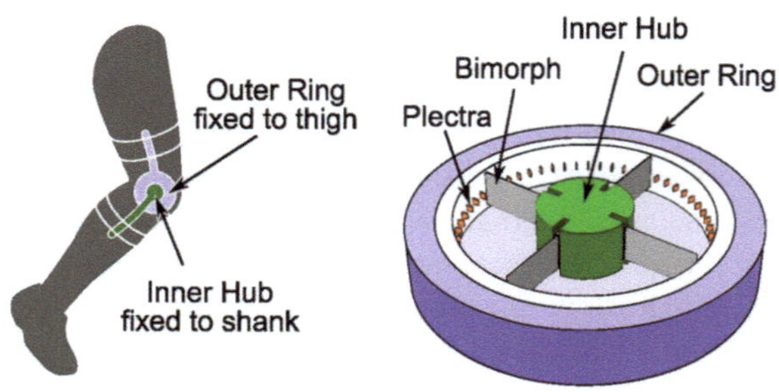

Fig. 2.37 Knee-joint piezoelectric harvester. It is worn on the external side of the knee and fixed by braces. Inside, a hub carries a number of bimorphs which are plucked by the ring-mounted plectra as the joint rotates during walking. Courtesy of Dr. M. Pozzi, Newcastle University. Reproduced with permission from [90]. Copyright @ 2011, IoP Publishing

Finite element (FE) modelling has been employed to analyze the harvester [91]. The power output is calculated from the voltage difference between the two electrodes, i.e., the voltage across the resistor. The boundary conditions include displacement constraints on the top and bottom surfaces in the clamped portion of the beam. The device this is modelled is a series bimorph with an internal brass shim and two sheets of PZT on either side, covering the entire surface, with given geometry and material properties. The geometry of the bimorph is determined by the acceptable dimensions of the overall harvester.

The model includes viscous damping with the damping matrix *C*, given by:

$$C = \beta_{\mathrm{PZT}} K_{\mathrm{PZT}} + \beta_{\mathrm{Brass}} K_{\mathrm{Brass}}, \tag{2.25}$$

where, β is stiffness matrix multiplier, *K* is portion of structure stiffness matrix. The β parameters are calculated from the tabulated mechanical quality factor Q_m and the resonant frequency of the device by:

$$\beta = \frac{1}{\omega Q_m}. \tag{2.26}$$

Two types of analyses are conducted: harmonic and transient. In the harmonic analysis, a sinusoidally varying force of 10 mN is applied at the tip and its frequency is scanned in the neighborhood of the first resonant frequency of the structure. This analysis is similar to the widely used conventional analyses, where the cantilever, with a large mass applied on its tip, is seismically excited by the vibration of its base. The force is not distributed along the cantilever, instead, it is localized on the nodes at the tip. The power output during the harmonic analysis is calculated as root-mean-square power using the amplitude voltage V_0 detected across the resistor *R*, as shown in Eq. (2.16).

The transient analysis consists of two steps to clarify the effects of different conditions applied for each step without interference of each other. The first is a loading step, where the bimorph is deflected to a predefined tip displacement at constant speed. The second step is the release phase, starting with a stepped removal of the tip constraints of a statically deflected beam. In the transient analyses, the instantaneous power $P(t_i)$ at substep *i* is calculated from the voltage across the resistor at the same substep, similar to Eq. (2.16):

$$P(t_i) = \frac{V^2(t_i)}{2R}. \tag{2.27}$$

From this equation, the average power $P_{\mathrm{avr.}}$ produced up to the time t_n can be calculated as:

$$P_{\mathrm{avr.}}(t_n) = \frac{1}{t_n}\sum_{i=0}^{n} P(t_i)\Delta t_i \quad (n > 0) \tag{2.28}$$

where Δt_i is the duration of each substep. The average power is of particular interest in the release phase, as it permits to optimize the interval between plucking actions and to predict the ensuing sustained power. Finally, the instantaneous power is also used to calculate the cumulative energy that is produced up to substep *n*, as a monotonic function of time:

$$E(t_n) = \sum_{i=0}^{n} P(t_i)\Delta t_i. \quad (2.29)$$

Harmonic response measurements with seismic excitations have been conducted to test the performance of the harvester. A Laser Doppler Vibrometer (Polytec CLV-2534) is used to measure the velocity of vibrations. The bimorphis made of PZT-5H that is poled for series operation, with a 130-μm-thick metal shim sandwiched between two layers of PZT, each of thickness of 125 μm. For the harmonic response test, the sample is mounted within copper-clad mechanical fixtures doubling up as pick-off electrodes, leaving a free length of about 22.5 mm. Seismic excitation is created with an acceleration of about 7.1 m s^{-2}. The plucking excitation is produced by loading the bimorph against a plectrum by displacing it over a given distance; release occurs upon retraction of the plectrum, mounted on a linear translation stage. Velocity data are collected with the Polytec LDV. A fixture is use to mount the bimorph test within the copper-clad plates leaving a free length of 26 mm. The plectrum is a 3 × 2 mm^2 rectangle 125 μm-thick Kapton® polyimide film. In both the harmonic and plucking excitation tests, the output of the bimorph is connected to the same circuit, with electrical loads ranging from 25 to 550 kΩ.

The harmonic analysis simulates the normal operating conditions of an environmental energy harvester. The power output curves for the piezoelectric bimorph correspond to maximum voltage amplitudes ranging from 2.7 mV at 300 Hz with 2 Ω up to 24 V at 320 Hz with 1 MΩ electrical load. The largest power output is found with an electrical load of 10 kΩ. As the electrical load increases, we observe a shift in the resonance frequency toward higher values; as is well known, this is due to the stiffening effect produced by the direct piezoelectric effect when the voltage on the electrodes is permitted to build up, which does not happen with lower resistors. The curve corresponding to 20 kΩ is particularly interesting as it has a maximum at frequencies intermediate between open and closed circuit and shows the largest area under the curve (it is lower but broader), indicating that it corresponds to maximum electrical damping, i.e., most efficient power transfer from the piezoelectric bimorph to the resistor, where it is dissipated. With 20 kΩ, a peak voltage amplitude of 6.1 V was detected at about 310 Hz. Meanwhile, the peak amplitude is the minimum in this case.

It has been found that the electrical loads with higher power are also accompanied by larger displacements. This is harmful to the PZT ceramic materials due to the potential failure. It is therefore important to determine the optimized value of the resistor that yields the maximum output power while limiting the strain of the material to a safe level. Therefore, it is necessary to define an appropriate "figure of merit" to figure out the optimal electrical load. This parameter should include both displacement (δ) and the electrical output. Noting that the voltage increases monotonically with increasing load and the power depends on the square of the displacement through the voltage, both V/δ and P/δ are not suitable. To address this problem, the quantity has been proposed:

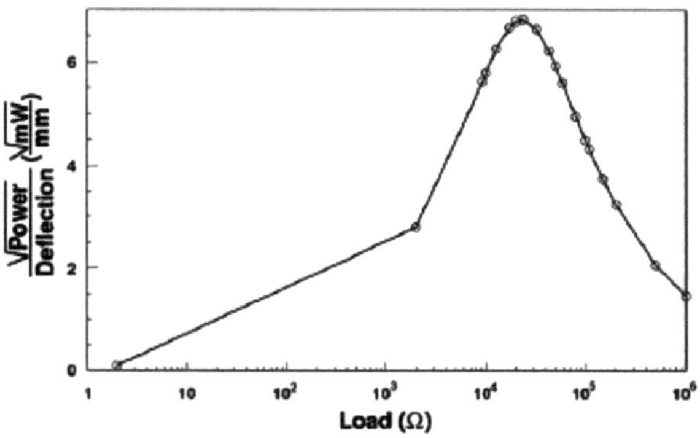

Fig. 2.38 Calculated "figure of merit" as defined in Eq. (2.30) to identify the optimal electrical load of the PZT-5H bimorph under harmonic excitations. Courtesy of Dr. M. Pozzi, Newcastle University. Reproduced with permission from [91]. Copyright @ 2011, IoP Publishing

$$f(R) = \frac{\sqrt{P}}{\delta}. \tag{2.30}$$

The values of "figure of merit" calculated with Eq. (2.30) for the devices are shown in Fig. 2.38 [91]. The best performance is obtained at 311 Hz with $f(R) = 6.8\ (\text{mW})^{1/2}\ \text{mm}^{-1}$ and electrical load of 20 kΩ. It has been found that there is a good agreement between the theoretical modeling and experimental results. Such an agreement is also observed in theoretical predictions of transient analysis (loading step and release step) and experimental measurements of plucking action.

A further study has been carried out to demonstrate the capability of energy harvesting of such harvesters for knee-joint wearable applications, by focusing on the effects of connection of the electrical load and excitation [92]. Two series of measurements are conducted at a speed of 5 s per revolution, as shown in Fig. 2.39. In one series, the bimorph is directly connected to a resistance, whereas in the other, a rectifying bridge is placed between the bimorph and potentiometer. Both curves have a maximum output energy, corresponding to an optimized value of resistance. This occurs when the impedance of the electrical load matches the impedance of the piezoelectric bimorph at the dominant operating frequency.

The optimal electrical load with the presence of a rectifying bridge (~100 kΩ) is higher than that without the bridge (~50 kΩ). This is because higher voltages are present across the bimorph electrodes if a higher resistance is connected. Therefore, the fixed forward voltage drop introduced by the diodes in the bridge becomes proportionally less important, i.e., the energy dissipated by the diodes becomes smaller fraction. As a result, when a rectifying bridge is used, higher electrical loads are beneficial and thus the maximum shifts toward higher values of resistance. In addition, when a bridge circuit is used, the amount of total output energy is decreased, with a magnitude of 50–80 μJ, which is due to the dissipation of the diodes in the bridge. Noting that the difference between the two maxima is only about 56 μJ, this loss due to the bridge is readily compensated by the higher electrical load.

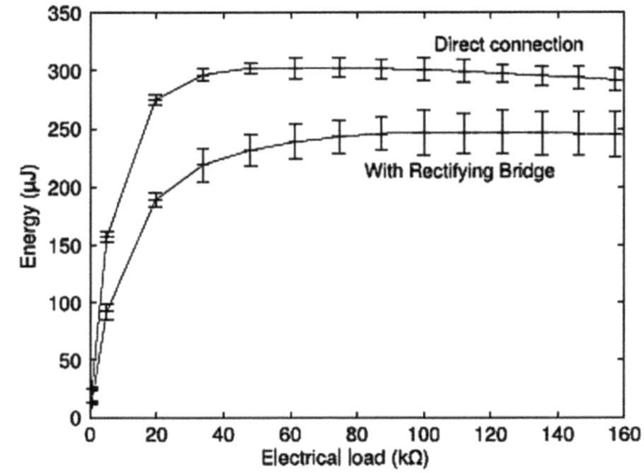

Fig. 2.39 Total energy produced by the bimorph during a full revolution of the outer ring, occurring in 5 s. The plot shows the effect of the resistive load with and without a rectifying bridge. The error bars represent the 90 % confidence level. Courtesy of Dr. M. Pozzi, Newcastle University. Reproduced with permission from [92]. Copyright @ 2012, IoP Publishing

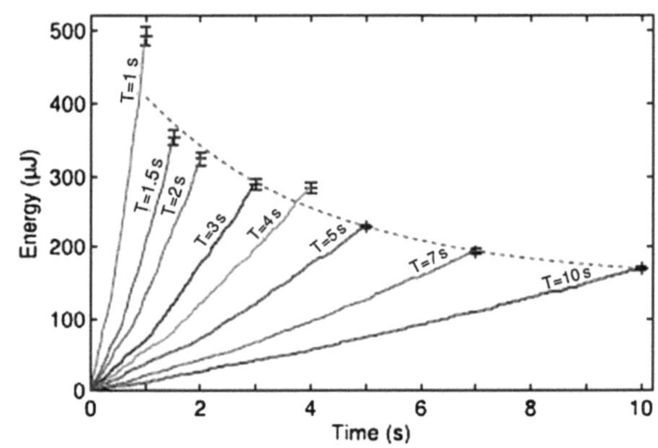

Fig. 2.40 Total energy produced by the bimorph during a full revolution of the outer ring. A 47.6 kΩ resistive load was directly connected to the bimorph. Courtesy of Dr. M. Pozzi, Newcastle University. Reproduced with permission from [92]. Copyright @ 2012, IoP Publishing

It has been reported that during normal gait of human being the angular speed of the knee-joint varies considerably, from 0 up to 6–7 rad s^{-1}. It is therefore important to characterize the harvester over a similarly wide range of angular speed in terms of practical applications. Two groups of measurements have been conducted to check the effect of the rotational speed: one with a directly connected electrical load (Fig. 2.40) and the other with a rectifying bridge interposed (Fig. 2.41) [92].

The results in the two figures indicate that the amount of energy that can be harvested by the harvester increases with increasing walking speed. This is simply because a higher walking speed means a larger total deflection of the bimorph, quicker initial deflection and clearer release, which are all favorable to a high energy output. The trend is likely to be exponential in both cases. This is a reflection of how the three factors combine. For example, larger deflections are observed only at the highest speed, while a quicker initial deflection gives a hyperbolic response.

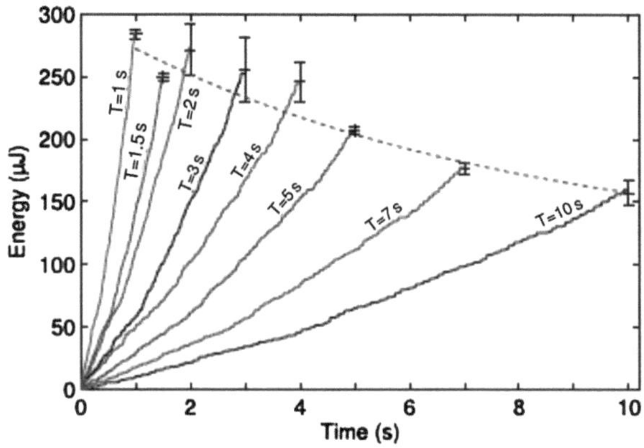

Fig. 2.41 Total energy produced by the bimorph during a full revolution of the outer ring. A rectifying bridge was connected between the bimorph and a 99.1 kΩ resistive load. Courtesy of Dr. M. Pozzi, Newcastle University. Reproduced with permission from [92]. Copyright @ 2012, IoP Publishing

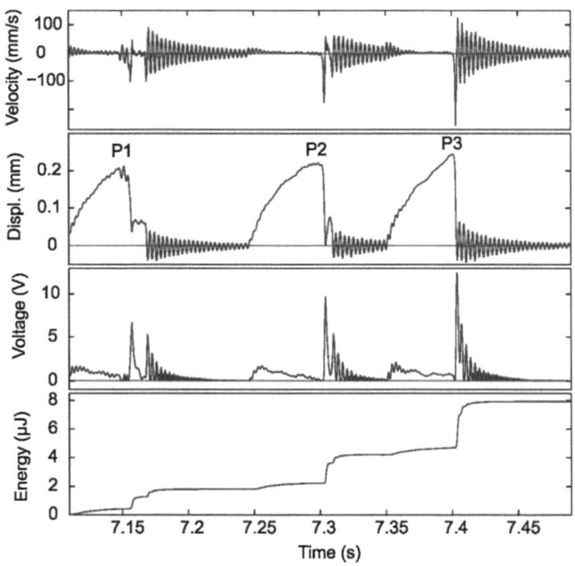

Fig. 2.42 Time domain data for a slow run ($T = 10$ s), from *top* to *bottom*: velocity, displacement (via integration of velocity and after baseline removal), voltage and energy. The first peak is the most common, showing an unclean release, with the plectrum slowly rubbing the bimorph; the second is cleaner and the third is almost ideal and a rare occurrence. Courtesy of Dr. M. Pozzi, Newcastle University. Reproduced with permission from [92]. Copyright @ 2012, IoP Publishing

The process of signal production has been analyzed in detail [92]. Time domain data for a slow run ($T = 10$ s) and fast run ($T = 1$ s) are shown in Figs. 2.42 and 2.43, respectively. Both figures are derived with a rectifying bridge. The displacement traces are extracted from the velocity via time integration, while the energy traces are obtained through time integration of the instantaneous power, calculated as $P(t) = V^2(t)/R$, with $R = 99.1$ kΩ and V is the rectified voltage. Figure 2.42 shows that, when the period is 10 s, all the plucking actions can be clearly identified. For instance, the bimorph is deflected by a plectrum over a time of about 60 ms to the maximum deflection before it is released and quickly returns to the initial state, but with the consequent vibrations about the rest position. The voltage signal already rises during the deflection, although it remains below 2 V in the 60 ms of the deflection. Much high voltages are produced just after the bimorph is released, with peaks of between 7 and 12 V. The energy curve exhibits a staircase-like character with three major steps, each of which is associated with a plucking action.

Although energy has been generated during the initial deflection, the greatest portion of energy is produced during the first several oscillations after the release. It is found that there are three main types of peaks. The first peak (P1) is characterized by a few of oscillations in the displacement before release and a fragmented step afterwards, implying an unclearn release. As a result, the voltage signal is relatively low and the energy step is small. The second peak (P2) has a cleaner release so that the output voltage is higher. The last peak (P3) is nearly an ideal release, with a sharp rise in displacement before the release point. Therefore, the output voltage and the energy step are both the highest. If the quality of the release is mainly determined by the plectra, there is always a fixed set of plectra that can provide the cleanest peaks.

As shown in Fig. 2.43, high speed ($T = 1$ s) leads to a difficulty in identifying the signals. The first plectrum (P1) causes decaying in the vibrations, due to the presence of a defect in the harvester. The following plectrum (P2) is too short to contact with the bimorph. As a consequence, there is a decaying voltage peak there. Also, the following plectrum (P3) is encountered by a bimorph almost at rest. However, there is still superposition of the fundamental vibration onto the displacement peak caused by the deflection. The next four plectra (P4 $\rightarrow$ P7) are met at short time intervals, yielding an uninterrupted oscillation of the bimorph, modulated by the contact with the plectra. As a result, the voltage signal shows a rapid succession of peaks with different heights and the energy trace is a wavy line instead of a sequence of steps. The outcome of this continuous vibration of the bimorph is an energy production that is significantly higher than that observed at lower speed.

Three major reasons have been identified to explain why higher energy is produced at faster rotational speeds. The first reason is a faster speed implies a faster initial deflection of the bimorph. The output voltage is proportional to the strain rate during deflection of the bimorph. This phenomenon can be understood from the fact that the energy produced by the harvester is approximately given by:

$$E = \frac{RQ^2}{\Delta t}, \tag{2.31}$$

where Q is the charge generated on the electrodes of the bimorphs, which is assumed to be independent of the strain rate, R is the electrical load and Δt is the deflection time [91]. At $T = 10$ s, the deflection occurs over about 60 ms (Fig. 2.42), at $T = 1$ s, it takes only 5–6 ms (Fig. 2.43). That is to say, a tenfold increase in the speed of revolution implies a tenfold increase in energy generation in the loading phase.

The second factor responsible for the high output energy is the cleanness of the release. As shown in Fig. 2.42, an unclean release, where the time of the contact

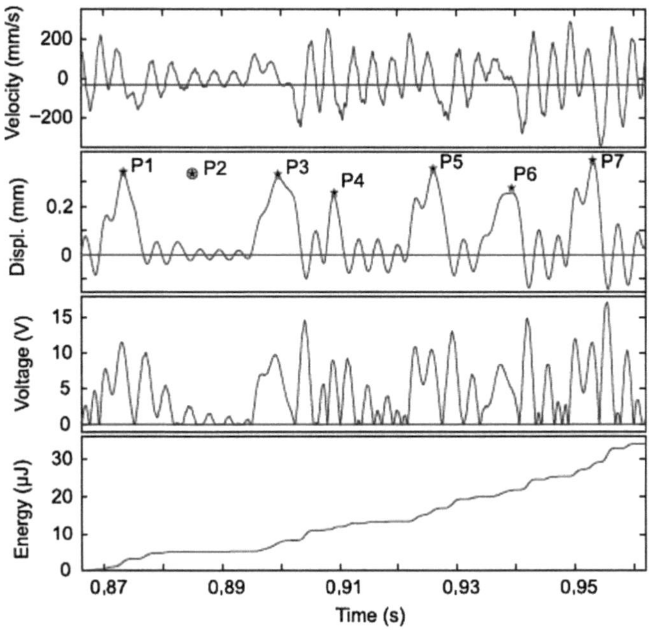

Fig. 2.43 Time domain data for a fast run ($T = 1$ s), from *top* to *bottom*: velocity, displacement (via integration of velocity and after baseline removal), voltage and energy. After the first plucking (*P1*), the bimorph has time to ring down as the immediately following plectrum (*P2*) is too short and does not touch the bimorph; the following plectrum (*P3*) is met by an almost still bimorph; the last peaks (*P3* → *P7*) are almost merged into each other and it can be seen that the bimorph continues to vibrate with considerable amplitude and generate large voltages. Courtesy of Dr. M. Pozzi, Newcastle University. Reproduced with permission from [92]. Copyright @ 2012, IoP Publishing

between the bimorph and the plectrum is too long, the bimorph is almost at the rest position, so that the deflection speed is very low. As a result, there is a lowlevel of energy production. At higher rotational speeds, however, the bimorph is released more quickly and sharply, so that high level of energy production is observed. The last factor is the increase in magnitude of the average maximum displacement, which is only effective at the highest rotational speed. In this case, the continuous encountering of the plectra and the bimorph makes the latter continuously vibrate.

It is also found that the ratio of achievable over measured energy increases with increasing speed, in both cases of with and without a bridge. This means that at higher speeds a larger portion of the measured energy is produced by the better performing plectra. In other words, at high speeds, the quality of the plectra is more important than at low speeds. This is because high quality plectra are less affected than poor performance plectra by speed. The use of a rectifying bridge makes the influence of the quality of the plectra less pronounced, due to the fact that a voltage drop is present.

Noting that the values of the maximum deflections are all well below 0.4 mm, while the displacement suggested by the supplier is 0.51 mm, there is much room to improve performance of the harvester. Other potential aspects include the use of high quality and multiple piezoelectric bimorphs, which is achievable when the PZT components made in laboratories are replaced by large-scale manufactured products.

A field test has been carried out to study the performance of the above mentioned knee-joint piezoelectric energy harvester [93]. Gait cycle data are collected with a marker-based motion capture system and are processed to control a custom-made knee simulator designed to test a piezoelectric energy harvester based on the plucking technique of frequency up-conversion. The performance of the harvester has been assessed by using the gait data collected with the human subject carrying three levels of backpack loads. The harvester can produce an average power of 2.06 $\pm$ 0.3 mW, during the normal walking of human being, which is sufficient for useful potential applications.

The harvester has been introduced earlier, which has four bimorphs mounted on the inner hub and 74 plectra mounted on the outer ring. The four PZT bimorphs are referred to as B1, B2, B3 and B4. The prototype harvester has a volume of 226 cm^3 and weight of 235 g. Kinematic data derived from marker-based motion capture systems is an accurate method for the characterization of human biomechanics. A healthy male subject is employed for the test, at a speed self-determined by the subject. A motion-analysis capturing system with six cameras (Qualisys Proreflex MCU240) is used, which captures at a frequency of 100 Hz.

Reflective markers are attached to lower limbs of the subject at the anterior superior iliac spine, posterior superior iliac spine, iliac crest, greater trochanter, fibula head, tibial tubercle, medial condyles, lateral condyles, calcaneus, lateral malleolus, and medial malleolus. The markers form the basis of anatomical reference frames and centers of rotations of the joints. Five rigid plates, each consisting of four noncollinear markers, are also secured on the antero-frontal aspect of the leg, thigh and around the pelvis, as shown in Fig. 2.44 [93]. The calibrated anatomical systems technique (CAST) is used to monitor and record the movement of these segments during the walking trials. The trial is repeated for five times, with three backpack load levels of 0, 12 and 24 kg. The backpack loads are evenly distributed to avoid bias and all of the markers are remained attached to minimize positional inconsistency in re-attachment.

The knee-joint angle displacement can be extracted from the main kinematics dataset. The joint kinematics is calculated by using an X–Y–Z Euler rotation sequence in which the center of the knee-joint is defined as the midpoint between the medial and lateral condyles markers. The angle between the thigh and shank in the sagittal plane is used in the knee-joint simulator, where a naturally standing extension is calibrated to 0° and all higher angular displacements represent flexion. A further extension is possible beyond the natural standing position, but giving negative angles. The angular displacement during the time interval of 0.01 s is averaged over all trials, offering a mean angular displacement sampled at 100 Hz.

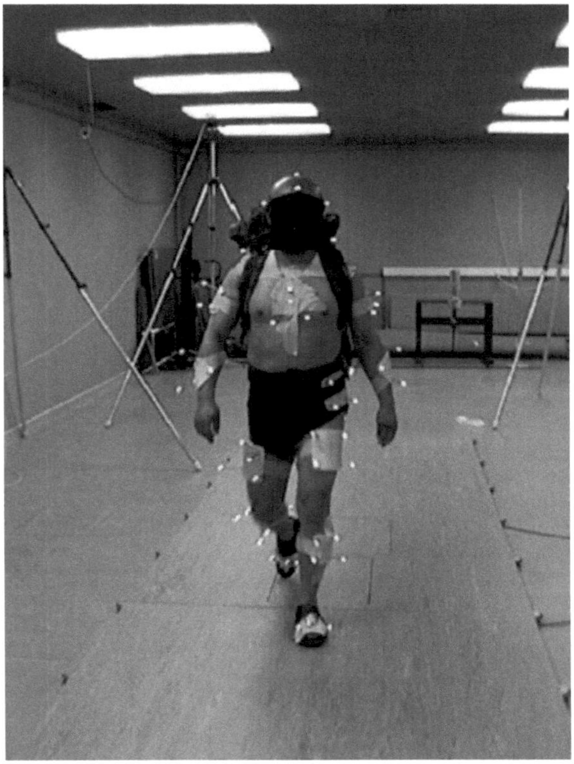

Fig. 2.44 The marker configuration and motion analysis system used for the calibrated anatomical systems technique (CAST). Courtesy of Dr. M. Pozzi, Newcastle University. Reproduced with permission from [93]. Copyright @ 2012, IoP Publishing

Figure 2.45 shows biomechanical data collected from the human subject carrying different levels of backpack loads [93]. It is found that the load indeed has influences on the gait pattern. There are two peaks in all three curves. The first peak, having an angle close to 20°, is attributed to the flexion of the knee-joint immediately following the heel-strike, when the leg is loaded with the weight of the body. The second one, slightly higher than 50°, is ascribed to lift-off, when the leg has left the ground and is carried forward before the next heel-strike. The gait patterns are almost the same for the 0 and 12 kg loads, while the highest load of 24 kg leads to a different gait pattern, with a delay of about 0.1 s, corresponding to a decrease in average step frequency from 0.95 to 0.88 Hz.

Figure 2.46 shows time-domain results of one of the four bimorphs (B4) tested with the kinematic data for the backpack load of 0 kg. In the first 0.5 s of motion, due to the low speed, the frequency of the plucking actions experienced by the bimorph is low, so that the plucking process can be clearly identified. The resonance vibration of the bimorph cannot be distinguished due to the relative high frequency of 300 Hz. The wide peak centered at about 0.4 s is due to the long time plectrum-bimorph contact. It means that plucking is not clear. Two groups of strong and dense peaks are observed in the second half of the gait cycle, with large angles experienced by the knee in short time duration. The appearance of the close peaks results in a continuous bimorph vibration [93].

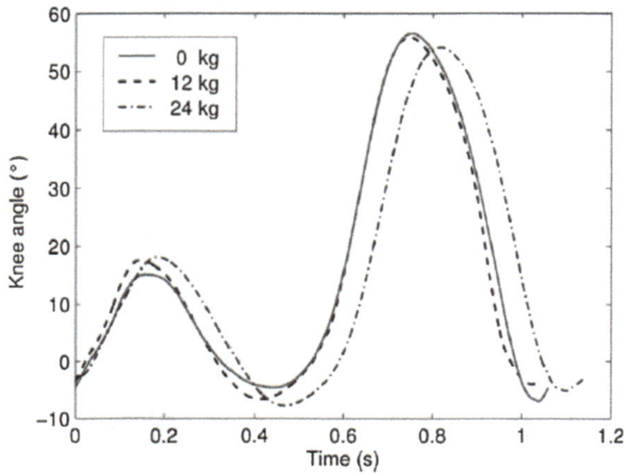

Fig. 2.45 Mean angular displacement covered by the knee-joint of the subject when carrying a selection of backpack loads. Courtesy of Dr. M. Pozzi, Newcastle University. Reproduced with permission from [93]. Copyright @ 2012, IoP Publishing

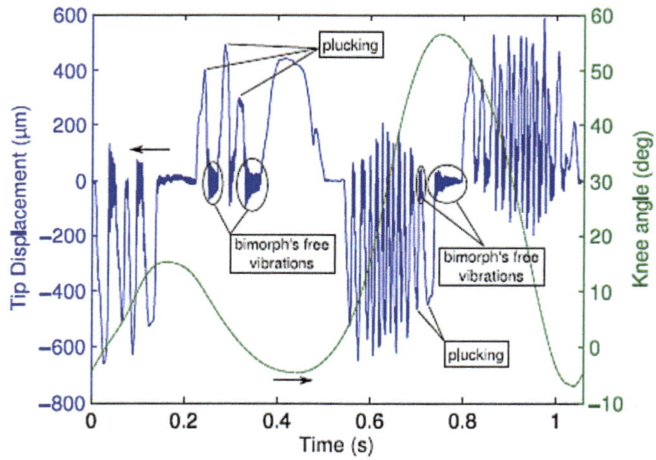

Fig. 2.46 Displacement of the tip of bimorph B4 during the gait cycle with 0 kg load. The angle covered by the knee-joint is also plotted (*dashed line*, *right* ordinate axis). Courtesy of Dr. M. Pozzi, Newcastle University. Reproduced with permission from [93]. Copyright @ 2012, IoP Publishing

The corresponding output power and voltage signals are shown in Fig. 2.47 [93]. In the first half of the gait cycle, the signal peaks are relatively low with large spaces, whereas two groups of dense peaks with much higher magnitude are observed in the second half. There are only two very low voltage peaks, with one in the beginning and the other at the end, for the wide displacement peak at about 0.4 s. This wide peak almost does not produce power, due to the slow deflection of the piezoelectric beams. Therefore, a clear plucking can produce higher power

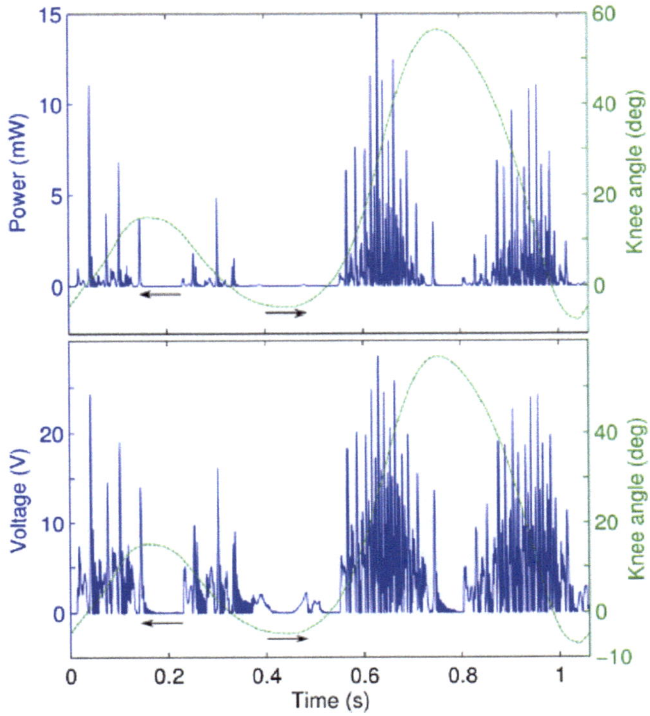

Fig. 2.47 Voltage and instantaneous power detected across a 52.9 kΩ equivalent resistor connected to bimorph B4 as a function of time during the gait cycle with a backpack load of 0 kg. The angle covered by the knee-joint is also plotted (*dashed line*, *right* ordinate axes). Courtesy of Dr. M. Pozzi, Newcastle University. Reproduced with permission from [93]. Copyright @ 2012, IoP Publishing

than a quasi-static deflection of the bimorph, although they may create the same amount of charges in open circuit. The maximum output power and voltage of the harvester are 20 V and 15 mW, respectively.

The energy versus time, *E*(t), during the gait cycle, calculated from the instantaneous power, *P*(t), by numerical integration via the trapezoidal method, $E(t) = \int_0^t P(t)\,\mathrm{d}t$, also indicates that the contribution of the first half gait cycle is only less than 20 %. In addition, the quality of the plectra also has a significant effect on the performance of the harvester. It is desired that all plectra have exactly the same dimension and same protrusion, so that they provide uniform and clear plucking to the bimorphs. This can be readily achieved in large-scale industrial production.

The final energy produced by the end of the gait cycle by the bimorphs has been evaluated [93]. In one round of test, three successive runs are performed for each of the three gait cycles, i.e., with three loads of 0, 12, and 24 kg, resulting in nine groups of measurements. In the next round, by reversing the direction of the motion, it is likely that the device is mounted on the other leg or connections to inner hub-shank and outer ring-thigh are reversed, corresponding another nine set of measurements. The four bimorphs have different energy harvesting capabilities. The energy produced by B3 is about 30 % higher than that by B2 in the first nine runs, while the difference in the amount of energy by B1 and B4 is less than 3 %. After the direction is reverse, bimorph B4 produces 33 % more energy than B3, while B1 and B2 produce the same energy (within 3 %). It means that the amount of energy generated by a bimorph can be different in different directions, implying that the exact shape of the plectra and the angle at which they are mounted have more pronounced effect than the bimorph. However, these effects have been balanced among the four bimorphs, so that the total average energy productions are almost the same in the first and second nine runs of experiments. In other words, the energy generated by the harvester presented here is not affected by the way that the device is mounted.

It is noticed that the gait cycle with the load level of 24 kg yields a slightly lower amount of energy than those of the cases with 0 and 12 kg [93]. According to the total angle, 166, 168, and 170°, for loads of 0, 12, and 24 kg, it would be expected that the gait cycle with the load 24 kg should facilitate more plucking actions and thus generate more energy. This means there is certain reason behind. It is found that the average bending speed of the knee-joint is affected by the level of the load. The average angle speeds are 161, 157, and 149° s^{-1} for 0, 12, and 24 kg, respectively. Therefore, it is further concluded that a high quality of plucking (quick and clear) ensures high energy generation [91, 92]. The average power produced by the bimorph is 2.06 ± 0.3 mW. Therefore, if more bimorphs are used, for example from the current 4 to 16–20, the output power can be over 30 mW.

2.4.3.2 Flexible Wearable Harvester

Different from the harvesters discussed above, in this subsection, a flexible energy harvester with a piezoelectric polymer PVDF in-shell structure that can generate high power from slow human motion [94]. Starting with description of working principle, fabrication and characterization of such flexible energy harvesters will be presented and discussed.

According the piezoelectric effects given by Eq. (2.12), if a linear relation between stress and strain is assumed, the total induced charge can be expressed as:

$$Q = \mathrm{d}YA\frac{\Delta l}{l_0}, \tag{2.32}$$

where Q is the total charge induced in the material, Y is Young's modulus, A is the surface area, and l_0 and Δl are the initial and differential lengths of the piezoelectric material, respectively. The effect of the electric field on the strain has been ignored. The output current and voltage over a resistive load can be obtained by differentiating both sides of Eq. (2.32) as:

$$I = \frac{V}{R_L} = \dot{Q} = \mathrm{d}YA\frac{l}{l_0}\dot{\Delta l}, \tag{2.33}$$

where I is the output current, V is the output voltage, and R_L is the load resistance. In addition, the energy generated by the materials in Eq. (2.32) can be expressed as:

$$W = \frac{Q^2}{2C} = \frac{tA}{2\varepsilon}d^2Y^2\left(\frac{\Delta l}{l_0}\right)^2, \tag{2.34}$$

where C is the capacitance of the piezoelectric layer and t is its thickness. Equations (2.33) and (2.34) indicate that a fast deformation of the piezoelectric material is required to have high output current and voltage, as can be found in the previous section. In addition, a large strain is necessary to generate high electrical energy and high output power, which is time derivative of the energy given in Eq. (2.34). Therefore, a pre-stressed shell structure is intentionally used to provide a large strain and fast transition, so as to achieve high output voltage and power.

The shell structure offers a fast transition from the initial state to the bending state, and vise-versa, when it is activated by a bending excitation [95, 96]. The proposed shell structure device consists of a PVDF layer attached to a curved polymer film, as shown in Fig. 2.48a as a concept [94]. In the initial state of the shell structure, no voltage is generated. When the shell structure is subject to a bending force at a threshold point, the film experiences a fast transition from its initial state to a folding state. As such, a tensile stress is applied to the PVDF layer, which thus produces an output voltage. When the structure is stretched, experiences a fast transition back to its initial state, which again leads to and output voltage, but in the opposite direction. With repeated folding and unfolding motion, the shell structure device converts mechanical energy into electrical energy continuously, although the motion might be irregular and slow. Figure 2.48b shows parameters of the shell structure in a cross-sectional view [94]. The curvature of the shell is defined by the following equation:

$$k = \frac{1}{r} \approx \frac{8h}{w^2 + 4h^2}, \tag{2.35}$$

where k is the laminate curvature, r is the radius of curvature, and w and h are the width and center height of the structure, respectively [97]. For a laminated piezoelectric film, as shown in Fig. 2.48c, the strain vector $\{S\}$ can be expressed as:

$$\{S\} = \{S_0\} + z\{k\}, \tag{2.36}$$

where $\{S_0\}$ is the mid-plane strain, $\{k\}$ is the mid-plane curvature and z is the distance variable in the direction perpendicular to the film surface. If the piezoelectric properties are isotropic in the given plane, the electrical displacement of the piezoelectric shell structure can be determined from Eq. (2.12) as:

$$D_{\text{shell}} = d_{31}\{110\}\{S\}. \tag{2.37}$$

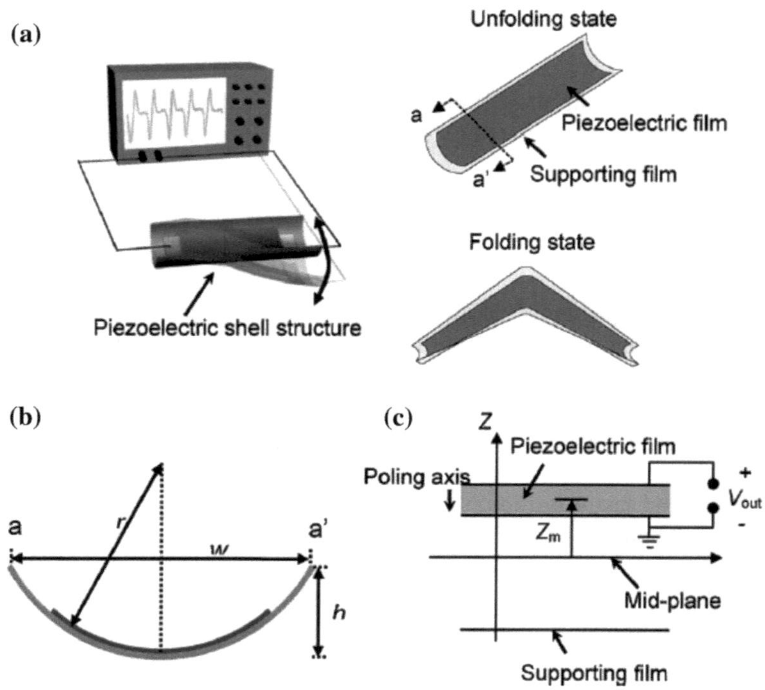

Fig. 2.48 **a** Schematic of the proposed shell structure device depicting the initial unfolding state and folding state. **b** Cross-sectional view of the piezoelectric shell structure. **c** A magnified partial cross section. Reproduced with permission from [94]. Copyright @ 2012, Elsevier

To find the charge induced by the shape transition of the piezoelectric shell structure, it is necessary to have its electrical displacements before and after the transition (Fig. 2.48a) [94]. However, the electrical displacement after the transition is negligible because the curvature in the transverse direction becomes almost flat in the folding state and the curvature in the longitudinal direction is very small as compared to the initial shell curvature. Therefore, the induced charge can be obtained by integrating the electrical displacement given by Eq. (2.37) over the surface deformation during the state transition:

$$Q = \int_{\Omega} D_{\text{shell}} \mathrm{d}\Omega = d_{31} \left(\int_{\Omega} S_0 \mathrm{d}\Omega + z_m \int_{\Omega} \kappa \mathrm{d}\Omega \right), \tag{2.38}$$

where z_m is the distance from the mid-plane of the shell film to the middle point of the piezoelectric layer. The first integral represents the contribution of the average membrane strains over the piezoelectric layer, while the second term stands for the charge generated by the bending of the piezoelectric layer. The above equation shows that the induced charge and consequently the generated energy are proportional to the initial curvature of the shell structure.

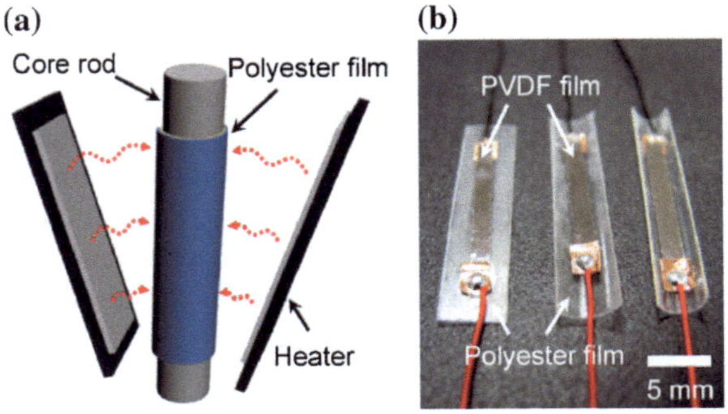

Fig. 2.49 **a** Schematic view of the fabrication method. **b** Photograph of the fabricated prototype devices. Reproduced with permission from [94]. Copyright @ 2012, Elsevier

The prototype shell structure device consists of a polyester film and a PVDF layer. A polyester film has a thickness of 127 μm and a dimension of 30 × 6 mm^2. Further, a thick commercial The PVDF film has a thickness of 110 μm, with metal electrodes on both sides over an area of 20 × 2 mm^2. Figure 2.49a shows a schematic view of the fabrication process [94]. The polyester film is rolled around a metal rod and fixed to the rod with a heatproof tape. It is then heat-treated at 150 °C for 3 min. The curvature of the polyester film is determined by the radius of the core rod. Metal rods with four different radii are used to produce shell structures with four curvatures ($\kappa = 200$, 300, 400, and 500 m^{-1}). The strains of the piezoelectric layers calculated with Eq. (2.36) are 2.36, 3.54, 4.72, and 5.90 %, respectively. After that, the PVDF films are attached to the concave surface of the polyester film. In this step, the PVDF film is positioned in such a way that positive voltage is generated on the upper surface when compressive stress is applied, as shown in Fig. 2.48c. Figure 2.49b shows a photograph of the prototype devices with curvatures of 0, 200, and 500 m^{-1}.

For the folding and unfolding experiment, a bending force is applied to the shell structure after it is mounted on a linear actuation system. The actuation system consists of a linear motion stage and step motor, with the moving distance, velocity and acceleration being accurately controlled. During the folding and unfolding process, the open-circuit voltages of the shell structures with different curvatures are measured. In addition, the effect of angular velocities or frequencies of the folding and unfolding operation has been tested, at a maximum bicep curl rate of 2 curls s^{-1} or an angular velocity of around 10 rad s^{-1}.

Experimental results have demonstrated that the shell structure generated much higher output voltage than the flat structure under the same conditions. This is because the shell structure experiences a faster transition between the two states. For the shell structure with a curvature of 500 m^{-1}, with a folding angle of 80° and folding/unfolding frequency of 2 Hz, the output voltages generated is negative during the folding and is positive during the unfolding. This is because the folding and unfolding correspond to tensile and compressive strain in the PVDF film, respectively.

The peak output voltage of the flat structure is always not over 10 V, which has only a very slight increase with angular velocity and the magnitude of displacement. In contrast, the output voltage of all the shell structures was almost linearly proportional to the angular velocity. The maximum output voltage is over 65 V at a folding angle of 80° and a folding and unfolding frequency of 3.3 Hz, when the shell curvature is larger than 300 m^{-1}. In addition, at low folding angles (40° and 60°), the output voltage is slightly dependent on the shell curvature. At high folding angles (80°), there is a critical curvature, 300 m^{-1}, output voltage has no change with increasing curvature.

The relationship between resistive load and output power has been studied to determine the optimal load for maximum output [94]. All measurements are conducted with a folding angle of 80° at a folding and unfolding frequency of 3.3 Hz, corresponding to an angular velocity of 9.3 rad s^{-1}. The optimal load resistance is 90 kΩ for the shell structure with a curvature of 500 m^{-1}. The maximum output power for the single shell structure is 0.87 mW, corresponding to a power density of 2.18 mW cm^{-2}, which is much larger than that of the flat structure (26 μW).

Durability of the shell structure has also been evaluated by measuring the output voltage with repeated folding and unfolding operations. The output voltage gradually decreases to 50 % of its initial value after 50 repeated operations. This degradation has been attributed to the reduction in curvature of the structure and peeling of the metal electrodes from the PVDF film. It is expected that these mechanical problems can be readily addressed by using more sophisticated fabrication technologies. For example, by using reactive metals such as chromium and titanium as an adhesion layer between PVDF and the metal electrode, the adhesion will be improved. High durable substrate materials with better mechanical strength can be used to replace the polyester.

An update version of the shell structure harvester, fabric device, has been developed, as shown in Fig. 2.50. In the fabric device, the piezoelectric shell structures are embedded in textiles. The fabrics are stitched and fitted to human body. The shell structures are inserted into the paths formed by stitching the fabric and the structures are connected in parallel. The fabrics are made in the form of a band for wearing either on elbow joints or on fingers. Three types of shell curvatures, i.e., 0 (flat), 200, and 500 m^{-1} have been tested. The numbers of the PVDF structures embedded in the fabrics are ten and four for the elbow joint and the finger, respectively.

Energy harvesting performances of the fabric devices have been measured. Figure 2.51 shows peak output voltages, as a function of frequency, of the fabrics with different curvatures, worn on elbow joint and finger. The bending angles are fixed at 130° and 70° for the elbow joint and finger motion, respectively. In all cases, the output voltage is proportional to the shell curvature and the bending speed.

The maximum output power generated by using the fabric harvester, made with the shell structures with a curvature of 500 m^{-1}, worn on the elbow, is 0.21 mW, corresponding to a power density of 0.05 mW cm^{-2}, for a resistive load of 90 kΩ

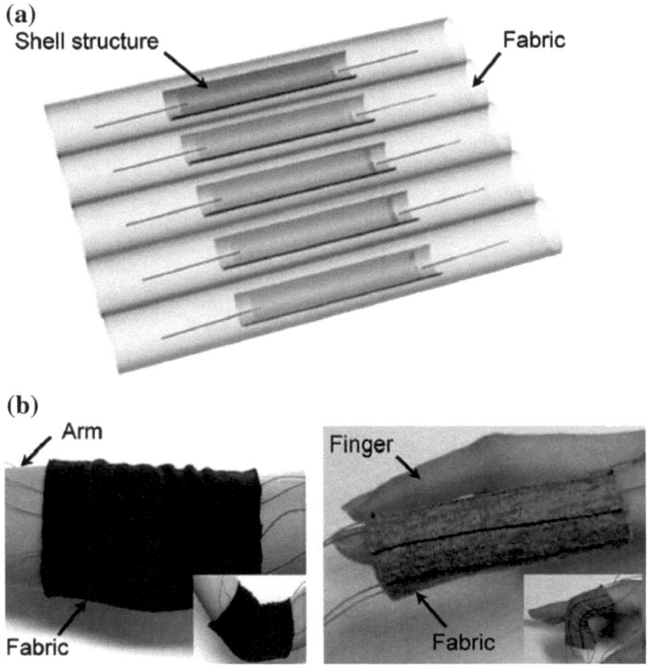

Fig. 2.50 **a** Conceptual view of fabricated fabric embedded with piezoelectric shell structures. **b** Photograph of fabrics worn on the elbow joint and the finger. Reproduced with permission from [94]. Copyright @ 2012, Elsevier

at a bending speed of 5 rad s^{-1}. The output powder is about 40 times that (~5.4 μW) offered by the fabric with flat piezoelectric structures. In summary, this device has obvious advantages, such as simple fabrication process, low cost and lightweight. It still has potential to further improve its performance.

2.4.4 Rotating Harvesters

2.4.4.1 Structure and Principle

A compact passively self-tuning piezoelectric energy harvester for rotation applications has been developed, which can be related to household activities [98]. The harvester rotates in a vertical plane, consisting of two beams: a relatively rigid piezoelectric generating beam and a narrow flexible driving beam with a tip mass mounted at the end. The mass impacts the generating beam repeatedly due to the influence of gravity. Centrifugal force from the rotation modifies the resonant frequency of the flexible driving beam and the frequency response of the harvester

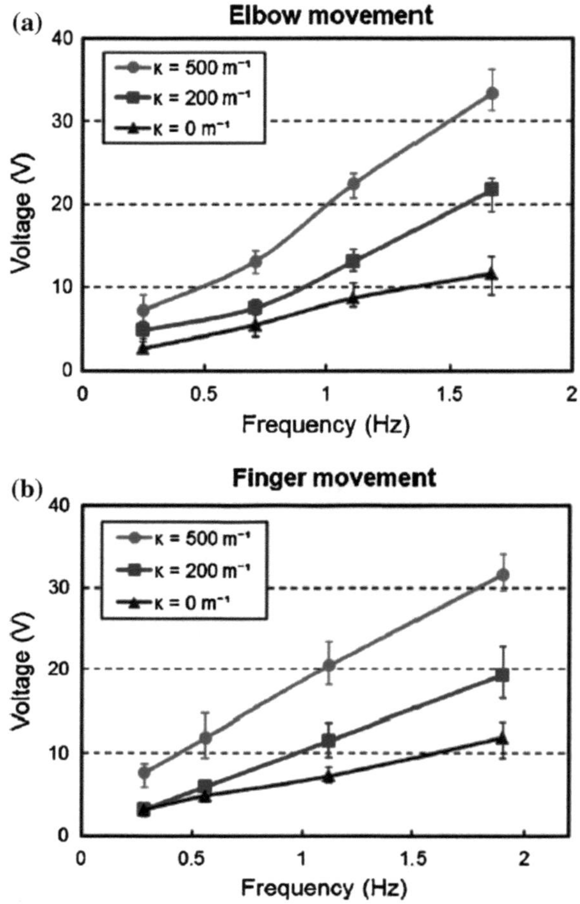

Fig. 2.51 Average output voltage generated when the fabrics are worn on **a** the elbow joint and **b** the finger. Reproduced with permission from [94]. Copyright @ 2012, Elsevier

beam. With an optimized design, resonant frequency of the harvester substantially matches the frequency of the rotation over a wide frequency range from 4 to 16.2 Hz. A prototype of the passive self-tuning energy harvester with a PZT generating beam exhibits a power density of 30.8 μW cm^{-3} and a bandwidth of more than 11 Hz. Passive tuning has also been demonstrated with PVDF as the generating beam.

Figure 2.52 shows schematic diagrams of structure and working principle of the rotatory piezoelectric harvester [98]. The harvester consists of a rigid piezoelectric generating beam located adjacent to a much more flexible driving beam with a steel ball at its tip. The walls of the harvester package can limit the motion of the ball to one direction, as shown in Fig. 2.52a. This ensures that motions have sufficiently large amplitude to apply an impact by the ball to the generating beam, so as to produce electrical output. The resonant frequency of the flexible driving beam is affected by the rotational frequency, thus affecting the natural frequency of harvester. The resonant frequency of the generating beam is almost not affected by

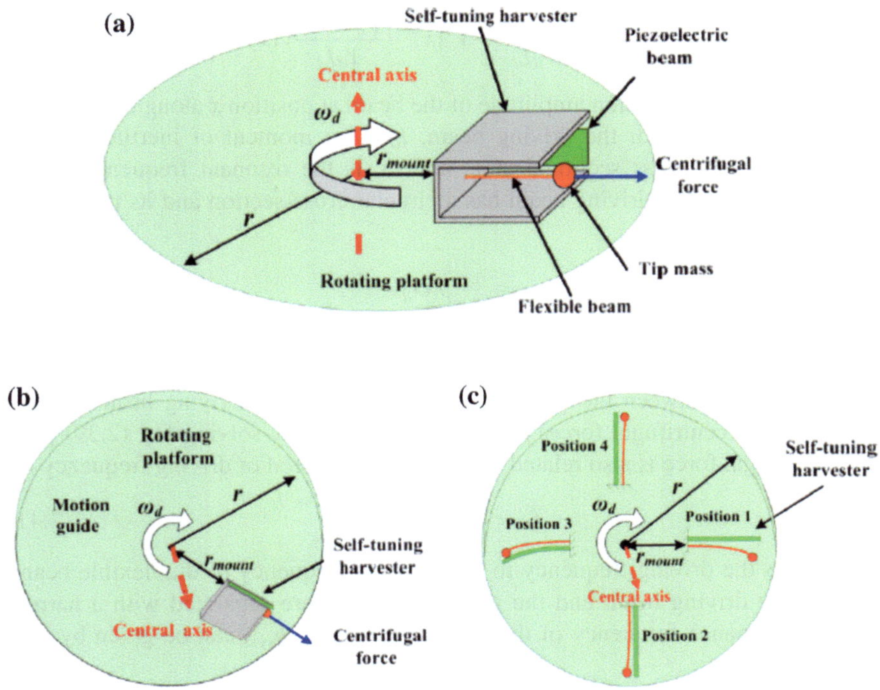

Fig. 2.52 **a** Schematic view of the impact harvester with frequency self-tuning. Although in reality the system rotates in the vertical plane, here it is drawn in the horizontal plane for clarity. **b** Schematic diagram showing the harvester mounted on a vertically rotating wheel. **c** Schematic diagram of the driving and generating beam positions as the wheel rotates through 360°. Beam positions are shown at 90° intervals. Reproduced with permission from [98]. Copyright @ 2012, IoP Publishing

the rotational motion. In operation, the harvester is mounted on a vertically oriented rotating platform such as a wheel or a fan (Fig. 2.52b and c).

When the flexible driving beam is mounted on a frictionless pivot at one end, the system can be described similar by using the conventional equations of a pendulum. However, the flexible driving beam in this harvester is mounted as a conventional cantilever beam, so that the displacement of the tip mass should be described by beam bending. According to classical beam theory, the stiffness of a cantilever beam can be tuned by applying an axial force [98]. Due to the tensioning of the frequency-dependent beam resonant frequency tracks the driving frequency in the rotational harvester. This approach has been used to ensure that the impact-driven harvester is more compact and can work at resonance over a wide frequency range.

For small deflections, the equation of motion for a cantilever beam with an applied centrifugal force f_c is given as by:

$$v''''(x) + \frac{f_c + \rho_d I_d \omega_r^2}{Y_d I_d} v''(x) - \frac{\rho_d A_d \omega_r^2}{Y_d I_d} v(x) = 0, \tag{2.39}$$

where $v(x)$ is the oscillation amplitude of the beam at position x along the beam, Y_d is Young's modulus of the driving beam, I_d is its moment of inertia, ρ_d is its density, A_d is its cross sectional area, and ω_r is the resonant frequency of the cantilever beam. The driving beam has a circular cross-section and its moment of inertia is given by:

$$I_d = \frac{\pi D_d^4}{64}, \tag{2.40}$$

where D_d is the diameter of the driving beam.

The relation between the resonant frequency ω_r of the driving beam and the applied force (centrifugal force) f_c can be established by solving Eq. (2.39) [98]. The centrifugal force is also related to the rotational speed or driving frequency as:

$$f_c = m(r + l_d)\omega_d^2, \tag{2.41}$$

which links the driving frequency to the resonant frequency of the flexible beam. Because the driving beam and the generating beam are separated with a narrow gap, the resonance frequency of the harvester system, ω_{sys}, can be given by:

$$\omega_{\text{sys}} = \frac{2\omega_r \omega_{\text{coup}}}{\omega_r + \omega_{\text{coup}}}, \tag{2.42}$$

where ω_r is the resonance frequency of the driving beam when it vibrates independently. The ω_{coup} is used to describe the time scale of the coupled motion of the driving beam and the generating beam after the generating beam is excited by the tip mass of the driving beam before they are separated.

During the coupled vibration, the tip mass of the driving beam remains in contact with the generating beam, so that they vibrate together. The spring constant of the generating beam is much larger than that of the flexible driving beam. According to Eq. (2.4), the resonant frequency, ω_{coup}, is given by:

$$\omega_{\text{sys}} = \sqrt{\frac{k_g + k_d}{m}}, \tag{2.43}$$

where k_g and k_d are the spring constants of the generating and driving beams, respectively. The effective mass of the generating beam is neglected. When $\omega_{\text{coup}} \gg \omega_r$, the resonant frequency of the harvester system can be approximated to $\omega_{\text{sys}} = 2\omega_r$, according to Eq. (2.42). Therefore, the resonance frequency of the harvester system can be related to the rotational speed of the system, which is the fundamental physics of the self-tuning rotational energy harvester. When the harvester works at its resonant frequency, the maximum open circuit voltage (V_{oc}) is given by:

$$V_{oc} = -\frac{d_{31} t_p \sigma_s}{\varepsilon}, \tag{2.44}$$

where d_{31} is the piezoelectric coefficient of the materials of the generating beam, t_p is the thickness of the piezoelectric beam, ε is dielectric constant of the piezoelectric layer and σ_s is the average stress on surface of the generating beam, which is express as:

$$\sigma_s = \frac{3k_g X_0 L_p}{b_p t_p^2}, \tag{2.45}$$

where k_g and X_0 are the spring constant and the deformation of the generating beam during the impact. L_p and b_p are the length and width of the piezoelectric beam. The deformation of the generating beam can be obtained by energy conservation, where the kinetic energy before impact is equal to the potential energy stored in the driving and generating beams at the maximum displacement [99]. Similarly to Eq. (2.27), the average electrical power generated by the harvester and delivered to the external load at resonant frequency can be evaluated by:

$$P_{\text{avr.}} = \frac{R_L}{2(R_S + R_L)^2} V_{oc}^2 = \frac{R_L}{2(R_S + R_L)^2} \left(\frac{-3d_{31} k_g X_0 L_p}{\varepsilon b_p t_p} \right)^2, \tag{2.46}$$

where R_L and R_S are the load resistance and the source impedance, respectively. It means that the output power of the harvester is proportional to the square of the magnitude of deflection by the generating beam, which is similar to that observed in a conventional harvester, but in a relatively complicated way. In this rotatory harvester, the deflection of the generating beam is dependent on the gap between the driving and the generating beams and the driving frequency. The gap determines the effective stiffness and therefore the amplitude of motion of the centrifugally tuned driving beam.

2.4.4.2 Parameter Studies

The design parameters have been optimized so that the harvester always works at its resonant frequency, while driving frequency range should be as wide as possible. The overall self-tuning performance of the harvester is determined by the driving beam. Frequency matching has been tested over a frequency range, with the highest drive frequency up to 20 Hz. The driving beam is made of nylon due to its low Young's modulus.

The first parameter is the diameter of the driving beam [98]. The resonant frequency of the harvester system, as a function of rotational frequency, for the flexible beam with various thicknesses (35, 50, 75, 100, and 150 μm), has been calculated. For thin beams (35 and 50 μm), the thickness has no effect on resonant frequency at given rotational frequencies. For thick beams (≥75 μm), the resonant frequency slightly increases with increasing thickness, because larger-diameter beams have larger zero-tension bending stiffness.

For each beam diameter, there is a specific rotational frequency that perfectly matches the resonant frequency of the harvester system. This frequency increases from 5 Hz for $D_d = 35$ μm to 19 Hz for $D_d = 100$ μm. Such an intersection means that the stiffness of the driving beams consists of two parts: a fixed zero-stiffness term and a frequency-tuned centrifugal tensioning term. This can be used as a guide to design a harvester with the perfectly-matched frequency to be within the frequency range of interest for specific applications. In this respect, it is easier to ensure a perfect matching between the resonant frequency of the harvester and the driving frequency, when using a thinner beam than a thicker beam, because thinner beam has smaller deviation.

The second parameter of interest is the proof mass [98]. It has been shown that the relationship between the resonant and driving frequency is relatively less sensitive to the proof mass. This insensitivity has been attributed to the fact that an increase in mass has two opposite effects. On the one hand, an increase in proof mass leads to an increase in centrifugal force, thus resulting in an increase in the resonant frequency of both the driving beam and the piezoelectric beam. On the other hand, it reduces the contribution of zero-tension to the resonant frequency of the driving beam and the harvester beam. Therefore, a larger proof mass is usually used to increase the displacement of the generating beam and thus the output power of the harvester.

The resonant frequency of the generating beam is also affected by the length of the driving beam (l_d) and the mounting radius (r_{mount}) [98]. The resonant frequency of the harvester, as a function of the rotational frequency, has been simulated, with the l_d varying from 15 to 25 mm, at a given r_{mount} of 5 mm. It is found that the matching between the rotational and resonant frequencies is quite sensitive to l_d, as compared to the driving beam thickness and the proof mass. This implies that the length driving of the driving has its effect through both the zero-tension stiffness and the centrifugal force. For a given harvester, the only parameter that can be changed is the r_{mount}. The simulated resonant frequency of the harvester, as a function of the rotational frequency, indicates that the quality of the frequency matching is very sensitive to the mounted radius r_{mount}. However, the frequency mismatch related to the length of the driving beam can be compensated by adjusting the mounting radius [98].

The last consideration is the optimization of the whole harvester system. With given optimal parameters, the predicted resonant frequencies of the harvester system and driving frequencies match well over a wide frequency range from 6 to 20 Hz with a maximum mismatching of only 0.5 Hz. The resonant frequency of the harvester system is almost the same as the driving frequency at the intersection frequency of 15 Hz, which has been actually treated as the resonant frequency of the self-tuning harvester. The predicted resonant frequencies of the harvester system also match quite well over a wide range of vehicle speed (up to 70 miles per hour (mph)).

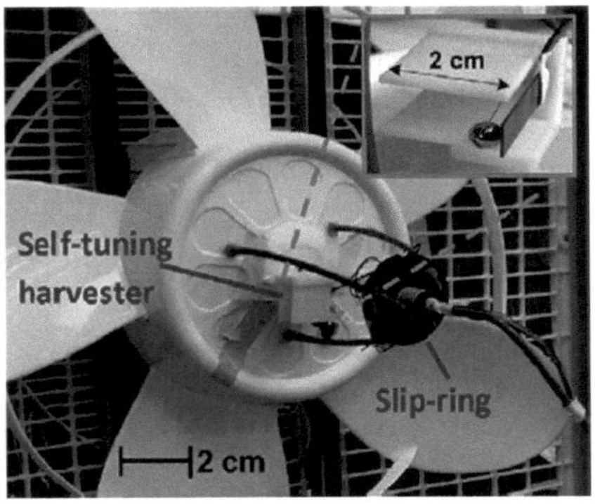

Fig. 2.53 Photograph of test bed with a harvester prototype mounted on an electrical fan. Reproduced with permission from [98]. Copyright @ 2012, IoP Publishing

Besides the advantage of the self-tuning capability, this harvester is able to harvest energy from low frequency sources, due to the frequency up-conversion. The impact vibration process involves two stages. The first stage begins when the driving beam's tip mass impacts the generating beam and thereafter they move together. If the mass of the generating beam is negligible as compared with the tip mass of the driving beam, the process is very similar to an inelastic impact. The second stage starts from the separation of the tip mass and the generating beam. During this stage, the generating beam vibrates alone with an exponentially decaying amplitude at its self-resonant frequency, while the driving beam is driven at the ambient driving frequency. The two stages are repeated at a frequency determined by the periodic rotational motion.

2.4.4.3 Characterization and Performance

Two types of materials, PZT and PVDF, have been used as generating beams of the rotational harvester [98]. Figure 2.53 shows photograph of a prototype of the harvester, which consists of a 0.4 g steel ball as tip mass mounted on a driving beam that is a 2 cm long and 75 μm-diameter nylon cable fixed to a frame holding a separate 25 mm long, 6.4 mm wide and 0.51 mm thick PZT generating beam. The gap between the steel ball and the generating beam is less than 1 mm, so that the amplitude of the ball's motion is larger than the gap. This is to ensure that the impact occurs at least at a driving frequency of up to 16.2 Hz, corresponding to a speed of 65 mph of a typical vehicle wheel. Two ABS plates are placed on the two sides of the driving beam to confine the motion of the steel ball within a path that impacts the generating beam. The harvester is attached to an electrical fan with the driving beam oriented parallel to the fan blade and along the radial direction. The root of the driving beam is offset from the center of the fan. The fan has been retrofitted to enable continuously varying speed and simulate the time-varying rotational motion of a wheel. The wires of the rotating harvester are connected to an optimal static load resistance through a slip-ring. A frequency range from 6.2 to 16.2 Hz is used to imitate the case of rotating tires at speeds ranging from 25 to 65 mph.

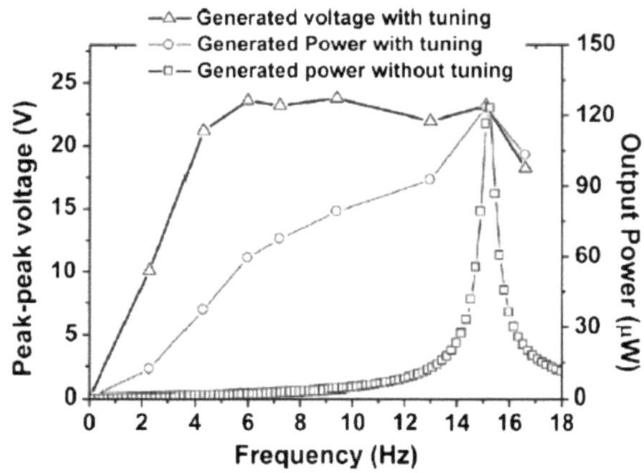

Fig. 2.54 Measured output power and generated voltage of the self-tuning energy harvester plotted versus frequency, along with output power of a comparable untuned harvester. Reproduced with permission from [98]. Copyright @ 2012, IoP Publishing

To evaluate the performance of the harvester, the optimal mounting radius r_{mount} is measured to be 7.5 mm, which is slightly larger than the theoretically predicted optimal radius, due to the effects of damping in the real system. It has been demonstrated that, at the optimal radius, the measured resonant frequency of the harvester matches the driving frequency very well, over a wide frequency range from 4 to 16.2 Hz. The maximum mismatching between the resonant and driving frequencies is less than 0.2 Hz.

Output voltage waveforms of the harvester have been measured with matched load at driving frequencies of 5.96, 9.44 and 15.2 Hz. The impact of the steel ball to the PZT beam has been characterized by the exponentially decayed signals. The observed peak–peak voltages are well within the range of 21.9–23.2 V, as the driving frequency is increased. The response of the harvester to the applied gravitational acceleration is constant over the frequency range. Although the stiffness of the driving beam increases with increasing rotational speed, which leads to decrease in amplitude, the potential energy stored in the driving beam remains constant at resonance. The potential energy transferred to the generating beam due to the impact can be considered to be equal to the maximum potential energy stored in the driving beam, which is responsible for the constant output voltage.

Figure 2.54 shows the output power and voltage of harvester with tuning and output power without tuning. The peak-to-peak voltage remains nearly constant from 4 to 15.2 Hz, indicating that the harvester works at resonance in this frequency range. The output power increases with impact frequency in this frequency range. The drop in the output power at high frequency can be attributed to the increase in the cable tensioning effects. The maximum output power is 123 μW at 15.2 Hz, corresponding to a maximum power density of 30.8 $\mu\text{W cm}^{-3}$. The self-tuning harvester shows a wider bandwidth 11 Hz as compared with that of the unturned harvester (0.8 Hz).

The optimal load resistance R_L is chosen to match the source resistance R_S, given by:

$$R_S = R_L = \frac{1}{\omega_g C_g} \tag{2.47}$$

where ω_g and C_g are the self-resonant frequency and the capacitance of the generating piezoelectric beam, respectively. Because the generating beam's bending stiffness is dominant over the effects of tensioning, its vibrational frequency remains essentially constant, so that the matched load resistance can be remained constant of 220 kΩ.

The power output of a harvester without tuning can be obtain analytically based on the measured peak output power when the harvester works at resonant frequency and the measured damping ratio. The damping ratio can be calculated from the exponentially decayed waveforms according to the experimental measurement of the resonant frequency of the harvester. The damping ratio ξ is given by:

$$\xi = \frac{1}{2\pi} \ln\left(\frac{p_1}{p_2}\right), \tag{2.48}$$

where p_1 and p_2 are consecutive peak amplitudes. The peak output power of the frequency-matched but harvester without can be considered to be the same as that of the tuned harvester at its resonant frequency. Thus, the semi-empirical plot can be evaluated by:

$$P_{\text{semi-impirical}} = \frac{4P_{\text{peak}}\xi^2(\omega/\omega_n)^2}{[1-(\omega/\omega_n)^2]^2 + (2\xi\omega/\omega_n)^2}, \tag{2.49}$$

where P_{peak} is the output power at resonant frequency, while ω and ω_n are the angular frequency and the resonant frequency of the untuned harvester, respectively.

There is always a potential problem of reliability when using PZT ceramics as active piezoelectric components in a harvester, due to its brittle characteristics. In comparison, PVDF is more flexible than PZT, therefore, the reliability problem is expected to be addressed by using the former to replace the latter. A 16 mm wide, 33 mm long, and 0.2 mm thick PVDF beam has been used as the generating beam in the above discussed self-tuned energy harvester. Since the PVDF beam is very flexible, the optimal mounting radius should be changed from 5 to 15 mm according to modeling results. Figure 2.55 shows measured output voltage and power of the harvester with the PVDF generating beam. With the matched load resistance of 586 kΩ, the PVDF-based harvester exhibits a maximum output power of 27 μW at 15 Hz with a working frequency bandwidth of more than 9.5 Hz. The low output power of the PVDF-based harvester is due to the lower piezoelectric coefficient of PVDF. It is expected that the output power can be further increased by using thicker PVDF beams.

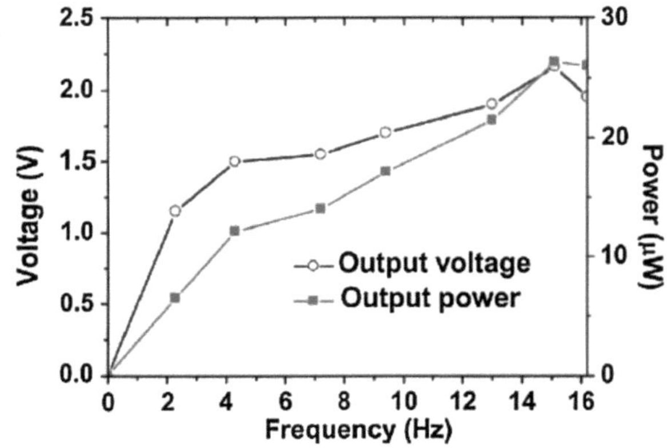

Fig. 2.55 Measured output power and generated voltage plotted versus frequency for a self-tuning energy harvester with a PVDF generating beam. Reproduced with permission from [98]. Copyright @ 2012, IoP Publishing

2.5 Energy from Civil Infrastructure and Transportation

2.5.1 Simulated Study of Bridge

This section is aimed to show the possibility of harvesting energy from civil infrastructure and transportation facilities by using piezoelectric harvesters. The examples are two case studies on bridges [100, 101]. In the first example, piezoelectric effects under various loading conditions on bridges are systematically studied [100]. A steel beam-slab type bridge specimen is fabricated and piezoelectric modules are attached at various positions, where different structural responses are expected under moving vehicles on the bridge. Traffic conditions, including vehicle weight and moving speed, and load conditions, such as load amplitude and loading frequency, have been tested. The performance of the harvester is evaluated by using the generated voltage. The test results are compared with the well-known analytical formulations of piezoelectric effects. It has been shown that the piezoelectric effects are sensitive to the rate of straining and peak strain of the piezoelectric materials, which are determined by the moving speed and weight of moving vehicle on the bridge. The second example is in situ study of a pre-stressed concrete highway bridge, with a piezoelectric cantilever harvester [101]. Although the bridge vibratins are characterized by small amplitude and low frequencies (<15 Hz), the mean output power is in the order of 0.03 mW, with voltages between 1.8 and 3.6 V. The experimental results have been simulated by using theoretical models.

As discussed earlier, when piezoelectric material is attached to a substrate, such as printed circuit board (PCB), the generated voltage of the piezoelectric module is determined by the stress and strain experienced by the piezoelectric material, which can be calculated with analytical formulations taking into account the stress and strain. There have been well-established analytical formulations to estimate the output voltages generated by the piezoelectric modules according to their dimensions, shape, and the attached modules. These analytical formulations

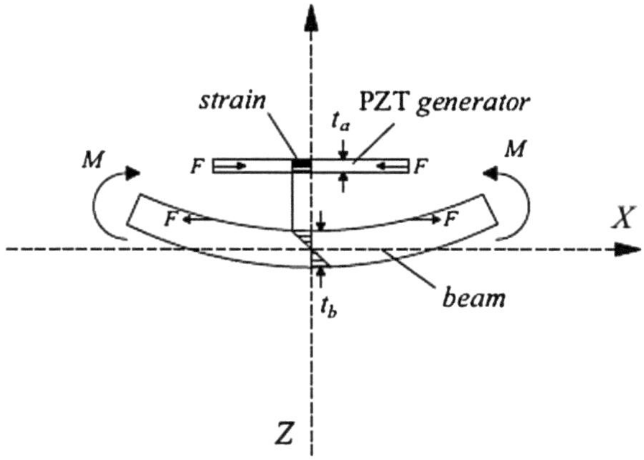

Fig. 2.56 Pin-force model of a unimorph-type piezoelectric module. Reproduced with permission from [100]. Copyright @ 2011, Elsevier

include the pin-force model, enhanced pin-force model, and Euler–Bernoulli model [100].

The pin-force model is able to estimate the output voltage of a piezoelectric harvester by considering the relationship of mechanical force between the piezoelectric material and the substrate. The piezoelectric material is attached to the substrate with a pin connection. Therefore, the shear force is intentionally designed to apply to the piezoelectric material at end of the pin connection. In this case, the shear stress exerted to the piezoelectric material is concentrated only within a small area at the end of the pin connection. The strain in the substrate is derived by Euler–Bernoulli beam theory, but the strain in the piezoelectric material is assumed to be constant. Because of this constant strain, the pin-force model does not consider the bending stiffness of the piezoelectric materials.

Figure 2.56 shows the pin-force model with a unimorph-type piezoelectric module, illustrating the strain distributions of the piezoelectric modules that consist of piezoelectric material and the substrate (PCB). When a pin-force is applied to the piezoelectric modules, an output voltage is produced, which is given by:

$$V = \frac{6g_{31}M}{bt_b(3 - Y_b t_b / Y_a t_a)}, \tag{2.50}$$

where g_{31} is piezoelectric constant, M is external moment, b is width and Y_b is Young's modulus of the piezoelectric material. Other parameters are shown in Fig. 2.56.

By considering the bending strength of the piezoelectric material, the pin-force model is modified to enhanced pin-force model. In the enhanced pin-force model, the stress and strain of piezoelectric materials are determined by the deformation shape of the substrate derived from the acting moments, as shown in Fig. 2.57. In this case, the output voltage can be estimated by:

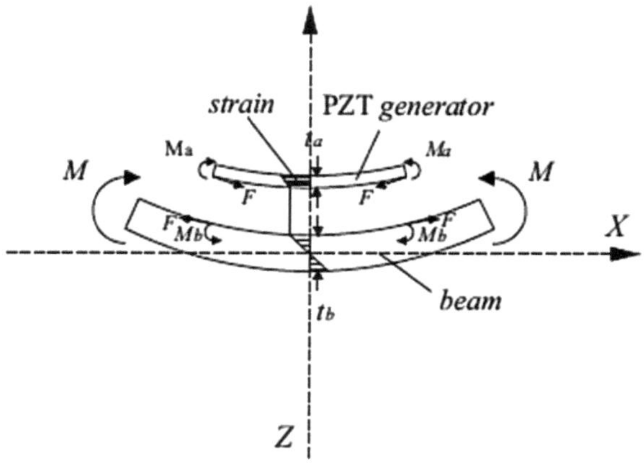

Fig. 2.57 Enhanced pin-force model of a unimorph-type piezoelectric module. Reproduced with permission from [100]. Copyright @ 2011, Elsevier

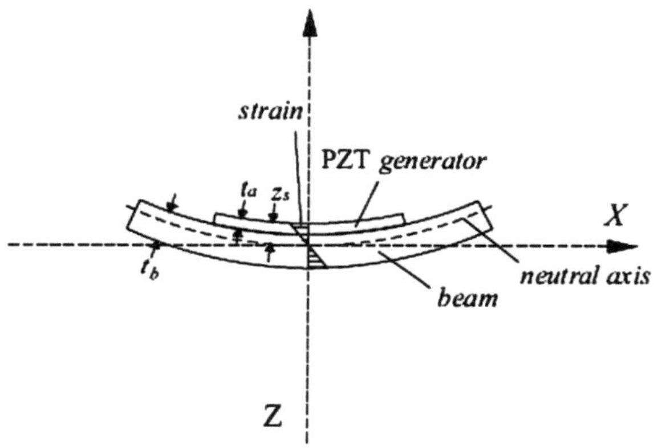

Fig. 2.58 Euler-Bernoulli model of the piezoelectric material and the substrate. Reproduced with permission from [100]. Copyright @ 2011, Elsevier

$$V = \frac{6g_{31}(t_b/t_a)M}{bt_a[3(t_b/t_a)^2 - 1 - (Y_b t_b/Y_a t_a)(t_b/t_a)^2]}, \tag{2.51}$$

Comparatively, the Euler–Bernoulli model is more accurate. This model assumes that the piezoelectric material and the substrate are deformed with the new neutral axis due to the perfectly composited piezoelectric material and the substrate, as shown in Fig. 2.58. The output voltage is given by:

$$V = \frac{6g_{31}M \cdot \frac{Y_b t_b}{Y_a t_a}\left(1 + \frac{t_b}{t_a}\right)}{bt_a\left[1 + \left(\frac{Y_b t_b}{Y_a t_a}\right)^2\left(1 + \frac{t_b}{t_a}\right)^2 + 2\frac{Y_b t_b}{Y_a t_a}\left(2 + 3\frac{t_b}{t_a} + 2\left(\frac{t_b}{t_a}\right)^2\right)\right]}. \tag{2.52}$$

The piezoelectric component is a unimorph, wording in 31 mode, with 8 mm width, 60 mm length, and 0.55 mm thickness. The unimorph is attached on a PCB with a width of 10 mm and a length of 97 mm, as shown in Fig. 2.59. The piezoelectric material is PZT ceramics prepared by using the traditional ceramic processing with a sintering temperature of 1,000 °C. After sintering, top and bottom sides of the bulk piezoelectric ceramics are coated with Ag electrode.

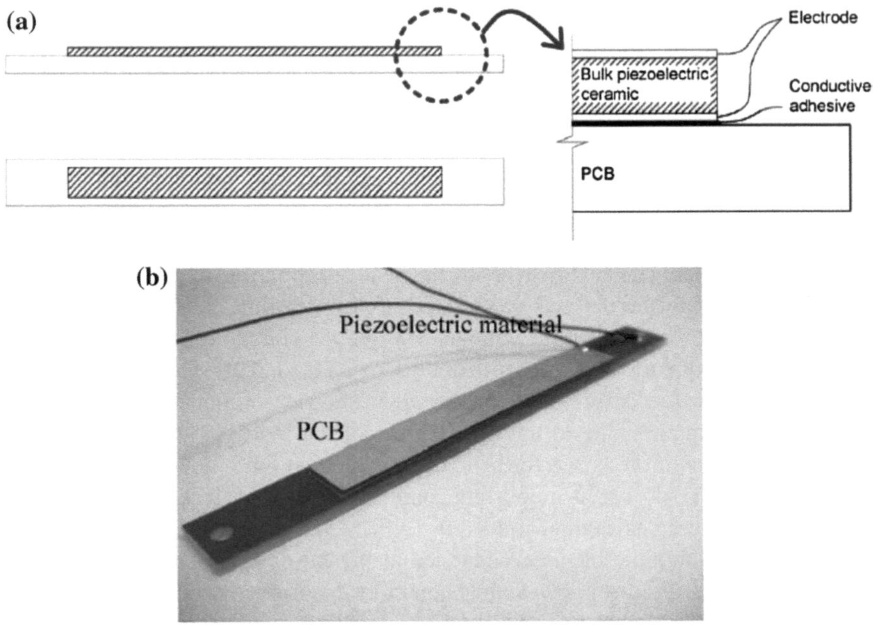

Fig. 2.59 PCB bonded piezoelectric material. **a** Detail of PCB bonded piezoelectric material. **b** Photograph of an actual PCB bonded PE material. Reproduced with permission from [100]. Copyright @ 2011, Elsevier

A steel beam-slab type bridge specimen with two steel plate girders are designed to examine the piezoelectric effect of the harvester under various loading conditions. The specimen is composed of two steel girders and a concrete slab, which is designed to have full shear connection for composite behaviors by installing shear connectors [100]. The steel girders have SS 400 grad hot rolled beams with a height of 294 mm and a width of 200 mm, while the concrete slab is 1,800 mm in width and 100 mm in thickness with a 30 MPa compressive strength. Natural frequency of the specimen is 21.0 Hz.

Ten piezoelectric modules are tested on the bridge specimen using an actuator of 250 kN capacity, with which cyclic loadings can be applied to simulate the loading conditions and effects on a real bridge. Strain gages and linear variable displacement transducers (LVDTs) are used to compare the generated piezoelectric effects with the behaviors of the structures. The output voltages of the piezoelectric modules are measured by using an oscilloscope, while the strains and displacements are measured by sing a data logger. Loading conditions include amplitude, frequency, and weight and speed of vehicles. A uniform loading test with constant velocity (PL-UV test) is conducted by changing the load amplitude from 10 to 100 kN, with constant loading frequency through the repeated process of loading and unloading. Load amplitude and the frequency variation test (PL-AP test) are carried out by varying the load amplitude and frequency. Loading frequency variation test (PL-AV test) is to change the frequency at two load amplitudes of 20 and 40 kN. There is also a loading effect test (PL-LE test) to check the piezoelectric effect caused by the loading effects of moving vehicles [101].

It is found that the output voltage of the piezoelectric modules has a peak-to-peak value of about 13.8 V under cyclic loading of 10 kN. The output voltage shows a distorted sine wave due to the distorted phenomenon of voltage, which must be calibrated in order to use the harvested energy as electric current. Theoretically, if the applied load to the piezoelectric material is constant, the voltage output should be constant. However, in the real deformation of the piezoelectric component could not be perfectly recovered after unloading, thus resulting in the distorted waveform of the output voltage. The output voltage is found to increase as the loading frequency is increased from 0.5 to 1 Hz, and after that, saturation is observed, because the deformation of the piezoelectric component is not further increase with frequency due to the overlapping of the loading deformation.

Output voltages of the harvester have been evaluated with uniform loading that ranges from 10 to 100 kN at a constant loading velocity of 10 kN s^{-1}. The uniform loading ranging is divided into 3.0–13.0, 3.0–23.0, …, and 3.0–113.0 kN. The piezoelectric module has been attached at the lower part of the lower flange and the lower part of the concrete slab at center of the bridge specimen, where both the tensile strain and compressive strain are at the maximum level. The output voltage of the piezoelectric module at the lower part of the lower flange is ±8.0 V, while that of the piezoelectric module at the lower part of the concrete slab is ±2.0 V. It is found that the output voltage is slightly increased as the loading level is increased, but the increment magnitude is not proportional to the level of the loading. In addition, the output voltage is not proportional to the magnitude of strain. It also does not have a clear relation to the strain range.

Because the output voltage does not correlate with simple variation of strain, it is of interest to test the effect of loading frequency. Therefore, cyclic loadings with a sine wave with frequencies from 1.0 to 3.0 Hz are applied on the bridge specimen. Their amplitudes are varied from 10 to 50 kN. The output voltage is up to ±30 V at the lower part of the lower flange in the center span, at the maximum tensile strain. The output voltage at the lower part of the concrete slab is only ±2.0 V. Generally, the output voltage increases almost linearly with increasing loading level. At a given loading level, the output voltage increases with increasing frequency. This is because an increase in loading frequency means an increase in straining rate.

It is expected that the loading condition on a bridge applied by moving vehicles continuously changes. Behaviors of a bridge, such as magnitude of displacement and rate of deformation, have close relation to vehicle speed, distance between moving vehicles and location of moving vehicles on the bridge, which also have influences on the magnitude of the output voltage.

The magnitude and wave shape of the output voltage of the piezoelectric module calculated by using the Euler–Bernoulli model are quite similar to the measurement results, while there are relatively large differences between experimental results and theoretical predictions when the other two models are used.

Fig. 2.60 The bridge for the in situ study. Reproduced with permission from [101]. Copyright @ 2013, IoP Publishing

2.5.2 In Situ Study of Bridge

The in situ study on energy harvesting from traffic-induced vibrations involves a prestressed concrete highway bridge located on a heavily trafficked itinerary in the north of France (Fig. 2.60) [101]. It has a 33 m long simply supported span carrying three one way lanes. The structure consists of five cross braced girders. The mean flow of heavy lorries is about 8,000 per day, with large variations according to the day of the week and time in the day.

Acceleration time series are measured at a sampled frequency of 300 Hz on girders 1 and 3 of the simply supported span at three locations labeled as 1–3 in Fig. 2.61. At a high traffic intensity, the root mean square value of the measured acceleration at mid-span on girder 1, which is more heavily loaded, is as low as 0:03 m s^{-2}. Significantly higher levels of vibration are measured on a water pipe fixed on girder 1, especially at the location labeled as 0 in Figs. 2.61 and 2.62. The root mean square value of the measured acceleration for a high traffic intensity is 0.3 m s^{-2}, which is ten times the estimated value on the girder. At low traffic intensity, the root mean square value of the acceleration decreased to 0.1 m s^{-2}. These estimates outline the expected range of acceleration to be considered for this ambient source of vibration.

The acceleration signal consists of successive short-time pulses, each pulse corresponding to an individual lorry crossing the bridge. This is especially obvious in nighttime, during which the traffic intensity is so low that the inter-arrival time exceed the time to cross the bridge. The calculated average duration of the pulse is 2 s, corresponding to the time for a five axle semi-trailer vehicle to cross the span of the bridge under free flowing traffic conditions. The average acceleration peak level is 0.58 m s^{-2}. At high traffic intensity, there is some overlap among the excitations of different vehicles, resulting in a signal that is more difficult to identify. Light vehicles produce a low-level signal (ambient noise) that cannot be distinguished.

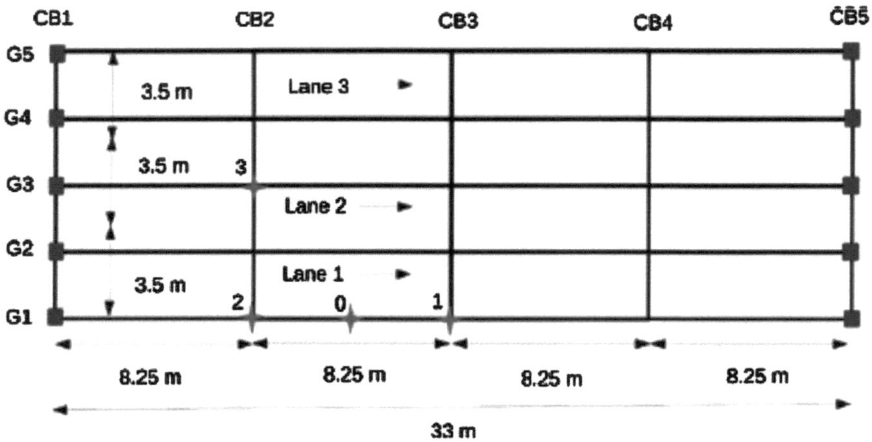

Fig. 2.61 Schematic of the bridge details. Reproduced with permission from [101]. Copyright @ 2013, IoP Publishing

Fig. 2.62 Locations of the sensors on the pipe under the bridge. Reproduced with permission from [101]. Copyright @ 2013, IoP Publishing

To determine the frequency characteristics of the vibration sources from the bridge, it is necessary to analyze the recorded signals. Smoothed power spectra of the acceleration responses to the traffic excitation are estimated by averaging 360 power spectra of 3,000 sample blocks. Three main resonant frequencies of the bridge deck are identified in the frequency range 0–50 Hz. Two close frequencies are in the vicinity of 4 Hz and another one is at 14.5 Hz. Two close frequencies are located at 3.9 and 4.3 Hz. These resonant frequencies correspond to the longitudinal bending, torsional and transverse bending deformations of the bridge deck. The uncertainty in the frequency estimation is about 0.1 Hz and is related to fluctuations in the traffic excitation conditions. The variation in the frequency of the first bending mode induced by the change in ambient temperature with 25 °C is not more than 0.1 Hz.

In addition, the frequency characteristics are dependent on the location on the bridge. Both the mode shape of the bridge and the fixation details of the pipe have

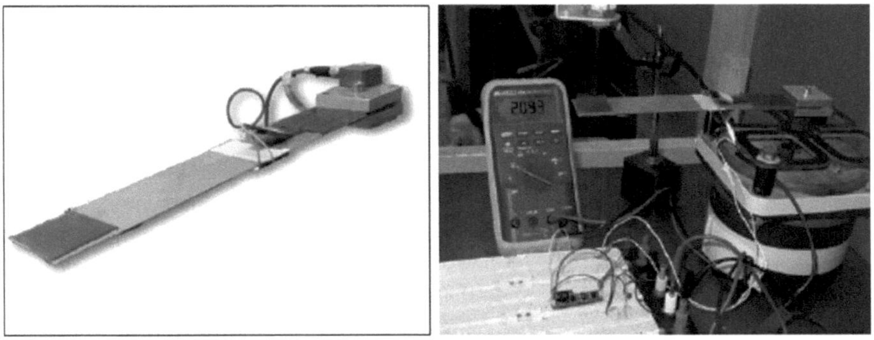

Fig. 2.63 The piezoelectric harvester designed for traffic-induced vibrations (*left*), and a view of the test set-up (*right*). Reproduced with permission from [101]. Copyright @ 2013, IoP Publishing

an influence on vibration behaviors of the pipe. The pipe usually vibrates at a resonant frequency of 14.5 Hz.

The piezoelectric energy harvesting device used in this study is a PZT cantilever, as shown in Fig. 2.63left. The resonant frequency of the harvester is tuned to be close to 14.5 Hz, thus matching the resonant vibration of the pipe discussed above. Two Mide QP20 W bimorph piezoelectric patches are bonded to the upper and lower surfaces of a 40 × 220 × 0.8 mm^3 steel plate, at the clamped end side. An additional 12 g concentrated mass is placed on the steel surface to tune the resonant frequency of the oscillator.

The experimental test setup is shown in Fig. 2.63right. An electromagnetic shaker is used to apply sinusoidal base excitations. To identify the parameters and get reference values of power conversion, tests are first carried out with a pure resistive load $R_L = 100\ k\Omega$ connected to the output of the piezoelectric electrodes wired in series. In these tests, the RMS amplitude of the harmonic excitation is set to 0.2 m s^{-2}. The frequency of the excitation is between 14 and 15 Hz. For each frequency, the RMS voltage (V_{RMS}) across the resistor is measured. According to the power-frequency curves, the device is found to have a resonant frequency close to 14.4 Hz, at which the maximum measured power is 0.19 mW. The harvester is subsequently tested by replacing the resistive load with an ALD EH300 circuit, and further characterized by using the excitation of the bridge. Under optimized conditions, the harvester produces a power of 0.029 mW. The half-power bandwidth is about 0.8 Hz for traffic-induced excitations, which is wider than that (0.2 Hz) for harmonic excitation. Field test results indicate that an average power of 0.02 mW can be achieved. The experimental results have been in a good agreement with the theoretical predictions [101].

2.6 Energy from Natural Sources

2.6.1 Wind Energy

Wind is a promising energy source that is already used on large scales, but is less developed at smaller scales, especially by using piezoelectric harvesters. Several approaches to harvest wind energy for low power systems have been proposed. In this subsection, a device for energy harvesting from wind flows by using flexible piezoelectric films will be presented. The piezoelectric films are sandwiched between two metallic electrodes to form a capacitor structure. The generated power from multiple wind flows by using various piezofilm geometries has been theoretically calculated and experimentally characterized [102].

2.6.1.1 Power Conversion

The available power (P_{wind}) of an incident wind over a cross section unit S is given by:

$$\frac{P_{\text{wind}}}{S} = \frac{1}{2}\rho_{\text{air}}v^3, \tag{2.53}$$

where ρ_{air} is the density of air (1.225 kg m^{-3} at room temperature), assuming laminar flow, for a given wind speed v. Power transfer from wind to mechanical wave depends on wind coupling and, in particular, the alignment. At normal incidence, wind impacts the piezoelectric films, while at parallel incidence, coupling occurs due to a turbulent flow caused by the Von Karman vortexes in a bluff body, which is similar to water flows [103]. Energy harvesting efficiency is dependent on electromechanical coupling, piezofilm geometry, and external stress.

2.6.1.2 Materials and Device

PVDF piezoelectric thin films are used due to the requirement of flexibility and slow air flows. The PVDF films are sandwiched between two metallic electrodes wired to external circuits. Two wind flows are employed: one is generated by using a hairdryer and the other is generated by using a wind tunnel. The wind tunnel is used provide a laminar flow. Wind speeds are measured by using a Pitot tube. The wind from the hair dryer is turbulent. In both cases, wind speeds are kept in the range of 2–18 m s^{-1}.

Three orientations of the piezofilm with respect to wind have been studied. As shown in Fig. 2.64, (a) is parallel incidence with two cylindrical 0.5 cm diameter bluff bodies, (b) is normal incidence of the wind to the surface film, which is attached at one boundary, and (c) is normal incidence of the wind to the film, which is fixed at both sides.

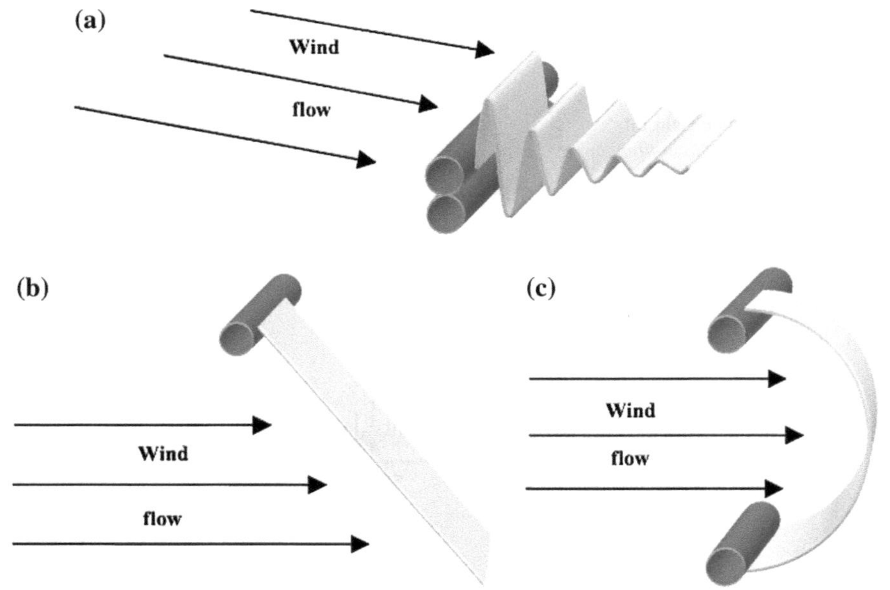

Fig. 2.64 Schematic illustration of wind incidence on the piezofilm with different oscillating behaviors. Wind incidence is parallel for setup (a) and normal for setups (b) and (c): **a** bluff body configuration, **b** one side fixed and **c** two sides fixed. Reproduced with permission from [102]. Copyright @ 2011, IoP Publishing

2.6.1.3 Characterization and Discussion

In setup (a), at low wind speed ($v < 2$ m s^{-1}), the piezofilm remains parallel to the wind flow. As the wind speed is increased to range of 2 m s^{-1}< v < 10 m s^{-1}, the film experiences small oscillations. Similarly, when the wind incidences perpendicularly (setup (b)), the film tends to align parallel to the flow but with small oscillations. When the film is fixed at both sides at normal wind incidence (setup (c)), oscillating waves are confined to the film. The oscillations of the three configurations seem to be similar one another. When the wind speed is further increased to $v < 15$ m s^{-1}, the oscillation amplitude of the piezofilm becomes larger, and stationary waves become chaotic movement. If the length of the film is increased, more vibrations will be observed, whereas films with smaller areas withstand stronger wind flows before becoming chaotic. Therefore, cantilever behavior is only observed at low speeds and for films with small sizes, otherwise, the films will experience traveling waves.

Theoretical relationship between wavelength and amplitude has been evaluated [102]. Wave amplitude u_0 is dependent on the configuration of the setup. For the parallel incidence (setup (a)), where stationary waves travel in a string with one end to be fixed, the piezoelectric film has a finite length, so that the relationship between the sinusoidal wave amplitude and its wavelength is given by:

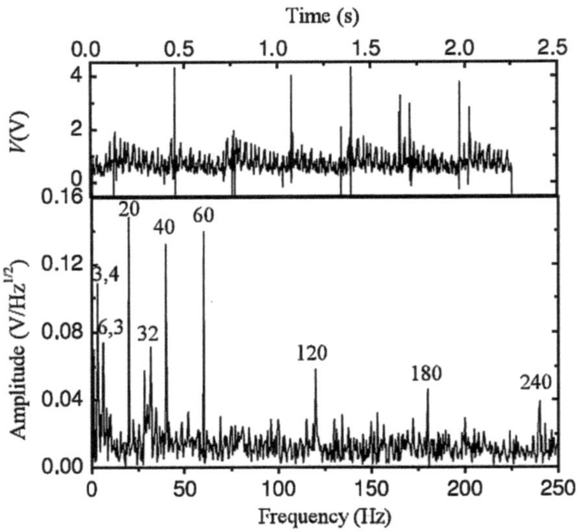

Fig. 2.65 Electrical response of a sample at 2.2 m s^{-1} in the wind tunnel excited with a bluff body. On the *top panel*, the signal is the open circuit output voltage. On the *bottom panel*, the signal is the frequency spectrum. Peaks are related to vibration modes at the indicated frequencies. Reproduced with permission from [102]. Copyright @ 2011, IoP Publishing

$$L = \int \sqrt{1 - \left(\frac{\mathrm{d}u}{\mathrm{d}x}\right)^2}\,\mathrm{d}x, \tag{2.54}$$

where, L is the total length of the film and du/dx is the derivative of the wave over the x-axis. This expression leads to elliptic integrals, which can be numerically solved for the films with specific lengths. The characteristics of the oscillations determine the stress applied to the piezofilms and thus charges produced by the harvester. This has been demonstrated by the frequency-dependent output voltage, as shown in Fig. 2.65, in which well-defined peaks are observed corresponding to for the various oscillating modes. The piezofilm of this device has a dimension of 1.9 cm × 15.6 cm × 40 μm.

Two parameters, speed and flow type, have been studied to characterize wind dependence energy harvesting efficiency. Experimental results are shown in Fig. 2.66. The amount of the stored energy in the capacitor increases with wind speed for a laminar flow measured in the wind tunnel. However, it does not follow the trend given by Eq. (2.53), which is mainly attributed to coupling between wind flow and the piezoelectric film, as can be seen later. The turbulent flow leads to a larger voltage than laminar flow. This is because the turbulent flow is rotational, so that it has additional contribution due to the Von Karman perturbation induced by the bluff bodies [27], which enhances the excitation effect.

It is also noticed that the energy harvesting efficiency increases with increasing thickness of the piezofilms. However, the increase in output voltage is not linear, because thickness affects mechanical damping, as discussed before. There is an optimal thickness that balances the volume of material with respect to the power increase due to the damping increase [104]. As for the effect of area of the piezofilms, only effective areas can be used for the evaluation.

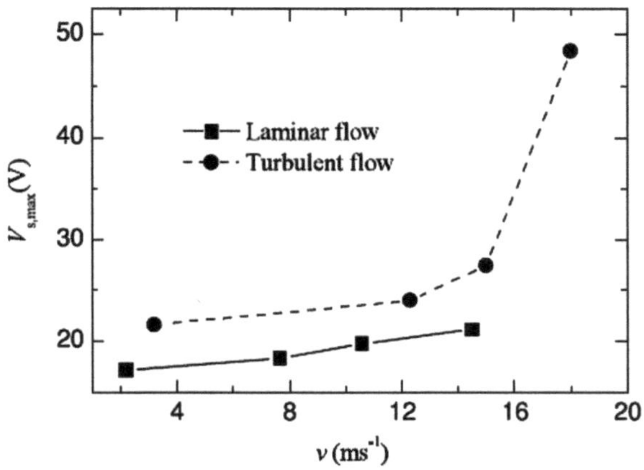

Fig. 2.66 Responses of the harvester in a bluff body configuration in the wind tunnel and by the fryer, for the piezoelectric film with a dimension of 1.2 cm × 3.0 cm × 40 μm, with a capacitor of 10 nF. Reproduced with permission from [102]. Copyright @ 2011, IoP Publishing

The power generated by the harvesters that can be delivered to capacitors with capacitance of 10 nF, 150 nF, and 1 mF have been evaluated. The current delivered by a harvester equals to the current flowing into the capacitor, which can be derived from the storage voltage as a function of time. This current is given by $I_C = C\ \mathrm{d}V_S/\mathrm{d}t$. The delivered power is thus calculated with the voltage at the storage capacitor and the current delivered to it, $P = V_S I_C$, where the losses during are neglected. The maximum available power transfer is $P_{\max} = V_{\max} I_{SC}/4$, where $V_{\max}$ is the open circuit voltage of the harvester and I_{SC} is the short circuit current. I_{SC} equals I_C when it is evaluated at the beginning ($t = 0$) when the storage capacitor voltage starts to rise. This simple evaluation has been validated by experimental results. The maximum amount of energy is produced by the harvester is 51.66 μJ, with piezofilm having an area of 7.44 cm^2 and a thickness of 64 μm, measured with a capacitor of 150 nF at a wind speed of 12.3 m s^{-1} by using a dryer in parallel configuration (setup (a)). The experimental results have a good agreement with the theoretical simulations [102].

2.6.2 Wind Energy Through Tree

Another example of using wind as an energy source is a micro-watt energy-harvester inspired by tree trunks swaying in wind [105]. A uniform flow vibrates a linear array of four cylinders affixed to piezoelectric energy transducers. Particular attention is paid to measuring the energy generated as a function of cylinder spacing, flow speed, and relative position of the cylinder within the array. Peak power is generated using cylinder center-to-center spacings of 3.3 diameters and flow speeds in which the vortex shedding frequency is 1.6 times the natural frequency of the cylinders. Using these flow speeds and spacings, the power generated by downstream cylinders can exceed that of leading cylinders by more than an order of magnitude. We visualize the flow in this system by studying the behavior of a dynamically matched flowing soap film with imbedded styrofoam disks. Our

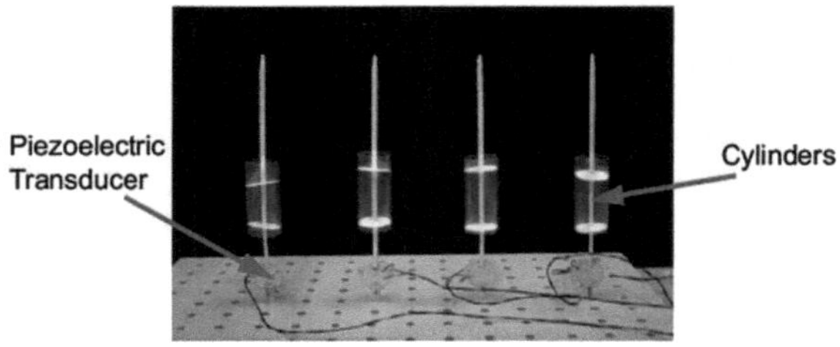

Fig. 2.67 Photograph of four piezoelectric devices arranged in series. Reproduced with permission from [105]. Copyright @ 2012, Elsevier

qualitative visualizations suggest that peak energy harvesting occurs under conditions in which vortices have fully detached from the leading cylinder.

2.6.2.1 Design and Analysis

Figure 2.67 shows a photograph of the harvester. It consists of cylinders affixed to the ground through a piezoelectric disk each. The piezoelectric disks are oriented in such a way that the cylinders can only tilt in the cross-flow direction, as demonstrated in Fig. 2.68 [105].

The cylinders in the in-line array have a geometry characterized by cylinder diameter *D*, height *H* and center-to-center distance *L*. Material properties of the cylinders include density ρ_c, elasticity *E* and dimensionless damping coefficient ζ. The fluid has density ρ_f, viscosity μ and far-field velocity *U*. The power generated by the harvester can be described as:

$$P = \rho_f U^3 DH \times g\left(\mathrm{Re}, \frac{L}{D}, \frac{H}{D}, \frac{f_s}{f_n}, \frac{\rho_c}{\rho_f}, \zeta\right), \tag{2.55}$$

where $\rho_f U^3 DH$ is the rate of the fluid on the cylinder and the function *g* depends on Reynolds number Re = UD/v, dimensionless spacing L/D and cylinder height *H/D*, dimensionless shedding frequency f_s/f_n, mass ratio ρ_c/ρ_f and damping coefficient ζ.

The Reynolds number is in the range of Re = 4,000–9,000, indicating that the viscosity effects are negligible. The cylinders are much heavier than the fluid, with a density ratio of $\rho_c/\rho_f \approx 85$. Because the amplitude of the motion of the cylinder is sufficiently small, it can be treated as two dimensional rather than three dimensional, although the motion is rotational (Fig. 2.68) [105]. The cylinders have dimensionless height of H/D = 2.2 and dimensionless pacing of L/D = 2–6. The height is kept at an intermediate value to increase the similarity between the piezoelectric system and the two-dimensional soap film visualizations.

The dimensionless group f_s/f_n is the ratio of vortex shedding frequency, f_s, to the natural frequency, f_n, of the elastic cylinders. In particular, resonance of the cylinder corresponds to $f_s/f_n = 1$. The natural frequency f_n of the cylinders is 11 Hz, and the cylinders under flow conditions continue to vibrate at 11–12 Hz. Because high energy generation occurs at resonance, the flow speed *U* should be tuned so that $f_s/f_n = 0.8$–2.0.

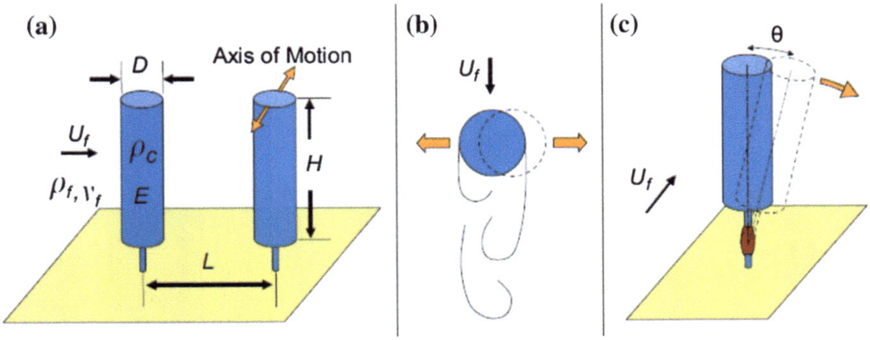

Fig. 2.68 Schematic diagrams illustrating the arrangement of the energy-harvesting cylinders. **a** *Side view* of the cylinders, along with variables characterizing the properties of the flow and the cylinder. **b–c** *Top* and *front views*. *Yellow arrows* indicate the direction of cylinder vibration, which is transverse to the incoming flow. Reproduced with permission from [105]. Copyright @ 2012, Elsevier

The resonant frequency of a circular beam in a fluid can be expressed as [106]:

$$f_n \sim \frac{D}{H^2}\left(\frac{E}{\rho_c}\right)^{1/2}\left(1+\frac{\pi\rho_f}{4\rho_c}\right)^{-1/2}. \quad (2.56)$$

Because $\rho_f \ll \rho_c$, f_s/f_n can be simplified as:

$$\frac{f_s}{f_n} \sim \frac{SU}{\sqrt{E/\rho_c}}\left(\frac{D}{H}\right)^2. \quad (2.57)$$

2.6.2.2 Fabrication and Characterization

High-speed video has been used to verify that the angular velocity of the cylinders is proportional to the output voltage. The tilting of the cylinders has been confirmed to in the cross-low direction, because the tilting in the cross-stream direction is 6°, as compared with the 1° tilting in the stream-wise direction. The output voltage of individual piezoelectric transducer is measured across a resistor of 1 MΩ, with which the output power can be calculated. The total output power of the array is given by $P_{\text{tot}} = \Sigma P_i$ (i = 1–4).

The cylinder walls are made of polyethylene storage tubes (35 mm inner in diameter, 0.4 mm in thickness and H = 78 mm in length). The tubes are attached to a bamboo rod (4.7 mm in diameter and 22 cm in length) with a pair of styrofoam spacers. A piezoelectric transducer is epoxied to the bamboo rod with a slit of 1 cm deep cut into the base of the rod. The piezoelectric transducers consist of disks for use as audio buzzers. They are composed of two ceramic layers sandwiching a 42 mm diameter brass plate between the layers. On the base of the transducer, a peg is epoxied, so as to enable the arrangement of several devices on

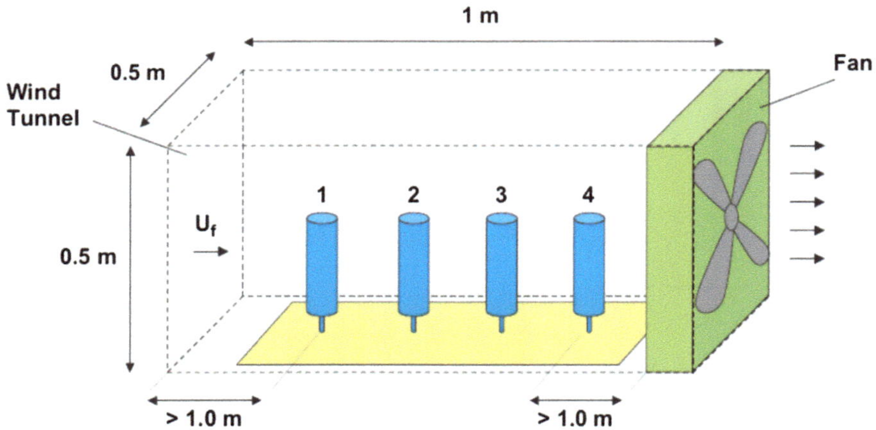

Fig. 2.69 Schematic diagram of the wind tunnel used for testing. Cylinders are numbered 1 (leading cylinder) to 4 (most downstream cylinder). Air is drawn through the tunnel, from the *left* to the *right*, by a box fan. To minimize edge effects, the first and last devices in the series are positioned at least 0.1 m from the ends of the 1 m long tunnel. Reproduced with permission from [105]. Copyright @ 2012, Elsevier

a wooden peg board, with center-to-center hole spacing of 1.5 in. The two layers of the transducer are connected in series. Wires are then run from each to a breadboard, from which voltage readings can be recorded. To tune the devices individually, such that they will resonant at 11 Hz, common metal paper clips are attached to the top of the bamboo rods, which can slid up and down the rod to change the rotational moment of inertia.

A schematic of the arrangement of four cylinders used in our tests is shown in Figs. 2.67 and 2.69. The pegboard base is designed in such a way that the arrangement of the devices can be adjusted. The cylinders are able to slide up and down the bamboo rod to adjust the mass distribution and thus the natural frequency of the devices. Figure 2.69 shows the experimental setup placed in a wind tunnel. The piezoelectric transducers and cylinders are labeled from 1 (the leading cylinder) to 4 (the most downstream cylinder). Because the devices are manually fabricated, there is a slight difference in the mechanical to electrical conversion efficiency from one device to the other. To eliminate this effect, the output power of each device has been normalized, by rotating the devices through each position, while keeping the other positions occupied, and measuring the average power output. Scaling factors of 1.26, 1.0, 1.08, and 1.30 are obtained for the devices 1–4.

Flow velocity in the wind tunnel is controlled by a combination of built-in speed controls on the fan and a 120 VAC Variac™ variable transformer. The flow speed calibrated is by velocity measurements with a hand held anemometer, as so to ensure that variation in velocity is less than 5 % over the area that the piezoelectric devices are placed. The damping ratio ζ is calculated for the piezoelectric devices and the soap film system, by measuring the logarithmic decrement. Devices in each system are manually deflected and released (i.e., a step input)

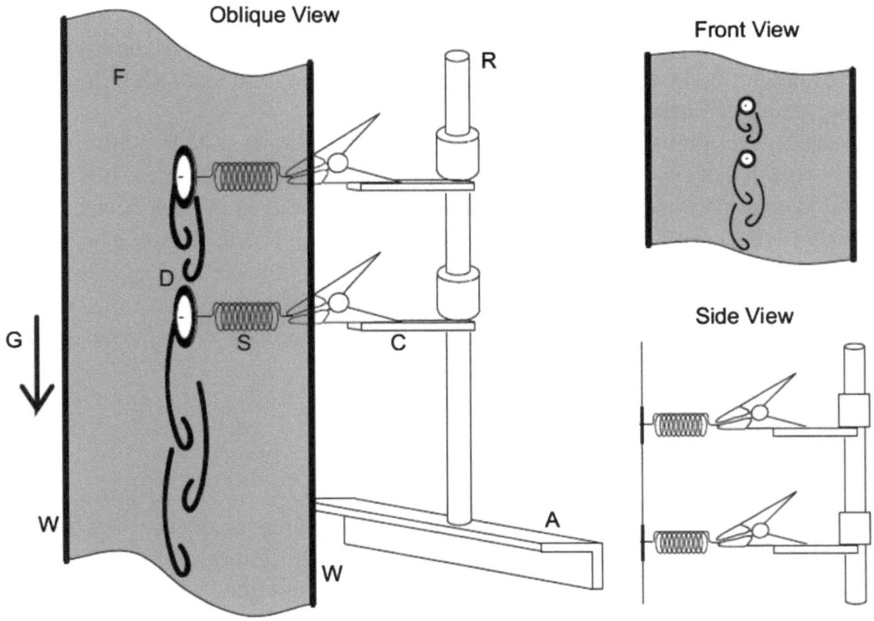

Fig. 2.70 Schematic diagram (oblique, *front* and *side* views) of the soap film apparatus. Disks (*D*) in the soap film (*F*) are suspended from linear springs (*S*) behind the film. The film flows from *top* to *bottom*, pulled by gravity (*G*). The clamps are vertically adjustable along a rod (*R*), which is fixed to an angled aluminum support (*A*). The wires (*W*) that support the film are also shown. Reproduced with permission from [105]. Copyright @ 2012, Elsevier

while the oscillating response is recorded. By measuring the decreasing motion over at least four oscillations, the damping ratio can be estimated. The springs have been verified to have a linear response in the range of flow speeds tested.

To understand the relation between vortex shedding and power output, a three-cylinder system is visualized in a dynamically matched two-dimensional flowing soap film, as shown in Fig. 2.70 [105, 107]. The styrofoam disks (diameter $D = 12.5$ mm) are suspended in the soap film with springs. The springs have a stiffness to provide dynamic similarity to the piezoelectric component. Dimensionless parameters for the wind tunnel and the soap film tests have been stated before. Two spacings, $L/D = 2$ and 5, and two flow speeds, corresponding to frequency ratios of $f_s/f_n = 0.5$ and 1.5, are used in the tests.

Systematic studies, including peak output power, relative power due to the variation in wind speed, spacing of the cylinder devices and so on, have been conducted to optimize the energy harvesting efficiency of the linear array harvesters [105]. There is an optimal wind speed for power production. Specifically, it is necessary to mention that the highest power has been found at $f_s/f_n = 1.6$, instead of $f_s/f_n = 1$, the common expectation. One of the explanations to this deviation could be related to vortex shedding frequency. The assumption that the vortex shedding frequency for the oscillating cylinders is the same as that for stationary cylinders does not take into account the effects of frequency lock-in. In addition, the assumption that vortex shedding would only occur in the horizontal plane may not be true, because the aspect ratio of the cylinders is fairly low. This aspect deserves further study.

Future generations of our device will likely use more than four cylinders, and more work is needed to understand the limitations of adding devices. It has been found that, at intermediate wind speeds, power is maximized in third device, while at higher wind speeds, the maximum is observed in the fourth device. This means that there should be certain degree of interactions among the devices when they experience the blow of the wind. Therefore, it is important to clarify this kind of interaction before large-scale applications of this approach when more devices have to be used.

2.6.3 Water Flow

The kinetic energy due to water flow is another source of energy that can be harvested by using piezoelectric harvesters [108–110]. A new energy-harvesting device based on flow-induced vibrations has been reported, which uses PVDF flexible piezoelectric materials [108]. The second example is to PZT-5H thin film to construct the energy harvester [109], that has a similar structure to that in Ref. [108]. Another energy harvester is assembled by using macro-fiber piezoelectric composite, which is a fish-like bimorph [110]. In this subsection, the first example of piezoelectric energy harvester will be discussed in detail [108].

Figure 2.71 shows a schematic operational principle of the piezoelectric energy harvest [108]. There is a flow channel with a flexible diaphragm that is connected to a flow source. The pressure in the chamber causes the diaphragm to deflect in the upward direction. As the pressure increases to the maximum level, the diaphragm reaches its highest position. When the pressure drops, the diaphragm moves downward. As the pressure decreases to the minimum level, the diaphragm reaches its lowest position. Therefore, by connecting the energy harvester to an ambient flow source, which is capable of providing the pressure change in the pressure chamber, the energy due to the oscillating movement of the diaphragm can be harvested by using the piezoelectric film attached to it.

Figure 2.71 shows the piezoelectric energy harvester. It consists of a flow channel with two glass tubes, a polydimethylsiloxane (PDMS) diaphragm bonded to the channel and a piezoelectric PVDF film glued to the PDMS diaphragm. The device harvests the energy due to the flow-induced vibration. The flow is bounded by the flexible structure and rigid walls. If the diaphragm has small inertia and is sufficiently flexible to be able to respond rapidly to the fluctuating pressure field set up by the flow, the diaphragm can oscillate at a frequency close to that of the flow. When the fluctuating pressure is applied to the piezoelectric film, it experiences lateral strains. The normal strain causes electrical charge to accumulate on the piezoelectric electrode, resulting in a voltage in the thickness direction of the piezoelectric film.

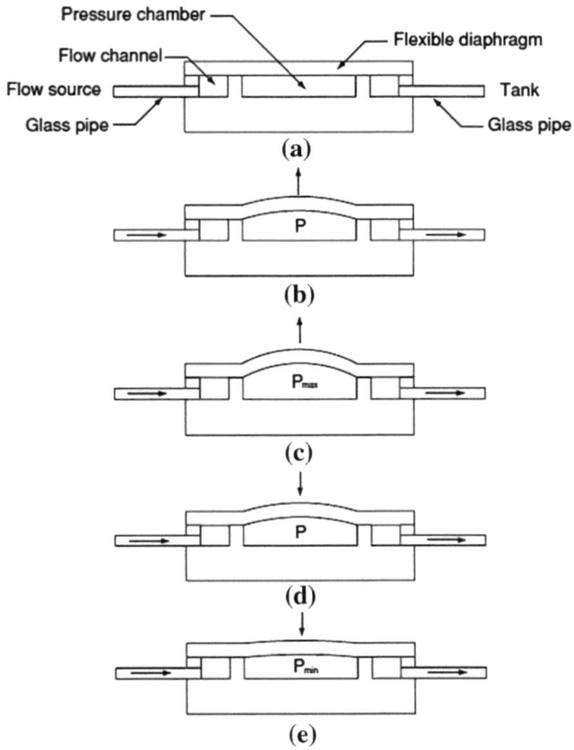

Fig. 2.71 Optional principle of the piezoelectric energy harvester due to flow caused vibration. Courtesy of Dr. D. A. Wang, National Chung Hsing University, Taiwan. Reproduced with permission from [108]. Copyright @ 2010, IoP Publishing

The piezoelectric film (LDT0-028 K/L, Measurement Specialties, Inc., USA) is a laminated film including a polyvinylidene fluoride (PVDF) film, two silver electrode layers and a polyester layer. The PDMS diaphragm has a thickness of 200 μm. The electrode layers with a thickness of 28 μm are attached to top and bottom surfaces of the PVDF film of 24 μm. A 125 μm polyester layer is laminated to top surface of the top electrode layer. When used in a bending mode, the laminated piezoelectric film develops much higher voltage output when flexed than a nonlaminated film. The neutral axis is in the polyester layer instead of in the PVDF film so the film is strained more when flexed. Because the behavior of the piezoelectric materials is assumed to be linear elastic in modelling, an almost linear increase in peak-to-peak voltage with increasing pressure difference is predicted by the theoretical simulation. For the pressure difference ranging from 1.790 to 2.392 kPa, the peak-to-peak output voltages are from 1.77 to 2.30 V.

Prototypes of the energy harvester have been fabricated for experimental evaluation. The PDMS flow channel is fabricated by a molding process with an acrylic mold. The acrylic mold is carved by a milling machine (PNC-3100, Roland DGA Co., Japan). Next, PDMS precursor material is poured over the mold to form the flow channel. The PDMS material is composed of two parts, a curing agent and the polymer. They are mixed with a volume ratio of 1:10. Before pouring into the

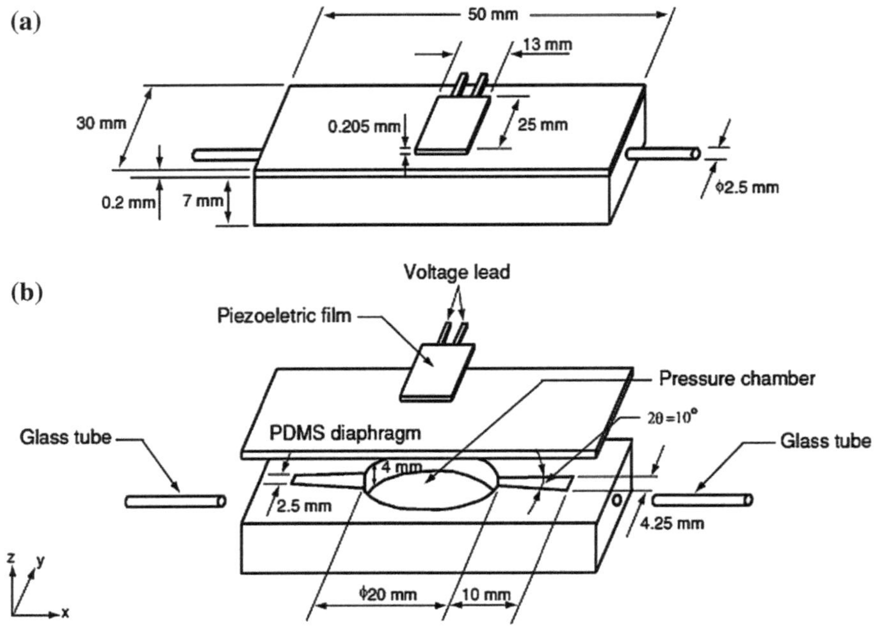

Fig. 2.72 Diagram of an assembled energy harvester (**a**) and components of the harvester (**b**). Courtesy of Dr. D. A. Wang, National Chung Hsing University, Taiwan. Reproduced with permission from [108]. Copyright @ 2010, IoP Publishing

mold, the mixture is degassed under vacuum until no bubbles appear. The PDMS is cured at 80 °C for 40 min. Then, the PDMS is peeled off from the mold (Fig. 2.72).

Figure 2.73 shows the steps to assemble the prototype of the piezoelectric energy harvester [108]. First, the inlet and outlet of the PDMS channel are prepared by piercing the front and back ends of the cured PDMS by using a copper tube with an outer diameter of 2.5 mm. After that, a PDMS diaphragm of 200 µm is glued to the flow channel with a thin layer of the liquid PDMS mixture. Two glass tubes are inserted into the inlet and outlet holes, respectively. The inlet and outlet are tightly sealed by using the PDMS mixture, after it is cured at 80 °C for 90 min. Finally, the piezoelectric film is glued to the flow channel by using an adhesive (3 M Scotch) to for the prototype. The thickness of the adhesive after the assembly step is nearly 50 µm.

Figure 2.74 shows schematic of the experimental apparatus for the characterization of the energy harvester [108]. The energy harvester is placed on the platform of a water tank. Tap water is pumped into the inlet of the energy harvester through a pulse pump to provide a periodic pressure in the pressure chamber of the energy harvester. The oscillating deflection of the piezoelectric film is monitored by using a Philtec D6 fiber optic displacement sensor. The generated voltage of the

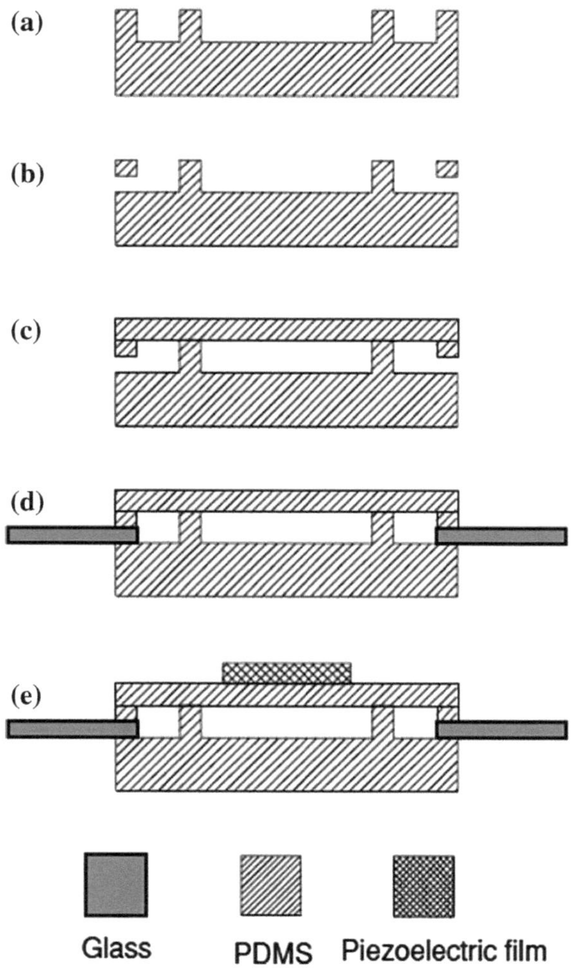

Fig. 2.73 Steps to assemble the piezoelectric energy harvester. Courtesy of Dr. D. A. Wang, National Chung Hsing University, Taiwan. Reproduced with permission from [108]. Copyright @ 2010, IoP Publishing

PVDF film is amplified and filtered by using a Stanford SR560 preamplifier, which is recorded and analyzed by using a data acquisition unit (PCI-5114, National Instruments Co., USA).

Experimental results indicate that the mean pressure and the pressure difference $P_{max} - P_{min}$ within the pressure chamber are 2.473 and 2.392 kPa, respectively. The measured deflection history of the center of the piezoelectric film shows that the film oscillates near a deflected position of 367 μm, with an amplitude of about 188 μm. The measured open circuit voltage generated by the PVDF film is shown in Fig. 2.75 [108]. The output peak-to-peak voltage is about 2.2 V, for an oscillation amplitude of 1.196 kPa and an excitation frequency of 26 Hz of the pressure in the pressure chamber. The experimental results do not show transient responses since they are recorded at steady state vibration. The output voltage has an almost

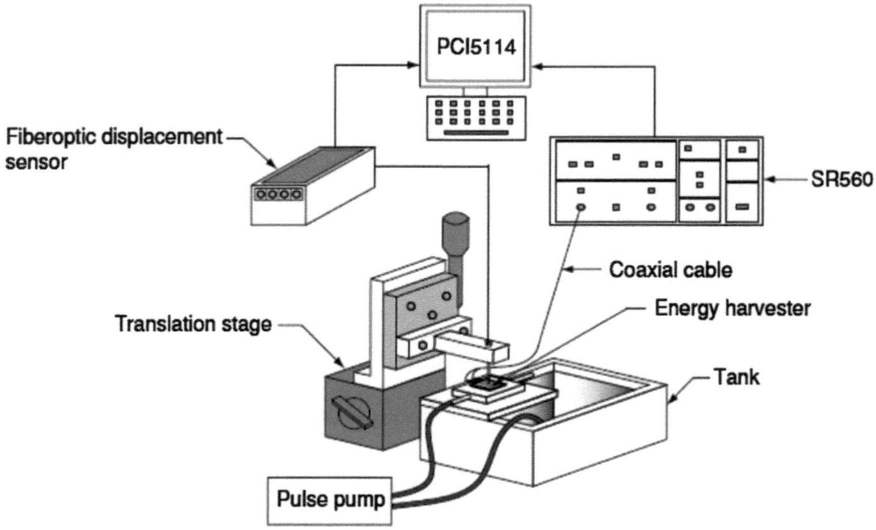

Fig. 2.74 Schematic of the experimental apparatus for the characterization of the energy harvester. Courtesy of Dr. D. A. Wang, National Chung Hsing University, Taiwan. Reproduced with permission from [108]. Copyright @ 2010, IoP Publishing

Fig. 2.75 Output voltage of the PVDF film. Courtesy of Dr. D. A. Wang, National Chung Hsing University, Taiwan. Reproduced with permission from [108]. Copyright @ 2010, IoP Publishing

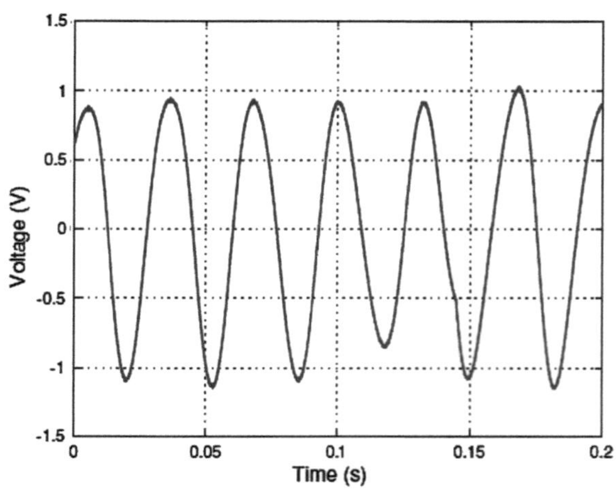

linear relationship with the pressure difference and is in good agreement with the result based on the finite element model, except the point of 2.392 kPa. The slightly lower output voltage at the pressure difference of 2.392 kPa could be attributed to the aging of the PDMS. The piezoelectric energy harvester has been operated continuously for 4 h, at a rate of 93,600 cycles per hour, with an oscillation amplitude of 1.196 kPa and an excitation frequency of 26 Hz of the pressure in the pressure chamber. The device is run for a total of 152 h without failure,

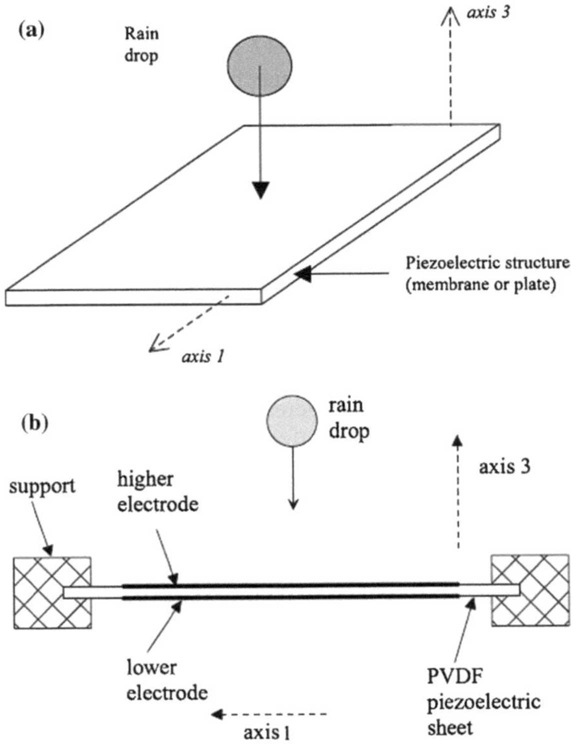

Fig. 2.76 Principle and system for raindrop energy harvesting: **a** 3D-view and **b** side-view. Reproduced with permission from [111, 112]. Copyright @ 2008, IoP Publishing

accumulating a total of 14 million cycles. The resonant frequency of the harvester is 26 Hz, which has been used to throughout the experiments.

2.6.4 Energy from Rain

Rain is liquid water in the form of droplets that have condensed from atmospheric water vapor and then precipitated, when they are heavy enough to fall under gravity. Piezoelectric effect has been proposed to collect the kinetic energy caused by rain drops. This idea has been explored theoretically, which is then validated experimentally [111, 112]. Figure 2.76 shows diagram of principle and system for energy harvesting from rain drops. In this case, PVDF should be selected as the piezoelectric component, due to the most important requirement of flexibility. Theoretical simulation results indicate that a 25 μm thick mono-stretched PVDF sheet with a piezoelectric strain coefficient $d_{31} = 2$ pJ N^{-1} is more effective than a 9 μm thick bi-stretched one with a piezoelectric coefficient of $d_{31} = 5$ pJ N^{-1} [111].

The experimental measurements conducted in various impact situations, such as different drop heights and drop sizes, have shown that the energy harvesting efficiency of the harvester is close to theoretical prediction, i.e., electrical energy of 1 nJ and instantaneous power of 1 μW collected from raindrops. If the harvester is used to harvest energy from downpour rain drops, the energy and power could be as high as 25 μJ and 12 mW, respectively. Therefore, this is a promising approach of piezoelectric mechanical energy harvesting.

2.7 Strategies to Enhance Energy Harvesting Efficiency

2.7.1 Brief Introduction

Most piezoelectric energy harvesters discussed above are based on vibration, which generate maximum power when the resonant frequencies of the generators are matched with the frequency of the vibrational sources. A slight deviation will result in a significant decrease in power generation efficiency. The resonant frequency of a harvester is fixed after its fabrication, while the energy sources could be different from one another. Therefore, mismatching between the resonant frequency of a harvester and the vibrational frequency of an energy source is easily encountered. Various strategies have been developed to address this problem, among which two main approaches have been widely adopted: frequency tuning and bandwidth widening. While these strategies have been well documented, a brief introduction will be given in this subsection [113].

2.7.2 Frequency Tuning

The power generation of a piezoelectric harvester has been given in Eq. (2.5) [6, 113]. The maximum power is generated when the resonant frequency of the harvester matches the frequency of the vibrational source. Therefore, a given harvester is only able to harvest energy from specific sources with vibrational frequencies matching with or close to its resonant frequency. If the resonant frequency can be tuned to match with the vibrational frequencies of different sources, the application range of a harvester will be widened.

The tuning of resonant frequency of harvester can be realized by changing the mechanical characteristics of the structure of or electrical loading on the generator [113]. There are two types of tuning through changing the mechanical characteristics of a harvester: passive tuning and active tuning. Passive tuning is a periodical tuning and thus only consumes power during the tuning until the frequency matching is achieved. In contrast, active tuning is a continuous tuning even though the frequencies are already matched. Because actually both approaches have active characteristics, they are called intermittent (passive) and continuous (active) tunings. Comparatively, intermittent tuning is advantageous over the continuous tuning, because the tuning is stopped once the frequency matching is achieved, so as to minimize the system energy consumption.

Several considerations should be taken into account when evaluating a tuning approach. The energy consumption should be as low as possible and should not exceed the energy produced by the harvester. The tunable frequency range should be sufficiently wide and the frequency resolution should meet the requirement for the give applications. The tuning should not increase the damping of the system, so as to maintain a high Q-factor for high power generation efficiency.

The resonant frequency of a generator can be tuned either mechanically or electrically. Mechanical tuning shifts the resonant frequency by changing mechanical properties of the whole structure. Electrical tuning is realized by adjusting the electrical loading.

2.7.3 Mechanical Tuning

For a cantilever with a mass at the free end, the resonant frequency is given by [113]:

$$f_r = \frac{1}{2\pi}\sqrt{\frac{Ywh^3}{4l^3(m + 0.24m_c)}}, \tag{2.58}$$

where similarly Y is Young's modulus of the cantilever material; w, h and l are the width, thickness and length of the cantilever, respectively, and m_c is the mass of the cantilever. Theoretically, the resonant frequency can be tuned by adjusting all these parameters. Therefore, for cantilever harvesters, mechanical tuning methods include: (i) changing dimension, (ii) shifting the center of gravity of the proof mass, (iii) varying the spring stiffness, and (iv) applying strain to the structure.

2.7.3.1 Changing Dimension

It is possible to change the length of a cantilever, while it is difficult to change its width and thickness. Furthermore, modifying length is suitable for intermittent tuning. The approach is just to change the clamping position of the cantilever, so as to adjust the effective length. Such an operation has no energy consumption. Moreover, as the resonant frequency is inversely proportional to $l^{3/2}$, a small adjustment in l can significantly change the resonant frequency f_r.

If l is the original length of the cantilever and l' is the newly adjusted length, there is $l' = l + \Delta$, where Δ is the difference between them. The mass of a cantilever is then changed to $m_c' = whl'\rho$, where ρ is the density of the material of the cantilever, while the original mass of a cantilever is $m_c = whl\rho$. Then the new resonant frequency becomes:

$$f_r' = \frac{1}{2\pi}\sqrt{\frac{Ywh^3}{4l'^3(m + 0.24m_c')}} = \sqrt{\frac{Ywh^3}{4(l + \Delta)^3\{m + 0.24[wh(l + \Delta)\rho]\}}}. \tag{2.59}$$

And the ratio of the frequency after tuning to the original frequency, which is called the normalized resonant frequency, is given by:

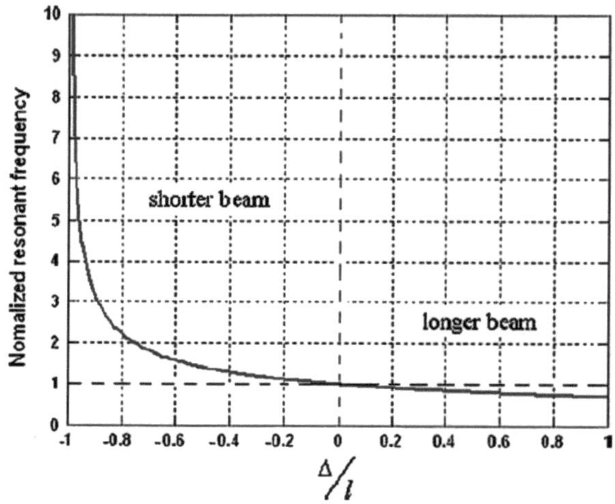

Fig. 2.77 Normalized resonant frequency as a function of the change in the length of the cantilever. Reproduced with permission from [113]. Copyright @ 2010, IoP Publishing

$$\frac{f_r'}{f_r} = \sqrt{\frac{l^3(m + 0.24m_c)}{l'^3(m + 0.24m_c')}} = \sqrt{\frac{l^3(m + 0.24\text{wh}l\rho)}{(l + \Delta)^3\{m + 0.24[\text{wh}(l + \Delta)\rho]\}}}. \tag{2.60}$$

Figure 2.77 shows the normalized resonant frequency as a function of the change in length of the cantilever. A negative Δ/l means that the new cantilever beam is shorter than its original length and thus has a higher resonant frequency, whereas a positive Δ/l means that the cantilever beam has been lengthened, thus giving a lower resonant frequency. It is found that shortening is more efficient than lengthening to tune the resonant frequency of the cantilever.

2.7.3.2 Shifting Gravity Center of the Proof Mass

It is difficult to add or reduce the mass once a generator is fabricated. However, the gravity center of the mass can be shifted to adjust the resonant frequency of the cantilever structure. The length of the cantilever without the mass is l and the proof mass on its free end is m. The gravity center of the proof mass is located at a position, with a distance to the end of the cantilever to be x. The tuned resonant frequency of this structure can be expressed as [113, 114]:

$$f_r' = \frac{1}{2\pi}\sqrt{\frac{Y\text{wh}^3}{12ml^3} \cdot \frac{r^2 + 6r + 2}{8r^4 + 24r^3 + \frac{21}{2}r^2 + 4r + \frac{2}{3}}}, \tag{2.61}$$

where w and h are the width and thickness of the cantilever, respectively, and $r = x/l$. If the mass of the cantilever beam is negligible as compared with the proof mass, the resonant frequency of a cantilever-based generator, given by Eq. (2.58), can be simplified as:

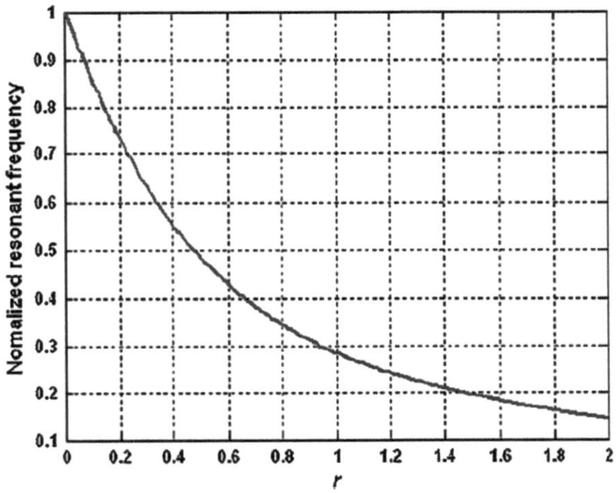

Fig. 2.78 Normalized resonant frequency as a function of the position of the gravity center of the proof mass. Reproduced with permission from [113]. Copyright @ 2010, IoP Publishing

$$f_r = \frac{1}{2\pi}\sqrt{\frac{Ywh^3}{4l^3m}}. \tag{2.62}$$

As a result, the ratio of the frequency after tuning to the original frequency is approximately given by:

$$\frac{f_r'}{f_r} = \sqrt{\frac{1}{3}\cdot\frac{r^2+6r+2}{8r^4+24r^3+\frac{21}{2}r^2+4r+\frac{2}{3}}}. \tag{2.63}$$

Figure 2.78 shows the normalized resonant frequency as a function of the position of the gravity center of the proof mass. The resonant frequency almost exponentially decreases with increasing distance between the gravity center of the proof mass and the end of the cantilever.

2.7.3.3 Varying Spring Stiffness

Softening spring stiffness is an effective method that has been widely used to tune the resonant frequency of cantilever harvesters. The principle is to apply a 'negative' spring in parallel to the mechanical spring. As a result, the system has a new spring constant, effective spring constant, k_{eff}, which is given by:

$$k_{\text{eff}} = k_m + k_a, \tag{2.64}$$

where k_m is the mechanical spring constant and k_a is the additional 'negative' spring stiffness. Therefore, the resonant frequency of the cantilever becomes:

$$f_r = \frac{1}{2\pi}\sqrt{\frac{k_{\text{eff}}}{m}} = \sqrt{\frac{k_m+k_a}{m}}. \tag{2.65}$$

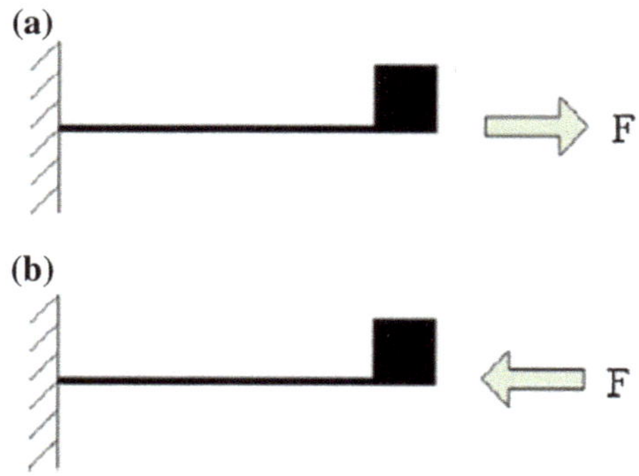

Fig. 2.79 Axial tensile (**a**) and compressive (**b**) load on a clamped-free cantilever. Reproduced with permission from [113]. Copyright @ 2010, IoP Publishing

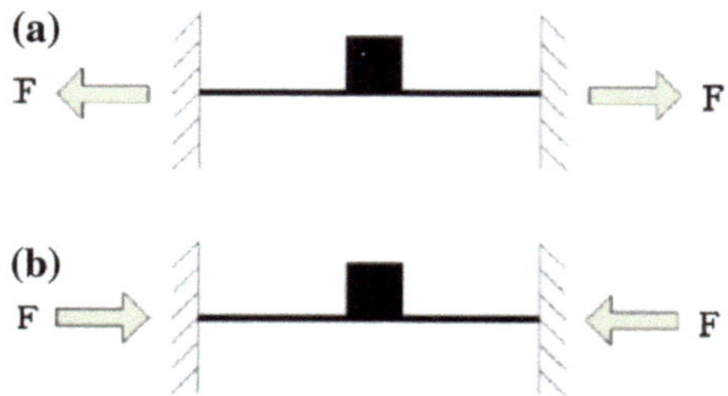

Fig. 2.80 Axial tensile (**a**) and compressive (**b**) load on a clamped–clamped cantilever. Reproduced with permission from [113]. Copyright @ 2010, IoP Publishing

The negative spring k_a can be applied electrostatically, piezoelectrically, magnetically or thermally.

2.7.3.4 Applying Straining

The effective stiffness of the structure can be varied by applying a stress to make it under certain degrees of strain. The resonant frequency of a cantilever structure can be tuned by applying an axial load. Examples of clamped-free and clamped–clamped cantilevers are shown in Figs. 2.79 and 2.80, respectively. An axial tensile load applied to the cantilevers (Figs. 2.79a and 2.80a) increases their resonant frequency, while an axial compressive load applied to the cantilevers (Figs 2.79b and 2.80b) decreases their resonant frequency.

An approximate formula for the resonant frequency of a uniform cantilever in mode i with an axial load, f_{ri}', is given by [113]:

$$f_{ri}' = f_{ri} \cdot \sqrt{1 + \frac{F}{F_b} \cdot \frac{\lambda_1^2}{\lambda_i^2}}, \tag{2.66}$$

where f_r is the resonant frequency in mode i without load, F is the axial load and F_b is the axial load required to buckle the beam, i.e., to make the fundamental resonant frequency to be zero. F is positive for a tensile load and negative for a compressive load. Variable λ_i is a dimensionless load parameter, which is a function of the beam boundary conditions applied to the cantilever for the ith mode of the beam. It is given by the ith positive solution of Eq. (2.67) for a cantilever and of Eq. (2.68) for a clamped–clamped beam:

$$\cos\lambda \cdot \cosh\lambda + 1 = 0, \tag{2.67}$$

$$\cos\lambda \cdot \cosh\lambda - 1 = 0. \tag{2.68}$$

The majority of cantilever-based piezoelectric energy harvesters work in the fundamental flexural mode (mode 1). The resonant frequency of a uniform cantilever in mode 1, with an axial load, f_{r1}', is given by:

$$f_{r1}' = f_{r1} \cdot \sqrt{1 + \frac{F}{F_b}}. \tag{2.69}$$

The ratio of the frequency after tuning to the original frequency is:

$$\frac{f_r'}{f_r} = \sqrt{1 + \frac{F}{F_b}}. \tag{2.70}$$

The buckling load F_b of a cantilever and a clamped–clamped beam is given by Eqs. (2.71) and (2.72), respectively:

$$F_b = \frac{\pi^2 \cdot Y \cdot w \cdot h^3}{48 \cdot l^2}, \tag{2.71}$$

$$F_b = \frac{\pi^2 \cdot Y \cdot w \cdot h^3}{3 \cdot l^2}, \tag{2.72}$$

where Y is Young's modulus of the material of the cantilever and w, h and l are the width, thickness and length of the cantilever, respectively.

Figure 2.81 shows the resonant frequency of a cantilever as a function of the axial load [113]. It is demonstrated that a compressive load is more efficient in tuning the resonant frequency than a tensile load. If the compressive force is larger than the buckling load, the cantilever beam will buckle and no longer oscillate in mode 1. If a very large tensile force is axially applied to the cantilever, i.e., much greater than the buckling load, the resonant frequency will approach that of a straight tensioned cable as the force associated with the tension in the cantilever becomes much greater than the beam stiffness [113].

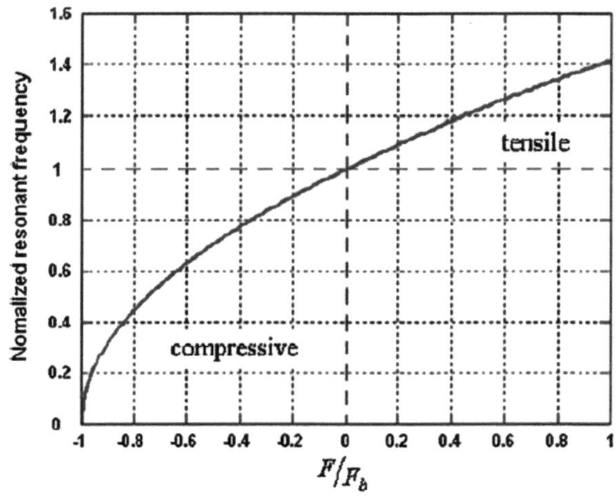

Fig. 2.81 Normalized resonant frequency as a function of the axial loads. Reproduced with permission from [113]. Copyright @ 2010, IoP Publishing

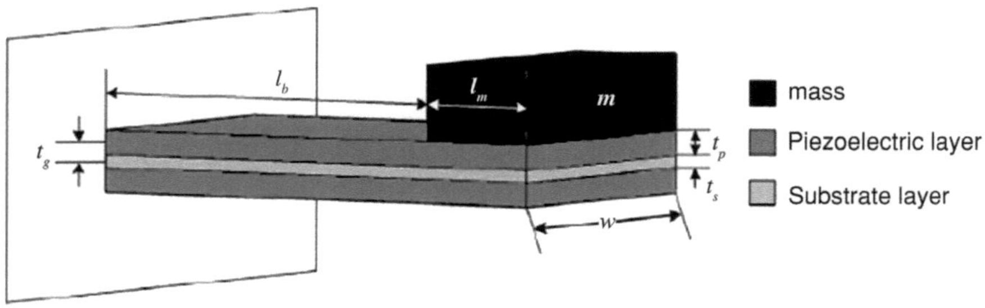

Fig. 2.82 Piezoelectric bimorph cantilever energy harvester. Reproduced with permission from [113]. Copyright @ 2010, IoP Publishing

2.7.4 Electrical Tuning

Besides mechanical methods, the resonant frequency of a cantilever can also be tuned by using electrical tuning methods. The basic principle of electrical tuning is to change the electrical damping of the cantilever by adjusting the load, so as to shift the power spectrum of the harvester. Electrical tuning is especially effective to piezoelectric harvesters. Because resistive loads usually reduce the efficiency of power transfer and it is difficult to vary the load inductances, electrical tuning is generally achieved through adjusting capacitive loads.

Figure 2.82 shows a schematic diagram of a bimorph piezoelectric generator with a mass, m, on the tip. l_b and l_m are the effective length of the cantilever and mass, respectively. w is the width of the cantilever. t_p and t_s are the thickness of the piezoelectric layer and substrate layer, respectively, and t_g is the distance from the

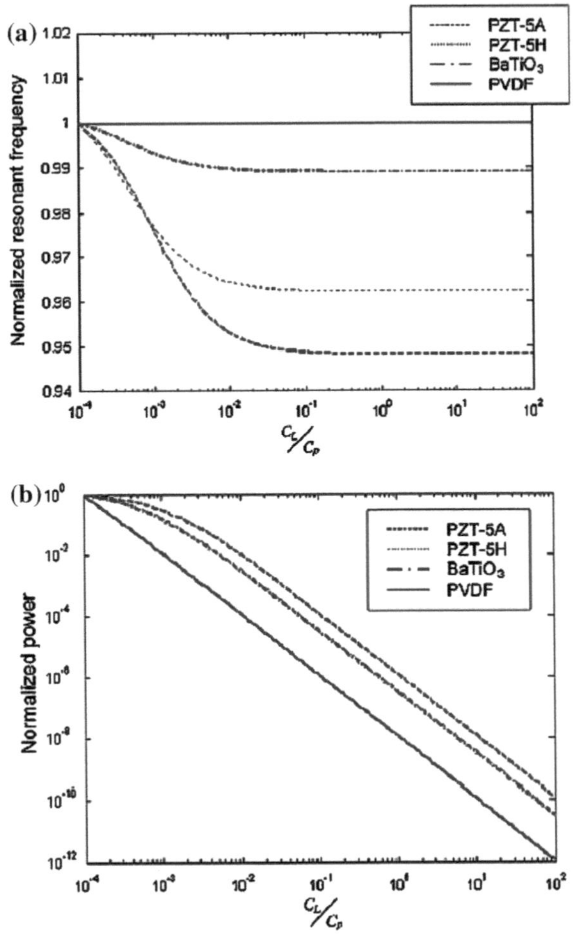

Fig. 2.83 Tuning effects on performances of the piezoelectric bimorph cantilever harvester: **a** normalized resonant frequency and **b** normalized output power. Reproduced with permission from [113]. Copyright @ 2010, IoP Publishing

center of the substrate layer to the center of the piezoelectric layer. Electrodes of the generator have been omitted in the figure. The resonant frequency of the harvester can be tuned by varying the capacitive load [113].

Figure 2.83 shows normalized resonant frequency and normalized output power of the piezoelectric harvesters, made with different piezoelectric materials, as a function of load capacitances, illustrating the effects of electrical tuning. Both the resonant frequency and the output power are decreased with increasing load capacitance. It is observed that PZT-5 H is the best among the four piezoelectric materials as an electrically tunable piezoelectric harvester.

Important aspects regarding the tunability of piezoelectric energy harvesters have been summarized [113]. The materials of the substrate layer and mass should have no effect on the tenability of the piezoelectric layer. A piezoelectric material with higher Young's modulus, strain coefficient and smaller permittivity has a

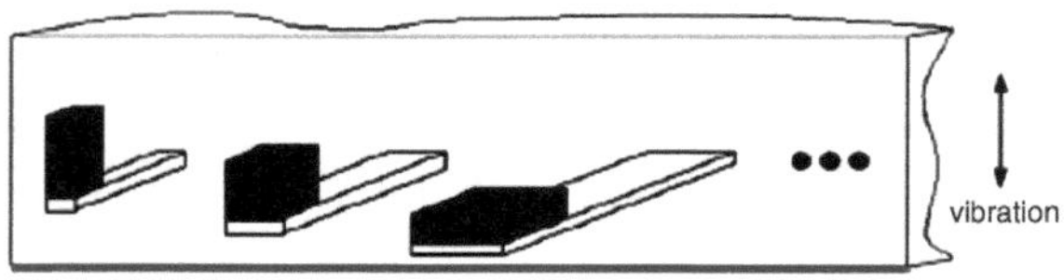

Fig. 2.84 A mechanical band-pass filter with a number of cantilever beams. Reproduced with permission from [113]. Copyright @ 2010, IoP Publishing

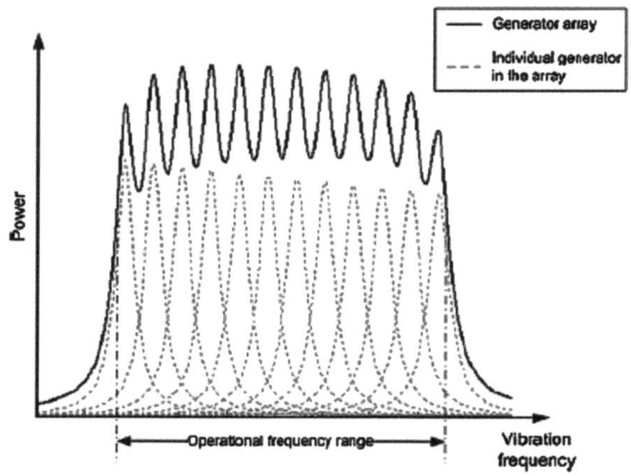

Fig. 2.85 Power spectrum of a generator array with resonant frequencies over a sufficiently wide range. Reproduced with permission from [113]. Copyright @ 2010, IoP Publishing

wider range of tunability. The ratio of the thickness of the piezoelectric layer to the thickness of the substrate layer should be small to have a wide range tuning. The capacitance of the piezoelectric layer should be minimized. If both piezoelectric layers are used for tuning, connection of these two layers in parallel will lead to a wider tuning range than connection in series. The total damping should be kept as low as possible.

2.7.5 Bandwidth Widening

The other widely adopted approach to increase the operational frequency range of a vibration energy harvester is to widen its bandwidth [113]. Various strategies have been effectively to widen the bandwidth of piezoelectric harvesters, include (i) using a harvester array consisting of numerous small generators with different resonant frequencies, (ii) introducing an amplitude limiter to the device, (iii) using coupled oscillators, (iv) employing nonlinear and bi-stable structures, and (v) designing a large generator with a large inertial mass and high degree of damping.

The effectiveness of this approach is demonstrated by the use of harvester arrays. A harvester array consists of numerous cantilever beams that are designed with different resonant frequencies. An example is shown in Fig. 2.84 [113]. Thus, the harvester has a wide operational frequency range, whereas the Q factor is not decreased. Figure 2.85 shows the power spectrum of the generator array which is a summation of the power spectra of all the small harvesters. The frequency band of the generator is obviously widened. The problem of this approach is the complexity of fabricating the arrays and the potentially large total volume of the device.

2.8 Summary and Perspectives

Small-scale mechanical energy harvesting has been studied intensively and extensively for more than a decade, with a number of potential applications in wireless sensors and electronic devices. Large-scale harvesting at 1 W–100 kW or even higher could lead to meaningful solution to the energy crisis. It can also find important applications in self-power active or semi-active vibration controls. Relatively, more challenges have been encountered in large-scale mechanical energy harvesting technology, with many questions to be answered. Although various sources of mechanical waste energy have been identified and studied for potential energy harvesting, it is important to further explore such sources.

For practical applications, efficiency of vibrational piezoelectric energy harvesting has become a concern. Addressing this problem requires more and higher efficient transducers, understanding of motion mechanisms and sophisticated power electronic circuits. While waiting for the emergence of new piezoelectric materials with higher performances might be unrealistic at the moment, it is highly potential to develop novel physical or geometrical configurations with the materials that are currently available.

For example, the optimization of electromagnetic motor should be paid more attention. If the back electromotive force coefficient k_e is increased, the efficiency can be significantly enhanced. Motion mechanism is also a very important factor to improve the efficiency. It is understood that partial energy is lost when the kinetic energy is transformed into electric energy, such as eddy current and friction. In addition, parasitic voltage drops, the inherent resistances of the electronic components and the power consumption by the controller, are also responsible for power loss. Strategies to reduce such voltage drops and the parasitic power consumption should be further studied. Furthermore, large-scale vibrations are usually in irregular ways at time-varying frequencies and in most cases at low velocities, which all make it difficult to harvest them at acceptably high efficiencies. Therefore, it still cannot be an option for efficient power takeoff technology. In this respect, the development of novel mechanical motion rectifiers that can convert irregular oscillatory vibrations into regular unidirectional rotations is of specific significance, which should be a direction of the future research in this area.

Vibration control is another specific challenge closely associated with the efficiency of mechanical energy harvesting, especially for large-scale applications. For small-scale energy harvesting, the vibration of the host structure is not a concern. However, when a vibration energy harvesting is at a large-scale, it is necessary to consider various issues, for instance, the protection of the mechanical systems, the human occupants and passengers during the vibration. As a result, many techniques for small-scale energy harvesting may not be applicable to the large-scale harvesting. For example, mechanic resonance can be used to magnify the vibration and thus improve the output power of the harvester in small-scale harvesting. In contrast, it becomes much difficult in large-scale applications, e.g., regenerative vehicle suspensions. Therefore, the ability of simultaneous vibration control and energy harvesting should be considered for each component and advanced control algorithms are required.

In addition, the motion mechanism should be efficient in driving and driven modes, the power electronics should be able to withdraw energy from the mechanical systems and also to inject energy into them if necessary, and the control algorithms should be able to regulate the electrical output voltages and control the regenerated electrical currents or voltages simultaneously. Because large-scale mechanical energy harvesting is not just a problem of design, it is multiple disciplinary problem and thus requires system level approaches, which involves structure dynamics and vibration, mechanical structures and design, power electronics, materials development and system controlling. Nevertheless, waste mechanical energy harvesting will be one of the most important energy technologies in the near future.

References

1. S. Roundy, R.K. Wright, J. Rabaey, A study of low level vibrations as a power source for wireless sensor nodes. Comput. Commun. **26**, 1131–1144 (2003)
2. S.P. Beeby, M.J. Tudor, N.M. White, Energy harvesting vibration sources for microsystems applications. Measur. Sci. Technol. **17**, R175–R195 (2006)
3. K.A. Cook-Chennault, N.Thambi, A.M. Sastry, Powering MEMS portable devices—a review of non-regenerative and regenerative power supply systems with emphasis on piezoelectric energy harvesting systems. Smart Mater. Struct. **17**, 043001 (2008)
4. S.R. Anton, H.A. Sodano, A review of power harvesting using piezoelectric materials (2003–2006). Smart Mater. Struct. **16**, R1–R21 (2007)
5. S. Priya, Advances in energy harvesting using low profile piezoelectric transducers. J.Electroceram. **19**, 167–184 (2007)
6. C.B. Williams, R.B. Yates, Analysis of a micro-electric generator for microsystems. Sens. Actuators A **52**, 8–11 (1996)
7. P. Glynne-Jones, M.J. Tudor, S.P. Beeby, N.M. White, An electromagnetic, vibration-powered generator for intelligent sensor systems. Sens. Actuators A **110**, 344–349 (2004)
8. D. Arnold, Review of microscale magnetic power generation. IEEE Trans. Magn. **43**, 3940–3951 (2007)
9. P. Mitcheson, P. Miao, B. Start, E. Yeatman, A. Holmes, T. Green, MEMS electrostatic micropower generator for low frequency operation. Sens. Actuators A **115**, 523–529 (2004)
10. Y.B. Jeon, R. Sood, J.H. Jeongand, S. Kim, MEMS power generator with transverse mode thin film PZT. Sens. Actuators A **122**, 16–22 (2005)
11. L. Wang, F.G. Yuan, Vibration energy harvesting by magnetostrictive material. Smart Mater. Struct. **17**, 045009 (2008)

12. W.J. Choi, Y. Jeon, J.H. Jeong, R. Sood, S.G. Kim, Energy harvesting MEMS device based on thin film piezoelectric cantilevers. J. Electroceram. **17**, 543–548 (2006)
13. G.H. Haertling, Ferroelectric ceramics: history and technology. J. Am. Ceram. Soc. **82**(4), 797–818 (1999)
14. D. Damjanovic, Ferroelectric, dielectric and piezoelectric properties of ferroelectric thin films and ceramics. Rep. Prog. Phys. **61**, 1267–1324 (1998)
15. L.B. Kong, J. Ma, H.T. Huang, W. Zhu, O.K. Tan, Lead zirconate titanate ceramics derived from oxide mixture treated by a high-energy ball milling process. Mater. Lett. **50**, 129–133 (2001)
16. L.B. Kong, T.S. Zhang, J. Ma, Y.C.F. Boey, Progress in synthesis of ferroelectric ceramic materials via high-energy mechanochemical techniques. Prog. Mater. Sci. **53**(2), 207–322 (2008)
17. L.E. Cross, R.E. Newnham, in *History of ferroelectrics, Ceramics and Civilization*, vol 3, ed. by W.D. Kingery. High-Technology Ceramics—Past, Present and Future (American Ceramic Society, Westerville, 1987), pp. 289–305
18. Y. Saito, H. Takao, T. Tani, T. Nonoyama, K. Takatori, T. Homma, T. Nagaya, M. Nakamura, Lead-free piezoceramics. Nature **432**, 84–87 (2004)
19. J. Rodel, W. Jo, K.T.P. Seifert, E.-M. Anton, T. Granzow, Perspective on the development of lead-free piezoceramics. J. Am. Ceram. Soc. **92**(6), 1153–1177 (2009)
20. G. Arlt, The influence of microstructure on the properties of ferroelectric ceramics. Ferroelectrics **104**, 217–227 (1990)
21. A.D. Polli, F.F. Lange, C.G. Levi, C G. Metastability of the fluorite, pyrochlore, and perovskite structures in the $PbO\text{-}ZrO_2\text{-}TiO_2$ system. J. Am. Ceram. Soc. **83**(4), 873–881 (2000)
22. C.A. Randall, N. Kim, J.P. Kucera, W. Cao, T.R. Shrout, Intrinsic and extrinsic size effects in fine-grained morphotropic-phase-boundary lead zirconate titanate ceramics. J. Am. Ceram. Soc. **81**(3), 677–688 (1998)
23. B. Jaffe, W.R. Cook, H. Jaffe, *Piezoelectric Ceramics* (Academic Press Limited, London, 1971)
24. I.N. Andryushinan, L.A. Reznichenko, V.A. Alyoshin, L.A. Shilkina, S.V. Titov, V.V. Titov, K.P. Andryushin, S.I. Dudkina, The PZT system ($PbZr_{1-x}Ti_xO_3$, $0.0 \leq x \leq 1.0$): specific features of recrystallization sintering and microstructures of solid solutions (part 1). Ceram. Int. **39**, 753–761 (2013)
25. B.W. Lee, Synthesis and characterization of compositionally modified PZT by wet chemical preparation from aqueous solution. J. Eur. Ceram. Soc. **24**, 925–929 (2004)
26. S.R. Shannigrahi, F.E.H. Tay, K. Yao, R.N.P. Choudhary, Effect of rare earth (La, Nd, Sm, Eu, Gd, Dy, Er and Yb) ion substitutions on the microstructural and electrical properties of sol-gel grown PZT ceramics. J. Eur. Ceram. Soc. **24**, 163–170 (2004)
27. L.B. Kong, J. Ma, T.S. Zhang, W. Zhu, O.K. Tan, $Pb(Zr_xTi_{1-x})O_3$ ceramics via reactive sintering of partially reacted mixture produced by a high-energy ball milling process. J. Mater. Res. **16**(6), 1636–1643 (2001)
28. N. Izyumskaya, Y.I. Alivov, S.J. Cho, H. Morkoc, H. Lee, Y.S. Kang, Processing, structure, properties, and applications of PZT thin films. Crit. Rev. Solid State Mater. Sci. **32**, 111–202 (2007)
29. L.B. Kong, A new solution method to deposit thick PZT films, unpublished works
30. B.M. Xu, Y.H. Ye, L.E. Cross, Dielectric properties and field-induced phase switching of lead zirconate titanate stannate antiferroelectric thick films on silicon substrates. J. Appl. Phys. **87**, 2507–2515 (2000)
31. B.M. Xu, L.E. Cross, D. Ravichandran, Synthesis of lead zirconate titanate stannate antiferroelectric thick films by sol-gel processing. J. Am. Ceram. Soc. **82**(2), 306–332 (1999)
32. B.M. Xu, Y.H. Ye, Q.M. Wang, N.G. Pai, L.E. Cross, Effect of compositional variations on electrical properties in phase switching (Pb, La)(Zr, Ti, Sn)O3 thin and thick films. J. Mater. Sci. **35**, 6027–6033 (2000)

33. Z.H. Wang, C.L. Zhao, W.G. Zhu, O.K. Tan, W. Liu, X. Yao, Processing and characterization of Pb(Zr, Ti)O3 thick films on platinum-coated silicon substrate derived from sol-gel deposition. Mater. Chem. Phys. **75**, 71–75 (2002)
34. W.G. Zhu, Z.H. Wang, C.L. Zhao, O.K. Tan, H.H. Hng, Low temperature processing of nanocrystalline lead zirconate titanate (PZT) thick films and ceramics by a modified solgel route. Jpn. J. Appl. Phys. **41**, 6969–6975 (2002)
35. Z.H. Wang, W.G. Zhu, C.L. Zhao, O.K. Tan, Dense PZT thick films derived from sol-gel based nanocomposite process. Mater. Sci. Eng., B **99**, 56–62 (2003)
36. C.L. Zhao, Z. Wang, W.G. Zhu, O.K. Tan, H.H. Hng, Microstructure and properties of PZT53/47 thick films derived from sols with submicron-sized PZT particle. Ceram. Int. **30**, 1925–1927 (2004)
37. Z.H. Wang, W.G. Zhu, C.L. Chao, X.F. Chen, Characterization of composite piezoelectric thick film for MEMS application. Surf. Coat. Technol. **198**, 384–388 (2005)
38. R.G. Kepler, R.A. Anderson, Ferroelectric polymers. Adv. Phys. **41**(1), 1–57 (1992)
39. V.V. Kochervinskii, Piezoelectricity in crystallizing ferroelectric polymers: poly(vinylidene fluoride) and its copolymers (a review). Crystallogr. Rep. **48**(4), 649–675 (2003)
40. T. Furukawa, Piezoelectricity and pyroelectricity in polymers. IEEE Trans. Electr. Insul. **24**, 375–394 (1989)
41. J.F. Tressler, S. Alkoy, A. Dogan, R.E. Newnham, Functional composites for sensors, actuators and transducers. Compos. Part A **30**, 477–482 (1999)
42. B. Hilczer, J. Kulek, E. Markiewicz, M. Kosec, B. Malic, Dielectric relaxation in ferroelectric PZT-PVDF nanocomposites. J. Non-Cryst. Solids **305**, 167–173 (2002)
43. K.A. Cook-Chennault, N. Thambi, A.M. Sastry, Powering MEMS portable devices-a review of non-regenerative and regenerative power supply systems with special emphasis on piezoelectric energy, harvesting systems. Smart Mater. Struct. **17**, 043001 (2008)
44. S.P. Beeby, M.J. Tudor, N.M. White, Energy harvesting vibration sources for microsystems applications. Meas. Sci. Technol. **17**, R175–R195 (2006)
45. Standards Committee of the IEEE Ultrasonics Ferroelectrics, and Frequency Control Society, *IEEE Standard on Piezoelectricity* (IEEE, New York, 1987)
46. A. Erturk, D.J. Inman, *Piezoelectric Energy Harvesting*, 1st edn. (Wiley, Chichester, 2011)
47. D. Zhu, in Vibration Energy Harvesting: Machinery Vibration, Human Movement and Flow Induced Vibration, Chapter 2, ed. by Y. K. Tan. *Sustainable Energy Harvesting Technologies—Past, Present and Future* (Intech, Rijeka, 2011)
48. H.A. Sodano, G. Park, D.J. Inman, Estimation of electric charge output for piezoelectric energy harvesting. Strain **40**, 49–58 (2004)
49. S.N. Chen, G.J. Wang, M.C. Chien, Analytical modeling of piezoelectric vibration-induced micro power generator. Mechatronics **16**, 387–397 (2006)
50. J. Ajitsaria, S.Y. Choe, D. Shen, D.J. Kim, Modeling and analysis of a bimorph piezoelectric cantilever beam for voltage generation. Smart Mater. Struct. **16**, 447–454 (2007)
51. G.K. Ottman, H.F. Hofmann, G.A. Lesieutre, Optimized piezoelectric energy harvesting circuit using step-down converter in discontinuous conduction mode. IEEE Trans. Power Electron. **18**, 696–703 (2003)
52. D. Guyomar, A. Badel, E. Lefeuvre, C. Richard, Toward energy harvesting using active materials and conversion improvement by nonlinear processing. IEEE Trans. Ultrason. Ferroelect. Freq. Control **52**, 584–595 (2005)
53. N. Kong, D.S. Ha, A. Erturk, D.J. Inman, Resistive impedance matching circuit for piezoelectric energy harvesting. J. Intell. Mater. Syst. Struct. **21**, 1293–1302 (2010)
54. S. Roundy, P.K. Wright, A piezoelectric vibration based generator for wireless electronics. Smart Mater. Struct. **13**, 1131–1144 (2004)
55. N.E. DuToit, B.L. Wardle, S. Kim, Design considerations for MEMS-scale piezoelectric mechanical vibration energy harvesters. Integr. Ferroelectr. **71**, 121–160 (2005)
56. L. Meirovitch, *Fundamentals of Vibrations* (McGraw-Hill, New York, 2001)
57. D.J. Inman, *Engineering Vibration* (Prentice Hall, Englewood Cliffs, 2007)

58. N.G. Stephen, On energy harvesting from ambient vibration. J. Sound Vib. **293**, 409–425 (2006)
59. A. Erturk, D.J. Inman, On mechanical modeling of cantilevered piezoelectric vibration energy harvesters. J. Intell. Mater. Syst. Struct. **19**, 1311–1325 (2008)
60. H.A. Sodano, G. Park, D.J. Inman, Estimation of electric charge output for piezoelectric energy harvesting. Strain **40**, 49–58 (2004)
61. N.E. duToit, B.L. Wardle, Experimental verification of models for microfabricated piezoelectric vibration energy harvesters. AIAA J. **45**, 1126–1137 (2007)
62. F. Lu, H.P. Lee, S.P. Lim, Modeling and analysis of micro piezoelectric power generators for micro-electromechanical-systems applications. Smart Mater. Struct. **13**, 57–63 (2004)
63. S.N. Chen, G.J. Wang, M.C. Chien, Analytical modeling of piezoelectric vibration-induced micro power generator. Mechatronics **16**, 387–397 (2006)
64. J.H. Lin, X.M. Wu, T.L. Ren, L.T. Liu, Modeling and simulation of piezoelectric MEMS energy harvesting device. Integr. Ferroelectr. **95**, 128–141 (2007)
65. A. Erturk, D.J. Inman, Issues in mathematical modeling of piezoelectric energy harvesters. Smart Mater. Struct. **17**, 065016 (2008)
66. A. Erturk, D.J. Inman, A distributed parameter electromechanical model for cantilevered piezoelectric energy harvesters. ASME J. Vib. Acoust. **130**, 041002 (2008)
67. A. Erturk, D.J. Inman, An experimentally validated bimorph cantilever model for piezoelectric energy harvesting from base excitations. Smart Mater. Struct. **18**, 025009 (2009)
68. C.J. Rupp, A. Evgrafov, K. Maute, M.L. Dunn, Design of piezoelectric energy harvesting systems: a topology optimization approach based on multilayer plates and shells. J. Intell. Mater. Syst. Struct. **20**, 1923–1939 (2009)
69. C. Jr De Marqui, A. Erturk, D.J. Inman, An electromechanical finite element model for piezoelectric energy harvester plates. J. Sound Vib. **327**, 9–25 (2009)
70. N.G. Elvin, A.A. Elvin, A coupled finite element—circuit simulation model for analyzing piezoelectric energy generators. J. Intell. Mater. Syst. Struct. **20**, 587–595 (2009)
71. Y. Yang, T. Tang, Equivalent circuit modeling of piezoelectric energy harvesters. J. Intell. Mater. Syst. Struct. **20**, 2223–2235 (2009)
72. A. Erturk, J. Hoffmann, D.J. Inman, A piezomagnetoelastic structure for broadband vibration energy harvesting. Appl. Phys. Lett. **94**, 254102 (2009)
73. S.C. Stanton, C.C. McGehee, B.P. Mann, Reversible hysteresis for broadband magnetopiezoelastic energy harvesting. Appl. Phys. Lett. **96**, 174103 (2009)
74. S.C. Stanton, C.C. McGehee, B.P. Mann, Nonlinear dynamics for broadband energy harvesting: investigation of a bistable inertial generator. Physica D **239**, 640–653 (2010)
75. S. Adhikari, M.I. Friswell, D.J. Inman, Piezoelectric energy harvesting from broadband random vibrations. Smart Mater. Struct. **18**, 115005 (2009)
76. N.S. Shenck, J.A. Paradiso, Energy scavenging with shoe-mounted piezoelectrics. IEEE Micro **21**, 30–42 (2001)
77. D. Fourie, Shoe-mounted PVDF piezoelectric transducer for energy harvesting (online version). http://web.vtc.edu/courses/el/elt2720/studentwork2012/KatieCloutier/index_files/shoe_mounted_piezo.pdf
78. J.G. Rocha, L.M. Goncalves, P.F. Rocha, M.P. Silva, S. Lanceros-Mendez, Energy harvesting from piezoelectric materials fully integrated in footwear. IEEE Trans Industr. Electron. **57**, 813–820 (2010)
79. L. Moro, D. Benasciutti, Harvested power and sensitivity analysis of vibrating shoe-mounted piezoelectric cantilevers. Smart Mater. Struct. **19**, 115011 (2010)
80. D. Benasciutti, L. Moro, Energy Harvesting with Vibrating Shoe-Mounted Piezoelectric Cantilevers (Chapter 6), in *Advances in Energy Harvesting Methods*, ed. by N. Elvin, A. Erturk (Springer Science+Business Media, New York, 2013)
81. W.R. Ledoux, H.J. Hillstrom, Acceleration of the calcaneus at heel strike in neutrally aligned and pes planus feet. Clin. Biomech. **16**, 608–613 (2001)

82. L. Mateu, F. Moll, Appropriate charge control of the storage capacitor in a piezoelectric energy harvesting device for discontinuous load operation. Sens. Actuators A **132**, 302–310 (2006)
83. W.M. Whittle, *Gait analysis, An Introduction*, 4th edn. (Elsevier, Philadelphia, 2007)
84. J. Perry, *Gait Analysis: Normal and Pathological Function* (SLACK Incorporated, Thorofare, 1992)
85. D.A. Winter, Foot trajectory in human gait: a precise and multifactorial motor control task. Phys. Ther. **72**, 45–53 (1992)
86. V.L. Giddings, G.S. Beaupre, R.T. Whalen, D.R. Carter, Calcaneal loading during walking and running. Med. Sci. Sports Exerc. **32**, 627–634 (2000)
87. B.K. Tripathy, A study on step distance and its relation with some morphometric features in adult male. Anthropologist **6**, 137–139 (2004)
88. P. Pillatsch, E.M. Yeatman, A.S. Holmes, A scalable piezoelectric impulse-excited energy harvester for human body excitation. Smart Mater. Struct. **21**, 115018 (2012)
89. M. Renaud, P. Fiorini, R. van Schaijkand, C. Van Hoof, Harvesting energy from the motion of human limbs: the design and analysis of an impact-based piezoelectric generator. Smart Mater. Struct. **18**, 035001 (2009)
90. S. Wei, H. Hu, S. He, Modeling and experimental investigation of an impact-driven piezoelectric energy harvester from human motion. Smart Mater. Struct. **22**, 105020 (2013)
91. M. Pozzi, M. Zhu, Plucked piezoelectric bimorphs for knee-joint energy harvesting: modeling and experimental validation. Smart Mater. Struct. **20**, 055007 (2011)
92. M. Pozzi, M. Zhu, Characterization of a rotary piezoelectric energy harvester based on plucking excitation for knee-joint wearable applications. Smart Mater. Struct. **21**, 055004 (2012)
93. M. Pozzil, M.S.H. Aung, M. Zhu, R.K Jones, J.Y Goulermas, The pizzicato knee-joint energy harvester: characterization with biomechanical data and the effect of backpack load. Smart Mater. Struct. **21**, 075023 (2012)
94. B. Yang, K.S. Yun, Piezoelectric shell structures as wearable energy harvesters for effective power generation at low-frequency movement. Sens. Actuators. A **188**, 427–433 (2012)
95. E. Kebadze, S.D. Guest, S. Pellegrino, Bistable prestressed shell structures. Int. J. Solids Struct. **41**(2004), 2801–2820 (2004)
96. S.J. Kim, J.S. Hwang, J. Mok, Sensor/actuator optimal design for active vibration control of shell structure. J. Intell. Mater. Syst. Struct. **11**, 848–856 (2000)
97. S. Daynes, K.D. Potter, P.M. Weaver, Bistable prestressed buckled laminates. Compos. Sci. Technol. **68**, 3431–3437 (2008)
98. L. Gu, C. Livermore, Compact passively self-tuning energy harvesting for rotating applications. Smart Mater. Struct. **21**, 015002 (2012)
99. M. Umeda, K. Nakamura, S. Ueha, Analysis of transformation of mechanical impact energy to electrical energy using a piezoelectric vibrator. Jpn. J. Appl. Phys. **35**, 3267–3273 (1996)
100. S.H. Kim, J.H. Ahn, H.M. Chung, H.W. Kang, Analysis of piezoelectric effects on various loading conditions for energy harvesting in a bridge system. Sens. Actuators. A **167**, 468–483 (2011)
101. M. Peigney, D. Siegert, Piezoelectric energy harvesting from traffic-induced bridge vibrations. Smart Mater. Struct. **22**, 095019 (2013)
102. V.J. Ovejas, A. Cuadras, Multimodal piezoelectric wind energy harvesters. Smart Mater. Struct. **20**, 085030 (2011)
103. J.J. Allen, A.J. Smits, Energy harvesting eel. J. Fluids Struct. **15**, 629–640 (2001)
104. P. Rakbamrung, M. Lallart, D. Guyomar, N. Muensit, C. Thanachayanont, C. Lucat, B. Guiffard, L. Petit, P. Sukwisut, Performance comparison of PZT and PMN-PT piezoceramics for vibration energy harvesting using standard or nonlinear approach. Sens. Actuators A **163**, 493–500 (2011)
105. W.B. Hobbs, D.L. Hu, Tree-inspired piezoelectric energy harvesting. J. Fluids Struct. **28**, 103–114 (2012)

106. C. Van Eysden, J. Sader, Resonant frequencies of a rectangular cantilever beam immersed in a fluid. J. Appl. Phys. **100**, 114916 (2009)
107. L. Ristroph, J. Zhang, Anomalous hydrodynamic drafting of interacting flapping flags. Phys. Rev. Lett. **101**, 194502 (2008)
108. D.A. Wang, H.H. Ko, Piezoelectric energy harvesting from flow-induced vibration. J. Micromech. Microeng. **20**, 025019 (2010)
109. D.A. Wang, N.Z. Liu, A shear mode piezoelectric energy harvester based on a pressurized water flow. Sens. Actuators A **167**, 449–458 (2011)
110. A. Erturk, G. Delporte, Underwater thrust and power generation using flexible piezoelectric composites: an experimental investigation toward self-powered swimmer-sensor platforms. Smart Mater. Struct. **20**, 125013 (2011)
111. R. Guigon, J.J. Chaillout, T. Jager, G. Despesse, Harvesting raindrop energy: theory. Smart Mater. Struct. **17**, 015038 (2008)
112. R. Guigon, J.J. Chaillout, T. Jager, G. Despesse, Harvesting raindrop energy: experimental study. Smart Mater. Struct. **17**, 015039 (2008)
113. D. Zhu, M.J. Tudor, S.P. Beeby, Strategies for increasing the operating frequency range of vibration energy harvesters: a review. Meas. Sci. Technol. **21**, 022002 (2010)
114. L. Roylance, J.B. Angell, A batch fabricated silicon accelerometer. IEEE Trans. Electron Devices **26**, 1911–1917 (1979)

Chapter 3
Waste Mechanical Energy Harvesting (II): Nanopiezoelectric Effect

3.1 Introduction

Recently, a new type of mechanical waste energy harvester—nanogenertor, based on nanopiezoelectric effect, has emerged. Nanogenerator is a device facilitated with modern nanotechnologies that can be used to convert mechanical or thermal energy as produced by small-scale physical change into electricity. Nanogenerators work is based on three typical effects: piezoelectric, triboelectric, and pyroelectric. Both the piezoelectric and triboelectric nanogenerators convert the mechanical energy into electricity. However, pyroelectric nanogenertors are used to harvest thermal energy from a time-dependent temperature fluctuation.

Piezoelectric nanogenerator is an energy harvesting device converting the external kinetic energy into an electrical energy based on the energy conversion by using the converse effect of nanostructured piezoelectric materials. Although its definition may include any types of energy harvesting devices with nanostructures converting various types of the ambient energy (e.g., solar power and thermal energy), it is used most of the time to specifically indicate the kinetic energy harvesting devices utilizing nanoscaled piezoelectric materials after its first introduction in 2006 [1]. Although still in the early stage of the development, it has been regarded as a potential breakthrough toward further miniaturization of the conventional energy harvesters, possibly leading to facile integration with the other types of energy harvesters that convert different types of energy and the independent operation of mobile electronic devices with the reduced concerns for the energy source consequently.

Triboelectric nanogenerator is an energy harvesting device that converts the external ambient mechanical energy into electricity based on the so-called nano-scale triboelectric effect. It is another mechanical energy harvesting technology in addition to the piezoelectric nanogenerator. This new type of nanogenerator was first demonstrated in Z. L. Wang at Georgia Institute of Technology [2]. Compared to piezoelectric nanogenerators, triboelectric nanogenerators have output voltages up to hundreds of volts [3]. The use of triboelectric nanogenerators has been proven to be a simple, cost-effective, robust, and an efficient approach to harvest

L. B. Kong et al., *Waste Energy Harvesting*, Lecture Notes in Energy 24,
DOI: 10.1007/978-3-642-54634-1_3, © Springer-Verlag Berlin Heidelberg 2014

mechanical waste energy [3, 4]. Both the piezoelectric and triboelectric nanogenerators are based on piezoelectric effect to harvest mechanical energy, while pyroelectric nanogenerators are using pyroelectric effect to convert thermal energy into electricity. As a result, this chapter will focus on piezoelectric (triboelectric) nanogenerators.

3.2 Crystal Characteristics of ZnO

Although theoretically all nanosized piezoelectric materials could be used to develop nanopiezoelectric nanogenerators, all results until now have shown that the most promising material for such specific applications is ZnO. Therefore, it is necessary to first introduce the crystal structure and properties of ZnO. ZnO has wurtzite crystal structure, having a hexagonal unit cell, with space group of *C6mc* and lattice parameters of $a = 0.3296$ nm and $c = 0.52065$ nm. The oxygen anions and Zn cations construct a tetrahedral unit. The crystal structure of ZnO lacks of central symmetry. The structure of ZnO can be simply described as a number of alternating planes that are composed of tetrahedrally coordinated O^{2-} and Zn^{2+} ions, which are stacked alternatively along the c-axis (Fig. 3.1a) [5].

Although the entire unit cell of ZnO is neutral, the distribution of the cations and anions could take specific configurations as determined by crystallography, so that some surfaces can be terminated entirely with either cations or anions, thus resulting in positively or negatively charged surfaces, which are called polar surfaces. The polar charge dominated surfaces are responsible for the unique growth behaviors of ZnO into nanowires. The most common polar surface is the basal plane. The oppositely charged ions produce positively charged Zn-(0001) and negatively charged O-$(000\bar{1})$ polar surfaces, resulting in a normal dipole moment and spontaneous polarization along the c-axis, as well as a divergence in surface energy. To maintain a stable structure, the polar surfaces generally have facets or exhibit massive surface reconstructions, but ZnO ±(0001) are exception, which are atomically flat, stable, and without reconstruction [6, 7]. Understanding the superior stability of the ZnO ±(0001) polar surfaces is a forefront research subject in surface physics [8].

Another polar surface is family of $\{01\bar{1}1\}$ planes. By projecting the structure along $[1\bar{2}10]$, as shown in Fig. 3.1b [5], beside the most typical ±(0001) polar surfaces that are terminated with Zn and O, respectively, $\pm(10\bar{1}1)$ and $\pm(10\bar{1}\bar{1})$ are also polar surfaces. The $\{10\bar{1}1\}$ surfaces are not commonly present in ZnO, but they have been observed in a nanohelical structure [9]. The charges on the polar surfaces are ionic charges, which are neither transferable nor mobile. Because the interaction energy among the charges depends on the distribution of the charges, the structure is arranged in such a configuration to minimize the electrostatic energy. This is the main driving force that polar surface dominated nanostructures can be grown.

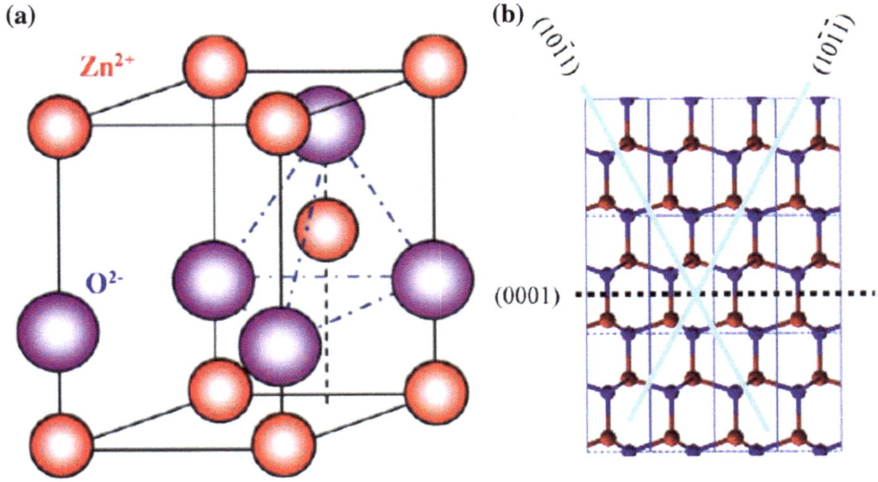

Fig. 3.1 **a** Wurtzite structure model of ZnO, which has noncentral symmetry and piezoelectric effect. **b** The three types of facets of ZnO nanostructures: ±(0001), $\{2\bar{1}\bar{1}0\}$ and $\{01\bar{1}0\}$. Courtesy of Dr. Z. L. Wang, Georgia Institute of Technology. Reproduced with permission from [5], Copyright @ 2009, Elsevier

ZnO has three types of fast growth directions: $\langle 2\bar{1}\bar{1}0\rangle$> (±$[2\bar{1}\bar{1}0]$, ±$[\bar{1}2\bar{1}0]$, ±$[\bar{1}\bar{1}20]$), $\langle 01\bar{1}0\rangle$ (±$[01\bar{1}0]$, ±$[10\bar{1}0]$, ±$[1\bar{1}001\ 1\ 00]$) and ±[0001]. Together with the polar surfaces due to atomic terminations, ZnO exhibits a wide range of novel structures that can be grown by tuning the growth rates along these directions. That is the reason why a huge number of morphologies have been synthesized and reported in the open literature for ZnO. The morphology adopted by ZnO under given synthesis conditions is determined by the relative surface activities of various growth facets. Macroscopically, different planes of a crystal have different kinetic parameters. Which parameter plays a determining role is dependent on the controlled growth conditions. Thus, after an initial period of nucleation and incubation, a crystallite will usually develop into a three-dimensional object with well-defined and low-index crystallographic faces. Figure 3.2 shows a few typical growth morphologies of 1D ZnO nanostructures [5]. These structures tend to maximize the areas of the $\{2\bar{1}\bar{1}0\}$ and $\{01\bar{1}0\}$ facets because of their relatively lower energy. The morphology shown in Fig. 3.2d is dominated by the polar surfaces, which can be grown by introducing planar defects parallel to the polar surfaces [10]. Planar defects and twins are usually observed parallel to the (0001) plane, while dislocations are rarely found [11].

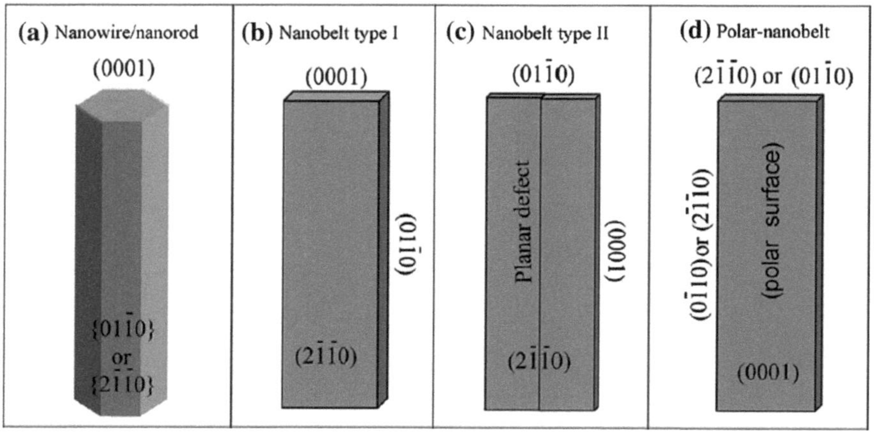

Fig. 3.2 Typical growth morphologies of one-dimensional ZnO nanostructures and the corresponding facets. Courtesy of Dr. Z. L. Wang, Georgia Institute of Technology. **a** Nanowire/nanorad. **b** Nanobelt Type I. **c** Nanobelt Type II. **d** Polar-nanobelt. Reproduced with permission from [5], Copyright @ 2009, Elsevier

3.3 Growth of Piezoelectric ZnO Nanostructures

The applications of ZnO nanostructures for nanogenerators have specific requirements on morphologies and properties of the materials, which are determined mainly by the synthetic methods used. Numerous physical and chemical methods have been developed to synthesize ZnO nanostructures with various dimensions and morphologies. It has been accepted that for nanogenerator applications ZnO should be grown as nanowire arrays. Currently, the most convenient techniques for the growth of high-quality ZnO nanowire arrays include vapor depositions, solution chemical deposition, hydrothermal reaction, and pulsed laser deposition (PLD). The applications of these methods will be discussed in this section.

3.3.1 Vapor Phase Deposition

There are two vapor phase deposition methods used to fabricate ZnO nanowire arrays: vapor–solid–solid (VSS) and vapor–liquid–solid (VLS). VSS is a simple and an effective method without the use of catalysts. It usually requires relatively higher temperatures, because pure ZnO has a very high melting point (1,975 °C) and thus relatively high evaporation temperature (>1,300 °C). VLS method uses catalysts, like gold (Au). Due to the formation of liquid intermediate phase for the growth of the ZnO nanowires, it is named as VLS. Both methods have their advantages and disadvantages, as discussed below.

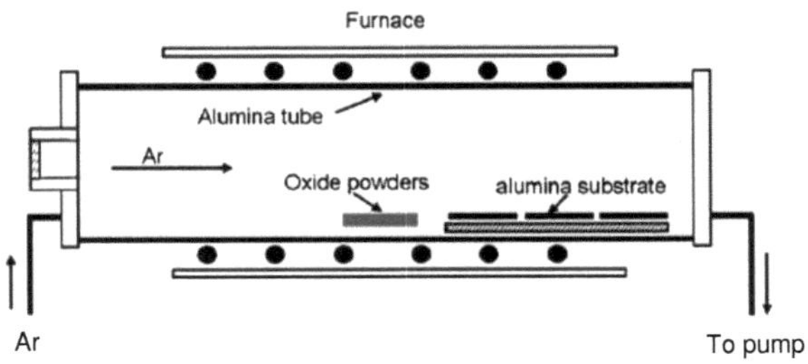

Fig. 3.3 Furnace system for the growth of ZnO nanowires through VLS and VSS process. Courtesy of Dr. Z. L. Wang, Georgia Institute of Technology. Reproduced with permission from [12], Copyright @ 2002, Cambridge University Press

Figure 3.3 shows a set up of the furnace system to grow oxides with various nanostructures through VSS [12]. The source materials are usually vaporized by using two approaches: thermal vaporization and laser ablation. Thermal vaporization means that source materials are heated at a given temperature to vaporize. The vapor phases are carried to lower temperatures zones by using a flow of carrying gas (Ar or N_2), where the vapors condense to form solid state materials. The condensation conditions can be controlled so as to control microstructures and thus properties of the final products. Experimental conditions include temperature, pressure, atmosphere, substrate, and so on. Morphology and phase composition of the synthesized products are closely related to the source materials used, growth temperature, temperature gradient, substrate, gas flow rate, and pressure and positions where samples are collected. In summary, VSS process is a simple yet effective technique to synthesize nanostructured materials. Figure 3.4 shows a collection of ZnO nanostructures with various morphologies and shapes [13]. Although nanostructured materials have been synthesized by using thermal vaporization technique like VSS, it is more desirable to control them for specific applications. To be used for nanogenerator applications, highly aligned ZnO nanowire arrays are preferred.

VLS process is an alternative process that has been demonstrated to be a feasible approach to grow ZnO nanowire arrays at relatively low temperatures. As such, the growth usually requires the use of substrates with catalyst or seeding layers. An example is shown in Fig. 3.5 [14]. Vertically aligned ZnO nanowire arrays are successfully grown on *a*-plane ((110) crystal surface) orientated single-crystal aluminum oxide (Al_2O_3, sapphire) substrates. Gold (Au) nanoparticles are used as catalyst layers. The growth of a ZnO nanowire is initiated and guided by the Au particle and the epitaxial relationship between ZnO and Al_2O_3 leads to the alignment (Fig. 3.5b).

To grow desirable-aligned ZnO nanowire arrays by using VSL process, the growth rate must be well controlled. Usually, the growth rate should be controlled in such a way that the catalyst particles have sufficient time to be molten, so as to form alloys with the materials to be grown and precipitate later to achieve epitaxial growth of ZnO on sapphire. An effective way to reduce the growth rate is to decrease the growth temperature, because the lower the temperature the lower the vapor concentration will be. By mixing ZnO with carbon (C) powder, which is

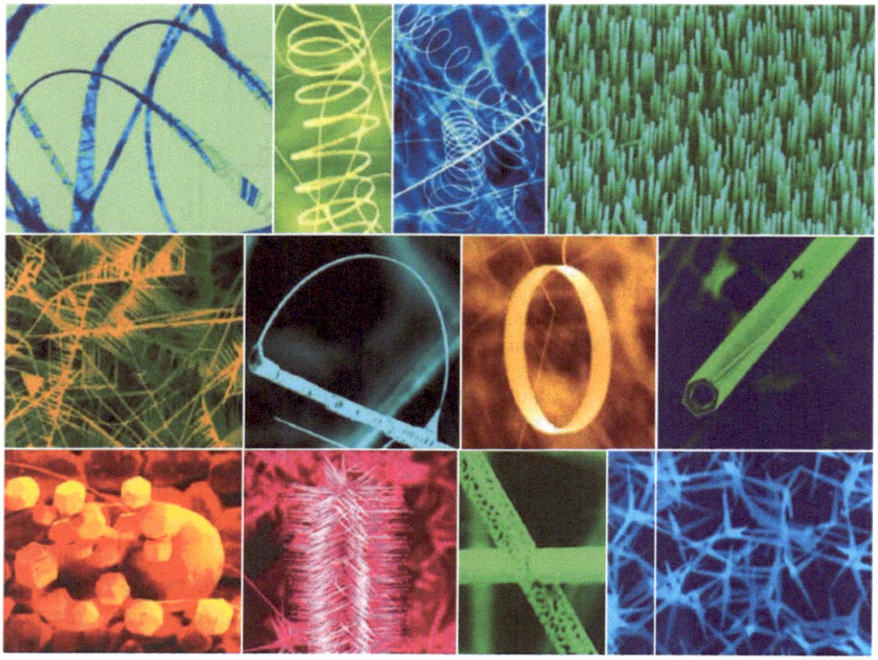

Fig. 3.4 A collection ZnO nanostructures with different morphologies and shapes. Courtesy of Dr. Z. L. Wang, Georgia Institute of Technology. Reproduced with permission from [13], Copyright @ 2004, Elsevier

known as carbon-thermal evaporation, the vaporization temperature can be reduced from >1,300 to 900 °C, due to the following reaction,

$$\mathrm{ZnO}\ (s) + \mathrm{C}\ (s) \leftrightarrow \mathrm{Zn}\ (v) + \mathrm{CO}\ (v) \tag{3.1}$$

Because this reaction is reversible at a relatively lower temperature, when the mixture of Zn vapor and CO is brought away from the high temperature center region of the furnace, condensation occurs in the substrate region and ZnO is formed. The ZnO particles could be interacting with the Au catalysts, which eventually facilitate the growth of ZnO nanowires from the liquid through the VLS growth process.

If the catalysts are prepatterned by lithography or self-assembly techniques, ZnO nanowires will be grown selectively in those areas with the catalysts remained [13, 15]. Therefore, well-aligned ZnO nanowire arrays with patterns could be fabricated. Figure 3.6 shows an example of ZnO nanowire arrays on substrates with prepatterned catalysts [13]. A sapphire substrate is covered with a two-dimensional, large area, self-assembled, and ordered monolayer of submicron polystyrene spheres (Fig. 3.6a). The layer of polystyrene spheres serves as a mask, through which a thin layer of Au particles is then deposited. After the polystyrene

Fig. 3.5 Epitaxial growth of ZnO nanowires on *a*-plane (110) sapphire.**a** Schematic illustration of ZnO *ab*-plane overlapping with (110) plane of the sapphire substrate and **b** ZnO nanowire arrays grown on the sapphire substrate. Reproduced with permission from [14], Copyright @ 2002, John Wiley & Sons

spheres are etched away, a hexagonal network of Au layer is retained (Fig. 3.6b). On this Au layer, well-aligned ZnO nanowire arrays are formed (Fig. 3.6c).

The degree of epitaxial growth is determined by the matching between ZnO and the substrates. Because ZnO has same crystal structures as sapphire and nitride [16, 17] and very small lattice mismatch, ZnO is very easily grown epitaxially on them. The reason why sapphire with $(11\bar{2}0)$ plane orientation is usually used as substrate is that there is the smallest lattice mismatch between the *c*-axis of Al_2O_3 and *a*-axis of ZnO. The epitaxial relationships between a ZnO nanowire and the *a*-plane sapphire substrate are $(0001)_{ZnO} \parallel (11\bar{2}0)_{Al2O3}$ and $[11\bar{2}0]_{ZnO} \parallel [0001]_{Al2O3}$. There is almost no difference in lattice parameter between $4[0110]_{ZnO}$ ($4 \times 3.249 = 12.996$Å) and $[0001]_{Al2O3}$ (12.99 Å), which is responsible for the epitaxial growth of the ZnO nanowires. In addition, because the $(11\bar{2}0)$ plane of Al_2O_3 is a rectangular lattice, whereas the (0001) plane of ZnO is a hexagonal one, there is only one direction that allows for the epitaxial growth to occur [13, 15].

Aligned ZnO NWs have been successfully grown on sapphire, GaN, AlGaN, and AlN substrates [16] through a VLS process, where the crystal structure of substrate is crucial for the orientation of NWs. The growth direction is controlled by the epitaxial relationship between the substrate and NWs (Fig. 2.7); while the

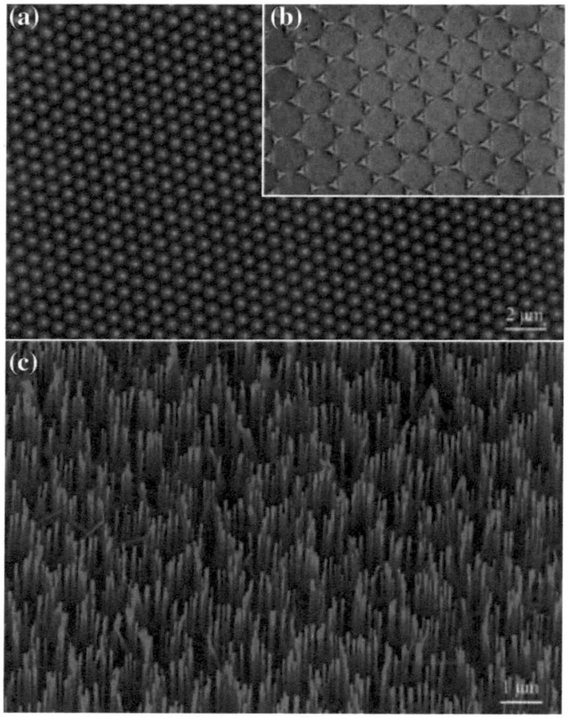

Fig. 3.6 Growth of patterned and aligned ZnO nanowire arrays. **a** Self-assembled monolayer of polystyrene spheres that serves as a mask. **b** Hexagonally patterned thin layer of Au catalyst on the substrate. **c** Aligned ZnO nanowires grown on a sapphire substrate with the honeycomb pattern defined by the catalyst mask. Courtesy of Dr. Z. L. Wang, Georgia Institute of Technology. Reproduced with permission from [13], Copyright @ 2004, Elsevier

quality of alignment is controlled by many other factors, such as chamber pressure, oxygen partial pressure, and thickness of catalyst layer [16, 17].

3.3.2 *Pulse Laser Deposition*

PLD has become one of the most widely used thin film fabrication techniques. Figure 3.7 shows a schematic diagram of a PLD set up, where a high power pulsed laser beam is focused inside a vacuum chamber to strike a target of the material that is to be deposited. This material is vaporized from the target (in a plasma plume) which deposits it as a thin film on a substrate. The experiment is usually conducted in ultra-high vacuum or in the presence of a background gas, such as oxygen which is commonly used when depositing oxides to fully oxygenate the deposited films. There are various parameters that have effects on properties of the films, such as substrate temperature, chamber pressure, laser energy density, frequency of pulsed layer, distance between target and substrate and rotating speed of target, and so on.

Figure 3.8 shows SEM images of the $Zn_{1-x}Mg_xO$ nanowires grown by using PLD [18], where a KrF excimer laser (Coherent Complex 205, wavelength of 248 nm) is used as the ablation source to focus on a ceramic target with the same

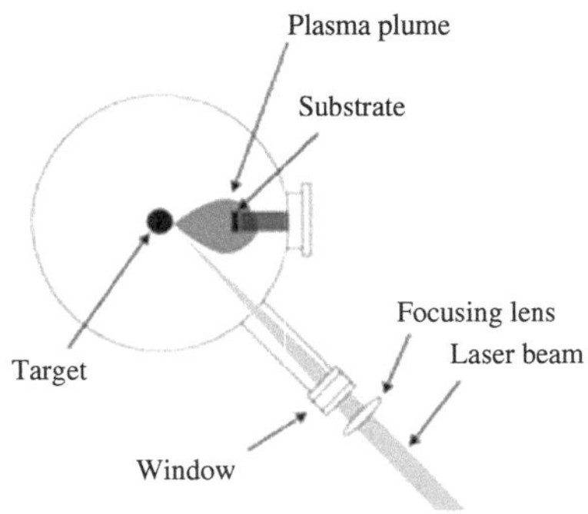

Fig. 3.7 Schematic diagram of PLD set-up

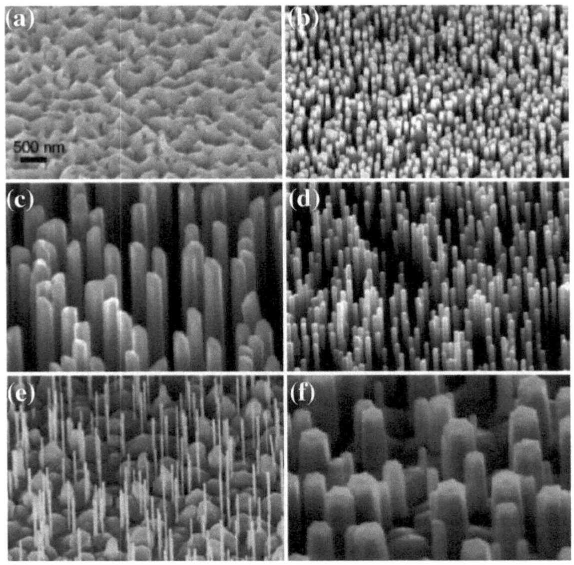

Fig. 3.8 SEM images with tilted views of the nanowires grown at **a** 700 °C **b** 750 °C **c** 800 °C **d** 850 °C and the growth pressure was 5 Torr. **e** and **f** Nanowire arrays grown at 6.0 and 4.0 Torr at 800 °C. All the images have a same scale bar of 500 nm. Courtesy of Dr. Z. L. Wang, Georgia Institute of Technology. Reproduced with permission from [18], Copyright @ 2009, American Chemical Society

composition, which is 0.5 at. % Li-doped $Zn_{0.95}Mg_{0.05}O$. The incident laser power density is about 3 J cm^{-2}. A textured ZnO buffer layer has been predeposited on a silicon substrate before nanowires are grown. Morphology and density of the nanowire arrays are related to four parameters: substrate temperature, chamber pressure, flow ratio of Ar and O_2,and laser repetition frequency [18]. Figure 3.8a–d shows SEM images of the nanowires deposited at different temperatures, at a laser repetition frequency of 5 Hz, a flowing mixture of Ar and O_2 with a ratio of 6:1 and a total pressure of 5.0 Torr. After deposition at 700 °C, no nanowires are formed, instead, only a film with a rough surface is deposited on the substrate (Fig. 3.8a). As the temperature is increased to 750 °C, nanowires, with a typical length of about 800 nm, are produced (Fig. 3.8b). Lengths of nanowires are increased to 2–3 μm after deposition at 800 °C (Fig. 3.8c). Further increasing temperature to 850 °C leads to merging of the nanowires (Fig. 3.8d). On the other hand, at given temperatures, morphologies of the nanowires can be controlled by controlling the growth pressure. For example, at 800 °C, as the growth pressure is increased to 6.0 Torr, the nanowires are very thin (Fig. 3.8e), whereas at 4.0 Torr, the nanowires are quite thick (Fig. 3.8f).

3.3.3 Chemical Processes

The above vapor deposition methods usually involve high temperatures and require crystal lattice matching between substrates and ZnO for epitaxial growth. By using chemical solution process, the problem of substrate dependence can be addressed. The available chemical techniques include solution deposition, catalyst facilitated deposition, electrochemical deposition, and hydrothermal reaction.

For chemical solution deposition, the strategy is the use of ZnO seeds in the form of either thin films or nanoparticles. Aligned growth of ZnO nanocrystals takes place on flat surfaces regardless of their crystallinity and surface chemistry. Currently, such substrates include ZnO and Al_2O_3 single crystals, transparent conducting oxides such as indium tin oxide (ITO), and fluorine-doped tin oxide (FTO), amorphous oxides including glasses and Si with native oxide and the oxide-free metals such as Au and Ti.

It has been reported that ZnO nanowires can be grown on polymeric substrate, such as polyimide, by using Au, Ag, or Pt as catalysts, through chemical process. The metallic particles are deposited onto polymeric substrates as thin films, on which ZnO nanowires are deposited. In this case, surface roughness of the catalyst thin films must be carefully optimized to promote nucleation of ZnO nanowire growth. In addition, the quality of the nanowire is largely determined by the surface morphology.

Electrochemical deposition has been found to be powerful technique to synthesize ZnO nanowire arrays with large area and uniform nanowire distribution [19, 20]. Electrochemical reaction usually occurs in strong acidic solutions, which requires substrates to be stable in such harsh environments. The substrates must also be conductive to carry out the electrochemical reactions. $ZnCl_2$/KCl has been widely used as electrolyte. The electrode solution is bubbled with O_2 to maintain saturation concentration of O_2. A standard three electrode setup, with a saturated Ag/AgCl reference electrode and a Pt counter electrode, is usually used for the electrochemical reactions. A potentiostat is used to control the current-voltage characteristics throughout the reaction process.

Recent studies indicated that nucleation and growth rate of ZnO nanowires can be further enhanced by using a solution of $Zn(NO_3)_2$ and hexamethylenetetramine (HMTA). Also, individual ZnO nanorod has been successfully grown in polymer substrates that have been prepatterned [19, 20]. In this case, the applied potential induces an increase in the concentration of OH^- ion due to the reduction in concentration of NO_3^- ion at the negative electrode. The reactions can be expressed as follows:

$$NO_3^- + H_2O + 2e- \rightarrow NO_2^- + 2OH^- \quad (3.2)$$

$$Zn^{2+} + 2OH^- \rightarrow ZnO + H_2O \quad (3.3)$$

The presence of HMTA promotes the growth of 1D nanowires by capping the nonpolar facets of ZnO and providing a continuous hydroxide source. Because there is no need to use seeds, this method is promising for the growth of nanowires on polymeric substrates.

Hydrothermal synthesis is another chemical method that has been widely used to synthesize ZnO nanowires, in which zinc nitrate hexahydrate ($Zn(NO_3)_2 \cdot 6H_2O$) and HMTA are used as starting chemicals [21, 22]. $Zn(NO_3)_2 \cdot 6H_2O$ is the source of Zn^{2+} ions to grow ZnO nanowires, while O^{2-} ions are from water. HMTA serves as a weak base and hydrolyzes in the water solution to supply OH^-. The slow hydrolysis of HMTA is critical to the growth of ZnO nanowires. Otherwise, too much OH^- ions are produced in a short time duration and thus raises the pH value of the solution. In this case, Zn^{2+} ions precipitate too fast to have an epitaxial growth. Chemical reactions can be expressed by:

$$(CH_2)_6N_4 + 6\ H_2O \leftrightarrow 4\ NH_3 + 6\ HCHO \quad (3.4)$$

$$NH_3 + H_2O \leftrightarrow NH_3 \cdot H_2O \quad (3.5)$$

$$NH_3 \cdot H_2O \leftrightarrow NH_4^+ + OH^- \quad (3.6)$$

$$Zn^{2+} + 2OH^- \leftrightarrow Zn(OH)_2 \quad (3.7)$$

$$Zn(OH)_2 \leftrightarrow ZnO + H_2O \quad (3.8)$$

All the five reactions are actually reversible and thus can be controlled by adjusting reaction parameters, including precursor concentrations, reaction temperature, and growth time, so as to push the reaction equilibriums forward or backward. In general, the density of the ZnO nanowires is determined by the precursor concentration in the solution. Growth time and temperature control the morphology and aspect ratio of ZnO nanowires.

An example is shown in Fig. 3.9 [22], in which a series of experiments are performed by varying the precursor concentration but keeping the ratio between the $Zn(NO_3)_2$ and HMTA to be constant. These experiments thus help to clarify the effect of concentration of the precursor on density of the ZnO nanowire arrays. Si substrates are taken as an example to show the growth of ZnO nanowires by using this method. Si substrate has been used as an example to describe the experimental procedure and to illustrate the effects of various experimental parameters on the growth. Si(100) wafers are cleaned following a standard cleaning progress, which is then coated with a 50-nm thick layer of Au. Between Si wafer and the Au layer, a 20 nm layer of Ti is predeposited as an adhesion layer

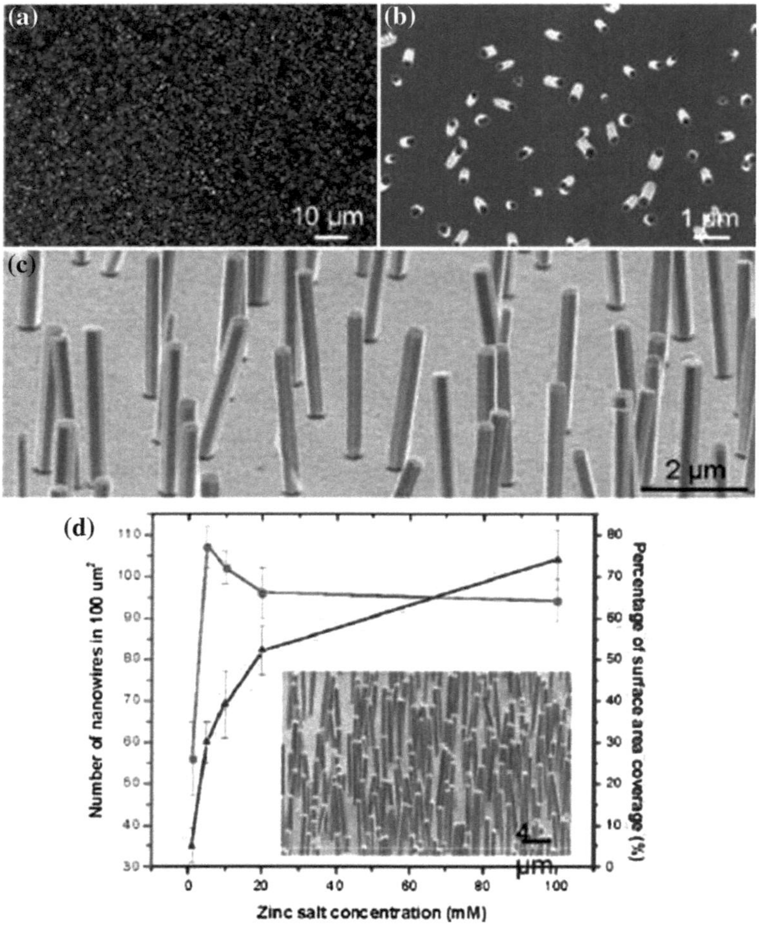

Fig. 3.9 A general view of the as-grown ZnO nanowire arrays from the 5 mM solution after growing for 24 h at 70 °C. **a** Top view **b** enlarged top view **c** image with a 60°tilt. **d** Density variation with concentration: plot of ZnO nanowire density in an area of 100 μm² (*red line*) and plot of area percentage covered by the ZnO nanowires (*blue line*). The inset is a representative image of the ZnO nanowires grown at 5 mM. Courtesy of Dr. Z. L. Wang, Georgia Institute of Technology. Reproduced with permission from [22], Copyright @ 2008, Cambridge Univ Press

to buffer the large lattice mismatch between Si(100) with native oxide on and Au(111) Then, the substrate is annealed at 300 °C for 1 h. The nutrient solution consists of $Zn(NO_3)_2$ and HMTA with a ratio of 1:1. The substrate is put face down, just contacting surface of the nutrient solution. The substrate is floated on the solution surface due to the surface tension.

Experimental results have demonstrated that the density of the NW arrays is closely related to the precursor concentration, as shown in Fig. 3.9d (red line) [22]. As the concentration of the solution is increased from 0.1 to 5 mM, the density of the ZnO nanowires, defined as number of the nanowires per 100 μm^2, increases significantly, which can be attributed to the following reasons. First, high concentration of zinc means high chemical potential, thus leading to more nucleation sites on the substrate surface. As a result, the density of ZnO nanowires in increased correspondingly. At low solution concentrations, this increasing trend is very pronounced. However, further increase in solution concentration will reach saturation at certain critical point. Second, those newly formed nuclei with sizes smaller than the critical size may not have chance to grow, instead, they are dissolved into the solution. Therefore, the density of the ZnO nanowires is mainly determined by the density of the nuclei formed in the early stage. In other words, there is saturation in nanowire density. Although the density of the ZnO nanowires is saturated at high precursor concentrations, the surface coverage increases slightly due to the lateral growth of the NWs (Fig. 3.9d, blue line).

3.3.4 Growth of Patterned Nanowire Arrays

3.3.4.1 Vertically Patterned Nanowire Arrays

Growth of patterned vertically aligned ZnO nanowire arrays is an important requirement for nanogenerator applications. To grow high quality-patterned ZnO nanowire arrays, there are three requirements. First, the growth temperature should be sufficiently low, so that normal substrates can be used. Second, the nanowire growth should be highly controllable, in terms of dimension, orientation, and shape. Finally, there should be no need of catalyst for integrating with silicon technology. A new approach that combines electron-beam lithography and a low-temperature hydrothermal synthesis has been developed to achieve patterned vertically aligned ZnO nanowire arrays at temperatures of <100 °C on inorganic substrates, such as Si and GaN, without using catalysts [23]. The Si doped *n*-type GaN substrate is a 2 μm thick GaN on *c* plane sapphire deposited by using metal organic chemical vapor deposition (MOCVD).

A 50-nm thick layer of PMMA is spun on the substrate, which is then baked on a hotplate at 180 °C for 2 min. The PMMA layer is patterned by an array of circle with 100 nm in diameter and 1 μm in pitch. After electron beam exposure and development, the substrate can be used to grow ZnO nanowire arrays. The nutrient solution is composed of 5 $mmol{\cdot}L^{-1}$ 1:1 ratio of $Zn(NO_3)_2$ and HMTA. Hydrothermal reaction is conducted at 70 °C or 95 °C for 24 h. Reaction at 70 °C leads to growth of multiple ZnO nanowires out of one single spot on Si wafer, while at 95 °C results in one ZnO nanorod in one single spot on Si wafer and GaN substrate.

Figure 3.10 shows SEM images of the ZnO nanowire arrays on Si substrates by using the hydrothermal synthesis [23]. After reaction at 70 °C, ZnO nanowire cluster is grown at every spot. The 95 °C samples consist of single nanowire at each spot. SEM images of the ZnO nanowires on GaN coated sapphire substrates

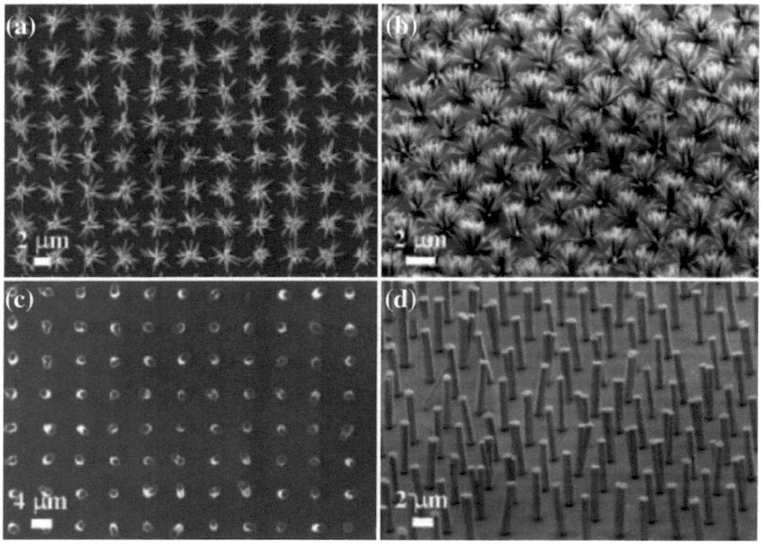

Fig. 3.10 **a** Top view and **b** 60° tilt view of the ZnO nanowire arrays on a Si substrate grown at 70 °C. **c** Top view and **d** 60° tilt view of the ZnO nanowire arrays on a Si wafer grown at 95 °C. Courtesy of Dr. Z. L. Wang, Georgia Institute of Technology. Reproduced with permission from [23], Copyright @ 2008, American Chemical Society

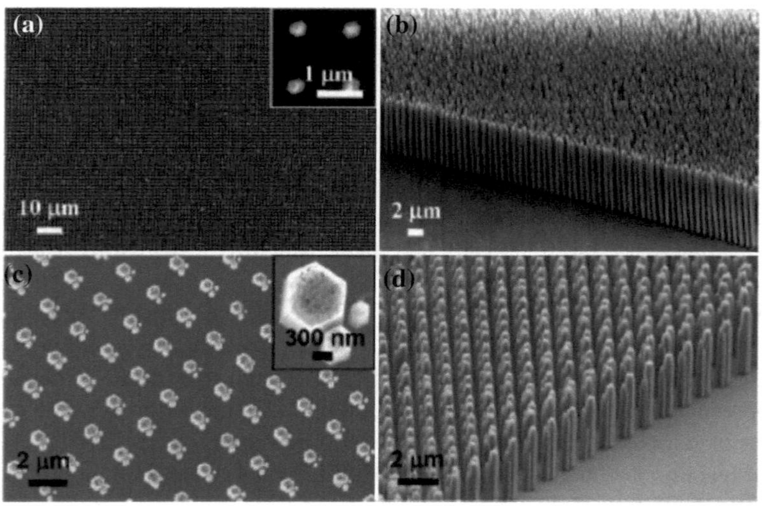

Fig. 3.11 **a** Top view and **b** 60° tilt view of the ZnO nanowire arrays on a GaN substrate. **c** Top view and **d** 60° tilt view of the nanowires growing out of 400, 200, and 100 nm sized holes. Insets are enlarged top view image. Insets are enlarged top view of the nanowires. Courtesy of Dr. Z. L. Wang, Georgia Institute of Technology. Reproduced with permission from [23], Copyright @ 2008, American Chemical Society

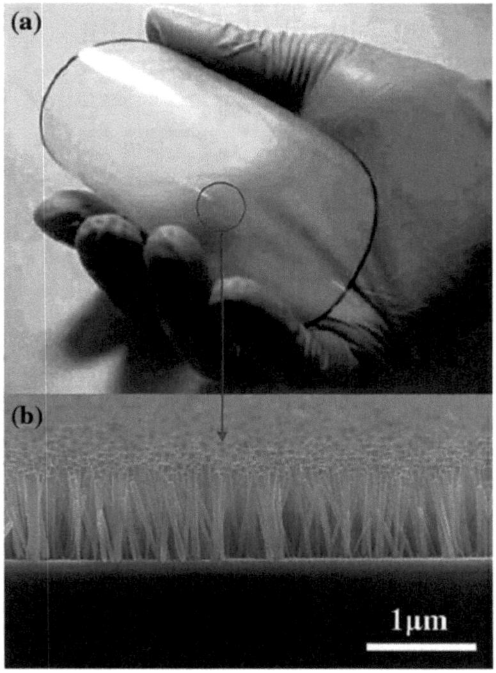

Fig. 3.12 **a** Photograph of the ZnO nanowires grown on a flexible 4-in. TPU substrate. **b** SEM image of the ZnO nanowires with a uniform length on the TPU substrate. Reproduced with permission from [24], Copyright @ 2006, American Chemical Society

are shown in Fig. 3.11 [23]. Because the lattice mismatch between the wurzite GaN (0001) plane and wurzite ZnO (0001) plane is only 1.8 %, the ZnO nanowires grown on GaN have better alignment.

3.3.4.2 Patterned Arrays on Flexible Substrates

It is also desirable to deposit ZnO nanowire arrays on flexible substrates, due some special requirements of nanogenerators for certain applications. Recent developments indicated that ZnO nanowires have been successfully synthesized on various organic substrates. The key to growth of high quality ZnO nanowires is the presence of a thin layer of ZnO seeds. Due to the use of polymer substrate, the deposition temperature should be as low as possible. Generally, the ZnO seeding layers can be deposited by using sputtering techniques at room temperature. Flexible polymer substrates that have been used for such a purpose include polyimide, polystyrene, polyethylene terephthalate and polyethylene naphthalate. Because ZnO nanowires are usually grown by using electrochemical method, the substrates should be conductive or precoated with a metal or a conducting oxides such as ITO. Figure 3.12 shows an example of ZnO nanowires grown on a 4 in. thermoplastic polyurethane (TPU) substrate [24]. The ZnO nanowires have a well-defined hexagonal morphology with an average diameter of about 70 nm.

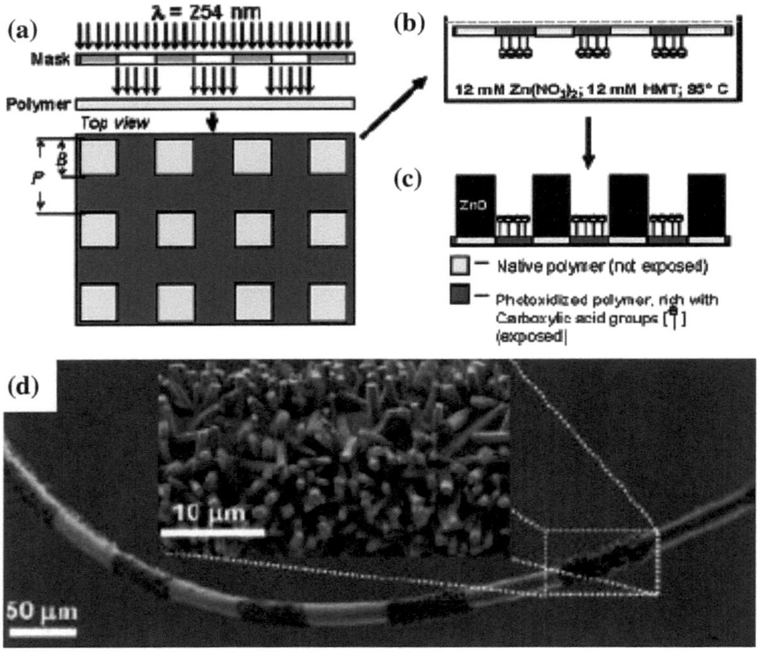

Fig. 3.13 Schematic illustration of the process to produce ordered microarrays of ZnO (**a–c**). SEM images of the patterned ZnO nanowires assembled on flexible PET fibers (**d**). Reproduced with permission from [25], Copyright @ 2007, American Chemical Society

Fibers with patterned ZnO nanowire arrays could be potentially integrated with woven textiles for wearable electronics. A simple catalyst-free process has been reported to coat ZnO nanowires on microfibers [25]. The polymer fibers are polyester (polyethylene terephthalate or PET). The main strategy is the use of UV radiation. It is found that in the regions that are radiated carboxylic groups are generated. These regions are covered by the protonated HMTA molecules, due to the chemical bonding with the carboxylic acid groups, thus blocking the nucleation of ZnO. Therefore, ZnO nanowire arrays are only grown selectively onto the regions without the carboxylic groups. The experimental procedure and representative SEM image is shown in Fig. 3.13 [25].

3.3.4.3 Laterally Aligned Nanowire Arrays

More recently, laterally aligned ZnO nanowire arrays have been successfully grown, as shown in Fig. 3.14 [26]. In this study, two materials are used: ZnO seeds for the growth ZnO nanowires and Cr layer for selective local growth. The first step is to fabricate a ZnO strip pattern covered with a thin layer of Cr (Fig. 3.14a). A cleaned (100) Si wafer is covered by layer of photoresist through spin-coating.

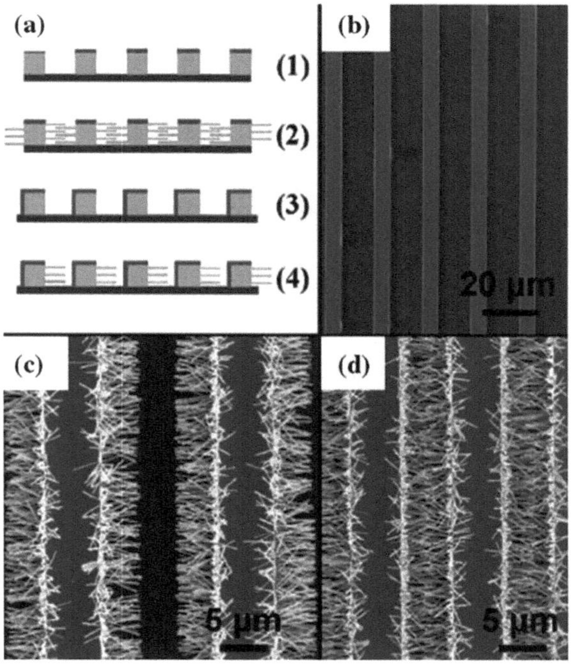

Fig. 3.14 **a** Schematic steps for growing patterned and laterally aligned ZnO nanowire arrays. **b** SEM image of the ZnO pattern covered with a 10 nm Cr shield layer prior to the nanowire growth. **c** and **d** SEM images of the ZnO nanowire arrays grown on Si substrates. Courtesy of Dr. Z. L. Wang, Georgia Institute of Technology. Reproduced with permission from [26], Copyright @ 2008, American Chemical Society

Then, a pattern of ZnO strips covered with Cr on top is prepared. Solution growth of ZnO nanowires onto the substrate is conducted for 12 h at 80 °C. The growth solution consists of 0.1878 g of $Zn(NO_3)_2 \cdot 6H_2O$ and 0.0881 g of HMTA in 250 mL water. The concentrations of $Zn(NO_3)_2$ and HMTA in the solution are both 0.0025 mol L^{-1}. The ZnO NW arrays grow from the lateral sides of the patterned strips. More than 70 % nanowires are parallel to the substrate. The ZnO nanowires have a diameter of about 200 nm and a length of about 4 μm. Hexagonal cross-section of the nanowires implies that *c* axis of the ZnO nanowires is along their length direction.

3.3.4.4 Laser Patterning Technique

A maskless pattern technique has been developed for the growth of ordered ZnO nanowire arrays by using laser interference lithography (LIL) [27, 28]. This method can be used to grow highly ordered and well-aligned ZnO nanowire arrays with high-throughput and low-cost at large scale. The patterned areas can be up to a whole 2-in. wafer. Perfectly aligned vertical ZnO NW arrays are grown at the predefined positions by using a low temperature hydrothermal method without using catalyst. Orientation, dimension, and location of the nanowire arrays can be well controlled. The ZnO nanowires are single crystal with growth direction along [0001].

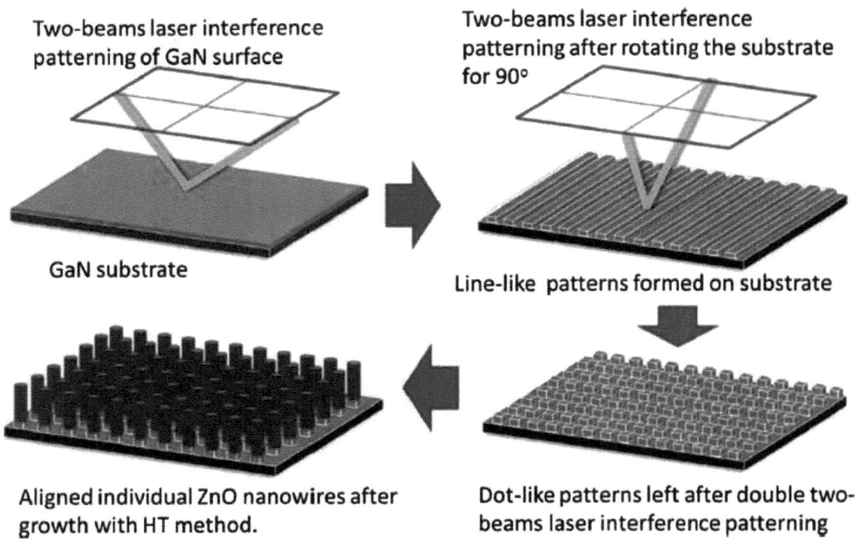

Fig. 3.15 Flow chart showing the fabrication steps of the ZnO nanowire arrays. GaN substrates are patterned with two-beam LIA to obtain line-like patterns on the surface, which are then rotated by 90° and patterned again to achieve dot-like array patterns. The patterned substrates are used to grow ZnO nanowires through a hydrothermal method. Courtesy of Dr. Z. L. Wang, Georgia Institute of Technology. Reproduced with permission from [27], Copyright @ 2010, John Wiley & Sons

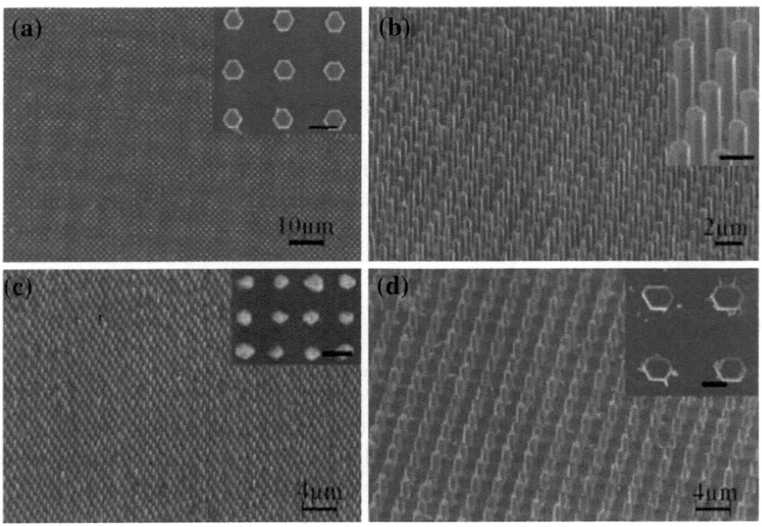

Fig. 3.16 SEM images of vertically aligned ZnO nanowire arrays. **a** Top and **b** 45° tiled views of the aligned ZnO nanowire arrays with uniform diameter, height and 2 μm period over a centimeter-scale area, where the insets are enlarged images, with scale bars of 1 μm. **c**, **d** ZnO nanowires grown on patterned GaN substrates with periods of 1 and 3 μm, respectively. Samples are tilted by 45°. Insets are enlarged top view SEM images, with scale bars of 1 μm. Courtesy of Dr. Z. L. Wang, Georgia Institute of Technology. Reproduced with permission from [27], Copyright @ 2010, John Wiley & Sons

The photoresist used is epoxy-based negative photoresist SU-8. Experimental steps are described in Fig. 3.15 in detail [27]. The surface of a GaN substrate is patterned through the interference of a two-beam laser. This step is repeated after rotating the substrates by 90° to achieve isolated mesas. The isolated mesas are areas without ablation, over which ZnO nanowires are grown. Size and periodicity of the areas are controlled by controlling the laser fluence and the angle between the interfering laser beams. A low temperature hydrothermal method, as discussed above, is used to produce patterned and aligned growth of ZnO nanowires at temperatures below 100 °C on the patterned GaN substrates, without using a catalyst. SEM images of representative samples are shown in Fig. 3.16 [27]. Almost all the nanowires have the same diameter and height. The aligned ZnO nanowire arrays are uniformly grown out of the patterned holes with high-fidelity at diameter of about 600 nm. All the NWs are perfectly aligned normal to the substrate and have the same height of abut 5 μm.

3.4 Concept and Principle of Nanogenerators

3.4.1 Piezopotential

Piezoelectric potential or piezopotential in short is an import concept and a quantitative measure of piezoelectric properties of ZnO nanowires in applications for nanogenerators, as shown in Fig. 3.17 [29]. As discussed above and in Chap. 2, piezoelectric phenomenon of a material is dictated by its crystal structure. ZnO is of the wurtzite structure. The lack of central symmetry is responsible for the piezoelectric property of ZnO. When an external strain is applied to ZnO, its structure is deformed and thus the charge centers of cations and anions separate which result in an electric dipole (Fig. 3.17c). Because the ionic charges are not free to move and the intrinsic free charge carriers can only partially screen them if the conductivity of the nanowires is sufficiently low, the piezoelectric field is preserved as long as the nanowire is strained. The potential created by the polar ions is called piezoelectric potential or piezopotential.

The presence of piezopotential is the physical fundamentals of the nanogenerators and piezotronics based on ZnO nanowires. When a strained ZnO crystal is subject to an external load, electrons in the circuit are driven to flow to partially screen the piezopotential, converting mechanical energy to electrical energy. This is the principle of nanogenerators. Because ZnO is usually a semiconductor, it is possible to use the piezopotential as a gate voltage to tune/gate the transport process of the charge carriers driven by an externally applied voltage (Fig. 3.17d). Devices based on this principle are called the piezotronic devices [30].

Piezopotentials of ZnO nanowires have been theoretically studied [31–33]. According to a perturbation theory, the maximum potential at the surface of a nanowire can be calculated as:

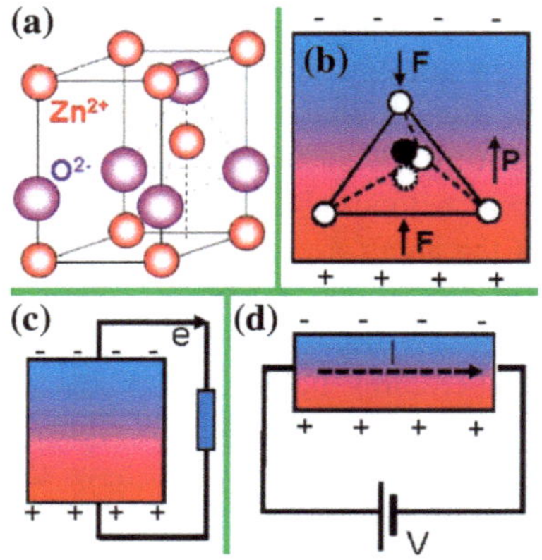

Fig. 3.17 **a** Atomic model of ZnO with the wurtzite structure. **b** Piezopotential of a stretched ZnO nanowire. **c** Power generation due to the piezopotential. **d** Current modulation in a semiconductor material from the piezopotential. Reproduced with permission from [29], Copyright @ 2010, Elsevier

$$\varphi_{\max}^{(T,C)} = \pm \frac{3}{4(\kappa_0 + \kappa_\perp)} [e_{33} - 2(1 + v)e_{15} - 2ve_{31}] \left(\frac{a}{l}\right)^3 v_{\max}, \qquad (3.9)$$

where κ_0 and $\kappa_\perp$ are dielectric constants of vacuum and the material, e is linear piezoelectric coefficient, v is Poisson ratio, $2a$ and l are diameter and length of the nanowire. This equation indicates that the piezopotential of a ZnO nanowire is directly related to its aspect ratio instead of its dimensionality. At a given aspect ratio, piezopotential of the nanowire is proportional to the maximum deflection at the tip. The analytical solution given under the first-order approximation produces a result that is within 6 % from the full numerical calculated result using a finite element method, confirming the accuracy and validity of the perturbation theory. The magnitude of piezoelectric potential for a ZnO nanowire with a diameter 50 nm and a length 600 nm is calculated to be about 0.3 V. This voltage is sufficiently high to drive the metal-semiconductor Schottky diode at the interface between AFM tip and the ZnO nanowire, which is supported by experimental results [31, 32].

The presence of piezopotential of ZnO nanowires has been confirmed by experimental measurement. By using a metal tip that to touch either at the tensile side or the compressive side of a nanowire, piezopotentials at the stretched and compressed sides can be measured, as shown in Fig. 3.18 [34]. During the measurement, the nanowire is deflected by air blowing. As the ZnO nanowire is subject to a periodic gas flow pulse, it is bent and a corresponding periodic negative voltage output, −25 mV, is detected by connecting the surface of the compressed side of the nanowire with the external measurement circuit (Fig. 3.18a). Similarly, a periodic positive voltage output is probed at the stretched side of the nanowire by using a Au-coated needle when the ZnO nanowire is periodically pushed by the Au-coated needle (Fig. 3.18b).

The ZnO nanowires used to construct nanogenerators are usually of a typical hexagonal dimension, grown along c-axis. The two ends and short segments adjacent to the ends of the nanowires are fully embraced by electrodes, where several types of forces, including tensile, compression, twisting, and combinations of them, could be present. The piezopotential distributions of a ZnO nanowire

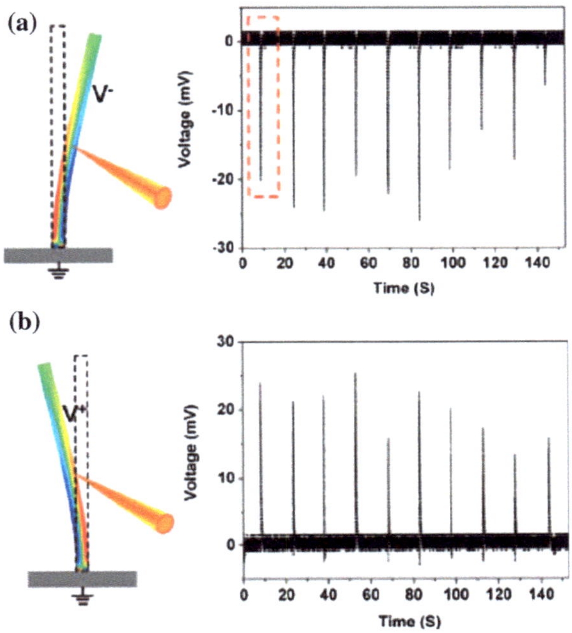

Fig. 3.18 Direct measurement of the asymmetric voltage distribution on the tensile and compressive side surfaces of a ZnO nanowire. **a** By placing a metal tip at the *right-hand side* and blowing Ar pulses at the *left-hand side*, negative voltage peaks of ~25 mV are observed once the pulse is on. **b** By quickly pushing and releasing the wire at the right-hand side by a metal tip, a positive voltage peak of ~25 mV is observed for each cycle of deflection. The deflection is repeated every 15 s. Courtesy of Dr. Z. L. Wang, Georgia Institute of Technology. Reproduced with permission from [34], Copyright @ 2008, American Chemical Society

under different mechanical strains have been theoretically calculated, as shown in Fig. 3.19 [35]. The nanowire is assumed to be nonconductive, so that there is no body force and no free charge in the nanowire for simplicity. Figure 3.19a shows a ZnO nanowire without any applied force. The axial length and the hexagonal side length of the nanowire are 1,200 and 100 nm, with a 100 nm length of contact domain at each end.

When a stretching force of 85 nN is applied to the nanowire in the *c*-axis direction, the length of the nanowire is increased by 0.02 nm, corresponding to a tensile strain of 2×10^{-5}. As shown in Fig. 3.19b, it creates a potential drop of approximately 0.4 V between the two end sides of the nanowire with the +*c* axis side showing higher potential. If the applied force is compressive, the piezoelectric potential is reversed accordingly. As shown in Fig. 3.19c, the nanowire length is decreased by 0.02 nm, with a compressive strain of -2×10^{-5}.

In both stretching and compressing, the piezoelectric potential always drops from one side to the other, implying a continuous increase in electron energy. Meanwhile, the Fermi level will be flattened all over the nanowire when an equilibrium state is arrived, because there is no external electrical field applied to

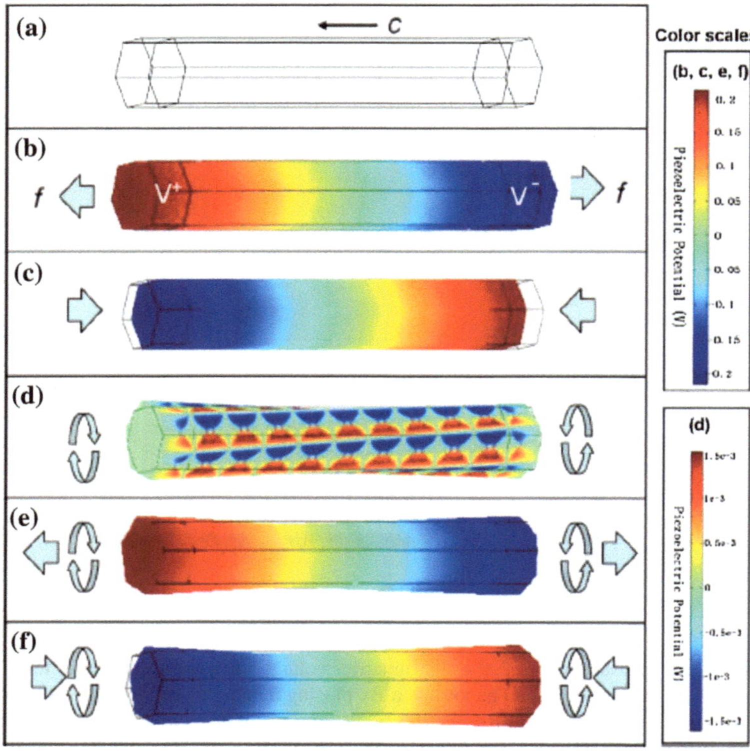

Fig. 3.19 Piezoelectric potential distributions of a ZnO nanowire without doping, subject to different mechanical strains, obtained by using numerical calculation. **a** An unstrained ZnO nanowire grown along *c*-axis with a length of 1,200 nm and a hexagonal side length of 100 nm. Two ends of the nanowire are assumed to be surrounded by electrodes for a length of 100 nm. Three-dimensional views of the piezoelectric potential distribution together with deformation shape for the nanowire **b** at a stretching force of 85 nN **c** at a compressing force of 85 nN **d** at a twist force pair of 60 nN **e** at a combination of 85 nN stretching force and 60 nN twist force, and **f** at a combination of 85 nN compressing force and 60 nN twist force. Courtesy of Dr. Z. L. Wang, Georgia Institute of Technology. Reproduced with permission from [35], Copyright @ 2009, Institute of Physics

the nanowire. As a result, the electron energy barrier between ZnO and metal electrode is raised at one side and lowered at the other side, which has been confirmed experimentally by the observation of an asymmetric *I*–*V* characteristic.

Figure 3.19d shows simulation results of piezopotential when the nanowire is twisted at two ends in opposite directions. In the growth direction, no electric potential drop is observed. The applied force can be a combination of twist and tension or twist and compression. In this case, there is a drop in piezopotential along the nanowire length direction, as shown in Fig. 3.19e, f. At the same time, the potential distribution over the cross-section of the nanowire is not uniform.

Similar theoretical studies have also been applied to calculate the piezopotentials of ZnO nanowires with semiconductive characteristics, doping concentration and doping type (*n*-type and *p*-type) [33–36]. Such studies have provided with useful guidance that can be used to design nanogenerators and nanogenerator facilitated devices with optimized performances.

3.4.2 Outputs of Nanogenerators

The anisotropic property of ZnO nanowires and the small output signals of the piezoelectric nanogenerators have made it potentially difficult to accurately measure the electrical signals. With the influence of measurement system accuracy and environment, it is highly possible to obtain artificial signals [37, 38]. How to identify whether a signal collected is truly generated by the nanogenerators or is an artifact is an important prerequisite in developing nanogenerators. Stringent criteria consisting of 11 tests that have been developed to rule out artifacts [39, 40]. These criteria include a Schottky behavior test (i) and switching-polarity tests (ii), as well as linear superpositions of current and voltage with eight configurations. A generator must pass each and all of the tests, the current and voltage collected are considered as true signals.

For the measurement of short-circuit current, the forward connection is measured first, where the positive probe and negative probe of the measurement system are connected to the positive and negative ends of the nanogenerator, respectively. Figure 3.20 shows the power outputs of short-circuit current of two single-wire generators (SWGs): SWG A and SWG B [39]. They have similar response behavior, but with different magnitudes. Their connection configurations are shown in the inserts. The positive peaks correspond to the state when the substrates are bent inward. When the substrates are released, the free state of the SWGs is recovered, resulting in the negative response peaks.

Switching-polarity test can be used to examine whether the signal is really due to the piezoelectric effect of the ZnO nanowire or it is an artificial one. To do this, the positive and negative probes of the current meter are connected to the negative and positive ends of the SWG, respectively. The result is shown in Fig. 3.20c and d. The output signals are opposite to those shown in Fig. 3.20a and b. Stretching the nanowire results in a negative response and a positive pulse is observed when the stretched nanowire is freed back to unstrained state. This switching-property test means that the signal is due to the possible error caused by the change of the system capacitor. The change in contact resistance could be a potential source of signal when the SWG is strained, but such signals have no change in their sign from positive to negative when the connecting is reversed. Moreover, there is a difference in magnitude between the signals of forward connection and the reverse connection. The nonsymmetric output of the SWG before and after the polarity switching is possibly related to the bias current of the measurement system. If this bias current adds to the current generated by the SWG with the forward connecting, it will be deducted from the current generated with the reverse connecting.

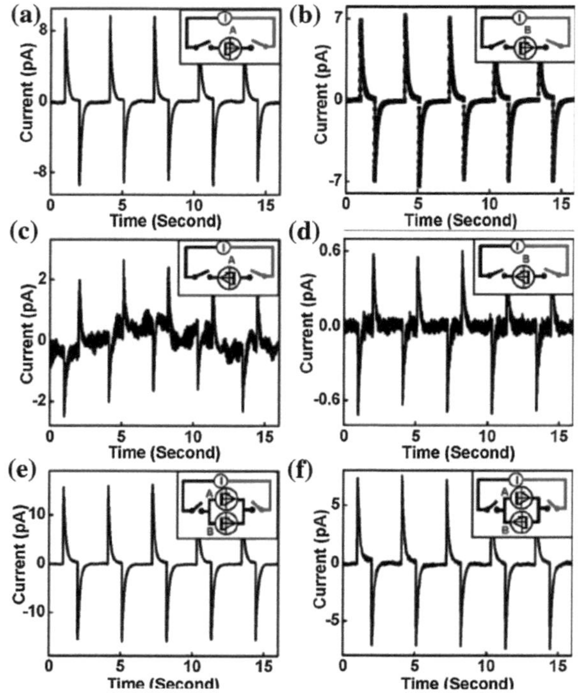

Fig. 3.20 **a** and **b** Short-circuit current power output from SWG A and SWG B, respectively, under forward connection. **c** and **d** Short-circuit current power output from SWG A and SWG B, respectively, under reverse connection. **e** and **f** Parallel connection of SWG A and SWG B demonstrates SWG's ability to "add up" or "cancel out." All insets illustrate the connection configuration of the two SWGs in reference to the measurement system. Courtesy of Dr. Z. L. Wang, Georgia Institute of Technology. Reproduced with permission from [39], Copyright @ 2010, Institute of Physics

The real signal generated by the SWG is an average in magnitudes of the ones observed with forward and reverse connecting configurations.

Due to the presence of a Schottky junction at one end of the SWG, the switching-polarity test is not sufficient to rule out all potential artifacts. Therefore, further test is necessary. Linear superposition is proposed for such a purpose. In this test, the two SWGs are connected in parallel to examine the liner superposition of currents. Since there is a Schottky contact at one end of the SWGs, the direction of their connection should be treated carefully. When the two SWGs are arranged in same direction, their output currents are added. In this case, the total current should be the sum of the signals shown in Fig. 3.20a and b. As seen in Fig. 3.20e, the resultant current output is slightly higher than 15 pA. However, when two SWGs are connected in antiparallel direction, as shown in Fig. 3.20f, the output current is decreased and approximately is equal to the sum of the signals shown in Fig. 3.20a and d, noting the off-set position of the signals with same signs. These experimental results prove that the current outputs of the two SWGs meet the linear superposition requirement, in addition to the switching-polarity testing.

Voltage output results of the two SWGs are shown in Fig. 3.21 [39]. SWG A and SWG B generate a positive voltage signal when the nanowires are stretched, which are accompanied by a negative signal when the nanowire are released, as shown in Fig. 3.21a and b. The voltage output also meets the requirement of witching-polarity test, as shown in Fig. 3.21c and d. For further testing, the voltage

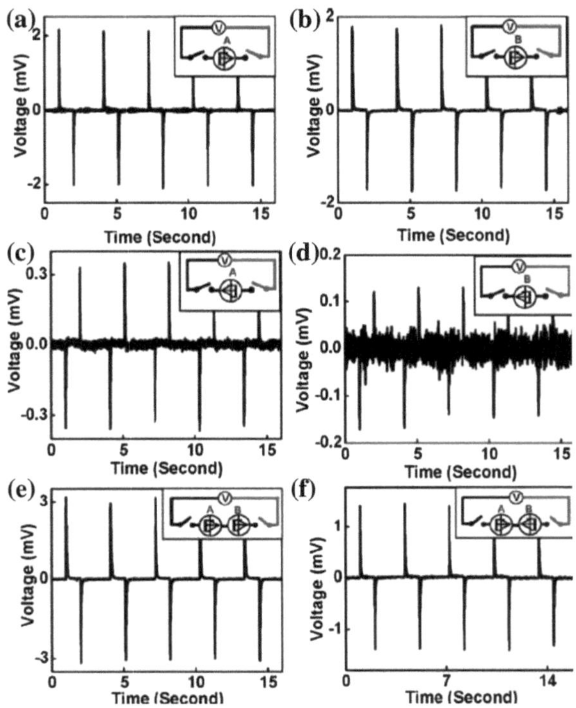

Fig. 3.21 **a** and **b** Open-circuit voltage output from SWG A and SWG B, respectively, under forward connection. **c**, **d** Open-circuit voltage power output from SWG A and SWG B, respectively, under reversal connection. **e**, **f** Serial connection of SWG A and SWG B demonstrates the add up or cancel out effect. All insets illustrate the connection configurations of the SWGs in reference to the measurement system. Courtesy of Dr. Z. L. Wang, Georgia Institute of Technology. Reproduced with permission from [39], Copyright @ 2010, Institute of Physics

Table 3.1 Averaged current and voltage outputs of the two SWGs [39]

	Current (pA)		Voltage (mV)	
	I^+	I^-	V^+	V^-
SWG A	9.54	−2.42	2.13	−0.36
SWG B	7.31	−0.70	1.81	−0.15
SWG A + SWG B	16.20	−3.76	3.18	−0.37
SWG A – SWG B	7.33	4.65	1.46	1.50

outputs of the two SWGs connected in serial are measured, as shown in 3.21e and f. When the two SWGs are connected in same direction, the total output is obviously increased. In contrast, when two SWGs are connected in opposite directions, the total output is reduced. In other words, the voltage outputs of the two SWGs also exhibit the linear superposition behavior, in addition to satisfying the switching-polarity test.

V^+ and $I^+$are used to represent the voltage and current for the forward connection and V^- and I^- are for the reverse connection. There are $V^- = -V^+$ and $I^- = -I^+$. The averaged output currents and voltages in Figs. 3.20 and 3.21 are listed in Table 3.1 [39]. Noting that a bias current is used during the measurement, the measured voltages/currents have difference in magnitude. However, a true electricity output of and SWG must have sign change before and after the change of connecting direction. The data for single device of two follow the requirement of switching-polarity test for both output current and voltage. The data also indicate that the SWGs meet the linear superposition criteria, with currents (parallel connection) and voltage (series connection) having the following relationships.

$$I_{A+B}^{+} = I_{A}^{+} + I_{B}^{+}, \quad I_{A+B}^{-} = I_{A}^{-} + I_{B}^{-}, \quad I_{A-B}^{+} = I_{A}^{+} + I_{B}^{-}, \quad I_{A-B}^{-} = I_{A}^{-} + I_{B}^{+}.$$
$$V_{A+B}^{+} = V_{A}^{+} + V_{B}^{+}, \quad V_{A+B}^{-} = V_{A}^{-} + V_{B}^{-}, \quad V_{A-B}^{+} = V_{A}^{+} + V_{B}^{-}, \quad V_{A-B}^{-} = V_{A}^{-} + V_{B}^{+}$$

3.5 Nanogenerators with Vertically Aligned ZnO Nanowires

3.5.1 Concept of Piezoelectric Nanogenerators

The concept of the nanogenerator is originated from a finding when measuring piezoelectric properties of ZnO nanowires by using an atomic force microscope (AFM) [1, 41, 42]. When an AFM tip scans across over an aligned nanowire array, both topography and lateral force image are produced, with which the elastic modulus of individual nanowires can be evaluated. This technique has been used to measurement mechanical properties of individual nanowires with different lengths nondestructively and directly.

Figure 3.22a shows the principle of the AFM measurement [42]. In contact mode, a constant normal force is applied constantly between the AFM tip and the sample surface. The tip scans over the top ends of the ZnO nanowires and the tip height is adjusted according to the surface morphology and local contacting force. Before the tip contacts a nanowire, a small lateral force is present (Fig. 3.22a(I). When the tip just contacts with the nanowire, the lateral force increases almost linearly, while the nanowire is elastically bent from its initial position (Fig. 3.22a(II) and (III)). As the bending approaches its maximum, as shown in Fig. 3.22a(IV), the tip crosses over the top end of the nanowire, after which point the nanowire is released and the lateral force drops abruptly (Fig. 3.22a(V)).

From the geometrical relationship shown in Fig. 3.22a(IV), when a vertical nanowire is subject to a lateral force f parallel to the scanning direction, the force f can be expressed as:

$$f = 3EI\frac{x}{L^3}, \tag{3.10}$$

where E and I are the elastic modulus and momentum of inertia of the nanowire, x is the lateral displacement perpendicular to the nanowire, and L is the length of the nanowire. According to Hook's law, the spring constant is $K = f/x$. The moment of inertia of a ZnO nanowire with a hexagonal cross-section is $I = (5(3^{1/2})/16)a^4$,

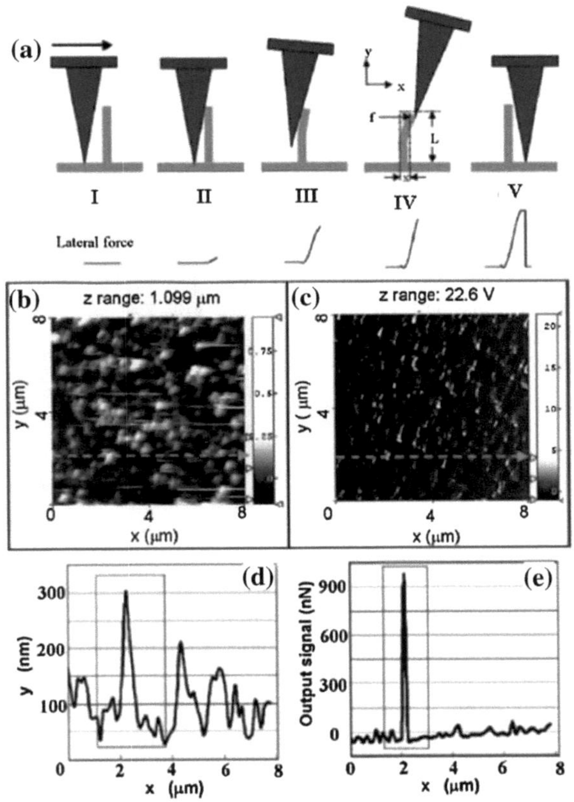

Fig. 3.22 **a** Procedure of measuring elastic modulus of a ZnO nanowire with AFM contact mode. **b** Topography image and **c** lateral force image of the aligned ZnO nanowire array derived with AFM contact mode. **d**, **e** Line scanning profiles along the dashed lines in (**b**) and (**c**), respectively. Courtesy of Dr. Z. L. Wang, Georgia Institute of Technology. Reproduced with permission from [42], Copyright @ 2007, The Royal Society of Chemistry

with a being the radius of the nanowire. Therefore, elastic modulus of the ZnO nanowire can be calculated by:

$$E = \frac{16L^3K}{15\sqrt{3}a^4}. \tag{3.11}$$

With the topography image and corresponding lateral force image, as shown in Fig. 3.22b and c, respectively, the displacement and lateral force can be calculated. The theoretical calculation requires the conical tip end to touch the center of the nanowire, so that the curves should be from the center of the nanowire, as indicated by the dashed lines in the images. Taking a line scan across the middle point of a spot in the topography image, a curve for the scanner retracting distance versus the nanowire lateral displacement was obtained, as shown in Fig. 3.22d. Likewise, the maximum lateral force for bending the nanowire was measured by taking a scanning profile at the corresponding line in the lateral force image (as indicated in Fig. 3.22e). With the measured x_m and f_m from the two line profiles, the spring constant $K = f_m/x_m$ can be calculated. For the ZnO nanowires grown on sapphire substrate with an average diameter of 45 nm, the elastic modulus is

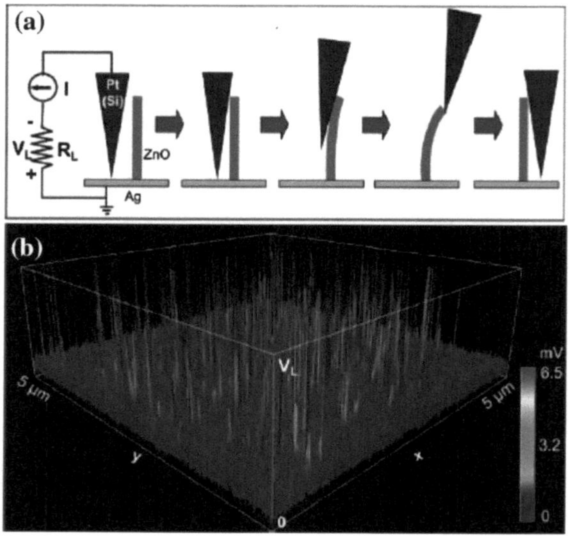

Fig. 3.23 **a** Experimental setup and procedures for generation of electricity by deforming a piezoelectric ZnO nanowire with a conductive AFM tip. **b** Output voltage image map of ZnO nanowire arrays. Courtesy of Dr. Z. L. Wang, Georgia Institute of Technology. Reproduced with permission from [42], Copyright @ 2007, The Royal Society of Chemistry

calculated to be 29 ± 8 GPa. Therefore, this is a technique of direct observation and measurement of the mechanical properties of aligned nanowires.

The application of the above technique to the measurement of electrical properties of ZnO nanowires leads to the finding of nanogenerators [16, 41, 42]. An example is shown in Fig. 3.23 [42]. To measure electrical properties, the ZnO nanowires are grown on conductive substrates. The AFM tip is Si coated with Pt film, which has a cone angle of 70°. A single ZnO nanowire is taken as a rectangular cantilever, with a calibrated normal spring constant of 0.76 N m^{-1}. In contact mode, a constant normal force of 5 nN applied between the tip and sample surface (top ends of the nanowires). The tip scans over the top end of every ZnO nanowire. Thermal vibration of the nanowires at room temperature is neglected. The bottom ends of the nanowires are connected to the measurement circuit by using. The output voltage across an outside load of resistance $R_L = 500$ MΩ is monitored as the tip scans over the nanowires. No external voltage is applied throughout the entire measurement process.

Both the topography (feedback signal from the scanner) and the corresponding output voltage images across the load can be recorded simultaneously as the AFM tip scans over the aligned nanowire arrays. In contact mode, as the tip scans over the array, all the vertically aligned nanowires are bent consecutively. The bending distance can be directly recorded in the topography image, from which the maximum bending deflection distance, elastic modulus, and density of the nanowires that have been scanned by the tip are directly obtained at the same time.

As shown in Fig. 3.23b [42], there are many sharp output peaks in the voltage output image for each contact position [1, 41]. The output voltage signal is negative with reference to the grounded end. Examination of the topological profile of a nanowire and its corresponding output potential indicates that there is a delay of the voltage output signal. This means that there is no electric output when the tip is just in contact with the nanowire and the sharp voltage peak is generated at the right moment when the tip is about to leave the nanowire. This delay is a key indication of the power output process. The output voltage V_L is converted from the current flowing through the external load R_L. This is the principle of nanogenerators.

The nanogenerators made with ZnO nanowires have several interesting experimental characteristics [31, 43, 44]. The output potential is a sharp peak that is negative with reference to the grounded end of the nanowire. No output current is detected as the AFM tip first touches and then pushes the nanowire. Electrical output is observed only when the tip is just to leave the nanowire. The power output is observable only when the tip touches the compressive side of the nanowire. Output signal is observed only when piezoelectric nanowires are tested. There is no electrical output if the nanowires are not piezoelectric, such as tungsten oxide, carbon nanotubes, silicon, or metals. The output is not due to friction or contact potential. The magnitude of the output signal is sensitive to the size of the nanowires. To generate electricity, the contact between the tip and the nanowire must be of characteristics of Schottky junctions and the contact between the nanowire and the grounded end should be Ohmic.

3.5.2 Schottky Barrier at the Electrode-Nanowire Interface

Two types of contact can be formed between a metal and a semiconductor: Ohmic contact and non-Ohmic. Ohmic contact does not impose a potential barrier at the interface so that electrons have a symmetric transport behavior at either forward or reversal bias. In this case, the *I–V* curve is a straight line. However, a potential barrier is established at the interface of a non-Ohmic contact, so that the transport of electrons across the interface is asymmetric. Figure 3.24a, b shows the configurations before and after a metal and an *n*-type semiconductor form a contact. A potential barrier of $\varphi_B = \Phi_M - \chi$, called Schottky barrier, is formed at the interface, where Φ_M is the work function of the metal and χ is electron affinity. The contact is Ohmic with $\Phi_M < \chi$ and it is Schottky with $\Phi_M > \chi$.

The interface of a Schotty contact behaviors like a "diode" (Fig. 3.24c and d). At forward bias with the metal having a higher potential, the current can flow through the barrier, e.g., electrons flow from the semiconductor side to the metal side. At reversal bias with the metal having a lower potential, electrons cannot overcome the barrier to reach the other side, there is no electrical current.

The presence of a Schottky contact at the interface between the metal and the ZnO nanowire is a key requirement to generate electrical current and power output for the nanogenerators. This has been confirmed experimentally, as shown in Fig. 3.25 [45]. Figure 3.25a shows that when a Si tip coated with 100 nm thick Pt film is used to scan the nanowires in contact mode, there is strong response, with voltage peaks are observed in Fig. 3.25b. The output voltage is about −11 mV. The negative sign of the signal is because the current flows from the grounded end

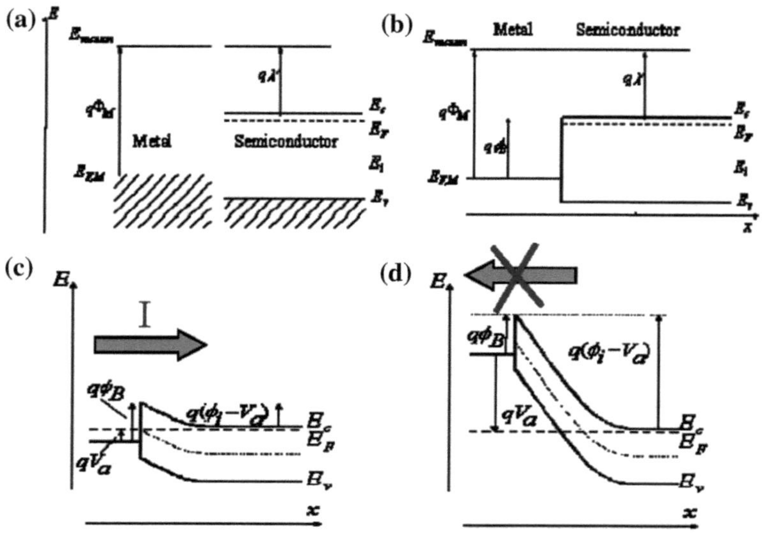

Fig. 3.24 **a**, **b** A metal and an n-type semiconductor before and after forming a contact. **c**, **d** The Schottky contact at forward and reverse bias. Courtesy of Dr. Z. L. Wang, Georgia Institute of Technology. Reproduced with permission from [31], Copyright @ 2011, Georgia Institute of Technology

into the external load. If a Si tip coated an Al-In (30 nm/30 nm) alloy layer is used, the ZnO nanowires have no piezoelectric response (Fig. 3.25c). The difference in performance between the two types of scanning tips is attributed to the different contact behaviors between the tips and the ZnO nanowires. They have *I–V* characteristics. A large piece of electrode is used to ensure the stability of the contact with the nanowires, as shown in Fig. 3.25d. The Pt–ZnO contact has a Schottky diode behavior (Fig. 3.25e), whereas the Al–In–ZnO is an Ohmic contact. Therefore, a Schottky contact between the metal electrode and ZnO nanowires is a prerequisite of nanogenerators [45].

3.5.3 Charge Generation and Output Processes

It is important to understand the underlying mechanism governing the creating, separating, accumulating, and outputting of the charges created by a nanowire generator Fig. 3.26 [43]. Figure 3.26a shows a vertical straight strain-free ZnO nanowire. As an AFM tip touches the top end of the nanowire, it is subject to a strain by the tip. Under such a strain, the nanowire experiences a deflection, with the outer surface being stretched and inner surface compressed, as shown in Fig. 3.26b. As a consequence, a piezoelectric potential is built up across the nanowire. Because the electrode at the base of the nanowire is grounded, the stretched side is positive and the compressed side is negative (Fig. 3.26c).

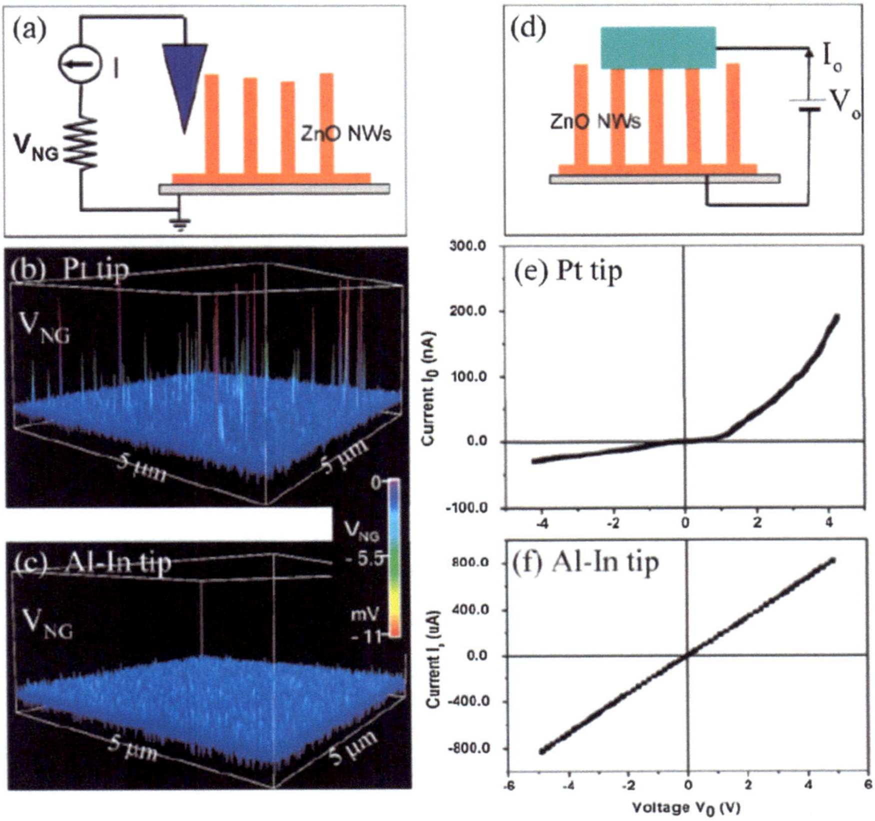

Fig. 3.25 **a** AFM-based measurement set-up to characterize the metal-ZnO contact behavior and the nanogenerator output. **b** Output potential generated by the ZnO nanowire array due to the scanning of a Pt coated Si tip. **c** No output potential is generated by the ZnO nanowire array scanned by an Al-In coated Si tip. **d** Experimental set-up to characterize *I–V* transport property of the metal-ZnO nanowire contact. **e** *I–V* curve of a Pt–ZnO nanowire contact, showing Schottky diode effect. **f** *I–V* curve of an alloyed Al/In–ZnO nanowire contact, showing Ohmic behavior. Courtesy of Dr. Z. L. Wang, Georgia Institute of Technology. Reproduced with permission from [45], Copyright @ 2008, American Chemical Society

The potential is produced as a result of the relative displacement of the Zn^{2+} cations with respect to the O^{2-} anions, owing to piezoelectric effect in the wurtzite crystal structure. These ionic charges are not freely movable and thus cannot recover if the strain is not released (Fig. 3.26c). The potential is maintained if the nanowire is in deflected state. Also, there are no foreign free charges, such as those from the metal contacts, injected in, if the doping level in the nanowire is sufficiently low. At this stage, electrical charges are created and separated.

The next step is regarding to the charge accumulation and releasing. As the conductive AFM tip that induces the deformation is still in contact with the stretched surface of positive potential V^T, there is a charge accumulation, as shown

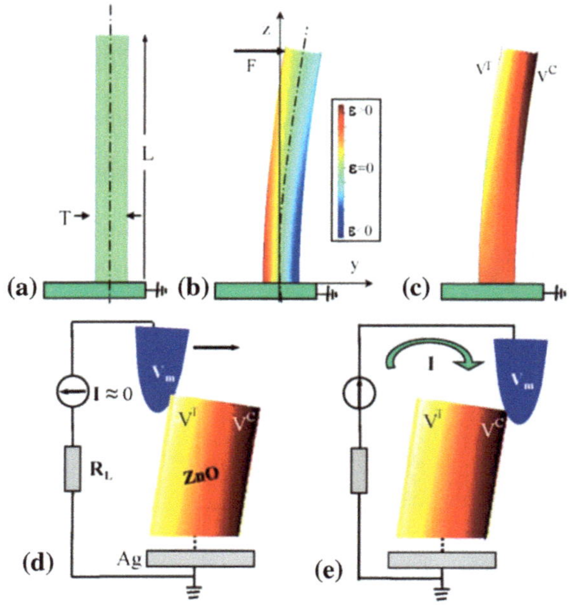

Fig. 3.26 Working mechanism of the power generation process of a piezoelectric ZnO nanogenerator, due to the coupling effect of piezoelectric and semiconducting properties of the ZnO nanowire, which forms a Schottky barrier at the AFM tip-semiconductor interface. **a** Schematic definition of a nanowire. **b** Longitudinal strain ε_z distribution in the nanowire that is deflected by an AFM tip from the left side. **c** Potential distribution in the nanowire as a result of piezoelectric effect, with the potentials on stretched and compressed side surfaces being positive and negative, respectively. **d**, **e** Metal and semiconductor contacts between the AFM tip and the semiconductor ZnO nanowire at two reversed local contact potentials (positive and negative), showing reverse and forward-biased Schottky rectifying behaviors, respectively. The process in **d** is to build up and preserve the charges/potential, while the process in **e** is to discharge the potential through a flow of electrons from the circuit. The negative output voltage on the load is a result of current flow from the AFM tip through the ZnO belt to the ground. Courtesy of Dr. Z. L. Wang, Georgia Institute of Technology. Reproduced with permission from [43], Copyright @ 2006, American Chemical Society

in Fig. 3.26c and d. The metallic Pt on the tip has a nearly zero potential, $V_m = 0$, so the metal tip-ZnO interface is negatively biased because $\Delta V = V_m - V^T < 0$. If the as-synthesized ZnO nanowires can be taken as *n*-type semiconductor, the Pt metal-ZnO semiconductor interface is similar to a reverse-biased Schottky diode (Fig. 3.26d), and thus little current can flow across it.

As the AFM tip sweeps over the top end and comes to in contact with the compressed side of the nanowire (Fig. 3.26e), the metal tip-ZnO interface becomes positively biased because $\Delta V = V_L = V_m - V^C > 0$. The metal-semiconductor interface in this case is like a positively biased Schottky diode, and thus a sudden increase in electrical current output takes place. The current is due to the flow of electrons from the semiconductor ZnO nanowire to the metallic tip, driven by ΔV. The free electrons flowing through the nanowire to the tip neutralize the ionic

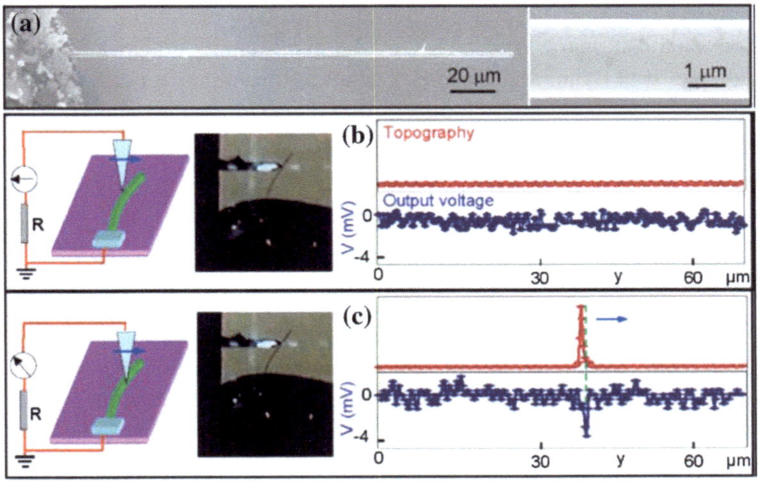

Fig. 3.27 In situ observation of the process as mechanical energy is converted into electric energy by a piezoelectric ZnO nanowire. **a** SEM images of a ZnO belt with one end fixed by silver paste onto a silicon substrate and the other end is free. **b**, **c** Three characteristic snapshots and the corresponding topography (*red curve*) and output voltage (*blue curve*) images when the tip scans across the middle section of the wire. The schematic illustration of the experimental condition is shown at the *left-hand side*, with the scanning direction of the tip indicated by an *arrowhead*. Courtesy of Dr. Z. L. Wang, Georgia Institute of Technology. Reproduced with permission from [43], Copyright @ 2006, American Chemical Society

charges distributed in the nanowire. As a result, the magnitudes of both the potential V^C and V^T are reduced.

This mechanism has been validated by experiments [43]. A sufficiently long ZnO nanowire that can be clearly identified under an optical microscope is used for the experiments. One end of the long ZnO wire is fixed on a Si substrate by silver paste, as shown in Fig. 3.27a. The substrate is intrinsic silicon which has a low conductivity. The wire has a small distance from the substrate to avoid friction. Both the topography (feedback signal) and the corresponding output voltage (V) images across can be obtained simultaneously by using the AFM tip to scan across the nanowire. The topography image is a result of the change in normal force applied perpendicular to the substrate, which shows a small jump only when the tip scans over the nanowire. The output voltage between the conductive tip and the ground is continuously monitored as the tip scans over the nanowire. No external voltage is applied throughout the experiment. The entire experimental process and the output images are captured by video recording, so that the electric generation process can be directly visualized.

The topography image is directly captured when if the tip passes over the belt. When the tip pushes the wire but does not go over and across it, according to the flat output signal in the topography image (Fig. 3.27b), no voltage output is generated, indicating that the stretched side has no piezoelectric response. Once the tip goes over the wire and touches the compressed side, there appears a peak in the topography image, corresponding to a sharp voltage peak (Fig. 3.27c). A close inspection on positions of the peaks in the topography image and the output voltage image, it is found that there is a delay in discharge. It occurs after the tip just leaves the wire. This clearly indicates that the negative piezoelectric discharge voltage is produced from the compressed side.

3.5.4 Principle of Nanogenerators with n-Type Materials

As mentioned above, the piezoelectric potentials created in the ZnO nanowires are due to the polarization of ions in the crystal. They are not movable charges. Because ZnO is usually an *n*-type semiconductor, the free carriers in the semiconductor nanowires might screen the piezoelectric charges. Moreover, the presence of oxygen vacancies and impurities, as well as a large portion of surface atoms (surface states), in the nanostructured materials, make the conductivity of the ZnO nanowire even higher. However, no matter how high the conductivity is, the free-mobile electrons cannot completely cancel or deplete the piezoelectric charges. Due to the presence of a Schottky barrier at the interface, those free-mobile electrons cannot across the boundary between the metal and the ZnO, thus, a "dipole" layer is formed between the accumulated electrons at the metal side and the piezoelectric charges at the ZnO side, but this is significantly different from the *p–n* junction in semiconductors.

The charge releasing process of a nanogenerator can be understood by using band structure model. The AFM tip (labeled as T in the figure) has a Schottky contact (with a barrier height Φ_{SB}) with the nanowire, while the nanowire has an Ohmic contact with the grounded side (labeled as G in the figure), as shown in Fig. 3.28a [38]. As the tip slowly pushes the nanowire, a positive piezoelectric potential V^+ is created at its tensile surface. If the tip continues to push the nanowire, thus bending the nanowire, there will be electrons slowly flowing from the grounded electrode through the external load to the tip. However, these electrons are blocked by the tip-nanowire interface, due to the presence of a reversely biased Schottky barrier at the contact (Fig. 3.28b). As a result, the accumulated free charges at the tip might have influences on the distribution of the piezoelectric potential in the nanowire, due to the screening effect of the charge carriers. The piezoelectric potential is created due to the rigid and immobile ionic charges in the nanowire. Therefore, it cannot be completely depleted by the free charge carriers. The local newly established potential V'^+ lowers the conduction band (CB) slightly.

When the tip comes in contact mode across the nanowire and reaches the center point of the nanowire (Fig. 3.28c), the local piezoelectric potential becomes zero. In this case, due to the sudden drop in the level of local potential, the originally accumulated electrons in the tip flow back through the load to the ground. This is a process faster than the charge accumulation one, as shown in Fig. 3.28b).

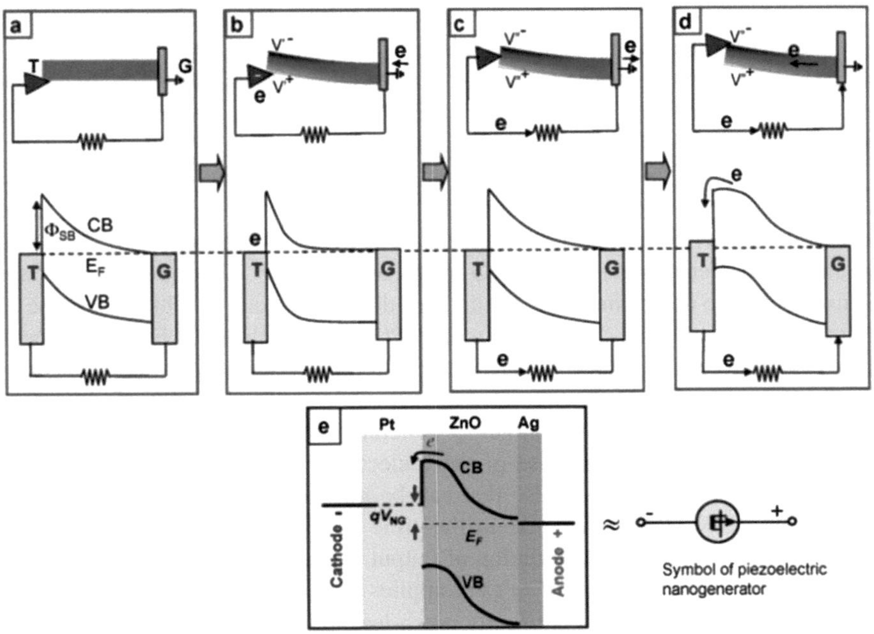

Fig. 3.28 Band diagram to explain the charge outputting and flowing processes in the nanogenerator. **a** Schematic and energy diagram of a nanowire with one end grounded (*G*) and the other being pushed by a conductive AFM tip (*T*). A Schottky barrier is present at the tip-nanowire interface. **b** As the nanowire is slowly deflected, the asymmetric piezoelectric potential in the nanowire changes the profile of the conduction band (*CB*). **c** When the tip scans across the nanowire and reaches its center point, a drop in local potential to zero results in a back flow of the accumulated electrons through the load into ground. **d** Once the tip reaches the compressive surface, a local negative piezoelectric potential raises the profile of the conduction band. If the piezoelectric potential is sufficiently high, electrons in the *n*-type ZnO nanowire can flow to the tip. This circular motion of the electrons in the circuit contributes to the output current. **e** Energy band diagram for the nanogenerator, showing the output voltage and the role of the piezoelectric potential. The diagram on the *right-hand side* is a symbol to represent a piezoelectric nanogenerator. Courtesy of Dr. Z. L. Wang, Georgia Institute of Technology. Reproduced with permission from [38], Copyright @ 2009, John Wiley & Sons

Alternatively, if the tip temporarily is left off the nanowire, the accumulated electrons will also flow back to the ground.

When the tip arrives at the compressive side of the nanowire (Fig. 3.28d), the local potential drops to V''^{-} (negative), which leads to a raise in height of the conduction band near the tip. If the increase in local potential energy, determined by the degree of bending of the nanowire, is sufficiently large, the local accumulated *n*-type carriers in the nanowire can quickly flow through the interface to the tip, which leads to a circular flow of electrons in the external circuit, which is electrical current. This process is much faster than the charge accumulation process. Therefore, the transient potential signal generated by the nanogenerator at the external load is strong enough to be detected out from the noises.

The presence of a Schottky junction at the tip-nanowire interface is mandatory for a nanogenerator, whose function is like a "gate" that can separate and slowly-accumulate the charges and then quickly release them. Without such a "gate," at e.g., an Ohmic contact, there is no charge accumulation and releasing. As a consequence, there are no detectable signals.

The magnitude of output voltage of the nanogenerator can be evaluated according to the energy band diagram shown in Fig. 3.28e. The piezoelectric potential has a tendency to drive the electrons from the ZnO nanowire to overcome the threshold energy at the metal-ZnO interface into the Pt electrode, but it cannot be used solely to determine the magnitude of the output voltage. This is because as more electrons are being "pumped" into the Pt electrode, the local Fermi surface is raised. Therefore, the output voltage is the difference between the Fermi energies for the Pt on the tip and the bottom Ag electrode.

Although total amount of charges Q generated by a single nanowire is small (1,000–10,000 e), a quick release of these electrons provides a sufficiently large electrical current or voltage pulse that can be measured with common facilities, because $V_L \approx R_L Q/\Delta t$, where Δt is the time interval of the charge releasing process. The conventional calculation of output voltage with the capacitance C of the nanowires and the system, $V = Q/C$, applies only to static process. The voltage predicted by this calculation is usually much less than the magnitude of the pulse observed experimentally.

In summary, the first process is a forward or back flow of the electrons through the load from the ground to the tip, while the second process is a circular flow of the electrons through the nanowire driven by the piezoelectric potential. The two processes generate the constructive currents flowing in same direction, resulting in a negative voltage over the load with reference to the grounded electrode. The output current can be significantly large if the charge releasing process is fast. In terms of the AFM tip contact, the second process is likely the dominant process. Therefore, the flow of external electrons driven by the piezoelectric potential is the key to the power output process of the nanogenerator.

3.5.5 Nanogenerators with p*-Type Materials*

3.5.5.1 Characteristic of Output Signals

As stated earlier, because ZnO is typically *n*-type semiconductor, the piezoelectric output of the ZnO nanogenerator measured by using an AFM usually is negative in sign and the potentials are detectable only when the tip touches the compressed side of the nanowire. It is therefore expected that a ZnO nanogenerator made with *p*-type nanowires should have a positive piezoelectric response. Figure 3.29 shows a comparison between two nanogenerators made with *p*-type and *n*-type ZnO nanowires [34]. As shown in Fig. 3.29a and b, the output voltage at the external load is always positive with reference to the grounded root of the nanowire. In

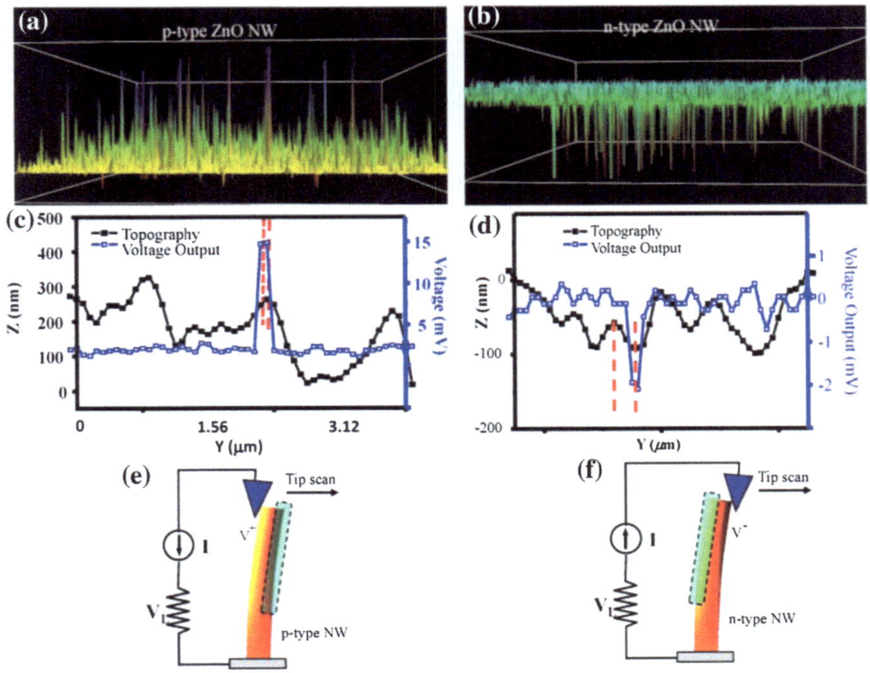

Fig. 3.29 A comparison of nanogenerators based on *n*-type and *p*-type ZnO nanowires. **a** 3D voltage output profile across the *p*-type ZnO nanowire array as the AFM tip scans in contact mode. **b** Comparison between the topological scan and output voltage scan, showing positive sign of the output voltage generated as the tip touches the stretched side of the nanowire. **c** Mechanism of the *p*-type nanowire nanogenerator. **d** 3D voltage output profile across the *n*-type ZnO nanowire array as the AFM tip scans in contact mode. **e** Comparison between the topological scan and output voltage scan, showing negative sign of the output voltage generated as the tip reaches the compressed side of the nanowire. **f** Mechanism of the *n*-type nanowire nanogenerator. Courtesy of Dr. Z. L. Wang, Georgia Institute of Technology. Reproduced with permission from [34], Copyright @ 2009, American Chemical Society

comparison, the output voltage of the *n*-type ZnO nanowire is negative, at the experimental conditions, as shown in Fig. 3.29d and e. In addition, the output voltage peak of the *p*-type nanowire appears in the first half of the whole contact process between the tip and the nanowire (Fig. 3.29c). It has been demonstrated earlier that the response of *n*-type nanowire is in the second half of the contact process (Fig. 3.29f). This is a reflection that the two types of nanogenerators have different mechanisms.

For *p*-type nanowires, the Fermi level is located closer to the edge of the valence band. The metal-semiconductor junction is forward biased if the semiconductor has a higher potential, otherwise it is reversely biased. As the tip is in contact with the nanowire, as shown in Fig. 3.29c, the negative piezoelectric charges created at the compressed side of the nanowire are partially screened by the positively charged holes. Once the magnitude of the positive piezoelectric

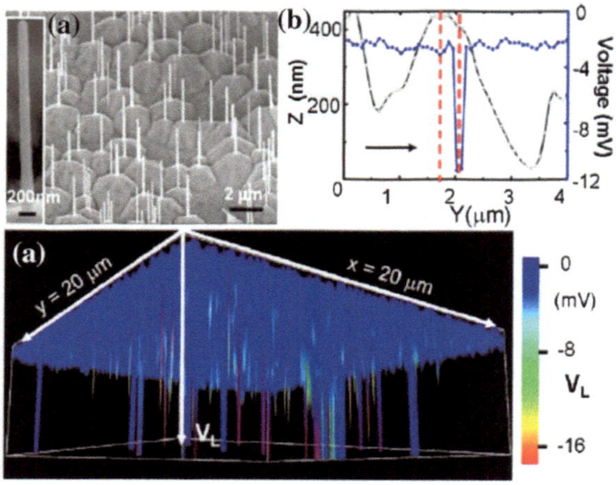

Fig. 3.30 **a** SEM image of the as-synthesized ZnO nanowire array by using PLD, with the inset showing a single nanowire whose diameter is about 150 nm. **b** Three-dimensional plot of output voltage of the as-synthesized *n*-type ZnO nanowire array. **c** A typical line scan profile from the AFM topography (*black*) and the corresponding output voltage (*blue*). Courtesy of Dr. Z. L. Wang, Georgia Institute of Technology. Reproduced with permission from [46], Copyright @ 2009, Institute of Physics

potential at the stretched side exceeds a value that can either drive the *p*-type carriers or the electrons in the metal tip across the junction, a positive output voltage can be observed in the external load. This scenario occurs in the first half of the scanning process of the metal tip over the nanowire.

For *n*-type nanowires, the metal-semiconductor interface is forward biased if the metal has a higher potential, otherwise the junction is reversely biased. When the tip starts to touch the nanowire, as shown in Fig. 3.29f, the positive piezoelectric charges created at the stretched side of the nanowire is partially screened by electrons. A potential different is built up between the tip and the nanowire once the tip touches the compressed side of the nanowire, which has a negative piezoelectric potential. This piezoelectric potential at the compressed side drives the electrons to flow across the junction, resulting in a negative voltage crossover the external load. This event is observed in the second half of the scanning of the metal tip over the nanowire.

3.5.5.2 Characterization of *p*- and *n*-Type Nanowires

The type of conduction of a ZnO film can be identified by using two methods: Hall Effect and Seebeck effect. Their application to thin films is very convenient and effective, because it is easy to fabricate electrodes for the measurements. However, these methods cannot be used for nanowires, due to their low dimensionality. Therefore, the carrier type of ZnO nanowires is identified by using single nanowire-based field effect transistor (FET).

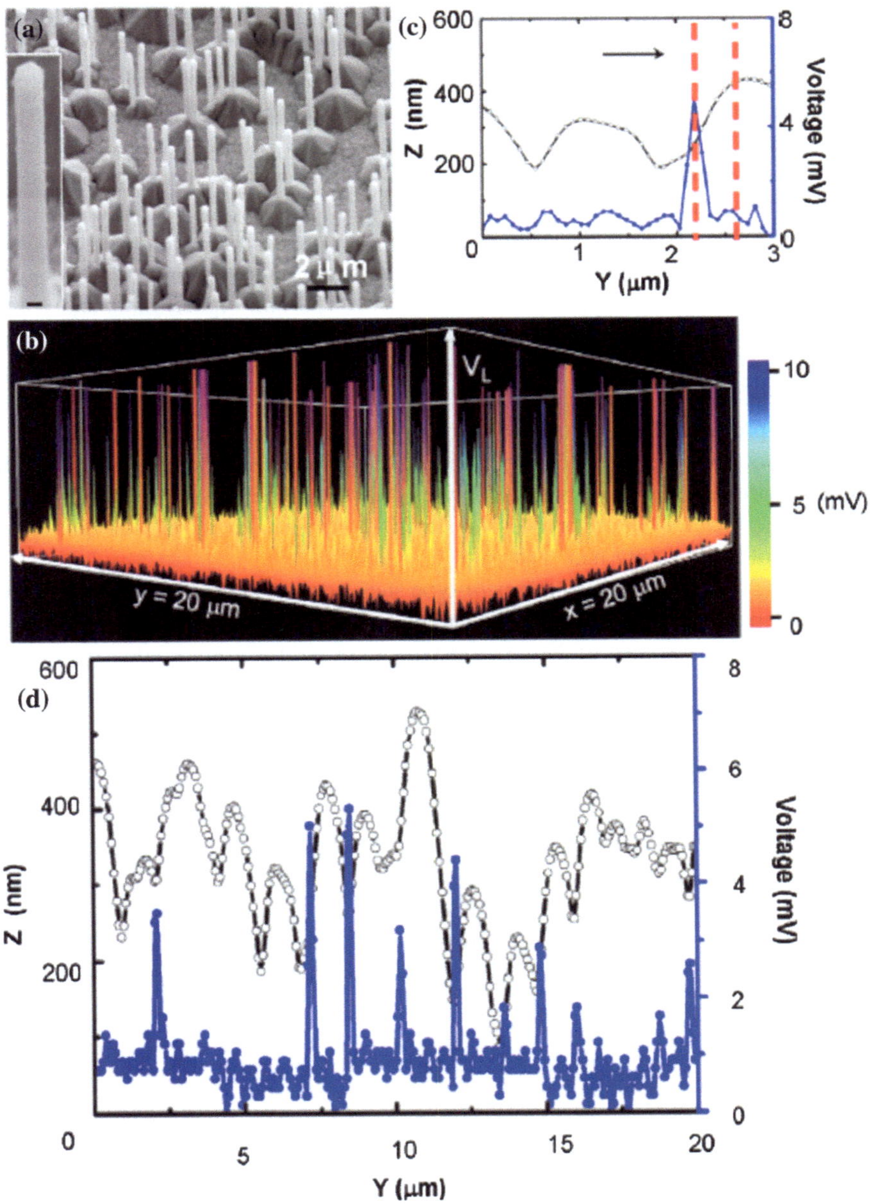

Fig. 3.31 **a** SEM image of the *n*/*p* core/shell ZnO nanowire array. The inset shows a single nanowire with a diameter of about 400 nm. **b** Three- dimensional plot of output voltages of the *n*/*p* core/shell ZnO nanowires. **c** A typical line scan profile from the AFM topography (*black*) and the corresponding output voltage image (*blue*). **d** A 20 μm line scan profile from the AFM topography (*black*) and the corresponding output voltage (*blue*) image. Courtesy of Dr. Z. L. Wang, Georgia Institute of Technology. Reproduced with permission from [46], Copyright @ 2009, Institute of Physics

A new criterion has been developed to identify the carrier type of individual ZnO nanowires, basing on the piezoelectric output of a nanowire when it is mechanically deformed by a conductive AFM tip in contact mode [46]. Experimental results indicate that ZnO nanowires with a *p*-type/*n*-type shell/core structure possess positive piezoelectric responses, while *n*-type nanowires exhibit negative piezoelectric behaviors. This is a simple and nondestructive method to identify the carrier type of ZnO nanowires.

Figure 3.30a shows a SEM image of the as-grown nanowires array by using PLD, with the inset showing the image of a single nanowire [46]. The as-synthesized nanowires have typical diameters of 100–150 nm. Figure 3.30b shows the corresponding three-dimensional (3D) voltage output image over a scanning area of $20 \times 20\ \mu m^2$ within the *n*-type ZnO nanowire array. The *n*-type ZnO nanowires exhibit negative voltage outputs when they are deformed by a Pt-coated AFM tip. A typical profile of voltage output is shown in Fig. 3.30c, demonstrating that the negative voltages are produced when the AFM tip touches the compressed side of the nanowires, as indicated by the delay of the voltage peak with reference to the peak in the morphological profile. As stated earlier, this delay is a unique characteristic of *n*-type ZnO nanowires.

The ZnO nanowires with a *p*-type/*n*-type shell/core structure can be synthesized via two steps. The PLD grown *n*-type nanowire array is used as a template. Then, a thin layer (300–400 nm) of N-doped *p*-type ZnO is uniformly coated on *n*-type nanowire array by using plasma-free MOCVD [46].

Figure 3.31a shows a SEM image of the nanowires after coating with the *p*-type layer. The *n*/*p* core/shell nanowires have typical diameters of about 400 nm, as shown in the inset. The larger diameter of the *n*/*p* core/shell nanowire is an evidence to confirm that the *p*-layer has been grown epitaxially on the *n*-type nanowire. 3D voltage output image over an area of $20 \times 20\ \mu m^2$ is shown in Fig. 3.31b. The *n*/*p* core/shell nanowires produce positive outputs over the entire scanned area. Detailed analysis indicates that the voltage peak is ahead of the peak in the morphological image with reference to the scanning direction of the tip, as shown in Fig. 3.31c. Figure 3.31d further shows that all the voltage peaks are positive.

3.5.6 Strategies Toward High-Performance Nanogenerators

The output power of nanogenerators using a single nanowire is too low for practical applications. Various strategies have been reported to obtain large-scale output power of ZnO nanogenerators. The latest develop of these strategies on the nanogenerators with vertically aligned ZnO nanowire arrays will be presented in this section.

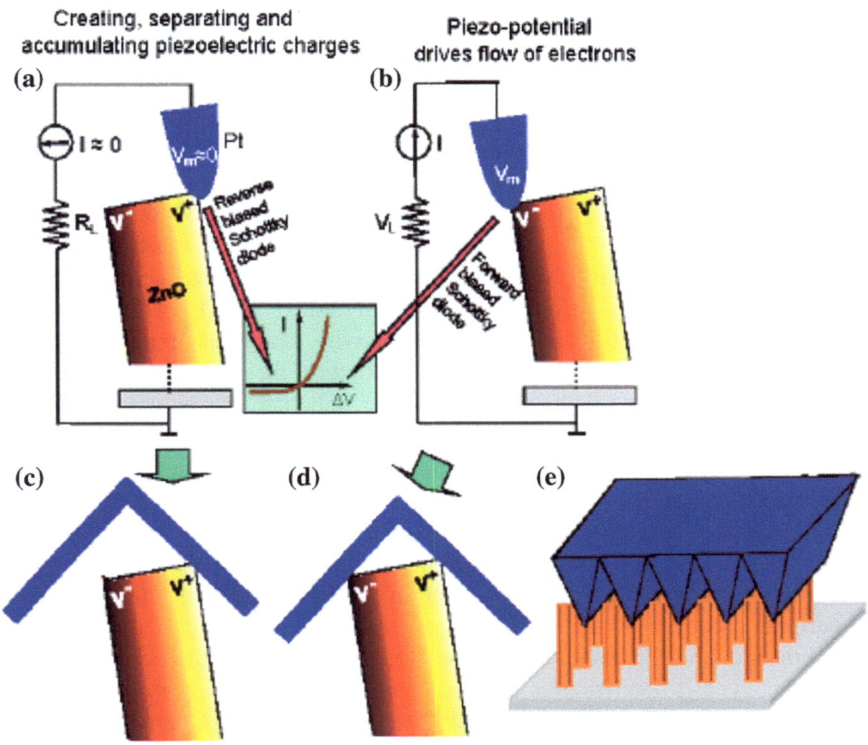

Fig. 3.32 **a**, **b** Current generation process of a piezoelectric nanowire due to the deformation by an AFM tip. **c–e** Schematic diagram showing the idea of using *zigzag-shaped* electrode to capture the electricity generated by a collect of nanowires rather than just a single nanowire. Courtesy of Dr. Z. L. Wang, Georgia Institute of Technology. Reproduced with permission from [31], Copyright @ 2011, Georgia Institute of Technology

3.5.6.1 Nanogenerators Driven by Ultrasonic Wave

The first example is the development of nanogenerators driven by ultrasonic wave [47, 48]. Three main achievements can be highlighted in this development. First, the use of AFM tip to produce mechanical deformation of the ZnO nanowires for power generation is replaced by a more adaptable, mobile, and cost-effective approach over a larger scale. Second, all the nanowires are used to generate electricity simultaneously and continuously, and the electricity can be effectively collected as an output. Finally, the energy to be converted from mechanical displacement into electricity can be realized by wave/vibration from the environment at very low frequencies.

Shown in Fig. 3.32a and b [31], the power output is due to the deformation of the single piezoelectric nanowire caused by an AFM tip, as discussed above. With the AFM tip, only nanowire can be used every time and it is also a discontinuous process. To replace AFM tip, an "inverted V-shape" (*I–V*) electrode has been developed [47, 48]. An example is shown in Fig. 3.32c. As the *I–V* electrode moves downward driven by an external excitation, the nanowire is deflected toward left-hand side. At the first point of contact, a positive potential is created due to the local tensile strain. In this case, the piezoelectric potential is preserved because the contact is a reversely biased Schottky barrier that blocks the charges. As the electrode further pushes the nanowire until it bends enough to touch the other side of the *I–V* electrode (Fig. 3.32d), a forward biased Schottky barrier is formed. As a result, the local potential drop drives the flow of electrons.

The idea of *I–V* electrode has been extended to zigzag electrode, as shown in Fig. 3.31e. The zigzag electrode is made of Si coated with Pt. The conductive metallic coating serves to enhance the conductivity of the electrode and form a Schottky barrier at the interface with ZnO. In practice, the metallic coating can be any conductive alloy, as long as it can form a Schottky barrier with ZnO. The zigzag electrode is like parallel AFM tip array. With a continuous excitation, the nanogenerator can produce electricity simultaneously and continuously.

Figure 3.33 shows the possible types of contact between a nanowire and the zigzag electrode [48]. Type I is deflected toward left-hand side. There is another possibility that the type I nanowire is deflected toward right-hand side, when they have similar height, a slight off-set in position. Regardless the deflection direction, the currents produced by the nanowires of Type I or similar will add up. Type II just vibrates due to the excitation by the ultrasonic wave. When the compressive side of the Type II nanowires is in contact with the electrode, they exhibit same discharge process as that of Type I, resulting in a flow of current from the electrode into the nanowires. The nanowire of Type III just touches with the tip of the electrode, whereas the one of Type IV is too short to reach the electrode. If nanowires of Type III are in compressive strain by the electrode without bending, the piezoelectric voltage created at the top of the nanowires is negative. Thus, a positively biased Schottky barrier is formed across the electrode-ZnO interface. As a result, electrons can freely flow across the interface. In this case, electrons flow from the nanowires into the top zigzag electrode as they are deformed. This discharging also contributes to the measured currents. The output current is a sum of those from all nanowires that are capable of contributing, while the voltage of the nanogenerator is determined by a single nanowire because all the nanowires are "in parallel" arrangement.

The nanogenerator is tested by immersing in water bath to measure the closed circuit current and open circuit voltage, under excitation of an ultrasonic wave with a frequency of 41 kHz. The ultrasonic wave is tuned on and off every 15 s. Figure 3.34a shows closed circuit current when the ultrasonic wave is turned on and off alternatively. The data clearly indicate that the current output is originated from the nanogenerator as a result of the excitation from the ultrasonic wave. A similar pattern in Open-circuit voltage output is also observed, as shown in Fig. 3.34b. The nanogenerator has high levels of current and voltage outputs, with a current of about 500 nA and voltage of about 10 mV. With the effective area of the nanowire array (6 mm^2), the nanogenerator achieves an equivalent current density of about 8.3 $\mu A/cm^2$ and a power density of about 83 nW/cm^2 [48].

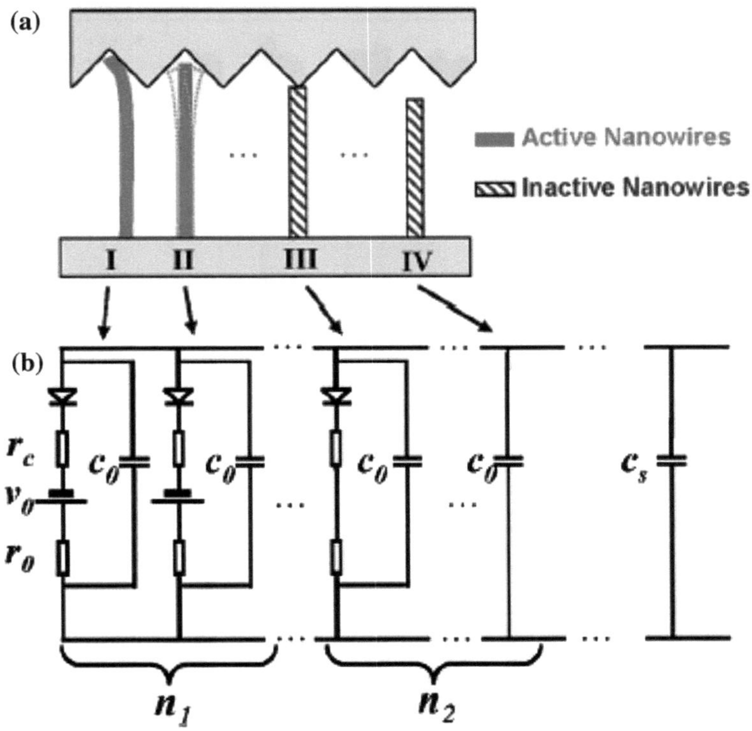

Fig. 3.33 Equivalent circuit analysis of a nanogenerator. **a** Three different types of nanowires are presented: Type I and Type II are active nanowires that output electricity; Type III is in contact with the top electrode but does not produce current; and Type IV is not in contact with the top electrode. **b** The corresponding circuit model, where c_s is added to represent the system capacitance. Courtesy of Dr. Z. L. Wang, Georgia Institute of Technology. Reproduced with permission from [48], Copyright @ 2008, American Institute of Physics

It is important to point out that the operation of the nanogenerators does not require mechanical resonance as required by those energy harvesting technologies discussed in previous chapter. Instead, mechanical deflection is the driving force for nanogenerators. Therefore, these nanogenerators can work in any range of frequencies and thus are wide range mechanical energy harvesters.

3.5.6.2 Integration of Nanotip-to-Nanowire

The advantages of the zigzag electrode have been elaborated. However, there is still room for further improvement of the nanogenerators. For example, the non-uniform heights and random distribution of the ZnO nanowires cannot ensure all the wires to have contribution to electricity generation. Also, there is a stringent requirement in integrating the electrode, because the gap between the zigzag

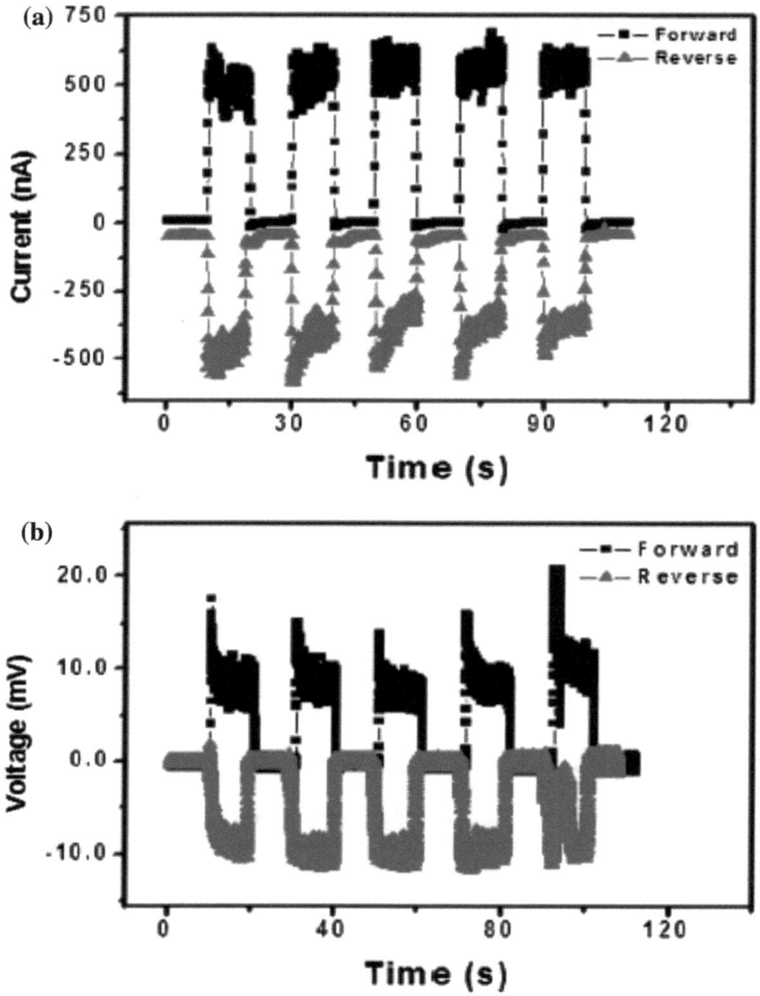

Fig. 3.34 Performance of a packaged NG when periodically excited by ultrasonic wave. **a** Closed-circuit current output and **b** Open-circuit voltage output measured at forward polarity (*dark line*) and reversed polarity (*red line*) connection with the measurement system. Courtesy of Dr. Z. L. Wang, Georgia Institute of Technology. Reproduced with permission from [48], Copyright @ 2008, American Institute of Physics

electrode and the nanowire arrays should be adjusted in such a way that it is sufficiently large to allow free bending or deflecting for the nanowires, but small enough to ensure an intimate contact between the nanowires and the electrode.

In this respect, a new approach has been developed to address such problems. With this approach, the nanogenerator is composed of integrated and paired nanobrushes made of pyramid-shaped metal-coated ZnO nanotip (labeled as NTP) arrays and hexagonal prism-shaped ZnO nanowire arrays [49]. All the ZnO nanotips and nanowires are synthesized by using a chemical route at temperatures of <100 °C on both surfaces of common substrates. A nanogenerator is assembled by stacking one piece of such structure in close proximity over another to form a layer-by-layer matched brush architecture. The nanogenerator can produce direct current upon excitation of ultrasonic waves. A nanogenerator with four-layer integration offers an output with power density of 0.11 $\mu W\ cm^{-2}$ at a voltage of 62 mV.

To assemble the nanogenerators, ZnO arrays with controlled orientation and shape are grown on double-side polished Si wafers. The wafers are coated with 100 nm thick Al_2O_3 film on both sides by using atomic layer deposition (ALD), which serves as an insulating layer to ensure independent function of the nanogenerators built in the adjacent layers. The wafer are also coated with a 20 nm thick Cr layer by direct current magnetron plasma sputtering on both surfaces, while the side surface must be kept uncoated. After that, a 50 nm thick ZnO layer is deposited by radio frequency magnetron sputtering in a similar fashion. The Cr layer has two functions: adhesion to bind the Al_2O_3 layer and the ZnO layer together and electrode to collect electricity.

A ZnO seeding layer is used to improve the quality of the aligned ZnO arrays. The seeding layer has a (0001) orientation by using diluted $Zn(CH_3COO)_2$ solutions in ethanol. Density of the ZnO seeds can be controlled by adjusting the concentrations of the $Zn(CH_3COO)_2$ ethanol solutions. Figure 3.35a shows a schematic diagram of the substrate, which is ready to grow the ZnO nanowires and the NTPs.

The growth of the ZnO NW/NTP arrays with controlled morphologies on different sides of the Si wafer is carried out in 5 mM solution with 1:1 ratio of $Zn(NO_3)_2$ and hexamethylenetetramine (HMTA), with the substrates floating on surface of the solutions. The ZnO morphologies can be controlled by growth temperature and time duration. Nanowires are grown at low temperature for longtime (Fig. 3.35b and g), whereas NTPs are formed at higher temperature for short time duration (Fig. 3.35c and h). Representative conditions for ZnO nanowires and NTPs are 70 °C for 48 h and 100 °C for 24 h, respectively. Finally, the side with NTPs is coated with a 100 nm thick Au to form a metal tip array [49].

A nanogenerator is assembled by stacking two pieces of the wafers with the designed structures, with the Au-coated NTPs partially interpenetrating into the nanowires, as illustrated in Fig. 3.35e [49]. Similarly, a multilayered nanogenerator is integrated by stacking layer-by-layer the multiple layers with such structures (Fig. 3.35f). The Au-coated NTPs acted as an AFM tip array, slightly interpenetrating into the spacing between the nanowire arrays, like two face to face brushes with a slight degree of overlap. This design has not a strict requirement on the height uniformity of the nanowires. There is also no need to maintain a specific gap between the layers. Therefore, this approach is more flexible as compared to that of zigzag electrode. Every two adjacent layers are insulated the Al_2O_3 film. The whole structure is sealed and packaged with epoxy resin and then is ready for testing. Positive electrode is the side with bare ZnO nanowire arrays, while negative electrode is the side with Au-coated NTP arrays.

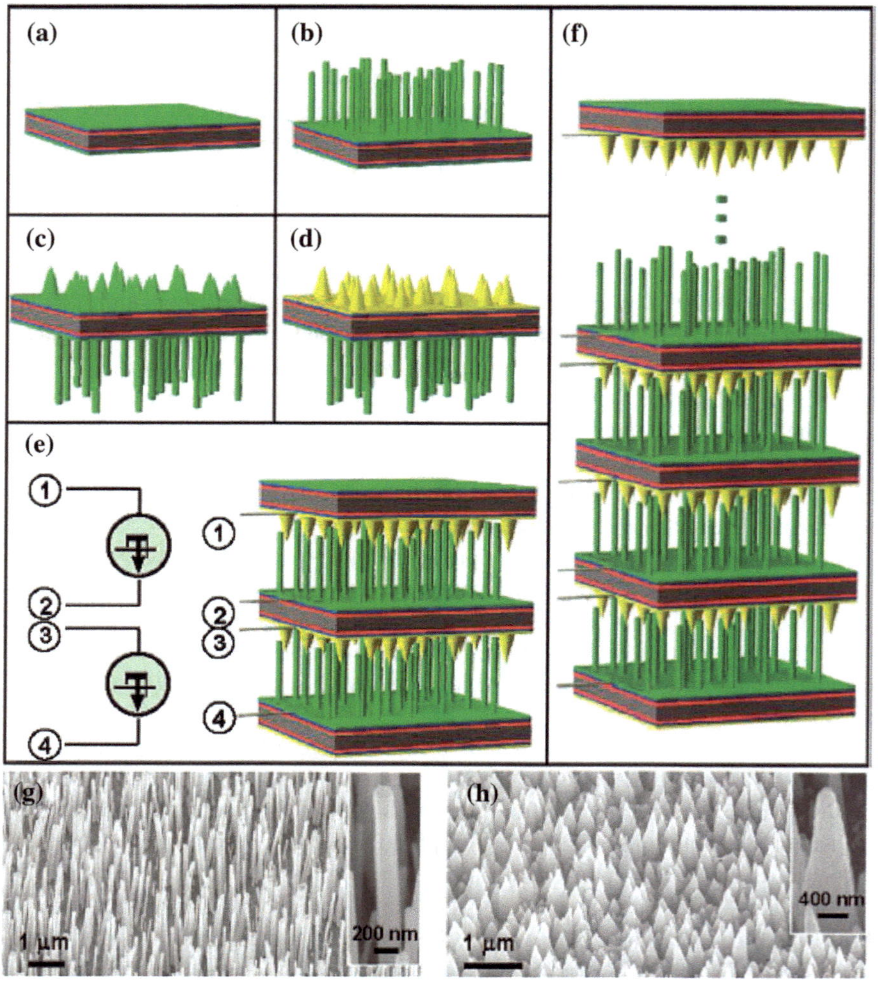

Fig. 3.35 Fabrication procedures of the multilayered nanogenerators and the growth of ZnO nanowires and NTP arrays. **a** As-fabricated substrate after coating with a seed layer. **b** Growth of hexagonal prism-shaped nanowire arrays by wet chemical method on one side. **c** Growth of pyramid-shaped NTP arrays on the other side. **d** Coating the NTP arrays with Au layer by magnetron plasma sputter. **e** Two-layered NG made by stacking three layers of the wafer structures shown in **d**, with the Au-coated NTP arrays facing and interpenetrating the bare nanowire arrays. The *right-hand side* is the designed symbol to represent the nanogenerators. **f** A multilayered three-dimensional nanogenerator integrated by stacking multiple layers of wafer structures. **g** 60° tilted view SEM image of the ZnO nanowire arrays. **h** 60° tilted view SEM image of ZnO NTP arrays. The inset SEM images are high magnification views of a single nanowire and NTP. Courtesy of Dr. Z. L. Wang, Georgia Institute of Technology. Reproduced with permission from [49], Copyright @ 2008, American Chemical Society

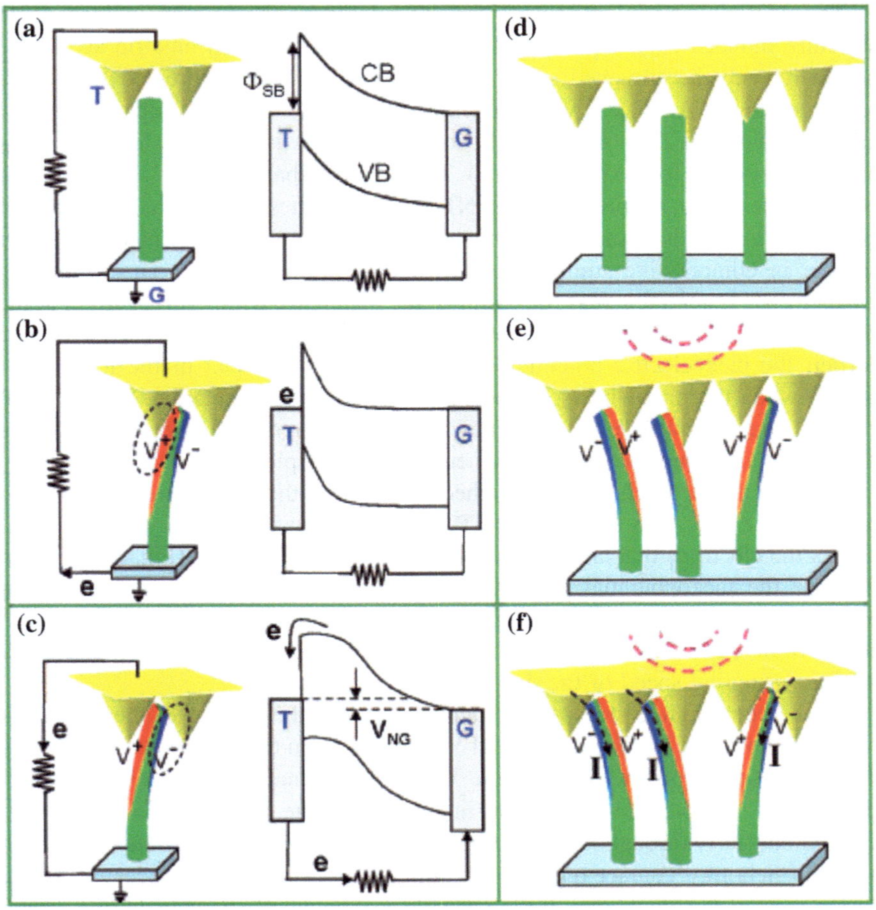

Fig. 3.36 Principle of a single-layered nanogenerator, in terms of band diagrams taken from the left/right boundaries of the nanowire. **a** Schematic diagram of the nanogenerator and the corresponding electron energy band diagram, where *T* is the Au-NTP and *G* is the grounded nanowire. **b** Under excitation of ultrasonic wave, there is a relative movement (in both *vertical* and *lateral directions*) between the NTP and the nanowire, and thus the NW is bent with the contact surface being stretched. The corresponding local piezoelectric potential (positive) creates a reversely biased Schottky barrier. This diagram represents the equilibrium status of the contact between the NTP and nanowire. **c** When the relative movement of the NTP and the nanowire is strong enough, the nanowire is bent enough to touch another adjacent NTP with its compressive side. Then, the local piezoelectric potential (negative) sets the Schottky barrier to be forward biased and drives the flow of electrons from the nanowire into the NTP. **d** Schematic diagram of multiple nanowire and NTP contacts with slight variation in heights and lateral distribution. **e** Regardless the NWs being deflected to left-hand or right-hand, all the first contacting surfaces with the tips are tensile surfaces with a positive piezoelectric potential. **f** Electric currents created by the driving force of the piezoelectric potential when the compressive surfaces of the nanowires touch the surfaces of adjacent tips. The contacts can be simultaneous or slightly delayed, but all the created transient currents flow in the same direction, from NTPs to nanowires. Courtesy of Dr. Z. L. Wang, Georgia Institute of Technology. Reproduced with permission from [49], Copyright @ 2008, American Chemical Society

The energy conversion process can be related to the change in local band structure induced by the piezoelectric potential, which is asymmetric across the nanowire with the stretched side being positive and the compressed side being negative. Since Au has a work function of 4.8 eV, which is higher than the electron affinity of ZnO (4.5 eV), a Schottky contact (with a barrier height of Φ_{SB}) at the interface is formed (Fig. 3.36a). The Schottky barrier exists as soon as the NTP touches the nanowire. When an NTP slowly pushes a nanowire, a strain field is created across the nanowire width, with its outer surface in tensile and inner surface in compressive strain. This asymmetric strain produces an asymmetric piezoelectric potential across the width of the nanowire, with V^- (negative) at the compressive surface and V^+ (positive) at the stretched surface. As mentioned earlier, the piezoelectric potential is due to the displacement of the ions in the crystal when the nanowire is subject to a mechanical deformation. This kind of charges cannot move freely, which may be partially screened by free carriers in the nanowire, but cannot be completely neutralized or depleted.

When a NTP touches the stretched surface of the nanowire, whose local potential is higher than that of the NTP, a reversely biased Schottky barrier (Φ_{SB}) blocks the electrons that are likely to flow through the interface (Fig. 3.36b) [49]. Because the piezopotential is built up quickly inside the nanowire, when it is pushed by the NTP, throughout pushing process, the whole system is in an equilibrium state. This is because the flowing speed of the charges is much faster than the scanning speed of the tip.

As the compressive sides of the nanowires touch the surface of their adjacent NTPs, the local piezoelectric potential V^- (negative) in the nanowires will modify the local shape of the conduction band of the nanowires near the NTPs (Fig. 3.36c). Theoretical calculation indicates that the piezoelectric potential is distributed nearly throughout the entire length of the nanowire, and the band shape near the Schottky contact region is modified most significantly, because of the height-raise of the local band. If the raise of local potential energy is large enough, which is dependent on the bending degree of the nanowire, the accumulated local *n*-type carriers in the nanowire can quickly flow through the contacting interface into the NTP, which creates an output current.

In terms of a single pair of nanowire and NTP, this process is transient. However, with the statistical contributions by thousands of NTP/NW pairs, the process is steady, outputting a stable and continuous electrical current. The role of the Schottky barrier is to prevent the flowing of electrons from the NTP into the nanowire, which is a key factor to preserve the piezoelectric potential and release the free electrons from the nanowire into the NTP. The role of the piezoelectric potential is to drive the electrons in the ZnO nanowire to overcome the threshold energy at the Au–ZnO interface and flow into the Au-NTP. The piezoelectric potential does not directly determine the magnitude of the output voltage.

In summary, although a single nanowire is in nonequilibrium or transient state, the statistical average of thousands of nanowires ensures to produce a stable and continuous output, which is similar to the process in solar cells. In practice, there

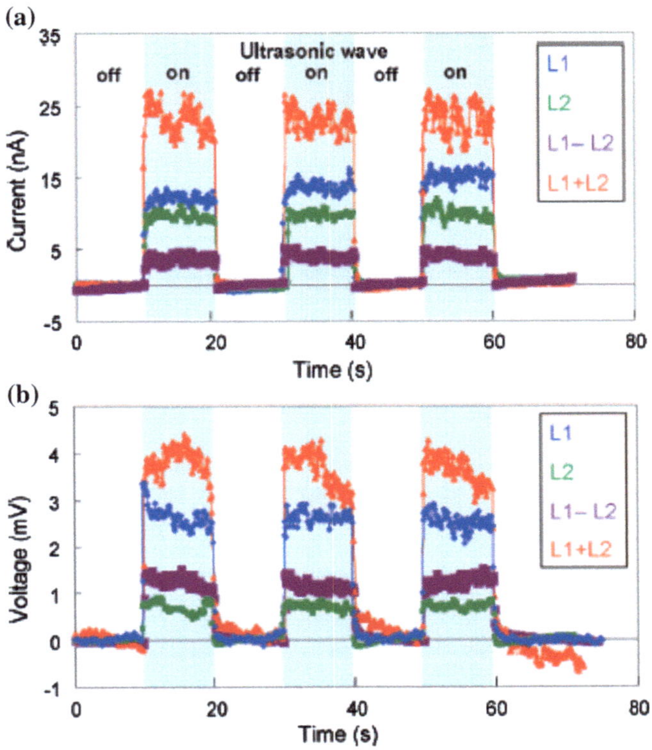

Fig. 3.37 Output current and voltage signals by connecting two single-layered nanogenerators in parallel and antiparallel, and serial and anti-serial. **a** Short-circuit current output measured from the nanogenerators when connected in parallel and antiparallel. **b** Open-circuit voltage output measured from the nanogenerators when connected in serial and anti-serial. The surface area of each nanogenerator is 6 mm^2. Courtesy of Dr. Z. L. Wang, Georgia Institute of Technology. Reproduced with permission from [49], Copyright @ 2008, American Chemical Society

are other factors, such as contact resistance, system capacitance and possible current leakage, which can lower the total voltage output.

The layer-by-layer integrated nanogenerators can be connected in different ways, to output current and voltage with different magnitudes [49]. By connecting two individual layers of nanogenerators in parallel, e.g., connecting electrodes 1 with 3 and electrodes 2 with 4 in Fig. 3.35e, the output current becomes a sum of the two nanogenerators. As shown in Fig. 3.37a, a single layer of nanogenerator L1 has short circuit output current of about 13 nA, and the other layer of nanogenerator has a value of about 10 nA under the same conditions. When they are connected in parallel, the output current is increased to about 22 nA (Fig. 3.37a, red curve). When they are connected in antiparallel, the total output current is decreased to 3 nA (Fig. 3.37a), purple curve), which is the difference of the two individual output currents. Also shown in Fig. 3.37a, the output signal of the L1 + L2 connection is more "unstable" than the L1 − L2 connection. This is

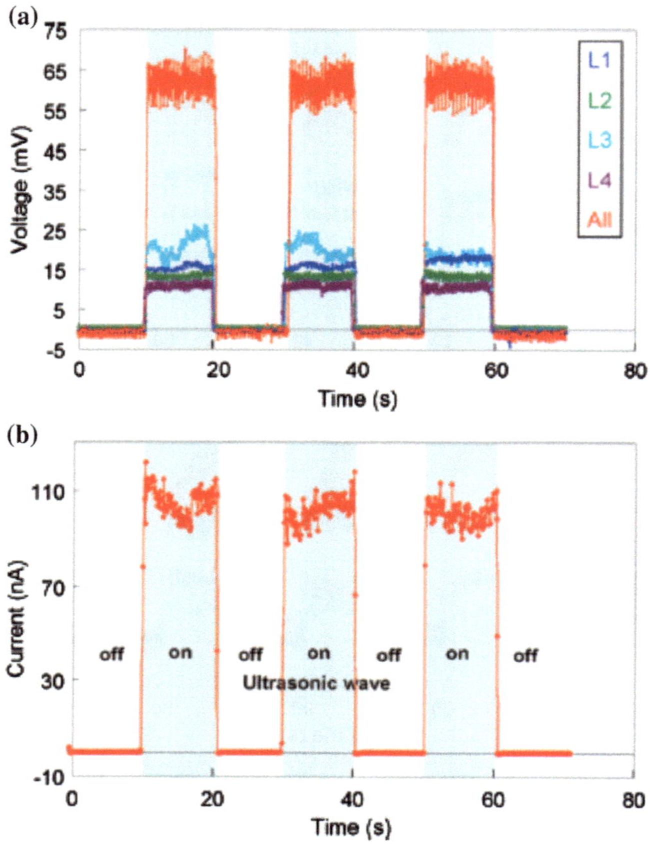

Fig. 3.38 Open-circuit voltage and Short-circuit current output measured from serially connected four-layer integrated nanogenerator. **a** Open-circuit voltage output measured from each individual layers and the serially connected four-layer integrated NG. **b** Short-circuit current output measured from the serially connected four-layer integrated nanogenerator. The regions when the ultrasonic wave was on and off are indicated. The surface area of the NG was 6 mm^2. Courtesy of Dr. Z. L. Wang, Georgia Institute of Technology. Reproduced with permission from [49], Copyright @ 2008, American Chemical Society

because the L1 + L2 connection double amplifies the instability introduced by the intensity and frequency of the ultrasonic wave source, while the L1 − L2 connection impairs the instability.

Similarly, by connecting two layers of nanogenerators in serial, e.g., connecting electrodes 2 with 3 in Fig. 3.35e, the output voltage is the sum of the output voltages of the two individual layers. As shown in Fig. 3.37b, L1 has an output voltage of about 2.6 mV, while L2 has voltage of about 0.8 mV. When those two layers of nanogenerators are connected in serial and tested under same conditions, the average output voltage is about 3.5 mV (Fig. 3.37b, red curve), which is the sum of their individual outputs. If two layers are connected in anti-serial, e.g., connecting electrodes 2 with 4 in Fig. 3.35e, the output voltage becomes 1.5 mV (Fig. 3.37b, purple curve), which is the difference of their individual output voltages.

Therefore, integrating these multilayer nanogenerators is able to increase the output voltages to be sufficiently high for practical applications or storages [49]. As shown in Fig. 3.38a, four individual layers of nanogenerators, L3, L4, L5, and L6 with 11, 14, 16, and 20 mV open-circuit output voltages, respectively, are connected in serial. The resultant total output voltage is about 62 mV, as expected. The corresponding short- circuit output current is about 105 nA (Fig. 3.38b). The maximum power output of the four-layer integrated nanogenerator is 6.5 nW, with each of them having a surface area of 6 mm^2, corresponding to a power density of 0.11 $\mu W/cm^2$ [49].

3.5.6.3 Integration of Nanogenerators with Firmly Contacted Ends

An alternating current (AC) nanogenerator (NG) (AC-NG), with high power outputs, has been developed, with vertically aligned ZnO nanowire arrays that are bonded at two ends with electrodes [50]. In this nanogenerator, the ZnO nanowires are excited by a periodic low-frequency uniaxial strain via the top electrode. The nanogenerator can be integrated to form multilayer structured harvesters with high output currents and voltages.

Figure 3.39 shows fabrication flow chart of the three-dimensionally integrated and packaged AC-NGs [50]. The vertically aligned ZnO nanowire arrays are grown on an Au-coated flat surface by using wet chemical method. Ti and Au thin layers are coated consecutively Si substrates by using magnetron sputtering (Fig. 3.39A(a)). ZnO nanowires with tip diameter of about 300 nm and length of about 4 μm are grown at <100 °C (Fig. 3.39A(b) and B(a)). The nanowire array is entirely filled with a layer of polymethyl methacrylate (PMMA) (Fig. 3.39A(c) and B(b)). The purpose is to increase stability and mechanical robustness of the structure and prevent possible short circuiting between the substrate and the top electrode. A thin layer of the PMMA film is etched away by using oxygen plasma etching to expose tips of the ZnO nanowires (Fig. 3.39A(d) and B(c)). A piece of Si wafer coated with 300 nm thick Pt film is placed on top of the ZnO nanowires (Fig. 3.39A(e)), so that a Schottky barrier is formed at the interface. The structure is encapsulated for measurement in a Faraday cage using a linear motor stimulator to generate mechanical strain at an impact speed of 0.1 m s^{-1}.

The working principle of the AC-NG is the coupling of piezoelectric and semiconducting properties of the ZnO nanowires. Because the wurtzite-structured ZnO nanowires are grown uniaxially parallel to *c*-axis, this crystallographic alignment ensures the alignment of their piezoelectric responses when deformed. Uniaxial straining produces a piezoelectric potential gradient along the *c*-axis of every nanowire (Fig. 3.39A(f)). The total response is the sum of the responses of all nanowires. When a uniaxial stress is applied onto the top electrode, the nanowires are under uniaxial compression, with negative piezoelectric potential at the tip, which is the Schottky contact side, and positive piezoelectric potential at

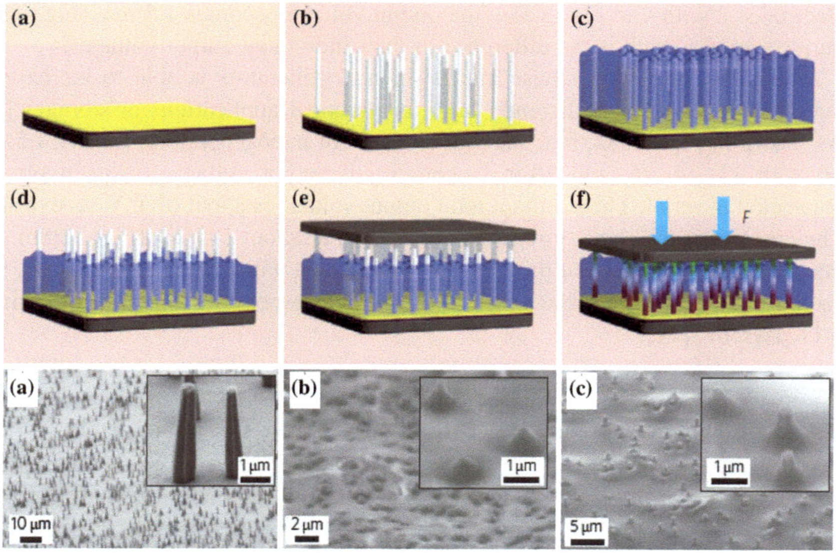

Fig. 3.39 **A** Fabrication flow chart of the AC-NG. **a** Au-coated Si substrate. **b** ZnO nanowire arrays on Si substrate. **c** Nanowire array filled with PMMA coating. **d** After oxygen plasma etching, tips of the nanowires are exposed. **e** A Pt coated flat electrode is placed on top of the ZnO nanowire array to form a firm Schottky contact. **B** SEM images of the **a** as-grown ZnO nanowire array **b** after spin coating with PMMA, and **c** after oxygen plasma etching. Courtesy of Dr. Z. L. Wang, Georgia Institute of Technology. Reproduced with permission from [50], Copyright @ 2010, Nature Publishing Group

the bottom, which is the Ohmic contact side. For simplicity, it is assumed that there is no shear strain component on the nanowires.

The negative piezoelectric potential raises the conduction band and the Fermi level at the tip side of the nanowires. Thus, electrons will flow from the tip side to the bottom side through external circuit. The Schottky barrier at the tip side blocks the electrons through the interface. Therefore, the electrons are accumulated at the bottom side of the nanowires. The Fermi level at the bottom side is consequently raised up until the piezoelectric potential is fully "screened," so that the Fermi levels of the two sides reach a new equilibrium state again. During this process, the flow of the electrons via external circuit is detected as a current pulse. As the external force is removed from the top electrode, the compressive strain of the nanowires is released, and the piezoelectric potential diminishes. Then, those electrons accumulated at the bottom side will flow back via the external circuit, creating a current pulse in the opposite direction.

The output voltage and current can be increased by using linear integration of the AC-NGs. When three AC-NGs, with individual output voltage of 80, 90 and 96 mV, respectively, are connected in serial; the output voltage is 0.243 V, which is nearly the sum of those of the three individuals. They have individual output current density of 6.0, 3.9 and 8.9 nA cm^{-2}, respectively. When they are connected in parallel, there is an output current density of 18.0 nA cm^{-2}, which is approximately the sum of those of the three individuals.

The maximum power density of the AC-NG can be estimated using the peak values of the output voltage and current. The AC-NG has a dimension of about 4 mm^2, outputting $V = 96$ mV and $I = 0.355$ nA. The average diameter of the nanowires is $D = 400$ nm and length is $L = 4$ μm. The number of nanowires is about 75,000, corresponding to an area density of 1.9×10^6 cm^{-2}. If 1/3 of the nanowires contribute actively to electricity generation in a perfectly synchronized process, the power density is estimated to be 2.7 mW cm^{-3}.

3.5.7 New Record of Power Output

3.5.7.1 20 V Nanogenerators

One example of increasing output performance of nanogenerators is to enhance electrical property of the ZnO nanowires. Various techniques have been used to improve the intrinsic properties of ZnO nanowires in order to achieve energy harvesting performance [51]. The techniques include oxygen plasma treatment, thermal annealing in air, and surface passivation with polymers. Theoretical calculation has shown that the performance of a nanogenerator is closely related to the carrier density of the ZnO nanowires.

It has been shown that the output voltage of a nanogenerator will be much lower if the ZnO nanowires with high level of doping. This is because the positive side of the piezopotential induced by the mechanical deformation is partially screened by the free electrons, which are the dominant charge carriers in *n*-type ZnO. The negative side of the piezoelectric potential can retained as long as the donor concentration is not too high, which is the case of as-grown *n*-type ZnO without unintentional doping. Therefore, it is expected that the strain-induced piezopotential can be remarkably increased if the carrier density in the ZnO nanowires is greatly decreased.

The first strategy is the use of oxygen plasma treatment. The as-grown ZnO films generally have several monolayers of hydroxide (OH) because of H adsorption on their surface, which leads to an electron accumulation. Oxygen plasma contains various oxygen ions and radicals, such as O^+, O_2^+, and O^*. These oxygen radical atoms, especially O^*, can combine with the surface-adsorbed H atoms and diffuse into the ZnO bulk to fill oxygen vacancies, which are doubly charged electron donors in ZnO. Therefore, oxygen plasma treatment can be used to efficiently suppress the surface accumulation layer of H and oxygen vacancies in ZnO. Experimental results have demonstrated that the measured output voltage is increased to 5 V and the output current is over 300 nA. The oxygen plasma treatment has very high reproducibility. However, after 2 weeks exposure in air, the performance of the nanogenerators is found to drop. It means that the treatment is not very stable, which requires a better packaging to retain the performance.

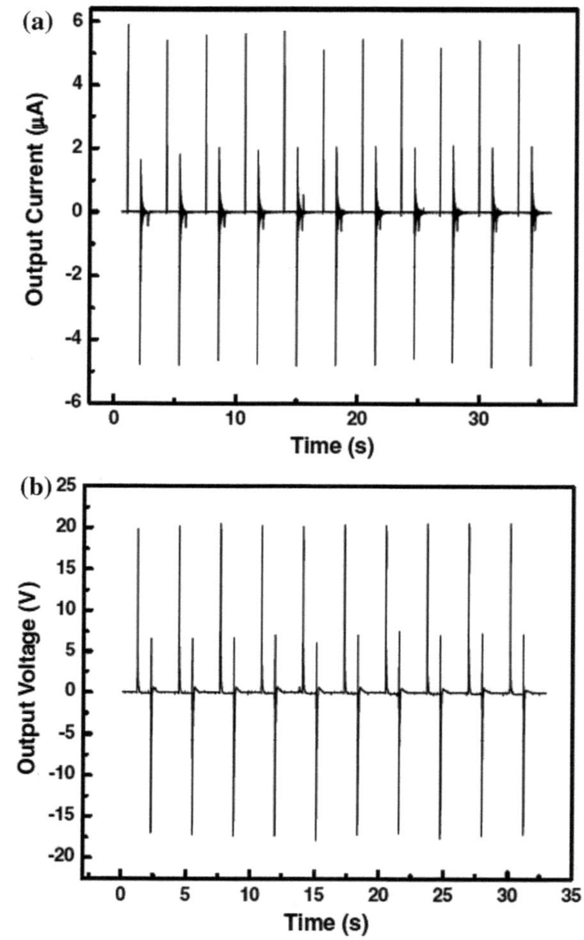

Fig. 3.40 Performance of the nanogenerators with the surface of the ZnO nanowire array being coated with PDAMAC and PSS through a layer-by-layer self-assembling method. Reproduced with permission from [51], Copyright @ 2012, John Wiley & Sons

The second technique is thermal annealing. Normally, ZnO nanowires grown with hydrothermal method have a high concentration of defects, although it is the most effective method to deposition high quality ZnO nanowires on flexible substrates due to its relatively low growth temperature. Thermal annealing at high temperatures has been found to be able to remove the oxygen-vacancy-related defects in as-grown ZnO crystals. It has been shown that a thermal annealing at 350 °C for 30 min can lead to significant increase in performance of ZnO nanowire nanogenerators. The output voltage is increased to 8 V, while the output current can be as high as 900 nA.

The third method is passivation of the surface of the ZnO nanowires by using polymers. Surface passivation can eliminate these various defects and make the surface to be more chemically inert. Positive charged poly (diallydimethylammonium chloride) (PDADMAC) and negative charged poly (sodium 4-styrenesulfonate) (PSS) have been used as passivation layers to improve the performance

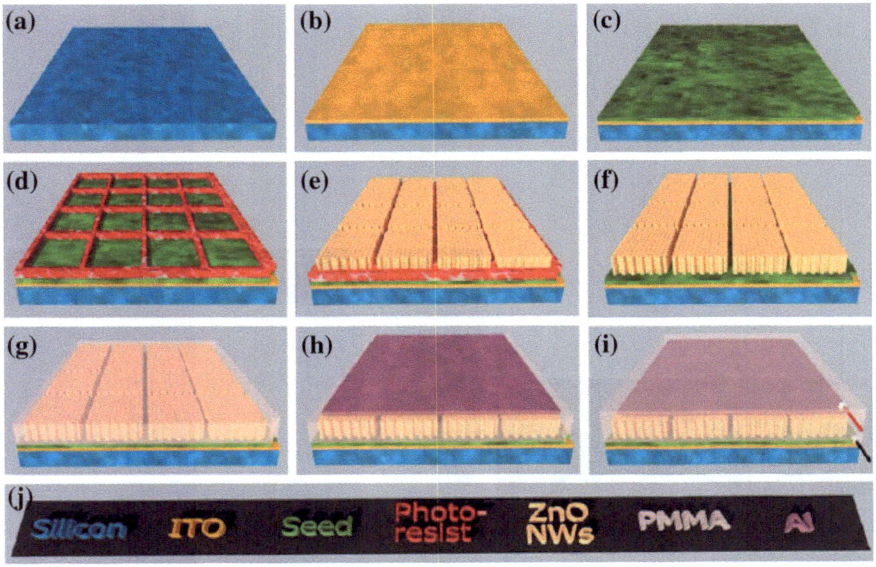

Fig. 3.41 Process flow to fabricate the newly designed nanogenerator. **a** Silicon substrate. **b** Deposition of an ITO layer by using sputtering. **c** Deposition of a ZnO seeding layer by using sputtering. **d** Window opening by photolithography. **e** Selective growth of ZnO nanowires on the window areas by using hydrothermal method. **f** Stripping off the residual photoresist. **g** Spin-coating of a PMMA layer to cover the structure. **h** Deposition of a metal layer as an electrode by e-beam evaporation. **i** Packaging with one more layer of PMMA through spin-coating. **j** Legend of the figure. Reproduced with permission from [52], Copyright @ 2012, American Chemical Society

of ZnO nanowire nanogenerators. Figure 3.40 shows the output performance of the nanogenerators after being applied with this pretreatment. The output voltage reaches 20 V and the output current is 6 μA, corresponding to a power density of 0.2 W/cm^3. In addition to the elimination of surface oxygen vacancies, this dramatic improvement of the output performance may be resulted from the introduction of a depletion layer at the surface of the ZnO nanowires by further transfer of electrons from the nanowires to the positively charged polymer PDADMAC [51]. The energy generated by the nanogenerator with 1,000 cycles of deformation can power an electronic watch for more than 1 min.

3.5.7.2 58 V Nanogenerators

The second example is the new technology to develop nanogenerators with high output. The technology is position-controlled vertical ZnO nanowires, producing a new record of output voltage of 58 V [52]. A thin insulating layer is used to replace the Schottky contact that can prevent current leakage. Figure 3.41 shows the flow of fabrication process of the newly designed nanogenerators. An ITO

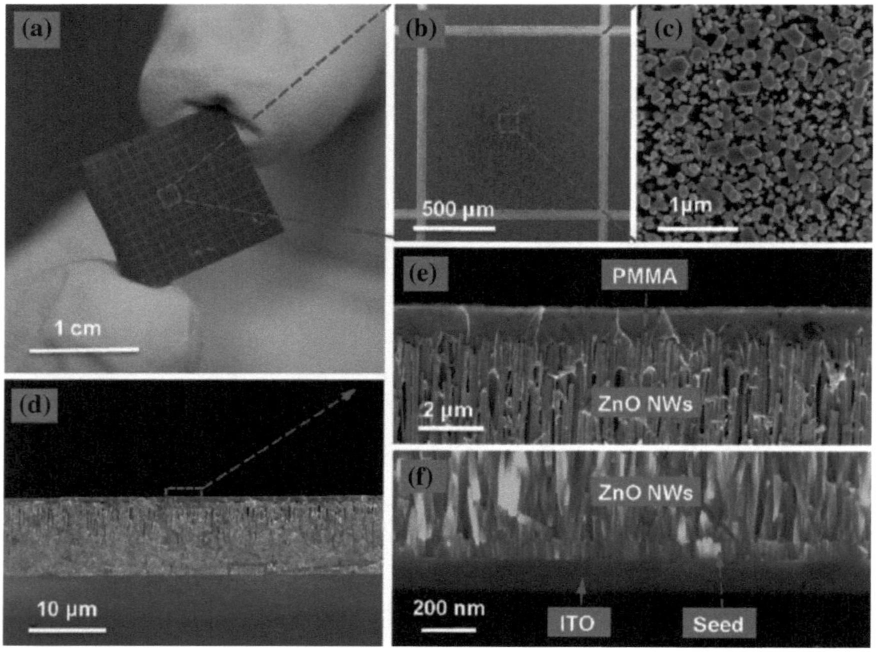

Fig. 3.42 Structure and microstructure of the nanogenerator. **a** Photograph of the nanogenerator with position-controlled grown ZnO nanowires. **b** SEM image of one segment of the nanogenerator defined by using photolithography. **c** Top-view SEM image of the ZnO nanowires. **d** Cross-sectional SEM image of the device with a spin-coated PMMA layer on top of the nanowire array. **e** High magnification SEM showing the nanowires covered by the PMMA layer. **f** SEM image of the seeding layer and the ITO layer. Reproduced with permission from [52], Copyright @ 2012, American Chemistry Society

layer and a ZnO seeding layer are deposited on Si substrates by using RF sputtering (Fig. 3.41a–c). The ITO layer is used as conductive electrode and also adhesion between the ZnO seeding layer and the substrate. An array of square windows, with narrow spacing, is patterned on photoresist (Fig. 3.41d). The ZnO nanowires are grown only from the exposed windows on the seeding layer surface, by using the wet chemical method discussed before (Fig. 3.41e). After the residual photoresist is removed (Fig. 3.41f), the ZnO nanowires are thermally annealed. After that, a layer of PMMA is applied to cover the nanowire arrays (Fig. 3.41g), on which a layer of aluminum is deposited as top electrode (Fig. 3.41h). Then, another layer of PMMA is used for packaging (Fig. 3.41i). The process of this design is compatible the silicon technology, and thus is scalable.

Figure 3.42a shows a piece of the nanogenerators, demonstrating the result of the position-controlled synthesis of the ZnO nanowires. All nanowires are grown only in designated areas (Fig. 3.42b). High magnification SEM image indicates that the ZnO nanowires are aligned vertically (Fig. 3.42c). The nanowires are grown in a very dense manner that most of them are connected or even merged

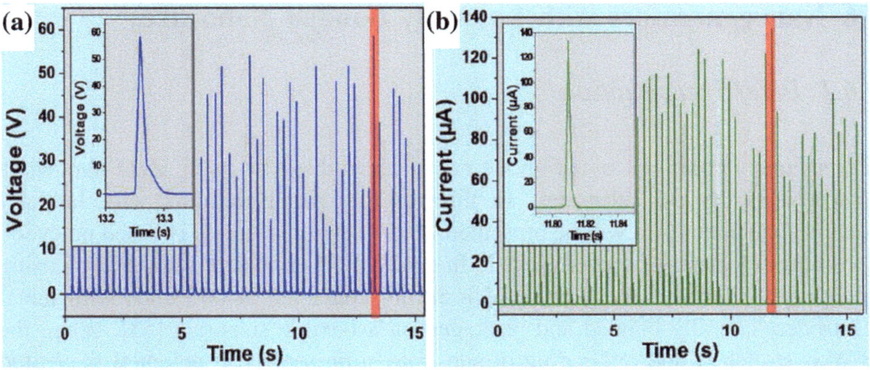

Fig. 3.43 **a** Open-circuit voltage of the nanogenerator pad rectified by using a bridge rectifier under impact by a human palm, with enlarged view of one cycle in the *shadowed region* (inset). **b** Short-circuit current of the nanogenerator pad rectified by using a bridge, with an enlarged view of one cycle in the *shadowed region* (inset). Reproduced with permission from [52], Copyright @ 2012, American Chemistry Society

together (Fig. 3.42d). Figure 3.42e shows cross-sectional SEM image of the nanowire arrays, with many nanowires being closely packed and submerged in PMMA

A nanogenerator with an effective dimension of 1 × 1 cm^2 × 10 μm of ZnO nanowires has an open-circuit voltage (V_{OC}) of 37 V and a short-circuit current (I_{SC}) of 12 μA, at a stress of 1 MPa. The experimental output voltage is lower than the theoretical predicted value of 45 V, which can be attributed to the presence of screening effect. The electric output can be further increased by scaling up with linear superposition. An energy-harvesting pad has fabricated with nine nanogenerators connected in parallel. Being it is punched by a human palm, the peak value of V_{OC} and I_{SC} are as high as 58 V and 134 μA, respectively (Fig. 3.43a and b). This value is new record of output voltage of the nanogenerators with vertically aligned ZnO nanowires. The electricity harvested by using this nanogenerator at palm impact for 20 times can be used to charge a capacitor of 2 μF to over 3 V.

The high performance and robustness of the new nanogenerators can be attributed to the use of a PMMA layer between the nanowires and the metal electrode. It acts as an insulating layer that prevents the leakage of the induced electrons through the ZnO-metal interface. It also replaces the Schottky contact in previous nanogenerators. These factors are beneficial to the improvement in electrical performance. The PMMA fills in the gaps/spaces among the nanowires and forms a capping layer at top of the nanowire array. Therefore, when a vertical force is applied to the nanogenerator, the stress can be effectively transmitted to all ZnO nanowires. In this respect, the presence of PMMA not only enhances the effectiveness of the nanogenerator, but also offers a protective function to both the ZnO nanowires and the whole structure.

3.6 Nanogenerators with Laterally Bonded Nanowires

3.6.1 Brief Introduction

The nanogenerators discussed above are all assembled by using ZnO nanowires grown vertically on substrates. In this section, nanogenerators with laterally bonded ZnO nanowires will be presented. The working principle is based on cyclic stretching-releasing of a piezoelectric fine wire (PFW), which is formed by using microwires or nanowires. The PFW is firmly contacted at two ends with metal electrodes, laterally bonded and packaged on a flexible substrate [53]. When the PFW is stretched due to bending of substrate, a piezoelectric potential is created along the PFW. The physics of this type of nanogenerator is the same as that of those with vertical ZnO nanowires.

By repeating stretching-releasing cycle at strain level of 0.05–0.1 %, a single PFW can create an AC output at a voltage of up to 50 mV. The energy conversion efficiency of the PFW is up to 6.8 %. The output is scalable by a serial connection to form multiple AC generators. This is a robust, nonsliding-contacting and packageable nanowire technology in polymer films. The nanogenerators can be used to harvest low-frequency energy due to mechanical deformations caused by vibration, air flow or wind, and so on. The flexible property of the nanogenerators makes them possible and feasible to be implanted in muscles, embedded in cloths, built in surface layers, and placed in shoe pads.

3.6.2 Fabrication Process

Figure 3.44 shows the fabrication process of the nanogenerators [53]. A ZnO PFW is laterally bonded on a Kapton polyimide film (Fig. 3.44a). The ZnO PFWs are synthesized by using a physical vapor deposition (PVD) method. They typically have diameters of 3–5 μm and lengths 200–300 μm. Long PFWs are used because they are manageable in terms of device assembling, but the principle is applicable to nanowires. The two ends of the ZnO PFW are fixed tightly on the flexible substrate by using silver paste. A current/voltage measurement meter is connected to two ends of the PFW without introducing any external power source in the circuit.

To measure the piezoelectric response of the PFW, a periodic mechanical bending is applied to the substrate by using a mechanical arm driven by a motor. The maximum bending radius of the substrate is 2 cm, which is much larger than the length of the PFW. In addition, the thickness of the Kapton film is much larger than the diameter of the PFW. As a result, the bending of the substrate film induces a tensile strain of 0.05–0.1 % in the PFW attached to its outer surface (Fig. 3.44b). Due to the piezoelectric property of ZnO, a piezoelectric potential field is created along the PFW, which drives the flow of electrons in the external circuit or other

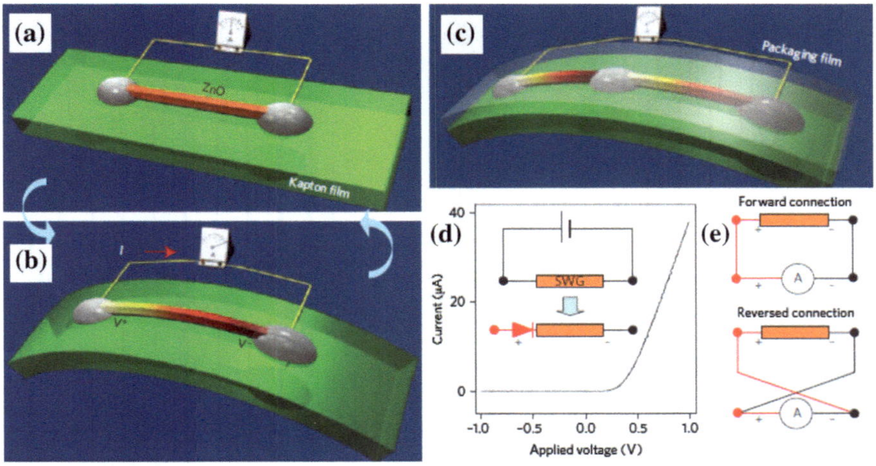

Fig. 3.44 Fabrication and characterization of the piezoelectric fine wire (PFW) generator on flexible substrates. **a** The PFW lays on a polymer Kapton film substrate, with two ends being tightly bonded to the substrate and outlet interconnects. **b** Mechanical bending of the substrate to create tensile strain and corresponding piezoelectric potential in the PFW. **c** Packaging of the generators made of multiple PFWs inside a thin and flexible polymer film, showing its integration with soft materials to improve mechanical stability and environmental adaptability. **d** *I–V* characteristic of a PFW generator with a single wire (SWG), showing a typical Schottky diode characteristic with a forward-bias threshold voltage of about 0.3 V. The end-contact of the SWG that has the Schottky behavior is defined to be positive, where a diode symbol is used to represent its presence at the interface (lower diagram). **e** Definition of the switching-polarity test. Courtesy of Dr. Z. L. Wang, Georgia Institute of Technology. Reproduced with permission from [53], Copyright @ 2009, Nature Publishing Group

interconnects as outputs. When the substrate is periodically bent and released, the PFW is periodically stretched and released, resulting in an alternating current (AC). Nanogenerators can be made with multiple PFWs so as to increase the output voltages (Fig. 3.44c). The physical stability of the nanogenerators can be ensured by packaged them with a thin layer of insulating wax or flexible polymer.

3.6.3 Electrical Characterization

Before the single-wire generator (SWG) is characterized for generating electricity, its transport property should be clarified to understand its performance and signal output. An effective SWG should have an *I–V* characteristic of Schottky behavior. To properly demonstrate the location of the Schottky barrier, a diode symbol is used at the left-hand side, with which the sign of the output signal is defined. The side with the Schottky diode is defined as the positive side of the SWG.

Short-circuit current (I_{SC}) and open-circuit voltage (V_{OC}) are used to characterize the performance of the nanogenerators. To ensure that the measured signals

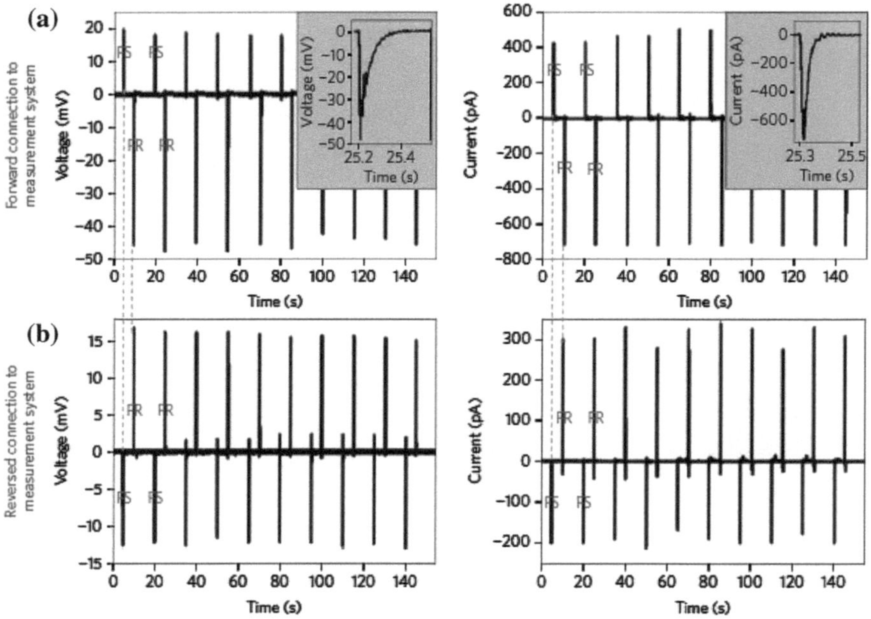

Fig. 3.45 Open-circuit voltage and short-circuit current outputs of a single-wire generator (SWG) and their stability. Alternating voltage and current generated by SWG when repeatedly fast-stretched (FS) and fast-released (FR), showing the reversal in signal output when the SWG is forwardly (**a**) and reversely (**b**) connected to the electrical measurement system, respectively, clearly ruling out the system artifacts. The insets in **a** and **b** are an enlarged output voltage and current peak, respectively. The area under the *I–t* curve is the total transported charges. Courtesy of Dr. Z. L. Wang, Georgia Institute of Technology. Reproduced with permission from [53], Copyright @ 2009, Nature Publishing Group

are generated by the nanogenerator rather than artificial signals from the measurement system, a switching-polarity test (as discussed above) is applied to all the measurements, as shown in Fig. 3.44e. When the current meter is forward connected to the nanogenerator, meaning that the positive and negative probes were connected to the positive and negative electrodes, respectively, a positive voltage/current pulse is detected for every fast stretching (FS) of the polymer substrate (Fig. 3.45a). A corresponding negative pulse is detected when the substrate is quickly released (FR). The angular bending rate of the substrate is $\sim 260^\circ\ s^{-1}$ at a radius of 2 cm. The output voltage/current for each cycle of FS and FR is a pair of positive and negative peaks. The output voltage is 20–50 mV and output current is 400–750 pA. The difference in heights of the current peaks for the FS and FR can be attributed to different straining rate. For each straining process, the total number of electrons transported is $\sim 10^8$.

When the current meter is reversely connected (Fig. 3.44e), i.e., the positive and negative probes are connected to negative and positive electrodes of the SWG, respectively, the voltage and current are reversed, shown in Fig. 3.45b, which confirms that output signals are generated by the nanogenerator.

The magnitude of the generated electricity depends on the deformation rate applied to the nanogenerator. To further verify this, slow stretch (SS) and slow release (SR) are used, with a rate of $7° s^{-1}$ for a radius of 2 cm [53]. Both SS and SR have lower outputs than FS and FR, but with broader responses. However, the areas under the signal peaks for both FS and SR or SS and FR are almost the same, implying that the total numbers of released charges are the same for a given maximum strain, regardless of the straining rate and straining process. Therefore, the total charges transported in the stretching and releasing processes are conservative and the SWG had little leakage current. At the slow rates of stretch and release, the nanogenerator also passes the switching-polarity test, which further proves that the signals are generated by the nanogenerator.

3.6.4 Principle of Nanogenerator

The working principle of the SWG is the same as that of the nanogenerators made with vertical ZnO nanowires. It is attributed to the presence of a Schottdy barrier and the creation of an inner piezoelectric field in the PFW due external tensile strain [53]. In first step, when tensile strain is applied to the PFW, a piezoelectric field is created in it. The piezoelectric charges can be present in the PFW for at least a few seconds or even much longer without "depletion" by the free carriers as long as the strain is preserved. If the piezoelectric potential is positive at the Schottky barrier side and is negative at the Ohmic side, the conduction band and Fermi level of the electrode at the left-hand side is raised. Thus, electrons will flow from the right-hand side electrode to the left-hand side through an external load. The electrons accumulate at the interfacial region between the left-hand side electrode and ZnO PFW due to the presence of the Schottky barrier. This process continues until the potential created by the accumulated electrons can balance the piezoelectric potential. As a result, the Fermi levels of the two electrodes reach a new equilibrium. This is corresponding to the first positive output current/voltage peak.

In second step, as the polymer substrate is released, the tensile strain in the PFW is released. An immediate disappearance of the piezoelectric potential lowers the Fermi surface of the right-hand side electrode, due to which the electrons originally accumulated at the interfacial region between the left-hand side electrode and the PFW flow back through the external circuit to the right-hand side electrode due to the presence of the Schottky barrier at the interface. This process ends when the Fermi levels at the two electrodes are at equilibrium and the system returns to the original status. As a result, negative output current/voltage peak is present. The total electrons transported in the two steps are the same regardless the straining process, provided there is no current leakage. A repeat of such two steps produces alternating outputs.

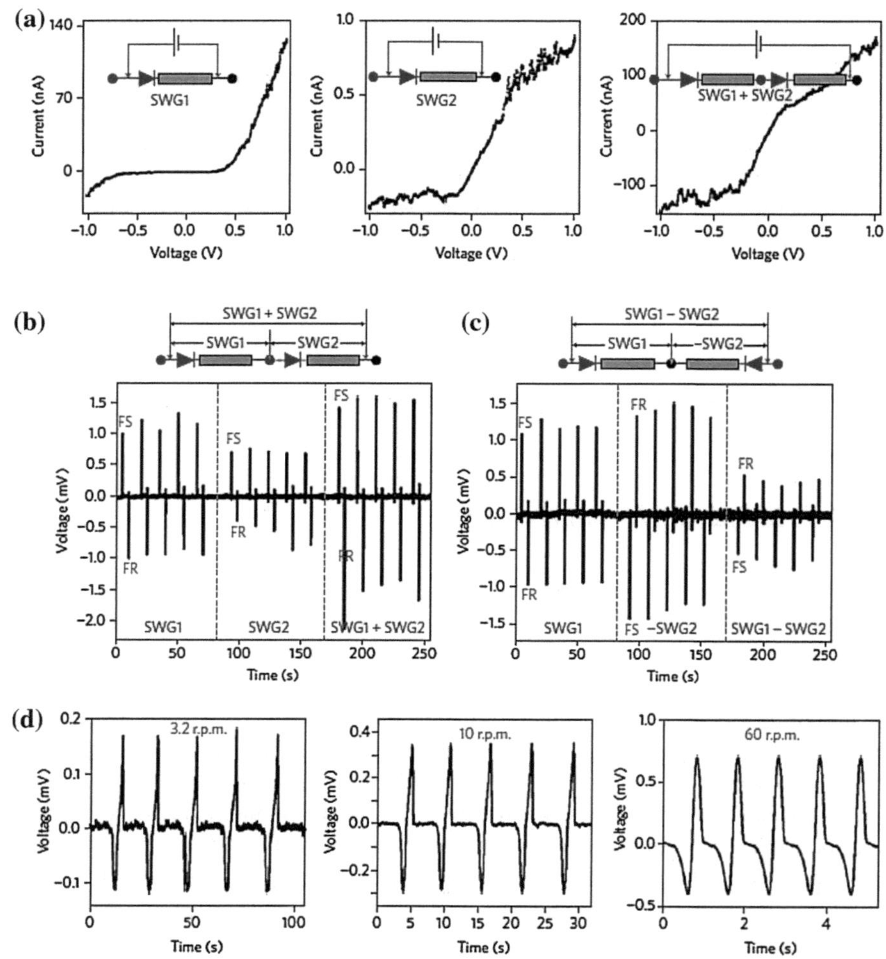

Fig. 3.46 Serial connection of the two SWGs to show their potential to scale up output power. **a** *I–V* characteristics of the two individual SWGs and their resultant transport properties after being connected in serial, so as to identify their polarity. **b** Voltage outputs of the two individual SWGs and those connected constructively in serial, showing a linear summation of the output voltages. **c** Voltage output of the two individual SWGs and those connected destructively in serial, showing the linear subtraction of the output voltages. **d** An increase in straining rate leads to an increase in output voltage. Courtesy of Dr. Z. L. Wang, Georgia Institute of Technology. Reproduced with permission from [53], Copyright @ 2009, Nature Publishing Group

3.6.5 Linear Connections

The nanogenerators can be connected in serial or parallel to achieve higher output voltage or current [53]. The polarity of the Schottky barriers at the ends of two SWGs in serial are determined by *I–V* measurement, as shown in Fig. 3.46a and b, so that they can connected in correct polarity and sequence as desired. By connecting the two in serial, the corresponding I–V characteristic still has a Schottky behavior. The measured voltage is approximately a sum of those generated by the two SWGs (Fig. 3.46b). If the two SWGs are reversely connected in serial, the output voltage is the difference of those generated by the two harvesters (Fig. 3.46c). This linear superposition test confirms that the signals are generated by the SWGs.

In addition, the output voltage and current can be significantly increased by increasing the straining rate, as shown in Fig. 3.46d. At a bending frequency of 60 rpm, the output voltage is close to sinusoidal, similar to real alternating current.

3.6.6 Power Conversion Efficiency

Efficiency of the SWG can be estimated by using thickness of the substrate and dimension of the piezoelectric wires. The Kapton substrate is assumed to be bent with a radius of R. The thickness of Kapton substrate is h. The ZnO fine wire has a length of L and diameter of D. Because $R \gg h$ and $h \gg D$, the strain of the ZnO fine wire is approximately equal to the strain of the outer surface of the Kapton film, $\varepsilon \approx h/2R$. The thickness of the Kapton substrate is 50 μm and the radius R is between 2.0 and 3.0 cm. Accordingly, the strain of the ZnO fine wires is $\varepsilon = 0.05$–0.1 %. The total mechanical deformation energy stored in the PFW after stretching is $W_m = \frac{1}{2}EA(L - L_0)^2/L_0$, where E is the elastic modulus of the PFW, L_0 and L are its original and final length after straining, respectively, and A is its cross-sectional area. The total electric energy generated can then be calculated. For a ZnO fine wire, with a diameter D of 4 μm, L_0 of 200 μm and E of 30–50 GPa, the energy conversion efficiency of the PFW is 6.8 %.

3.6.7 Applications-Harvesting Biomechanical Energies

3.6.7.1 In Vitro Applications

Living species have abundant sources of mechanical energy, such as walking, breathing, and heart beating. A wide range of fluctuation in frequency and intensity is a challenging issue, when it is attempted to harvest these energies. This is the reason why the progress of energy harvesting by using traditional techniques is very limited. The nanogenerators have been shown to be a great potential technique to harvest the mechanical energy from irregular motion with a magnitude at very gentle level without adding significant burden.

One example of biomechanical energy harvesting is demonstrated by using a SWG to harvest energy due to the motion of a human finger [54]. The SEG is assembled with fine wire of 100–800 nm in diameter and 100–500 μm in length. A flexible polymer layer is used to improve robustness and adaptability of the SWG.

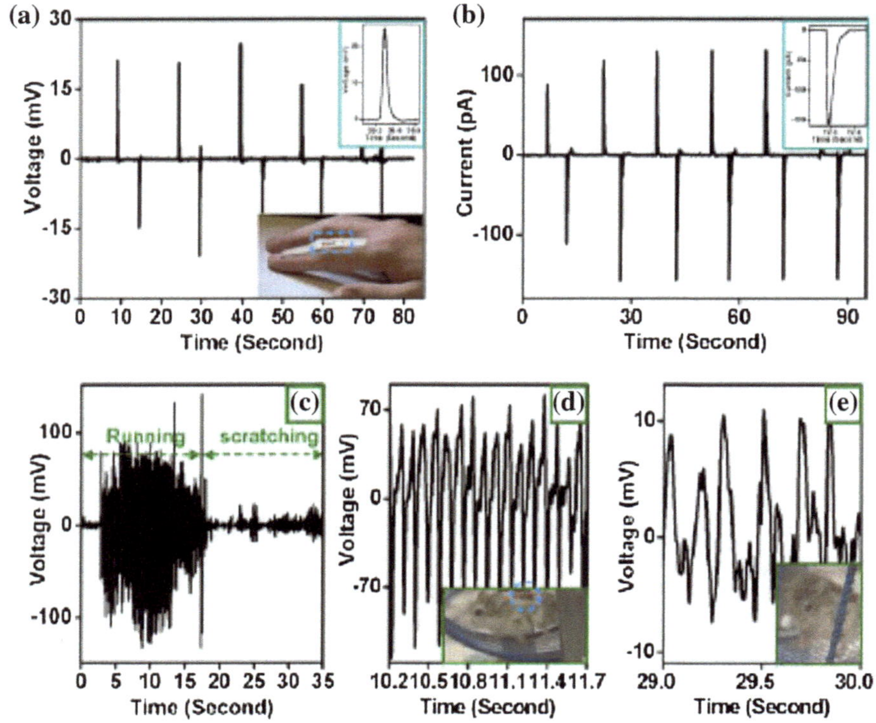

Fig. 3.47 Energy harvesting from the motion of human finer with the PFW nanogenerator. **a** Open-circuit voltage and **b** short-circuit current output of the generator fixed of top of the index, as shown in the inset. **c** Open-circuit voltage output from a nanogenerator fixed to the jacket worn by a hamster while it is running or scratching. **d**, **e** Enlarged voltage output corresponding to the running hamster and scratching hamster, respectively. The insets in (**d**) and (**e**) are the corresponding snapshots of the hamster with the nanogenerator on its back. Courtesy of Dr. Z. L. Wang, Georgia Institute of Technology. Reproduced with permission from [54], Copyright @ 2009, American Chemical Society

The SWG is fixed at the joint position at the top of an index finger, as shown in the inset in Fig. 3.47a. The tapping of the finger induces a deformation of the ZnO wire in the SWG with a strain rate about 4–8 $\times$ 10^{-3} s^{-1} and a maximum strain of $\sim$0.2 %. As discussed above, the strain results in a piezoelectric polarization along the fine wire and thus a potential difference between its two ends, which drive the electrons to flow in the external circuit. This is a very gentle and slow motion, which can be found in human activities.

Measured open-circuit voltage and closed circuit current are presented in Fig. 3.47a and b, respectively. The periodic peaks are corresponding to the periodic tapping of the finger. Figure 3.47a shows that the voltage output is up to 25 mV while the current shown in Fig. 3.47b is higher than 150 pA, as a single SWG harvesting device is used.

Another example is demonstrated by using a living hamster. The hamster belongs to the Campbell's dwarf type, which can produce both regular and irregular motion, like running or scratching. A special jacket is made for the hamster to wear, so that the hamster can move freely in and around the cage to drive the SWG on the jacket at the same time. Electric output signal from the SWG is measured without disturbing the movement of the hamster, as shown in Fig. 3.47c–e. With motion, running or scratching, the hamster provided mechanical energy with distinct pattern and energy intensity to be harvested by the SWG. The output electricity from the SWG well reflects the movements. The enlarged voltage output signal in Fig. 3.47d from the running hamster shows a clear periodic pattern with magnitude of 50–100 mV and frequency of 10–11 Hz. The periodicity coincides with the running of the hamster. The running hamster applies a much faster strain rate than the human finger to the SWG. In comparison, the voltage output due to scratching of the hamster, as shown in Fig. 3.47d, is irregular with lower magnitude, due to the irregular movement and less intense activity. Similar observations are found in the results of closed circuit current measurement. A running hamster produces a periodic alternating current of up to 0.5 nA and a scratching hamster produces a lower irregular current.

3.6.7.2 In Vivo Applications

In in vivo applications, due to the presence of biofluids, it is necessary to encapsulate entire device with a flexible polymer to isolate it from the surrounding medium and to improve its robustness [55]. The nanogenerator has been used to convert the mechanical deformation to electricity, due to the periodic expansion and contraction of the diaphragm of a rat, as shown in Fig. 3.48. The adult rats used for the experiment are SD (Hsd: Sprague Dawley, male, 200–224 g). The anesthesia procedure of the rat starts with the intake of isoflurane gas (1–3 % in pure medical grade oxygen), followed by the injection of Ketamine cocktail (intraperitoneal, ketamine 50–90 mg kg^{-1}, xylazine 5 mg kg^{-1}) and buprenorphine (intraperitoneal, 0.03 mg kg^{-1}) for anesthesia induction and maintenance, respectively. The tracheotomy is first operated; and a tracheal intubation is connected to a respirator, which provides artificial respiration and sustains the life of the rat through the entire experiment. Then the rat's abdominal cavity is opened to make the ventral-side of the diaphragm reachable for implanting a SWG (Fig. 3.48). Part of the liver was cut off after proper ligation for easy observation and operation. The output of the SWG is typically less than 50 mV and 500 pA in most cases.

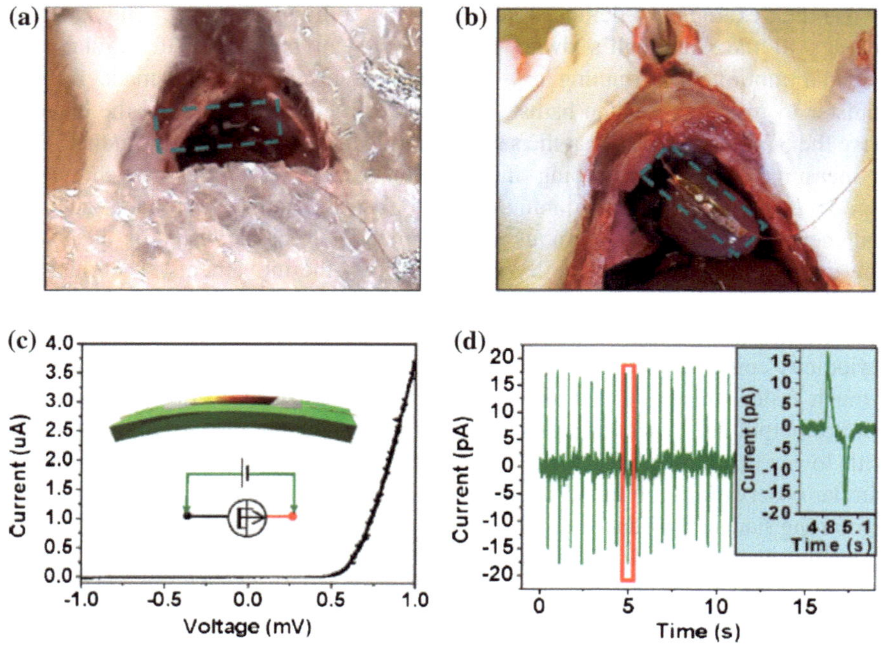

Fig. 3.48 Energy harvesting from the breathing and heart beating of a live rat with the SWG. A SWG attached to a live rat's diaphragm (**a**) and heart (**b**), which drives the SWG to periodically bend and produce AC power output. **c** *I–V* characteristic of the SWG. The inset illustrates the schematic of the SWG and its connection configuration with reference to the measurement system. **d** Typical current output of the SGW under in vivo condition from (**a**). Courtesy of Dr. Z. L. Wang, Georgia Institute of Technology. Reproduced with permission from [55], Copyright @ 2010, John Wiley & Sons

3.6.8 High-Output Nanogenerators with Lateral Nanowire Arrays

Although the single-wire (PFW) generator has been shown to be a robust device to harvest low-frequency energy generated by in vitro human or animal motion and in vivo animal activity, its output power is not sufficiently high for practical applications. Therefore, it is necessary to scale up the design of the AC-NGs discussed above. One way is to integrate the contributions of sufficiently huge number of nanowires so as to achieve high-output powers.

3.6.8.1 Lateral Integrated Nanogenerator

A lateral integrated nanogenerator (LING) has been proposed and developed to achieve high power output for potential practical applications [50]. There are

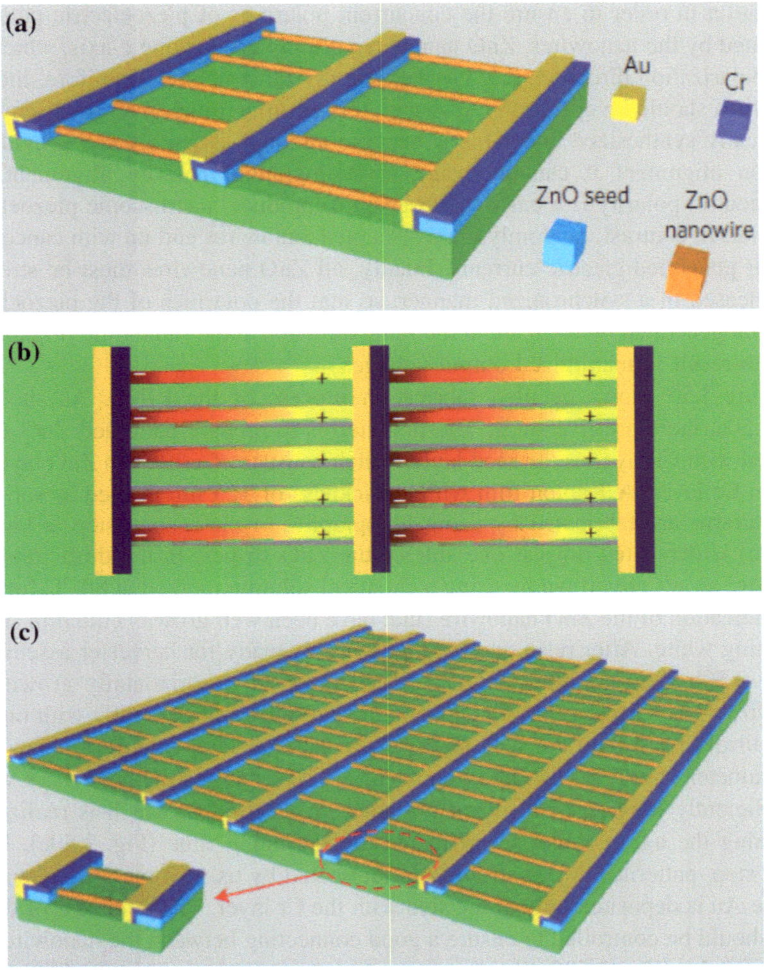

Fig. 3.49 Concept of a lateral nanowire-integrated nanogenerator (LING) array. **a** Schematics of the LING structure, where Au and Cr are used to create Schottky and Ohmic contacts, respectively, at the two ends. **b** Working mechanism of the LING when it is subjected to a mechanical deformation. The "+, −" signs indicate the polarity of the local piezoelectric potential created in the ZnO nanowires. **c** Schematics of a LING array made of rows of nanowires. Courtesy of Dr. Z. L. Wang, Georgia Institute of Technology. Reproduced with permission from [50], Copyright @ 2010, Nature Publishing Group

several factors that have to be considered to assemble a nanogenerator with a huge number of single-wire. First, the metal contacts at the two ends of the nanowire arrays must be nonsymmetric to ensure a Schottky contact at one side and an Ohmic contact at the other side (Fig. 3.49b), which has been discussed earlier. Second, the contacts at the two ends of the nanowires must be robust, so that mechanical excitation can be more effectively transmitted from the electrodes to the nanowires. Third, all ZnO nanowires should have same crystallographic orientation in order to ensure the concurrent polarities of piezoelectric potentials generated by the nanowires. ZnO nanowires usually grow along *c*-axis, which also their polarization direction, due the wurtzite crystal structure. Therefore, the ZnO nanowires should be epitaxially grown on the substrate rather than by assembly of chemically synthesized one. This is because although assembly can provide orientation alignment it cannot ensure crystallographic polarity alignment. The alignment in polarity is a basic requirement to produce macroscopic piezoelectric potential. In contrast, randomly *c*-axis-oriented nanowires end up with cancelation in their generated electric currents. Finally, all ZnO nanowires must be stretched and released in a synchronized manner, so that the polarities of the piezoelectric potential generated by them are in the same direction at the same time (Fig. 3.49c), so as to result in maximized output voltages.

Figure 3.50 shows detailed fabrication process of the LING. As shown in Fig. 3.50a, there are five steps. The first step is to prepare patterned ZnO stripes covered with Cr layer as the seed layer, which consists of a 300 nm ZnO layer and a 5 nm Cr layer. A Kapton film with a thickness of 125 μm is used as substrate. The patterns are fabricated by using lithography. In the second step, a layer Cr (10 nm) is deposited only at one side of the ZnO stripes. In the third step, ZnO nanowire arrays are grown by using a chemical solution method at 80 °C for 12 h. One indication of the ZnO nanowires that have been well grown is the film surface becoming white. After resin and baking, they are ready for harvester assembling. Figure 3.50b shows a representative SEM image of a horizontally grown ZnO nanowire arrays. Most of the ZnO nanowires are grown horizontally with one end being firmly fixed onto the seed layer. The nanowires have length of about 5 μm and diameter of several hundred nanometers. The length of the nanowires should be sufficiently long so as to reach the adjacent electrodes, which is realized by refreshing the aging solution and increasing the aging time (Fig. 3.50e). In the fourth step, patterned Au electrodes are fabricated by using a masking technique, and the Au is deposited only at the side with the Cr layer. The thickness of the Au layer should be controlled to ensure a good connecting between the nanowires and the electrodes (Fig. 3.50f). Finally, the entire structure is packaged with an insulative soft polymer, such as a photoresist (MicroChem PMMA 950 K A2). This packaging layer is to fix the ZnO nanowires firmly on the substrate and ensures all the nanowires to be synchronized in mechanical stretching or releasing. Optical microscopy images of the fabricated LING are shown in Fig. 3.50c and d. The ZnO NW arrays are connected with each other end-to-end by the patterned electrodes, as shown by the SEM images in Fig. 3.50e and f. A fully packaged LING with large size is shown in Fig. 3.50g and its flexibility is demonstrated in the inset.

A periodic external force is applied to the Kapton substrate so that it makes the nanowires to experience a cyclic stretching-releasing process. With this, energy harvesting performance of the LING can be tested. Since the thickness of the Kapton substrate is much larger than the diameter of the nanowires, a push to the substrate at its middle by using a mechanical motor results in a stretch at its outer surface, which produces a tensile strain across the rows of the nanogenerators on

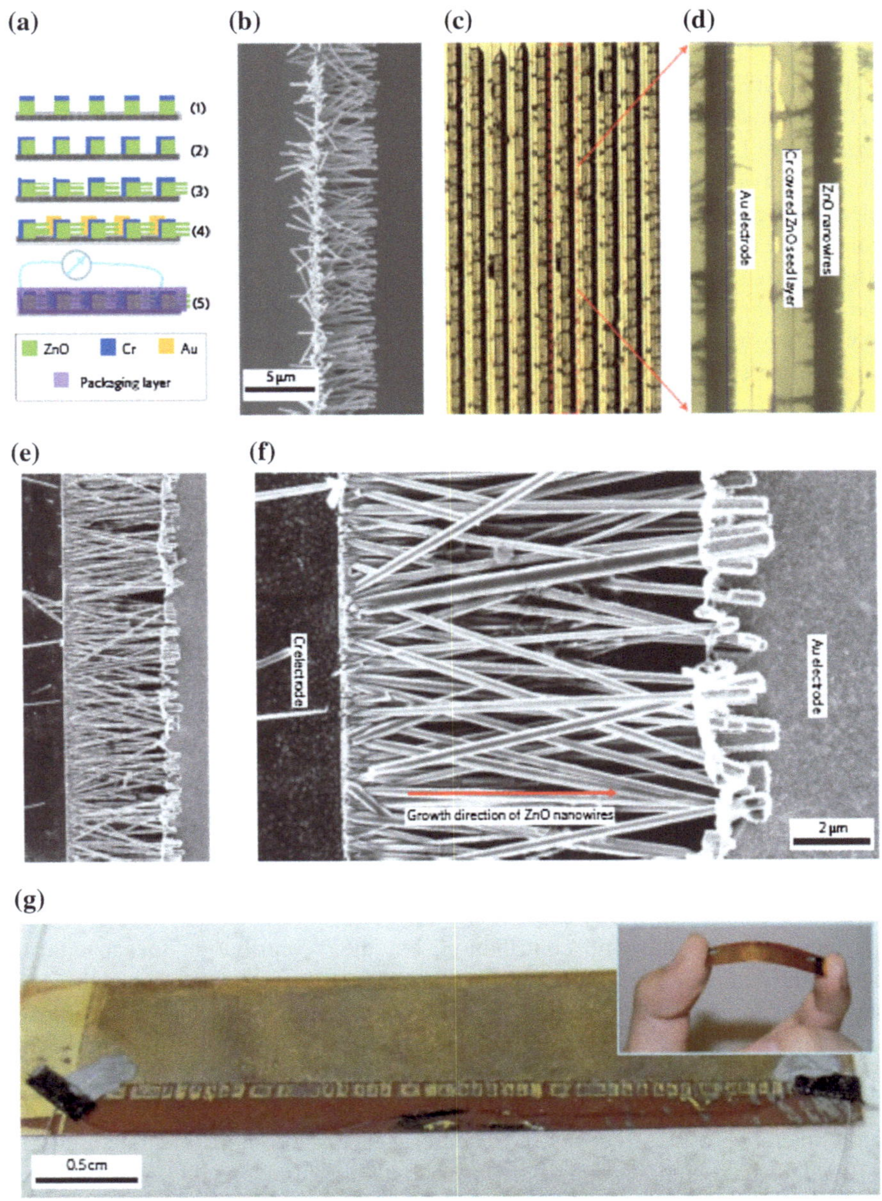

Fig. 3.50 Fabrication process and structure characterization of the LING. **a** Schematics showing the growth of the orientation aligned nanowire arrays parallel to the substrate surface. **b** SEM image of a row of laterally grown ZnO nanowire array. **c**, **d** Optical microscopy images of the LING structure using a number of rows of the nanowire arrays. **e**, **f** SEM images of a single row LING structure. **g** An low-magnification optical image of the LING and a demonstration of its flexibility (inset). Courtesy of Dr. Z. L. Wang, Georgia Institute of Technology. Reproduced with permission from [50], Copyright @ 2010, Nature Publishing Group

the top. The strain can be estimated basing on the curvature of the bending and thickness of the substrate film, as presented above.

The LING is composed of 700 rows of nanowires, with each row having about 20,000 nanowires. The magnitudes of the current/voltage peaks are slightly different as the LING is stretched and released, because of the difference in straining rate in the two processes. The stretching is driven by an external force, while the releasing occurs naturally. Upon stretching, a piezoelectric potential is created along the ZnO nanowires at the Au contact side, which drives the electrons in external circuit to flow from low potential side to high potential side and accumulated at the contact interface. As the nanogenerator is released, the piezoelectric potential disappears, as a result, the accumulated electrons flow back in the opposite direction.

The LING has an average output voltage pulse of 1.2 V and a current pulse of 26 nA, at a straining rate of 2.13 % s^{-1} and strain of 0.19 %. The average output voltage peaks were greater than 1.2 V, with a maximum voltage reaching 1.26 V. This result demonstrates the potential of scaling up the output voltage by serially integrating rows of nanowires. Most of the output current peaks are greater than 25 nA, with a maximum current peak of 28.8 nA. By assuming that 1/3 of the nanowires are actively contributing to the current output, the average current generated by one nanowire is estimated to be 4.3 pA, which is comparable to the output current of $\sim$10 pA when a nanowire is triggered by an AFM tip. By taking out the area occupied by the electrodes, the peak output power density reaches 70 nW cm^{-2}.

This is a significant progress toward practical applications of the nanogenerators. There are still rooms to further improve the performance of nanogenerators. For example, the orientation alignment of the lateral-nanowires is not perfect. Some of the nanowires are sticking out the electrodes, so that only a fraction of them are in contact to the Au electrodes. Among the nanowires that are in good contact with the electrodes, only a fraction of them contribute actively to electricity output. Those inactive nanowires have no contribution, but their contribution to capacitance reduces the output voltage. Also, the bonding between the electrode Au and the ZnO nanowires can be stronger, because the repeated mechanical stretching can destroy weak bonding. The weak bonding prevents the application of larger strain to the nanowires. The maximum strain is 0.19 % currently. However, theoretical simulation indicates that the maximum tensile strain that a ZnO nanowire can withstand before fracture is 6 %. Moreover, the inner resistance of the entire integrated sheet is 1–10 MΩ, which is the reason of the low output current that can be achieved.

3.6.8.2 Flexible High-Output Nanogenerator

More recently, a simple and effective approach, scalable sweeping-printing-method, to fabricate flexible high-output nanogenerator (HONG), has been developed [56]. The HONG can harvest mechanical energy, which can be used to drive small commercial electronic components. A single layer of HONG offers an

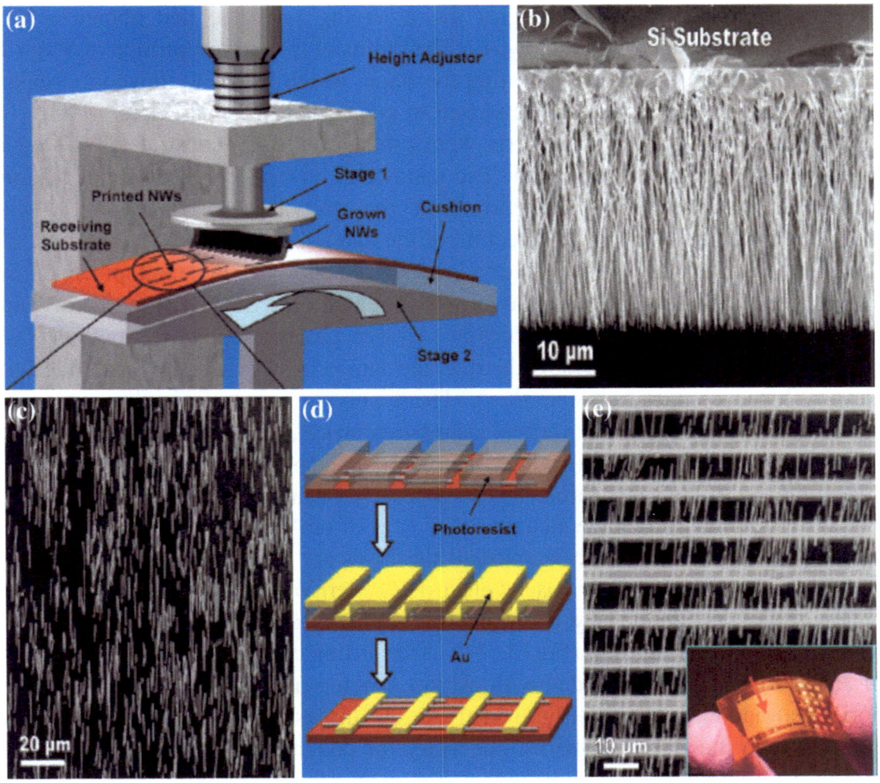

Fig. 3.51 Fabrication process and structure characterization of the flexible high-output nanogenerator (HONG). **a** Experimental setup for transferring vertically grown ZnO nanowires to a flexible substrate to make horizontally aligned ZnO nanowire arrays with crystallographic alignment. **b** SEM image of the as-grown vertically aligned ZnO nanowires by using a physical vapor method on Si substrate. **c** SEM image of the as-transferred horizontal ZnO nanowires on a flexible substrate. **d** Process of fabricating Au electrodes on the horizontal ZnO nanowire arrays, including photolithography, metallization and lift-off. **e** SEM image of the ZnO nanowire arrays bonded by the Au electrodes. The inset is an as-fabricated HONG, with the arrowhead indicating the effective working area. Courtesy of Dr. Z. L. Wang, Georgia Institute of Technology. Reproduced with permission from [56], Copyright @ 2010, American Chemical Society

open-circuit voltage of up to 2.03 V and a peak output power density of about 11 mW cm^{-3}. The generated electric energy can be stored in capacitors, which can be used to light up a commercial light-emitting diode (LED).

The fabrication of the HONG consists of two main steps, as shown in Fig. 3.51 [56]. Firstly, vertically aligned nanowires are transferred to a substrate to form horizontally aligned arrays. The major components used for the transfer are two stages (Fig. 3.51a). Stage 1 has a flat surface that faces downwards and holds the vertically aligned nanowires, while Stage 2 has a curved surface and holds the receiving substrate. Polydimethylsiloxane (PDMS) film on surface of Stage 2 is used as a cushion layer to support the receiving substrate and enhances the alignment of the transferred nanowires. The radius of the curved surface of Stage 2 equals the length of the rod supporting the stage, which is free to move in circular motion. Second, electrodes are deposited to connect all the nanowires.

Vertically aligned ZnO nanowires were synthesized on Si substrates by using a physical vapor deposition method. The nanowires have a length of about 50 μm and a diameter of about 200 nm, with the growth direction along *c*-axis (Fig. 3.51b). The same growth direction of the nanowires ensures the alignment of the piezoelectric potentials in every individual nanowire. A small piece of Si substrate with the grown ZnO nanowires is mounted onto Stage 1 (Fig. 3.51a) and a piece of Kapton® film with a thickness of 125 μm is attached to Stage 2. The distance between the receiving substrate and nanowires is precisely controlled to form a loose contact between them. The receiving substrate is then counter-clockwise swept across the vertical nanowires arrays, so that they are detached from the Si substrate and aligned on the receiving substrate along the direction of sweeping due to the shear force. The as-transferred nanowires are demonstrated in Fig. 3.51c, with an average density of $1.1 \times 10^6 cm^{-2}$. There is a variation in length of the nanowires, because not all the NWs are broken off at their roots.

In the second step, evenly spaced electrode pattern over the horizontally aligned nanowires is first defined by using photolithography and then followed by sputtering 300 nm thick Au film (Fig. 3.51d). After lifting off the photoresist, 600 rows of stripe-shaped Au electrodes with 10 μm spacing are deposited on top of the horizontal nanowire arrays (Fig. 3.51e). Au electrodes are used to form Schottky contacts with the ZnO nanowires, necessary for collecting the piezoelectric electricity. Approximately, 3.0×10^5 nanowires in an effective working area of 1 cm^2, as shown in the inset of Fig. 3.51d, are in contact with the electrodes at both ends. Finally, the harvester is encapsulated with a layer of PDMS to further enhance mechanical robustness and protect the device from invasive chemicals.

Working principle of the HONG is demonstrated by the schematic diagrams in Fig. 3.52a, b [56]. All nanowires are connected in parallel, collectively contributing to the current output. The nanowires in different rows are connected in serial, constructively improving the voltage output. The same growth direction of all the ZnO nanowires, combined with the sweeping printing method, ensures the alignment in crystallographic orientations of the horizontally grown nanowires, so as to achieve the alignment in polarity of the induced piezopotentials. This leads to a macroscopic potential contributed constructively by all the nanowires.

A linear motor is used to periodically deform the HONG in a cyclic stretching-releasing agitation (0.33 Hz). The open-circuit voltage (V_{OC}) and the short-circuit current (I_{SC}) are measured with caution to rule out potential artifacts. At a strain of 0.1 % and strain rate of 5 % s^{-1}, peak voltage and current reach up to 2.03 V and 107 nA, respectively. Assuming that all the integrated nanowires actively contribute to the output, the current generated by a single nanowire is averaged to be about 200 pA, while the voltage from each row is about 3.3 mV in average. With the size of the working area of the nanogenerator (1 cm^2) (Fig. 3.52e, inset), the peak output power density is estimated to be about 0.22 μW cm^{-2}. Noting that the

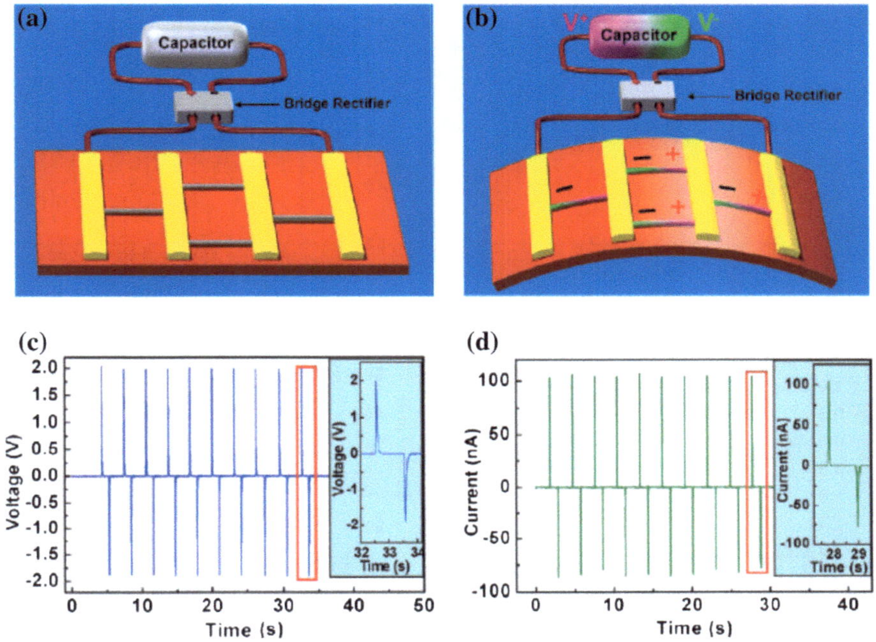

Fig. 3.52 Working principle and output measurement of the HONG. **a** Schematic diagram of its structure without mechanical deformation, with Au being used to form Schottky contacts with the ZnO nanowire arrays. **b** Demonstration of the output scaling up when mechanical deformation is induced, where the "±" signs indicate the polarity of the local piezoelectric potential created in the nanowires. **c** Open-circuit voltage of the HONG. **d** Short-circuit current measurement of the HONG. The measurement is performed at a strain of 0.1 % and strain rate of 5 % s^{-1}, at a deformation frequency of 0.33 Hz. The insets are the enlarged views of the boxed area for one cycle of deformation. Courtesy of Dr. Z. L. Wang, Georgia Institute of Technology. Reproduced with permission from [56], Copyright @ 2010, American Chemical Society

nanowires have a diameter of about 200 nm, the power volume density is calculated to be about 11 mW cm^{-3}.

Further scaling up the power output is still possible. For instance, if the nanowires can be uniformly and densely packed as a monolayer over the entire working area, and all of them actively contribute to the output, the maximum power area density can reach 22 μW cm^{-2}. The power volume density can be improved to 1.1 W cm^{-3}. With 20 layers of such nanowire arrays stacked together, the power area density will be 0.44 mW cm^{-2}.

The performance of the HONG is affected by the magnitude of strain and straining rate [53]. At a given strain rate (e.g., 5 % s^{-1}), the output increases with increasing magnitude of strain (Fig. 3.53a, b). Similarly, at a constant strain (0.1 %), the output is proportional to the straining rate (Fig. 3.53c, d). However, beyond certain levels of strain and straining rate, there will be saturation in outputs. This is probably attributed to the converse piezoelectric effect, which is a

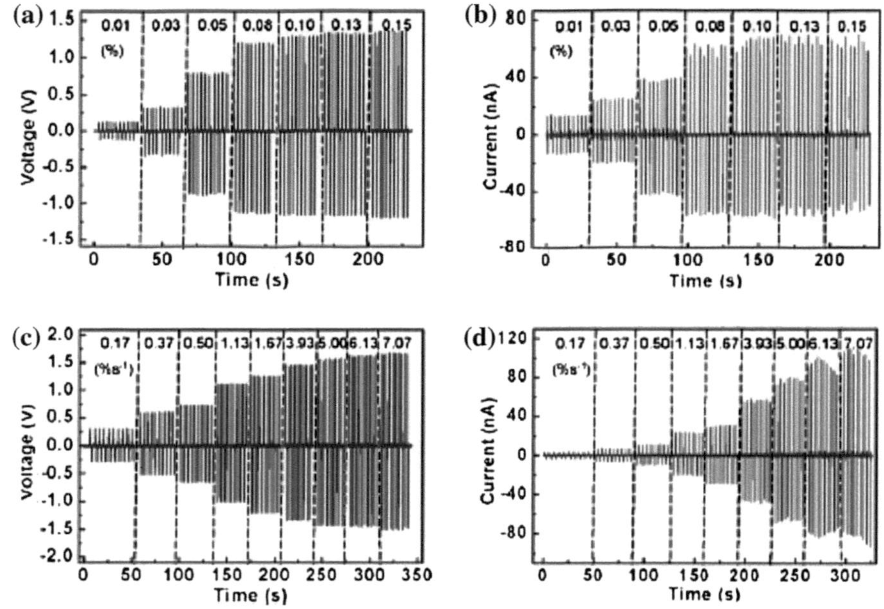

Fig. 3.53 Performance characterization of the HONG with increasing strain and straining rate. **a** Open-circuit voltage of the HONG with increasing strain at a given straining rate of 5 % s^{-1}. **b** Short-circuit current of the HONG with increasing strain at a given straining rate of 5 % s^{-1}. **c** Open-circuit voltage of the HONG with increasing strain rate at a constant strain of 0.1 %. **d** Short-circuit current of the HONG with increasing strain rate at a constant strain of 0.1 %. For all measurements, the mechanical deformation frequency is fixed at 0.33 Hz. Courtesy of Dr. Z. L. Wang, Georgia Institute of Technology. Reproduced with permission from [56], Copyright @ 2010, American Chemical Society

strain created by the piezopotential and it is opposite to the externally induced strain.

3.6.8.3 Storage and Application the Harvested Energy

The energy generated by the nanogenerators should be stored and can be used to power devices, which are important steps toward practical applications. Storing the generated energy and driving functional devices are extremely important steps toward practical applications of the nanogenerator. This is demonstrated in this section, by using a charging-discharging circuit with two consecutive steps (Fig. 3.54). The function of the circuit is controlled by the status of a switch (Fig. 3.54a, inset). The switch is at position A for energy storage by charging the capacitors. Once the charging is completed, the switch is switched to position B for energy discharging to power a device, e.g., light-emitting diode (LED).

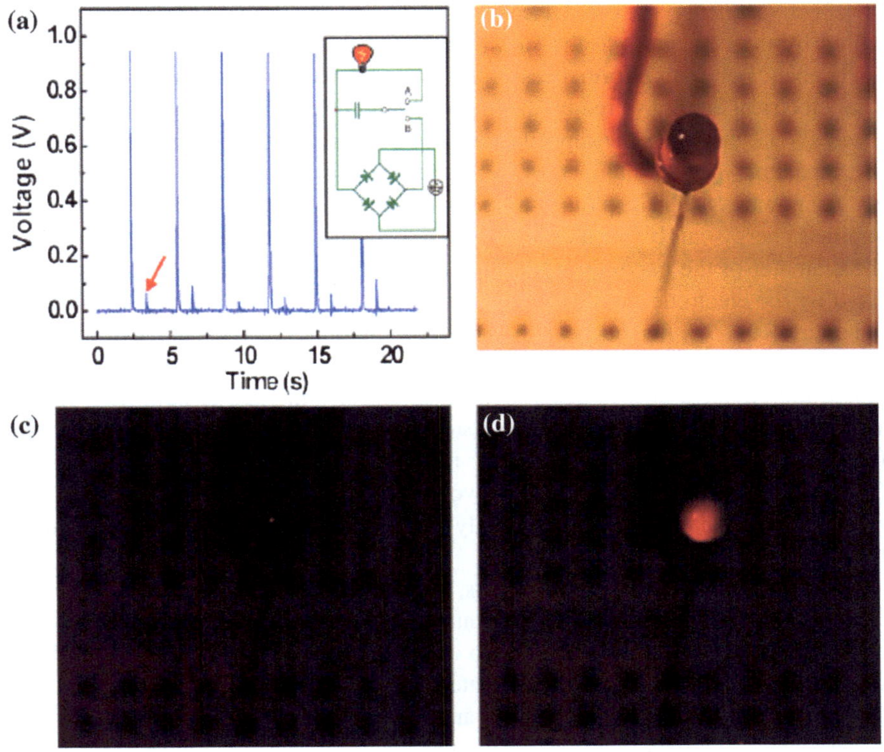

Fig. 3.54 Storage and application of the electric energy generated by the HONG. **a** The electric output measured through a full wave rectifying bridge. Signals with negative signs are reversed, as pointed by the *arrowhead*. The inset is schematic of the charging–discharging circuit for storing and discharging the energy generated by the HONG, respectively. **b** Image of a commercial LED, which is incorporated into the circuit. **c** Image of the LED in dim background before it is lit up. **d** Image of the LED in dim background at the moment when it is lit up. Courtesy of Dr. Z. L. Wang, Georgia Institute of Technology. Reproduced with permission from [56], Copyright @ 2010, American Chemical Society

To store the energy as an output with alternating characteristics, it is necessary to use an integrated full wave rectifying bridge between the HONG and capacitors. The output of the HONG measured through the bridge becomes positive signals (Fig. 3.54a). Full wave rectification achieved by the bridge ensures energy storage at an enhanced efficiency, although the rectified signal, as shown by the arrowhead in Fig. 3.54a, has a largely reduced magnitude due to the reverse current leakage of the diodes in the bridge. This effect is especially notable when output current is too low. In order to have an effective charging process, the output frequency of the HONG is tuned to 3 Hz by reducing the periodicity of the mechanical excitation. Ten capacitors are connected in parallel to be charged simultaneously. The voltage across every capacitor finally reaches 0.37 V.

Effective energy generation efficiency is defined as the ratio of the energy stored by the capacitors to the strain energy input to all the active nanowires. It also depends on the performances of the electronic components in the circuit. The total electrical energy stored by the capacitor can be calculated by $W_{stored} = CV^2n/2 = 1.37$ J, where C is the capacitance of a single capacitor, V is the voltage across the capacitor, n is the number of the capacitors. Since the dominant strain in the ZnO nanowires is tensile strain, shear strain can be neglected, therefore, the totally strained energy can be estimated by $W_{strain} = \pi D^2 L_0 E \varepsilon^2 f t n_0 / 8 = 30$ J, where D is the diameter of the nanowire (200 nm), L_0 is its original length (10 μm) which is fixed by the electrodes, E is the Young's modulus (30 GPa), ε is the strain of the nanowires (0.1 %), f is the frequency of deformation (3 Hz), t is the total charging time (7,200 s), and n_0 is the number of integrated nanowires (300,000) [56]. As a result, the effective energy generation efficiency is estimated to be ~4.6 %. This value is lower than the energy conversion efficiency of a single nano/micro wire (~7 %), which is defined as the ratio of the generated electric energy ($W_{generated} = \int VI\mathrm{d}t$, where V is the voltage and I is the current) to the input mechanical strain energy. This is mainly attributed to the energy dissipation due to the rectifying bridge and capacitors.

After finishing the charging process, the capacitors are reconnected from parallel connection to series connection, thus becoming a source with total voltage of 3.7 V. The stored electricity is used to drive a commercial red LED (Fig. 3.54b), which has an emission spectrum centered at 635 nm. Its turn-on voltage and forward-biased resistance are 1.7 V and 450 Ω, respectively. The discharging process leads to a maximum discharging current of 4.5 mA and the LED is lit up. The emitted light lasts 0.1–0.2 s, so that it can be clearly captured in the dim background (Fig. 3.54c, d. During the whole charging-discharging process, no other power sources are involved. The entire circuit is essentially a complete self-powered system, which consists of three components: an energy harvester (the HONG), storage units (capacitors) and a functional device (the LED). Therefore, it is expected that such a nanogenerator- based self-powering technology will find potential applications in mobile electronics, health monitoring, environmental inspection, cargo shipping tracking system, infrastructure monitoring, and defense technology [56].

3.7 Flexible Fiber Nanogenerators

Nanogenerators can be fabricated on a range of substrates, from polymer, semiconductor to metal, and from flat substrate to even fiber-based substrate, because ZnO nanowires can be grown at low temperature on substrates of different materials and different shapes by using the chemical solution growth. It has been successfully demonstrated that nanogenerators can be fabricated by using fibers on which ZnO nanowires are grown [57, 58].

3.7.1 Microfiber-Nanowire Hybrid Nanogenerators

3.7.1.1 Structure of the Nanogenerators

ZnO NWs are grown radially around Kevlar 129 fibers by using a hydrothermal method. The as-grown ZnO nanowires are then chemically bonded to surface of the fiber and to one another by coating with tetraethoxysilane (TEOS). A double-fiber nanogenerator is assembled by entangling one fiber covered with the as-grown nanowires around the other fiber covered with Au-coated nanowires. By holding the two ends of one fiber and sliding the other fiber back and forth, a relative brushing motion between the two fibers produces output current due to the coupled piezoelectric-semiconducting properties. Short-circuit current and Open-circuit voltage can be recorded when the two fibers slide with respect to each other.

The Kevlar 129 fibers have a diameter of 14.9 μm. A 100 nm-thick ZnO seeding layer is uniformly coated on surface of the fiber by using magnetron sputtering. ZnO nanowires are then grown radially on the fiber by using a hydrothermal method, as described before. The fibers coated with the ZnO nanowires are immersed in 99.9 % tetraethoxysilane (TEOS) for 2–3 min. Because the ZnO seed layer usually has cracks due to its incompatibility with the fiber, due to capillary force, TEOS flows through the cracks to reach the interface between the ZnO seed layer and the fiber, as well as the roots of the nanowires. Thus, two layers of TEOS, above and below the ZnO seeding layer, are formed.

A representative SEM image of a ZnO NW-covered Kevlar fiber is shown in Fig. 3.55a. Along the entire length of the fiber, ZnO nanowires are grown radially, with a very uniform coverage and well preserved cylindrical shape. Some splits on the nanowire arrays can be identified (Fig. 3.55b), which are caused by the surface tension in the seeding layer during the growth. All the ZnO nanowires are single crystalline and have a hexagonal cross-section, with a diameter of 50–200 nm and a typical length of about 3.5 μm. Their top and side surfaces are smooth and clean, meaning that they are able to form reliable metal-semiconductor junctions that are necessary for a nanogenerator. The space between the nanowires is in the order of a few hundreds nanometers, which is sufficiently large for them to be bent. The tips of the nanowires are separated from one another due to their small tilting angles ($<\pm 10^\circ$), but their bottom ends are tightly connected (inset of Fig. 3.55b). As a result, a continuous ZnO film at the roots of the nanowires can serve as a common electrode. It has been shown that the solution grown ZnO nanowire has output voltage of up to 45 mV.

Figure 3.55c shows the two layers of tetraethoxysilane (TEOS) infiltrated above and below the ZnO seed layer as a binding agent. The Si–O bonds in TEOS are highly reactive with the OH^- groups on the ZnO surface. The organic chains in TEOS firmly bind to the aromatic polyamide fiber. As a result, the ZnO seed layer and the fiber core are tightly bound with each other due to the presence of the thin layer of TEOS. Furthermore, since TEOS can easily form cross-linked chains, the nanowires can be firmly bundled and bound together at their roots and fixed on the

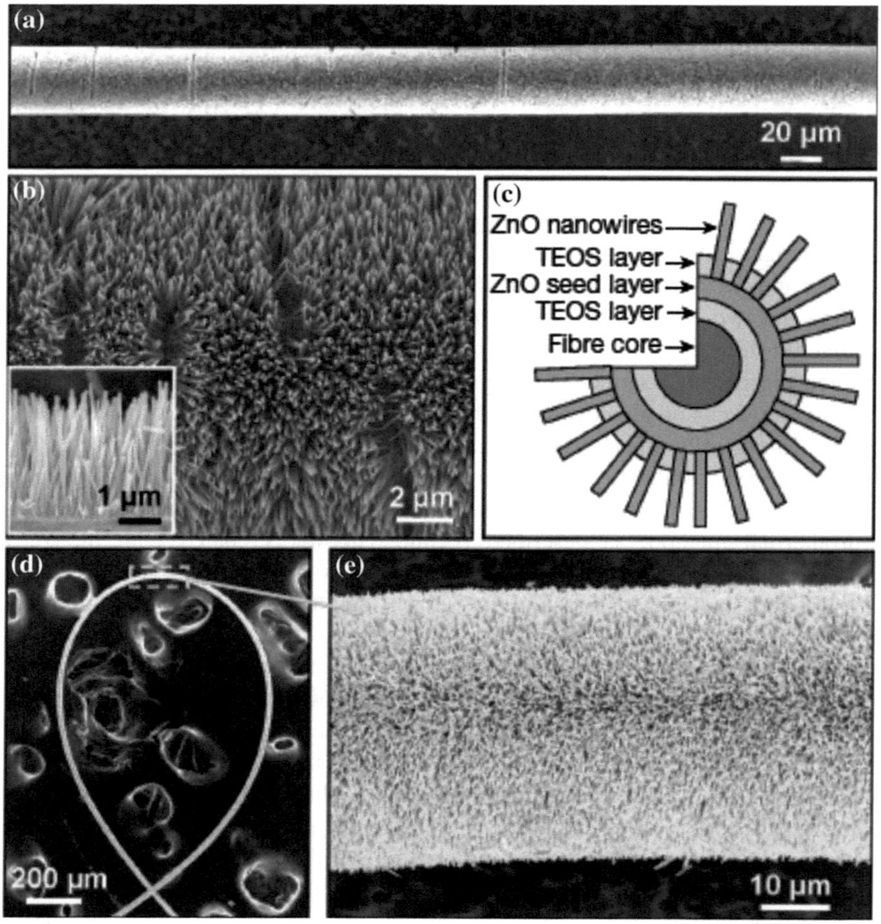

Fig. 3.55 ZnO nanowire-coated Kevlar fibers. **a** SEM image of a Kevlar fiber covered with ZnO nanowire arrays along the radial direction. **b** Higher magnification SEM image and a cross-section image (inset) of the fiber, showing a uniform distribution of the nanowires. **c** Schematic showing the cross-sectional structure of the TEOS-enhanced fiber designed for improving its mechanical strength. **d** SEM image of a looped fiber showing the flexibility and strong binding capability of the nanowire layer. **e** Enlarged section of the looped fiber showing the distribution of the ZnO nanowires at the bending area. Courtesy of Dr. Z. L. Wang, Georgia Institute of Technology. Reproduced with permission from [57], Copyright @ 2008, Nature Publishing Group

ZnO seeding layer, so that the nanowires are not scratched/stripped off during mechanical brushing/sliding. The fiber can be bent into a circle of ~1 mm in diameter, without cracks, loosing pieces or peeling-offs in the ZnO nanowire coating layer (Fig. 3.55d). In the bending area, the radial aligning morphology of the ZnO nanowires is also very well preserved (Fig. 3.55e), demonstrating the toughness of the structure.

3.7.1.2 Fabrication of the Nanogenerator

A double-fiber nanogenerator is assembled with two 3 cm ZnO nanowire-coated fibers, one is coated with Au and the other is freshly grown. The Au layer is coated by using a DC sputtering system. During the sputtering, the fiber is fixed on the stage at one end and its body is left freestanding. Therefore, a uniform Au layer is coated along the entire fiber by rotating the sample stage. The thickness of the Au coating is 300 $\pm$ 20 nm. Both ends of the fiber with as-grown ZnO nanowires are fixed on a glass substrate with one end grounded and electrically connected to an external measurement circuit as the nanogenerator output anode. One end of the gold-coated fiber is attached to a small spring, which is affixed to the substrate; while the other end is attached to a pulling string, so that it can freely move back and forth. The gold-coated fiber is electrically connected to the external circuit as the nanogenerator output cathode. The effective length of the fiber nanogenerator is 4–5 mm and typically has nine cycles, with each cycle to be about 500 μm in length.

The double-fiber NG is fixed on a stationary stage, with the movable gold-coated fiber being attached to a pulling bar. The pulling bar is driven by a speed control motor with an extrusion on its spin axis, which transfers its rotary motion into a back-forth motion of the pulling bar at a chosen frequency. In each round, the gold-coated fiber first moves to the pulling direction when the pulling bar is lifted up; then is retracted back to its original position by the spring attached to the other end. The back and forth brushing movement between the two fibers is thus achieved with a controlled frequency.

A double-fiber model system has been designed, as shown in Fig. 3.56a [57]. The pulling force ensures a good contact between the two fibers, as shown by an optical microscopy image in Fig. 3.56b. In this design, the gold-coated ZnO nanowires act as an array of scanning metal tips that deflected the ZnO nanowires rooted at the other fiber. The gold coating completely covered the ZnO nanowires, forming a continuous layer along the entire fiber. Once the two fibers are firmly entangled together, some of the gold-coated nanowires penetrate slightly into the spaces among the uncoated nanowires rooted at the other fiber, as shown in Fig. 3.56c. Thus, when there is a relative sliding/deflection between them, the bending of the uncoated ZnO nanowires produces a piezoelectric potential across their width, and the Au-coated nanowires act as the “zigzag” electrode as discussed earlier.

3.7.1.3 Mechanism

Figure 3.56d shows the mechanism of charge generation by the fiber nanogenerators. It is analogous to that of deflecting a ZnO nanowire by an AFM tip, as described earlier. Briefly, when the top fiber moves to the right-hand side, the gold-coated nanowires bend the uncoated ZnO nanowires to right-hand side, with an assumption that the Au-coated NWs are much stiffer and suffer little bending.

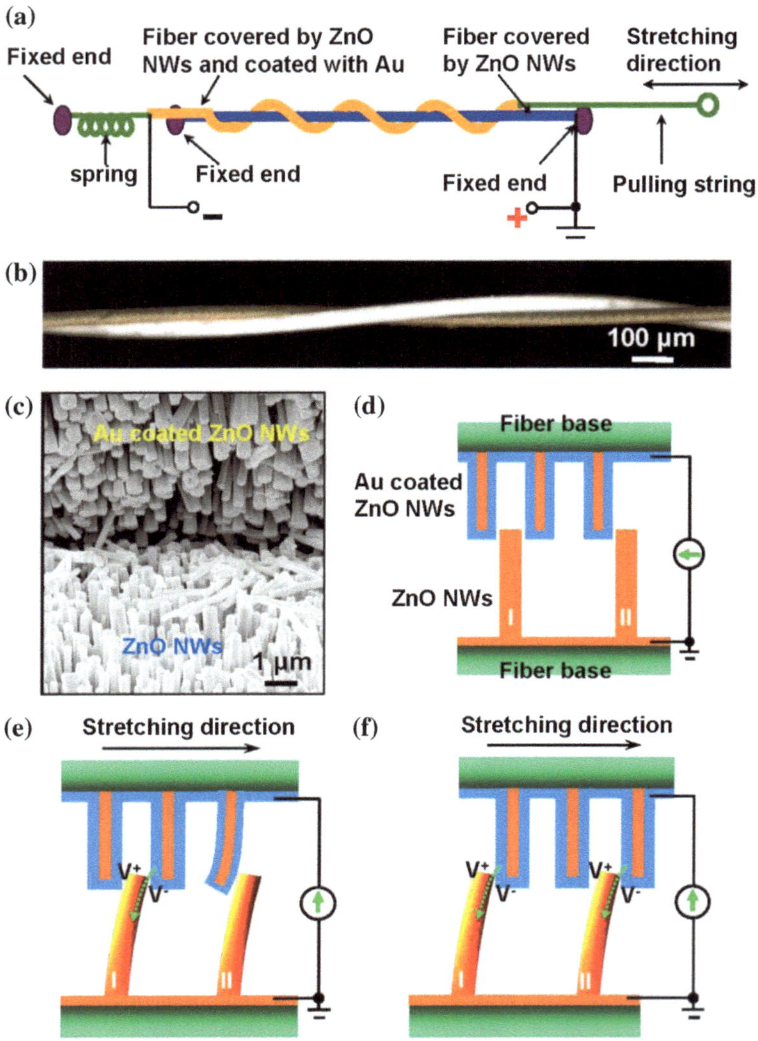

Fig. 3.56 Design and mechanism of the fiber-based nanogenerator. **a** Schematic experimental setup of the fiber-based nanogenerator. **b** An optical micrograph of a pair of entangled fibers, one of which is coated with Au (in darker contrast). **c** SEM image at the "teeth-to-teeth" interface of two fibers covered by the nanowires, with the top one coated with Au. The Au-coated nanowires at the top serve as the conductive "tips" that deflect/bend the nanowires at the bottom. **d** Schematic illustration of the teeth-to-teeth joining between the two fibers covered by the nanowires. **e** Piezoelectric potential created across the nanowires I and II under the pulling of the top fiber by an external force. The side with positive piezoelectric potential does not allow the flow of current due to the presence of a reversely biased Schottky barrier. Once the nanowire is pushed to bend sufficiently to reach the other Au-coated nanowire, electrons in the external circuit will be driven to flow through the uncoated nanowire due to the forward-biased Schottky barrier at the interface. **f** When the top fiber is further pulled, the Au-coated nanowires scrub across the uncoated nanowires. At the moment that the two types of nanowires are in the final contact, the interface is a forward biased Schoottky barrier, resulting in further output of electric current, as indicated by *arrowheads*. The output current is a sum of all those contributed by the individual nanowires, while the output voltage is determined by only one nanowire. Courtesy of Dr. Z. L. Wang, Georgia Institute of Technology. Reproduced with permission from [57], Copyright @ 2008, Nature Publishing Group

Piezoelectric potential is thus generated across the uncoated NW due to its piezoelectric effect, with the stretched surface to positive (V^+) and the compressed surface to be negative (V^-). The positive potential side has a reversely biased Schottky contact with Au and thus prevents the flow of current, while the negative potential side has a forward biased Schottky contact with the Au and it allows the current to flow from the Au to the NW.

Since the array has high density of nanowires (Fig. 3.55b), a bent nanowire on the uncoated fiber touches the backside of another gold-coated nanowire after the bending (such as nanowire I in Fig. 3.56e). In this case, the negative potential surface of the ZnO nanowire contacts the gold layer, so the Schottky barrier at the interface is forward biased, resulting in a current flowing from the gold layer into the ZnO nanowire. Then, when the top fiber keeps moving further toward the right-hand side (Fig. 3.56f), the gold-coated nanowires scan across the ZnO nanowire tips and reach their negatively charged side (nanowires I and II in Fig. 3.56f). Therefore, more current will be released through the forward-biased Schottky barrier. The currents from all the nanowires will add up constructively, regardless the nanowires being deflected to which direction even in same cycle of pulling.

The output voltage is defined by the characteristic of one nanowire, and the sign of the voltage does not change in responding to the deflection configuration of the nanowire due to the rectifying effect of the Schottky barrier at the Au–ZnO interface. The same effect is expected if the top fiber is driven to retract toward the left-hand-side. Due to the similar mechanical property of the top and bottom nanowires, the gold-coated nanowires are also possibly bent by the nanowires rooted at the uncoated fiber, but this does not affect the mechanism presented in Fig. 3.56. For the Au-coated fiber, all the nanowires are completely covered by a thick gold layer, and they can be considered as an equal potential electrode connected to the external measurement circuit. Thus, the role played by the ZnO nanowires is only a template for supporting the Au coating and no piezoelectric charges are preserved inside the gold-coated nanowires.

3.7.1.4 Performance Improvement

To further improve the performance of such nanogenerators, a yarn nanogenerator is assembled [57]. It is made of six fibers, among which three are covered with nanowires and coated with Au and the other three have only nanowires. All the three Au-coated fibers are movable. At a frequency of 80 rpm, an average current of about 0.2 nA is achieved, which is 30–50 times larger than the output signal from a signal fiber nanogenerator, due to the largely increased surface contact area among fibers. At the same time, the width of each pulse is broadened, due to the unsynchronized movement and a relative delay in outputting current among the fibers.

It is also found that the output current can be enhanced by reducing the inner resistance of the fibers and the nanowires. By depositing a conductive layer directly onto the fiber before depositing the ZnO seeding layer, the inner resistance of the nanogenerator can be reduced from ~1 GΩ to ~1 kΩ, so that the output current of a double-fiber nanogenerator is increased from ~4 pA to ~4 nA. It means that the output current is approximately inversely proportional to the inner resistance of the nanogenerator. In addition, the output voltage and current can be increased by connecting nanogenerators in serial and parallel.

The textile fiber-based nanogenerators have the following advantages. First, they are flexible, foldable, wearable, and robust, so that the nanogenerators made with them can be power sources in any shape, such as power shirt. Second, the output electricity can be dramatically enhanced by using fiber bundles in the form of yarns, which are the basic units for fabrics. The optimum output power density can be 4–20 mW m^{-2}. Third, the nanogenerator can be operated at low frequencies, in the range of conventional mechanical vibration, foot stepping and heart beating. Finally, since the ZnO nanowire arrays are grown at a low temperature of 80 °C, it can be applied a wide range of materials and applications.

3.7.2 Flexible Fiber Nanogenerators Driven by Pressure

3.7.2.1 Brief Introduction

Air/liquid flow induces dynamic pressure change and thus can be important source of mechanical energy harvested by using nanogenerators. Fluctuation in air pressure is rather irregular, with a wide range of amplitude and frequency, which makes it difficult to be harvested directly by using the conventional technologies. For example, when using the piezoelectric cantilever method as energy harvester, it has maximum energy harvesting efficiency only when the external mechanical stimulating frequency matches the resonance frequency of the cantilever. If the stimulating frequency is too low, the required size of the resonator has to be large, which can only harvest the energy if the magnitude of the applied force is sufficiently strong enough. If the applied frequency is too high, the size of the resonator will be very small, but high frequency mechanical signals are not as popular as the low-frequency excitations around us, especially in biological systems, such as breathing and heart beating.

To address this problem, a flexible fiber nanogenerator (FNG) has been developed [58]. Such FNGs can be used for applications in smart shirt, flexible electronics, and medical. The FNGs are based on carbon fibers which are covered cylindrically with textured ZnO thin films. When a uni-compression is applied to the cylindrical ZnO thin film, it will experience a compressive strain, which results in a macroscopic piezopotential across its inner and exterior surfaces, due to the textured structure of the film. This is the driving force for generating electricity. The harvester has an output peak voltage of 3.2 V and the average current density of 0.15 n μA cm^{-2}. The FNG can be driven by air pressure, so that it can work in a noncontact mode, such as rotating tire, flowing air/liquid and blood vessel. The pressure-driven FNG in a syringe shows its potential to harvest energy in a blood vessel, gas pipe, and oil pipe, as long as there is a fluctuation in pressure or turbulence. The heart pulse-driven FNG can serve as ultrasensitive sensor for monitoring human heart behavior, which may possibly be applied to medical diagnostics as sensors and measurement tools.

3.7.2.2 Textured ZnO Thin Film and Nanogenerator Assembling

Figure 3.57a shows the design of the FNG with radial-textured ZnO thin film grown on carbon fibers. In principle, any fiber can be used for this purpose as long as it is conductive. The fiber is used not only as a flexible soft substrate to grow ZnO thin film at high temperature but also as an electrode for charge transport. The key point is to produce textured ZnO film that is made of densely packed nanowires with [0001] orientation. The ZnO film covers the surface of the fiber to form a radially textured, cylindrical, and shelled structure. The density of the nanowires can be controlled by experimental conditions without using ZnO seeds.

The ZnO film is coated on carbon fibers by using the low temperature (960 °C) VSS method, as discussed earlier. ZnO powder and activated carbon particles with equal ratio in mass are used as the source materials. The carbon fibers are freely suspended in the tube of the furnace on the way of carrier gas. The carrier gas consists of O_2 and N_2 with a ratio of 1:4 at a flow rate of 40 sccm.

The as-grown thin film is composed of nearly parallel aligned ZnO nanorods to form a textured film ($\sim$250 nm in thickness) with the normal direction along [0001]. As shown in Fig. 3.57c, the densely packed ZnO nanorods have hexagonal flat tops. These nanorods can have a random orientation in the film plan, but a good alignment in the normal direction so that the entire film has a unipolar structure along the *c*-axis. The distribution of the nanorods around the fiber can be clearly seen if the density of the nanorods is not high (Fig. 3.57d).

To assemble a fiber nanogenerator (FNG), the ZnO thin film at one side of the carbon fiber is etched off with NaOH solution to expose the fiber as electrode. The other electrode of the FNG is contacted to the top surface of the ZnO thin film by using a silver tape/paste (Fig. 3.57e). As shown in Fig. 3.57, the wurtzite structured ZnO nanorods form a densely packed and polar direction aligned thin film, with the polar direction pointing radial outward. The FNG is constructed by aligning the ZnO coated carbon fibers in parallel, with one electrode being connected to the carbon fibers and the other electrode being connected to top surface of the ZnO film. A plastic substrate is used to support the aligned fibers, by fixing all the fibers to the substrate.

3.7.2.3 Mechanism of the Fiber Nanogenerator

The textured ZnO film can be treated as a "single-crystal" in the FNG. When the film is compressively strained due to the external pressure through air/liquid, piezoelectric potential gradient is produced along the *c*-axis of ZnO. The textured

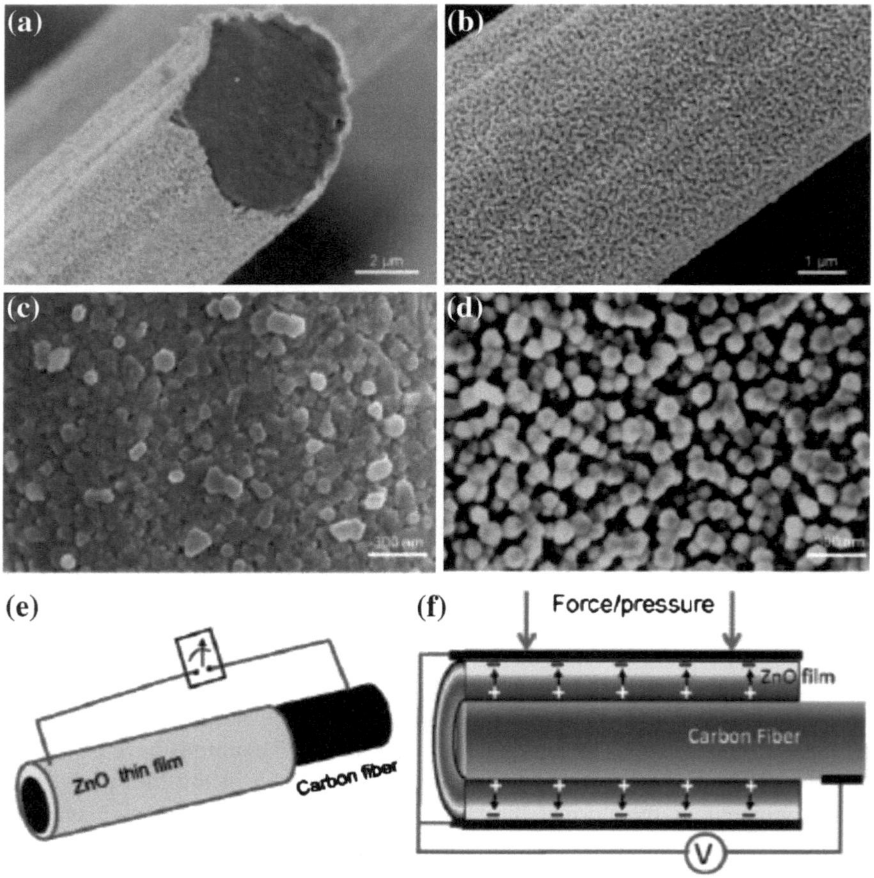

Fig. 3.57 Textured ZnO thin films grown on surface of a carbon fiber. **a**, **b** Low-magnification SEM images of the ZnO/fiber structure. **c** A ZnO film made of densely packed ZnO nanorods with a uniaxial direction of *c*-axis. **d** A ZnO film made of sparely aligned ZnO nanorods with their *c*-axes closely in parallel. **e** Schematic structure of the fiber nanogenerator-based ZnO thin film-coated carbon **f** Working principle of the fiber nanogenerator, where the "±" signs indicate the polarity of the local piezoelectric potential created on the inner and outer surfaces of the ZnO thin film. Courtesy of Dr. Z. L. Wang, Georgia Institute of Technology. Reproduced with permission from [58], Copyright @ 2011, John Wiley & Sons

structure of the thin film results in a macroscopic piezoelectric potential across the thickness of the film (Fig. 3.57f). If the *c*-axis points outward, the negative side of the piezopotential is the external surface of the film, which raises the conduction band and the Fermi level at the electrode, forcing the electrons to flow from this side to the other electrode through the external load. As the external force is removed and the compressive strain is released, the piezoelectric potential inside the film diminishes. The electrons accumulated at the other electrode flow back via the external circuit, creating an electric pulse in the opposite direction. A cycled strain results in an AC output. The key role played by the fiber is to effectively use the pressure from all directions.

The external excitation leads to an angular bending rate of about 260° s^{-1} with a radius of 2 cm for the substrate. A FNG made of 150 carbon fibers with a diameter of ~10 μm and a length of about 20 mm offers a maximum output voltage of 2.0–2.2 V and output current of 60–120 nA.

3.7.2.4 Air Pressure Experiment

Air pressure can apply to every part of an object without a specific direction. The ZnO film-coated carbon fibers are just loosely packaged on a flexible substrate, as shown in Fig. 3.58c, which is similar to the connection as shown in Fig. 3.57e. A uni-radial compression due to the air pressure on the fiber creates a piezoelectric potential, as shown in Fig. 3.57f. A dynamic variation in pressure creates a compression-releasing cycle to the cylindrical shaped ZnO film. The generated piezoelectric potential leads to back and forth flows of electrons in the external load as a response to the fluctuation of the air pressure.

By placing the FNG inside a syringe (Fig. 3.58c) and applying a periodic pressure through the pushing/pulling of the piston, an AC output is created (Fig. 3.58d). The generated electricity output is not quite symmetric because the magnitude of the pressure and rate of the pressure are different when pushing and pulling the piston. The output peak voltage is up to 3.2 V and the average current density is 0.15 μA cm^{-2}. The surface area used for calculation is the area of the working carbon fibers in the nanogenerator.

3.7.2.5 Exhalation Experiment

Gentle air flow can always change its direction. The relatively small force of the air flow may too weak to excite a nanogenerator packaged with materials with high stiffness and damping. Also, the frequency of the air flow is not constant. To cater for these characteristics of gentle breeze, a new assembling has been proposed, by using long fibers, as shown in Fig. 3.59a. A ring shaped FNG is fixed on a glass substrate. One electrode is connected to the carbon fibers and the other one is joined to the ZnO film. When blowing the FNG with a gentle exhalation, the breezing wind applies mechanical force to the FNG, so that it will change its shape and vibrate, which results in electricity output (Fig. 3.59). Repeated exhalation results in an AC output, with an average output voltage of 1.5 mV and an average output current of 0.5 nA. The low output is due to the low level strain by exhalation. This low output may not be useful as energy harvesting, but it can be used as a breathing sensor in potential biomedical and health care applications.

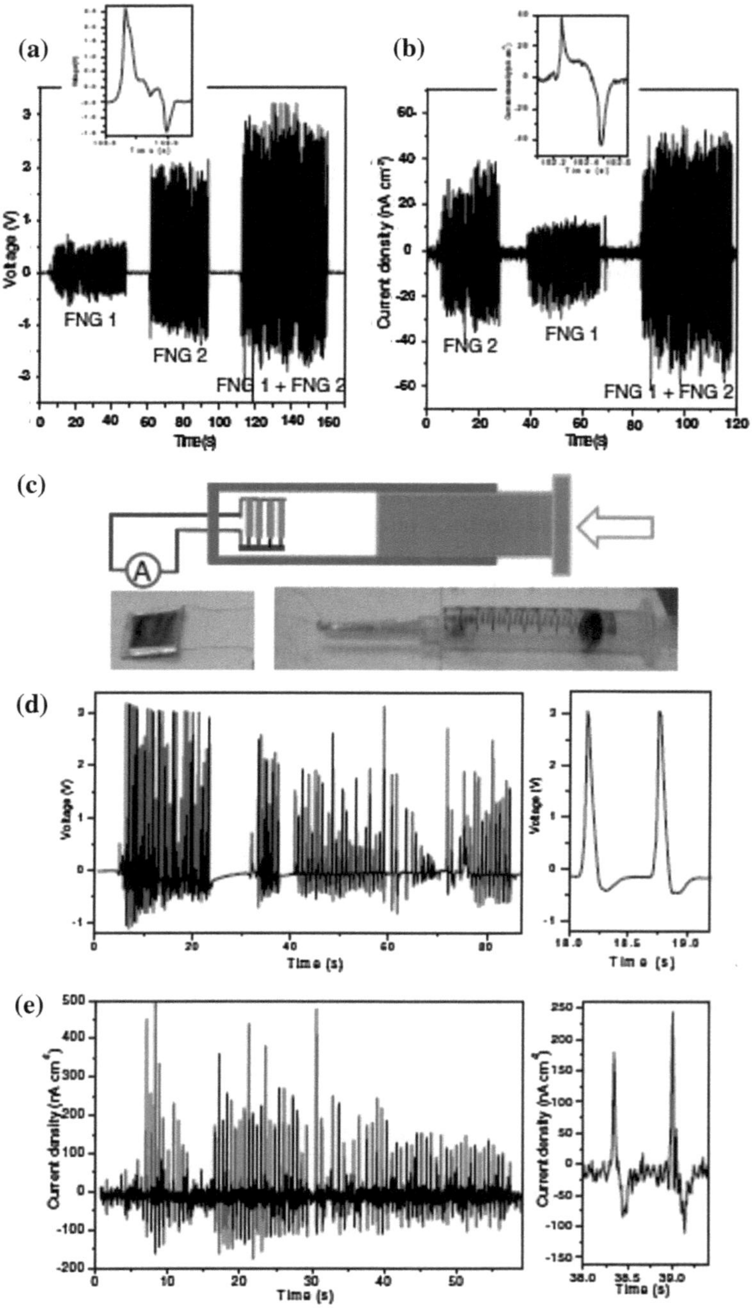
(a)
Voltage (V)
FNG 1
FNG 2
FNG 1 + FNG 2
Time(s)
(b)
Current density (nA cm⁻²)
FNG 2
FNG 1
FNG 1 + FNG 2
Time(s)
(c)
A
(d)
Voltage (V)
Time (s)
(e)
Current density (nA cm⁻²)
Time (s)

◀ **Fig. 3.58** Two normal FNGs for a "linear superposition" test of outputs current density **a** and voltage **b**. The output voltage of the two FNGs (FNG1+FNG2) is the sum of those of FNG1 and FNG2 when they are connected in serial. The output current density also adds up when the two FNGs are connected in parallel. The insets are the details of the output signals. **c** Schematics of an air pressure driven FNG when it is placed inside a syringe. The variation of air pressure is created by pushing/pulling the piston of the syringe. Open-circuit output voltage **d** and short-circuit output current **e** of the FNG with 100 carbon fibers. The maximum output voltage peak reaches 3.2 V. The insets at the *right-hand side* are one cycle of the output voltage and current. Courtesy of Dr. Z. L. Wang, Georgia Institute of Technology. Reproduced with permission from [58], Copyright @ 2011, John Wiley & Sons

3.7.2.6 Wrist Pulse Experiment

Pulse pressure (PP) is one of the most important sources of mechanical energy from human activity. These sources, such as signals of a human heart system, are complex, time-dependent, and nonlinear, reflecting the fluctuation of motion activities and health conditions. Quantitative data of such pulse signal scan be used as important information for medical monitoring and diagnostics.

One example is to use the pulse of human wrist as a mechanical energy sours. A FNG made with relatively long fibers of $\sim$4 cm is bonded to the wrist with cloth athletic tape. In order to monitor the in situ original blood pressure signal, a medical blood pressure gauge is fixed on top of the FNG. The output of the FNG is a reflection of its responses to the heart beating behavior. The heart beating behaviors can be used as references to diagnose the health states of patients.

3.7.2.7 Summary

Flexible fiber-based nanogenerator is a potential mechanical energy harvester that can be used for fabrication of smart textiles and flexible electronics. This type of nanogenerators does not need a thick layer of polymers for protection, so as to address the problems of damps and thus lead to high outputs. Also, it works in a noncontact mode and can be incorporated on all the available surfaces, making it highly feasible for practical applications, such as three-dimensional sensors in rotating tire, flowing air/liquid, and even in blood vessel. The energy harvested can be used to monitor the status and operation of the fluid/gas. The heart pulse-driven nanogenerator can be used as ultrasensitive sensors to monitor human heart behaviors for medical diagnostics.

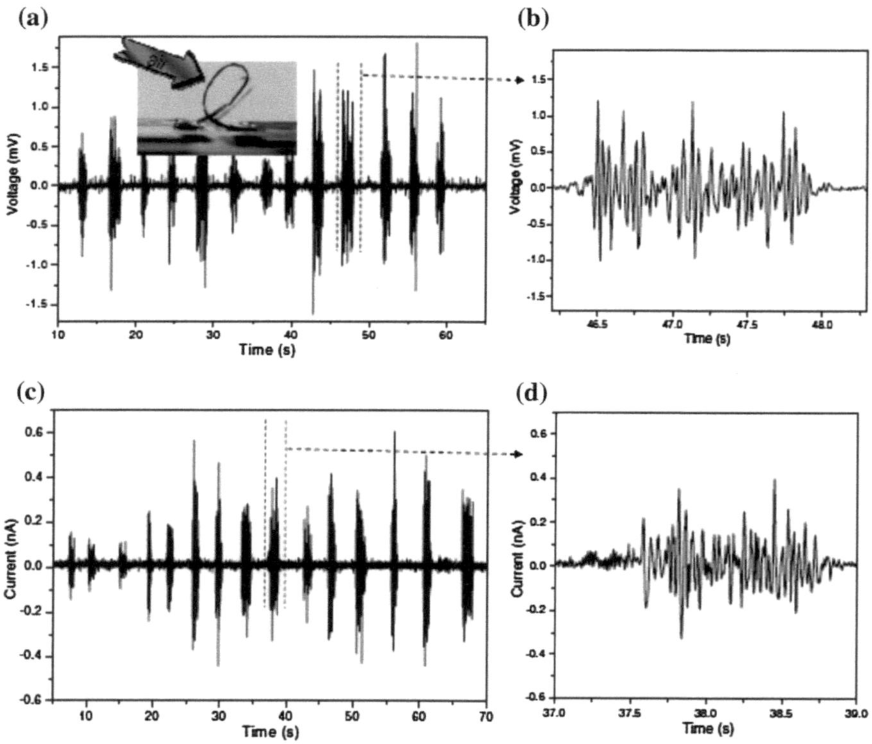

Fig. 3.59 Performance of a fiber nanogenerator driven by air flow. The inset in **a** is a loop shape FNG fixed to a glass substrate, which is made with 50 carbon fibers. When breezing the air by exhalation, the open-circuit output voltage (**a**, **b**) and short-circuit output current (**c**, **d**) for the FNG are detected. Courtesy of Dr. Z. L. Wang, Georgia Institute of Technology. Reproduced with permission from [58], Copyright @ 2011, John Wiley & Sons

3.8 Multifunctional Nanogenerators

3.8.1 Nanogenerators with Noncontact Nanowires

3.8.1.1 Brief Introduction

The ZnO nanowires, used in all the nanogenerators discussed above, no matter whether they are vertically or laterally oriented, are directly contacted with electrodes. A new type of nanogenerators made with noncontact ZnO nanowires has recently been reported [59]. The method to fabricate this nanogenerator is simple, cost-effective, robust, and scalable, with outputs that are sufficiently strong to continuously power a commercial liquid crystal display (LCD). To assemble such nanogenerators, ZnO nanowires with conical shape are dispersed in polymer matrix to form composite. The geometry-induced unipolar assembly of the conical

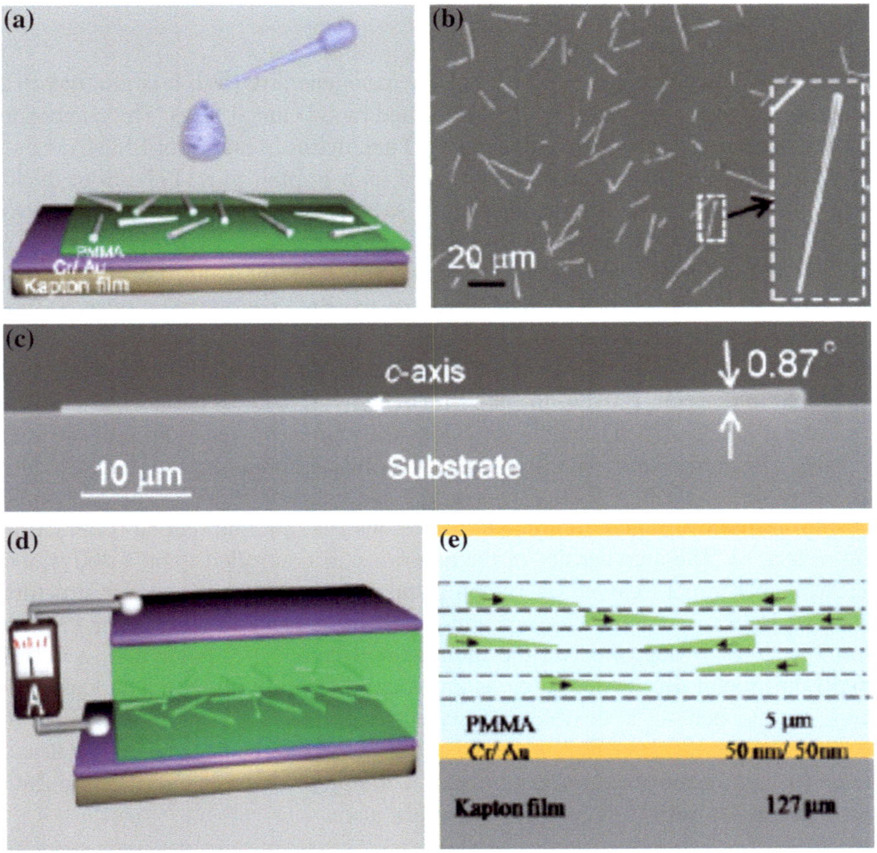

Fig. 3.60 Fabrication process of the nanogenerator with noncontact ZnO nanowires. **a** Schematic diagram showing the fabrication process of the nanogenerator. **b** SEM image showing that the conical ZnO nanowires are fairly uniformly distributed on the substrate with a random lateral orientation. The insert high magnification image showing the *conical shape* of the nanowires. **c** Cross-sectional SEM image of a conical nanowire laying on a flat substrate. The bottom side surface of the nanowire tightly attached to the substrate, with a conical angle of 0.87°. The *white arrowhead* indicates that the *c*-axis of the nanowire is pointing downward into the substrate. **d** Schematic image of the fabricated device structure. **e** Schematic diagram showing the idea and working principle of the nanogenerator. Courtesy of Dr. Z. L. Wang, Georgia Institute of Technology. Reproduced with permission from [59], Copyright @ 2010, American Chemical Society

ZnO nanowires in the composite results in a macroscopic piezoelectric potential across its thickness when subject to mechanical deformation. A compressive strain of 0.11 % at a straining rate of 3.67 % s^{-1} produces an output voltage of up to 2 V. It has potential to power small size personal electronic devices.

3.8.1.2 Design and Fabrication

Figure 3.60 shows fabrication process of the nanogenerator [59]. It is multilayered structure, with the composite being sandwiched by two metal films. The composite is made of unipolar assembly of conical ZnO nanowires in matrix of PMMA. First, a Cr/Au (50/50 nm) metal layer is deposited on a Kapton film (127 μm in thickness) by using electron beam evaporation. A layer of PMMA ($\sim$2 μm in thickness) is then coated on the metal film by using spin-coating method.

The ZnO nanowires are grown on a solid substrate by using the vapor deposition method discussed before. The length of the ZnO nanowires is over 30 μm. The nanowires have a conical shape due to the fastest growth rate along the *c*-axis and a much slower growth rate in the basal plane. The conical shape of the ZnO nanowires has been proven to be critical for the nanogenerator to harvest mechanical energy. Then, conical ZnO nanowires are removed from the substrate by using ultrasonic wave. By dispersing the solution containing the conical ZnO nanowires on the PMMA film, a layer of the nanowires is formed. The nanowires are uniformly distributed on the substrate with a random lateral orientation (Fig. 3.60a, b). The area density of the nanowires is controlled to be 1,400–1,500 nanowires per mm^2) to avoid overlapping and aggregation. A multilayer structure with five layers is obtained by repeating the two steps. Then, a thicker PMMA layer ($\sim$2 μm) is deposited on the multilayered composite. After that, a Cr/Au (50/50 nm) metal film is deposited as an electrode (Fig. 3.60d, e). The nanogenerator is attached to a flexible polystyrene substrate ($\sim$1 mm in thickness). External forces can be applied to the nanogenerator from the back of the substrate. In this case, the nanogenerator experiences a compressive strain and thus the ZnO nanowires are subject to a compressive strain also.

3.8.1.3 Energy Harvesting Mechanism

The conical ZnO nanowires are randomly distributed within the composite, with lateral orientations (Fig. 3.60b), while their bottom surfaces are tightly attached to the flat surface of the substrate (Fig. 3.60c). The [0001] growth direction of the ZnO nanowires is their polar direction. The conical shape of the nanowires results in a constructive alignment in the projected polar directions for all the nanowires, which is in the direction perpendicular to the substrate and pointing downward into the substrate, as shown in Fig. 3.60e, in which the dashed line indicates one cycle of deposition of the ZnO nanowires. The conical-shaped nanowires and their corresponding *c*-axes are indicated by the arrowheads, which are the symmetry axes of the nanowires [59].

Owing to the geometrical shape, the component of the *c*-axis of each nanowire in the normal direction of the substrate is $c \cdot sin(\alpha/2)$, where α is the conical angle of the nanowires. This projected component of all the nanowires along the vertical direction constructively adds up, which is the source of pizoelectric polarization across the thickness of the composite to create piezopotentials. The unipolar

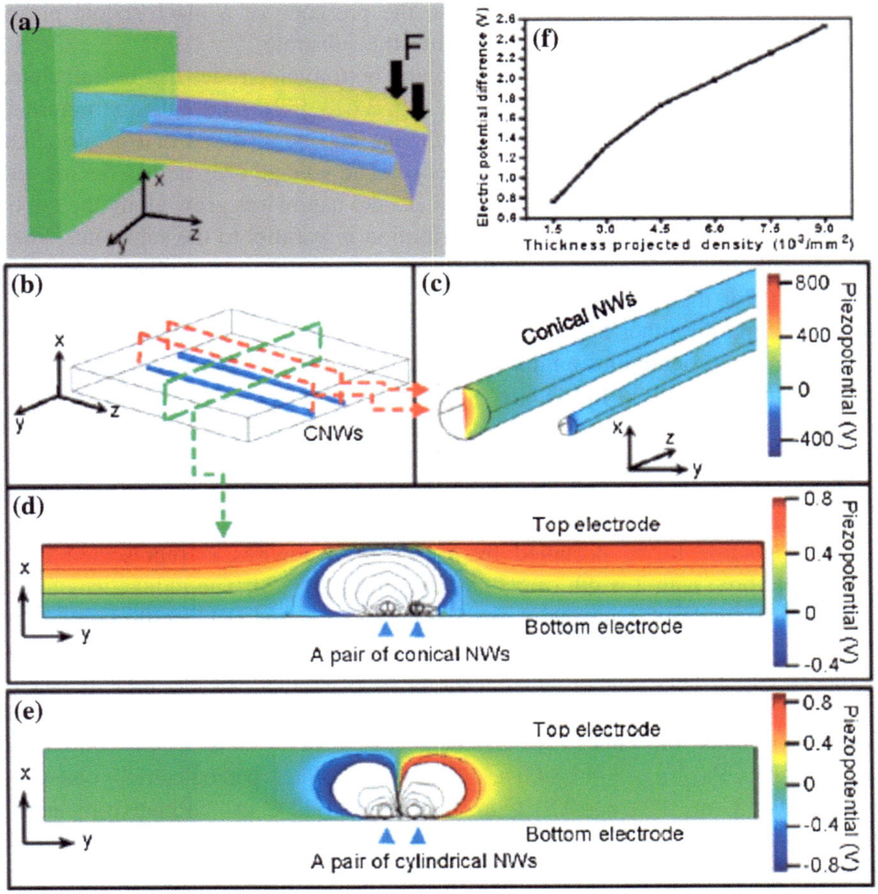

Fig. 3.61 **a** Schematic model showing the set up to measure the energy conversion of the nanogenerator. The conical ZnO nanowires are subject to a compressive strain during the deformation. **b** The unit cell and model used for calculating the potential distribution across the *top* and *bottom* electrodes of the nanogenerator with a pair of nanowires, The corresponding cross-sections at which the potential distributions are exhibited are indicated by *dashed lines*. The results are shown in (**c, d**), respectively. Owing to the large magnitude variation in the potential distribution across the cross-section, both color grade and equal potential lines are used to present the local potential. The blank region close to the nanowires is the region with the calculated piezopotential of smaller than -0.4 V, beyond the range selected for the color plotting in order to show the detail in this region, only equal potential lines are used. The nanowires are positioned close to the bottom of the unit cell in **b** to ensure that they are subject to compressive strain once a transverse force is applied. **e** The calculated potential of a perfect cylindrical nanowire. The result indicates that there is no potential difference at the two electrodes. **f** Calculated potential difference between the *top* and *bottom* electrodes of a nanogenerator as a function of the thickness of the projected conical nanowire density. The distance between the top and bottom electrodes is kept constant (5 μm). The density required for a uniform, fully packed, monolayer coverage of the substrate is ~90,000 mm^{-2}. Courtesy of Dr. Z. L. Wang, Georgia Institute of Technology. Reproduced with permission from [59], Copyright @ 2010, American Chemical Society

assembly of the nanowires is the key to the presence of a macroscopic piezoelectric potential in the direction normal to the substrate.

The potential generated across the top and bottom electrodes can be calculated by using a simple model, as shown in Fig. 3.61a [59]. The entire structure is considered as a free standing beam, with one end being fixed and a periodical transverse force being applied at its top edge. The voltage drop across the top and bottom electrodes is calculated. The conical ZnO nanowires grow along the c-axis and they have a random orientation distribution in parallel to the substrate. When an external strain is created along the z-axis in parallel to the substrate, statistically, 50 % of the nanowires whose c-axis projections in parallel to z are along $+z$ axis, while the other half are along $-z$ axis. A pair of the nanowires, having opposite c-axis orientations on the substrate in parallel to z-axis, is selected for the calculation, as shown in Fig. 3.60e.

With the density of the conical ZnO nanowires on the substrate, each pair of nanowires occupies an average volume. Because of the random orientation of the nanowires on the substrate and their projected length along the z-axis, a factor of 2/p is multiplied to the real density to account for this equivalent effect. Therefore, the effective volume to be occupied by a pair of nanowires is represented by a rectangular box, with a width of 50 μm and a height 5 μm. The radius, length and semi-conical angle of a nanowire are 500 nm, 45 μm and 0.4°, respectively. The externally shear stress is 40 MPa (equivalent to a total force of F = 0.01 N applied to the top edge). This force produces a compressive strain of 0.12 % at the fixed end of the nanowire. The material constants used in the calculation are: anisotropic elastic constants of ZnO: c_{11}= 207 GPa, c_{12} = 117.7 GPa, c_{13} = 106.1 GPa, c_{33} = 209.5 GPa, c_{44} = 44.8 GPa, c_{55} = 44.6 GPa, piezoelectric constants $e_{15} = -0.45$ C m^{-2}, $e_{31} = -0.51$ C $m^{-2}$and e_{33} = 1.22 C m^{-2}. Relative dielectric constants of ZnO are $k = 7.77$ and $k_{//} = 8.91$. The PMMA Young's modulus, Poisson ratio and relative dielectric constants are E = 3 GPa, v = 0.4 and k = 3.0, respectively.

In Fig. 3.61e, the left-hand side of the structure is affixed and the right-hand side is free, on which the vertical shear stress is applied at the top end surface. The top and bottom surfaces are electrodes so that the top and bottom surfaces are planes with equal potential. The bottom surface electrode is grounded. In case of open circuit, the total charges on the top and bottom surface must be zero.

This model is a capacitor-like plate structure with ZnO nanowires and PMMA as the dielectric components in the composite. By fixing one end of the plate and applying a transverse mechanical force (Fig. 3.61a), the mechanical deformation of the plate can be calculated. Under such a deformation, the distribution of the piezoelectric field in the nanowires is calculated by using the paired-nanowire model. The nanowires are placed in the region that is subject to a compressive strain, to reflect the experimental condition where the entire nanogenerator is excited by a compressive strain from the backside of the substrate. The distributions of the induced charges in the electrode plates at the top and bottom surfaces are then calculated with proper boundary conditions, so that the electric potential difference between the two plates can be obtained.

The coupling between the piezoelectric field and the inductive charges in the electric plates is ignored under the first-order approximation. The ZnO nanowires are assumed to be intrinsic without doping. To represent the random distribution of the nanowires on surface of the PMMA film, two nanowires with opposite *c*-axes directions are selected for the simulation, as shown in the model in Fig. 3.61b. The presence of the induced electric potential difference across the two plate electrodes is the driving force for the flow of electrons in the external load. Although the magnitude of the potential difference depends slightly on the relative depth of the nanowires in the nanogenerator due to variation in local strain, the physical principle is valid. If the strain in the nanwires is released, the piezoelectric field will disappear and the inductive charges in the electrode plates have to flow back. Thus, an AC output current is produced.

Due to the conical shape of the nanowires and the opposite *c*-axis, the piezopotentials inside the two nanowires are opposite in sign under compressive strain, but with a small off-set in centers of the charge in the direction normal to the substrate surface. This is the fundamental mechanism of creating the inductive charges at the top and bottom electrodes. By adjusting the display scale of the piezopotential in the space outside the nanowires, a 0.8 V induced potential difference across the two electrodes is clearly shown (Fig. 3.61d), which is generated by a compressive strain of 0.12 % at the fixed end of the nanowire. This is the driving force of the nanogenerator to produce the AC current.

To demonstrate that the conical shape of the nanowires is the key to the presence of the piezopotentials in nanogenerator, a cylindrical nanowire with zero conical angle is calculated as a comparison. No potential difference across the two electrodes has been observed (Fig. 3.61e). It is also found that the voltage across the top and bottom electrodes is approximately proportional to the thickness projected density of the nanowires, if the projection of all the nanowires in the composite cannot form a dense monolayer (Fig. 3.61f) [59].

3.8.1.4 Experimental Output

The nanogenerator with only one cycle of deposition of the ZnO nanowires, corresponding to an area density of 1,400–1,500 nanowires per mm^2, is tested. In this case, the density of the nanowires is rather much lower than that of a uniform and fully packed monolayer, which is about 90,000 per mm^2. The measured output voltage is about 0.25 V and the output current is about 5 nA with a strain of 0.11 % at a straining rate of 3.67 % s^{-1}. The output of the nanogenerator is approximately linearly proportional to the thickness projected density of the nanowires (Fig. 3.62a), in agreement with the theoretical prediction (Fig. 3.61f). For the nanogenerator with five cycles of deposition of the nanowires, corresponding to a thickness projected density of 7,000–7,500 nanowires per mm^2, the output voltage is increased to 1.5 V and the current to 30–40 nA (Figs. 3.62b, c). The output current can be rectified by using a diode (Fig. 3.62d) and stored for later use. The device has a high stability (Fig. 3.62e).

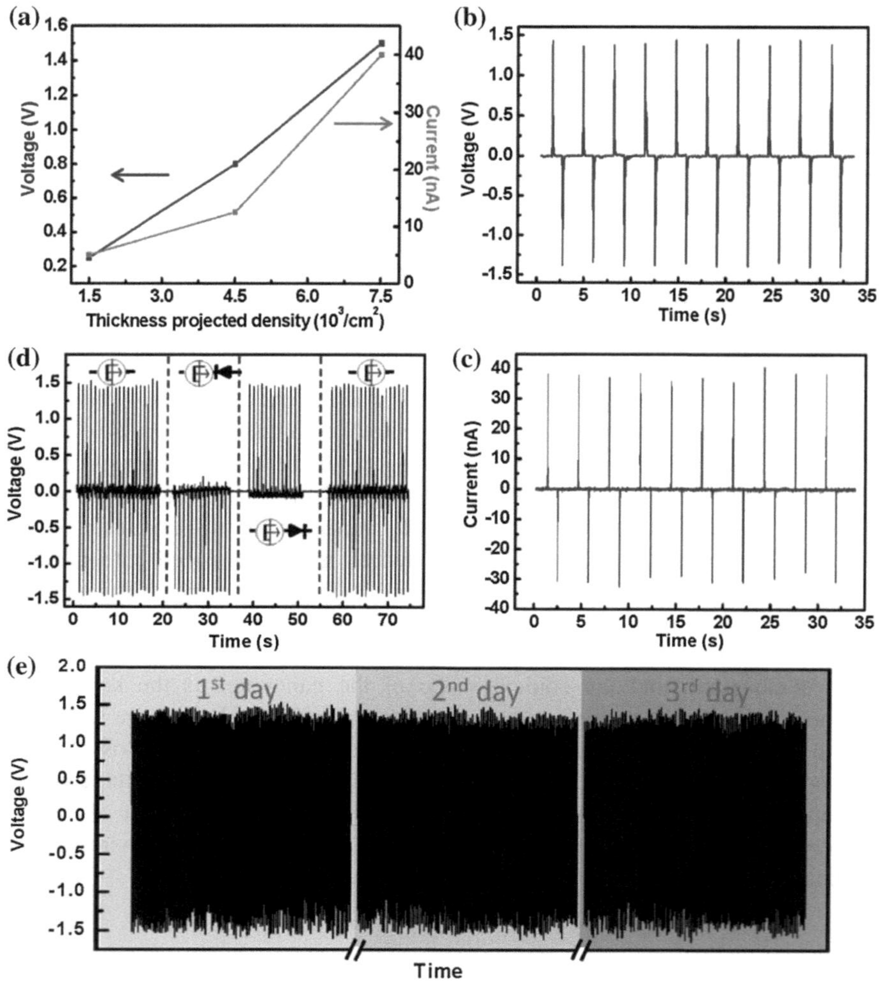

Fig. 3.62 Performance of the nanogenerator with noncontact ZnO nanowires. **a** Measured output of the nanogenerator as a function of thickness projected density of the conical ZnO nanowires. **b** Measured output voltage and **c** output current of the nanogenerator with a thickness projected density the nanowires to 7000–7500/mm^2. **d** Output voltage of the nanogenerator before and after being rectified with a diode. **e** Long-term stability of the nanogenerator. Courtesy of Dr. Z. L. Wang, Georgia Institute of Technology. Reproduced with permission from [59], Copyright @ 2010, American Chemical Society

3.8.1.5 Applications

The output power of the nanogenerator is sufficient to drive a LCD screen. LCD is a nonpolar device that can be driven by an AC power as long as its output potential exceeds a threshold value. A LCD screen taken from a calculator has been used to test the nanogenerator. The LCD screen is directly connected to the nanogenerator without involving any other external sources or measurement meters. The LCD blinks corresponding to the application of an external mechanical excitation to the nanogenerator.

The nanogenerator can continuously power a LCD. LCD screen is a capacitive device that can release the input charges within a certain period of time, during which it continuous lights up if the electric field generated by the remaining stored charges is above the threshold. This is possible if the LCD discharging time is longer than the time interval between two consecutive mechanical excitations.

3.8.1.6 Summary

The nanogenerators with noncontact conical ZnO nanowires can be constructed on any types of substrates. The composited structure of such nanogenerator makes them much more flexible than those with ZnO thin films. Because the nanowires are entirely enclosed with polymer matrix, there is no direct contraction with the electrodes. This noncontact design therefore offers robust nanogenerators with high stability. In addition, the fabrication process is simple and cost-effective and thus highly potential to be scale up for large-scale applications.

3.8.2 Hybrid Nanogenerators

3.8.2.1 Introduction

There is an abundance of energies, including light, thermal, mechanical (such as vibration, sonic wave, wind and hydraulic), magnetic, chemical and biological, surrounding us. Harvesting these energies is of critical importance for long-term energy consumption and sustainable development of our society. Various materials and technologies have been developed to convert solar and mechanical energies into electricity. Photovoltaic cells have been constructed by using materials and structures, such as inorganic *p–n* junctions, organic thin films and organic–inorganic hetero junctions. Mechanical energy harvesters have been designed based on principles of electromagnetic induction and piezoelectric effect. These approaches are used as independent technologies, because of their drastically different physical principles in harvesting specific types of energy. A solar cell works only under sufficient light illumination. A mechanical energy generator relies on sufficient mechanical movement/vibration. It is desired to develop hybrid devices that can harvest different types of energies, so that the energy resources can be effectively and complimentary utilized whenever and wherever one or all of them are available. Such hybrid nanogenerators have been developed to simultaneously harvest solar, mechanical and chemical energies. One of their applications is to develop wireless self-powered systems. A nanowire-based hybrid cell has been demonstrated to harvest both solar and mechanical energies [60].

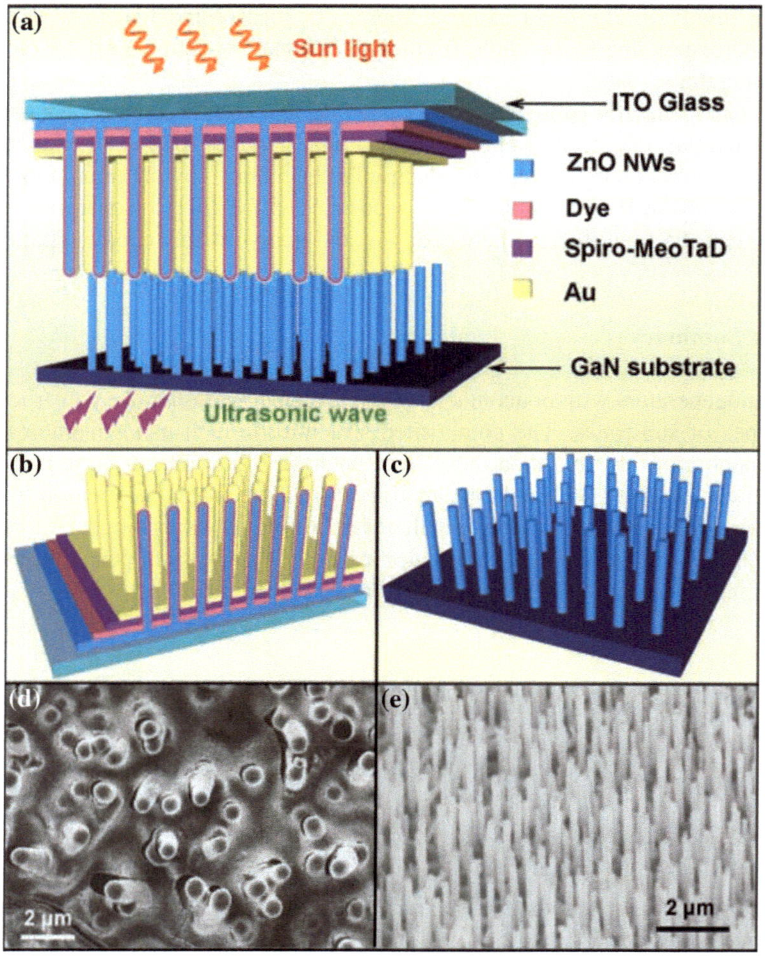

Fig. 3.63 Design of the compacted hybrid cell (CHC) structure composed of a dye-sensitized solar cell (DSSC) and a nanogenerator (NG). **a** Schematic illustration of a CHC, which is shined by sunlight from the top and excited by ultrasonic wave from the bottom. The ITO layer on the DSSC part and GaN substrate are used as the cathode and anode of the CHC. **b** Schematic illustration of a solid state DSSC. **c** Schematic illustration of a vertical aligned ZnO nanowire array grown on GaN substrate. **d** Top view SEM image of the DSSC. **e** SEM image of the ZnO nanowire array for the nanogenerator. Courtesy of Dr. Z. L. Wang, Georgia Institute of Technology. Reproduced with permission from [60], Copyright @ 2011, John Wiley & Sons

3.8.2.2 Structure Design

It consists of a dye-sensitized solar cell (DSSC) and a piezoelectric nanogenerator. The nanogenerator is constructed by using aligned ZnO nanowire arrays grown on surfaces of a flat substrate, on which a DSSC cell is integrated. The DSSC harvests solar energy irradiating on the top and the nanogenerator harvests ultrasonic wave energy from the surroundings. They can work simultaneously or individually. In addition, they can be integrated in parallel or in serial to have higher output current and voltage, respectively. Under a light illumination of a simulated sun emission (100 mW cm^{-2}), the optimum power is enhanced by 6 % after incorporating the contribution of the nanogenerator. This research sets a platform toward multimode energy harvesting as practical power sources.

The design of the compacted hybrid cell (CHC) is to convolute the roles of the nanowire arrays to simultaneously perform their functionalities in a nanogenerator and a DSSC, as shown in Fig. 3.63 [60]. The main frame of the structure is made of two sets of ZnO nanowire arrays that are configurated teeth-to-teeth (Fig. 3.63a), similar to the fiber-based nanogenerators discussed previously. The DSSC cell is built with the top nanowire array as a template. A thin metallic layer deposited on the nanowire array as the electrode for the nanogenerator. The bottom nanowire array is used as piezoelectric structure to convert mechanical energy into electricity. Solar energy illuminates the device from the top and the ultrasonic wave is applied from the bottom.

To fabricate the solid state DSSC, vertically aligned ZnO nanowires are grown on an ITO-coated glass substrate, by using the hydrothermal method described before. After a standard cleaning process, the substrate is coated with a thin film of ZnO with 200 nm in thickness by using magnetron RF sputtering. Then, ZnO nanowires are synthesized by using the chemical route discussed above at 80 °C for 24 h. After rinsed with acetone, the nanowires are immersed in a 0.5 mmol L^{-1}sensitized solution of $(Bu_4N)_2Ru(dcbpyH)_2(NCS)_2$ (N719 dye) in ethanol for 1 h to load the dye. After that, an amorphous organic hole-transport material, 2,2′,7,7′-tetrakis(*N*,*N*-di-*p*-methoxyphenyl-amine) 9,9′-spirobifluorene (OMeTAD) is on the dye-sensitized ZnO nanowires by using spin-coating at a spin rate of 2,000 rpm for 60 s, which is then baked at 100 °C to get rid of the organic contents. As shown schematically in Fig. 3.63b, a cone shaped surface is thus created. A continuous gold (Au) thin film (60 nm in thickness) is coated to form a cone shapes electrode, as required for the nanogenerator. Figure 3.63d shows SEM image of the cone shape surfaces, with a spacing between the cones to be 1–2 μm.

ZnO nanowires (Fig. 3.63c) on the GaN (0001) surface are 2–3 μm in length, with an interspacing between the nanowires to 400–700 nm. The two sets of nanowire arrays are stacked interdigitatively face-to-face. The Au-coated cone shaped electrode on the top serves as a “zig-zag” electrode to mechanically trigger the nanowires in the array at the bottom (Fig. 3.63a). The nanogenerator is excited by ultrasonic wave.

The connection of the DSSC and nanogenerator in serial to form a CHC is presented in Fig. 3.64a, in which the ITO serves as the cathode while the silver (Ag) paste in contact with the GaN substrate serves as the anode. After connecting to output wires, the entire CHC is encapsulated and packaged with epoxy resin to prevent infiltration of liquids, while the window of the DSSC is open to solar illumination.

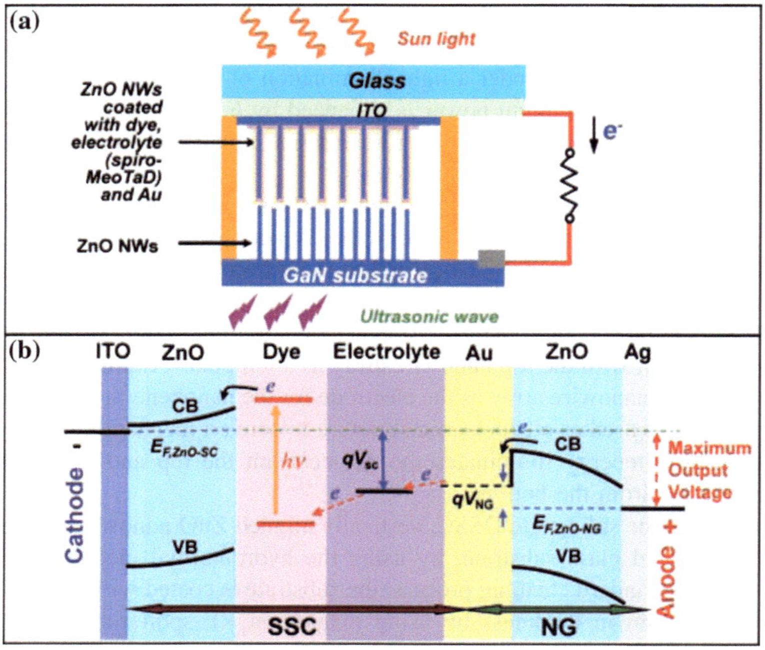

Fig. 3.64 Design and physical principle of the CHC. **a** Schematic structure of the CHC. **b** Electron energy band diagram of the CHC, showing that the maximum output voltage is a sum of those produced by the DSSC and the nanogenerator. The abbreviations are: conduction band (*CB*), valence band (*VB*), and Fermi level (*EF*). Courtesy of Dr. Z. L. Wang, Georgia Institute of Technology. Reproduced with permission from [60], Copyright @ 2011, John Wiley & Sons

3.8.2.3 Mechanism

The working principle of the CHC is described by using the electron energy band diagram shown in Fig. 3.64b. The electrons are driven by the piezoelectric potential and photovoltaic potential consecutively through the two devices. The maximum achievable output voltage is the difference between Fermi level of the ZnO nanowires in the DSSC (EF, ZnO–SC) and that of the ZnO nanowires in the nanogenerator (EF, ZnO-NG). The output voltage of the whole device is the sum of the output voltages of the nanogenerator and the DSSC.

In the nanogenerator part, the gap between the Fermi level of the ZnO nanowires (EF, ZnO-NG) and that of the Au determines the maximum voltage output of the nanogenerator (VNG). The Au–ZnO junction forms a Schottky contact because Au has a work function of 4.8 eV that is higher than the electron affinity of ZnO (4.5 eV), which serves as a “gate” to block the back flow of electrons. When the Au electrode slowly pushed the nanowires to apply a strain field across the nanowire width, outer surface is in tensile strain and the inner surface is in compressive strain. The piezoelectric potential at the compressive side of the nanowire sets the Schottky contact as forward biased and drives the electrons to across the Au–ZnO junction. Through an electron-transfer process, these charge carriers continuously transport in the solid state electrolyte into the DSSC.

In the DSSC part (Fig. 3.64b), the maximum voltage output (VSC) is determined by the difference between the ZnO Fermi level (EF, ZnO-SC) and the electrochemical potential of the electrolyte. Visible-light absorption by the dye sensitizer excites electron transfer to the conduction band of ZnO. The electron injection from the excited sensitizer into ZnO is followed by regeneration of the dye sensitizer by the electron from OMeTAD. The electrons in the conduction band of ZnO and the holes in the electrolyte are separated and subsequently transported to the contact electrodes.

3.8.2.4 Characterization

Both the DSSC and the nanogenerator units of the CHC can work independently and concurrently. The CHC is characterized by affixing it on the water surface in an ultrasonic generator cavity, with the transparent DSSC side facing the source of stimulated sunlight and the nanogenerator side in direct contact the water, where an ultrasonic generator with a frequency of 41 kHz is applied from the bottom side (Fig. 3.63a). J–V curves of the CHC, DSSC and NG are recorded individually. Short-circuit current is measured by serially connecting the CHC to a DS345 30 MHz synthesized function generator, with a resistance of 50 Ω sweeping from −1 to 1 V as an external load. All the signals are converted through a BNC-2120 analog-to-digital converter. Photovoltaic performance of the DSSC is characterized under a simulated sunlight illumination without applying the ultrasonic wave. The Open-circuit voltage (UOC-SC) is 0.42 V and the Short-circuit current density (JSC-SC) is 0.25 mA cm^{-2}. The fill factor of the DSSC reaches 30.6 %, corresponding to an overall energy conversion efficiency of 0.03 %.

The nanogenerator is characterized by applying the ultrasonic wave without the illumination of sunlight. The J–V curve indicates that the nanogenerator has an output voltage of about 19 mV and an output current density of about 0.3 pA cm^{-2}. No matter whether the sunlight is turned on and off, the corresponding output of the nanogenerator has no change, indicating that there is no contribution from the DSSC to the nanogenerator, when only the nanogenerator is tested. Furthermore, an *I–V* curve of the NG has also been recorded at near-zero-point region when the device is in dark without applying ultrasonic wave. There results further confirm that the DSSC and the nanogenerator units in the CHC can work independently, when only one type of energy source is available.

Technological feasibility of the CHC has been demonstrated by simultaneously performing the harvesting of solar and mechanical energies. Under the shinning of sunlight without the ultrasonic wave, the CHC has an output voltage of 0.415 V and a current density of 252 μA cm^{-2}. When both the ultrasonic wave and sunlight are turned on, the output voltage is increased to 0.433 V, while the current density is still 252 μA cm^{-2}. The difference in the output voltage of the CHC by 19 mV, before and after turning on the ultrasonic wave, is the output voltage of the nanogenerator without the illumination of sunlight. This has been confirmed by the measurement of J–V between the cathode and anode of the DSSC unit without including the nanogenerator unit.

It has been shown that the CHC has an enhanced the energy harvesting performance, as compared with those of either one of the subunits. When only the DSSC component is in operation under full illumination, the optimum output power density is 32.5 μW cm^{-2} at short-circuit current density $J_{SC} = 140$ μA $cm^{-2}$and open-circuit voltage $U_{OC} = 0.231$ V. When both the DSSC and the nanogenerator are operating in serial connection, the output power density becomes 34.5 μW cm^{-2} at $J_{SC} = 141$ μA cm^{-2} and $U_{OC} = 0.243$ V. An increment (ΔP_{HC}) of 2 μW cm^{-2} in power density is thus achieved after turning on the ultrasonic wave, which is 6 % enhancement in optimum power. Therefore, the CHC has an added-up total power outputs from both the DSSC and the nanogenerator.

3.8.3 Nanogenerators for Self-Powering

3.8.3.1 Concept of Self-Powering

An integrated system can has more than one function, which consists of multiple components, such as sensors, transducers, data processors, control units and communication systems. As the size of components shrinks to the micro or even nanometer scale, the power consumption also drops significantly. Such a drop in power consumption level makes it possible to powder the devices by using the energy harvested from the surrounding sources, such as airflow, vibration, sonic wave, solar, chemical and thermal energy. This is called self-powered system. Therefore, a self-powered system can work independently, sustainably and wirelessly without using energy storage devices, such batteries and supercapacitors.

An integrated self-powered system can be simply described in Fig. 3.65a [61]. The power source in this system includes energy harvesting and storage units. The harvester scavenges certain kind of energy (solar, thermal, mechanical or chemical) from the surrounding environment and stores it in the energy storage unit. Thus the collected power can be used to power other parts of the system. The sensors detect the changes in the environment, while the data processor and controller analyze the information. Then, the signal is sent out by the data transmitter, and simultaneously the response is received. Therefore, it is self-sustainable system. Significant progress has been made in developing self-powered systems by using nanogenerators with piezoelectric ZnO nanowires. An example of such self-powered systems is demonstrated in Fig. 3.65b [61]. It consists of a nanogenerator to harvest mechanical energy, integrated with a low-loss full-wave bridge rectifier, a capacitor for storing the energy, an infrared photodetector and a wireless data transmitter.

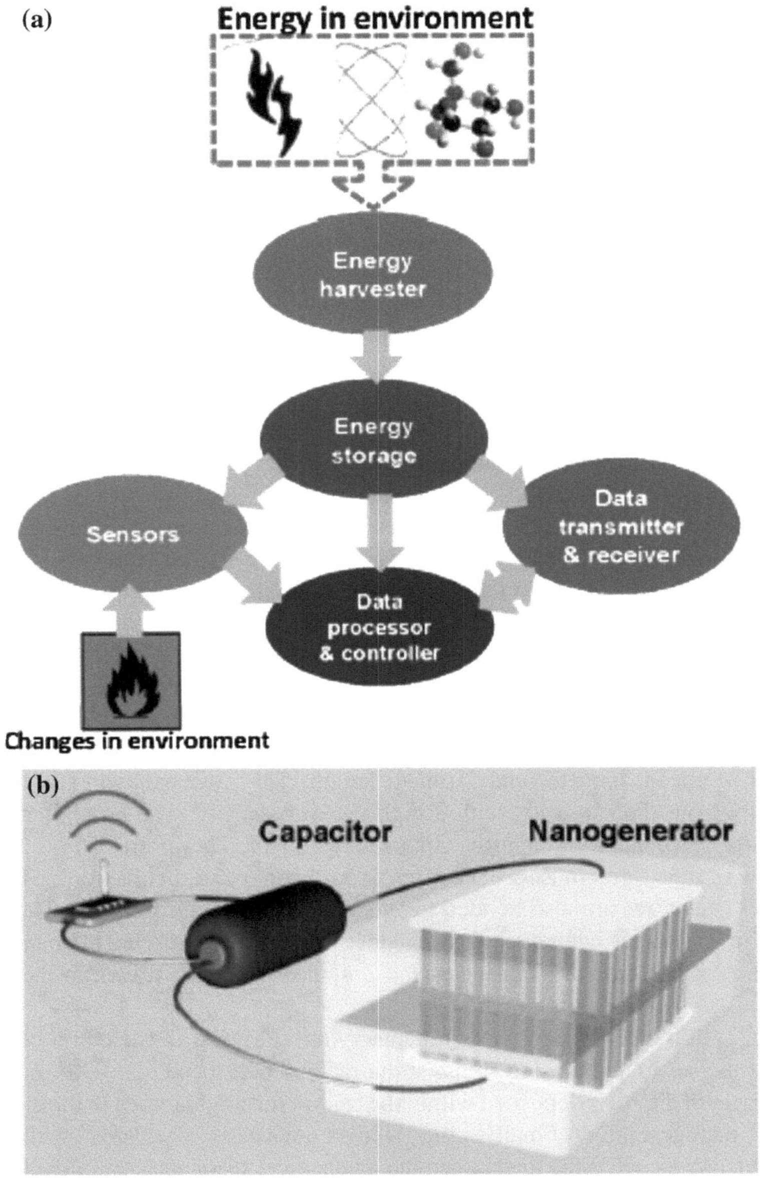

Fig. 3.65 Schematic diagram of the integrated self-powered system. **a** Components of a self-powered system: energy harvester, energy storage, sensors, data processor and controller, and data transmitter and receiver. **b** Prototype of an integrated self-powered system by using a piezoelectric ZnO nanogenerator as the energy harvester. Courtesy of Dr. Z. L. Wang, Georgia Institute of Technology. Reproduced with permission from [61], Copyright @ 2011, American Chemical Society

The power generated by a nanogenerator may not be sufficiently high to continuously power a device, but the accumulation of charges generated over a period of time can be sufficient to power the device to work for a few of seconds. This is very suitable for applications in the fields of sensing, infrastructure monitoring and sensor networks. An integrated system can have many sensors, and each sensor is expected to work independently and wirelessly, but all of them will be linked through a network or internet. Each sensor is not required to work continuously and simultaneously, instead, it is only required to work instantaneously with a standby status and active status. The standby mode is normally longer, while the active mode is shorter. The energy scavenged and stored during the standby status can be used to drive it in the active mode. Therefore, nanogenerators can be used for such purposes.

3.8.3.2 The Design of Nanogenerator

The nanogenerator used in the integrated system is a free-standing cantilever beam made of a five-layer of densely packed ZnO nanowire textured films [61]. A piece of flexible polyester (PS) is used as substrate (220 μm in thickness), as shown in Fig. 3.66a. Firstly, a 5 nm thick Cr adhesion layer and a 50 nm thick ZnO seeding layer are deposited within the selected rectangular region of $1 \times 1\ cm^2$ on the top and bottom surfaces of the substrate. The ZnO seeding layer is used to induce the growth of densely packed ZnO nanowires by using a wet chemical method. Figure 3.66b shows SEM image of the as-grown ZnO nanowires. The nanowires are about 150 nm in diameter and 2 μm in length. The ZnO nanowires are grown vertically from the substrate, with a high packing density. The bottoms of these nanowires are bonded through the ZnO seeding layer. The insert of Fig. 3.66b is top view of the as-grown ZnO nanowires. It has been found that the top surfaces of the ZnO nanowires are also bonded together tightly as a uniform film. Therefore, the entire ZnO layer is a textured film, consisting of fully packed ZnO nanowires. According to the growth mechanism, the ZnO nanowires are grown in the *c*-axes direction, as shown in Fig. 3.66a.

Then, a thin layer of polymethyl methacrylate (PMMA) is coated on both surfaces of the substrate by using spin-coating method at a speed of 3,000 rpm. After that, a layer of Cr/Au is deposited within the central rectangular area as the electrode of the nanogenerator. Finally, the whole device is packaged with polydimethylsiloxane (PDMS) to enhance the mechanical robustness and flexibility. A photo of the nanogenerator is shown in the lower right corner of Fig. 3.66a. The size of the effective working area of the nanogenerator is $1 \times 1\ cm^2$. Two leads are connected to the top and bottom electrodes, respectively. Because of the low processing temperature (<100 °C), this process can be used for flexible electronics.

The distributions of piezopotentials in the nanowire films have been calculated in order to understand the working mechanism of the nanogenerator. The structure of the nanogenerator is modeled as a cantilever beam, with textured nanowire films on a common substrate. The films on the top and bottom surfaces have [0001]

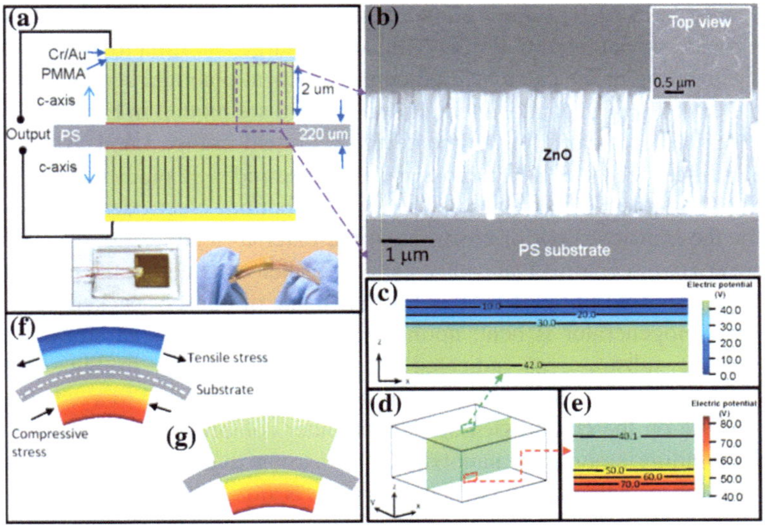

Fig. 3.66 Nanogenerator fabrication and its working mechanism. **a** Fabrication process of the nanogenerator. The *lower right* parts are photos of a nanogenerator after packaging. The bending of the nanogenerator shows its good mechanical flexibility. **b** Cross-sectional SEM image of the as-grown ZnO nanowire textured film on the substrate. The insert is top view of the nanowire film. **d** The model for simulation of working mechanism of the nanogenerator. It is a triple-layer *rectangular box*, containing the flexible substrate and the ZnO films on the top and bottom surfaces. The calculated local potentials across the top electrode and bottom electrode are presented in **c** and **e**, respectively. **f** Distribution of piezopotentials in the structure if the ZnO nanowires form a densely packed solid film that behaves like a continuous medium. The *white dashed line* in the substrate indicates the strain neutral plane. **g** Distribution of piezopotentials in the structure if the ZnO nanowires form a densely packed film but with small gaps or interwire sliding, so that the nanowires on the stretched side of the substrate may contribute nothing to the output voltage, while the nanowires on the compressed side always creates a piezopotential drop. Courtesy of Dr. Z. L. Wang, Georgia Institute of Technology. Reproduced with permission from [61], Copyright @ 2011, American Chemical Society

uniaxial texturing. The electric potential difference across the top and bottom electrodes is calculated when the structure is bent. Because the ZnO nanowires on the substrate are close-packed with top and bottom ends being tightly bonded together, they can be treated as a thin film. By doing this, the calculation can be simplified. The device is represented by using a rectangular box with a dimension of $500 \times 500 \times 224\ \mu m^3$, as shown in Fig. 3.66d. It is a triple-layer structure, including two single crystal ZnO thin films (2 μm thick each) on the top and bottom surface of the polymer substrate (220 μm in thickness). Since the thickness of the electrodes is 100 nm, they can be ignored in the calculation.

The strain distribution in the cantilever is nonuniform. The mean strain in the beam applied in parallel to the substrate along the *y*-axis (Fig. 3.66d) is about 0.2 %. To represent the metal electrodes in the nanogenerator, the top and bottom surfaces of the structure are set as equal potential planes, with the bottom one being grounded. In the case of open circuit, the total charge on the top and bottom surface must be zero. The ZnO films are assumed to be intrinsic without doping. The piezoelectric potential in the ZnO films has a large variation in magnitude. For easy visualization of the calculated potential, the local potential across the top and bottom electrodes of the nanogenerator under strain are plotted in Fig. 3.66c and e, respectively. The potential inside the device is plotted in Fig. 3.66d. An inductive potential difference of 83.8 V across the two electrodes is predicted by the calculation. This potential drop across the top and bottom electrodes is the driving force for the transient flow of electrons in the external load.

There are two scenarios that can be expected when considering the piezoelectric response of the ZnO films. If the ZnO film is made of densely packed nanowires, when the nanogenerator is bent, noting that the neutral plane of strain is at the center of the substrate, as indicated by white dashed line in Fig. 3.66f, the nanowire film on the stretched surface of the substrate is subject to a tensile stress, while the one on the compressed surface is under compressive stress.

If the bonding among the nanowires is so strong that a solid film is formed, noting that the growth direction of the nanowires is along the *c*-axis, a tensile stress perpendicular to the nanowires results in a compressive strain along the *c*-axis direction, thus creating a piezopotential drop from the roots of the nanowires to their top-ends, as shown in Fig. 3.66f. At the same time, a compressive stress is applied to the nanowires on bottom surface of the substrate, resulting in a tensile strain along the *c*-axis. As a result, the top-ends of the nanowires have a higher piezopotential than their roots. Therefore, the piezopotential drops in the top and bottom ZnO films have the same polarity, so that they add up constructively. This piezopotential distribution will produce induced charges in the top and bottom electrodes and consequently generate the output voltage.

Alternatively, if the bonding among the nanowires is very weak, with the presence of interwire sliding/gaps, no piezoelectric potential is produced in the film on top surface of the substrate that is under tensile stress. However, the nanowires still can squeeze one another due to their full packaging, a piezopotential drop is thus created, but with a certain degree of degradation, in the film at bottom surface of the substrate that is subject to a compressive stress (Fig. 3.66g). Therefore, a potential drop is also present between the top and bottom electrodes, but with a decreased magnitude by more than half as compared to that in the first case. The actual bonding condition among nanowires should be between the two scenarios. In addition, the as-grown ZnO nanowires usually are *n*-type semiconductor, which can significantly screen the higher piezopotenital side, with almost no effect on the lower piezopotential side. Because of these, the observed output voltage of the nanogenerator will be smaller than the theoretically predicted value.

3.8.3.3 Charge Storage with Supercapacitor

It has been shown the energy generated by using the nanogenerators can be stored with supercapacitors [62]. The supercapacitor is fabricated by using multiwalled carbon nanotubes (MWNT) coated ITO electrodes and PVA/H_3PO_4 electrolyte.

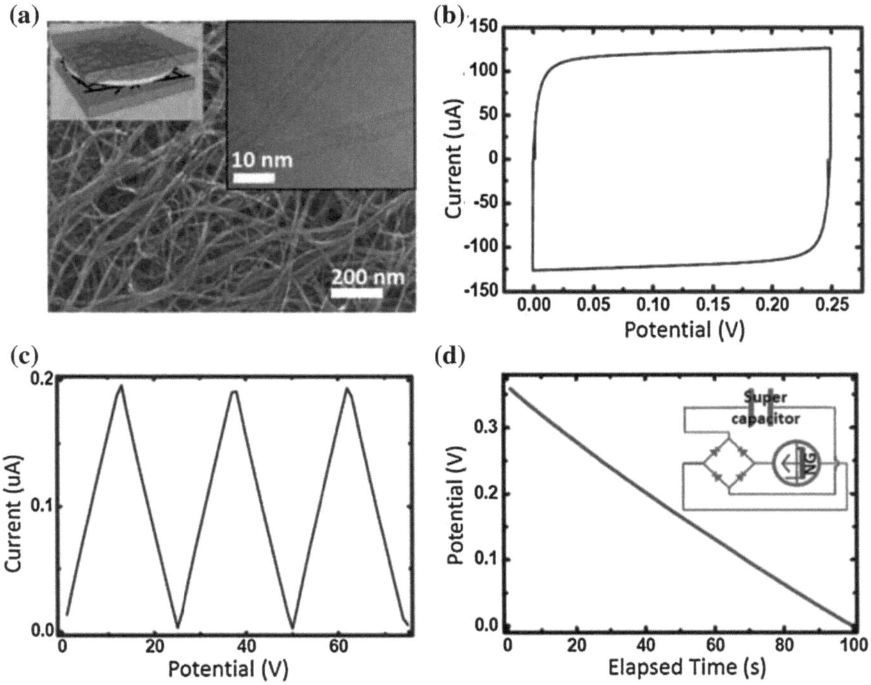

Fig. 3.67 Characteristics of multi-walled carbon nanotube-based supercapacitor and storage test combined with a nanogenerator. **a** Schematic diagram and SEM/TEM images of the carbon nanotube network-based supercapacitor. **b** Cyclic voltammetry of the supercapacitor. **c** Galvanostatic charge–discharge characteristics of the supercapacitor. **d** Discharging graph of the supercapacitor, which has been charged by a ten-layer integrated nanogenerator for about 1 h. Courtesy of Dr. Z. L. Wang, Georgia Institute of Technology. Reproduced with permission from [62], Copyright @ 2011, The Royal Society of Chemical

Direct spray coating method has been applied to make uniformly dispersed MWCNT layers on ITO glass substrates (Fig. 3.67a). The MWCNT coated ITO glass substrates are used as working and counter electrodes of the supercapacitor. PVA/H_3PO_4 polymer like solution is used as electrolyte. Right upper inset in Fig. 3.67a shows schematic diagram of the structure of the supercapacitor. Electrochemical properties of the supercapacitor are characterized by using a potentiostat/galvanostat. Figure 3.67b shows cyclic voltammetry, indicating that it has high electrochemical stability and high capacitance. The scan range is between 0 and 250 mV, with a scan rate of 100 mV s^{-1}. Figure 3.67c shows galvanostatic charge–discharge measurement of the MWCNT-based supercapacitor. To store the generated charge generated by the nanogenerators, a supercapacitor and 10-layer-integrated nanogenerators are connected with rectifying bridge (inset in Fig. 3.67d). The generated AC outputs from the nanogenerators are rectified into direct current (DC) signals, which can be stored in the supercapacitor.

3.8.3.4 Sensor Applications

Applications of self-powering by using nanogenerators have been demonstrated [62]. One example is wireless data transmission by using the energy harvested with a nanogenerator. A single transistor radio frequency (RF) transmitter is used to send out the detected electric signal. The oscillation frequency is tuned to be 90 MHz, while a commercial portable AM/FM radio (CX-39, Coby) is used to receive the transmitted signal. When the transmitter is triggered by the stored energy harvested with the nanogenerator, there is a disturbing noise received by the radio. Due to the low power consumption (<1 mW) of the transmitter, the energy generated and stored during three straining cycles of the nanogenerator is high enough to transmit the signal. The maximum transmission distance is 5 m.

It is further demonstrated that the self-powered system can work independently and wirelessly, by using a signal transmitting and receiving experiment [62]. To use the nanogenerators as power supplier, three factors are important to determine the power output performance: length of the nanowires, thickness of the substrate and mechanical strain applied to the nanogenerator. In practical applications, there are two ways to trigger the nanogenerators, depending on the form of the mechanical energy to be scavenged by the nanogenerators from environment. If the mechanical energy source has a constant stress, such as air flow, the piezopotential between the two electrodes increases with increasing length of nanowires and decreasing thickness of the substrate. If applied strain is at constant, such as the vibrations of a bridge which is rigid, the piezopotential varies in the opposite trend. In other words, by adjusting the two competing factors, thickness of the substrate and length of the nanowires, the nanogenerators can be optimized to have maximized power harvesting efficiency, according to the character of the scavenged energy. Increasing strain can also significantly enhance the output voltage. In addition, ZnO is a biocompatible and environmentally friendly material. The nanowire films can be grown at very low temperature (<100 °C) on various substrate with different shapes. Therefore, the ZnO nanowire-based nanogenerators have high potential industrial applications in flexible/stretchable electronics and many other areas.

3.8.3.5 Environmental Applications

Another example is to use nanogenerators to self-power an environmental sensor to detect Hg^{2+} in water [62]. A single-walled carbon nanotube (SWNT)- based FET is used as Hg^{2+} sensor and ZnO nanowire array- based nanogenerator is used to harvest mechanical energy from the environment. The concentration of Hg^{2+} ion can be quantitatively evaluated according to the intensity of a LED.

A circuit with two independent loops has been designed to accomplish the self-powered environmental sensor with nanogenerators. The circuit has two loops, one to connect the nanogenerator and the rectifying bridge to generate and store energy and the other to connect the detector. When detecting Hg^{2+} ion, the LED is

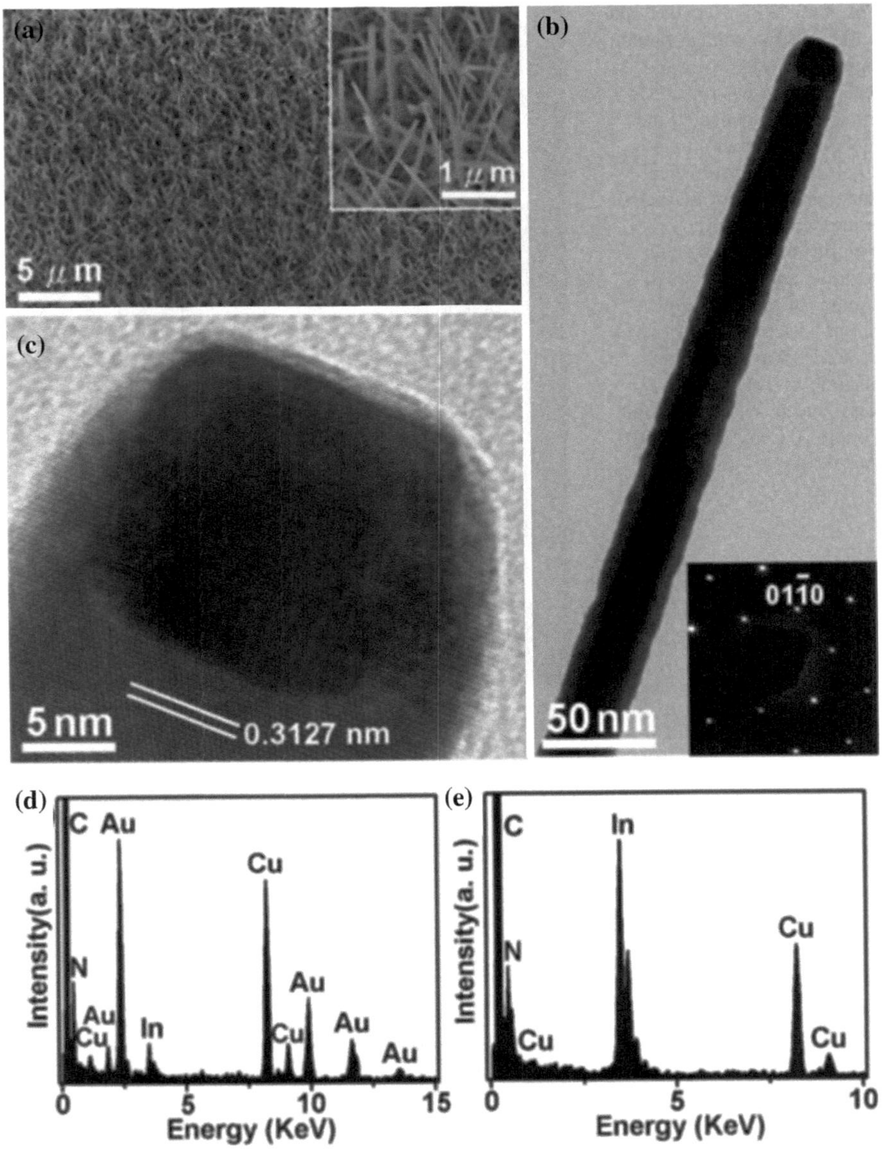

Fig. 3.68 Crystal structures and compositions of the InN nanowires. **a** 30° tilted view of the InN nanowires. The inset is an image at high resolution. **b** Low-magnification TEM image of the InN nanowire. The inset is the corresponding SAED of InN nanowire with [0001] zone axis. Growth direction of the InN nanowire is along [01$\bar{1}$0]. **c** Atomic resolution TEM image of the InN nanowire with a catalyst embedded in its tip. **d**, **e** EDS spectra of the tip region and body region of the InN nanowire in (**b**). Courtesy of Dr. Z. L. Wang, Georgia Institute of Technology. Reproduced with permission from [65], Copyright @ 2010, John Wiley & Sons

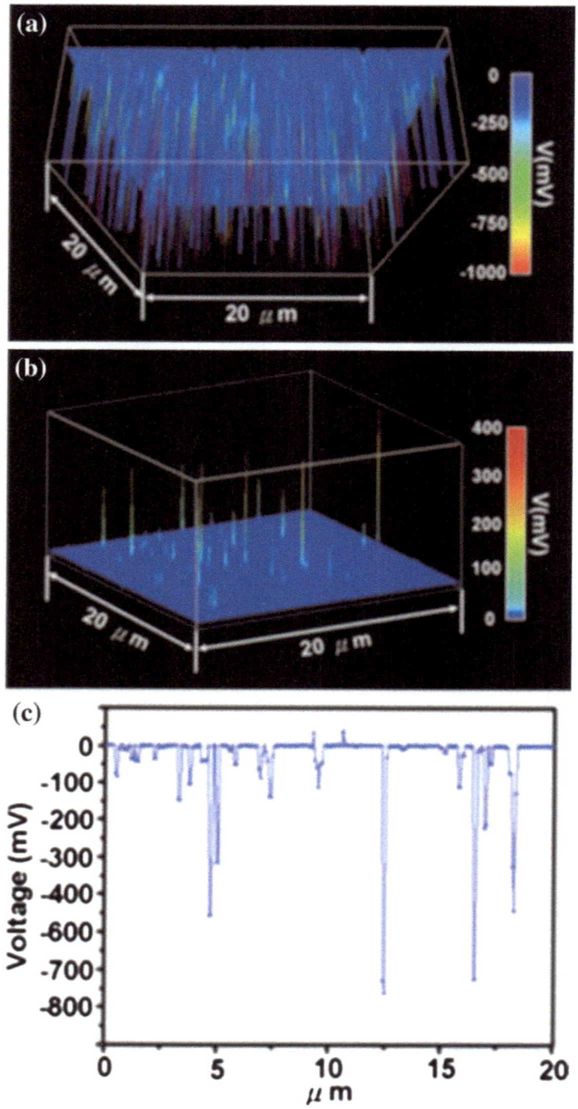

Fig. 3.69 **a** 3D Negative and **b** 3D positive voltage output signals recorded over an external load when the AFM tip scans and deflected the InN NWs with an area of 20 × 20 μm^2. **c** A typical line scan profile of the output voltage, revealing the majority of output signals are negative and only a small number of positive signals can be observed. Courtesy of Dr. Z. L. Wang, Georgia Institute of Technology. Reproduced with permission from [65], Copyright @ 2010, John Wiley & Sons

lightened. The intensity of the LED increases with increasing concentration of Hg^{2+} in the tested water. In addition, the LED can be replaced by an RF unit so that the signal can be detected wirelessly over a longer distance.

3.9 Nanogenerators with Other Nanowires

3.9.1 Wurtzite Nanowires

Besides ZnO, nanowires of CdS [63], GaN [64] and InN [65] in the wurtzite family have also been synthesized and characterized as nanogenerators. Piezoelectric outputs of the GaN and InN nanowires are particularly encouraging. Figure 3.68 shows microstructure and composition of the InN nanowires, deposited on Si substrates [65], with indium ingot and NH_3 as sources of In and N, by using VLS method. Au layer is used as catalyst. The nanowires are high quality single crystal. Almost all of the output signals are negative, with a maximum output voltage of about -1 V, as shown in Fig. 3.69. The mechanism of the charge output process of these non-ZnO nanowires is the same as that of ZnO nanowire.

3.9.2 Ferroelectric Nanowires

Although ferroelectric materials have much higher piezoelectric properties than ZnO, it is very difficult to synthesize nanowires with perfect crystal structure, as compared to ZnO. This is the reason why nanogenerators made with ferroelectric nanowires have rarely reported. However, recent development has demonstrated that ferroelectric nanowires or nanoribbons can be fabricated with various techniques and strategies. As a result, nanogenerators with ferroelectric nanowires or nanoribbons start to appear in the open literature. This section is intended to demonstrate the progress in fabrication and application of nanosized ferroelectric materials for nanogenerators. The ferroelectric materials include $PbZr_xTi_{1-x}O_3$ (PZT) [66–69], $BaTiO_3$ (BT) [70], 0.5 $Ba(Zr_{0.2}Ti_{0.8})O_3$–$0.5(Ba_{0.7}Ca_{0.3})TiO_3$ (BZT-BCT) [71] and $KNbO_3$ (KN) [72].

PZT nanowires (nanoribbons) have been fabricated by using a method, called photolithography and etching for nanoscale lithography (PENCiL) [66]. The fabrication process is shown in Fig. 3.70a. Ni thin film with a thickness of 100 nm is first deposited on certain types of substrates. An array of 1–2 μm wide resist lines is patterned by using the standard photolithography. Then, the Ni thin film is etched by using electrochemical method in concentrated phosphoric acid by undercutting the resist windows to form the nanowire structures. The shape and size of the nanowires can be well controlled by adjusting the applied voltage, patterns and heights of the photoresist and etching time. Nanowires of sub-40 nm in diameter can be fabricated by using this method, as shown in Fig. 3.70b. Because only photolithography and etching are involved, nanowires can be produced at large scale, as shown in Figs. 3.70c and 3.71. The Ni nanowires can be further polished by using laser annealing technique, which is called self-perfection by liquefaction (SPEL).

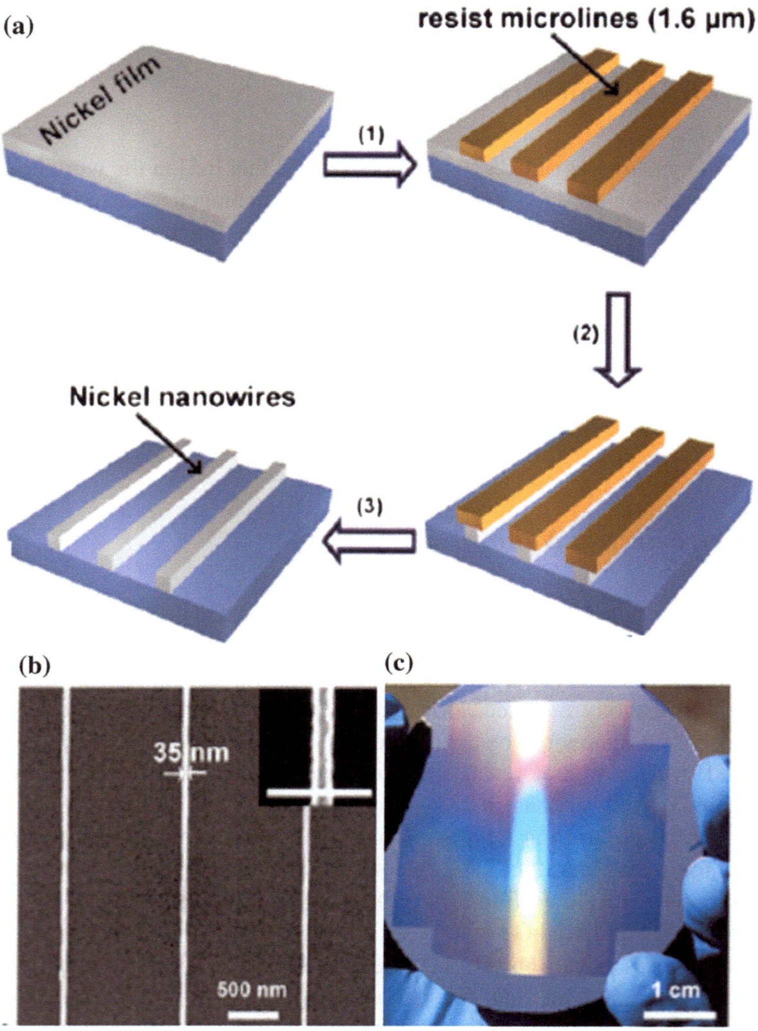

Fig. 3.70 Wafer-scale nanomask patterned by using PENCiL. **a** Schematic showing nanopatterning by using PENCiL: step 1, a Ni thin film is deposited on a substrate and photolithography is carried out on top; step 2, the Ni film is electrochemically etched to yield Ni nanowires undercut from the resist pattern; step 3, the resist is removed, revealing a Ni nanomask. **b** SEM image of the Ni nanowire arrays with diameters of 35 nm. The inset shows a single nanowire at a higher magnification. Scale bar is 200 nm. **c** Nanowires patterned by using PENCiL across an entire wafer form a grating which diffracts natural light. The horizontal lines are monolithically patterned contact pads. Reproduced with permission from [66], Copyright @ 2010, American Chemical Society

The formation of PZT nanowires is shown in Fig. 3.72 [66]. PZT thin film, with a composition of 52/48 Zr/Ti and thickness of 300–500 nm, is first deposited epitaxially on Pt/MgO or Si substrates, by using RF-sputtering. After the deposition, the

Fig. 3.71 SEM of the Ni nanowire arrays (*vertical*) patterned over a large scale. The *horizontal lines* are monolithically patterned contact pads. Reproduced with permission from [66], Copyright @ 2010, American Chemical Society

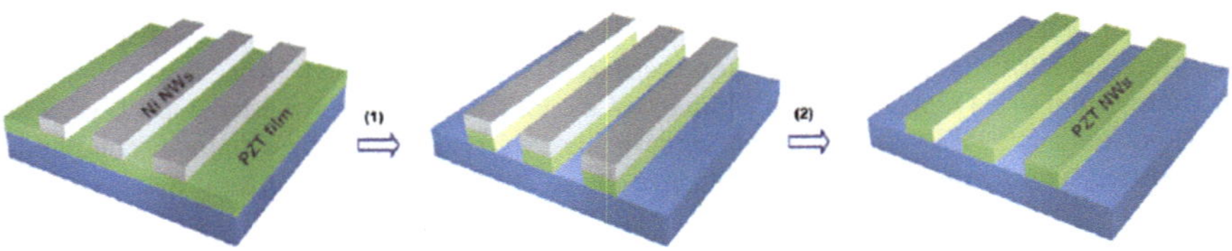

Fig. 3.72 Schematic showing the process to translate PZT nanowires from the Ni nanowire mask at large scale: Step 1, PENCiL is performed on top of a PZT thin film, and RIE is used to selectively etch the exposed PZT; step 2, the Ni nanowire mask is completely removed to reveal PZT nanowires. Reproduced with permission from [66], Copyright @ 2010, American Chemical Society

thin film is thermally annealed at 700 °C for 15 min to form a perovskite structure. Then, PENCiL is applied to the PZT thin film, which is followed by dry reactive ion etching (RIE) to remove the exposed part of the film, thus leading to PZT nanowire array. After removing the Ni nanowire mask by using 63 % nitric acid at 60 °C for 10 min, PZT nanowire arrays can be obtained. The highly ordered PZT nanowires fabricated at large scale are similar to the Ni nanowires shown in Fig. 3.63. The average diameter of PZT nanowires is 75 μm. EDS result indicates that the composition of the PZT nanowires has no significant change as compared to their thin film counterpart. The PZT nanowires have effective piezoelectric coefficient (d_{eff}) of 114 $pm\ V^{-1}$ and 145 $pm\ V^{-1}$ before and after poling at 10 V for 30 m.

The PZT nanoribbons prepared by using this method have been further made with wave geometry. Figure 3.73 shows fabrication process of wavy PZT nanoribbons [67]. PZT ribbons (5–10 μm wide and 250–500 nm thick) are patterned on a MgO host substrate and subsequently released from the mother substrate with phosphoric acid (85 % concentration, 75 °C, for ~50 s). A slab of PDMS

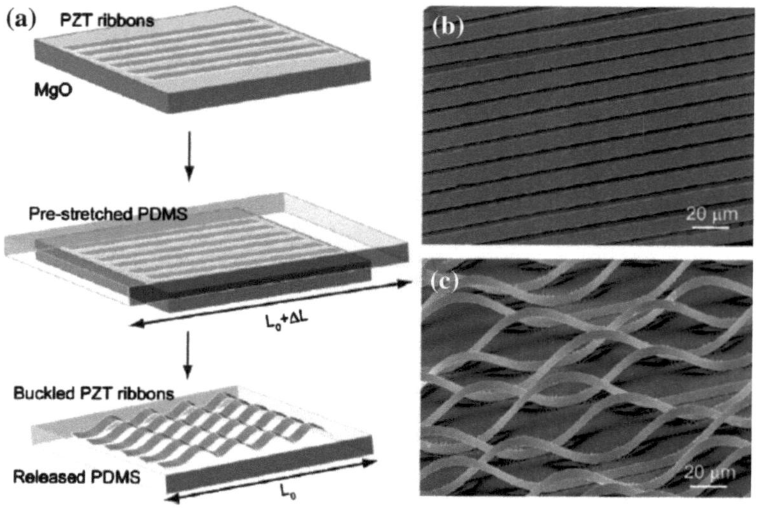

Fig. 3.73 Formation process of wavy/buckled piezoelectric PZT ribbons. **a** From *top* to *bottom*: PZT ribbons are patterned on an MgO substrate and undercut etched to release them from the mother substrate; a slab of prestrained PDMS is laminated against the ribbons and peeled off quickly; the retrieved PZT ribbons are transferred onto PDMS and formed wavy/buckled structures upon strain relaxation. **b** SEM image of the PZT ribbons transfer printed to PDMS with zero prestrain. **c** PZT ribbons spontaneously are buckled under prestrained conditions. Reproduced with permission from [67], Copyright @ 2011, American Chemical Society

($\sim$2 mm thick) is then elastically stretched and attached to the ribbons. The PZT ribbons are then transferred to the elastomer via adhesive van der Waals forces in the surface dominated ribbons, by peeling off the PDMS. Finally, the releasing of prestrain in the PDMS results in a compressive force in the PZT ribbons as the PDMS is relaxed to zero strain, leading to periodic de-adhesion and buckling. Due to the transfer of mechanical compressive energy into bending energy, the PZT ribbons become wavy. Figure 3.73b shows representative SEM image of the PZT ribbons transferred with unstrained PDMS, whereas Fig. 3.73c shows the PZT ribbons with a wavy/buckled structure induced by the prestrained PDMS. The buckled PZT ribbons have an enhancement in piezoelectric effect by 70 %. Nanogenerator based on the wavy ribbons exhibits an output current density of 2.5 $\mu A\ mm^{-2}$ at a strain of 8 % [67].

A similar "top-down" method has been used to fabricate PZT nanowires to assemble nanogenerators with graphene transparent electrodes [68]. PZT thin film, with composition of 52/48 Zr/Ti, is deposited on Si substrates by using a sol–gel method. The substrates coated with electrode, Pt (150 nm)/Ti (10 nm)/SiO_2 (300 nm)/Si wafer. The PZT solution, with a concentration of 0.4 M, is prepared with lead acetate trihydrate, titanium-isopropoxide and zirconium-*n*-propoxide. Spin-coating method is used to deposit the PZT thin film, at 4,000 rpm for 20 s, which is repeated for 6 times. Thermal treatment includes baking at 300 °C for 5 min and annealing at 650 °C for 30 min.

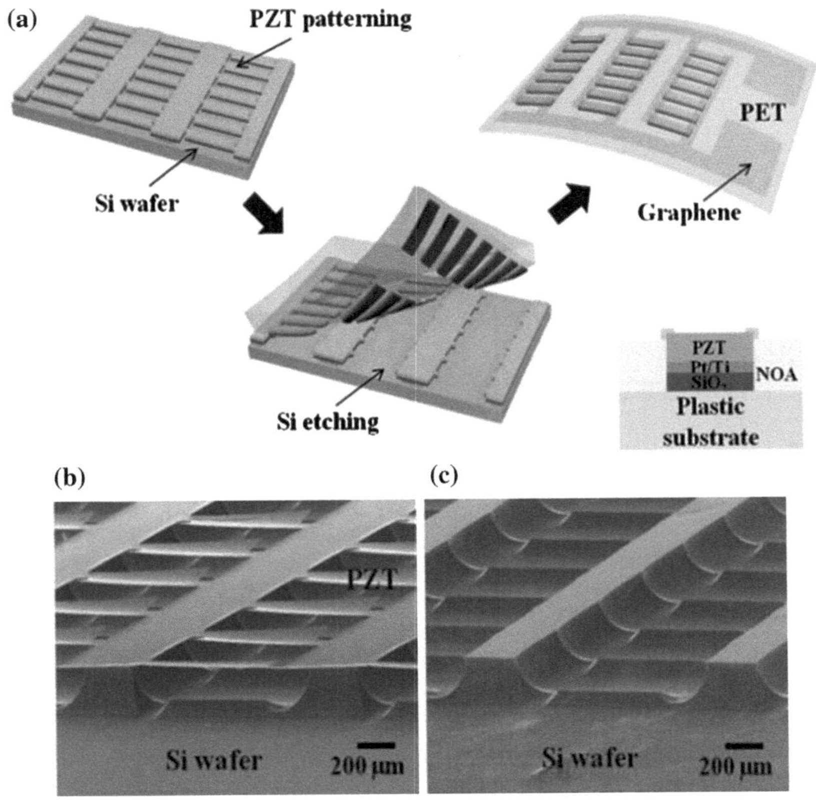

Fig. 3.74 Schematic illustration of the fabrication steps of a flexible PZT nanogenerator with graphene electrodes and optical images. **a** Overview of the fabrication process of the PZT nanogenerators and schematic cross-sectional image. **b** PZT nanoribbons after the lateral etching of the sacrificial layer (silicon). **c** The silicon wafer with the PZT nanoribbons being transferred. Reproduced with permission from [68], Copyright @ 2012, The Royal Society of Chemistry

Figure 3.74 shows the fabrication process of the PZT ribbons. The PZT thin film is first patterned by using the conventional photolithographic method, with dimension of 500 μm by 100 μm. PZT is etched out by using diluted HF solution, while Pt, Ti and SiO_2 layers are etched out by using inductively coupled reactive ion etching (RIE), with Ar, Cl_2 and CHF_3 gases, respectively. Lateral etching of Si is carried out by using a XeF4 dry method. After the etching process, PZT nanoribbons are transferred to Norland optical adhesion (NOA) coated polyethylene terephthalate (PET) flexible substrate with a PDMS stamp. Graphene sheets were patterned on PZT ribbons. Graphene sheets are used as electrodes, which are grown on Cu foil by using CVD method. To form multilayered graphene film, poly(methyl methacrylate) (PMMA) is coated on the graphene/Cu foil. After etching the Cu with ammonium persulfate (APS) solution, the PMMA-graphene film is transferred to another graphene/Cu foil. The step is repeated to produce

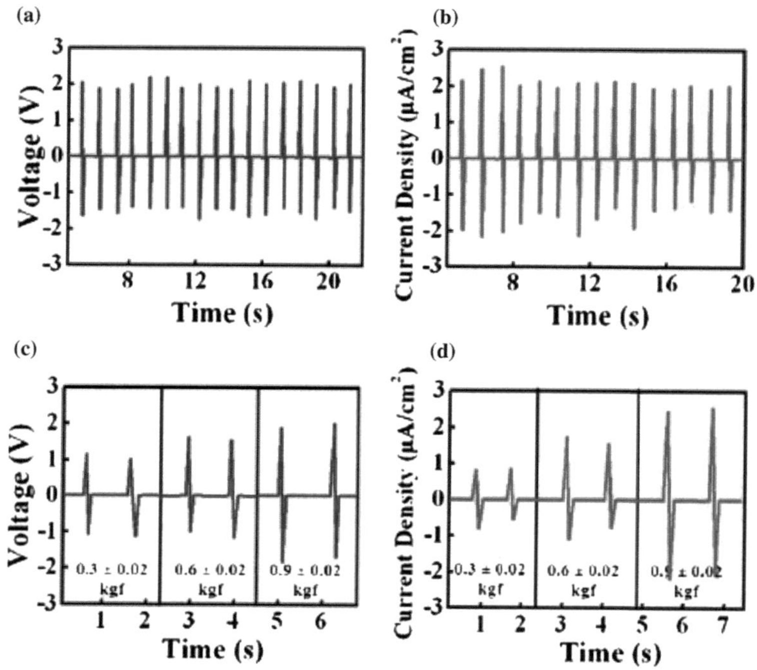

Fig. 3.75 Outputs of the PZT nanogenerator. **a** Voltage and **b** current density of the PZT nanogenerator with doped graphene electrodes when a dynamic load is applied on top of the generator by touch. **c** Voltage and **d** current density signals, at different pressures. Reproduced with permission from [68], Copyright @ 2012, The Royal Society of Chemistry

graphene film with desired thickness. After that, it is transferred onto plastic substrates. The PMMA support is removed by using boiled acetone up to 80 °C.

Flexible nanogenerator has been assembled with the PZT ribbons and the graphene electrodes. Voltage and current responses of the nanogenerator, when it is subjected to a cyclic compressive/releasing external force with respect to time, are shown in Fig. 3.75a and b, respectively. The generator has an output voltage ~2 V and an output current density ~2 mA cm^{-2}. The voltage and current responses of the nanogenerator to the change in the compressive force are shown in Fig. 3.75c and d. The voltage and current outputs increase with increasing compressive force. In addition, the generator exhibits good stability.

Before energy harvesting testing, the PZT ribbons are polled at a strong electric field at a temperature below the Curie point, by applying a voltage between the graphene interdigitated electrodes. When an external mechanical stress is applied on top surface of the nanogenerator, a nonuniform charge distribution due to the displacement of the positive-negative charge centers in the PZT ribbons is created. This generates net charges of opposite nature at the PZT/graphene interfaces. These charges are harvested through the interdigitated graphene electrodes, leading to the output voltage and current signals. When the external mechanical

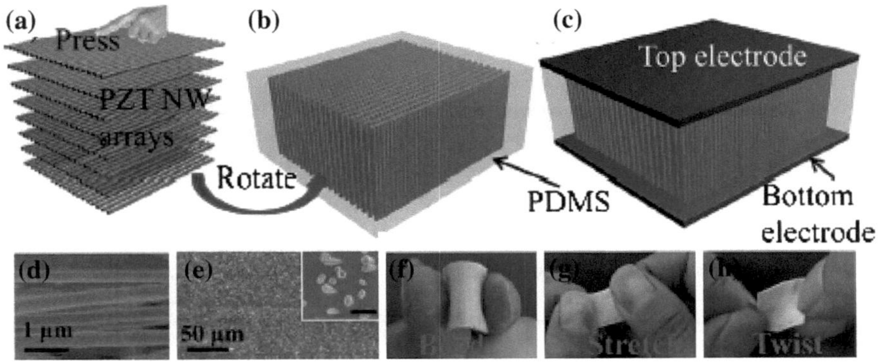

Fig. 3.76 Fabrication process and structure characterization of the nanogenerator. **a–c** Experiment setup to fabricate the high-output nanogenerator by using regionally orientated electrospun PZT nanofibers. **d** SEM image of the regionally oriented electrospun PZT nanofibers. **e** Top-view SEM images of the fabricated nanofibers. The inset is a magnified image with a scale bar of 2 μm. The *white spots* are the exposed tops of the PZT nanowires. **f–h** Optical photographs of nanowire arrays under different deformations, showing its flexibility and robustness. Reproduced with permission from [69], Copyright @ 2013, American Chemical Society

stress is released, charge carriers start to flow in the reverse direction, resulting in negative voltage and current responses, while piezopotential in the PZT ribbons disappears. The energy harvested by using the nanogenerator can be used to power a liquid crystal display (LCD) [68].

An alternative method to fabricate high-performance nanogenerator with PZT nanofiber array has been reported [69]. The PZT nanofibers with an average crystal size of 16 nm are synthesized by using electrospinning. Figure 3.76 shows fabrication process of the nanogenerator and characterization of the PZT nanofibers. Orientated PZT nanofiber films are stacked layer by layer to form a multilayer film, as shown in Fig. 3.68a. Then, diluted polydimethylsiloxane PDMS ($V_{PDMS}/V_{diluent}$ = 1:1) is filled into the PZT nanofiber films. The multilayered film is gently pressed to extrude the redundant PDMS without damaging its structure. The whole structure is heated at 80 °C for 1 h to solidify the PDMS. By doing this, a laterally aligned PZT multilayer film can be obtained. The samples are cut into small cuboids, which are rotated by 90° and then bonded laterally with PDMS to form vertically aligned ultra long PZT nanowire arrays, as shown in Fig. 3.76b. Nangenerator can be assembled by applying electrodes on both sides of the fiber arrays (Fig. 3.76c). Microstructure of the PZT nanofibers and mechanical strength of the nanofiber arrays are demonstrated in Fig. 3.76d–h. This kind of nanogenerators has potential application in flexible energy harvesting and self-powered systems.

After being poled at 5 kV mm^{-1}at 130 °C for 15 min, the nanogenerator can be tested for energy harvesting. A periodic impact is applied by using a heavy object. The working mechanism is shown in Fig. 3.77a, b. The heavy object produces an impact pressure is applied on the NG. The application of the pressure generates a piezoelectric potential due to the compressive strain of nanofiber array, leading to

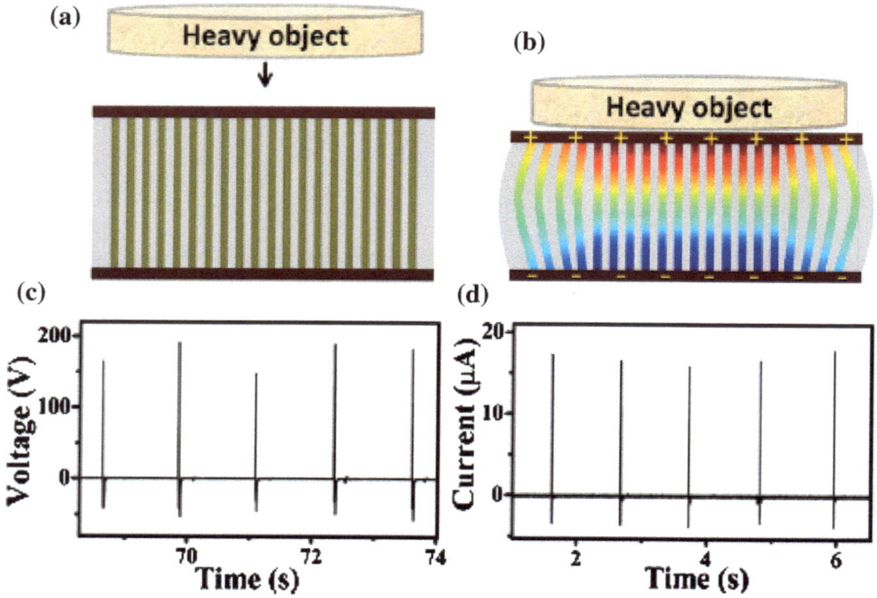

Fig. 3.77 Working principle and performance of the nanogenerator. **a** Schematic diagram of nanogenerator structure without deformation. **b** Schematic picture of piezoelectric potential distribution in a deformed nanogenerator when it is subject to an impact force. The *red* and *blue* color regions indicate a positive and negative piezoelectric potential, respectively. **c** Output voltage and **d** output current of the nanogenerator at a periodic pressure of 0.53 MPa. Reproduced with permission from [69], Copyright @ 2013, American Chemical Society

a flow of electrons through the external load. When the object is taken away from the generator, the pressure is released and the accumulated electrons flow back through the load in the external circuit. As a result, positive and negative current and voltage signals are detected, respectively. The nanogenerator with an area of 2.25 cm^2 shows a maximum peak output voltage of 198 V with an external load of 100 MΩ (Fig. 3.77c) and a peak output current of 17.8 μA with an external load of 100 Ω (Fig. 3.69d). The output voltage and current can be further increased to 209 V and 53 μA, which is new record of output of nanogenerators [51, 52].

The instant output of the nanogenerator has been used to stimulate frog's sciatic nerve to excite a synchronized contraction of frog's gastrocnemius. The nanogenerator can directly powder a commercial LED, without the need of energy storage, showing strong potential for practical applications [69].

Although PZT has best piezoelectric properties among ferroelectric materials, due to the toxicity of Pb, it will be replaced gradually by Pb-free materials. Therefore, more and more attentions have been paid to Pb-free ferroelectric nanosized materials for nanogenerator applications. Three examples will be discussed in this section [70–72].

$BaTiO_3$ is a typical ferroelectric material with strong piezoelectric properties. A nanogenerator, made with $BaTiO_3$ nanoparticles rather than nanowires, has been

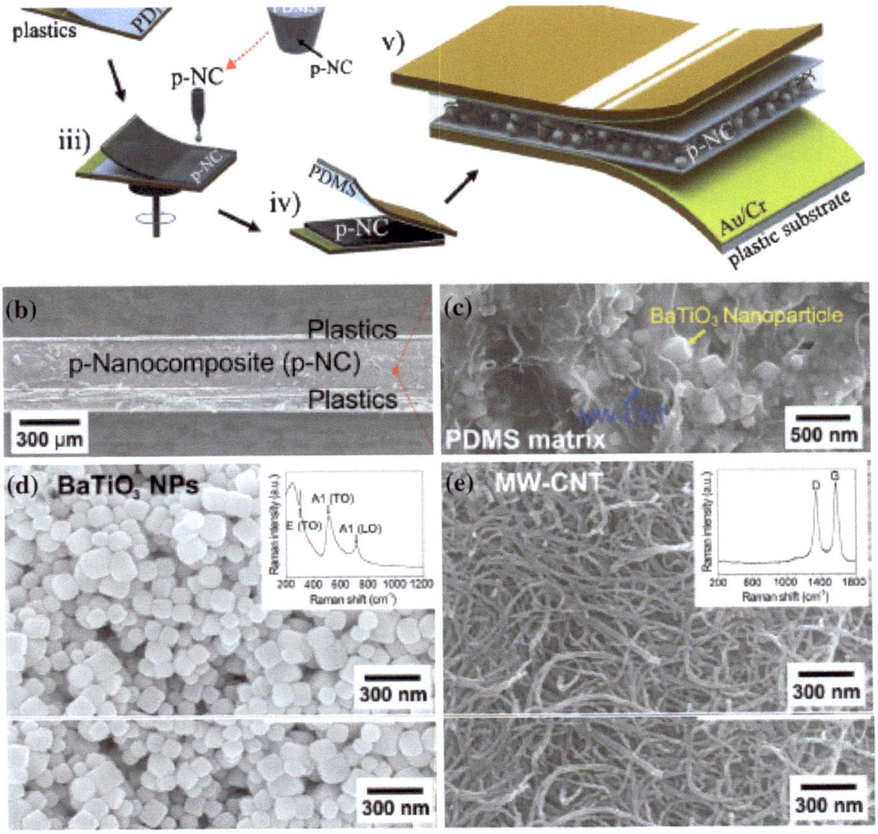

Fig. 3.78 **a** Schematic illustration of the process to fabricate the nanogenerator. **b** A cross-sectional SEM image of the nanogenerator. **c** A magnified cross-sectional SEM image of the piezoelectric nanocomposite. **d** SEM image of the $BaTiO_3$nanoparticles, with the inset showing a Raman spectrum of the nanoparticles. **e** SEM image of the MWCNTs used to make the nanocomposite. The inset shows a typical Raman shift of the MWCNTs with large D bands. Reproduced with permission from [70], Copyright @ 2012, John Wiley & Sons

reported recently [70]. Figure 3.78 shows schematic diagram of fabrication process of the nanogenerator and characterization of $BaTiO_3$ nanoparticles used to assemble the nanogenerator. The $BaTiO_3$ nanoparticles used to fabricate the nanogenerator have tetragonal perovskite structure, with spherical like particles of about 100 nm, which are synthesized by using a hydrothermal method. The particles are mixed with multi-walled carbon nanotube (MWCNT) as stress reinforcing agent, at a low concentration, so as not to affect the dielectric property of the nanocomposite that is formed by dispersing the MWCNT-$BaTiO_3$ in PDMS. The nanocomposites are very flexible, bendable and even stretchable. A nanogenerator is assembled with a 250 μm thick nanocomposite sandwiched between the top and bottom metal-coated plastic structures, as electrodes.

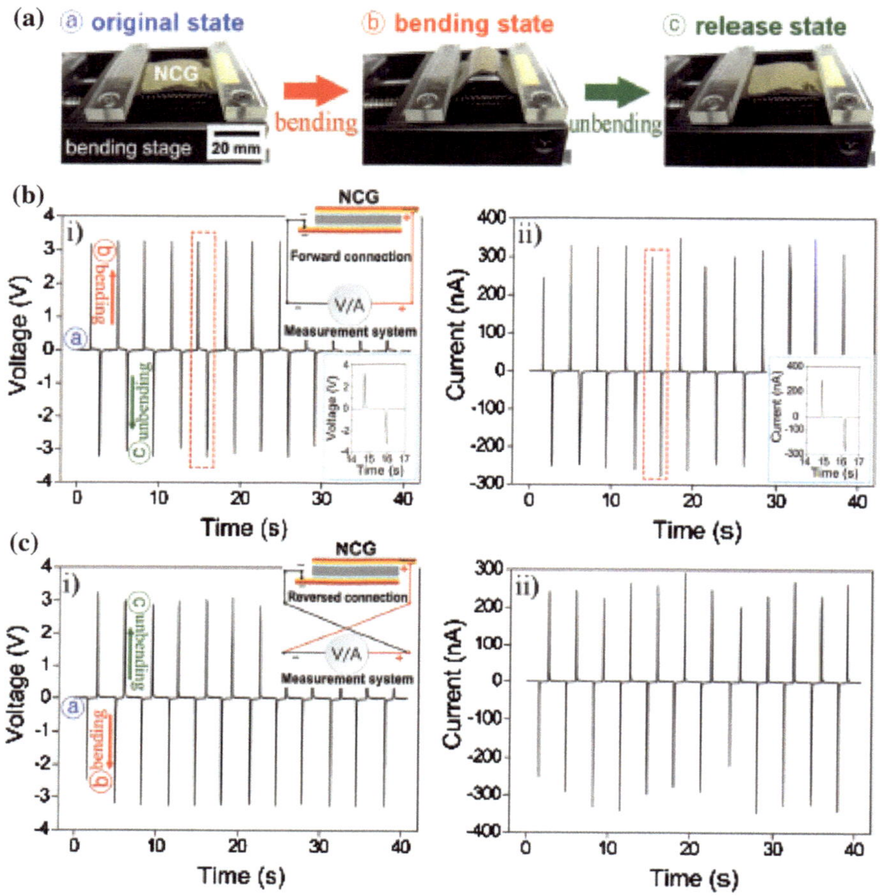

Fig. 3.79 **a** Optical images of nanogenerator in original, bending and release states. **b** Measured output voltage and current signals of the nanogenerator in the forward connection during the periodic bending and unbending movement. **c** Open-circuit voltage and short-circuit current signals generated in the reverse connection. Reproduced with permission from [70], Copyright @ 2012, John Wiley & Sons

The nanogenerator has been tested for energy harvesting, when it is subject to a periodic bending and unbending motion on a bending stage. Experimental results are shown in Fig. 3.79. Switching-polarity test has indicated that measurement results are due to the piezoelectric effect of the $BaTiO_3$ nanoparticles (Fig. 3.79b, c). A continuous bending and unbending cycle is created by extending the nanogenerator from 4 cm to 5 cm, at a deformation rate of 0.2 m s^{-1}. Under such a bending and unbending cycle, the nanogenerator produces an open-circuit voltage of ~3.2 V and a short-circuit current of 250–350 nA. The nanogenerator also shows very good stability.

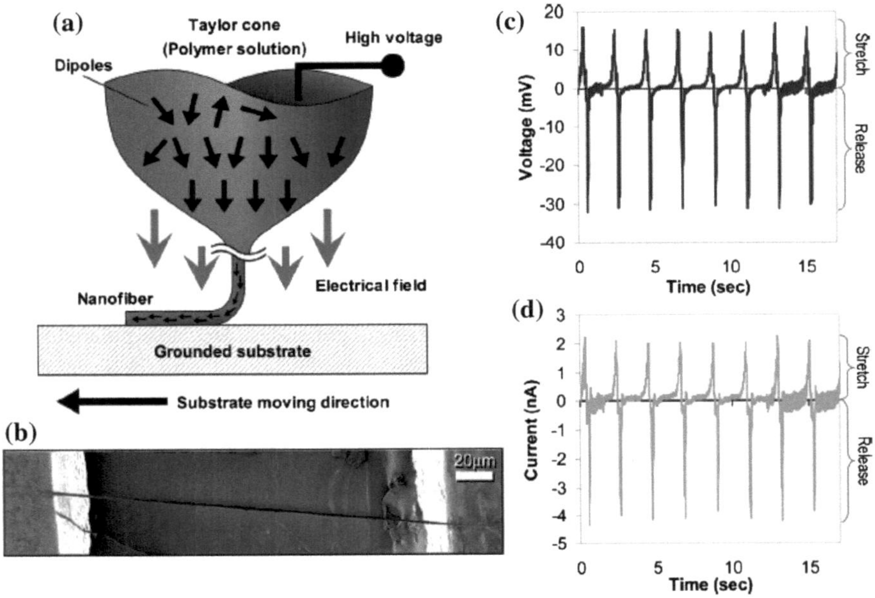

Fig. 3.80 Fabrication and performance of nanogenerator with piezoelectric PVDF nanofibers. **a** Near-field electrospinning (NFES) combining direct-write, mechanical stretching and in situ electrical poling to create and place piezoelectric nanogenerators onto a substrate. **b** SEM image of a nanogenerator made with a single PVDF nanofiber, two contact electrodes and a plastic substrate. **c** Output voltage measured with respect to time at a strain at 2 Hz. **d** Output current measured with respect to time at a strain at 2 Hz. Reproduced with permission from [73], Copyright @ 2010, American Chemical Society

Nanogenerators with similar energy harvesting capability have also been fabricated by using $0.5Ba(Zr_{0.2}Ti_{0.8})O_3$–$0.5(Ba_{0.7}Ca_{0.3})TiO_3$ nanowires [71] and $KNbO_3$ nanowires [72]. The $0.5Ba(Zr_{0.2}Ti_{0.8})O_3$–$0.5(Ba_{0.7}Ca_{0.3})TiO_3$ nanowires are prepared by using electrospinning method, which lead to nanogenerator with output voltage, output current, and power density of 3.25 V, 55 nA, and 338 μW cm^{-3}, respectively. The $KNbO_3$ nanowires are synthesized by using a hydrothermal method. It has been found that monoclinic $KNbO_3$ nanowires have much better energy harvesting capability than orthorhombic nanowires.

3.9.3 Poly(Vinylidene Fluoride) (PVDF)

3.9.3.1 Direct-Write PVDF Nanogenerators

A direct-write technique by means of near-field electrospinning (NFES) has been used to synthesize PVDF nanofibers for nanogenerator applications [73]. The as-produced PVDF nanofibers are directly placed on working substrates with in situ mechanical stretching and electrical poling (Fig. 3.80a). Due to the strong electric fields (>107 V m^{-1}) and stretching force from the electrospinning process, the dipoles of nanofiber crystals are aligned during the electrospinning, in which the nonpolar α phase (with randomly orientated dipoles) is transformed into polar β phase, determining the polarity of the electrospun nanofiber. The as-spun PVDF nanofibers have diameters ranging from 500 nm to 6.5 μm with variable lengths defined by the separation distance (100–600 μm) between two metallic electrodes (Fig. 3.80b).

When an axial stress is applied by bending the plastic deposition substrate, a piezoelectric potential in the PVDF nanofibers is generated. Because piezoelectric constant, d_{33}, of PVDF is negative, stretching the PVDF nanofibers along their poling axis generates a voltage with the polarity opposite to the electric field direction (the poling axis), during the electrospinning process. Therefore, a positive voltage output is expected by connecting the positive port of the multimeter to the distal end of the electrospun nanofiber. When the substrate is stretched and released repeatedly, voltage and current outputs can be detected, respectively (Fig. 3.80c, d). The typical electrical outputs of more than 50 tested nanogenerators are 5–30 mV and 0.5–3 nA. It is found that both nonpiezoelectric poly(ethylene oxide) nanofibers and randomly distributed PVDF nanofibers fabricated by using the conventional electrospinning process have no measurable electrical outputs.

The PVDV nanofiber nanogenerators have a maximum energy conversion efficiency of 21.8 %, with an average of 12.5 %. This is much higher than the power generators made with PVDF thin films and commercial PVDF thin films, which has been attributed to several reasons. For example, thin films may have internal defects, while the nanofibers made with the NFES process have high degree of crystallinity and chain orientation and thus much less defects. Nevertheless, such nanogenerators have been shown to possess high energy conversion efficiency, manufacturability and potential to integrate with other devices [73].

3.9.3.2 Hybrid Cells with PVDF Nanogenerators

By using PVDF nanogenerators, various hybrid cells have been designed and fabricated to harvest biomechanical and biochemical energies. These hybrid energy harvesters may find a number of potential in vivo applications. The piezoelectric PVDF nanofibers used to assemble the nanogenerators are synthesized by using the conventional electrospinning method. The potential biomechanical energies to be harvested include breathing, heart beating, biofluid and so on. The two energy harvesting approaches can work simultaneously or individually.

Figure 3.81 shows fabrication and patterning, characterization and poling of the PVDF nanofibers, by using the conventional electrospinning method [74]. PVDF

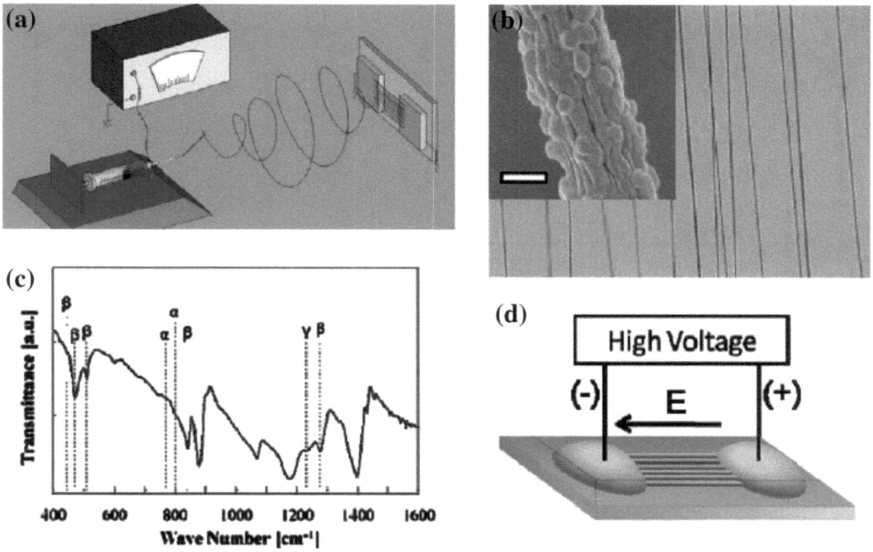

Fig. 3.81 **a** Schematic of the electrospinning process to fabricate aligned PVDF nanofibers. **b** Optical image of the aligned PVDF nanofibers on Kapton film, high resolution SEM of the nanofiber surface morphology shown in the inset (400 nm). **c** FTIR transmission spectrum of the PVDF nanofibers. **d** Schematic of high field in-plane poling of the PVDF nanofibers. Courtesy of Dr. Z. L. Wang, Georgia Institute of Technology. Reproduced with permission from [74], Copyright @ 2010, American Chemical Society

powder (MW 534,000) and DMF are mixed in acetone to form homogeneous solution as precursor of the PVDF nanofibers. The electrospun fibers are collected onto two grounded copper pieces from the needle, with the fibers being electrostatically aligned across the electrode gap.

The nanofiber has a textured morphology on surface (Fig. 3.81b, inset), due to the formation of small crystallites. The as-spun PVDF nanofibers are a mixture of the polar β phase and the nonpolar α and γ phases (FTIR, Fig. 3.81c). The random dipole orientation of the polar β phase has been oriented after the nanofibers are encased in PDMS and in-plane poled at high electric field ($\sim$0.2 MV cm^{-1}) in-plane poling.

A piece of Kapton@Dupont thin film, with dimension of 1×2 cm^2 $\times$ 30 μm, is used as the supporting substrate to assemble nanogenerators. A 50 nm thin layer of Au is used as electrodes, on the Kapton substrate, with a 1 mm gap in the middle. The electrospun fibers are then transferred across the Au electrodes and electrically bonded with silver paste. A 0.5 mm layer of PDMS is deposited over the device for dielectric protection and biocompatibility. The final device is then poled in paraffin oil at 20 kV for 15 min.

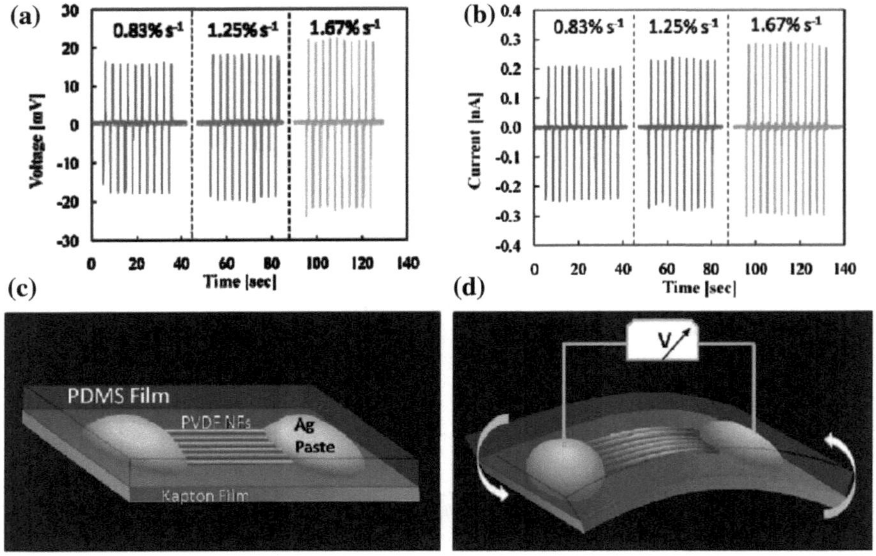

Fig. 3.82 Open-loop voltage **a** and short-circuit current of a PVDF nanogenerator at different strain rates with a maximum strain of 0.05 %. **c** The PVDF nanofibers on a Kapton substrate, with both ends bonded with silver paste and the entire device encapsulated with PDMS. **d** Mechanical bending of the substrate creates tensile strain and a corresponding piezoelectric field along the fiber that drives the electrons through an external load back and forth following the cycled mechanical action. Courtesy of Dr. Z. L. Wang, Georgia Institute of Technology. Reproduced with permission from [74], Copyright @ 2010, American Chemical Society

The working principle of the PVDF nanogenerator is based on the insulating property of the PVDF nanofiber and the creation of an inner piezoelectric field when it is subject to an external tensile strain. As the device is deformed by alternating compressive and tensile forces (Fig. 3.82c, d), the NF acts like a "capacitor" and "charge pump" to drive flow of electrons back and forth through the external circuit. The charging and discharging process leads to an AC power output.

Figure 3.82a, b shows the open-loop voltage and short-circuit current of the PVDF nanogenerator, when it is subject to cyclic mechanical loading at different strain rates with a fixed strain of 0.05 %. With increasing the strain rate, the open-loop voltage is increased from 15 mV to 20 mV and the Short-circuit current is increased from 0.2 to 0.3 nA, which is consistent with piezoelectric theory presented before. The output voltage is determined by the performance of individual nanowires, while the output current is the sum of those of all the active nanowires. The output of the PVDF nanogenerator could be improved by replacing PDMS with a material having higher breakdown voltage, so that it can be poled at higher fields to achieve higher remnant polarization. In addition, nanofibers can be further integrated like that of ZnO nanowires discussed earlier.

One example of hybrid cells is to combine with a biofuel cell (BFC) [74]. An enzymatic BFC is used to harvest the chemical energy of glucose and oxygen in

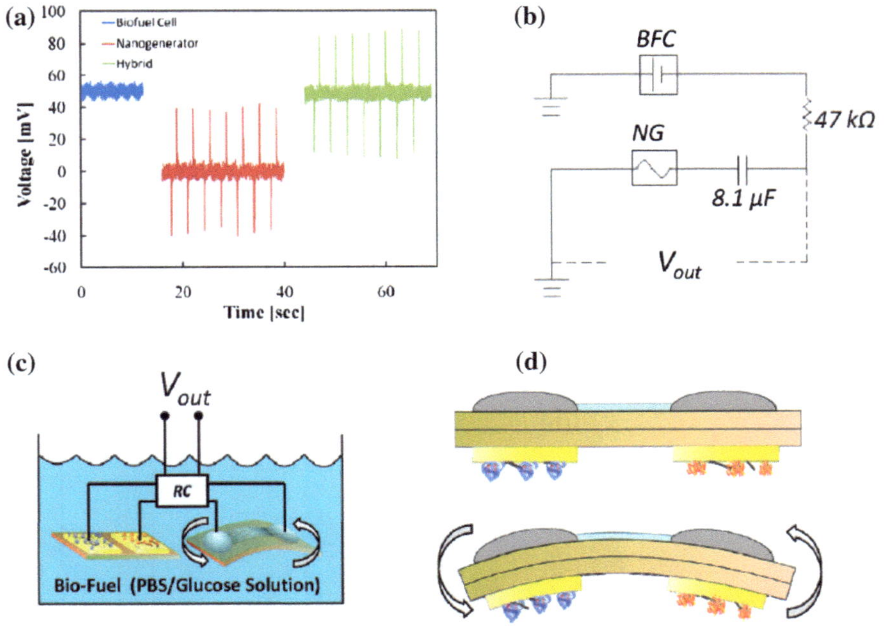

Fig. 3.83 **a** Open-circuit voltage of the independent and combined devices. **b** Circuit diagram used to integrate the hybrid BFC-nanogenerator device. Due to the inner resistance of the nanogenerator, the time required to charge the capacitor is much longer than the duration of the strain applied to the nanogenerator, so that the output voltage is a sum of those of the BFC and the nanogenerator. **c** Schematic of the hybrid device integration. **d** Conceptual design of a BFC-nanogenerator hybrid device. Courtesy of Dr. Z. L. Wang, Georgia Institute of Technology. Reproduced with permission from [74], Copyright @ 2010, American Chemical Society

biofluid. To fabricate the BFC, electrodes are patterned on Kapton film and coated with MWCNTs, followed by immobilization of glucose oxidase (GO_x) and formation of the anode and cathode. The use of the MWCNTs is beneficial to electron transfer between the enzymes and the electrodes. When the device is in contact with a biofluid that contains glucose (such as blood), glucose is electro oxidized to gluconolactone at the anode and the dissolved O_2 is electro-reduced to water at the cathode.

A hybrid cell of the BFC and a PVDF nanogenerator can harvest biochemical and biomechanical energies [74]. The independent and integrated operations of the PVDF nanogenerator, the BFC and the hybrid BFC-nanogenerator, are shown in Fig. 3.83a–c. To integrate the AC voltage of the PVDF nanogenerator with the DC voltage of the BFC, a simple RC high pass filter (Fig. 3.83b) is used to block the DC voltage of the BFC in one direction while passing the AC voltage of the nanogenerator. By integrating the two devices, the peak voltage is nearly doubled from 50 mV to 95 mV. Furthermore, due to packaging of the PVDF nanogenerator with PDMS, it can operate inside biofluid and in vivo environments. In addition,

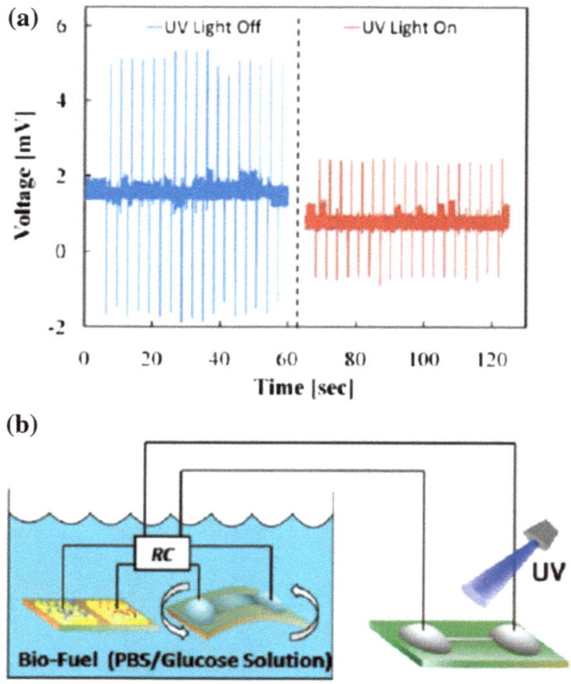

Fig. 3.84 Integration of the BFC-nanogenerator hybrid cell with a UV nanosensor to demonstrate "self-powered" nanosystem. **a** Voltage drop across the ZnO nanowire UV light sensor when UV light is off and on. **b** Schematic illustration of the self-powered hybrid nanosystem. Courtesy of Dr. Z. L. Wang, Georgia Institute of Technology. Reproduced with permission from [74], Copyright @ 2010, American Chemical Society

due to the use of a flexible Kapton film substrate, the BFC can be integrated back-to-back, as shown in Fig. 3.83d.

The power output of the hybrid device is the sum of those of the BFC and the nanogenerator. The output voltage of the BFC is V_{BFC} and the output AC voltage of the nanogenerator is $\pm V_{NG}$. Due to the large resistance of the PVDF nanofibers, the output voltage of the nanogenerator is $\pm V_{NG}$ even with the presence of a load. The inner resistance of the BFC is relatively low since it is determined by electron transfer between the active center of the enzymes and the MWCNT electrodes. Therefore, the voltage applied to an external load R is $V_{BFC} \pm V_{NG}$, corresponding to an output power of $(V_{BFC} \pm V_{NG})^2/R$. The average peak output power for each cycle of the mechanical action is $[(V_{BFC} + V_{NG})^2/R + (V_{BFC} - V_{NG})^2/R]/2 = (V_{BFC})^2/R + (V_{NG})^2/R$. It is expected the AC output of the nanogenerator can be rectified, so as to obtain DC output of both the BFC and the nanogenerator.

Another example is to use the hybrid BFC-nanogenerator to power a ZnO nanowire-based UV light sensor (Fig. 3.84b). When there was no UV light, The resistance of the ZnO nanowire is about 7 MΩ without UV light, corresponding toa peak voltage drop on the nanosensor to be ~5 mV, as shown in Fig. 3.84a. Upon irradiation with UV light, the nanowire resistance drops to ~800 kΩ while the peak voltage drop across the nanowire device is decreased to ~2.5 mV. This is a demonstration of the potential of a fully "self-powered" nanosystem for in vivo applications.

3.10 Summary and Prospects

The development of nanogenerators and the recent advances in power generation by using nanogenerators, as well as the power generation mechanisms, have been presented. These novel piezoelectric nanogenerators are versatile power generation strategy to harvest ambient waste energy for potential practical applications. The nanogenerators are especially promising for the miniaturization of power packages and self-power nano-systems than may find applications in implantable biosensing, environmental monitoring and personal electronics. Particularly, flexible and foldable nanogenerators are useful in those areas that have special mechanical and structural requirements, such as implanted biosensors in the muscle or joint. They also have the potential to directly harvest biomechanical or hydraulic energy in human body, such as flow of body fluid, blood flow, heartbeat and contraction of blood vessels, muscle stretching or eye blinking. Energy harvested in these ways can be used instantaneously to power implanted medical biodevices. Sound-driven nanogenerators can convert sound energy from speech, music or even noise into electrical energy. For example, speech vibration during conversations can be harvested to charge mobile phones.

Considerable efforts in fundamental research and engineering development have been made in order to enhance the efficiency of energy harvesting and to commercialize the applications of nanogenerators. There have been demonstrations that strong enough electrical energy harvested by using nanogenerators can continuously powder commercial LCDs and light up commercial LEDs and LDs, confirming the feasibility of using nanogenerators for powering mobile and personal microelectronic devices. More efforts are needed to output into the development and fabrication of powerful integrated amplifiers with the nanogenerators to amplify the electricity for large-scale applications. It is also necessary to explore the possibility and capability to harvest multiplex energy through the nanogenerators for electricity generation with even higher efficiencies. Also, more study on DC power generation through nanogenerators is deserved, so that the output can be feasibly used for the miniaturization of power packages. Other aspects include development of deposition and fabrication of large-scale nanosized ZnO nanowire arrays, with desired dimensions, on various substrates of interest, application of nanogenerator mechanism to other materials with higher piezoelectric performances.

References

1. Z.L. Wang, J.H. Song, Piezoelectric nanogenerators based on zinc oxide nanowire arrays. Science **312**, 242–246 (2006)
2. F.R. Fan, Z.Q. Tian, Z.L. Zhong, Flexible triboelectric generator. Nano Energy **1**(2), 328–334 (2012)
3. G. Zhu, C. Pan, W. Guo, C.Y. Chen, Y. Zhou, R. Yu, Z.L. Wang, Triboelectric-generator-driven pulse electrodeposition for micropatterning. Nano Lett. **12**(9), 4960–4965 (2012)

4. F.R. Fan, L. Lin, G. Zhu, W. Wu, R. Zhang, Z.L. Wang, Transparent triboelectric nanogenerators and self-powered pressure sensors based on micropatterned plastic films. Nano Lett. **12**(6), 3109–3114 (2012)
5. Z.L. Wang, ZnO nanowire and nanobelt platform for nanotechnology. Mater. Sci. Eng. R **64**, 33–71 (2009)
6. O. Dulub, L.A. Boatne, U. Diebold, STM study of the geometric and electronic structure of ZnO(0001)-Zn, (000$\bar{1}$)-O, (10$\bar{1}$0), and (1120) surfaces. Surf. Sci. **519**, 201–217 (2002)
7. B. Meyer, D. Marx, Density-functional study of the structure and stability of ZnO surfaces. Phys. Rev. B **67**, 035403 (2003)
8. A. Wander, F. Schedin, P. Steadman, A. Norris, R. McGrath, T.S. Turner, G. Thornton, N.M. Harrison, Stability of polar oxide surfaces. Phys. Rev. Lett. **86**, 3811–3814 (2001)
9. R.S. Yang, Y. Ding, Z.L. Wang, Deformation-free single-crystal nanohelixes of polar nanowires. Nano Lett. **4**, 1309–1312 (2004)
10. Y. Ding, X.Y. Kong, Z.L. Wang, Doping and planar defects in the formation of single-crystal ZnO nanorings. Phys. Rev. B **70**, 235–408 (2004)
11. Y. Ding, Z.L. Wang, Structures of planar defects in ZnO nanobelts and nanowires. Micron **40**, 335–342 (2009)
12. Z.L. Wang, Z.W. Pan, Z.R. Dai, Structures of oxide nanobelts and nanowires. Microsc. Microanal. **8**, 467–474 (2002)
13. Z.L. Wang, Nanostructures of zinc oxide. Mater. Today **7**, 26–33 (2004)
14. P. Yang, H. Yan, S. Mao, R. Russo, J. Johnson, R. Saykally, N. Morris, J. Pham, R. He, H. Choi, Adv. Func. Mater. **12**, 323–331 (2002)
15. X.D. Wang, C.J. Summers, Z.L. Wang, Large-scale hexagonal-patterned growth of aligned ZnO nanorods for nano-optoelectronics and nanosensor arrays. Nano Lett. **3**, 423–426 (2004)
16. X. D. Wang, J. H. Song, P. Li, J. H. Ryou, R. D. Dupuis, C.J. Summers and Z. L. Wang, Growth of uniformly aligned ZnO nanowire heterojunction arrays on GaN, AlN, and $Al_{0.5}Ga_{0.5}N$ substrates,*J.Am. Chem. Soc.*, **127**, 7920-7923 (2005)
17. J.H. Song, X.D. Wang, E. Riedo, Z.L. Wang, Systematic study on experimental conditions for large-scale growth of aligned ZnO nanwires on nitrides. J. Phys. Chem. B, **109**, 9869–9872 (2005)
18. S.S. Lin, J.I. Hong, J.H. Song, Y. Zhu, H.P. He, Z. Xu, Y.G. Wei, Y. Ding, R.L. Snyder, Z.L. Wang, Phosphorus doped $Zn_{1-x}Mg_xO$ nanowire arrays. Nano Lett. **9**, 3877–3882 (2009)
19. B. Weintraub, Y.L. Deng, Z.L. Wang, Position-controlled seedless growth of ZnO nanorod arrays on a polymer substrate via wet chemical synthesis. J. Phys. Chem. C **111**, 10162–10165 (2007)
20. B. Weintraub, Z. Zhou, Y. Li, Y.L. Deng, Solution synthesis of one-dimensional ZnO nanomaterials and their applications. Nanoscale **2**, 1573–1587 (2010)
21. L. Vayssieres, Growth of arrayed nanorods and nanowires of ZnO from aqueous solutions. Adv. Mater. **15**, 464–466 (2003)
22. S. Xu, C. Lao, B. Weintraub, Z.L. Wang, Density-controlled growth of aligned ZnO nanowire arrays by seedless chemical approach on smooth surfaces. J. Mater. Res. **23**, 2072–2077 (2008)
23. S. Xu, Y.G. Wei, M. Kirkham, J. Liu, W.J. Mai, R.L. Snyder, Z.L. Wang, Patterned growth of vertically aligned ZnO nanowire arrays on inorganic substrates at low temperature without catalyst. J. Am. Chem. Soc. **130**, 14958–14959 (2008)
24. T.Y. Liu, H.C. Liao, C.C. Lin, S.H. Hu, S.Y. Chen, Biofunctional ZnO nanorod arrays grown on flexible substrates. Langmuir **22**, 5804–5809 (2006)
25. S.A. Morin, F.F. Amos, S. Jin, Biomimetic assembly of zinc oxide nanorods onto flexible polymers. J. Am. Chem. Soc. **129**, 13776–13777 (2007)
26. Y. Qin, R.S. Yang, Z.L. Wang, Growth of horizonatal ZnO nanowire arrays on any substrate. J. Phys. Chem. C **112**, 18734–18736 (2008)
27. D. Yuan, R. Guo, Y. Wei, W. Wu, Y. Ding, Z.L. Wang, S. Das, Heteroepitaxial patterned growth of vertically aligned and periodically distributed ZnO nanowires on GaN using laser interference ablation. Adv. Funct. Mater. **20**, 3484–3489 (2010)

28. Y. Wei, W. Wu, R. Guo, D. Yuan, S. Das, Z.L. Wang, Wafer-scale high-throughput ordered growth of vertically aligned ZnO nanowire arrays. Nano Lett. **10**, 3414–3419 (2010)
29. Z.L. Wang, R.S. Yang, J. Zhou, Y. Qin, C. Xu, Y.F. Hu, S. Xu, Lateral nanowire/nanobelt based nanogenerators, piezotronics and piezo-phototronics. Mater. Sci. Eng., R **70**, 320–329 (2010)
30. Z.L. Wang, Piezopotential gated nanowire devices: piezotronics and piezo-phototronics. Nano Today **5**, 540–552 (2010)
31. Z.L. Wang, *Nanogenerators for self-powered devices and systems* (Georgia Institute of Technology, Atlanta, 2011)
32. Y.F. Gao, Z.L. Wang, Electrostatic potential in a bent piezoelectric nanowire The fundamental theory of nanogenerator and nanopiezotronics. Nano Lett. **7**, 2499–2505 (2007)
33. Y.F. Gao, Z.L. Wang, Equilibrium potential of free charge carriers in a bent piezoelectric semiconductive nanowire. Nano Lett. **9**, 1103–1120 (2009)
34. M.P. Lu, J.H. Song, M.Y. Lu, M.T. Chen, Y.F. Gao, L.F. Chen, Z.L. Wang, Piezoelectric nanogenerator using p-type ZnO nanowire arrays. Nano Lett. **9**, 1223–1227 (2009)
35. J. Zhou, P. Fei, Y.F. Gao, Y.D. Gu, J. Liu, G. Bao, Z.L. Wang, Mechanical-electrical triggers and sensors using piezoelectric micowires/nanowires. Nano Lett. **8**, 2725–2730 (2008)
36. Z.Y. Gao, J. Zhou, Y.D. Gu, P. Fei, Y. Hao, G. Bao, Z.L. Wang, Effects of piezoelectric potential on the transport characteristics of metal-ZnO nanowire-metal field effect transistor. J. Appl. Phys. **105**, 113707 (2009)
37. M. Alexe, S. Senz, M.A. Schubert, D. Hesse, U. Gosele, Energy harvesting using nanowires? Adv. Mater. **20**, 4021–4026 (2008)
38. Alexe et al., Energy harvesting using piezoelectric nanowires-a correspondence on "Energy harvesting using nanowires?"Adv. Mater. **21**, 1311–1315 (2009)
39. R.S. Yang, Y. Qin, C. Li, L.M. Dai, Z.L. Wang, Characteristics of output voltage and current of integrated nanogenerators. Appl. Phys. Lett. **94**, 022905 (2009)
40. R.S. Yang, Y. Qin, L.M. Dai, Z.L. Wang, Power generation with laterally packaged piezoelectric fine wires. Nat. Nanotech. **4**, 34–39 (2009)
41. Z.L. Wang, Oxide nanobelts and nanowires-growth, properties and applications. J. Nanosci. Nanotech. **8**, 27–55 (2008)
42. X.D. Wang, J.H. Song, Z.L. Wang, Nanowire and nanobelt arrays of zinc oxide from synthesis to properties and to novel devices. J. Mater. Chem. **17**, 711–720 (2007)
43. J.H. Song, J. Zhou, Z.L. Wang, Piezoelectric and semiconducting coupled power generating process of a single ZnO belt/wire. A technology for harvesting electricity from the environment. Nano Lett. **6**, 1622–1656 (2006)
44. P.X. Gao, J.H. Song, J. Liu, Z.L. Wang, Nanowire piezoelectric nanogenerators on plastic substrates as flexible power sources for nanodevices. Adv. Mater. **19**, 67–71 (2007)
45. J. Liu, P. Fei, J.H. Song, X.D. Wang, C.S. Lao, R. Tummala, Z.L. Wang, Carrier density and Schottky barrier on the performance of DC nanogenerator. Nano Lett. **8**, 328–332 (2008)
46. S.S. Lin, J.H. Song, Y.F. Lu, Z.L. Wang, Identifying individual n- and p-type ZnO nanowires by the output voltage sign of piezoelectric nanogenerator. Nanotechnology **20**, 365703 (2009)
47. X.D. Wang, J.H. Song, J. Liu, Z.L. Wang, Direct-durrent nanogenerator driven by ultrasonic waves. Science **316**, 102–105 (2007)
48. J. Liu, P. Fei, J. Zhou, R. Tummala, Z.L. Wang, Toward high output-power nanogenerator. Appl. Phys. Lett. **92**, 173105 (2008)
49. S. Xu, Y.G. Wei, J. Liu, R. Yang, Z.L. Wang, Integrated multilayer nanogenerator fabricated using paired nanotip-to-nanowire brushes. Nano Lett. **8**, 4027–4032 (2008)
50. S. Xu, Y. Qin, C. Xu, Y. Wei, R. Yang, Z.L. Wang, Self-powered nanowire devices. Nat. Nanotech. **5**, 366–373 (2010)
51. Y.F. Hu, L. Lin, Y. Zhang, Z.L. Wang, Replacing a battery by a nanogenerator with 20 V output. Adv. Mater. **24**, 110–114 (2012)
52. G. Zhu, A.C. Wang, Y. Liu, Y.S. Zhou, Z.L. Wang, Functional electrical stimulation by nanogenerator with 58 V output voltage. Nano Lett. **12**, 3086–3090 (2012)

53. R.S. Yang, Y. Qin, L.M. Dai, Z.L. Wang, Power generation with laterally packaged piezoelectric fine wires. Nat. Nanotech. **4**, 34–39 (2009)
54. R.S. Yang, Y. Qin, C. Li, G. Zhu, Z.L. Wang, Converting biomechanical energy into electricity by a muscle-movement-driven nanogenerator. Nano Lett. **9**, 1201–1205 (2009)
55. Z. Li, G. Zhu, R.S. Yang, A.C. Wang, Z.L. Wang, Muscle-driven in vivo nanogenerator. Adv. Mater. **22**, 2534–2537 (2010)
56. G. Zhu, R.S. Yang, S.H. Wang, Z.L. Wang, Flexible high-output nanogenerator based on lateral ZnO nanowire array. Nano Lett. **10**, 3151–3155 (2010)
57. Y. Qin, X.D. Wang, Z.L. Wang, Microfibre-nanowire hybrid structure for energy scavenging. Nature **451**, 809 (2008)
58. Z.T. Li, Z.L. Wang, Air/liquid-pressure and heartbeat-driven flexible fiber nanogenerators as a micro/nano-power source or diagnostic sensor. Adv. Mater. **23**, 84–89 (2011)
59. Y.F. Hu, Y. Zhang, C. Xu, G. Zhu, Z.L. Wang, High-output nanogenerator by rational unipolar assembly of conical nanowires and its application for driving a small liquid crystal display. Nano Lett. **10**, 5025–5031 (2010)
60. C. Xu, Z.L. Wang, Compact hybrid cell based on a convoluted nanowire structure for harvesting solar and mechanical energy. Adv. Mater. **23**, 873–877 (2011)
61. Y.F. Hu, Y. Zhang, C. Xu, L. Lin, R.L. Snyder, Z.L. Wang, Self-Powered System with Wireless Data Transmission. Nano Lett. **11**(6), 2572–2577 (2011)
62. M.B. Lee, J.H. Bae, J.H. Lee, C.S. Lee, S.H. Hong, Z.L. Wang, Self-powered environmental sensor system driven by nanogenerators. Energy Environ. Sci. **4**, 3359–3364 (2011)
63. Y.F. Lin, J.H. Song, D. Yong, S.Y. Lu, Z.L. Wang, Alternating the output of a CdS nanowire nanogenerator by a white-light-stimulated optoelectronic effect. Adv. Mater. **20**, 3127–3130 (2008)
64. C.T. Huang, J.H. Song, W.F. Lee, Y. Ding, Z.Y. Gao, Y. Hao, L.J. Chen, Z.L. Wang, GaN nanowire arrays for high-output nanogenerators. J. Am. Chem. Soc. **132**, 4766–4771 (2010)
65. C.T. Huang, J.H. Song, C.M. Tsai, W.F. Lee, D.H. Lien, Z.Y. Gao, Y. Hao, L.J. Chen, Z.L. Wang, Single-InN-nanowire nanogenerator with up to 1 V output voltage. Adv. Mater. **22**, 4008–4013 (2010)
66. T.D. Nguyen, J.M. Nagarah, Y. Qi, S.S. Nonnenmann, A.V. Morozov, S. Li, C.B. Arnold, M.C. McAlpine, Wafer-scale nanopatterning and translation into high-performance piezoelectric nanowires. Nano Lett. **10**, 4595–4599 (2010)
67. Y. Qi, J. Kim, T.D. Nguyen, B. Lisko, P.K. Purohit, M.C. McAlpine, Enhanced piezoelectricity and ptretchability in energy harvesting devices fabricated from buckled PZT ribbons. Nano Lett. **11**, 1331–1336 (2011)
68. J. Kwon, W. Seung, B.K. Sharma, S.W. Kim, J.H. Ahn, A high performance PZT ribbon-based nanogenerator using graphene transparent electrodes. Energy Environ. Sci. **5**, 8970–8975 (2012)
69. L. Gu, N. Cui, L. Cheng, Q. Xu, S. Bai, M. Yuan, W. Wu, J. Liu, Y. Zhao, F. Ma, Y. Qin, Z.L. Wang, Flexible fiber nanogenerator with 209 V output voltage directly powers a light-emitting diode. Nano Lett. **13**, 91–94 (2013)
70. K.I. Park, M. Lee, Y. Liu, S. Moon, G.T. Hwang, G. Zhu, J.E. Kim, S.O. Kim, D.K. Kim, Z.L. Wang, K.J. Lee, Flexible nanocomposite generator made of $BaTiO_3$ nanoparticles and graphitic carbons. Adv. Mater. **24**, 2999–3004 (2012)
71. W. Wu, L. Chenga S. Bai, W. Dou, Q. Xu, Z. Wei, Y. Qin, Electrospinning lead-free 0.5Ba$(Zr_{0.2}Ti_{0.8})O_3$-0.5$(Ba_{0.7}Ca_{0.3})TiO_3$ nanowires and their application in energy harvesting. J. Mater. Chem. A, **1**, 7332–7338 (2013)
72. S. Kim, J.H. Lee, J. Lee, S.W. Kim, M.H. Kim, S. Park, H. Chung,Y.I. Kimand W. Kim, Synthesis of monoclinic potassium niobate nanowires that are stable at room temperature, J. Am. Chem. Soc. **135**, 6–9 (2013)
73. C. Chang, V.H. Tran, J. Wang, Y. Fuh, L. Lin, Direct-write piezoelectric polymeric nanogenerator with high energy conversion efficiency. Nano Lett. **10**, 726–731 (2010)
74. B.J. Hansen, Y. Liu, R.S. Yang, Z.L. Wang, Hybrid nanogenerator for concurrently harvesting biomechanical and biochemical energy. ACS Nano **4**, 3647–3652 (2010)

Chapter 4
Waste Thermal Energy Harvesting (I): Thermoelectric Effect

4.1 Overview on Waste Thermal Energies: Definition, Identification, and Classification

Waste thermal energy (heat) is the second type of energy to be discussed. Usually, it is generated in a process by way of fuel combustion or chemical reaction, and then "dumped" into the environment even though it could still be reused for some useful and economic purpose. The essential quality of heat is not the amount but rather its "value." The strategy of how to recover this heat depends in part on the temperature of the waste heat gases and the economics involved.

Waste thermal energy is one of the largest sources of inexpensive, clean, and fuel-free energy available. The vast amount of heat that is discharged into the atmosphere everyday is one of the best sources of clean, fuel-free, and inexpensive energy. According to the US Department of Energy (DOE), up to 50 % of all fuels burned in the US goes unused into the atmosphere as waste heat is released to the atmosphere. Research indicates that the energy currently wasted by the industrial facilities in the U.S. could produce as much as 20 % of the total US electrical output with the associated 20 % reduction in greenhouse gas emissions. Various sources that can be classified as waste thermal energy (heat) with their properties are listed in Tables 4.1, 4.2, 4.3, and 4.4 [1].

Among the large quantities of waste heat that have been directly discharged into the Earth's environment, much of it is at temperatures which are too low to recover by using the conventional electrical power generators. Thermoelectric power generation, also known as thermoelectricity, has been demonstrated as a promising technology in the direct conversion of low-grade thermal energy, such as waste heat energy into electrical power. Probably, the earliest application is the utilization of waste heat from a kerosene lamp to provide thermoelectric energy to power a wireless device. Thermoelectric generators have also been used to provide small amounts electricity to remote regions, for instance, Northern Sweden, as an

L. B. Kong et al., *Waste Energy Harvesting*, Lecture Notes in Energy 24,
DOI: 10.1007/978-3-642-54634-1_4,

Table 4.1 Source and quality waste thermal energy [1]

No.	Source of waste heat	Quality of waste heat
1	Heat in flue gases	The higher the temperature, the greater the potential value for heat recovery
2	Heat in vapor streams	As above but when condensed, latent heat also recoverable
3	Convective and radiant heat lost from exterior of equipment	Low grade—if collected may be used for space heating or air preheats
4	Heat losses in cooling water	Low grade—useful gains if heat is exchanged with incoming fresh water
5	Heat losses in providing chilled water or in the disposal of chilled water	High grade if it can be utilized to reduce demand for refrigeration Low grade if refrigeration unit used as a form of heat pump
6	Heat stored in products leaving the process	Quality depends upon temperature
7	Heat in gaseous and liquid effluents leaving process	Poor if heavily contaminated and thus requiring alloy heat exchanger

Table 4.2 Typical waste heat temperature at high temperature range from various sources [1]

Types of devices	Temperature (°C)
Nickel refining furnace	1,370–1,650
Aluminum refining furnace	650–760
Zinc refining furnace	760–1,100
Copper refining furnace	760–815
Steel heating furnace	925–1,050
Copper reverberatory furnace	900–1,100
Open hearth furnace	650–700
Cement kiln (Dry process)	620–730
Glass melting furnace	1,000–1,550
Hydrogen plants	650–1000
Solid waste incinerators	650–1,000
Fume incinerators	650–1,450

Table 4.3 Typical waste heat temperature at medium temperature range from various sources [1]

Types of devices	Temperature (°C)
Steam boiler exhaust	230–480
Gas turbine exhaust	370–540
Reciprocating engine exhaust	315–600
Reciprocating engine exhaust (turbo charged)	230–370
Heat treatment furnace	425–650
Drying and baking ovens	230–600
Catalytic crackers	425–650
Annealing furnace cooling systems	425–650

Table 4.4 Typical waste heat temperature at low temperature range from various sources [1]

Source	Temperature (°C)
Process steam condensate	55–88
Cooling water from: Furnace doors	32–55
Bearings	32–88
Welding machines	32–88
Injection molding machines	32–88
Annealing furnaces	66–230
Forming dies	27–88
Air compressors	27–50
Pumps	27–88
Internal combustion engines	66–120
Air conditioning and refrigeration condensers	32–43
Liquid still condensers	32–88
Drying, baking, and curing ovens	93–230
Hot processed liquids	32–232
Hot processed solids	93–232

alternative to costly gasoline-powered motor generators. In this waste-heat-powered thermoelectric technology, because it is unnecessary to consider the cost of the thermal energy input, the low conversion efficiency of thermoelectric power generators is not a critical problem. Recent development indicates that they can be used in more and more cases, such as those used in cogeneration systems, to improve overall efficiencies of energy conversion systems by converting waste heat energy into electrical power.

This chapter is aimed to provide a detailed summary on the progress in thermoelectric effect and materials as a potential technique to harvest waster thermal energy. Principles of thermoelectric effect will be discussed first, which is followed by a brief description on efficiency of thermoelectric energy conversion by using thermoelectric devices. After that, a thorough list, together with detailed discussion on structure, properties, and performances, of thermoelectric materials that have been developed and reported in the open literature. Theoretical considerations from thermoelectric physics point of view on how to enhance the performance of modern thermoelectric materials will be systematically presented. Practical strategies according to the theoretical considerations that have been applied to currently available thermoelectric materials will be demonstrated. While more detailed description on each section can be found in respective excellent review [2–9], this chapter serves as a thorough reference to the latest progress in thermoelectric effect and thermoelectric materials. The chapter will be wrapped up with some concluding remarks and the potential direction of future research in this important area.

4.2 Principle of Thermoelectric Effect

4.2.1 Thermoelectric Effect

Thermoelectric effect is defined as the direct conversion of temperature differences to electric voltage and vice versa. A thermoelectric device creates a voltage when there is a different temperature applied on each side. Conversely, when a voltage is applied to such a device, it creates a temperature difference. At the atomic scale, an applied temperature gradient causes charge carriers in the material to diffuse from the hot side to the cold side, thus inducing a thermal current, which is similar to a classical gas that expands when heated, leading a flux of the gas molecules.

This effect can be used to generate electricity, measure temperature, or change the temperature of objects. Because the direction of heating and cooling is determined by the polarity of the applied voltage, thermoelectric devices are also efficient temperature controllers.

The term "thermoelectric effect" encompasses three separately identified effects: the Seebeck effect, Peltier effect, and Thomson effect. In most textbooks, it is known as the Peltier–Seebeck effect. This name is given due to the independent discoveries of the effect by French physicist Jean Charles Athanase Peltier and Estonian–German physicist Thomas Johann Seebeck. Joule heating, a heat that is generated whenever a voltage is applied across a resistive material, is related, though it is not generally termed as thermoelectric effect. The Peltier–Seebeck and Thomson effects are thermodynamically reversible, whereas Joule heating is not.

4.2.2 Seebeck Effect

The Seebeck effect is the conversion of temperature differences directly into electricity, which is named after German–Estonian physicist Thomas Johann Seebeck. In 1821, Seebeck discovered that a compass needle was deflected by a closed loop formed by two metals joined in two places, if there was a temperature difference between the junctions. This is because the metals responded differently to the temperature difference, creating a current loop and a magnetic field. Seebeck did not recognize that there was an electric current involved, so he called the phenomenon the "thermomagnetic effect". Danish physicist Hans Christian Ørsted rectified the mistake and coined the term "thermoelectricity." The voltage created by this effect is of the order of several microvolts per kelvin difference. For example, copper–constantan has a Seebeck coefficient of 41 $\mu V\ K^{-1}$ at room temperature.

Figure 4.1 shows a diagram of the circuit of Seebeck effect. The voltage V developed can be derived by using the following equation:

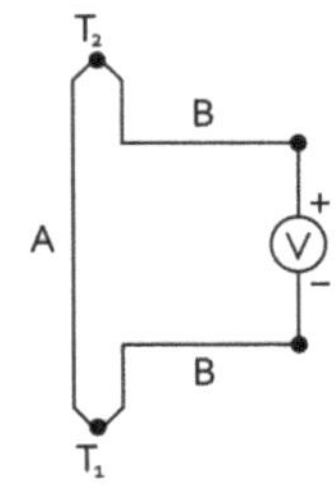

Fig. 4.1 Diagram of the circuit of Seebeck effect, with two different materials (A and B)

$$V = \int_{T_1}^{T_2} [S_B(T) - S_A(T)]\mathrm{d}T, \tag{4.1}$$

where S_A and S_B are the thermopowers (Seebeck coefficients) of metals A and B as a function of temperature, while T_1 and T_2 are the temperatures of the two junctions. The Seebeck coefficients are nonlinear as a function of temperature, and depend on absolute temperature of the conductors, as well as properties of the materials. If the Seebeck coefficients are independent of temperature in the measured temperature range, the above formula can be approximated as:

$$V = (S_B - S_A)(T_2 - T_1) \tag{4.2}$$

The Seebeck effect can be used in a thermocouple to measure a temperature difference. It also can be use to measure absolute temperature if the temperature of one end is known. A metal of unknown composition can be classified by its thermoelectric effect if a metallic probe of known composition, kept at a constant temperature, is held in contact with it. Industrial quality control instruments use this as thermoelectric alloy sorting to identify metal alloys. Because the voltage induced over each individual couple could be very small, it is necessary to use thermocouples connected in series to form a thermopile, in order to increase the overall output voltage. Thermoelectric generators are used for creating power from heat differentials.

4.2.3 Thermopower

The thermopower or Seebeck coefficient of a material, represented by S, measures the magnitude of an induced thermoelectric voltage in response to a temperature difference across that material and the entropy per charge carrier in the material. S has units of V K^{-1} or μV K^{-1} in most cases. Typical thermoelectric materials have values of S in hundreds of μV K^{-1}. An applied temperature difference causes charged carriers in the material to diffuse from the hot side to the cold side. Mobile charged carriers migrating to the cold side leave behind their oppositely charged nuclei at the hot side, thus creasting a thermoelectric voltage. Since a separation of charges creates an electric potential, the buildup of the charged carriers onto the cold side eventually reaches a maximum value when the electric field is at equilibrium. An increase in the temperature difference resumes a buildup of charge carriers on the cold side, leading to an increase in the thermoelectric voltage, and vice versa. The value of S is closely related to the properties of the materials. Generally, metals have small thermopower because of their half-filled bands caused by the equal number of negative and positive charges. In contrast, semiconductors can be doped with excessive electrons or electron holes, thus leading to relatively high magnitude of S. The sign of the thermopower reflects which charged carriers dominate the electric transport.

If the temperature difference, ΔT, between the two ends of a material is sufficiently small, the thermopower of a material is defined approximately as:

$$S = -\frac{\Delta V}{\Delta T}, \tag{4.3}$$

and a thermoelectric voltage of ΔV can be observed at the terminals.

This can be written in relation to the electric field E and the temperature gradient ∇T by the following approximate equation:

$$S = \frac{E}{\nabla T}. \tag{4.4}$$

The absolute thermopower of a material of interest is rarely practically measured because electrodes attached to a voltmeter must be placed onto the material in order to measure the thermoelectric voltage, which induces a thermoelectric voltage across one leg of the measurement electrodes. The measured thermopower then includes the thermopower of the material of interest and the material of the electrodes, which is represented by:

$$S_{AB} = S_B - S_A = \frac{\Delta V_B}{\Delta T} - \frac{\Delta V_A}{\Delta T}. \tag{4.5}$$

This allows a direct measurement of absolute thermopower of a material of interest. In addition, thermopower can also be derived from Thomson coefficient, μ, through the following relation:

$$S = \int \frac{\mu}{T} \mathrm{d}T \tag{4.6}$$

4.2.4 Charge Carrier Diffusion

The Seebeck effect of a material is caused by two factors: charge carrier diffusion and phonon drag. Charge carriers in the materials will diffuse when their two ends are at different temperatures. Hot carriers diffuse from the hot end to the cold end, because there is a lower density of hot carriers at the cold end of the materials, and vice versa. When a thermodynamic equilibrium is reached, the heat distribution will be evenly throughout the materials. The movement of heat, in the form of hot charge carriers, from one end to the other is a heat current and an electric current as charge carriers is moving.

In a system where both ends are kept at a constant temperature difference, there is a constant diffusion of carriers. If the rates of diffusion of hot and cold carriers in opposite directions are equal, there is no net change in the number of charge. The diffusing charges can be scattered by impurities, imperfections, and lattice vibrations or phonons. If the scattering is energy dependent, the hot and cold carriers will diffuse at different rates, creating a higher density of carriers at one end of the materials and thus an electrostatic voltage.

This electric field opposes the uneven scattering of carriers and an equilibrium is reached where the net number of carrier diffusing in one direction is canceled by the net number of carrier moving in the opposite direction. This means the thermopower of a material depends greatly on impurities, imperfections, and structural changes that vary with temperature and electric field. Therefore, the thermopower of a material is a collection of many different effects.

4.2.5 Phonon Drag

Phonons are not always in local thermal equilibrium and they move against the thermal gradient. They lose momentum by interacting with electrons, or other carriers, as well as imperfections in the crystal. If the phonon–electron interaction is predominant, the phonons will tend to push the electrons to one end of the material, thus losing momentum and leading to the thermoelectric field. This contribution is most important in the temperature region where phonon–electron scattering is predominant, which is described as:

$$T \approx \frac{1}{5}\Theta_D, \tag{4.7}$$

where Θ_D is the Debye temperature. At lower temperatures, there are fewer phonons available for drag, whereas they tend to lose momentum in phonon–phonon scattering instead of phonon–electron scattering at higher temperatures. This region of the thermopower versus temperature function is highly variable at externally applied magnetic fields.

4.2.6 Peltier Effect

The Peltier effect is the presence of heat at an electrified junction of two different metals, which was discovered by French physicist Jean–Charles Peltier in 1834. When a current flows through a junction composed of different materials, A and B, heat is generated at the upper junction at T_2 and absorbed at the lower junction at T_1. The Peltier heat, $\dot{Q}$, absorbed by the lower junction per unit time is equal to:

$$\dot{Q} = \Pi_{AB} I = (\Pi_B - \Pi_A) I, \tag{4.8}$$

where Π_{AB} is the Peltier coefficient for the thermocouple composed of materials A and B, while Π_A and Π_B are the Peltier coefficients of material A and B, respectively. Π varies with temperature and is determined by the specific composition of the materials. For example, p-type silicon typically has a positive Peltier coefficient of <550 K, whereas Peltier coefficient of n-type silicon is typically negative.

The Peltier coefficients represent the amount of heat current that is carried per unit charge through a material. Because charge current must be continuous across a junction, the associated heat flow will develop a discontinuity, if Π_A and Π_B are different. Depending on the magnitude of the current, heat must accumulate or deplete at the junction, due to a nonzero divergence caused by the carriers attempting to return to the equilibrium that exists before the current is applied by transferring energy from one connector to another. Individual couples can be connected in series to enhance the effect. This phenomenon has been used in either thermoelectric heat pumps or cooling devices such as refrigerators.

4.2.7 Thomson Effect

Thomson effect was predicted and subsequently observed by Lord Kelvin in 1851. It describes the heating or cooling of a current-carrying conductor with a temperature gradient. Any current-carrying conductor, except for a superconductor, with a temperature difference between two points either absorbs or emits heat, depending on the properties of the materials. If a current density, J, is passed through a homogeneous conductor, the heat production, q, per unit volume is given by:

$$q = \rho J^2 - \mu J \frac{\mathrm{d}T}{\mathrm{d}x}, \tag{4.9}$$

where ρ is resistivity of the material, $\mathrm{d}T/\mathrm{d}x$ is the temperature gradient along the wirem and μ is the Thomson coefficient. The first term is Joule heating, which does not change in sign, while the second term is the Thomson heating, which changes sign following J.

In some metals, such as zinc (Zn) and copper (Cu), whose temperature is directly proportional to their potential, when current moves from the hotter end to the colder end, there is a generation of heat and positive Thomson effect is observed. Conversely, in other metals, such as cobalt (Co), nickel (ni), and iron (Fe), whose temperature is inversely proportional to their potential, when current moves from the hotter end to the colder end, there is an absorption of heat and negative Thomson effect takes place.

If the Thomson coefficient of a material is measured over a wide range of temperature, it can be integrated by using the Thomson relations to determine the absolute values for the Peltier and Seebeck coefficients. This needs to be done only for one material, since the values of others can be determined by measuring pairwise Seebeck coefficients in thermocouples containing the reference material and then adding back the absolute thermopower of the reference material.

4.2.8 Thomson Relations

Thomson coefficient is unique among the three main thermoelectric coefficients, because it is the only one directly measurable for individual materials. Because both Peltier and Seebeck coefficients can only be determined for pairs of materials, no direct methods are available to determine absolute Seebeck or Peltier coefficients for an individual material. In 1854, Lord Kelvin found relationships among the three coefficients, implying that only one could be considered unique.

The first Thomson relation is

$$\mu = T\frac{\mathrm{d}S}{\mathrm{d}T}, \tag{4.10}$$

where T is the absolute temperature, μ is the Thomson coefficient, and S is the Seebeck coefficient. The second Thomson relation is:

$$\Pi = S \cdot T, \tag{4.11}$$

where Π is the Peltier coefficient.

4.3 Criteria of Thermoelectric Materials for High Efficiency

4.3.1 Figure of Merit

One key parameter that determines the efficiency of thermoelectric effect is called figure of merit, ZT, which is defined by the following equation. The figure of merit Z for thermoelectric devices is defined as:

$$Z = \frac{\sigma S^2}{\kappa}, \tag{4.12}$$

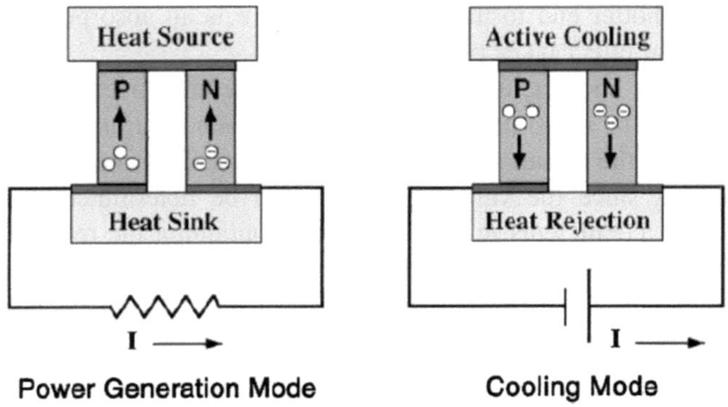

Fig. 4.2 A thermoelectric module illustrating the versatility of these materials for use in solid-state thermoelectric refrigeration or in power generation. The thermoelectric module is composed of an *n*-type and a *p*-type semiconducting material connected electrically in series through metallic electrical contact pads and thermally in parallel between ceramic ends

where σ is electrical conductivity, κ is thermal conductivity, and S is Seebeck coefficient. The dimensionless figure of merit, ZT, is formed by multiplying Z with average temperature:

$$ZT = \frac{\sigma S^2 T}{\kappa}, \tag{4.13}$$

$$T = \frac{T_2 + T_1}{2}. \tag{4.14}$$

A larger ZT indicates a greater thermodynamic efficiency, subject to certain provisions, particularly that the two materials in the couple have similar Z. ZT is therefore a method for comparing the potential efficiency of devices using different materials. Values of one are considered good; values in the 3–4 range are essential for thermoelectric to compete with mechanical devices in efficiency. To date, the best reported ZT values are in the range of 2–3. Currently, this goal of high ZT values is referred to as: "high-figure-of-merit" (Fig. 4.2).

4.3.2 Device Efficiency

The efficiency of a thermoelectric device for electricity generation is given by η, defined as:

$$\eta = \frac{\text{Energy provided to the load}}{\text{Heat energy absorbed at the hot junction}}. \tag{4.15}$$

The maximum efficiency, $\eta_{\max}$, is defined as:

$$\eta_{\max} = \frac{T_H - T_C}{T_H} \frac{\sqrt{1 + ZT} - 1}{\sqrt{1 + ZT} + \frac{T_C}{T_H}}, \tag{4.16}$$

where T_H is the temperature at the hot junction and T_C is the temperature at the surface being cooled. ZT is the modified dimensionless figure of merit, which takes into consideration the thermoelectric capacity of both thermoelectric materials being used in the device and is defined as:

$$\eta_{\max} = \frac{\left(S_p - S_n\right)^2 T}{\left[\left(\rho_n k_n\right)^{1/2} + \left(\rho_p k_p\right)^{1/2}\right]^2}, \tag{4.17}$$

where ρ is electrical resistivity, T is average temperature between the hot and cold surfaces, and the subscripts n and p denote the properties related to the n- and p-type semiconducting thermoelectric materials, respectively. Because thermoelectric devices are heat engines, their efficiency is limited by the Carnot efficiency. Regardless materials, the coefficients of performance of currently available commercial thermoelectric refrigerators are in the range from 0.3 to 0.6, which are only one-sixth the value of traditional vapor-compression refrigerators. Therefore, the key to the applications of thermoelectric materials is to increase their figure of merit.

4.4 Thermoelectric Materials

Numerous thermoelectric materials systems have been developed and reviewed previously [2–9]. Our goal herein is to give an overview and most recent progress in thermoelectric materials. First, single-phase bulk materials will be discussed with particular attention to the chemistry, crystal structure, physical properties, and optimization of thermoelectric performance. These systems will be described in sequence based on materials class. Second, bulk nanostructure composite materials will be examined. The opportunities for enhanced performance in nanostructure materials will be discussed as well as our current understanding of this class of materials.

4.4.1 Single-Phase Materials

4.4.1.1 Skutterudites

The skutterudites derive their name from a naturally occurring mineral, skutterudite or $CoAs_3$, first found in Skutterud, Norway. The structure is cubic and contains 32 atoms per unit cubic cell. Skutterudites have been acknowledged to be

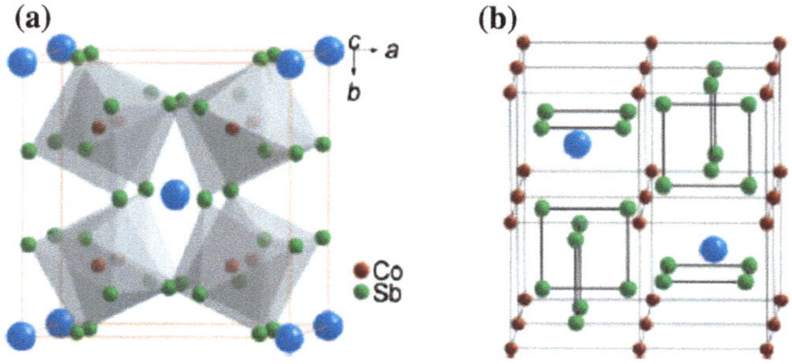

Fig. 4.3 Two model structures of the skutterudite, $CoSb_3$; the void cages are filled with *blue spheres* for clarity. **a** The unit cell of skutterudite structure. The transition metals (Co) are at the center of octahedra formed by pnicogen atoms (Sb). **b** The model shifted by the fractional coordinates (¼, ¼, ¼) from the unit cell. The Co atoms are connected for clarity. The only chemical bonds in this model are those of the Sb squares. Reproduced with permission from [3], Copyright @ 2009, John Wiley & Sons

promising candidates at high-performance thermoelectric materials and thus widely and extensively studied. They possess the $CoAs_3$-type structure with cubic space group of $Im\overline{3}$. The structure is composed of eight corner-shared XY_6 (X = Co, Rh, Ir; Y = P, As, Sb) octahedra. The $CoAs_3$ type structure is a severely distorted perovskite AB_3 type structure. As shown in Fig. 4.3a and b, the linked octahedral produce a void at center of the $(XY_6)_8$ cluster, with the void space taking a body-centered position of the cubic lattice. This void is so large that many large metallic atoms can be hosted in it to form filled skutterudites. They are characterized by the presence of the square anionic rings of the pnicogen atoms, such as $[P_4]^{4-}$, $[As_4]^{4-}$, etc., which link the transition metal ions to form the cubic structure. Therefore, the composition can be written as $\square_2X_8Y_{24}$ as illustrated in Fig. 4.3a or $\square_2X_2[Y_4]_6$ as illustrated in Fig. 4.3b [3].

A simple semiconductor transport model has predicted that *ZT* values of skutterudites could be as high as 0.3–1.4 over 300–1000 K [10]. Although the original skutterudite $CoSb_3$ has a high power factor, its too high lattice thermal conductivity (~ 10 W m^{-1} K^{-1} at room temperature) leads to a low *ZT*, so that it is not useful as thermoelectric materials. To address this problem, chemical approach has been developed by void-filling in the structure with various elements, including lanthanide, actinide, alkaline-earth, alkali, thallium, and Group IV elements [11]. Skutterudite antimonides have largest voids and thus are potential thermoelectric materials for practical applications.

Because the void-filling atoms can be either donor or acceptor, the electron concentration of the materials can be effectively optimized. At the same time, the presence of these atoms provides with strong phonon scattering centers to greatly reduce the lattice thermal conductivity of the materials. Therefore, this is called the rattling effect of the void-filling atoms in skutterudites [11]. The smaller and heavier the ions in the voids, the larger the disordering that is produced and, as a result, the larger the reduction in the lattice thermal conductivity. Although it has been hypothesized that the reduction in thermal conductivity is attributed to the rattling effect of the large-sized void-filling atoms, the evidence of scattering phenomenon in filled skutterudites has not been experimentally confirmed. This is because there are other factors, such as lattice disordering and the presence of point defects, which should be taken into account.

Various materials, such as $La_{0.9}Fe_3CoSb_{12}$, $Ce_{0.9}Fe_3CoSb_{12}$, $Yb_xCo_4Sb_{12}$, and $Ce_yFe_xCo_{4-x}Sb_{12}$, either *p*-type or *n*-type, have been reported to have high *ZT* values [10, 12–14]. It has been found that only a low concentration of La or Ce [15] in the voids of $CoSb_3$ results in a significant reduction in thermal conductivity. In some cases, the partially filled materials have shown high power factors [13, 16]. It has been found that only partial filling can result in higher *ZT* values. For example, $Yb_{0.19}Co_4Sb_{12}$ has a $ZT \approx 1$ at 600 K [10, 13]. Recently, higher *ZT* values have been reported in partially filled skutterudites with a small amount of Ni for Co, $Ba_{0.30}Ni_{0.05}Co_{3.95}Sb_{12}$ ($ZT \approx 1.25$ at 900 K) [17], and $Ca_{0.18}Ni_{0.03}Co_{3.97}Sb_{12.4}$ ($ZT \approx 1$ at 800 K) [18], as compared with those without Ni, $Ba_x Co_4Sb_{12}$ [19] ($ZT \approx 0.8$ at 800 K), and $Ca_xCo_4Sb_{12}$ [20] ($ZT \approx 0.45$ at 800 K).

The understandings on the thermal transport processes and phonon scattering mechanisms in skutterudite have stimulated great interest to search other materials with similar thermoelectric properties. Although it is still an open question whether these partially filled skutterudite compounds are of PGEC characteristics, the concept of the rattling effect [21] has led to various skutterudite-structured materials with *ZT* values to be higher than 1.

4.4.1.2 Clathrate

Clathrate are characterized by open frameworks composed of tetrahedrally coordinated Al, Ga, Si, Ge, or Sn, thus having low thermal conductivity. Within the framework structure, there are various cages that can incorporate large electropositive atoms. There are two main types of structure, Type I and Type II, with Type I being more commonly encountered. The Type I structure can be represented by a general formula $X_2Y_6E_{46}$ (Fig. 4.4a, Na_8Si_{46} for example), where X and Y are "guest" atoms encapsulated in two different polyhedral cages E_{20} and E_{24}, with E representing tetrahedrally coordinated framework atoms [3]. The Type II structure is composed of E_{20} and E_{28} cages.

It has been accepted that the presence of the guest atoms in these cages that can rattle and scatter lattice phonons, together with the open nature of the framework, effectively decrease the lattice thermal conductivity [22, 23]. In the last decade, research in this relatively large class of materials has accelerated. According to electronic band structure calculations with density function theory (DFT) [24], the Type I clathrate can be optimized in compositions to have high *ZT* values. For instance, *ZT* values of $Sr_8Ga_{16}Ge_{30}$ and $Ba_8In_{16}Sn_{30}$ can reach 0.5 at room temperature and as high as 1.7 at 800 K [25]. With measured Seebeck coefficient and resistivity of polycrystalline $Ba_8Ga_{16}Ge_{30}$, $Ba_8Ga_{16}Si_{30}$, $Ba_8Ga_{16}Sn_{30}$, $Sr_8Ga_{16}Ge_{30}$ and $Ba_8Ga_{16}Ge_{30}$ at high temperatures, combined with estimated high

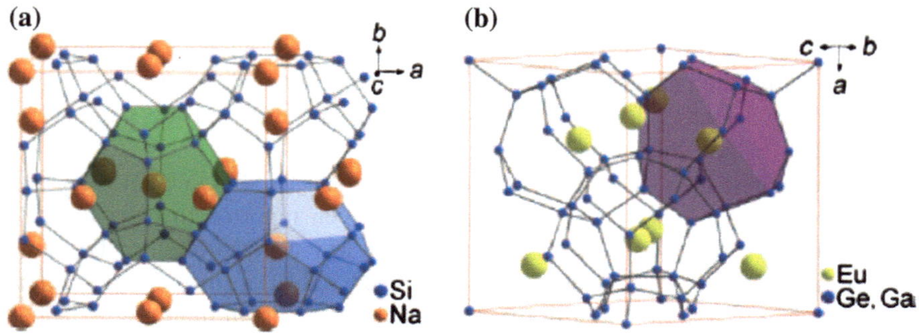

Fig. 4.4 **a** Crystal structure of the Type I clathrate, Na_8Si_{46}. Framework composed of Si atoms (*blue*) and two different cages with guest Na atoms, the tetrakaidecahedral cage ($[5^{12}6^2]$; *blue*) and the pentagonal dodecahedral cage ($[5^{12}]$; *green*). **b** Crystal structure of the Type VIII clathrate, $Eu_8Ga_{16}Ge_{30}$. The framework ($[3^34^35^9]$; *violet*) is composed of Ge and Ga atoms. ($[A^x]$: A = number of vertices, x = number of faces). Reproduced with permission from [3], Copyright @ 2009, John Wiley & Sons

temperature thermal conductivity of $Ba_8Ga_{16}Ge_{30}$ and $Ba_8Ga_{16}Si_{30}$ from open literature, high values of figure of merit $ZT = 0.7$ at 700 K and $ZT = 0.87$ at 870 K are expected [26].

Recent efforts in optimizing thermoelectric properties of the Type I clathrate at high temperatures have achieved significant progress. For example, a $Ba_8Ga_{16}Ge_{30}$ crystalline ingot synthesized by using the Czochralski method has a Seebeck coefficient of -45 to -150 μV K^{-1} and electrical conductivity of 1,500–600 S cm^{-1} at temperatures of 300–900 K. Its thermal conductivity decreases from 1.8 W m^{-1} K^{-1} at 300 K to 1.25 W m^{-1} K^{-1} at 900 K, which corresponds to a ZT of 1.35 at 900 K, without the presence of a maximum [27]. An even higher $ZT \approx 0.9$ at 1,000 K has been reported for the same material [3], which means that the $Ba_8Ga_{16}Ge_{30}$ prepared by using the Czochralski method is a promising candidate thermoelectric materials for high-temperature applications. However, the high cost of Ga and Ge could be a limiting factor for this specific material for large-scale commercial or industrial applications. Therefore, it is highly recommended to extend this special processing technique to other materials containing low-cost elements.

The Type III structure of clathrate with a formula of $X_{24}E_{100}$ is composed of three kinds of cages, E_{20} pentagonal dodecahedra, open dodecahedra and distorted cubes. As an example, n-type clathrate $Ba_{24}Ga_xGe_{100-x}$ ($x = 15$) has a high figure of merit of $ZT = 1.25$ at 943 K, with a power factor of 1.15×10^{-3} W m^{-1} K^{-2} and a temperature-independent thermal conductivity of about 0.85 W m^{-1} K^{-1} [28].

It is also of interest to explore other structure types of clathrate, such as Type VIII clathrate $Eu_8Ga_{16}Ge_{30}$ (Fig. 4.4b). The Type I structured Eu8Ga16Ge30 has ZT $\approx$ 0.4 at 400 K for both n-type and p-type doping [29], whereas the Type VIII analog $Eu_8Ga_{16}Ge_{30}$ also has a figure of merit of $ZT \approx 0.3$ at 400 K for n-type doping [30]. However, theoretical analysis has predicted that a $ZT = 1.2$ can be achieved by using this Type VIII material if it is p-type doped [31]. There is still possibility for further improvement in thermoelectric performance of the materials.

4.4.1.3 Half-Heusler Intermetallic Compounds

Another representative class of potential thermoelectric materials for high-temperature applications are the half-Heusler (HH) intermetallic compounds, with a formula of MNiSn (M = Ti, Hf, Zr). HH phases have crystal structure similar to that MgAgAs, which consists of three filled interpenetrating *fcc* sublattices and one vacant sublattice.

In the general formula of XYZ, X and Y are transition metals and Z is a main-group element [32]. One important advantage is that they can be easily synthesized relatively. Another advantage of these compounds is their high melting points of 1,100–1,300 °C. They also have high chemical stability, with almost no sublimation at temperatures of up to 1,000 °C, enabling their potential applications at high temperatures. In the unit cell of TiNiSn, Ti and Sn occupy a NaCl lattice and Ni takes an *fcc* sublattice. The Heusler intermetallic compounds with fully filled sublattices have metallic characteristics (full-Heusler alloys). In contrast, those HH compounds with vacant sites of Ni atom are characterized by narrow bands, with d-orbital hybridization and thus belonging to semiconductors [32–35]. Due to the presence of the rather narrow bands, these compounds possess a large effective mass, which is responsible for their large thermopower [36].

The three filled sublattices of the compounds can be independently modified in terms of chemical composition to improve their thermoelectric properties. For example, Sn site doping usually increases the concentration of charge carriers, while doping at Ti and Ni sites leads to high degree of mass fluctuation that is beneficial to reducing thermal conductivity. Most importantly, HH alloys have relatively large room temperature Seebeck coefficient of about 100 mV K^{-1} and high electrical conductivities of 1,000–10,000 S cm^{-1}, so as to be promising thermoelectric materials [36–39].

Among the HH alloys that have been studied to improve their thermoelectric properties, ZnNiSn is the most intensively investigated [36–41]. Sb-doped TiNiSn alloys have power factors of as high as 70 mW cm^{-1} K^{-2} at 650 K [38]. Despite the large power factor, However, due to the high thermal conductivity of about 10 W m^{-1} K^{-1}, their ZT is only 0.45 at 650 K.

It has been reported that n-type $Zr_{0.5}Hf_{0.5}Ni_{0.8}Pd_{0.2}Sn_{0.99}Sb_{0.01}$ has a high figure of merit of $ZT = 0.7$ at 800 K [41], while the ZT value of n-type $(Zr_{0.5}Hf_{0.5})_{0.5}$ $Ti_{0.5}NiSn_{1-y}Sb_y$ is up to 1.4 at 700 K [40]. Although these results have not been reproduced, similar HH compounds have drawn considerable attention as promising thermoelectric materials. Recently, electronic structure and ab initio calculations have been applied to predict the thermoelectric performances of about 30 HH compounds with the 18 valence-electron count [42]. For instance, Co, Rh-, and Fe-based HH compounds could be promising p-type materials, whereas LaPdBi and several compounds are potential *n*-type thermoelectric materials with large power factors.

One way to reduce the lattice thermal conductivity of HH alloys is to increase the atomic disordering at the transition metal sites with other elements to induce mass fluctuations and strain field effects [41, 43]. For example, room temperature thermal conductivity of $Hf_{0.75}Zr_{0.25}NiSn_{0.975}Sb_{0.025}$ is $\kappa = 7.1$ W m^{-1} K^{-1}, as compared with 11.14 W m^{-1} K^{-1} of ZrNiSn. A further reduction in thermal conductivity to 6.4 W m^{-1} K^{-1} has been observed in $Hf_{0.60}Zr_{0.25}Ti_{0.15}$-$NiSn_{0.975}Sb_{0.025}$, in which Ti is introduced to the transition metal site [44]. However, at temperatures over 1,000 K, the thermal conductivity is increased due to the onset in ionic thermal conduction within the open crystal structure.

It has also reported that doping the Sn site with Sb is able to increase *ZT* value of MNiSn HH alloys. For example, $Hf_{0.75}Zr_{0.25}NiSn_{0.975}Sb_{0.025}$ has *ZT* values of 0.81 at 1,025 K and 0.78 at 1,070 K. This is mainly because an increase in Sb content shifts the maximum of Seebeck coefficient to higher temperature, thus leading to larger power factors. There are other members of the HH family with semiconducting properties, such as the (RE)MSb [45] (M = transition metal, RE = rare-earth metal), which deserve further studies.

4.4.1.4 β-Zn_4Sb_3

β-Zn_4Sb_3 is a *p*-type semiconductor, which exhibits relatively low thermal conductivity at moderate temperatures [46]. The crystal structure of β-Zn_4Sb_3 is shown in Fig. 4.5 [3]. It has one Zn site and two independent Sb sites, where the Sb1 and Sb2 sites are Sb^{3-} and Sb_2^{4-} dimers. The site that is disordered with Zn is the reason to cause the controversy in stoichiometry of the compound.

X-ray single-crystal diffraction and powder synchrotron radiation diffraction results indicate that, in the crystal structure of β-Zn_4Sb_3, there are three different interstitial sites in addition to the Zn site. It is found that only 90 % of the Zn site is occupied, so that the composition is refined as $Zn_{12.8}Sb_{10}$ [47, 48]. The *p*-type characteristic of the crystal structure has been confirmed by both theoretical calculation and experimental verification [48]. According to the calculation results, the Zn interstitial atoms act as electron donors and play an important role in enhancing the thermopower.

Although β-Zn_4Sb_3 has been shown to have a very high *ZT* value of 1.3 at 670 K, it is not a promising thermoelectric material, because it decomposes into ZnSb and Zn as the approaches to its melting point of 841 K [49]. When compared with that of HH alloys (70 mW cm^{-1} K^{-2} near 400 °C) [38], the power factor of β-Zn_4Sb_3 is relatively low (13 mW cm^{-1} K^{-2} at 400 °C). This could be one of the problems for β-Zn_4Sb_3 as a high *ZT* material [49]. In this respect, the high *ZT* value of β-Zn_4Sb_3 is mainly attributed to its "phonon-glass" behavior, which is responsible for its low thermal conductivity of about 0.9 W m^{-1} K^{-1} at room temperature [47]. It is the interstitial atoms that result in the glass-like phonon damping to suppress the lattice thermal conductivity of the material.

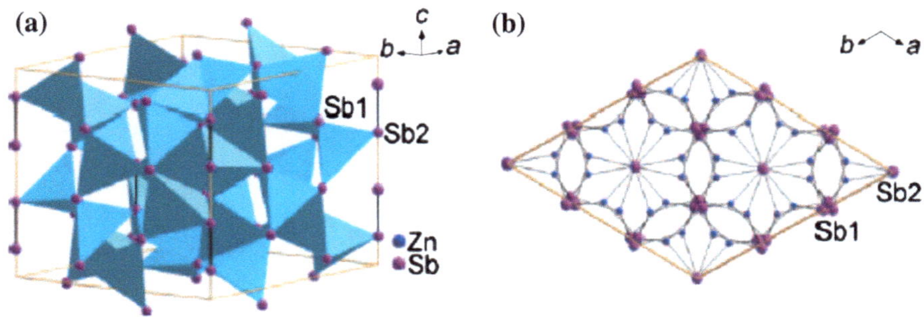

Fig. 4.5 Crystal structure of β-Zn_4Sb_3 consisting of (**a**) three-dimensional corner-sharing tetrahedra of $[ZnSb_4]$ units and (**b**) Sb2 dimers that are formed in the octahedral holes within the distorted hexagonal Sb1 channels (view down to the c-axis). Reproduced with permission from [3], Copyright @ 2009, John Wiley & Sons

The thermal conductivity of β-Zn_4Sb_3 can be further decreased by doping with other elements [47, 50]. However, it does not allow for high doping concentrations. For example, when Cd_4Sb_3 is used to form solid solution $(Zn_{1-x}Cd_x)_4Sb_3$ with Zn_4Sb_3, x cannot be higher than 6 mol% at 400 °C, even though they are isostructural compounds [51]. This could be the main reason that there is no report on further improvement in ZT of for this material that excesses the record high value until now.

4.4.1.5 Zintl Phase $Yb_{14}MnSb_{11}$

Intermetallic $Yb_{14}MnSb_{11}$ has been considered to be a promising thermoelectric material for applications at very high temperatures. The compound belongs to the Zintl family, $A1_4MPn_{11}$, where A is an alkaline-earth or rare-earth metal, M is a transition or main-group metal and Pn is a pnicogen. Several members with this structure have been extensively studied in terms of magnetic properties, among which the Yb analog appears to be an excellent p-type thermoelectric material [52]. Figure 4.6 shows the cubic structure of $Yb_{14}MnSb_{11}$, with a similar complex structure of $Ca_{14}AlSb_{11}$ [52]. It contains one $[AlSb_4]^{9-}$ tetrahedron, one $[Sb_3]^{7-}$ polyatomic anion, four Sb^{3-} anions located between the $[AlSb_4]^{9-}$ and $[Sb_3]^{7-}$ units, and $14Ca^{2+}$ per formula. $Yb_{14}MnSb_{11}$ is a valence precise semiconductor, according to the classical concept of Zintl, in which the strongly electropositive Yb atoms donate electrons to Sb atoms. Actually, the material exhibits weakly metallic or semimetallic behavior, according to its conductivity that is measured as a function of temperature.

Electrical conductivity and Seebeck coefficient of $Yb_{14}MnSb_{11}$ synthesized by using hot-pressing are measured at 1,200 K to be 185 S cm^{-1} and 180 mV K^{-1}, respectively. Although it has relatively low power factor of about 6 mW cm^{-1} K^{-2}, its thermoelectric figure of merit is pretty high, $ZT \sim 1.0$ at

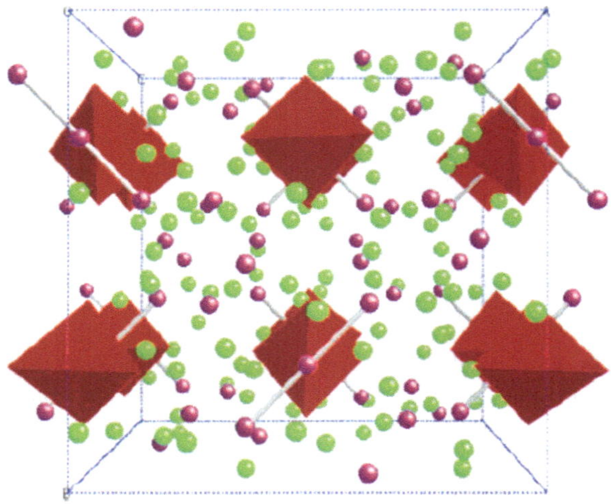

Fig. 4.6 Body-centered, $I4_1/acd$ crystal structure of $Yb_{14}MnSb_{11}$. The *green* and *purple spheres* represent Yb and Sb, respectively. The $MnSb_4$ tetrahedron is shown as a filled *red* polyhedron. Reproduced with permission from [52], Copyright @ 2006, American Chemical Society

1200 K, due to the significantly low thermal conductivity of the material, which is in the range of 0.7–0.9 W m^{-1} K^{-1} at temperatures of 200–1,275 K [52]. This low thermal conductivity has been found to be associated with its large lattice constant, structural complexity and the ionic character of the bonding in the lattice. The *ZT* value of $Yb_{14}MnSb_{11}$ is nearly twice at high temperatures (975–1,275 K) that of *p*-type SiGe based materials (with maximum $ZT \sim 0.6$ at 873 K) [53]. Si–Ge alloy has been an important thermoelectric material used in the radioisotope thermoelectric generator (RTG) for deep-space probes, due to its high *ZT* at 1,200 K. $Yb_{14}MnSb_{11}$ is currently the only material that can replace *p*-type Si–Ge alloys for this application.

Significant efforts have been made to improve the thermoelectric performance of $Yb_{14}MnSb_{11}$ based material systems through the substitutions with other alkaline-earth, rare-earth metals and transition/main-group metals [54, 55]. The effect of pressure on charge transport properties of $Yb_{14}MnSb_{11}$ has been systematically studied. It is found that its conductivity is decreased and thermopower is increased with increasing pressure [56]. For instance, room temperature conductivity and Seebeck coefficient of single-crystal samples are 689 S cm^{-1} and 47 mV K^{-1}, at ambient pressure, while they are 645 S cm^{-1} and 55 mV K^{-1} at 2.3 GPa.

4.4.1.6 $FeSb_2$

$FeSb_2$ has a record high Seebeck coefficient (-4.5×10^4 mV K^{-1}) at low temperatures ($\sim$10 K), which leads to the largest power factor ever reported in open literature ($\sim 2.3 \times 10^3$ mW cm^{-1} K^{-2}) [57]. $FeSb_2$ possesses marcasite crystal structure and is one of the two phases in the Fe–Sb system.

According to magnetic susceptibility, Mossbauer spectra, resistivity and Seebeck coefficient, $FeSb_2$ has been acknowledged as a strongly correlated electron system. Band structure calculations have demonstrated that there are localized Fe *d*-states in the valence band of the compound, which could be the main contributor to the large experimental Seebeck coefficient, because *S* is dependent on the variation of the density of state (DOS) near the Fermi energy [58]. However, $FeSb_2$ has a large lattice thermal conductivity, which leads to its low value of *ZT*. For example, its *ZT* is only 0.005 at 12 K [57]. Therefore, it is a critical challenge for this material to be used for thermoelectric applications.

It has been proposed that there is a very narrow band at temperatures of below 10 K, which is formed due to the weak hybridization of Fe 3d states with Sb 5p state. The colossal Seebeck coefficient is observed within the "very narrow band." If this kind of narrow band can be created at high temperatures or in other materials, it is possible to develop new thermoelectric materials with high performances.

4.4.2 Anisotropic Chalcogenide Compounds

Another group of potential thermoelectric materials is chalcogenide compounds, which are semiconductors with high stability in air and high melting points. Because of their strong flexibility in hosting other elements and the small difference in electronegativity among sulfur, selenium, and tellurium, it is highly potential to synthesize chalcogenide semiconductors with energy gaps (e.g. 0.1–0.8 eV) suitable for thermoelectric applications over a wide range of temperature.

Among various chalcogenide compounds, Bi_2Te_3, together with its solid solutions, such as *p*-type $Bi_{2-x}Sb_xTe_3$ and *n*-type $Bi_2Te_{3-x}Se_x$, is the most widely used thermoelectric materials for large-scale cooling applications [59]. PbTe is another important example of chalcogenide compounds, with a maximum *ZT* of about 0.8 at 770 K, which is thus used for power generation at intermediate temperatures. Recent studies have shown that germanium-based TAGS (Te–Ag–Ge–Sb) has higher performance than PbTe, but its high sublimation rate, high cost, and the presence of a low-temperature phase transition could be obstacles for their practical applications.

4.4.2.1 Tl_9BiTe_6, Ag_9TlTe_5, and Tl_2SnTe_5

Thallium chalcogenide have very low thermal conductivities and high Seebeck coefficients, but relatively low electrical conductivity. It has been predicted that their power factor can be increased by controlling the carrier concentration through doping [60], but no attempt has been made in this regard. Potential thallium chalcogenide as thermoelectric materials include Tl_9BiTe_6, Tl_2SnTe_5, and Ag_9TlTe_5.

Tl_9BiTe_6 is a derivative of the isostructural compound Tl_5Te_3, which is formed by replacing Tl^{3+} with Bi^{3+} in $2[Tl_5Te_3] = \left[Tl_8^{+}\left(Tl^{+}Tl^{3+}\right)Te^{2-}{}_3\right]$. The nearest neighbors of the Te atoms are exclusively Tl atoms and the central position surrounded by Te atoms in the octahedral pocket is equally occupied by Tl and Bi. Tl_9BiTe_6 has been reported to exhibit a thermoelectric figure of merit of *ZT* of about 1.2 at 500 K, which is mainly attributed to its remarkably low lattice thermal conductivity of about 0.39 W m^{-1} K^{-1} at 300 K [61].

Tl_2SnTe_5 has a tetragonal structure with infinite chains of $[SnTe_5]^{2-}$ that are parallel one another, with eightfold coordinate Tl^+ ions between the channels. One of the main reasons for the very low lattice thermal conductivity (0.5 W m^{-1} K^{-1}) of this compound is the relatively long Tl–Te bonds, which correspond to very low frequency phonons. This compound could be optimized to have a *ZT* of about1 at 500 K [62]. Ag_9TlTe_5 has a same crystal structure as Ag_2Te, which has an even lower lattice thermal conductivity than Tl_2SnTe_5 [63, 64]. The extremely low thermal conductivity of 0.22 W m^{-1} K^{-1} and relatively low electrical resistivity Ag_9TlTe_5 make it to have a relatively high figure or merit *ZT* = 1.23 at 700 K.

Another interesting thallium-containing compound is $TlIn_{1-x}Yb_xTe_2$. The solid solutions of $TlInTe_2$–$TlYbTe_2$ are *p*-type semiconductors, which have a notably high *ZT* value over 500–700 K. Single-crystal $TlIn_{0.94}Yb_{0.06}Te_2$ grown by using float-zone melting have promising physical properties at 700 K, including $Z = 2.61 \times 10{-}3$ K^{-1}, thermopower of about 630 mV K^{-1}, electrical conductivity of approximately 39.5 S cm^{-1} and lattice thermal conductivity of about 0.61 W m^{-1} K^{-1} [65]. However, because of the toxicity and environmental concerns, thallium-containing compounds are not suitable for large-scale practical applications, but they can be good examples to provide useful information on how to reduce thermal conductivity and optimize the performance of thermoelectric materials.

4.4.2.2 Alkali-Metal Bismuth Chalcogenide

Bismuth chalcogenide have drawn significant attentions during the last decade, because of their high potential as thermoelectric materials and the abundance of new compounds that can be synthesized in the type of materials [66–82]. Some of the new compounds include KBi_3S_5 [66], $KBi_{6.33}S_{10}$ [67, 68], $K_2Bi_8S_{13}$ [67, 68], α-, β-$K_2Bi_8Se_{13}$ [69, 70], $K_{2.5}Bi_{8.5}Se_{14}$ [70], $A_xBi_4Se_7$ [71] (x = 1, 2), $BaBiTe_3$ [72], $CsBi_4Te_6$ [73], $ALn_{1-x}Bi_{4-x}S_8$ [74] (A = K, Rb and Ln = La, Ce, Pr, Nd), BaLa–Bi_2Q_6 [75] (Q = S, Se), α-, β-$APbBi_3Se_6$ [76] (A = K, Rb, Cs), $K_{1-x}Sn_{5-x}Bi_{11+x}Se_{22}$ [77], $A_{1+x}M'_{4-2x}Bi_{7+x}Se_{15}$ [78] (A = K, Rb and M' = Sn, Pb), $Sn_4Bi_2Se_7$ [79], $SnBi_4Se_7$ [80], $CdBi_2S_4$, $CdBi_4S_7$, $Cd_{2.8}Bi_{8.1}S_{15}$, $Cd_2Bi_6S_{11}$ [81], $Ba_3Bi_{6.67}Se_{13}$ and $Ba_3MBi_6Se_{13}$ [82] (M = Sn, Pb). These types of compounds have shown low thermal conductivity, high thermopower, and often high electrical conductivity, with $CsBi_4Te_6$ having a *ZT* value of 0.8 at 225 K [73].

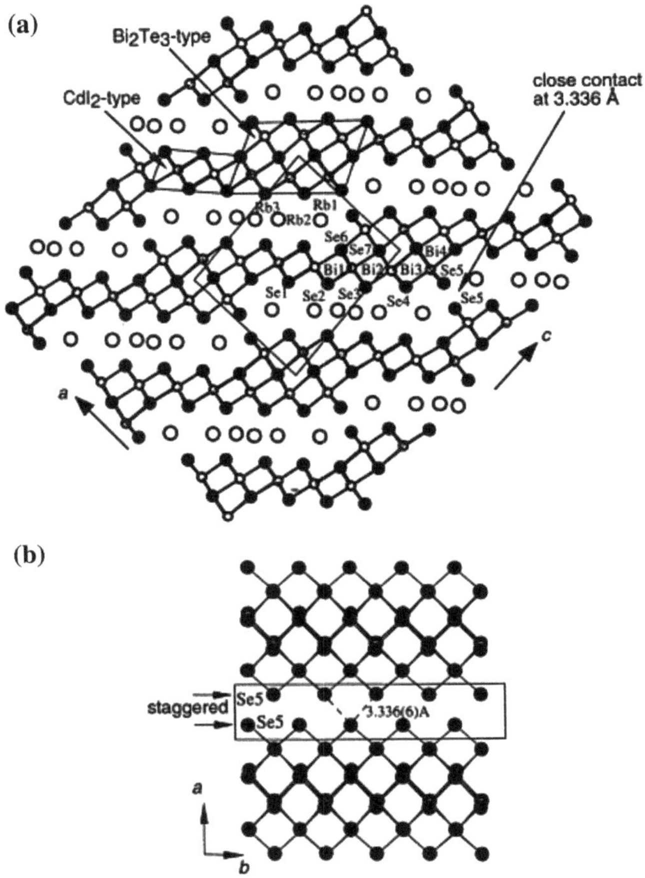

Fig. 4.7 2D-layered structure (**a**) and the Se5 atoms in each layer (**b**). Projection of structure of $Rb_2Bi_4Se_7$ down the *b*-axis. The Bi1 and Bi2 octahedra are distorted with Bi–Se bond lengths between 2.870(3) and 3.061(a) Å. In each Bi octahedron there is a short bound *trans* to a long bond. For example, in the Bi4 octahedra, there are two short Bi–Se bonds of 2.792(2) Å trans to two long bonds of 3.205(3) Å. The bond angles for Bi1 and Bi2 octahedra are more regular, varying between 86.74(8)° and 92.58(9)° for Bi1 and 86.55(8)° and 93.47(9)° for Bi2, while they are less regular in the Bi3 and Bi4 octahedra, varying between 82.06(8)° and 94.64(9)° and 82.05(8)° and 97.80(1)°, respectively. Reproduced with permission from [71], Copyright @ 2000, Wiley John and Sons

The common feature of these materials is that they have a high structural anisotropy. The bismuth chalcogenide frameworks are composed of condensed Bi–Q octahedra through sharing their edges to form blocks that have the NaCl-, Bi_2Te_3-, CdI_2- and Sb_2Se_3-type structures. Representative crystal structures are shown in Fig. 4.7 [71] and Fig. 4.8 [81]. These octahedral blocks possess various shapes and different sizes. They are usually connected one another either directly or through metal atoms of high coordination number (>6). These compounds are members of large homologous series that are defined by the adjustable blocks

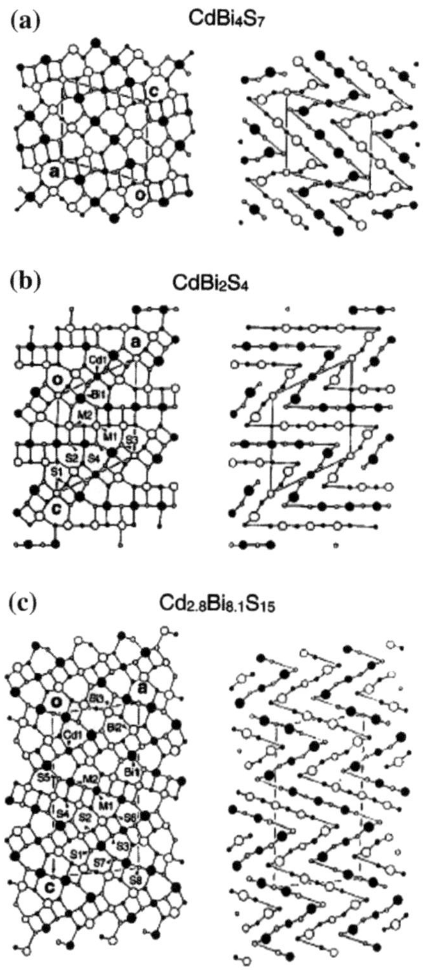

Fig. 4.8 The [010] projections of the $Cd_xBi_2S_{x+3}$ structures. We emphasize metal-sulfur bonds on the left side of the figures while we show the infinite zigzag line segments on the right. The open and closed circles are at $z = 0$ and $z = 0.5$, respectively. **a** The $CdBi_4S_7$ structure viewed down the [010] axis. $CdBi_4S_7$ is isotypic with Y_5S_7 structure type, with the atomic coordinates being taken from the Y_5S_7 crystal structure. This structure has a (4, 3) zigzag sequence. The Y1 site (which is possibly a Cd site in $CdBi_4S_7$) is located at the origin of the unit cell. **b** The $CdBi_2S_4$ structure viewed down the [010] axis, which has a (5, 3) zigzag sequence. **c** The $Cd_{2.8}Bi_{8.1}S_{15}$ structure viewed down the [010] axis, a (5, 3, 4, 3) zigzag sequence. Reproduced with permission from [81], Copyright @ 1997, American Chemical Society

following the same assembly principle. These characteristics bring out huge potential to synthesize new phases as thermoelectric materials with chemical and structural complexity, diversity, and disordering characteristics that are desirable to achieve high values of *ZT* [3, 83].

4.4.2.3 β-$K_2Bi_8Se_{13}$

β-$K_2Bi_8Se_{13}$ has been shown to have promising thermoelectric properties, due to its very low thermal conductivity and relatively high power factor [70]. Further studies indicate that the *ZT* value of this system can be substantially improved. β-$K_2Bi_8Se_{13}$ has a low-symmetry monoclinic structure, which consists of two

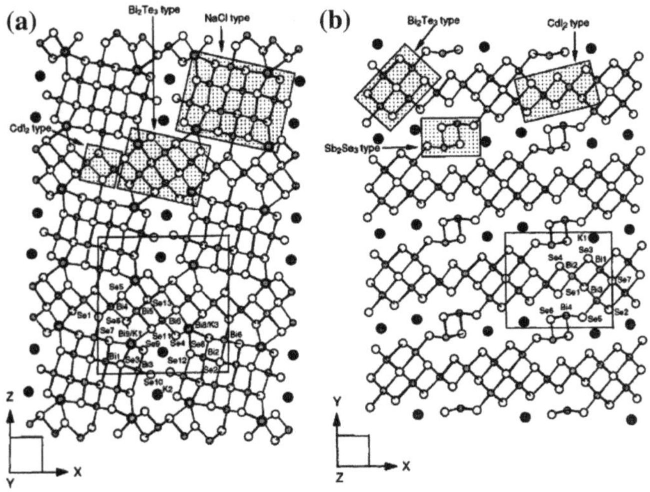

Fig. 4.9 **a** Projection of the structure of β-$K_2Bi_8Se_{13}$ viewed down the *b*-axis. NaCl-, Bi_2Te_3-, and CdI_2-type fragments are found in this framework, which are highlighted by the *shaded areas*. **b** Projection of the structure of α-$K_2Bi_8Se_{13}$ viewed down the *b*-axis. Sb_2Se_3-, Bi_2Te_3-, and CdI_2-type building blocks in the structure are highlighted by the *shaded areas*. Reproduced with permission from [70], Copyright @ 1997, American Chemical Society

different interconnected Bi/Se building blocks, the so-called NaCl (100)- and NaCl (111)-type, and K^+ ions in the channels, as shown in Fig. 4.9 [70]. These two Bi/Se blocks are infinitely extended along the crystallographic *b*-axis, which are connected to each other at special mixed occupancy K/Bi sites. Its structure is highly anisotropic, so that needle-shaped crystals are usually obtained.

According to experimental results, β-$K_2Bi_8Se_{13}$ single crystals have a low thermal conductivity of ~ 1.3 W m^{-1} K^{-1} and a relatively high power factor ($S^2\sigma \approx 10$ mW cm^{-1} K^{-2}) at room temperature, thus leading a figure of merit, $ZT = 0.22$ [70]. The *n*-type character of the compound is confirmed by its negative Seebeck coefficient. It is a highly degenerate semiconductor. Based on β-$K_2Bi_8Se_{13}$, solid solutions with the isostructure, such as $K_2Sb_8Se_{13}$, $Rb_2Sb_8Se_{13}$, and $K_2Bi_8S_{13}$, have been reported [84–89]. For example, polycrystalline 0.2 % Sn doped $K_2Bi_{8-x}Sb_xSe_{13}$ ($x = 1.6$) has been demonstrated to have a substantial improvement in power factor at high temperatures, with a slight reduction in thermal conductivity ($\kappa \approx 1.08$ W m^{-1} K^{-1}) [84]. It is predicted that this system of materials could offer a *ZT* of about 1 at 700–800 K. Their charge transport properties have also been studied at high pressures, where a significant increase in power factor and a peak in Seebeck coefficient are observed, suggesting the materials experience an electronic topological transition upon being subject to compression [90].

The alkali-metal bismuth selenide system is usually characterized by a mixed occupancy K/Bi disordering at the sites bridging the two different structural units. According to electron diffraction, charge transport properties and Hall coefficients of a series of solid solutions, such as $K_2Bi_{8-x}Sb_xSe_{13}$ and $K_{2-x}Rb_xBi_8Se_{13}$, it is suggested that the degree of disordering at the mixed occupancy K(Rb)/Bi(Sb) sites has played a very important role in affecting the thermoelectric properties of these materials [85]. This suggestion is highly supported by the results of ab initio density functional theory band structure calculations [91]. Therefore, specific attention should be paid to these sites for further optimizing the thermoelectric properties of these systems.

4.4.2.4 $CsBi_4Te_6$

$CsBi_4Te_6$ is a promising thermoelectric material for low-temperature applications. This compound has a strong anisotropic structure, containing both Bi^{3+} and Bi^{2+} centers. The reduced Bi^{2+} centers lead to Bi–Bi bonds of 3.238 Å. The compound has a lamellar structure with slabs of $[Bi_4Te_6]^-$, alternating with layers of Cs^+ ions, as shown in 4.10 [92]. $CsBi_4Te_6$ is so sensitive to the type and level of doping elements that both *p*-and *n*-type materials can be synthesized at their corresponding conditions. The presence of Bi–Bi bonds in the structure results in a very narrow energy gap of about 0.08 eV, which is almost half of that of Bi_2Te_3 [92–94]. The narrower band gap is responsible for the maximum *ZT* value in $CsBi_4Te_6$ to appear at lower temperatures than that of Bi_2Te_3. With different dopants, *p*-type $CsBi_4Te_6$ has high values of power factor of >30 mW cm^{-1} K^{-2} at 100–220 K. Materials doped with Sb, Bi, SbI_3, and BiI_3 have power factors of 40–60 mW cm^{-1} K^{-2} at 150–180 K. *n*-type $CsBi_4Te_6$ doped with In_2Te_3 and Sn, exhitits power factors of about 25 mW cm^{-1} K^{-2} at 100–150 K [73].

Band calculations have predicted that the electronic structural features of $CsBi_4Te_6$ make it highly potential in developing thermoelectric materials with desirable performances for practical applications [95]. The most important feature leading to the high *ZT* value of $CsBi_4Te_6$ is the large anisotropy in its effective mass. This hypothesis has been confirmed in *p*-type $CsBi_4Te_6$ by angle-resolved photoelectron spectroscopy studies (ARPES) [93]. Another report [94] on the band structure of $CsBi_4Te_6$ suggests that the carrier concentration of the best *p*-type $CsBi_4Te_6$ material ($ZT \approx 0.8$) is close to the optimal value achievable, but the thermoelectric properties of *n*-type $CsBi_4Te_6$, when optimally doped, may exceed those of the *p*-type doped materials. It would be interesting to experimentally investigate these predictions.

4.4.2.5 Bi_2Te_3

Bi_2Te_3 is a narrow-gap semiconductor with an indirect gap of about 0.15 eV. It has rhombohedral crystal structure, with space group of $R\overline{3}m$. The structure comprises plates, which are made up of five atomic layers (Te1–Bi–Te2–Bi–Te1) and stacked through van der Waals interactions along the *c*-axis in the unit cell. The currently widely used Bi_2Te_3 materials with $ZT \approx 1$ are synthesized by alloying with Sb to form p-type and Se to form n-type components. To assemble devices, the p-type "legs" are generally $Bi_{2-x}Sb_xTe_3$ ($x \approx 1.5$) pellets, prepared by using hot-pressing and thermal annealing, so as to have required mechanical properties. The n-type counterpart is usually $Bi_2Te_{3-y}Se_y$ ($y \approx 0.3$) ingot that is grown by using zone melting techniques.

Significant progresses in enhancing the ZT value of Bi_2Te_3-based bulk materials have been made in recent years (Fig. 4.19) [96–98]. A quasi-layered nanotube n-type Bi_2Te_3 with a holey one-dimensional structure, synthesized by using hydrothermal method, has been reported to have $ZT \approx 1.0$ at 450 K [96]. By imbedding the nanotubes in a Bi_2Te_3 matrix, grown by using the zone melting technique, structural composite thermoelectric materials with n-type properties can be obtained, which exhibit a ZT value of 1.25 at 420 K. In contrast, a combined operation of melt spinning followed by spark plasma sintering (SPS) results in p-type Bi_2Te_3 ingots with a ZT value of 1.35 at 300 K [97]. Within the composite structure, there are 25 nm wide ribbons composed of nanostructure layers of Bi_2Te_3 crystals with an interplanar distance of about 1 nm.

The highest ZT value for was reported recently. It has been reported more recently that a p-type Bi_2Te_3 bulk material has a $ZT \approx 1.4$ at 100 °C. It is fabricated by using ball milling followed by direct current hot-pressing [98]. The enhancement in ZT value of this material system has been attributed to the reduction in phonon thermal conductivity while retaining a comparable power factor to that of the bulk p-type $Bi_{2-x}Sb_xTe_3$. It is still a single-phase material, but a mixture of nanograins and macrograins. A significant concern is the long-term stability of the nanograined structures during application at high temperatures.

It has been expected that superlattice structures with low dimensionality can be used to greatly enhance the thermoelectric figure of merit values of ZT, because they have confinement effects on the electronic DOS. For example, a p-type Bi_2Te_3/Sb_2Te_3 superlattice thin film grown by using chemical-vapor deposition (CVD) method can reach a record high value of $ZT \approx 2.4$ at room temperature [99], which is much higher the long-standing $ZT \approx 1$ record of Bi_2Te_3 based alloys. This high record of ZT has not been validated separately. In addition, this high ZT value of 2.4 is available only in the direction perpendicular to the artificially constructed Bi_2Te_3 layers, which have been attributed to two factors: (i) exceedingly low lattice thermal conductivity (0.24 W m^{-1} K^{-1}) and (ii) a cross-plane electron mobility that is equal to the in-plane electron mobility. This observation is very surprising because the cross-plane mobility is less than half that of the in-plane mobility in pure Bi_2Te_3 thin film. It has been suggested that the first factor-low lattice thermal conductivity is due to the phonon back reflection at the Bi_2Te_3/Sb_2Te_3 interfaces. One of the reasons for the second factor is the very small band offsets between the Bi_2Te_3 and Sb_2Te_3 respective layers. However, this explanation is still in argument. If it is true, the cross-plane mobility should also be equal to that of in-plane mobility in pure Bi_2Te_3, because the valence band offsets between Bi_2Te_3 layers is zero. Furthermore, thermoelectric materials in the form of thin films will encounter difficulties in large-scale applications, as compared with bulk materials [6].

4.4.3 Isotropic Chalcogenide Compounds

4.4.3.1 PbTe

PbTe is a unique thermoelectric material for applications at intermediate temperatures (600–800 K). It has the NaCl crystal structure, with Pb atoms as the cations and Te atoms as anions. With a band gap of 0.32 eV, PbTe can be readily made to be either *n*- or *p*-type semiconductor through appropriate doping. PbTe has been reported to maximum *ZT* values of 0.8–1.0 at a temperature of near 650 K. The underlying reasons for the high thermoelectric performance of PbTe have been well clarified [100]. The lattice thermal conductivity of PbTe is about 2.2 W m^{-1} K^{-1} at room temperature and decreases with a $1/T$ dependence at higher temperatures. Strategies to further improve the thermoelectric properties of PbTe include the development of nanostructures and the modification of DOS, so as to create resonance states in the conduction band, thus increasing the Seebeck coefficient. In addition, mechanical properties of PbTe have also been investigated for potential device applications [101–103].

The electronic states produced in a semiconductor due to the doping of elements that can resonate with the semiconductor matrix [104–107]. According to the theoretical prediction on the effect of Group 13 dopants on the electronic structure of PbTe, corresponding experiments have been carried out on PbTe and $Pb_{1-x}Sn_xTe$ by introducing various impurities [107]. Some elements have no such effect, while others are very promising in creating a resonance. For example, when $Pb_{1-x}Sn_xTe$ is doped with in, the resonance level of In is located neither in the conduction nor in the valence band, therefore, no enhancement in thermoelectric properties has been observed [108]. In contrast, is PbTe doped with Tl, the resonant state is located in the valence band. This technique has been confirmed in a PbTe:Tl system, which has a large *ZT* value of 1.5 at 773 K. due to the increase in Seebeck coefficient [109]. An increase in the DOS has been evidenced by the experimental results of specific heat measurements by using optical spectroscopic technology (Fig. 4.10).

However, the above mentioned strategies are not very effective in reducing the thermal conductivity of this system. Therefore, if the resonant state technique could be coupled with a reduction in thermal conductivity, further increase in *ZT* value should be achievable. It has been shown that nanostructure PbTe have different thermal and charge transport properties, as compared with those of the bulk counterparts, which will be discussed later. While nanostructuring has been acknowledged to an important technique to increase the performance of PbTe-based materials, solid solution alloying is still of great significance in terms of large-scale applications, as shown by the examples in Ref. [110].

Figure 4.11 shows powder *X*-ray diffraction (XRD) patterns of $Pb_{9.6}Sb_{0.2}Te_{10-x}Se_x$ (x = 0 to 10) samples [110]. All samples are of single phase with NaCl-type structure ($Fm\bar{3}m$). The lattice parameter decreases almost linearly with increasing content of Se (Fig. 4.11b). This is because Se atom is smaller than Te atom.

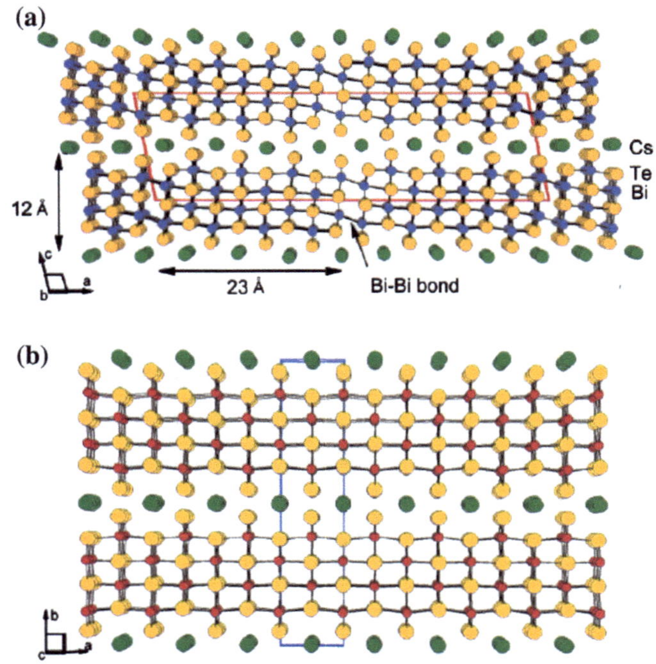

Fig. 4.10 Perspective view of the structures of **a** $CsBi_4Te_6$ along the *b*-axis, with Bi atoms in blue and Te atoms in *yellow*. **b** $CsPbBi_3Te_6$ composed of NaCl-type layers. The red atoms are Bi or Pb atoms in mixed occupancy. In $CsBi_4Te_6$, the [Bi_4Te_6] slabs have a finite width, with 12×23 Å^2 cross-section area, which are interconnected side by side and are linked by Bi–Bi bonds at 3.238(1) Å. In $CsPbBi_3Te_6$, the slabs have infinite width and extend continuously along the crystallographic *a*-axis. Reproduced with permission from [92], Copyright @ 2004, American Chemical Society

According to single-crystal XRD patterns, a sample of $Pb_{9.6}Sb_{0.2}Te_{10}$ has lower symmetry ($Pm\overline{3}m$) and a composition of $Pb_{3.82}Sb_{0.12}Te_4$. It means that the $Pb_{9.6}Sb_{0.3}Te_{10}$ exhibits a pseudo NaCl structure type, so that the symmetry is reduced to $Pm\overline{3}m$, thus causing the positions of Na and Cl to split into two crystallographically independent metal positions and two anion positions. It is also found that there is Pb/Sb atomic ordering in the structure, as shown in Fig. 4.12a. Figure 4.12b indicates that the M1(1/2, 1/2, 1/2) position is occupied by Pb and Sb with a total occupancy factor of 94 % (82 %Pb + 12 %Sb + 6 % vacancy), while the M2(1/2, 0, 0) position is fully occupied by Pb.

According to TEM image of $Pb_{9.6}Sb_{0.2}Te_7Se_3$ shown in Fig. 4.13a, nanocrystals with varying sizes and shapes are embedded and dispersed inside the PbTe-rich matrix [110]. The wide size distribution and the uniform dispersion of the nanocrystals are effective to scatter phonons over mid-to-long wavelength, which is beneficial to the reduction of lattice thermal conductivity. The Sb-rich nanocrystals are formed during the cooling process. Due to the slight lattice mismatch

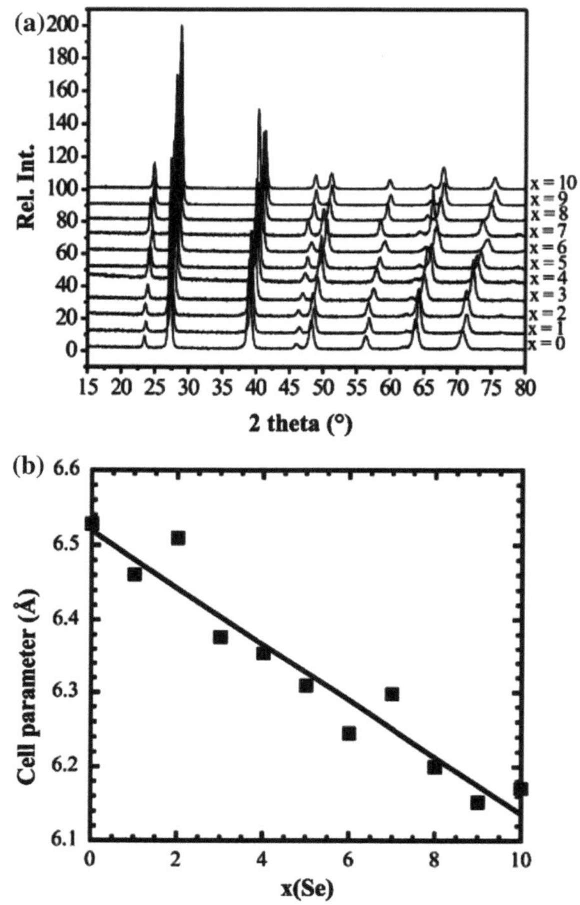

Fig. 4.11 **a** Powder XRD patterns of the $Pb_{9.6}Sb_{0.2}Te_{10-x}Se_x$ (x = 0 to 10) samples, showing lattice contraction with increasing x (Se). **b** Variation in cell parameter (refined from the power patterns) as a function of the content of Se (x). Reproduced with permission from [110], Copyright @ 2006, American Chemical Society

between regions with different compositions, crystal boundaries are formed, as shown in Fig. 4.13b and c. Different domains also have different periodicity, as demonstrated in Fig. 4.13d and e. These features further lead to decreased thermal conductivity. Maximum *ZT* of about 1.20 at 650 K has been achieved in $Pb_{9.6}Sb_{0.2}Te_3Se_7$.

4.4.3.2 $AgSbTe_2$

$AgSbTe_2$ is a *p*-type thermoelectric material with a *ZT* value of as high as 1.3 at 720 K [2]. This material is the only ternary phase that is found in the Ag–Sb–Te ternary phase diagram. Although its simple ternary stoichiometry seems to quite simple, the compound actually has very complex structures, as revealed by the recent studies. According to the phase diagram of the Ag–Sb–Te system, indicates

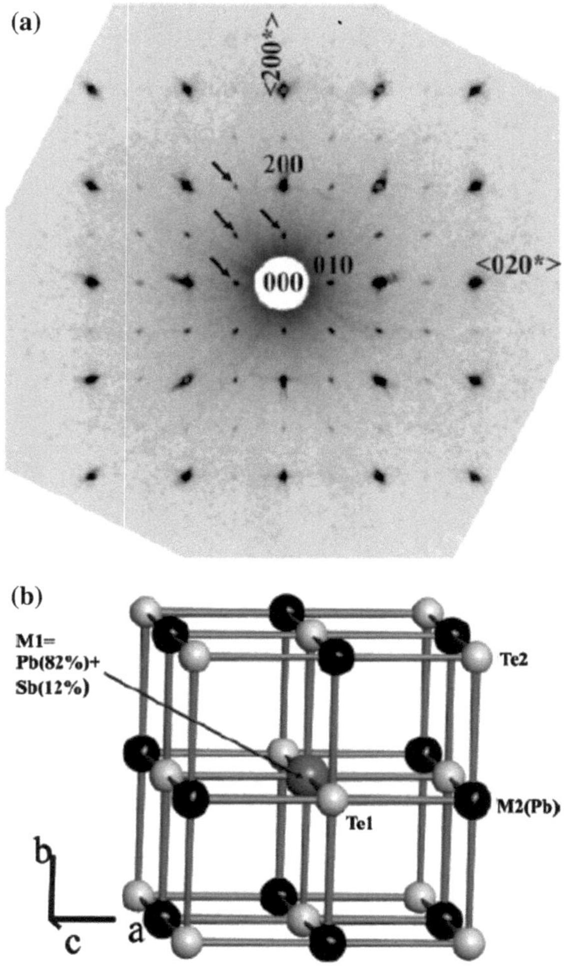

Fig. 4.12 **a** Precession photograph of *hk*0 zone simulated from image plate diffraction data of the $Pb_{9.6}Sb_{0.3}Te_{10}$ single crystal. The arrows indicate supplementary Bragg diffraction spots breaking the $Fm\overline{3}m$ symmetry. **b** Average cubic structure of $Pb_{9.6}Sb_{0.3}Te_{10}$ in the space group $Pm\overline{3}m$. The metal position M1 is mixed occupied by Pb and Sb with 6 % vacancy. Reproduced with permission from [110], Copyright @ 2006, American Chemical Society

$AgSbTe_2$ is metastable at low temperatures, which is prone to decompose into Ag_2Te and Sb_2Te_3 [111, 112]. Therefore, most $AgSbTe_2$ samples reported in the open literature could be in fact three-phase mixtures of $AgSbTe_2$, Ag_2Te, and Sb_2Te_3. This hypothesis has been confirmed by experimental results [113].

$AgSbTe_2$ is of the rock salt structure, in which Ag and Sb atoms are supposed to be randomly distributed in the Na sites. However, it has been shown that there is ordering Ag and Sb atoms in the structure. The presence of the ordering is expectable, because random distribution of Ag and Sb in the Na sites would create second-neighbor contacts of Ag^+–Ag^+ and Sb^{3+}–Sb^{3+}, which are energetically unfavorable when compared with the contacts of Ag^+–Sb^{3+}. Due to the very similar scattering lengths of the three atoms, it is difficult to identify the localized ordering in $AgSbTe_2$ by using the common diffraction techniques, such as *X*-ray,

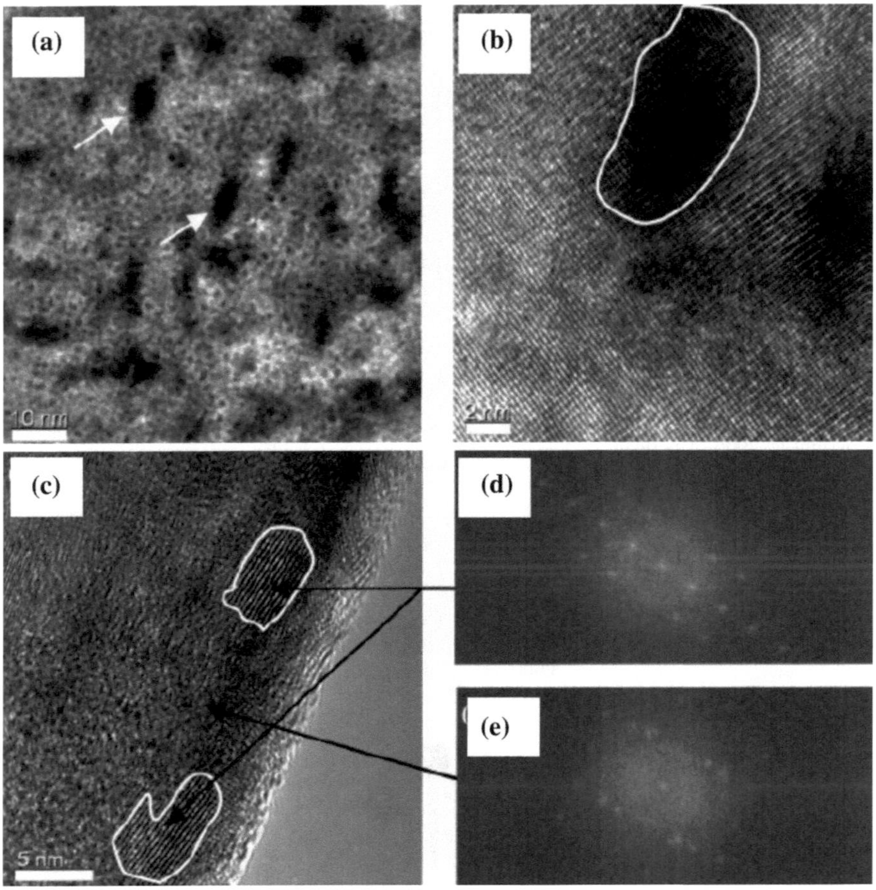

Fig. 4.13 **a** Low magnification TEM image of the $Pb_{9.6}Sb_{0.2}Te_7Se_3$, showing Sb-rich nanocrystals of various sizes and shapes embedded inside the PbTe-rich matrix. **b** High-resolution image (HRTEM) of a selected portion of the $Pb_{9.6}Sb_{0.2}Te_7Se_3$ sample, showing an about 5 nm nanocrystal embedded inside the PbTe-rich matrix. **c** High-resolution image (HRTEM) showing the coexistence of domains with different features. The fast Fourier transform (FFT) of both domains showed smaller periodicity for the islands compared with the bulk matrix. Reproduced with permission from [110], Copyright @ 2006, American Chemical Society

neutron and electron. As a result, unique ordering model based on structure refinement has not been established, although evidence of ordering has been confirmed. Therefore, it is reasonable to suggest that there should be ordered and disordered regions of Ag and Sb atoms in $AgSbTe_2$.

From these discussions, it is now readily understandable that there are significant complexities in the chemistry of $AgSbTe_2$, which have been the reasons for the argument on the nature of the structure, composition and conductive properties (semiconductor or semimetal) of it [114, 115]. Temperature dependence of its resistivity shows that there are energy gaps with values ranging from 0.6 to 0.2 eV. Galvanomagnetic and thermoelectric studies indicate that Hall coefficients of the samples prepared under similar conditions can be either positive or negative, while Seebeck coefficient is always positive, which suggests strong variation from one sample to the other [116]. Noting the thermodynamic instability of $AgSbTe_2$ as observed in its phase diagram, it is expected that the processing conditions of synthesis and crystal growth could have strong effects on the properties and thus performances of these thermoelectric materials.

First principle calculations have been employed to study the electronic, optical, and lattice vibrational properties of $AgSbTe_2$, through the use of the generalized gradient approximation (GGA), which indicate that it is semimetal because there is a negative band gap [117]. However, this prediction is not in agreement with experimental observations. This problem has been addressed by using the screened-exchange local-density approximation (sx-LDA) method. The sx-LDA results predict a vanishing density of states at the Fermi level, which is consistent with the semiconducting behavior of $AgSbTe_2$ [118]. Various optical properties, including the dielectric characteristics, absorption coefficient and refractive index, as a function of the photon energy, have also been theoretically studied by using the sx-LDA method, with all predictions being in a very good agreement with experimental results.

According to calculated phonon spectra, the optical modes of $AgSbTe_2$ have a very low frequency and thus would scatter strongly with acoustic modes during heat transport. The scattering of acoustic phonons by optical modes and the possible presence of the Ag/Sb disordering can be used to explain the extremely low lattice thermal conductivity of $AgSbTe_2$. However, the effects of the presence of Ag_2Te and Sb_2Te_3 in the matrix of $AgSbTe_2$ as mentioned above have not been taken into account during these theoretical studies. Therefore, these conclusions still need further clarifications.

Experimental studies have confirmed that $AgSbTe_2$ is a semiconductor with a very narrow energy gap of only 7 meV. It possesses highly mobile electrons to dominate the Hall measurements and holes in a heavy band to dominate the thermoelectric power [119, 120]. Because this gap energy is comparable to the thermal energy, k_BT, as the temperature is above 100 K, it is suggested that $AgSbTe_2$ can be considered as an indirect zero-gap material above 100 K.

4.4.3.3 $AgSbTe_2$/GeTe

$AgSbTe_2$ and GeTe can be used to form alloys $(AgSbTe_2)_{1-x}(GeTe)_x$, which are usually called TAGS-*m*, with *m* representing the mole percent of GeTe [121]. They are intrinsically *p*-type semiconductors, which are typically used together with PbTe as *n*-type leg in thermoelectric modules. The materials with compositions of $m = 80$ and 85 have *ZT* values of 1.4 and 1.5 at 750 K, respectively, which stimulated the subsequent interests in the materials with (GeTe)-rich compositions.

This is one of the materials to possess figure of merit $ZT > 1$ at high temperatures that have been used for practical applications.

There is a polymorphic transformation at 510 K, from a low-temperature polar rhombohedral ($R3\,m$) to a high temperature NaCl-type cubic ($Fm\overline{3}m$) structure, in the TAGS system [122]. The transition temperature is determined by the ratio of GeTe to $AgSbTe_2$ and the ratio of Sb to Ag. According to the results of high-temperature *X*-ray diffraction (XRD) and TEM observation, no evidence has been found to show the presence of any second phases, either coherent or incoherent, in this material [122, 123]. This implies that the GeTe-$AgSbTe_2$ system is a proper solid solution [124].

As the composition is varied from $AgSbTe_2$ to GeTe in the solid solution, the transport properties vary transiently, with the presence of a double minimum in thermal conductivity for the samples with 80 and 85 % GeTe. These minima are corresponding to the extremely low lattice thermal conductivity (0.3–1.0 W m^{-1} K^{-1}) of the two samples. Meanwhile, the materials possess large carrier mobilities, which is in the range of 100–200 cm^2 V^{-1} s^{-1}. Therefore, *ZT* values of as high as 1.7 can be achieved in the sample of TAGS-80 [125].

4.4.4 Oxide Thermoelectric Materials

Significant progress has been achieved in developing thermoelectric materials. However, the stability and toxicity of the above mentioned compounds have become a critical issue for some applications. Therefore, it is desired to develop oxide based thermoelectric materials due to their less toxic property and relatively high stability. Until now, it is still a challenge to find an oxide based thermoelectric material with performance similar or close to the best nonoxide counterparts. There are also *n*-type and *p*-type oxide theremoelectric materials. The strageties that have been used to improve the performances of oxide thermoelectric materials are mainly doping of other elements. Since oxide thermoelectric materials have been well reviewed [3, 126, 127], a brief introduction will be presented in this subsection.

4.4.4.1 *p*-Type Thermoelectric Oxides

Alkali and alkaline-earth cobaltite compounds have been considered to the most promising *p*-type oxide thermoelectric materials, because of their large Seebeck coefficients, as evidenced by the changes in oxygen content, have been reported to exhibit a significant influence on thermoelectric properties of $Ca_3Co_4O_9$ [128]. $Ca_3Co_4O_9$ has a layer structures containing CoO_2 planes, thus forming a path for *p*-type electronic conduction, while the interfaces between every two adjacent layers and the other structural components scatter lattice phonons to reduce heat

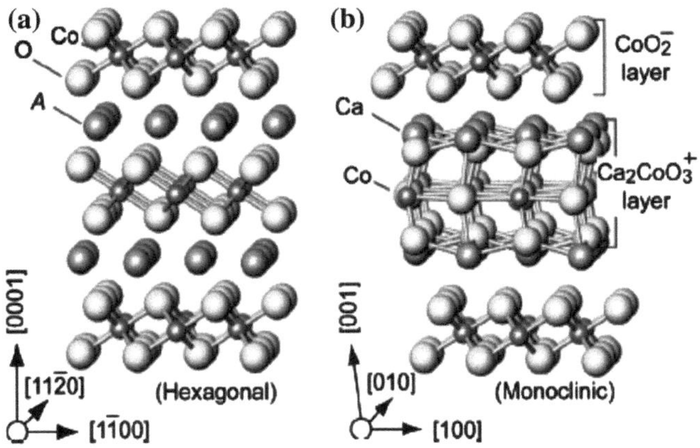

Fig. 4.14 Schematic structures of A_xCoO_2 (A = Li, Na, Ca, Sr, etc.) (**a**) and $Ca_3Co_4O_9$ (**b**). Reproduced with permission from [129], Copyright @ 2008, American Chemical Society

conductivity. Figure 4.14 shows the schematic structures of $Ca_3Co_4O_9$ [129–132] and Na_xCoO_2 [133–135], which are two of the most promising thermoelectric oxides reported in the open literature. The CoO_2 planes in $Ca_3Co_4O_9$ are separated by a layer of Ca_2CoO_3 to form a rock salt type structure (Fig. 4.14b), while those in Na_xCoO_2 are separated by a layer of sodium ions (Fig. 4.14a). Similar compound can also be formed with calcium ion (Ca^{2+}), Ca_xCoO_2 [136]. There is also another calcium cobaltite, $Ca_3Co_2O_6$ [137]. $Ca_3Co_2O_6$ has a larger Seebeck coefficient but a low electrical conductivity, so that $Ca_3Co_4O_9$ is more suitable candidate for thermoelectric applications.

Thermoelectric properties of $Ca_3Co_4O_9$ have been systematically reported and well documented, which can be found in open literature [138–144]. Typical properties of $Ca_3Co_4O_9$ include: conductivity of about 104 S m^{-1}, a Seebeck coefficient of about 150 $\mu V\ K^{-1}$ and a thermal conductivity of about 2 W K^{-1} m^{-1}. Its thermoelectric properties can be further improved through doping of other elements. Bismuth (Bi) has been the most frequently used as dopant to enhance the thermoelectric performance of $Ca_3Co_4O_9$. The doping of Bi leads to an increase in both the electrical conductivity and Seebeck coefficient, and a decrease in thermal conductivity simultaneously.

The increase in conductivity is attributed to the increase in carrier mobility, rather than carrier concentration, which is generally desired to increase the figure of merit of thermoelectric materials. This is because an increase in carrier concentration leads to a decrease in the Seebeck coefficient. In addition, the improvement in conductivity has also been attributed to the change in the microstructure caused by the doping. The decrease in thermal conductivity has been ascribed to the larger size and heavy mass of Bi as compared with those of Ca. *ZT* values of the $Ca_3Co_4O_9$ samples doped with Bi and other ions have been well available in the open literature [127]. Although doing of Bi can enhance all the three important properties that are required to increase the *ZT*, the experimental values of *ZT* still need to be further improved, as compared with that of pure $Ca_3Co_4O_9$.

Silver (Ag) is the element that has the greatest impact on the figure of merit of $Ca_3Co_4O_9$. Ag can be introduced as a dopant or as a second (metallic) phase. As a dopant, Ag has been shown to increase the electrical conductivity [145, 146], which is attributed to an increase in carrier concentration and mobility. The Seebeck coefficient can be either increased or decreased due to the addition of Ag. The introduction of Ag is helpful in decreasing the thermal conductivity, which has been attributed to its large mass. Silver has also been shown to increase electrical conductivity as a codopant with barium (Ba) or lutetium (Lu). The presence of silver as a second phase also increases the electrical conductivity, but, in contrast to its use as a dopant, leads to a decrease in the Seebeck coefficient. When combined with a dopant, however, the silver second phase can lead to an increase in both electrical conductivity and thermopower.

Transitional metallic elements have also been widely used to increase the properties of $Ca_3Co_4O_9$ [143, 147, 148]. The replacement of Co with copper (Cu) always increases electrical conductivity of $Ca_3Co_4O_9$, but at the same time Seebeck coefficient is also decreased. Cu occupies the sites in the Ca_2CoO_3 rock salt layer instead of the CoO_2 layer. Cu has also been used to replace Ca, which however, results in an increase in Seebeck coefficient. However, a second phase is detected by using XRD, implying that Cu is not stable at the Ca sites. Iron (Fe) and manganese (Mn) are found to occupy Co sites in the CoO_2 layer. Some transitional elements have shown to be able increase electrical conductivity, while most of can increase thermopower of $Ca_3Co_4O_9$ only. Heavier transitional metals have different effects. For example, rhodium Rh) increases electrical conductivity but decreases thermopower, whereas and tantalum (Ta) increases thermopower, but decreases electrical conductivity.

Figure 4.15 shows XRD patterns of $Ca_3Co_2O_6$ ceramics doped with various transitional metallic elements, forming $Ca_3Co_{1.8}M_{0.2}O_6$ (M = Mn, Fe, Co, Ni, and Cu), together with $Ca_{2.7}Na_{0.3}Co_2O_6$ as a comparison [148]. Except for Ni, all samples are of nearly single-phase $Ca_3Co_2O_6$. It means that it is difficult to replace the Co site with Ni. Both *a* and *c* lattice parameters are increased when Co site is partially replaced with Mn, while only *c* parameter is increased when Cu is used. The variation in lattice parameter is negligible for other elements due to their similar ionic radius to that of Co. In addition, the presence Mn and Fe prohibits grain growth and that of Cu and Na promotes grain growth of $Ca_3Co_2O_6$.

Resistivities ($\log\rho$) of the samples are shown in Fig. 4.16. A thermally activated behavior is observed in the samples over the entire temperature range. However, the plots of log ρ versus 1,000/*T* are not straight line, especially in the high temperature region above 500 K. This behavior is not due to neither the phase transition nor the oxygen loss, according to thermal analysis experiments. Instead, it can be attributed to the complicated band structure due to the highly anisotropic crystallographic structure of $Ca_3Co_2O_6$. In addition, grain boundary effect and the

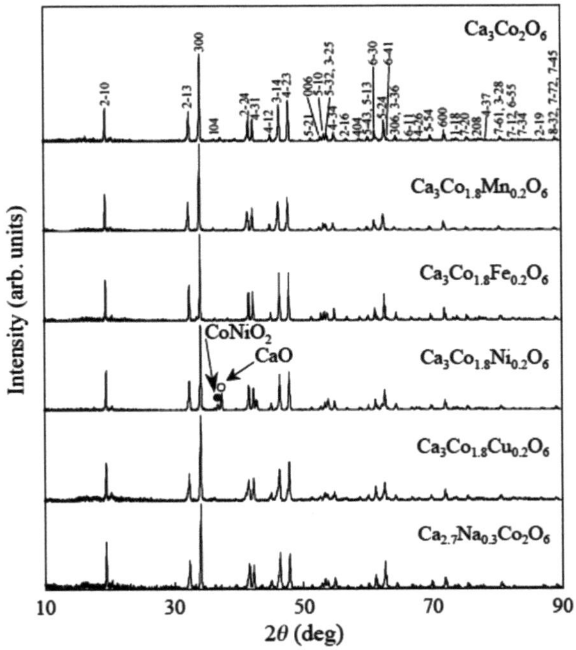

Fig. 4.15 XRD patterns of the $Ca_3Co_{1.8}M_{0.2}O_6$ (M = Mn, Fe, Co, Ni, and Cu) and $Ca_{2.7}Na_{0.3}Co_2O_6$ ceramics. Reproduced with permission from [148], Copyright @ 2005, Elsevier

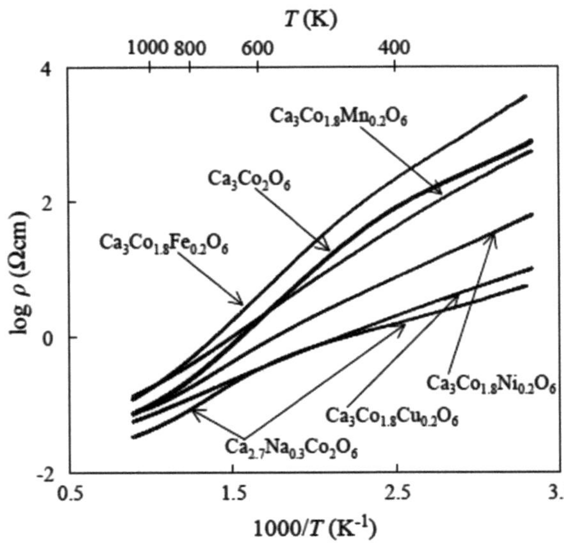

Fig. 4.16 logρ versus 1000/T of the $Ca_3Co_{1.8}M_{0.2}O_6$ (M = Mn, Fe, Co, Ni, and Cu) and $Ca_{2.7}Na_{0.3}Co_2O_6$ ceramics. Reproduced with permission from [148], Copyright @ 2005, Elsevier

impurity phase could also be contributors to the complex nature of resistivity. The value of ρ decreases rapidly with increasing temperature. Since a high ZT requires low ρ, this property is desired for high temperature applications of thermoelectric materials.

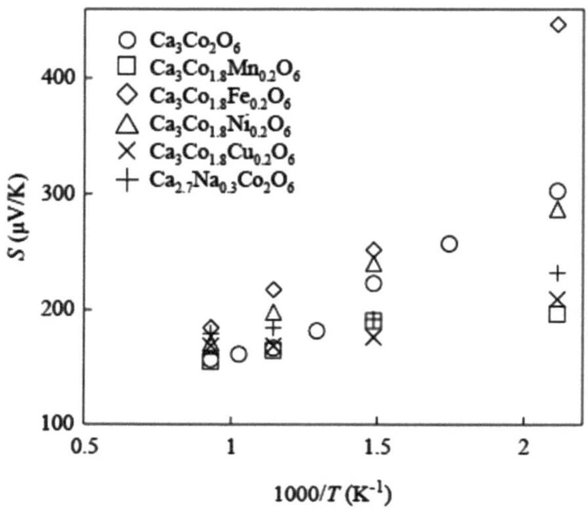

Fig. 4.17 Seebeck coefficient (*S*) versus 1000/T of the $Ca_3Co_{1.8}M_{0.2}O_6$ (M = Mn, Fe, Co, Ni, and Cu) and $Ca_{2.7}Na_{0.3}Co_2O_6$ ceramics. Reproduced with permission from [148], Copyright @ 2005, Elsevier

Figure 4.17 shows temperature dependence of Seebeck coefficients (*S*) of the samples. Similarly, a thermally activated behavior is present. However, the activation energy values for *S* are much smaller than those ρ [148]. This suggests the presence of a polaron conduction and the difference in magnitude corresponds to the hopping energy, which contributes to the change of mobility by temperature. The Fe substitution increases the value of *S*, those of Cu, Mn, and Na decrease *S* at temperatures below 673 K. At the same time, the Ni has almost negligible effect on the value of *S*. Although the value of *S* decreases with increasing temperature, all samples exhibit large *S* values of >160 mV K^{-1}, even at a temperature of as high as 1,073 K. Together with that of ρ, this variation of *S* with temperature makes the materials for high temperature applications.

The thermoelectric power factors (PF = S^2/ρ) of the samples, derived from was calculated from measured ρ and *S* values, are shown in Fig. 4.18. The PF values increase with increasing temperature. This temperature dependence is mainly caused by the significant reduction of ρ at high temperatures. Although the sample doped with Fe has an increased value of *S*, the increase in ρ does not result in a large increase in PF. The Cu and Na substitutions lead to an enhancement in PF due to the reduction in ρ without serious degradation in *S*. The PF value of the Na-substituted sample at temperatures of above 873 K is more than two times that of the pure sample. The enhancement in PF of the samples could also be related to their microstructures as a result of the substitution.

Lanthanide elements have been considered as suitable doping elements to improve the performance of $Ca_3Co_4O_9$ and $Ca_3Co_2O_6$ thermoelectrical oxides, due to their large ionic radius, heavy atomic mass and special electronic structure [149–151] have been reported. The introduction of lanthanide elements, including neodymium (Nd) [152], europium (Eu) [149], holmium (Ho), dysprosium (Dy), erbium (Er) [153], ytterbium (Y) [131] and lutetium (Lu) [153], increases

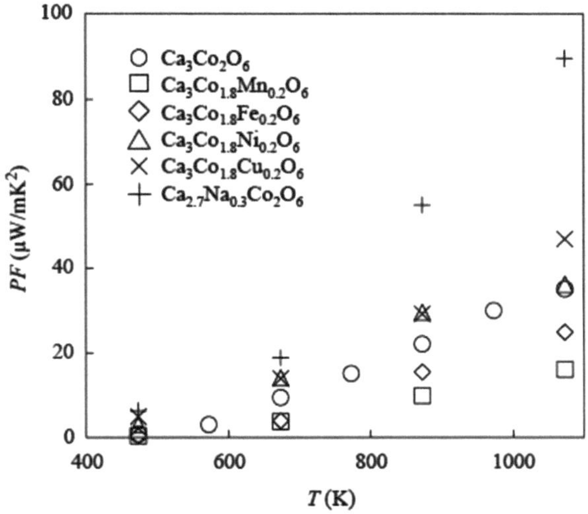

Fig. 4.18 Temperature dependence of Power factor (PF) as a function of temperature of the $Ca_3Co_{1.8}M_{0.2}O_6$ (M = Mn, Fe, Co, Ni, and Cu) and $Ca_{2.7}Na_{0.3}Co_2O_6$ ceramics. Reproduced with permission from [148], Copyright @ 2005, Elsevier

thermopower and reduces electrical conductivity. Controversial results have been reported in different occasions, in terms of its effect on thermopower and electrical conductivity [134, 135]. Addition of gadolinium (Gd), dysprosium (Dy), holmium (Ho), and ytterbium (Y) is able to decrease thermal conductivity.

Both sodium (Na) and potassium (K) have been used to relapse Ca of $Ca_3Co_4O_9$ and $Ca_3Co_2O_6$ [144, 148, 154, 155]. Na is also used as a codopant with other elements, such as Mn and Nd, because single-doping with Na has very weak effect in increasing the Seebeck coefficient of $Ca_3O_4O_9$ [144, 155–157]. The replace of Ca with K can increase electrical conductivity of $Ca_3O_4O_9$, together with a slight increase in Seebeck coefficient but an increase in thermal conductivity. Therefore, the increase in the figure of merit due to the doping is moderate. Codoping of with La increases electrical conductivity, without significant effect on Seebeck coefficient [154].

Due to their same chemical valency, barium (Ba) and strontium (Sr) can be used to replace Ca of $Ca_3O_4O_9$ without charge disruption. The effect of Ba is controversial, which can be attributed to microstructural difference between the samples prepared by using different methods, in different reports. Sr addition can reduce thermal conductivity of $Ca_3Co_4O_9$, which leads a slight increase in the figure of merit.

Na_xCoO_2 is another important layered cobaltite based thermoelectric oxide [158–164]. The variation in properties of Na_xCoO_2 is large than that of $Co_3Co_4O_9$, Na content can be varied in a wide range. In general, however, the electrical conductivities of Na_xCoO_2 are higher than those of $Co_3Co_4O_9$ family. Figure 4.19 shows XRD patterns of the Na_xCoO_2 (x = 0.65–0.85) samples, prepared by using the conventional solid-state reaction method [163]. All samples can be indexed as hexagonal γ-Na_xCoO_2 phase with x up to 0.80. The sample with x = 0.85 has a trace of Na_2CO_3. The lattice parameters a and c are found to increase and decrease, respectively, with increasing content of Na, without noticeable change in

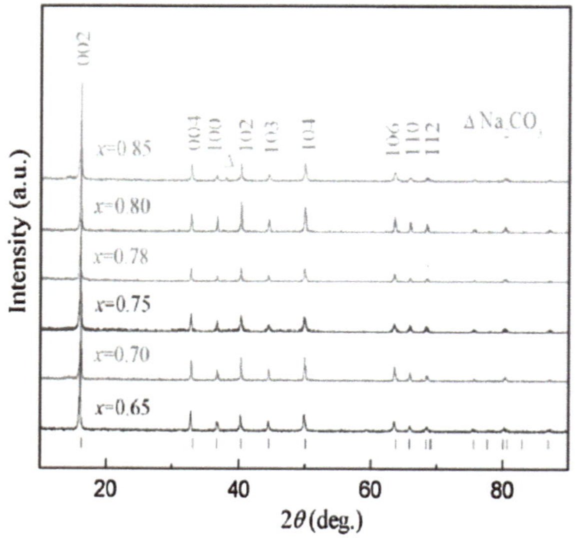

Fig. 4.19 XRD patterns of the as-synthesized Na_xCoO_2 (x = 0.65–0.85) samples. Reproduced with permission from [163], Copyright @ 2008, Elsevier

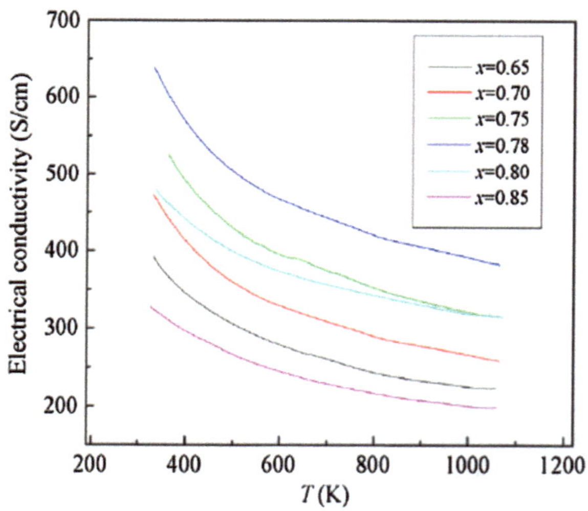

Fig. 4.20 Temperature dependences of electrical conductivity of the as-synthesized Na_xCoO_2 (x = 0.65–0.85) samples. Reproduced with permission from [163], Copyright @ 2008, Elsevier

cell volumes. The increase in a could be attributed to the gradual increase in coulomb repulsion of Na ions in the same layer, while the shrinkage in c is most likely due to the increased coulomb attraction between the CoO_2 layers and sandwiched Na ion layers.

Figure 4.20 shows electrical conductivities of the samples as a function of temperature [163]. All samples exhibit metallic conduction behavior. The magnitude of electrical conductivity at ambient temperature is in the range of 300–650 S cm^{-1}. The electrical conductivity is strongly dependent on the concentration of Na ion. Up to x = 0.78, the electrical conductivity increases monotonically with

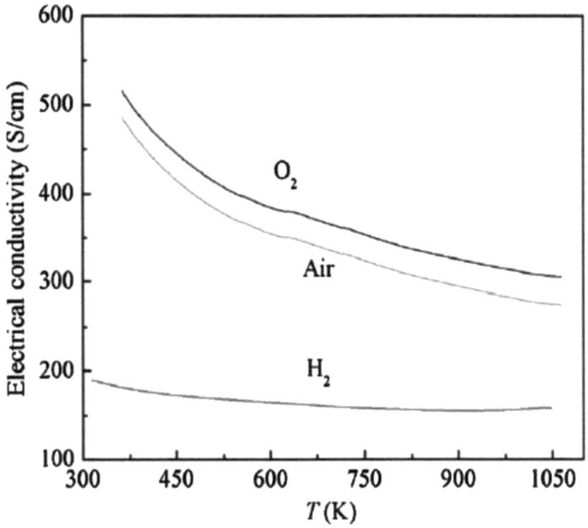

Fig. 4.21 Temperature dependences of the electrical conductivity of the sample with $x = 0.78$ calcined in different atmospheres. Reproduced with permission from [163], Copyright @ 2008, Elsevier

increasing content of Na. The samples with high contents of Na ($x = 0.80$ and 0.85) start to decrease.

It is found that electrical conductivity of Na_xCoO_2 can be altered by annealing in different atmospheres, as shown in Fig. 4.21 for $Na_{0.75}CoO_2$ [163]. Obviously, the electrical conductivity increases with increasing oxygen partial pressure. The sample sintered in H_2 has much lower electrical conductivity the samples sintered in pure O_2 or air. Because Na_xCoO_2 is *p*-type semiconductor, reductive atmosphere decreases the concentration of carriers by pumping in electrons. Figure 4.22 shows thermoelectric powers of the Na_xCoO_2 samples, as a function of temperature. The thermoelectric power increases with increasing content of Na, which is different from the variation trend of electrical conductivity. The large thermoelectric power observed in Na_xCoO_2 has been attributed to both the large degeneracy in spin states of cobalt oxides and the strong correlation of the 3d electrons. The increase in the content of Na would lead to a decrease in the concentration of Co^{4+} ion in the CoO_2 layers.

Similar to those of $Co_3Co_4O_9$, thermoelectric properties of Na_xCoO_2 can also be modified by doping of other elements. Silver doping increases electrical conductivity and thermopower and decreases thermal conductivity [165, 166]. If Ag is doped as second phase, it can increase electrical conductivity and thermopower, but also increase thermal conductivity, so its effect on figure of merit is canceled out [167]. The effects of tansitional metallic elements on the properties of Na_x-CoO_2 are diversed. Such elements include Cu, Zn, and Ni [160, 168–172]. Lanthanide elements that have been used are Nd, Sm, and Y [160, 161, 173]. Comparatively, although Na_xCoO_2 has higher ZT values, but $Co_3Co_4O_9$ is more stable in terms of composition variations [174, 175].

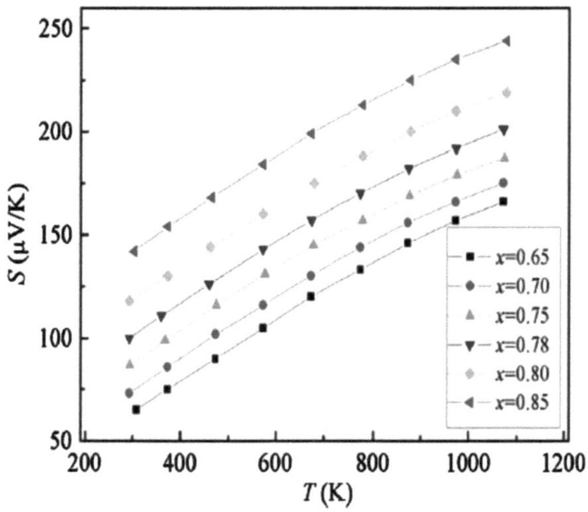

Fig. 4.22 Temperature dependences of thermoelectric power of the as-synthesized Na_xCoO_2 (x = 0.65–0.85), with the thermoelectric power increasing with the content of Na. Reproduced with permission from [163], Copyright @ 2008, Elsevier

4.4.4.2 *n*-Type Thermoelectric Oxides

Strontium titanate, $SrTiO_3$ (or ST in short), is a good electronic conductor when doped with ions of higher valence (i.e., electron donors). Both the thermal and electrical conductivities of $SrTiO_3$ are higher than those of *p*-type layer cobaltite discussed above. Similarly, the thermoelectric properties of $SrTiO_3$ can be modified or improved by using doping [176–179].

$SrTiO_3$ is of perovskite crystal structure, but when it is doped properly, a superlattice Ruddlesden–Popper structure will be formed, which has a similar mechanism to that in the layered cobaltite in reducing thermal conductivity [180]. Figure 4.23 shows a shematic Ruddlesden–Popper structure $(SrO)(SrTiO_3)_n$ or $Sr_{n+1}Ti_nO_{3n+1}$, with n = 2, which consists of perovskite layers separated by SrO layers [127]. Although the formation of the Ruddlesden–Popper phase results in a decrease in thermal conductivity, but it also leads to a decrease in electrical conductivity at the same time. Therefore, such a structure is not desired to have a significant effect on the values of *ZT* [181]. Varius elements have been used to dope $SrTiO_3$, attempting to increase its thermoelectric performances. A La-doped $BaTiO_3$–$SrTiO_3$ solid solution, $Ba_{0.3}Sr_{0.6}La_{0.1}TiO_3$, has been shown to possess promising thermoelectric properties [182]. Further improvement is necessary for this material to be used for practical applications.

Another promising *n*-type thermoelectric oxide is $CaMnO_3$. $CaMnO_3$ can be doped at either Ca or Mn site [183–186]. *Y* and Nb are among the most widely used dopants to replace Ca and Mn, respectively. *Y* doping leads to higher electrical conductivity and Seebeck coefficient, which results in a larger figure of merit. The introduction of *Y* is especially effective in reducing the thermal conductivity, due to the decrease in phonon contributions to the thermal conduction mechanism. Nb is to increase the electrical conductivity and decrease the thermal

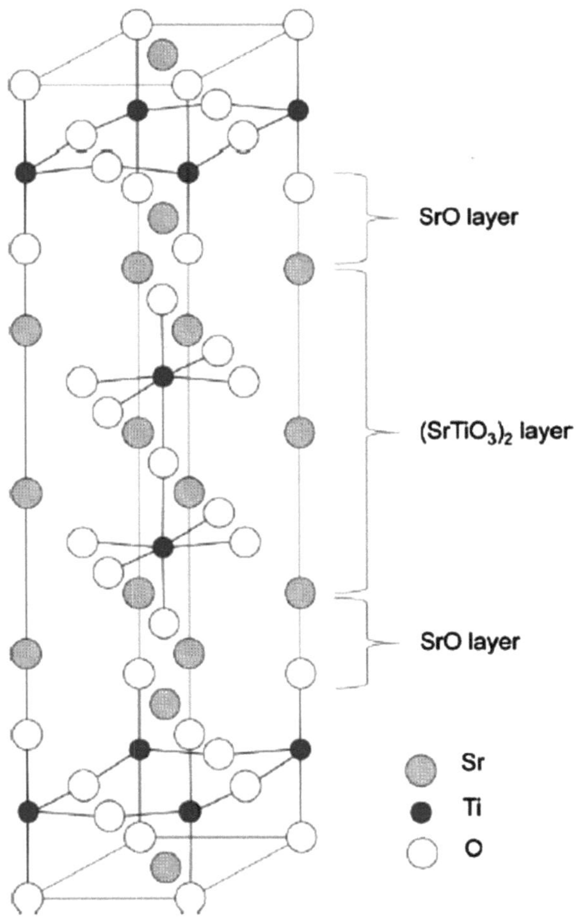

Fig. 4.23 Schematic structure of Ruddlesden–Popper (SrO) $(SrTiO_2)_n$ (n = 2). Reproduced with permission from [127], Copyright @ 2012, Elsevier

conductivity, but it all decrease the Seebeck coefficient. There are also reports on the doping effects of other elements on thermoelectric properties of $CaMnO_3$ [127, 187].

ZnO is another promising thermoelectric material that requires doping for adequate n-type conduction [188–193]. ZnO has a close conductivity of doped $SrTiO_3$, but its conductivity is higher at high temperatures than $SrTiO_3$. One of the problems of ZnO is its high thermal conductivity. This problem can be partly addressed by doping with Al. The beneficial effect of Al has been attributed to the decrease in c/a ratio of ZnO. Many other elements have also been applied to improve the performance of ZnO [192, 193].

In summary, both p-type and n-type thermoelectric oxides have been available to develop thermal energy conversion devices. Layer structured cobaltite compounds are the best p-type thermoelectric oxides. Na_xCoO_2 based materials have better thermoelectric performances, while $Co_3Co_4O_9$ based materials have higher stability and are more widely used for practical applications currently. There are three n-type oxides for thermoelectric applications. $CoMnO_3$ has lower thermal conductivity, whereas $SrTiO_3$ and ZnO have higher electrical conductivity. Thermoelectric performances of the oxides are relatively lower than those of their nonoxide counterparts. Therefore, further efforts have to be made for these materials to be used with high efficiencies.

4.5 Physics and Strategies of Thermoelectric Materials

4.5.1 Electrical Transport

In this subsection, electron transport coefficients relevant to thermoelectric materials, including diffusive Seebeck coefficient (S), electrical conductivity (σ), and thermal conductivity (κ_e), will be elaborated with the expressions derived from the Boltzmann transport equation (BTE). Their physical origin and evolution, as a function of temperature, carrier concentration and the form of materials (from bulk to quantum well), as well as their interrelationship, will be described in a detailed way. With this understanding, the optimal conditions for maximized power factor ($S^2\sigma$) of a material can be identified and the various methods to further enhance the performance will be understood. Detailed descriptions and discussion with related materials can be found in Ref. [4], a brief summary will be given below.

4.5.1.1 Boltzmann Transport Equation

The electron transport coefficients can be derived from the solution to the BTE. The formal expression of the BTE [194] is:

$$\left(\frac{\mathrm{d}f}{\mathrm{d}t}\right)_{sc} = \frac{\mathrm{d}f}{\mathrm{d}t} + \frac{\mathrm{d}\kappa}{\mathrm{d}t}\nabla_{\kappa}f + \frac{\mathrm{d}r}{\mathrm{d}t}\nabla_{r}f, \tag{4.18}$$

where t is time, k and r are the wave and position vectors of electrons, and f is the nonequilibrium distribution function. The BTE describes the evolution of particles in a system due to the redistribution of the momentum and position of the particles, caused by a perturbation. For electrons, external forces, such the application of an electric field or the presence of temperature gradient, can facilitate such perturbations. By using the relaxation time, where electrons are in an equilibrium state through the effect of random scattering within a relaxation time (τ) $(\mathrm{d}f/\mathrm{d}t)_{sc} = -(f-f_0)/\tau$, the BTE can be solved to obtain f. The solution is then stated in term of the equilibrium distribution function (f_0). At equilibrium, electron distribution follows the Fermi–Dirac statistics through:

$$f_0(E) = \frac{1}{\exp[(E - E_F)/k_B T] + 1}, \tag{4.19}$$

where E is electron's energy level, E_F is the ''Fermi energy'', and k_B is Boltzmann constant. In a more precise form, the Fermi–Dirac function should be described in term of chemical potential [194]. If Fermi energy is still used, it is also called Fermi potential or Fermi level. At low electron concentration, the equilibrium distribution of electrons can also be approximated by using the Maxwell–Boltzmann function [195]. However, to obtain accurate results for thermoelectric materials, it is necessary to use Fermi–Dirac statistics. In this case, the first order steady state ($(\mathrm{d}f/\mathrm{d}t) = 0$) solution to the BTE can be expressed as:

$$f(E) - f_0(E) = -\tau(E)v(E)\frac{\mathrm{d}f_0}{\mathrm{d}E}\left[\mp e\varepsilon - \left(\frac{E - E_F}{T}\right)\frac{\mathrm{d}T}{\mathrm{d}x}\right]. \tag{4.20}$$

Subsequently, Eq. (4.20) is used to determine the magnitude of charge density (J) and heat current density (Q) as:

$$J \equiv \mp nev \mp e \int_{-\infty}^{+\infty} g(E)v(E)[f(E) - f_0(E)]\mathrm{d}E, \tag{4.21}$$

$$Q \equiv n(E - E_F)v = \int_{-\infty}^{+\infty} g(E)(E - E_F)v(E)[f(E) - f_0(E)]\mathrm{d}E, \tag{4.22}$$

where n is the carrier concentration, v is the carrier velocity and $g(E)$ is DOS of the material. An electron always carries the same amount of electrical charge (e), whereas it may carry any amount of thermal energy, which is determined by ($E-E_F$). The transport coefficients, S, σ, and κ_e, can then be defined in term of J and Q.

Figure 4.24 shows the characteristic feature of the Fermi–Dirac distribution function, which means that the probability of finding an electron with E far below and far above the E_F is 100 and 0 %, respectively. Moreover, if the ratio of the change in distribution ($\mathrm{d}f_0$) over the incremental change in energy ($\mathrm{d}E$) for electrons near E_F is quite high, the probability is virtually zero for those electrons with energy far greater or far lesser than E_F, i.e., $(\mathrm{d}f_0/\mathrm{d}E) = 0$ for $|E|\rangle\rangle E_F$. It is then obvious from Eqs. (4.20–4.22) clearly indicate that the external disturbances are applied to only those electrons close to Fermi level, which have contribution to electrical conductivity.

This semi-classical BTE approach has been successfully applied in the studies of thermoelectric materials and has been validated by experimental results. It is also validate for nanostructure materials that have quantum mechanical effects. The quantum effects, such as electron confinement and tunneling, can be readily accounted through the modification of the DOS. The predictions given by the simple BTE-based methods have been in a good agreement with those by those

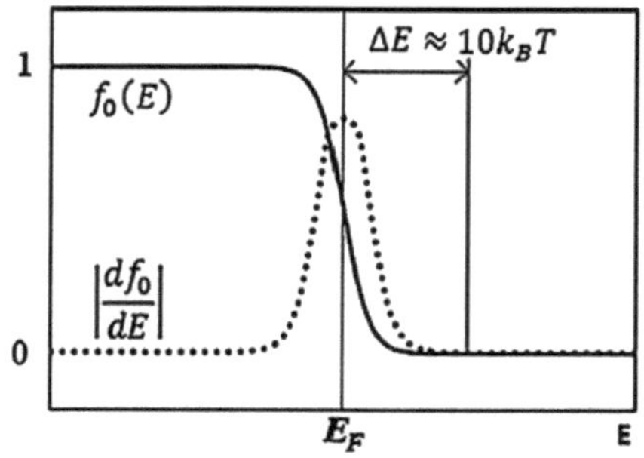

Fig. 4.24 The Fermi–Dirac distribution function, $f_0(E)$ (*solid line*) and its first derivative, df_0/dE (*dotted line*). The implication of this diagram is that only those electrons with energy close to E_F contribute to electrical/thermal transports. Reproduced with permission from [4], Copyright @ 2010, Elsevier

sophisticated modern approaches, such as nonequilibrium Green's function (NEGF) methods, in which the results are obtained by directly solving the Schrodinger equation [196, 197].

4.5.1.2 Electronic Band Structure

J and *Q* are determined by several important parameters, such as the DOS and carrier velocity, which can be derived from the electronic band structure of the e-material. That is, electronic band structure of a material determines its transport coefficients.

The complete electronic band structure of any material is very complex. However, only the portions close to E_F have significant influence on thermoelectric properties. For example, the thermoelectric properties of *p*-type Si are determined by the characteristics of the local band structure of the valence band (VB) maximum at the *G*-point, while those of *n*-type Si are governed by the conduction band (CB) minimum located at the position of about 80 % of the way from the *G*-to the *X*-points. These electrically active portions of the band, which are close to the E_F, i.e., the *G*-point VB maximum in *p*-type Si and the CB minimum in *n*-type Si, are usually known as conduction valleys and carrier pockets, respectively.

The location in *k*-space, in terms of the crystal of the structure material, determines the degeneracy factor (*N*) of the conduction valleys. For instance, the CB minimum also has a sixfold degeneracy ($N = 6$), because Si has an *fcc* crystal structure with sixfold symmetry. On the other hand, because the *G*-point is the center of the Brillouin zone, the VB maximum has a single degeneracy ($N = 1$). Most thermoelectric materials, such as Bi_2Te_3, SiGe and PbTe, have such ''multi-valley'' band structures ($N > 1$).

The transport coefficients may be determined for each conduction valley. In particular, each valley's transport coefficient is distinguished by the curvature of the band, which is inversely proportional to the effective mass, i.e., $m^* = \hbar^2 (d^2E/dK^2)^{-1}$. The curvature and effective mass of a parabolic band are constant. However, the curvature of nonparabolic bands is not constant, while the effective mass is more complicated to define because it is a function of energy. Moreover, an anisotropic band could have multiple distinct effective masses. Due to this anisotropy of the effective masses, there will have anisotropic transport properties in the materials.

The degenerate valleys may have the same band structure, but they could have different orientations, in terms of the directions of electrical conduction and the confinement characteristics of the electrons, such as nanostructures. These asymmetric characteristics can result in the breaking of degeneracy. In the cases of different directions of electrical conduction, the asymmetry can be eliminated by properly averaging the effective masses in the relevant directions. However, in the case of electron confinement, it is necessary to treat the valleys distinctively and assume that each one has its distinct transport coefficient.

The transport coefficients in a given band are the summation of those of the constituent conduction valleys by using a parallel conductor model. In this model, each conduction valley is assumed to have no interaction one another, given by:

$$S = \frac{\sum_{i=1}^{N} S_i \sigma_i}{\sum_{i=1}^{N} \sigma_i}, \tag{4.23}$$

$$\sigma = \sum_{i=1}^{N} \sigma_i, \tag{4.24}$$

$$\kappa_e = \sum_{i=1}^{N} \kappa_{e,i}, \tag{4.25}$$

where the transport coefficient of a valley is represented by the ith term. The transport coefficients of the material can be similarly obtained from those of the bands. However, if the electrical conductivities of all other bands far away from the Fermi level are much lower than that of the band closest to the Fermi level, the materials are single-band materials, such as metals and degenerate semiconductors. In this case, such rules are no longer necessary. In contrast, for semimetals and intrinsic semiconductors, whose conduction valleys in both VB and CB must be taken into account, it is necessary to use Eqs. (4.23–4.25). For instance, because the signs of S for conduction valleys in VB and CB are opposite, semimetals usually have relatively low values of $|S|$ according to Eq. (4.23).

If the size and alignment of the band gap (direct or indirect) does not directly affect the conduction transport coefficients of the valleys, it still has role in determining the overall behavior of a material, but not in a very strong way. For example, materials with small direct band gaps, such as Bi_2Te_3 and PbTe, have a strong VB–CB coupling, thus leading to nonparabolic conduction and valence bands, with small effective masses. In contrast, indirect band gap materials, such as

SiGe, have parabolic and broader conduction bands, with larger effective masses. However, the valence bands are still prone to be nonparabolic, which is because of the heavy-and light-hole band interaction. Furthermore, the temperature limit is determined by the width of the band gap, with the doped semiconductor becoming intrinsic owing to the thermal excitation of minority carriers.

The density of states and carrier velocity of a material are determined by two important parameters: the effective masses (m^*) and the degeneracy factor (N), derived from the electronic band structure. Together with relaxation time, they can be used to determine the charge and heat current densities through Eqs. (4.20–4.22) and thus the transport coefficients.

4.5.1.3 Density of States

The local band structure near the maxima or minima can be described by a parabolic dispersion function, which defines the energy–momentum (E–k) relationship of electrons in that band. For bulk materials (three-dimension or 3D, with x, y and z), the energy dispersion function at the Brillouin zone center is given by:

$$E_{3\mathrm{D}}(\kappa) = \frac{\hbar^2}{2}\left(\frac{\kappa_x^2}{m_x} + \frac{\kappa_y^2}{m_y} + \frac{\kappa_z^2}{m_z}\right), \tag{4.26}$$

where $\hbar$ is the reduced Planck constant, $k_{(x,\ y,\ z)}$ is the wave number and $m_{(x,\ y,\ z)}$ is the principal effective masses in x-, y-, and z-directions, respectively.

Similarly, holes in the valence band can also be described by a dispersion function, when a consistent energy scale is used. The energy scale for holes is inverted with respect to that of electrons, as illustrated in Fig. 4.25 [4]. For consistency, the E_F must always be measured with respect to the ground state defined in each case, where the ground state energy levels are set preferably to be zero for both the CB and the VB, i.e., $E_{3\mathrm{D}}(k = 0) = 0$.

For quantum wells (2D), due to the discretization of carrier momentum in the directions of the confined quantum, this dispersion needs to be modified. For instance, if the confinement is in the z-direction, the electron dispersion function for the nth sub-band will be expressed by:

$$E_{2\mathrm{D}}(\kappa) = \frac{\hbar^2}{2}\left(\frac{\kappa_x^2}{m_x} + \frac{\kappa_y^2}{m_y}\right) + E_n, E_n = \frac{\hbar^2\pi^2 n^2}{2a^2 m_z}, \quad n = 1, 2, 3\ldots \tag{4.27}$$

where a is the thickness of the quantum well and E_n is the confinement energy. In this case, the ground state energy in this band has been increased by E_1 relative to the same band in the counterpart of bulk material. Similarly, the dispersion function for nanowire (which is 1D structure) can be modified with the carrier confinement in two directions (y and z). For the nmth sub-band, it is given by:

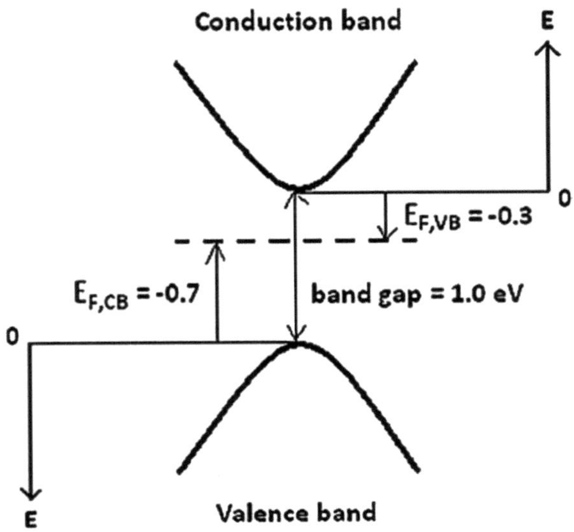

Fig. 4.25 The energy dispersion and scale defined for electrons in the upper CB and holes in the lower VB. The E_F is measured from the minima of the CB and the maxima of the VB. As an example, the E_F is equal to −0.3 and −0.7 eV, when considering carriers in the CB and VB, respectively. Reproduced with permission from [4], Copyright @ 2010, Elsevier

$$E_{1\mathrm{D}}(\kappa) = \frac{\hbar^2}{2}\left(\frac{\kappa_x^2}{m_x}\right) + E_{nm}, E_{nm} = \frac{\hbar^2\pi^2}{2a^2}\left(\frac{n^2}{m_y} + \frac{m^2}{m_z}\right), \quad n, m = 1, 2, 3\ldots \quad (4.28)$$

where a is the width of the nanowire with a square cross-section. Besides the physical length scales (a), the confinement potential energies for sub-bands of both quantum wells (E_n) and nanowires (E_{nm}) are also related to the principal effective masses in the directions of confinement. The corresponding effective masses are called ''confinement effective mass'' (m_c), i.e., $m_c = m_z$ for quantum wells, while $m_{c1} = m_y$ and $m_{c2} = m_z$ for nanowire.

The DOS per unit volume, for a single band/sub-band of three-, two-and one-dimensional systems, together with a spin-degeneracy factor of two, can be obtained through the dispersion functions of Eqs. (4.26–4.28) as [198]:

$$g_{3\mathrm{D}}(E) = \frac{1}{2\pi}\left(\frac{3m_d}{\hbar^2}\right)^{3/2} E^{1/2}, E \geq 0, \quad m_d = \sqrt{[3]m_x m_y m_z}, \quad (4.29)$$

$$g_{2\mathrm{D},n}(E) = \frac{m_d}{a\pi\hbar^2}, E^{1/2}, E \geq E_n, \quad m_d = \sqrt{[2]m_x m_y}, \quad (4.30)$$

$$g_{1\mathrm{D},nm}(E) = \frac{1}{a^2\pi}\left(\frac{2m_d}{\hbar^2}\right)^{1/2} (E - E_{nm})^{1/2}, E \geq E_n, \quad m_d = m_x. \quad (4.31)$$

The total DOS is the summation of all individual sub-bands, $g(E) = \sum_{n=1}^{+\infty} g_n(E)$. Figure 4.26 shows the staggered patterns of DOS of the materials with different dimensionalities, with considering the multiple sub-bands.

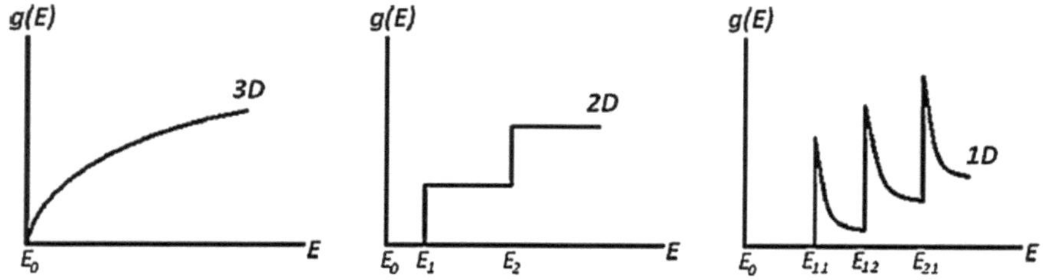

Fig. 4.26 Density of states, $g(E)$, for 3D, 2D and 1D electron gases in bulk materials, quantum wells and nanowires, respectively. The staggered pattern is due to the superposition of multiple sub-bands, each of which is shifted up from the bulk ground state energy (E_0) by the confinement energy (E_1, E_2, etc.). Reproduced with permission from [4], Copyright @ 2010, Elsevier

In practice, it is not necessary to take into account the higher sub-bands ($n > 1$), because consider only the lowest sub-band ($n = 1$) is sufficient in most cases, which is closest to the E_F. In this case, the total DOS for N number of degenerate conduction valleys can be described only by the power law function:

$$\begin{aligned} g(E) &= \frac{N}{g_D a^{3-D}} \left(\frac{2m_d}{\hbar^2}\right)^{D/2} (E - E_0)^{D/2-1}, E \geq E_0, \\ g_D &= \begin{Bmatrix} 2\pi^2 & D = 3 \\ D\pi & D = 2, 1 \end{Bmatrix}, \end{aligned} \tag{4.32}$$

where E_0 is the ground state energy level and the exponent, D = 3, 2, 1, is called the ''dimensionality'' factor for bulk materials, quantum wells and nanowire, respectively. In these expressions, a new term, called density of states effective mass (m_d), is introduced. It is the geometric mean of the principal effective masses, i.e., $m_d = \sqrt[D]{\prod_{i=1}^{3-D} m_i}$. Therefore, in the systems with lower dimensionality, there is no contribution to m_d from the effective mass in the directions of confinement. Because the conduction valleys can have different orientations with respect to the confinement, they can have different values of m_d and confinement potentials or energies at the same time.

The DOS and electron confinement are significantly important in determining transport parameters of materials, it is necessary to mention several aspects for nanostructures. First, the energy dependence of the DOS is changed from $D/2-1 = +1/2$ for bulk materials to 0 from quantum wells and to $-1/2$ for nanowire. Second, there is an inverse proportionality of the DOS to the thickness of quantum wells or nanowire, with the DOS increasing with decreasing a. Thirdly, the effective band gap due to upward or downward shift of the CB or VB ground states will be increased, corresponding to the confinement potential energy which is inversely proportional to m_c and a. All these factors have been successfully used to enhance the power factor by developing nanostructure thermoelectric materials, as have been discussed in the previous section.

4.5.1.4 Carrier Velocity and Relaxation Time

At low electric fields, the diffusive carrier motion is assumed to be along the direction perpendicular to the direction of confinement, the carrier velocity is given by:

$$v^2(E) = \frac{2(E - E_0)}{Dm_\sigma}, \tag{4.33}$$

where m_σ is the conductivity of inertial effective mass, defined through the harmonic average of the effective masses of degenerate conduction valleys in the direction of conduction, i.e., $1/m_\sigma = \sum_{i=1}^{N} 1/m_{\sigma,i}$. For example, the conductivity effective mass for the Δ-point valleys in *n*-type Si for conduction along ⟨100⟩ is $m_\sigma = 3(2/m_t + 1/m_l)^{-1}$, with m_t and m_l being the transverse and longitudinal effective mass, respectively. With the use of the conductivity effective mass, it is able eliminate the asymmetry in orientation of each valley with respect to the direction of conduction. The dimensionality factor, *D*, in Eq. (4.33) is used to describe the equipartition of energy among the available degrees of freedom.

The relaxation time is determined by the type and magnitude of carrier scattering within the materials. It is difficult to derive the carrier relaxation time, $\tau(E)$, from first principle calculations, because it involves simultaneous carrier interaction among various scatterers, such as acoustic and optical phonons, plasmon, defects, interfaces and even other carriers. In most cases, the relaxation time can be modeled by using the power law function, as:

$$\tau(E) = \tau_0(E - E_0)^r, \tag{4.34}$$

where the exponent *r* is the characteristic ''scattering constant'' that is specific for a particular scattering process and t_0 is a constant that is related to both the scattering processes and material properties. The value of *r* can be derived from experiments, inducing (i) measurements of mobility as a function of temperature (ii) the variation in the Seebeck coefficient as a function of carrier concentration and (iii) the measurements of the Nernst coefficient, which are also used to determine the sign of *r* [199].

The values of *r* can also be analytically determined [200]. For example, for scattering by weakly screened ionized impurities, $r = 3/2$ and for neutral impurity scattering, $r = 0$. However, if lattice deformation potential scattering by acoustic and optical phonons (ADP and ODP, respectively) is involved or screened ionized impurity scattering is very strong, *r* cannot be determined, because it is not a constant. It is also because τ is inversely proportional to the DOS. When there are multiple scattering processes, *r* also cannot be determined.

It is also possible that *r* is determined by more than one scattering process and thus the calculation of the relaxation time becomes complicated. In this case, if the scattering processes can be considered to be independent of one another, the Matthiessen's rule [194] can be applied to solve the problem. Therefore, the total relaxation time can be derived from those of all the scattering processes by using $(1/\tau) = (1/\tau_1) + (1/\tau_2) + \cdots + (1/\tau_n)$. Obviously, if these processes have different values of γ, results will be inaccurate [200].

4.5.1.5 Electrical Conductivity (σ)

With the standard definitions of the charge and heat current density, described in Eqs. (4.20–4.22), the three important transport coefficients, σ, S, and κ_e, for conduction perpendicular to directions of confinement, can be expressed as:

$$S \equiv \frac{\varepsilon}{\frac{dT}{dx}}\bigg|_{J=0} = \mp\frac{1}{eT}\left[\frac{X_1}{X_0} - E_F\right], \tag{4.35}$$

$$\sigma \equiv \frac{J}{\varepsilon}\Big|_{dT/dx \neq 0} = e^2 X_0, \tag{4.36}$$

$$\kappa_e \equiv \frac{Q}{\frac{dT}{dx}}\bigg|_{J=0} = \frac{1}{T}\left[X_2 - \frac{(X_1)^2}{X_0}\right], \tag{4.37}$$

where $X_i = -\int_{E_0}^{+\infty} g(E)\tau(E)v^2(E)E^i(df_0/dE)dE$. Therefore, by incorporating Eqs. (4.32–4.34), with assumption of a single band/sub-band with parabolic energy dispersion, together with a power law relaxation time, the integral X_i can be readily given by:

$$X_{i,D} = \frac{N\tau_0}{m^\sigma g_D a^{3-D}}\left(\frac{2m_d}{\hbar^2}\right)^{D/2}(k_BT)^{r+D/2+i}\left(\frac{2(r+i)}{D}+1\right)F_{r+D/2+i-1}(\eta), \tag{4.38}$$

where $F_j(\eta) = \int_{E_0}^{+\infty} x^j/[\exp(x-\eta)+1]dx$ is the jth order Fermi integral and $\eta = (E_F - E_O)k_BT$, which is the reduced Fermi level. The constant g_D has values of $2\pi^2$, 2π, and π, for D = 3, 2, and 1, respectively.

The reduced Fermi level is a useful variable because it can be used to represent both the changes in temperature of the materials and the Fermi level. It also allows X_n to be expressed in terms of Fermi integrals. However, in the case of zeroth-order integral, where $F_0(\eta) = \ln(1 + e^\eta)$, $F_j(\eta)$ can only be evaluated numerically.

The electrical conductivity is a measure of the number (n) of carriers (electrons or holes) can contribute to the electrical current along with their mobility (μ), which is given by: $\sigma = ne\mu$. Both n and μ can vary as a function of temperature (T). They can be measured experimentally through the characterization of resistivity and Hall coefficient. The magnitude of power factor is very sensitive to n and a suitable concentration is crucial for the optimization thermoelectric performances. The enhancement in power factor of nanostructure materials is solely determined by the relation of n, E_F and the DOS.

With the assumption of parabolic energy dispersion with power law relaxation time, the equilibrium carrier concentration, n, in a set of degenerate conduction valleys, is given by:

$$n = \int_{E_0}^{+\infty} g(E)F_0(E)\mathrm{d}E = \frac{N}{g_D a^{3-D}} \left(\frac{2k_B T m_d}{\hbar^2}\right)^{D/2} F_{D/2-1}(\eta). \tag{4.39}$$

In some cases, when the degeneracy factor (N) is combined with the density of states effective mass, m_d, it is possible to derive the density of states effective mass for all valleys, as $m_{d,\text{ total}} \equiv N^{2/D} m_d$. However, it is better to list N and m_d separately. As n is closely related to the material parameters and is proportional to temperature, T, the factors can be extracted out to plot the normalized carrier concentration, $\tilde{n} = n/\theta$, as shown in Fig. 4.27a. The normalization factor is given by:

$$\theta = \frac{N}{(a')^{3-D}} \left(\frac{m_d}{m_o} \frac{T}{300\,\mathrm{K}}\right)^{D/2}, \tag{4.40}$$

where a' is equal to the thickness of the quantum wells or nanowire in nanometer (nm), with $\theta = 1$ for $m_d = m_0$, $N = 1$, $a = 1$ nm, and $T = 300$ K. Due to the normalization, Fig. 4.27a is thus applicable to any materials at any temperature, as long as the basic assumptions discussed above are satisfied. It can be used to determine n at a given E_F.

Similarly, the E_F can be variable or a constant depending on the properties of materials. E_F is also dependent on n and T. For example, for extrinsic semiconductors, because n may not increase significantly with T, E_F must decrease with increasing T, according to Eq. (4.39). In contrast, for intrinsic semiconductors, E_F is in the middle of the band gap and remains relatively constant, and here fore n increases with increasing T. This relationship is also affected by the changes in DOS. For example, because a quantum well of 2 nm has larger DOS than a quantum well of 5 nm due to smaller a of the former, which is present in the denominator of $g(E)$, the former has smaller E_F than the latter, given that n and T are the same. Similarly, if two different materials have the same n and T, the one with larger m_d and N has larger DOS and thus smaller E_F.

In addition, in many cases, it is very difficult to model how the mobility, μ, of nanostructure materials differs from that of bulk materials. From the previous descriptions for σ and n, the following expression can be derived:

$$\mu = \left(\frac{e}{m_\sigma}\right) \frac{\int_{E_0}^{+\infty} g(E)\tau(E)v^2(E)(\mathrm{d}f_0/\mathrm{d}E)\mathrm{d}E}{\int_{E_0}^{+\infty} g(E)v^2(E)(\mathrm{d}f_0/\mathrm{d}E)\mathrm{d}E} = \frac{e\langle\langle\tau\rangle\rangle}{m_\sigma}, \tag{4.41}$$

where $\langle\langle\tau\rangle\rangle$ represents an averaged relaxation time involving the nonequilibrium terms of the distribution ($\mathrm{d}f_0/\mathrm{d}E$) [200]. Because there is $v^2(E) \propto E$, $\langle\langle\tau\rangle\rangle$ can be simplified by using integration-by-parts, as:

$$\langle\tau\rangle = \frac{\int_{E_0}^{+\infty} g(E)\tau(E)f_0(E)\mathrm{d}E}{\int_{E_0}^{+\infty} g(E)f_0(E)\mathrm{d}E} = \frac{e\langle\langle\tau\rangle\rangle}{2r/D+1} = \tau_0(k_B T)^r \frac{F_{r+D/2-1}(\eta)}{F_{D/2-1}}, \tag{4.42}$$

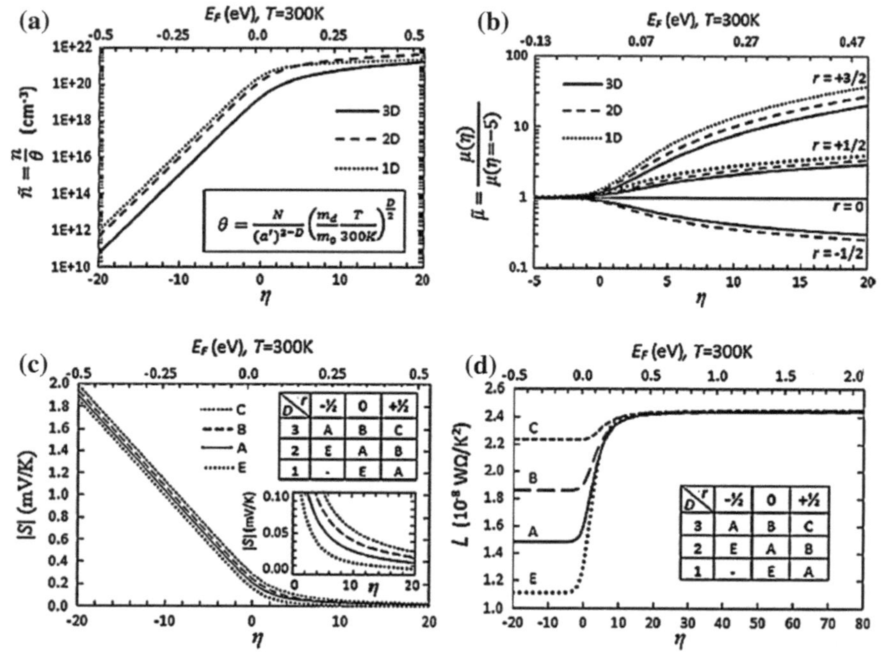

Fig. 4.27 **a** Normalized carrier concentration, **b** normalized mobility, **c** $|S|$, and **d** L as a function of η. Equivalent value of E_F at 300 K is labeled on the top horizontal axis. n is normalized to the constant θ as defined in the inserted box, where T is temperature in Kelvin, and a_0 is the width of quantumwells or nanowires in nm. $|S|$ and L are degenerate for several combinations of the r and D, as listed in the inserted table. All plots made with an assumption of single sub-bandwith parabolic energy dispersion and power law variation of the relaxation time. Reproduced with permission from [4], Copyright @ 2010, Elsevier

where $\langle\tau\rangle$ is the average over the equilibrium carrier distribution.

Although it is easy to calculate the variation of μ as a function of n through their mutual relationship with η, it is difficult to predict which scattering processes are dominant if the conditions of the materials are changed. Moreover, τ_0 is determined by a number of parameters, which cannot be covered in detail due its complexity [200]. The ideal relationship between μ and η can be demonstrated by using the normalized mobility, $\tilde{u}(\eta) = \mu(\eta)/\mu(\eta = -5)$, as shown in Fig. 4.27b. In this case, μ increases with η if $r > 0$, and vice versa. If $r = 0$, then $\langle\langle\tau\rangle\rangle = <\tau> = \tau_0$ and $\mu = e\tau_0/m_\sigma$ are both constants.

Because μ may change with increasing η, σ may also be variable. For example, σ of metal usually decreases with increasing η, because the decrease in μ due to the acoustic phonon scattering ($r = -1/2$) dominates the slighter increase in n when η is sufficiently large. In contrast, σ of nondegenerate semiconductors generally increases with increasing η, due to the large increase in n when η is sufficiently small.

4.5.1.6 Seebeck Coefficient (*S*)

Until now, the Seebeck coefficient is not as well understood as electrical conductivity. Most likely, the Seebeck coefficient can only be described in a incomplete and nonquantitative way. It is dependent on several factors, including the symmetry characteristic of the electronic band [201] or the energy dependence of the DOS in the range of the Fermi level [109]. Also, it is only accepted that a "high and steep" DOS is required to have large Seebeck coefficient [202].

It has been acknowledged that the net Seebeck effect is determined by more than one distinct physical process, with the most prominent and common process arising from the diffusion of electrons. Therefore, it is only necessary to derive the diffusive Seebeck coefficient, as described in its most general form in Eq. (4.35). If *g*(*E*) and *t*(*E*) are assumed to be power law functions, the expression of *S* for a single band can be simplified as:

$$S = \mp\frac{1}{eT}\left(\frac{\int_{E_0}^{+\infty} g(E)\tau(E)Ev^2(E)(\mathrm{d}f_0/\mathrm{d}E)\mathrm{d}E}{\int_{E_0}^{+\infty} g(E)v^2(E)(\mathrm{d}f_0/\mathrm{d}E)\mathrm{d}E} - E_F\right) = \mp\frac{1}{eT}\left(\frac{\langle\tau E\rangle}{\langle\tau\rangle} - E_F\right), \tag{4.43}$$

where $\langle x\rangle = \int_{E_0}^{+\infty} g(E)x(E)f_0(E)\mathrm{d}E / \int_{E_0}^{+\infty} g(E)f_0(E)\mathrm{d}E$ is the average value of an arbitrary function *x*(*E*) over the equilibrium carrier distribution. If *g*(*E*) and *t*(*E*) are not power law functions, for example, in the case of nonparabolic bands, the more complicated average over nonequilibrium carrier distribution should be used.

It is clearly demonstrated in Eq. (4.43) that the *S* of a single band is proportional to the difference in the average energy between the electrons weighted by the relaxation time and the E_F. If the relaxation time is constant, $\langle\tau E</>\tau\rangle$ can be simplified to be $\langle E\rangle$. Therefore, the diffusive Seebeck effect is the result of the tendency of the electrons in that band, which have an average energy, $\langle E\rangle$, to return to the lowest energy state represented by the Fermi level, E_F. The behavior of the Seebeck coefficient can thus be easily understood by considering the changes in the average energy of the electrons. With Eqs. (4.35) and (4.38), the diffusive Seebeck coefficient can be expressed as:

$$S = \mp\frac{k_B}{e}\left(\frac{(r+D/2+1)F_{r+D/2}(\eta)}{(r+D/2)F_{r+D/2-1}(\eta)} - \eta\right). \tag{4.44}$$

The behaviors of |*S*|, as a function of the reduced Fermi level, η, for different energy dependences of the DOS and relaxation time (*D* and *r*, respectively), can be plotted, as shown in Fig. 4.27c.

It is thus found that |*S*| always decreases with increasing η, i.e., increasing E_F or decreasing *T*. Because an increase in Fermi level means an increase in carrier concentration, the average energy is increased also. However, since the average energy increases relatively slowly, |*S*| always decreases as the Fermi level is increased. An increase in temperature also leads to an increase in carrier concentration and average energy, which means that |*S*| will increase and vice versa.

For an extrinsic semiconductor, the increase in temperature does not mean that there is a significant increase in the carrier concentration. Instead, it is a fixed value, depending on the concentration of dopants. Therefore, there is also no increase in the average energy. However, due to the decrease in the Fermi level, $|S|$ is still increased. If the semiconductors become intrinsic at sufficiently high temperatures, $|S|$ will decrease with increasing temperature. This is because both the conduction and valence bands exhibit their effects. At the same time, the Seebeck coefficients of opposing sign in each band result in a relatively small net value, according to Eq. (4.23).

If it is assumed that there is a nonchanging Fermi level, the value of $|S|$ also decreases because the energy dependence of the DOS is decreased. Generally, the shape or profile of the DOS is considered to have a significant influence on the Seebeck effect. For example, it is commonly believed that a sharp peak in the DOS, such as those found in nanowire, corresponds to a large Seebeck coefficient, but this is not always true. The influence of the shape of DOS on the value of $|S|$ is easy to understand, if its effect on the average energy is taken into account. The parabolic profile of the DOS in bulk materials actually corresponds to a larger $|S|$ than the step-like or peak-like profile of the DOS in quantum wells or nanowire, if Fermi levels of all three cases are the same. Figure 4.26 shows that the DOS of bulk materials is desired to achieve high energy states, while the DOS of nanowire tends to provide lower energy states, which means the average electron energy will be slightly larger in the bulk materials than in the nanowire, if both of them have the same Fermi level.

Because the average energy should be weighted by the relaxation time, according to Eq. (4.45), the increase in the scattering constant r will similarly result in an increase in $|S|$. Therefore, for example, a weakly ionized impurity scattering (with $r = +3/2$) will give a larger $|S|$ than a phonon scattering (with $r = -1/2$), if both of them are dominant. The relaxation time can be manipulated through the use of resonant scattering, which occurs when electrons in the allowed energy band of the host material are trapped in the quasi-local states of impurity atoms before they are ejected after a finite period of time [203]. The total relaxation time near the range of resonant energy can be affected subsequently, leading to either increase or decrease in the energy dependence of the total relaxation time and $|S|$. For example, it has been found that if the resonant states are below the Fermi level (low energy states), an increase in $|S|$ is likely observed. Otherwise, $|S|$ will decrease if the resonant states are above the Fermi level (high energy states). In practice, the effect of resonant scattering can be detected by the measurement of r. If resonant scattering is really responsible for the increase in $|S|$, an increase in r should be observed. However, experimental results on Tl-doped PbTe have shown that resonant scattering is not the reason for the increased $|S|$, because there is no change in r [109].

According to Eq. (4.44), $|S|$ is independent on several factors, including effective masses, the valley degeneracy, the thickness of the quantum wells or nanowire and τ_0. When the Fermi level is a constant and effective mass is increased, the magnitude of the DOS and carrier concentration will be increased.

However, because the DOS, with the assumption of the parabolic band, is a power law function and is scale invariant with respect to effective mass Eq. (4.32), the average energy and |*S*| will not be changed. This feature has been the basis of the proposal that there is an optimal Seebeck coefficient at which the power factor can be maximized.

If the Seebeck coefficients are estimated by using the approximate expressions, their limitations should be taken into account. According to Fig. 4.27c, three regimes can be identified, in terms of the behavior of *S*. In the regime of nondegenerate semiconductor ($\eta < -2$), |*S*| is large and decreases almost linearly with increasing η, which can be approximated by using the classical Maxwell–Boltzmann statistics, given by [195]:

$$S_{nd-SC} \cong \mp \frac{k_B}{e}\left(r + \frac{D}{2} + 1 - \eta\right), \eta < -2. \tag{4.45}$$

In the regime of metallics ($\eta \gg 0$), a smaller |*S*| is observed, which is inversely proportional to η and can be expressed by using the Mott formula [204]:

$$\begin{aligned} S_{\text{metal}} &\cong \mp \frac{\pi^2 k_B}{3e\eta}\left(\frac{\partial \ln(g(E))}{\partial lun(E)} + \frac{\partial \ln(\tau(E))}{\partial lun(E)} + \frac{\partial \ln(v^2(E))}{\partial lun(E)}\right) \\ &= \mp \frac{\pi^2 k_B}{3e\eta}\left(r + \frac{D}{2}\right), \quad \eta > 30. \end{aligned} \tag{4.46}$$

It is necessary to mention that the limits of η of -2 and 30 are only loosely defined and thus should not be taken to be applicable to any case. Because Eqs. (4.45) and (4.46) are derived from Eq. (4.44), they are subject to the same assumptions of parabolic band and power law relaxation time, and so on. However, they cannot be used to accurately describe the transport coefficients of degenerate semiconductors. The typical values of |*S*| are about 1 mV K^{-1} for nondegenerate semiconductors, about 10 μV K^{-1} for metals and about 100 μV K^{-1} for degenerate semiconductors.

In addition, these expressions alone are also not sufficient to accurately describe the Seebeck coefficients of materials, when multiple bands are involved. Such materials include intrinsic semiconductors and semi-metals. This is because, in these materials, the Fermi level is too close to both the CB and VB bands, leading to significantly high conductivity in both bands. Therefore, it is still necessary to calculate all individual *S* values of the conduction and valence bands first by using Eq. (4.44), and then the total valued of *S* can be obtained by using Eq. (4.35).

4.5.1.7 Electronic Thermal Conductivity (k_e)

It is understandable that the electronic thermal conductivity (k_e) is very similar to the electrical conductivity, except that it accounts for thermal energy, instead of electrical energy (charge), both of which are carried by electrons. Therefore, k_e

also increases with n and μ. The ratio of k_e to σ is known as the Lorentz number (L), given by:

$$L = \frac{\kappa_e}{\sigma T} = \left(\frac{k_B}{e}\right)^2 \left[\frac{(r+D/2+2)F_{r+D/2+1}(\eta)}{(r+D/2)F_{r+D/2-1}(\eta)} - \left(\frac{(r+D/2+2)F_{r+D/2}(\eta)}{(r+D/2)F_{r+D/2-1}(\eta)}\right)^2\right]. \tag{4.47}$$

In the regime of metals ($\eta \gg 0$), the Lorentz number approaches the value specified by the empirical Wiedemann–Franz law $L = \pi^2/3(k_B/e)^2 = 2.44 \times 10{-}8$ ($W\Omega/K^2$), as shown in Fig. 4.27d. L has a strong variation with D, r, and η, in the regime of degenerate semiconductors ($\eta \approx 0$), and similarly reaches a constant in the regime of nondegenerate semiconductor ($\eta < 0$). Similar to S, L is also independent of material parameters.

The dependences of L on D and r in the nondegenerate and degenerate semiconductor regimes are due to the greater sensitivity of κ_e to the energy dependence of the DOS and $\tau(E)$, respectively. This sensitivity is caused by the fact that hot electrons from higher energy states carry more thermal energy than cold electrons, while the σ is rather insensitive to the energy distribution of the carriers since all electrons carry the same amount of charge, regardless of their energy levels. For instance, with same values of n and σ, κ_e of nanowire is smaller than that of bulk material, because a greater proportion of the electrons in the nanowire are ''cold''electrons and have lower energy states. Similarly, κ_e is reduced in the cases of $r < 0$, because the hot electrons have a smaller $\tau(E)$, which means that they possess stronger scattering effect than cold electrons.

4.5.2 Higher Z Through Increasing Power Factor

It is well known that Z is maximized when the power factor $S^2\sigma$ is maximized. According to the general tentative variations of $S^2\sigma$, κ_e, κ_L, and Z with η, the power factor exhibits a peak due to the opposing variations of S and σ, while κ_e increases monotonically with increasing η. Because κ_L is usually not so related to electrical properties, it is assumed to be a constant, independent of η.

In the metallic regime where the η is very large, Z tends to be small because of (i) the inherently small $|S|$ and (ii) the large κ_e due to the large σ. Because κ_e is usually much larger than κ_L in metals, therefore, for metals, $\kappa \approx \kappa_e$ and $Z = S^2/LT$. Obviously, as both the $|S|$ and $1/L$ are increased with increasing η (Fig. 4.27), a higher Z is always achievable if η is decreased, or in other words, if the material is of less metallic characteristics. Interestingly, a very large σ does not mean an even high Z, because κ_e is also very high.

Of course, a very small σ will definitely bring out a small Z, such as the case of insulating nondegenerate semiconductors, although $|S|$ is relatively large, because κ_e is also very small, $\kappa \sim \kappa_L$ and $Z = S^2\sigma/\kappa_L$, i.e., Z decreases with decreasing σ.

As a result, the optimal reduced Fermi potential, η_{opt}, where the power factor is maximized, is always in the regime of degenerate semiconductors. This is the reason why promising thermoelectric materials, such as Bi_2Te_3, PbTe and SiGe are all heavily doped degenerate semiconductors.

4.5.2.1 Optimization of Power Factor

The $S^2\sigma$ is significantly affected by the carrier concentration and η. Because $|S|$ decreases and σ increases with increasing η, the product $S^2\sigma$ is expected to experience a maximum at an optimal value of η, which can be determined by using Eqs. (4.35) and (4.36), with given material parameters and temperature. Noting that S is independent of η and σ can be normalized, $S^2\sigma$ can be written as:

$$S^2\sigma = [BT^{r+D/2}]S^2\tilde{\sigma}(\eta, r, D), \tag{4.48}$$

where $\tilde{\sigma}$ is the normalized conductivity and $B = \tau_0 N(2m_d)^{D/2}/m_\sigma a^{3-D}$ is similar to the figure of merit of the material [205]. From this equation, it has been shown that there are an optimal reduced Fermi level and Seebeck coefficient, which always leads to the maximum power factor, and it is independent of the effective masses (m_d and m_σ) of the materials, degeneracy factor (N), thickness (a) of quantum wells and nanowire, τ_0 and temperature [206].

According to Eq. (4.48), the power factor is scale invariant with the common factor, $BT^{r+D/2}$, and thus its maxima is only determined by η, r and D. As a result, there are a universal and optimal reduced Fermi level (η_{opt}) and a corresponding optimal Seebeck coefficient (S_{opt}), as determined by Eq. (4.44). The optimal carrier concentration also corresponds to η_{opt}, but its value varies according to the properties of the materials (m_d, N and a) and temperature (T). The values of η_{opt} and S_{opt} have been well studied and reported [206, 207].

When $r > 0$, where weakly screened ionized impurity scattering is dominant, there is a possibility that the increase of power factor is unlimited, due to the combined increase in n and μ that are larger than the reduction in S^2, as η is increased. However, this is not observed in reality as strongly screened scattering (e.g., $r = -1/2$) dominates only at high values of η and n. Because the scattering rates of carriers through the acoustic and optical phonon deformation potential (ADP and ODP, respectively) and strongly screened ionized impurities are proportional to the DOS [200], r is not a constant and changes with D in these occasions. When multiple scattering processes are present, r will take certain intermediate values. Generally, as the Fermi integral changes gradually with incremental changes in its indices, the value of η_{opt} will also change gradually, and should follow a specific trend. The values of η_{opt} indicate that the maximum $S^2\sigma$ always occurs in the degenerate semiconductor regime with $\eta \approx 0$.

Theoretical calculations have demonstrated that when the $|S|$ is in the optimal range of 130–187 μV K^{-1}, it is expected to have a maximum $S^2\sigma$, for any material at any temperature. This maximum power factor has been observed at an optimal value, $|S| \approx 0.167$ mV K^{-1}, for a wide variety of materials and temperatures, including bulk PbTe at 300 K [208], *p*-type $Si_{0.7}Ge_{0.3}$ at 300 K [209], $Sr_{1-x}La_x$-TiO_3 at 300 K [210] and $SrNbTiO_3$ at 1,000 K [211]. In other words, the calculated optimal *S* can be used as a guidance to optimize the power factor of thermoelectric materials. For bulk materials, if there is $|S|$ = 0.130–0.167 mV K^{-1}, the power factor will be maximized.

Besides the fact that the $S^2\sigma$ is always maximized at η_{opt}, its magnitude can freely increase with the variations in material parameters and temperature. As shown by Eq. (4.48), high-performance thermoelectric materials should have multivalley band structure (large *N*), large effective masses (m_d), and high mobility (large τ_0 and small m_σ). The $S^2\sigma$ also be increases with increasing operating temperature, *T*. In this case, E_F must also increase to maintain the η_{opt}, which is a constant. However, E_F usually decreases with increasing *T*, so the optimal carrier concentration must increase with increasing temperature.

The $S^2\sigma$ is eventually suppressed by the onset temperature of bipolar conduction. It has been estimated that the maximum temperature beyond which *Z* starts to decrease is $T_{max} \sim E_g/10k_B$T, where E_g is the band gap [212]. As a result, the absolute maximum power factor of a material is expected at T_{max} and η_{opt}. For bulk materials, the maximum power factor is then limited by the intrinsic properties of the materials (such as m_d and *N*) at η_{opt} and T_{max}.

Fortunately, theoretical and experimental studies have demonstrated that quantum confinement is an effective strategy to enhance $S^2\sigma$ in nanostructure materials to a level that is beyond the value of their bulk counterparts. The presence of the thickness of quantum wells or nanowire, *a*, through the *B* factor, in Eq. (4.48), has indicated that it is possible to increase $S^2\sigma$ by reducing *a*. In fact, the requirements of large *N* and m_d and small *a* are equivalent to having a large magnitude of the DOS. Most methods to increase *Z* via electron confinement are aimed at increasing the DOS.

4.5.2.2 Power Factor Beyond Bulk Value

The $S^2\sigma$ can be increased by tuning the Fermi level through changing the carrier concentration and increasing temperature, but there is an inherent limit that is achieved at η_{opt} and T_{max}. However, there has been alternative way to increase $S^2\sigma$ by using confinement effects in nanostructures, such as quantum wells and nanowire [213]. With the expression $S^2\sigma = (S^2n)e\mu$, it is very clear that the enhancement of $S^2\sigma$ by using the confinement effect is attributed completely to the increase in the magnitude of the DOS, which in turn increases the value of S^2n. This concept is illustrated in the case of Si nanowire in the following.

The increase in S^2n as a result of the increase in E_F has been demonstrated in bulk materials [4]. According to a carrier distribution in the conduction band of bulk Si, there are parameters: $m_d = 0.32m_0$, $N = 6$, with an arbitrary $E_F = -0.05$ eV, which is equal to $n = 3.8 \times 10^{18}$ cm^{-3} and η = (−0.05 eV)/[(8.6 × 10–5 eV K^{-1})(300 K)] = −1.93 at 300 K. With $r = 0$, Eq. (4.44) indicates that

$\langle E\rangle$ = +0.07 eV and $S = (1/e)/[(0.07 + 0.05)$ eV/300 K] = −0.4 mV K^{-1}. If the concentration of dopant is increased such that E_F = +0.1 eV (η = 3.86), a larger $n = 1.7 \times 10^{20}$ cm^{-3} is achieved. Because the corresponding increase in $\langle E\rangle$ to +0.13 eV is smaller than the increase in E_F, the value of $|S|$ is reduced by four fold to −0.1 mV K^{-1}. However, S^2n has been increased by a factor of three, from 6.1×10^{23} to 17×10^{23} μV^2 K^{-2} cm^{-3}. Because μ has been assumed to be a constant (r = 0) with respect to η, the value of $S^2\sigma$ will be increases by a factor of three at the same time. Since the η is smaller in one case and greater in the other than the predicted optimal value of 2.47, it is reasonable to suggest that the optimal n is between 3.8×10^{18} cm^{-3} and 1.7×10^{20} cm^{-3}.

To show the mechanism responsible for the enhancement of power factor by using quantum confinement, the DOS of a 2 nm diameter Si nanowire has been calculated. It is found that the nanowire has a much larger DOS than its bulk counterpart, over the range of occupied energy states (up to ~0.1 eV). With same $n = 1.7 \times 10^{20}$ cm^{-3}, the E_F of the nanowire is −0.01 eV (as measured from lowest sub-band edge of the nanowire), as compared with +0.10 eV of the bulk Si. The electron distribution in the nanowire is also skewed toward the lower energy states due to the shape of the DOS, which results in a lower average energy, $\langle E\rangle$, of +0.05 eV as compared with +0.13 eV of the bulk Si. As a consequence, S of the Si nanowire has been doubled to 0.2 mV K^{-1}, due to the large decrease in E_F. The value of S^2n is greatly increased by four fold [4].

If mobility remains the same or is not significantly reduced, the increase in S^2n will lead to increase in power factor. However, it is difficult to predict the difference in the mobility between quantum wells or nanowire and their bulk material counterparts, because too many contributing factors are involved. Firstly, the DOS is changed, which may lead to either increase or decrease in the Fermi level. The dominant scattering mechanism may also be changed. For instance, interface scattering, which is not dominate in bulk materials, but can be dominate in nanostructures that have a high surface-to-volume ratio. As shown in Fig. 4.27b, all these variations (η, r ,and D) have influence on μ.

The enhancement in S^2n is primarily caused by the increase in the magnitude of the DOS, but not by the change in its shape. For example, with similar shapes of the DOS, a 5 nm nanowire has smaller magnitude as compared to that of the bulk Si, except for a narrow range of energy below ~0.01 eV. Therefore, given the same $n = 1.7 \times 10^{20}$ cm^{-3}, the Fermi level of the quantum wire (+0.15 eV) has to be higher than that of bulk material (+0.1 eV). Consequently, S^2n is decreased by almost one order of magnitude as compared with that of the bulk value. Therefore, the enhancement in S^2n can only be available when the naowire are sufficiently small in diameter, so that the DOS becomes larger than that of the bulk material.

According to the enhancement in S^2n by using confinement effects, the DOS is represented by the width of each shape, with larger magnitude of the DOS at the bottom of the band when the dimensionality is decreased [4]. At the same time, the ground state energy increases gradually due to the increased confinement potential at lower dimensions. The increase in ground state energy has no effect on S, σ or κ_e, when only a single band is considered. Consequently, with the same n, the E_F

(relative to the bottom of the band) may progressively decrease as the materials are changed from 3D (bulk) to 1D (nanowire). Although the average energy of the electrons, $\langle E \rangle$, also decreases, the magnitude of the decrease is relatively smaller as compared with the E_F, so that $|S|$ is increased for the same n, as the dimensionality is decreased from 3D to 1D. It has been predicted that the power factor can be enhanced by one order of magnitude, if μ is not decreased [213].

There is a maximum thickness, $a_{\max}$, below which the $S^2\sigma$ starts to be increased over bulk value. This thickness can be obtained by solving the inequality: $S^2\sigma(\eta_{\text{opt},\ r,\ D\ =\ 2,\ 1}) > S^2\sigma(\eta_{\text{opt},\ r,\ D\ =\ 3})$. Assuming a constant relaxation time, $r = 0$ and τ_0 to be a constant regardless of D, this maximum thickness for quantum wells or nanowire is given by:

$$a_{\max} = \sqrt{\frac{\hbar^2}{2\kappa_B T}} \left[\frac{1}{\sqrt{\prod_{i=1}^{3-D} m_{c,i}}} \frac{2\pi}{D} \frac{F_{D/2-1}\eta_{\text{opt},D}}{F_{1/2}(2.47)} \left(\frac{S_{\text{opt},D}}{130\,\text{mV/K}} \right)^2 \right]^{1/3-D}, \tag{4.49}$$

where $m_{c,i}$ is the effective mass in the direction of the confinement. The values of $a_{\max}$ for some theremoelectric materials have been estimated by assuming the effective mass to be isotropic. The larger the effective mass, the higher the enhancement in power factor will be. Due to the limitation of technology, fabrication of materials with very small length scales is still challenger. This is one of the reasons why the effect of very high effective mass on power factor is less studied. This also suggests that the confinement direction should be the one that has the smallest effective mass, which in turn will reduce the requirement for fabrication. In addition, this ensures that m_d will be largest that thus results in higher value of $S^2\sigma$.

The location of the conduction valleys is also an important factor, when the orientation is optimized to achieve the highest $S^2\sigma$. It has been found that the symmetry of previously degenerate conduction valleys could be broken by the confinement [214]. For example, the conduction valleys in PbTe are located at the L-point along the $\langle 111 \rangle$ axes with a degeneracy factor of 4, together with anisotropic longitudinal and transverse effective masses, m_l and m_t, respectively. If PbTe is fabricated as quantum well along the [111] direction, the three oblique (to the growth/confinement direction) valleys have a greater confinement potential than the single longitudinal valley, whose $m_c = m_l$ is largest and which leads to smallest confinement energy. The lifting of degeneracy of the three oblique valleys from the single longitudinal valley results in a decrease in the DOS at the bottom of the conduction band and thus a reduction in $S^2\sigma$. To remain the valley degeneracy, the PbTe quantum wells should be grown in the [100] orientation [202]. However, for n-type Si quantum wells, whose conduction valleys are located in the $\langle 100 \rangle$ axes, the preferred growth orientation should be along the [111] direction [215].

4.5.2.3 Carrier Pocket Engineering

Carrier pocket engineering is a strategy to take further advantage of confinement effects by shifting the energy levels of the carrier pocket/conduction valleys, so as to enhance the $S^2\sigma$ due to the increase in the number of active conduction valleys [216]. In bulk materials, different carrier pockets have different confinement energies; electrons will occupy those with lower energies. By shifting the carrier pocket energies, it is possible to make them to be close one another, so that electrons can occupy more pockets.

This accomplishment has been demonstrated by the carrier pockets of a Si/Ge multilayer structure, in which the X-valley is related to Si and the L-valley is to Ge. Without confinement, all electrons occupy the X-valleys, because they have lower energy as compared with the L-valleys. In this case, the properties of the material/structure are dominated by the characteristics of Si, since the Ge layers have electrical contribution. If the thicknesses of both the Si and Ge layers are reduced sufficiently to form quantum wells, the electrons in Si will be confined to the X-pockets, while those in Ge confined will be confined to the L-pockets. Because the L-pockets have a smaller shift due to their larger effective mass as compared with the X-pockets, both X- and L-pockets have similar ground state energy levels. In this case, both types of pockets are occupied by electrons. Electrons are confined locally since the intervalley transfer of electrons between X-pockets and L-valleys requires substantial change in momentum. Therefore, both X-pockets and L-pockets contribute significantly to the overall properties of the superlattices, according to the parallel conductor model discussed earlier.

An enhancement in $S^2\sigma$ is then observed, due to (i) the local increase in the DOS for both the X-pockets and the L-pockets and (ii) the global increase in the DOS due to the addition contribution of the L-pockets, which are inactive in the bulk-like structure. If it is assumed that the mobility is not changed, the power factor of the superlattices can be high than that of the bulk materials.

The carrier pocket engineering requires the materials to have similar properties, especially compatible crystal structure, so as to ensure defect-free microstructure and matched electronic band structures with favorable energy configuration. Such materials include Si/Ge and GaAs/AlAs [215, 216]. The band edges of Γ-valleys and L-valleys in bulk GaAs and X-valleys in bulk AlAs are within 0.28 eV of each other. In a GaAs (2 nm)/AlAs (2 nm) superlattice, these valleys become much closer, with an energy difference of only 0.05 eV with the $\langle 111 \rangle$ orientation. The value of ZT of the GaAs/AlAs superlattice is expected to be increased to $\sim$0.1 from the bulk value of $\sim$0.01 due to the confinement effect on the G-pocket and further to 0.4 owing to the addition of L-pockets and X-pockets through the carrier pocket engineering. However, the Z value of the GaAs/AlAs superlattice is not very high, due to the low Z of the bulk GaAs and AlAs.

Comparatively, n-type Si has higher intrinsic power factor. Moreover, the lattice parameter mismatch of Si and Gem makes it possible to use strain effect to further manipulate the band structure. For example, when Ge is grown on Si substrates, an in-plane compressive strain is created, which induces an energy splitting that causes an increase in energies of the longitudinal carrier pocket and a decrease in that of transverse carrier pocket. The Si (2 nm)/Ge (2 nm) superlattices grown on (111) $Si_{0.5}Ge_{0.5}$ substrate have been predicted to possess a *ZT* of $\sim$0.96 at 300 K and $\sim$1.75 at 700 K [212].

Therefore, carrier pocket engineering is to increase the number of conduction valleys by controlling favorable growth orientations of materials with similar properties to form confined nanostructures. This method requires sophisticated material design processing. Also, it may not be applicable to the materials with conduction valleys only at one point in the Brillouin zone, such as *p*-type Si and Ge, which have the valence band maximum at the Γ-point.

4.5.2.4 Semimetal-Semiconductor Transition

Metallic Bi has been considered to be potential thermoelectric materials because of two factors. Firstly, it has a large μ, due to its large relaxation time and low effective mass, with $\mu = 3.5\ \text{m}^2\ \text{V}^{-1}\ \text{s}^{-1}$, as compared with $0.14\ \text{m}^2\ \text{V}^{-1}\ \text{s}^{-1}$ of Si. Secondly, it has a low κ_L, owing to its heavy atomic mass, with $\kappa_L \approx 0.9\ \text{W m}^{-1}\ \text{K}^{-1}$ [217]. However, because Bi is a semimetal, its CB and VB overlap slightly. Therefore, its $|S|$ is very low, due to the competing contributions of the two bands.

To increase the power factor of Bi, it is necessary to reduce the overlapping of the two competing contributions. It is expected that confinement potential can be used to separate the CB and the VB, so as to create a band gap in Bi, by using quantum wells [218]. The quantum wells can be further doped to form *p*-type or *n*-type semiconductors to maximize $S^2\sigma$. The semimetal (SM)–semiconductor (SC) transition Bi has been successfully employed to achieve higher $S^2\sigma$ of Bi [217] and Bi–Sb alloys [219].

4.5.2.5 Barrier Layers in Superlattice

The quantum confinement is usually realized by using superlattice materials with similar properties. In superlattice structures, multiple quantum wells are separated by a barrier material with wide band gap. It is also attempted to include nanowire embedded in the matrix. The transport coefficients of superlattice can be described by treating the quantum wells and nanowire, as well as the barrier materials, as parallel conductors. For example, the properties of quantum well superlattice (subscript "SL") are the summation of the properties of the quantum well (subscript "*a*") and barrier (subscript "*b*") materials as:

$$S_{\text{SL}} = \frac{S_a\sigma_a a + S_b\sigma_b b}{\sigma_a a + \sigma_b b}, \tag{4.50}$$

$$x_{\mathrm{SL}} = \frac{x_a a + x_b b}{a + b}, \quad x = \sigma, \kappa_e(\kappa_L), \tag{4.51}$$

$$Z_{\mathrm{SL}} = \frac{S_{\mathrm{SL}}^2 \sigma_{\mathrm{SL}}}{\kappa_{e,\mathrm{SL}} + \kappa_{L,\mathrm{SL}} b} \overset{\substack{\sigma_a a \gg \sigma_b b \\ \kappa_{e,a} a \gg \kappa_{e,b} b}}{=} \frac{S_a^2 \sigma_a}{\kappa_{e,a} + \kappa_{L,a} + (\kappa_{L,b})^{b/a}}. \tag{4.52}$$

The barrier material should be electrically insulating, because the charge carriers are confined in the quantum well layers. It is therefore assumed that $\sigma_a a \gg \sigma_b b$, i.e., the sheet resistance of the barrier layers is much greater than that of the quantum well layers, and $\kappa_{e,a} \gg \kappa_{e,b} b$. However, it is not able to assume that there always is $\kappa_{L,a} a \gg \kappa_{L,b} b$. With these assumptions, the figure of merit of the superlattice (Z_{SL}) become to be dependent on the lattice properties of both materials, while the electrical properties are only determined by materials of the quantum wells, as shown in Eq. (4.52). In such superlattice, there are several issues concerning the effect of barrier material [214, 220–223].

To enhance the power factors ($S_a^2\sigma_a$), the thickness of the quantum wells or nanowires (*a*) have still to be as small as possible, while the thickness of the barrier layer (*b*) should be optimized to balance the effect of electron tunneling and parasitic heat conduction on *Z*. Electron tunneling is a quantum mechanical effect, during which electrons can penetrate the potential barrier, thus escaping the confinement in the quantum wells nanowires. However, the tunneling electrons will not increase carrier concentration or electrical conductivity of the barrier material themselves, because they are not present within the barrier layers. The coupling of nearby quantum wells or nanowires due to the tunneling will result in additional energy states, which make the DOS to be broadened and decreased in the quantum wells or nanowires. The reduction in DOS thus results in a decrease in power factor and *ZT*, when $a = b = 2.5$ nm [221]. To prevent the electron tunneling from occurring , it is necessary to either increase the thickness of the barrier layers or enlarge the band offset between the quantum wells or nanowire and the barrier materials.

However, if thicker barrier layers are used, parasitic heat conduction will be increased, thus reducing *ZT*. As meant by Eqs. (4.51) and (4.52), it is obvious that an increase in *b* (or *b*/*a*) results in an increase $\kappa_{L,\mathrm{SL}}$ and a decrease in Z_{SL}. Parasitic heat conduction means that the barrier layers will take much heat away from the source to the sink directly without contributing to electrical power generation. Similarly, parasitic heat flow counters the heat-pumping action in a refrigeration device, leading to significant reduction of efficiency. Because materials are usually selected as barrier layers because they are similar to those of the quantum wells or nanowire, to facilitate heterostructure growth, their thermal conductivities are comparable and both must be considered. The significant reduction in *ZT*, from the value for $b = 0$ to that for $b = a$, has been attributed to this kind of parasitic heat conduction in superlattice structured thermoelectric materials.

If wide band gap materials are used as the barrier layers, the tunneling effect can be reduced due to the increase in the band offset [202]. However, an increase in band offset will increase the scattering of electrons at the interfaces, which in turn will reduce mobility and thus electrical conductivity of the whole structures [214]. For example, if the band offset is 0.2 eV, the mobility is almost not affected, wheres a band offset of 1 eV causes a decrease in mobility by up to 40 %.

The definition on quantification of the $S^2\sigma$ or Z of superlattice has been found to be inconsistent in the open literature [220–222]. For instance, the carrier concentration or electrical conductivity have been defined using either the volume of the quantum wells (i.e., σ_a) or the volume of the superlattice [220–222] (i.e., $\sigma_{SL} = \sigma_a(b/a + b)$, which is equivalent to that of Eq. (4.51) if it is assumed that there is ($\sigma_a a \gg \sigma_b b$). Moreover, with same data, S^2n of PbTe has been reported to be higher in quantum wells of superlattice than in bulk when the carrier concentration is normalized with the volume of just the quantum wells [224], but the result is inversed if the volume of the whole structure without excluding the volume of the barrier layers [202]. In other words, the PbTe quantum wells have superior performance over the bulk PbTe, but the thermoelectric effect of the entire structure have been diluted due to the presence of the barrier layers. Therefore, the use of thick barriers should be avoided, because a large b reduces S^2n of the superlattice (S^2n_{SL}).

However, it seems that the apparent reduction in S^2n_{SL} has been compensated by the simultaneous reduction in $\kappa_{e,SL}$. As shown by Eq. (4.52), Z_{SL} is equal to $Z_a = S_a^2\sigma_a/(\kappa_{e,a} + \kappa_{L,a})$, without the term b, if $\kappa_{L,b}$ is ignored. Therefore, the use of a thick barrier only leads to an increase in parasitic heat conduction, through the relation of $\kappa_{L,b}(b/a)$. Therefore, it is necessary to carefully optimize the barrier thickness in the structures of quantum wells or nanowire, so that the adverse effects of tunneling, parasitic heat conduction and loss of carrier mobility due to the interface scattering can be minimized.

4.5.3 Lattice Thermal Conductivity ($\mathbf{k_L}$)

The total thermal conductivity of thermoelectric materials consists of two parts: carriers/electrons (k_e) and the lattice (k_L). k_e is directly proportional to the electrical conductivity (σ) through the Wiedemann–Franz law, which is usually smaller in magnitude than k_L. In terms of maintain a high ZT, it is necessary to maintain a high electrical conductivity. Therefore, there is no much room to reduce thermal conductivity through reducing k_e. In other words, the key to the reduction of thermal conductivity is to reduce k_L, as that observed in nanostructure thermoelectric materials. Reducing k_L means to increase the proportion of heat transport by electronic carriers and thus to enhance efficiency of heat-to-electrical energy conversion. In this section, we will first review the phenomenology behind k_L in order to get a better appreciation of the extant ideas and experimental efforts aimed at reducing k_L.

4.5.3.1 Semi-Classical Theory

According to elementary principles of the kinetic theory of gases, with consideration of energy transport through noninteracting particles, the lattice thermal conductivity is $k_L = C_L vl$, where C_L is the heat capacity of the lattice per unit volume, v is the average particle velocity and l is the particle mean free path [225, 226]. The "particles" that are responsible for heat transport in solid materials are lattice modes or phonons. The formula for k_L can also be obtained with the semi-classical form of the Boltzmann transport equation, Eq. (4.18). At steady state ($\mathrm{d}f_P/\mathrm{d}t = 0$) and neglecting the momentum dispersion of the phonons ($\nabla_k f = 0$), there is $f_P = f_{P,o} - \tau v \nabla_k f$, where $v = \mathrm{d}r/\mathrm{d}t$. With the definition of heat flux, $Q(r,t) = \sum_k v f_p E(k)$, it can be derived as:

$$Q(r,t) = \int v[f_{P,o} - \tau v \nabla_r f_p] E(k) g(E) \mathrm{d}E, \tag{4.53}$$

where the summation over all the modes is replaced by an integral involving a continuous DOS, $g(\mathrm{E})$. The integral over $f_{P,o}$ vanishes, together with $\nabla_r f_p = \nabla_T f_p \nabla T$, there is:

$$Q(r,t) = \nabla T \int \tau v^2 \nabla_T f_p E(k) g(E) \mathrm{d}E. \tag{4.54}$$

The k_L is defined as the ratio of the heat flux $Q(r,t)$ to the temperature gradient (∇T), given by:

$$\kappa_L = \int \tau v^2 \nabla_T f_p E(k) g(E) \mathrm{d}E. \tag{4.55}$$

This expression can be further simplified to $\kappa_L = \tau v^2 C_L$, by using the definition of the lattice heat capacity

$$C_L = \mathrm{d}U/\mathrm{d}T = d\left[\int f_p E(k) g(E) \mathrm{d}E\right]/\mathrm{d}T. \tag{4.56}$$

It is the temperature derivative of the total energy (U). With the definition of phonon mean free path, $l = v\tau$, the lattice thermal conductivity can then be written as:

$$\kappa_L = C_L v l. \tag{4.57}$$

Therefore, the reduction of κ_L can be achieved by decreasing the individual components, C_L, v, and l, which will be discussed as follows.

4.5.3.2 Phonon Energy Dispersion

As stated before, there have been two assumptions: (i) the particles or phonons have no interaction and (ii) their momentum dispersion is neglected. According to Eq. (4.57), these assumptions do not change the dependences of k_L on C_L, v, and l, qualitatively, but they can help to deeply understand the phonon interactions and dispersion, thus enabling an elucidation of k_L. The interaction between atoms at their discrete lattice sites in response to an external applied force can be described by using a simple model, by assuming that the interactions between nearest neighbors are harmonic, so that follows Hooke's law can be applied, in which the displacement is linearly proportional to the force. The resultant modes of vibration of the lattices—phonons, can be correlated to the vibrational energy. It is necessary to mention that there is no longer an analogy to noninteracting particles because the number and momentum conservation laws are not strictly valid in this case [194]. As a result, phonon dispersion relationships, i.e., phonon frequency (ω) versus wave vector (κ), can be given by:

$$\omega = \sqrt{4\,\mathrm{K/m}}\,\sin(\kappa a_0/2), \tag{4.58}$$

where a_0 is the lattice point spacing and K is an indicator of the strength of coupling, related to a spring constant-like parameter, between the points.

The phonon group velocity, v, can then be determined, which is include in the expression of κ_L with v = dω/dk. It is noted that the continuum mechanics formula $v = \sqrt{Y/\rho}$, with Y being the elastic modulus and ρ the material density, is strictly valid for $\lambda \to \infty$ or $\kappa \to 0$, e.g., for Si, with $Y \approx$ 180 GPa and $\rho \approx$ 2330 kg cm^{-3}, $v \approx 9 \times 10^3$ m s^{-1}. In addition, the number of phonons per unit energy, i.e., the DOS, is inversely proportional to v, as:

$$\mathrm{DOS} = \frac{\mathrm{d}N}{\mathrm{d}\omega} = \frac{\mathrm{d}N}{\mathrm{d}\kappa}\frac{\mathrm{d}\kappa}{\mathrm{d}\omega} \tag{4.59}$$

A single-valued and multivalued basis at a lattice site are corresponding to lower energy acoustic phonon modes and higher energy optical phonon modes, because they respond to acoustic (sound) disturbances and electromagnetic radiation, respectively. The acoustic phonons are caused by the interunit cell response of the lattice to external stimulus, such as gradient of temperature and electric field, while the optical phonons are attributed to intracell interactions [90]. The acoustic phonon modes have a larger energy dispersion, corresponding to a wider distribution of velocities of phonon, according to v = dω/dk. Comparatively, the energy dispersion of the optical phonon modes is much less pronounced, implying smaller dω/dk and lower velocities. It is easier to reduce κ_L when v is low, corresponding to features with the ω–k dispersion or phonon band structure having smaller slopes. The presence of internal structural defects, such as variation of charge and difference of mass, at lattice sites will provide additional low group velocity optical modes, and thus have no contribution to k_L. Moreover, the increased entropy of the system, due to the presence these additional modes, is also favorable to have a lower k_L.

The acoustic and optical modes can be further decomposed into one longitudinal and two transverse branches, in terms of polarization. By extrapolating the phenomenology of lattice sites to unit cells, for p number of primitive cells with q atoms or unit cell, the total number of degrees of freedom is $3pq$, out of which only $3p$ degrees can be accommodated through the acoustic modes. As a result, the rest $3p(q-1)$ degrees of freedom are accommodated through the optical branches. Therefore, structures with a larger number of unit cells (larger p) and with more complex crystal structures (larger q) will have a larger number of optical modes and thus a lower k_L. In other words, materials with higher crystal complexity have lower k_L. A complexity factor, CF, defined as the number of atoms per primitive unit cell, has been used to represent structural characteristics of materials. It has been found that the complexity factor has a direct effect on k_L, with a relation of $k_L \propto \mathrm{CF}^{-2/3}$ [227, 228]. The values of CF have a very wide range, for example, CF = 1 for rare gas crystals, such as Ar, Kr, etc., CF = 2 for Si, Ge, GaAs, PbTe, etc., CF = 14 for $MgAl_2O_4$, CF = 105 for β-boron and CF = 414 for YB_{68}. As the k_L varies over five orders of magnitude, from 10^3 to 10^{-2} W m K^{-1}, the minimum k_L can correspond to a maximum CF of up to 10^7 [93].

In a similar way, the crystal structures, through the acoustic and optical phonon energies, also contribute to lattice heat capacity, C_L. According to Debye model approximation to evaluate the energy (U), the velocity (v) of the acoustic modes is assumed to be constant, which is independent on polarization. The C_L is found to be proportional to $(T/\Theta_D)^{-3}$, where Θ_D is the Debye temperature. Θ_D has the same importance for the phonons as E_F for electrons. It can be used to describe the energy of the acoustic phonons through $k_B\Theta_D$. Θ_D is inversely proportional to the density of materials, which thus can be influenced through tensile and compressive strains [194]. However, much less effort has been made to reduce C_L through the manipulation of Θ_D. The contribution of the optical phonons to C_L has been evaluated by using the Einstein model, in which harmonic oscillators are assumed to be independent at a constant frequency, ω_0 [194]. The C_L evaluated with this model has the Dulong–Petit value, 24.9 J mol^{-1} K^{-1}, at high temperatures, at which the assumption of noninteracting oscillators is more valid.

The variation of k_L as a function of temperature is similar to that of C_L, i.e., $k_L \propto T^3$. Such a dependence has been found in certain types of materials, especially at low temperatures, e.g., for Si [229] and Ge [230], at $T < 20$ K. At higher temperatures, anharmonic interactions, e.g., phonon–phonon, phonon-carrier, and phonon-defect interactions, which involve redistributions of energy and momentum, should be taken into account to evaluate the variation of k_L. For phonon–phonon interactions, it has been found that the dominant interactions belong to the three-phonon collision type [231, 232], which can be considered by using the N-process (Normal) and U-process (Umklapp).

An example of a momentum conserving N-process involves a phonon–phonon collision of the type: $k + k' \rightarrow k''$, where k and k' are the wave vector of the incident phonons, while k'' is the wave vector of the resultant phonon. In the

N-process, the phonon energy is only redistributed from the incident phonons to the transmitted phonons and an infinite k_L is implied. On the other hand, momentum nonconserving U-processes, represented by $k + k' \rightarrow k'' + G$, result in excess momentum from the phonon collisions, represented by using G, which is the reciprocal lattice wave vector. The presence of $G = 2\pi/a$, where a is the lattice parameter, indicates a net phonon backscattering and a resistance to phonon and thermal transport, thus resulting in a finite k_L. The tendency of the occurrence of U-processes is found to be dependent on temperature, with a relationship of $k_L \infty e^{-\Theta_D/2T}$, which is the basis of Eucken's law [233], where k_L decreases with increasing temperature above Θ_D. For thermoelectric materials, working at room temperature and above, generally follows the latter relationship, i.e., k_L scaling as $k_L \infty 1/T$, which has to be considered in the evaluation of the figure of merit.

The reduction of k_L can also be achieved by the reduction of the phonon mean free path $l = v\tau$, which is the underlying reason why nanostructure materials exhibit improved thermoelectric performances, such as superlattice and nanowire. Although both elastic and inelastic scattering processes are possibly present, the inelastic scattering length is usually much greater than the elastic scattering length and consequently we only consider the latter limiting process. The length scales involved in phonon transport can be classified according to the phonon wavelengths. With the phonon energy dispersion, there are a wide range of wavelengths and wave vector (λ and k), from the spacing between individual lattice sites ($\lambda = 2a$, $k = \pi/a$) to the continuum limit ($\lambda = \infty$, $k = 0$). Additionally, important phonon wavelengths (λ_{ph}) can be associated with dominant interaction forces, e.g., when the temperature is taken as a generalized force to drive the vibrations of the lattice, then the corresponding λ_{ph} can be calculated through $E = hv/\lambda_{ph} = k_B T$, which is ~1 nm at 300 K [234]. In addition, if the corresponding wave vector is close to the edge of the Brillouin zone, there will be a large number of such thermal phonons, due to the increased DOS. For a given C_L and v, k_L will reach its minimum value, as l is of the order of the smallest λ_{ph}, i.e., $\sim 2a$.

If the transport of thermal energy is random from one oscillator/atom to another throughout the solid over an average time, $t = 1/f$, there will be $l = v/f = \lambda_{ph}$ [235]. In this case, a minimum value of k_L is given by:

$$\kappa_{L,\min} = \left(\frac{\pi}{6}\right)^{1/2} k_B n^{2/3} \sum_i v_i \left(\frac{T}{\Theta_D}\right)^2 \int_0^{\Theta_D/T} \frac{x^3 e^x}{(e^x - 1)^2} \mathrm{d}x. \quad (4.60)$$

Equation (4.60) has been widely used to predict $k_{L,\min}$ for a given mode velocity (v_i) and number density (n). For example, $k_{L,\min}$ of Si is ~1 W m^{-1} K^{-1} [236]. Until now, values of $<k_{L,\min}$ have not been observed, which can be used to justify the validation of the model. Therefore, once the minimum $k_{L,\min}$ has been reached, further increasing ZT is only possible through the increment of power factor.

In nanostructure thermoelectric materials, such as nanoparticles, nanorods, nanowire, and superlattice, their sizes are comparable to or even smaller than l. As a consequence, temperature gradient cannot be well-defined across the nanostructure, so that it is necessary to modify the basic Fourier heat conduction theory [237, 238]. When the gradient cannot be well defined [239], the heat flux (Q) between two temperatures, T_1 and T_2 and across an area (A) should be considered in terms of radiative transport [240], by using a Stefan-Boltzmann type relationship, $Q = A\sigma_B(T_1^4 - T_2^4)$ where $\sigma_B = (\pi^2/120)(k_B^4/\hbar^3)\Sigma 1/c_i^2$ is a constant. In this case, the thermal conductivity is derived from the concomitant equation of phonon radiative transport (EPRT) [240], where the l for determining k_L is replaced by an effective mean free path, $l_{\text{eff}} = l/[1 + (4\ l/3d)]$, with d being the size of the materials. An increasing ratio of l to d means a decreased k_L.

The upper limit of λ_{ph} is determined by the size, d, of the thermoelectric materials. For materials with large d, it is related to the continuum limit and the center of the Brillouin zone. Another length scale involved in the phonon transport is called coherence length (ξ_{coh}). Below this length, the effects of phonon wave interference must be considered, otherwise diffusive particle-like motion becomes dominant [241]. The ξ_{coh} bulk GaAs has been estimated to be about 1 nm for the transverse phonons and about 2 nm for the longitudinal phonons [242]. As a result, if there is $l < \xi_{\text{coh}} < d$, phonons can be treated as particles, whereas if $l < d < \xi_{\text{coh}}$, phonon wave interference will be enhanced, which leads to the onset of localization of phonon. This localization reduces the propagation of phonon and thus decreases thermal conductivity. There is also localization of phonon when λ_{ph} is comparable to l, i.e., $k\ l \approx 2\pi$. By treating phonons as waves, it also allows to clarify the minimum thermal conductance due to phonons, $K = k_L A/L$, for heat transport through a length L and cross-sectional area A. The understanding of possible smallest quantity of heat transport in a material will provide an alternative way to understand the possible lowest k_L, which has led to comprehensive studies of thermal transport behaviors by using quantum mechanics as an example.

4.5.3.3 Quantum of Thermal Conductance

As a macroscopical quantity, k_L has been defined according to the kinetic theory and the Boltzmann transport equation. Alternatively, the total thermal conductivity or conductance of a material or structure is the summation of the contributions by all discrete phonon modes with a number of M. The maximum amount of heat that can be transported by an individule mode, at a certain temperature, T, is represented by the quantum of thermal conductance, g_o [243]. The total thermal conductance of a material will thus be proportional to Mg_o. When phonons (with energy of $\hbar\omega$) propagate along a one-dimensional nanowire, for i longitudinal and transverse branches, the total energy flux is given by [108]:

$$J_{\text{th}} = \sum_i \int_0^{\infty} \frac{\mathrm{d}\kappa}{2\pi} \hbar\omega n(\omega) v. \tag{4.61}$$

The thermal conductance, G_{th} can be defined as the ratio of J_{th} to ΔT, given by:

$$G_{th} = \sum_i \frac{k_B^2 T}{h} \int_{x_m}^{\infty} \frac{x^2 e^x}{(e^x - 1)^2} dx, \quad \text{with} \quad x = \frac{\hbar\omega}{k_B T}. \tag{4.62}$$

For the modes with the smallest energy, i.e., $\hbar\omega(k = 0) = 0$, there is $G_{th} = g_o = \pi^2 k_B^2 T/3\ h = 9{:}456 \times 10^{-13}$ W K^{-2} (T), with $k_B T < \hbar\omega$, In this case, g_o represents the possible maximum value of the thermal conductance for a given mode at a given temperature [244]. The value of g_o has been determined experimentally, by using carefully controlled geometries in one-dimensional ballistic phonon channels [244].

4.5.3.4 Reducing k_L with Nanostructures

Nanostructuring has been considered to be an effective technique to reduce the values of k_L of the corresponding bulk thermoelectric materials. Such nanostructuring strategies that have been successfully used to achieve low values of k_L include selecting materials with high degree of crystal complexity, fabrication of nanocomposite by embedding nano-sized inclusions in bulk materials, increasing interface effects, using supperlatice in thin films and confinement with one-dimensional or quasi one-dimensional structures, which will be elaborated as follows.

Reducing Thermal Conductivity of Bulk Materials

According to its definition, k_L is a reflection of the properties of the lattice. Therefore, properties related the lattice, such as the compressibility (χ), thermal expansion coefficient, melting point (T_m), density (ρ) and the atomic weight (A) of the constituent atoms, all have certain degree of effects on k_L [245]. The concurrent effect of these parameters on k_L can be well described by using a unified empirically derived expression, Keyes' relationship, $k_L T = B(T_m^{3/2} \rho^{2/3} / A^{7/6})$, where B is a constant of proportionality, which is related to the Gruneisen parameter, a measure of the change in the phonon dispersion spectrum for a given dilatation that is inversely related to the χ [245]. This relationship has been validated by experimental results in various materials. Therefore, soft materials (with high porosity and thus low density—ρ), constituted with atoms of atomic high weight (A), together with low melting points (T_m), will have a low k_L. In this respect, it can also be deduced that materials of ionic bonding have lower k_L than those of covalent bonding, because the fomers have lower melting point than the latter. The presence of charged offers additional scattering mechanisms, due to the creation of additional optical modes, related to the formation of the sublattices of the positive and negative ions. It is also understandable that k_L decreases with increasing cation-to-anion mass ratio.

This explains why high-performance thermoelectric materials are those that have low density and open crystal structures, such as skutterudites, clathrates, and HH intermetallic compounds, as discussed previously [21]. In these materials, loosely bound metal guest atoms are dispersed into the lattice of the matrix, acting as rattlers to perturb phonon from transporting thermal energy. Therefore, phonon scattering is enhanced at the atomic level while the periodicity of the matrix lattice is unchanged. The materials behave like glasses in terms of phonon transport and crystalline in terms of electrical carrier transport, which has been well known as "phonon glass electron crystal" (PGEC). Once the phonon frequency equals to the rattler frequency, a maximal scattering to the propagating acoustic phonons occurs due to resonance.

However, this principle cannot be applied to simple monatomic substitutions, it requires ''random non-central distortions of the lattice'' [235]. There are other requirements to produce PGEC behavior. For example, the rattler atoms should be uncorrelated one another, without forming lattice with well-defined positions, i.e., no long range ordering and phase coherence, with respect to the vibrational motion [233, 246]. The concentration of the rattler atoms should be higher than 3 % of the total mass of the crystal.

The presence of PGEC behavior in thermoelectric materials has been confirmed by experimental measurement of k_L, specific heat and electrical resistivity of the Tl filled skutterudites, such as $Tl_xCo_{4-y}Fe_ySb_{12}$ and $Tl_xCo_{4-y}Sn_ySb_{12-y}$ [247]. A decrease in k_L, by a factor of as high large as 5, has been reported. According to the measurement of specific heat, Tl atoms behave as independent harmonic oscillators [248]. The phonon mean free path (l) corresponds to the average distance between adjacent Tl atoms, which is proportional to $x^{1/3}$. Based on crystallographic analysis and powder diffraction measurements, it is found that 80 % of the voids in $CoSb_3$ have been filled by Tl atoms, resulting in a large atomic displacement parameter (ADP) of 0.05 Å^2, which measures the mean square displacement amplitude of an atom about its mean position. The ADP is nearly one order of magnitude larger than that in terms of nominal static disorder, which strongly supports the rattler hypothesis.

It has been reported more recently that there is a quasi-harmonic coupling between the inserted rattler and the host lattice in La- and Ce-filled Fe_4Sb_{12} skutterudites, which is revealed by using neutron spectroscopy [246]. This implies that the increased complexity of the filled crystals through the *CF* factor and the decreased group velocity of the optical phonon modes, rather than the rattling effect discussed above, can be more responsible for the reduced k_L. Additional evidence to support the presence of such coupled modes has been found in clathrate thermoelectrics, $Ba_8Ga_{16}Ge_{30}$, as shown in Fig. 4.28 [249]. According to neutron triple axis spectroscopy, the phonon lifetimes are estimated to be about 2 ps, which is one order of magnitude longer than the 0.18 ps expected according to the inelastic scattering of the acoustic phonons with the rattler atoms [249]. Such elongated lifetime has been attributed to the absence of scattering caused by an avoided band crossing between the acoustic phonon mode of the host lattice and a flat quasi-localized low energy mode due to the rattler atom. Therefore, it is necessary to re-elavulate the PGEC type of mechanism.

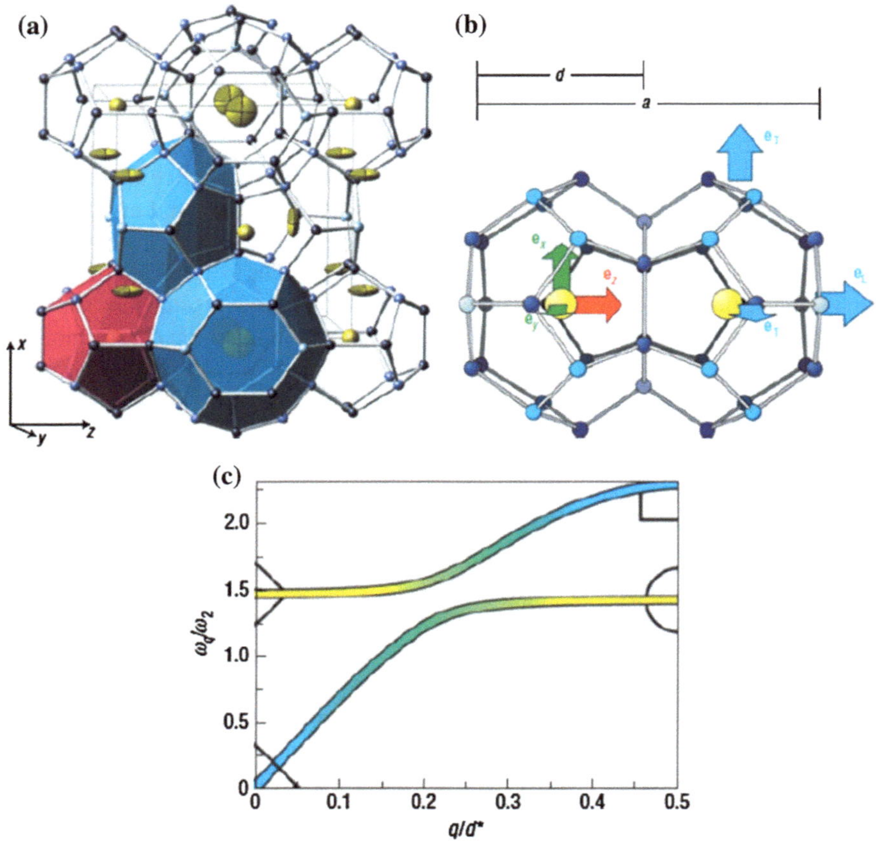

Fig. 4.28 **a** Illustration of a typical open cage-like clathrate structure of $Ba_8Ga_{16}Ge_{30}$ with guest (rattler atom: Ba)-host (cage framework: Ga and Ge atoms) interactions. **b** The rattling motion of the guest can be decomposed into longitudinal (e_L) and transverse (e_T) modes. **c** The interaction of the guest mode phonon dispersion with the host mode is postulated to result in an avoided band crossing, at which the phonon lifetimes are increased. Reproduced with permission from [249], Copyright @ 2008, Macmillan Publishers Limited

Bulk Nanocomposite Thermoelectrics

Superlattice and nanowire, with lower dimensionalities, possess promising thermoelectric performance, due to the great reduction in k_L, which is realized by reducing the mean free path, l. However, in terms of real applications, these nanostructure thermoelectric materials have a serious problem in large-scale production with low cost. In this respect, bulk nanomaterials with embedded nanoscale inclusion are more advantageous for practical large-scale applications.

In nanostructure materials, the radiative transport of phonons across the various nanostructures with length scales smaller than l is the main reason that is responsible for the reduction of k_L. Monte–Carlo simulations have predicted that

k_L of a material can be greatly reduced, by forming nanocomposite with nanoparticles embedded in a matrix [250]. However, these simulations use bulk phonon dispersion parameters and have not considered phonon polarization or phonon dispersion [251]. They also have the assumption that l is independent of phonon frequency [252]. As a result, a good agreement between the theoretical prodictions and experimental results has not been reported until now [242].

The above discussion implies that the phonon transport can be modulated by controlling the orientation and shape of the nanoparticles (Fig. 4.26), so as to reduce k_L. It has been shown that a random orientation with a hierarchical distribution of the nanoparticles is more effective in decreasing k_L than an ordered orientation with a uniform distribtuion. However, the effects of degree of randomness and the influence of interfaces on k_L have not been quantified.

As discussed in previous section, reduced k_L and improved *ZT* have been observed in PbTe or Bi_2Te_3, by embedding nanostructural inclusions in bulk materials, through suitable materials processing techniques. The nanoscale inclusions are lower dimensional structures, which are similar to zero-dimensional quantum dots and thin films. One of such examples is the class of LAST alloys, $AgPb_mSb_nTe_{2+m}$ [253]. The reasons that these exhibit high thermoelectric performance is that they meet most of the requirements as discussed above, such as (i) complex crystal structure (ii) low intrinsic k_L (1.3 W m^{-1} K^{-1}) and (iii) tunable electrical carrier concentrations and electrical conductivity (σ) over a wide range, through the generation of a large number of compositions, by varying m and n [253]. The LAST alloys have a NaCl lattice structure, where the Ag, Pb, and Sb atoms are distributed over the "Na" sites while the Te atoms are located at the "Cl" sites. The isoelectronic substitution of Pb^{2+} ions for the Ag^+ and Sb^{3+} creates structural and electronic distortions, which reduce the k_L and increase the Seebeck coefficient at the same time. Compositional modulations arising from the proximity and high concentrations of the cations have been observed in these alloys. This morphology has the features of two- and quasi zero-dimensional nanostructures, respectively.

Spinodal decomposition in $Pb_{1-x}Sn_xTe$ based systems has facilitated the nucleation of such nanostructural features, where Sn is introduced to enhance the electrical conductivity σ [254]. For instance, the spontaneous separation of PbS and PbTe phases can cause structural modulations of about 2 nm in wavelength in $(Pb_{0.95}Sn_{0.05}Te)_{1-x}(PbS)_x$ (x = 0.04–0.3) mixtures, as shown in Figs. 4.37 and 4.38 [139, 140-need to determined later]. Additionally, nucleation and growth of nanocrystals, 3–10 nm in size, have been observed in the samples with lower values of x (Fig. 4.39) [140-need to determined later]. The nucleated nanocrystals or nanodots have an intrinsic lattice parameter mismatch by 2–5 % with the surrounding matrix, with endotaxial orientation [200]. This endotaxy is a process of crystallographically oriented segregation of a phase from a matrix, such as Ag–Sb rich nanocrystals from PbTe-rich matrix, which occurs due to the disintegration of the original native solid solutions [255].

Dislocations at the nanocrystal-matrix interface and the large number of interfaces serve to scatter these phonons and thus decrease the thermal conductivity, which almost negligibly affects the electrical conductivity σ, if the doping leves are not too high. Reduced k_L has been observed in various material systems, such as $AgPb_mSb_nTe_{2+m}$ solid solutions, with $k_L \approx 0.43$ W m^{-1} K^{-1} and ZT of 1.45 at 620 K) [256], $Pb_{9.6}Sb_{0.2}Te_{10-x}Se_x$ bulk materials, with $k_L \approx 0.4$ W m^{-1} K^{-1} and ZT of 1.2 at 650 K) [110] and Ag free $Na_{1-x}Pb_mSbyTe_{m+2}$, with $k_L \approx 0.85$ W m^{-1} K^{-1} and ZT of 1.7 at 650 K [257]. It has been found in $(Pb_{0.95}Sn_{0.05}Te)_{1-x}(PbS)_x$ that, an increase in the content of PbS in the range of $0 < x < 0.3$, leads to an enhancement in the one-dimensional spatial modulations but a decrease in the nucleation of the nanocrystals [258]. This results in a three-fold reduction in k_L as compared with the unmodified solid solutions. Lower values of k_L have also been observed in the samples with $x = 0.08$ (0.38 W m^{-1} K^{-1}) and $x = 0.16$ (0.4 W m^{-1} K^{-1}), as compared with the one with $x = 0.3$ (1 W m^{-1} K^{-1}), which implies that zero-dimensional features are more effective in reducing the k_L than the one-dimensionally oriented stripes.

The introduction of nanostructural inclusions in p-type $Bi_xSb_{2-x}Te_3$ decrease its k_L from 1.4 W m^{-1} K^{-1} of the bulk ingot to about 1.1 W m^{-1} K^{-1}, with corresponding increase in ZT from about 1 to 1.2 at 300 K and 1.4 at 373 K [235]. The materials consist of grains and Te precipitates of 10–50 nm in size, with embedded Sb-rich nanodots of 2–10 nm. The phonon scattering at various length scales in the ''nano-bulk'' materials have contributed to the reduced k_L. However, the surprisingly increased σ can be attributed to the fact the grain boundaries are devoid of charge and act as Schottky barriers [259]. At equilibrium, as the Fermi energy (E_F) is constant through the material, the grain boundaries serve as a source of electrical holes which are injected into the grains and hence enhance the σ. Consequently, grain boundaries which serve as interfaces for enhancing phonon scattering can also have positive effect on the σ.

Interface Engineering

Once the length scales of materials are below or approaching the phonon mean free path, l, a thermal boundary resistance (TBR) caused by the interfaces will be formed, which reduces the thermal conductivity or conductance [260]. This corresponds to an additional thermal resistance, which is represented by R_{TBR} and calculated as per unit area. Together with the intrinsic resistance, $R_{\mathrm{int}} = L/k_L$, the net thermal resistance, $R_{\mathrm{net}} = L/k_{\mathrm{net}} = R_{\mathrm{TBR}} + R_{\mathrm{int}}$. The heat transport across the interface of two materials is significantly affected by the TBR, caused by the reflection, transmission, and absorption of phonons at the interfaces. The difference in distribution of phonons at the two sides results in a temperature drop across the interface, corresponding to a thermal resistance given by the ratio of the temperature drop to the transmitted heat flux. This is usually called Kapitsa resistance, which is dependent on the ratio of the Debye temperatures of the materials [261]. It has been found in Si/Ge interface a ratio of about 1.7 results in a better match of the phonon wavelengths and thus enhanced phonon transmission, for a Pb/diamond interface with a ratio 0.05.

Two models have been developed to understand the thermal conduction across an interface or boundary: the acoustic mismatch model (AMM) and the diffuse mismatch model (DMM) [262]. In the AMM, the constituents on the two sides of the interface are treated as a continuum, each with an acoustic impedance, $A = \rho v$, where ρ is density of the bulk material and v is the associated acoustic velocity. The heat flux flowing across the interface due to the longitudinal phonon waves, from medium 1 to medium 2, is determined by the product of the number of incident phonons and the transmission probability, $T_{1\to2}$ [263]. With incidence angle of θ_i and refraction angle of θ_t, it can be derived as [263]:

$$T_{1\to2} = \frac{4(A_2/A_1)(\cos\theta_t/\cos\theta_i)}{(A_2/A_1 + \cos\theta_t/\cos\theta_i)^2}. \tag{4.63}$$

It means that the maximum transfer of heat occurs when the two materials have similar values of density and specific heat. If the two layers have identical acoustic impedances, i.e., $A_1 = A_2$, the AMM predicts a $T_{1\to2}$ to be unity, for normal phonon incidence ($\theta_i = \theta_t = 0°$). The wave-like propagation also implies that there is a critical angle (θ_c), above which all the incident phonons are totally internally reflected and do not contribute to the heat transfer, thus leading to a reduced TBR. Because the materials have assumed to be continua, the detailed nature of the interface is ignored in the AMM approximation, meaning that there is λ_{ph} is larger than the interface roughness. However, with increasing phonon energy and decreasing λ_{ph}, the l can be comparable to the scale of the surface roughness and imperfections and the AMM becomes inappropriate [262, 263].

In comparison, the DMM takes into account the details of the interface by considering individual phonon traversal. The phonons impinging on the interface lose their original state, such as direction, polarization, etc., subsequent to scattering. The transmission of any phonon is thus determined by whether there is a corresponding phonon, on the other side, with the same energy to which scattering can occur, i.e., by the phonon density of states $g(\omega)$. When the phonon dispersion is taken into account, all the incident phonons now scatter elastically from the interface, with the $T_{1\to2}$ being determined by the degree of the mismatching in phonon DOS at the two sides. For both longitudinal and/or transverse mode phonons, expressed through the index j, the transmission is determined by the ratio of the in-and out-fluxes, product of the phonon velocities and number, as:

$$T_{1\to2}(\omega) = \sum\nolimits_j v_{2,j}g_{2,j}(\omega,T) \Big/ \sum\nolimits_j v_{1,j}g_{1,j}(\omega,T). \tag{4.64}$$

Both the AMM and the DMM have been applied to simulate experimental results, but their simplified formulation generally cannot provide a precise prediction [262]. For example, for SiO_2 and SiN_x films deposited on Si substrates, although the DMM has shown a good agreement at low temperatures (<20 K), its prediction at room temperature deviates the experimental result by nearly one order of magnitude [264]. Because the DMM considers elastic scattering of a large number of phonons, it usually gives higher thermal conductance or conductivity than the AMM [262, 265]. However, there are also inelastic scattering processes, due to the presence of point defects [266], interfacial roughness, tunneling, excitation of surface phonons, phonon down conversion [267], influence of electrical carriers, and so on, have not been taken into account by the models, which makes them inaccurate in predicting the more complicated reaslity. Also, the presence of strain at the interfaces is similar to the case of increase in the number of interfaces, which has been overlooked in the two models.

To address these problems, molecular dynamics (MD) simulation has been used, which serves to examine the diffusive approximations inherent in the Boltzmann transport formalism [268]. In MD, mutual atom–atom interactions are modeled through Lennard–Jones [225, 268] or other interatomic potentials [269] and the heat flux is derived from the product of the atomic forces and velocities. According to MD simulations, two orders of magnitude reduction in thermal conductivity can be obtained by using superlattice [270] and nanowire [269], as compared with their bulk counterparts.

The TBR has also been used in modeling the reduction of k_L with decreasing grain size of polycrystalline materials, with a Kapitza thermal resistance to the grain boundaries [271], $R_{gbdy} = R_{TBR}$. The net thermal conductance can be calculated by assuming two parallel conduction processes through the grain and grain boundary, respectively. It is thus expected that, a decrease in grain size implies a reduction in l, and thus an increase in the contribution of R_{gbdy}. When there is a lack of direct proportionality between the grain size and the measured thermal resistance or conductivity, k_{net} can be used as an evidence of the contribution of R_{gbdy} [147, 148]. The TBR models do not consider the effects of the mean free path and uses a Kapitsa resistance to the grain, which should be further studied for clarification [148, 149].

Superlattice Structures

The reduction in k_L in nanostructure thermoelectric materials, such as thin films and nanowire, has been attributed to the reduction in l. However, these k_L values are anisotropic due to the anisotropic microstructures. For example, in GaAs/AlGaAs thin film superlattice, the l in the superlattice growth direction is smaller, which means that a smaller value of k_L in the direction will be expected, as compared with the value in the plane of the film [272]. It is fount that the k_L of AlAs/GaAs superlattice [273], consisting of an equal thickness of AlAs and GaAs, decreases with decreasing layer thickness and superlattice period (d). However, the measured k_L is higher than that of $Al_{0.5}Ga_{0.5}As$ bulk alloy, alghough the k_L of $Al_{0.5}Ga_{0.5}As$ is lower than those of AlAs and GaAs. The lowest k_L in the superlattice is limited by that of the alloy, which means that alloy scattering is the strongest k_L suppression mechanism. Two samples of GaAs/AlAs monolayer (ML) superlattice, $[GaAs\ (3\ ML)]_{159}/[AlAs\ (3\ ML)]_{159}$, and (ii) $[GaAs\ (12\ ML)]_{27}/[AlAs\ (14\ ML)]_{27}$, have k_L values of 3.1 and 5.8 W m^{-1} K^{-1}, both of which are lower than that of the bulk GaAs (46 W m^{-1} K^{-1}) [274]. The lower k_L of first sample indicates that the presence of a larger number of interfaces is helpful in reducing k_L.

Similar decrease in k_L has also been observed in other superlattice, such as $Si/Si_{0.7}Ge_{0.3}$ [275, 276]. A decrease in d, corresponding to an increase in the number of interfaces or unit length for given thicknesses, results in a decrease in k_L. Sometime, such variations in k_L with d are not as significant as expected, which has been attributed to the smaller acoustic impedance mismatching. For example, the acoustic impedance mismatching of $Si_{0.84}Ge_{0.16}/Si_{0.74}Ge_{0.26}$ alloy superlattice is 1.03, whereas that of the $Si/Si_{0.7}Ge_{0.3}$ superlattice is 1.15. However, a k_L value lower than that of the bulk alloy and higher than that of amorphous Si is found in Si_m/Ge_n superlattice, with m and n being the number of the monolayers. When 3 nm $< d <$ 7 nm, the k_L decreases with decreasing d [265]. Interesting observation in this material is that when $d >$ 13 nm an even smaller k_L has been obtained. This special observation can be used to rule out the role of mini-Umklapp scattering in modulating the thermal conductivity, but the corresponding detailed information on the number and nature has not been available. Noting that the critical thickness for the pseudomorphic growth of Ge on Si has been shown to be 1–3 nm, it is also necessary to monitor the possible presence of morphology variations and static impfections such as dislocations, which can provide with additional phonon scatterings, when they have sufficiently high concentrations (about 10^{11} cm^{-2}).

Another example is the symmetrically strained Si/Ge (2/2 nm) superlattice, with k_L values in the range of 2.9–4.0 W m^{-1} K^{-1}, which are lower than the values (8.8 W m^{-1} K^{-1} at 300 K) of the equivalent composition bulk $Si_{0.5}Ge_{0.5}$ alloys [277]. The strains of these samples have been compensated by using a 1-2 mm thick buffer layer with lattice parameter being equal to the average lattice parameters of Si and Ge. According to TEM results, the dislocation density is about 10^4 cm^{-2}, which is too low to significantly influence the k_L. It has been reported that k_L decreases with increasing carrier concentration (10^{18}–10^{19} cm^{-3}) in p-doped and n-doped superlattice. Also, it decreases with increasing period. However, an increase in k_L with increasing doping concentration in n-doped samples has been observed, which is unexpected.

Reduction in cross-plane k_L has also been achieved in Bi_2Te_3/Sb_2Te_3 superlattice, Bi_2Te_3 (1 nm)/Sb_2Te_3 (5 nm) and Bi_2Te_3 (3 nm)/Sb_2Te_3 (3 nm), with $k_L \approx 0.25$ W m^{-1} K^{-1} [278]. The effect of the mean interface roughness (η) on the reduction of k_L has been reported for small values of d, which can be attributed to additional scattering of phonons. Theoretical simulations indicate that a random roughness can cause phonon backscattering and thus reduce k_L [279]. The effective mean path (l_{eff}) is given by: $l_{\text{eff}} = [(1+p)/(1-p)]d$, where p is a specularity parameter, $p = \exp[(-16\pi^3/\lambda_{\text{ph}}^2)\eta^2]$, with a Gaussian distribution for the surface roughness [280]. $p = 1$ with purely specular scattering, while $p = 0$ represents purely diffusive scattering that leads to the smallest k_L.

The effect of zero-dimensional nanostructured inclusions on k_L has been studied by using quantum dots (QD) made by alternative deposition of two thin films with different lattice parameters. When a critical thickness for maintaining registry of a thin film with an underlying layer is exceeded, the strain energy minimization will induce the spontaneous break down of the overlaid film into randomly distributed islands or QDs. For example, in PbSe/PbTe superlattice, due to the 5 % lattice parameter mismatch of PbSe with PbTe, 8000 periods of 13 nm thick PbTe/$PbSe_{0.98}Te_{0.02}$ QDs can be formed on a film of 0.1 mm thick [281]. With a low value of $k_L \approx 0.33$ W m^{-1} K^{-1}, the QD incorporated PbSe/PbTe superlattice have values of *ZT* of 1.6 at 300 K and 3.5 at 570 K, which is mainly attributed to the reduced k_L. Even lower k_L values can be obtained by using quaternary alloys, such as lattice mismatched PbTe/$Pb_{1-x}Sn_xSe_yTe_{1-y}$ QD superlattice, where the addition of Sn is to enhance the σ and the *S*, due to the creation of larger DOS effective mass.

It is also possible to reduce k_L by engineering the phonon dispersion in superlattice, with strategies including (i) increasing phonon interactions leading to phononic band gaps to modulate phonon and thermal transport and (ii) increasing the number of phonon modes and flattening the phonon energy dispersion to reduces the overall group velocity.

For superlattice, consisting of thin layers with spacing $d = na$, with *n* being an integer and *a* being the spacing of the lattice points, a new periodicity corresponding to *d* is introduced, in the direction perpendicular to the growth direction of the films. Translated to reciprocal space, the imposed periodicity intersperses a mini-Brillouin zone of extent $2\pi/d$ into the original Brillouin zone of span $2\pi/a$. A larger *d* corresponds to smaller mini-Brillouin zones, and vice verse, The periodic interruption of the band structure, with a flattening of the bands or energy dispersion at the edges of the mini-Brillouin zones, reduces the phonon group velocity and thus k_L. In addition, the enhanced scattering through mini-Umklapp processes makes it possible through a smaller backscattering mini-reciprocal lattice vector, which also contributes to increasing the thermal resistance [280]. For a superlattice of two constituents with masses m_A and m_B, it has been predicted that k_L can be decreased by as much as 25 % if there is $\Delta m/m_{avg} = 1$, where $\Delta m = m_A - m_B$ and m_{avg} is the average mass.

As mentioned before, phonons can be treated as particles, when $d > l$, while wave propagation should be considered, when $d < l$. In the latter case, the interaction of phonons with wavelength of λ_{ph}, of arbitrary propagation direction and polarization, in the superlattice, can be governed through a form of Bragg's law:

$$n\lambda_{ph} = 2d \cos \theta_k, \quad (4.65)$$

where θ_k is the angle between the superlattice growth direction and the phonon propagation direction, e.g., when a phonon travels perpendicular to an atomically perfect interface, complete backscattering due to constructive interference occurs when $\lambda_{ph} = 2d$ [282]. Such phonon wave interference effects will be clearly manifested, if the interface roughness is smaller than the phonon coherence length (ξ_{coh}) of about 1 nm. Such interferences can be used to selectively backscatter or transmit phonons, thus leading to the possibilities of phonon filters, as observed in a (111) GaAs/AlGaAs superlattice [283–285].

In a binary multilayer system, 1—GaAs and 2—AlGaAs, where $d = d_1 + d_2$ is the sum of thicknesses of the respective layer, wave interference between different phonon propagation velocities can also produce a phononic band gap, with magnitude corresponding to the difference in the acoustic impedances, $A = \rho v$, of the layers. Such a phenomenon can be similar to the electronic band gaps. Electronic band gaps, with magnitude being proportional to the Fourier component of the crystal potential, are formed due to electron interference. Photonic band gaps, with magnitude being corresponding to the dielectric constant contrast, are formed due to electromagnetic wave reflection at the edges of the Brillouin zone [225].

For InGaAs/AlAs superlattice, when the complete interaction between all the phonon polarization modes, high-energy longitudinal mode and two lower energy transverse modes, is considered, intrazone frequency gaps have been observed experimentally in the phonon dispersion spectrum, which can be attributed to the mixing and subsequent anticrossing of phonon energies. Such an energy level mixing and the consequent flattening of the phonon dispersion, leading to a reduced group velocity, is especially pronounced for phonon modes propagating oblique to the superlattice growth direction. For example, the slow transverse (ST) and the longitudinal acoustic (L) phonon modes, being in the same plane, are coupled to each other and propagate independently of the orthogonally polarized fast transverse (FT) mode. With increasing phonon frequency, near the edge of the Brillouin zone, the ST mode is folded back into the main Brillouin zone, and through an avoided crossing type interaction with the L mode, it induces the formation of a forbidden frequency gap [282]. Quantitatively, the band gap is formed through satisfying an intermode Bragg reflection with wave vector conservation, i.e., $k_{\perp}^{\mathrm{L}} + k_{\perp}^{\mathrm{ST}} = 2n\pi/d$ for the normal component, and $k_{||}^{\mathrm{L}} = k_{||}^{\mathrm{ST}}$ for the parallel component, along with $\omega_L = \omega_{\mathrm{ST}}$. The formation of the band gaps also offers a possibility of $L \leftrightarrow T$ mode conversion, which implies that at a certain interface the transmitted and reflected waves can carry equivalent amounts of energy flux, corresponding to a zero net energy transport and thus a minimized k_L.

Structures with phononic band gaps, allowing selective phonon frequencies to transmit, i.e., no phonon propagation in the band gap, which have been used to fabricate devices, such as phonon mirrors and phonon resonators, can also be applied to minimize k_L through changing the thickness of the layer to modulate the frequency of transmission and selecting the constituent materials to varying the acoustic impedance [163]. Another possibility is to localize the phonons to be near the phononic band edges, so that a decreased or even zero phonon group velocity can be realized, so as to reduce k_L [286, 287]. The increased confinement, due to the localization, could also increase the scatterings due to Umklapp, boundary and impurity. According to simulations, it has been found in 10 nm Si freestanding quantum wells that the increased impurity scattering of the confined acoustic phonons results in a strong reduction in the group velocity, which leads to a decrease in k_L by seven folds [288].

The onset of phonon localization can be represented through a reduced value of kl, approaching 2π, with k being the magnitude of the wave vector ($2\pi/\lambda_{\mathrm{ph}}$). For example, a kind of atomically smooth p-type Bi_2Te_3/Sb_2Te_3 superlattice has found to have $kl < 1$. In this case, the $l = l_{\mathrm{mfp}}$ can be calculated from $k_L = C_L vl$, with C_L and v being assumed to have bulk values and the k being estimated from the phonon phase velocity. Reduced k_L values have been obtained, which leads to a ZT of about 2.4, the largest value in thermoelectric thin films [99, 278]. Because the measurement temperature (300 K) is much higher than the Debye temperatures of Bi_2Te_3 and Sb_2Te_3, it is concluded that the major contributors are high frequency low wavelength phonons. It is also found that the superlattice period, d, is more critical than the thicknesse of individual layers of Bi_2Te_3 and Sb_2Te_3. For example, the k_L values of the 6 nm period superlattice, with different individual layer thicknesses of Bi_2Te_3/Sb_2Te_3, which are 3/3 nm, 1/5 nm and 2/4 nm, are almost the same [278].

Another issue that can be exploited for the reduction of k_L is the anisotropic characteristic of heat transport in thin films. In this case, phonon transport occurs preferentially in directions with smaller elastic constants, corresponding to lower velocities and thus lower energies, which is called phonon focusing [289]. Such concept has been applied Si, leading to a difference in k_L between different directions to be as high as 90 %, which is expected to be applicable to thermoelectric superlattices.

Therefore, the reduction in k_L of bulk thermoelectric materials through length scaling, such as superlattices, has been found to be quite effective and is believed to be research topic in the near future. The involvement of various parameters, such as strain, impurities and defects at the interface and compositional uniformity, has made it difficult to deep understand their effects, with the scattered data available in the open literature. Nevertheless, engineering the phonon dispersion and transport through the creation of structures with phononic band gaps will be definitely one of the research directions in reducing k_L.

Nanowires

It has been proposed that systems comprising nanowires (NWs) and nanotubes (NTs) have high degree of confinement, which could lead to significant modulation of phonon energy dispersion [290, 291]. The reduced phonon mean free path in two dimensions and enhanced phonon-surface, phonon–phonon, and phonon-carrier scatterings contribute to anisotropy and thus low k_L. The thermal boundary resistance can also be modified by such scattering [262], such as those observed experimentally in single-walled [292, 293] and multiwalled carbon nanotube (CNT) bundles, which have a lower k_L than individual CNT [294].

The factors as discussed above, including surface roughness, degree of specular or diffuse scattering at the interfaces, have also influence in terms of reducing k_L. For instance, the k_L values of the Si NWs with a sawtooth morphology and core-shell structures are two orders of magnitude lower than that of bulk Si, which is attributed to the increased number of diffuse reflections and inelastic scatterings [269, 295, 296]. Similarly, by embedding Si NWs in Ge bulk matrix, the k_L

decreases with decreasing diameter of the nanowire, due to the decrease in phonon scattering at the Si–Ge interface [297].

Furthermore, the k_L of NWs can be minimized by embedding them in a matrix with smaller acoustic impedance (A). According to the mechanism of elastic energy minimization, a redistribution of the phonon modes from the acoustically harder nanowire to the acoustically softer matrix, will result in phonon depletion in the NWs, which can minimize the thermal energy transport [298]. The increase in surface roughness has been found to be one of the major contributing factors for the great reduction of k_L of Si NWs, with a value of about 1.2 W m^{-1} K^{-1}, which is very close to the value of amorphous SiO_2 and the $k_{L,\ min}$ of Si ($\sim$1 W m^{-1} K^{-1}) [299].

A comparison study has been on bulk Si and Si NWS prepare by using electroless etching (EE) and vapor-liquid-solid (VLS) deposition methods [300]. The EE NWs with rought surface exhibit a reduction in k_L by a factor of five, as compared to that of Si VSL NWs whose surface is smooth. It is also found that the peak of the k_L–T curve shifts toward high temperature with increasing roughness. Such peaks are indicative of the magnitude of Debye temperature (Θ_D) and onset of Umklapp scattering [300]. Therefore, it can be concluded that the nanostructures have relatively higher stiffness and the phonon scattering is increased with increasing roughness. The effect of roughness has been demonstrated by using the ratio of k_L values between the bulk and NW, which is about 100 at 300 K and reaches 25,000 at 25 K, as shown in Fig. 4.29 [299]. With decreasing temperature, wavelength of the phonons increases. The dominant phonon wavelength is calculated using $\lambda_{ph} = hv/3k_BT$ together with $v = 9 \times 10^3$ m s^{-1} at $T = 25$ K to about 6 nm, which is comparable to the surface roughness of the Si NWs [299].

In summary, the principle of reducing the k_L of NWs is to increase the phonon scattering at different length scales, with strategies of decreasing the nanowire diameter to increase boundary scattering, increasing surface roughness to induce scattering of phonon with longer wavelength, and promoting atomic scattering to cater to phonons with short wavelength.

Phonon Drag of Nanostructures

As mentioned earlier, the Seebeck coefficient, S, includes contributions from both the diffusion of the carriers and the component due to the influence of carriers on the phonons. The second component is called phonon drag. The first component determines the measured S at room temperature, because the second one is an impurity and size-dependent phenomena at low temperatures. However, the results of Si NWs indicate that phonon drag could have contributed to an enhanced S and the reduced k_L, thus leading to k_L values of even lower than $k_{L,min}$ [301]. Such phenomena open a new venue to increase the thermoelectric ZT.

The phonon drag is characterized by producing a thermoelectric voltage due to the sweep of the electrons by the phonons, which is increased when the respective heat flow contributions or energies are comparable. Usually, due to their larger group

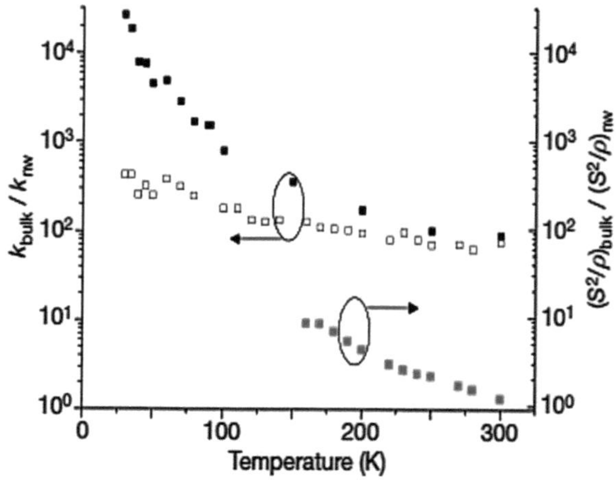

Fig. 4.29 A sharp increase in the ratio of the bulk to nanowire thermal conductivity with decreasing temperature, which is attributed to the increasing influence of surface roughness of the nanowires. The $S^2\sigma$ is adversely affected by the roughness. Reproduced with permission from [299], Copyright @ 2008, Macmillan Publishers Limited

velocities, the longitudinal phonon modes have more contribution to the phonon drag than the transverse modes. The magnitude of the voltage due to phonon drag is small in metallic materials (<5 mV K^{-1}) and is evident at $T < \Theta_D$. In contrast, it can be orders of magnitude higher in semiconductors [302], at doping levels $\leq 10^{15}$ cm^{-3}, typically with 5–20 mV K^{-1} at T < 20 K in Ge and Si [302–304].

The thermopower contributed by phonon drag, $S_{ph\text{-}e}$, is directly proportional to the ratio of the phonon relaxation time ($\bar{\tau}$) and the carrier relaxation time ($\bar{\tau}_e$) along with the fraction (f_r) of electronic carrier momentum lost to the phonons through Im $\overline{3}$ and decreases with increasing carrier concentration and temperature [3, 187]. In the equation, m^* and v are the effective mass and acoustic velocity, respectively. A saturation of $S_{ph\text{-}e}$ occurs as the carrier concentration is increased to a certain level, e.g., at 10^{18} cm^{-3} in p-type and n-type Si, due to the increased interactions of the electrical carriers with the phonons, and a consequent decrease of the mutual drag. As a result, below a critical free path $l_p = v\tilde{\tau}$, which is estimated to be about 1 μm for Si, $S_{ph\text{-}e}$ is greatly decreased, which is called quenching [183].

It has been reported that the k_L value of 0.76 ± 0.15 W m^{-1} K^{-1} of a 10 nm Si NW lower than the theoretical limit of $k_{L,\,min} \approx 1$ W m^{-1} K^{-1}. This additional reduction in k_L has been attributed to the contribution of phonon drag [301]. One of the arguments is that when originally define the $k_{L,\,min}$ a minimum phonon mean free path of $\lambda_{ph}/2$ has been assumed, which is valid for three-dimensional transport but may not be valid in the case of one-dimension. In this regard, a mechanism involving one-dimensional phonon propagation has been developed, where the NW boundaries are incorporated into the phonon mode, so that there is a new mode of heat transport. The l_p, below which $S_{ph\text{-}e}$ is reduced, can be extended to mean the length of the NWs, instead of the diameter. In this case, phonon drag phenomena can be used to explain the results.

As a result, there is a phonon mode propagating along the NW, with positive and negative dilatations, as shown in the inset of Fig. 4.30, corresponding to

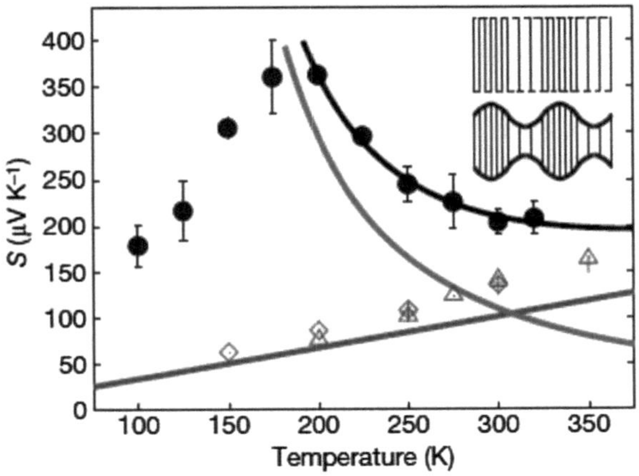

Fig. 4.30 Variation of the Seebeck coefficient (S) with temperature, with both the phonon drag component ($S_{ph\text{-}e}$) and diffusive electron transport component (S_e). While S_e is directly proportional to temperature T through Eq. (4.46), $S_{ph\text{-}e}$ varies as exp(Θ_D/T)—1, with Θ_D being the Debye temperature. The $S_{ph\text{-}e}$ component relies on the one-dimensional propagation of phonons along the nanowire, which is also favored by a reduced thermal conductivity. Reproduced with permission from [301], Copyright @ 2008, Macmillan Publishers Limited

alternate heating and cooling and heat transport along the NW [301]. For longitudinal acoustic (LA) modes propagating along the axis of the NW, which is the main contributor to the $S_{ph\text{-}e}$, a large l (low v) corresponds to a large distance between temperature fluctuations and an adiabatic situation, whereas a small l (large v) implies an isothermal behavior. Because the phonon drag is maximized when the phonon energy (wavelength—l_{ph}) is comparable to the electrical carrier energy (wavelength—l_e), the dominant l_{ph} can be calculated with a given carrier concentration, n. With a three-dimensional carrier distribution, l_{ph} is estimated with $\lambda_e = 2\pi/\sqrt[3]{3\pi^2 n}$, with $n = 3 \times 10^{19}$ cm^{-3}, to be about 6 nm, which is larger than the thermal phonon wavelength ($l_{th} = hv/k_B T$) of about 1.5 nm, assuming a bulk v of 10^4 m s^{-1} at $T = 300$ K. Therefore, an adiabatic condition is achieved.

In this adiabatic case, a lack of contact with the ambient leads to a longer phonon relaxation time, $\bar{\tau}$, and thus a larger $S_{ph\text{-}e}$. The transverse acoustic modes, which experience less scattering as compared with the LA modes, due to the confinement perpendicular to the length of the NW, are also considered to have an adiabatic character, with contribution to the thermopower [305]. Therefore, the total S is thus the sum of the contributions S_e and $S_{ph\text{-}e}$. Since the scattering rate, related to $(\bar{\tau})^{-1}$, depends on the number of phonons through the Bose–Einstein distribution function, $f_{BE} = 1/[\exp(\Theta_D/T)-1]$, the $S_{ph\text{-}e}$ is proportional to $\exp(\Theta_D/T)-1$. The S_e is proportional to the temperature through the Mott formula, $S_e \sim aT$. The data in Fig. 4.39 can be well fit to the expression, $S = S_e + S_{ph\text{-}e} = aT + b[\exp(\Theta_D/T)-1]$, with a and b being constants of proportionality used to justify the hypotheses.

Therefore, a larger $\bar{\tau}$ only applies to the phonons that have participated in the phonon drag, so that the k_L which is due to the overall contributions of all the phonon modes is not affected. However, a reduction in k_L can be achieved by reducing the phonon group velocity, peculiar to the phonon propagation in the one-dimensional nanostructures, as shown in the inset of Fig. 4.30. Although k_L values of lower than the $k_{L,\ \mathrm{min}}$ have been observed and explained, much details regarding phonon propagations and their contributions to a reduced k_L still need to be further clarified, such as appropriate geometries from three-dimensional to one-dimensional crossover for the phonon propagation and the incorporation of the NW boundaries into the propagation mode, the mechanism of spectral distribution of the phonons contributing to the $S_{\mathrm{ph}\text{-}e}$ along with the range of carrier concentrations for which $S_{\mathrm{ph}\text{-}e}$ can be tailored through an increased $\bar{\tau}$ and the validity of this model to other materials.

4.6 Nanostructured Thermoelectric Materials

4.6.1 Heterogeneous Nanoscomposite

According to previous discussion, it has been confirmed that the significantly increased *ZT* values in nanoscale thin film thermoelectric materials are attributed to the tremendous reduction in thermal conductivity. However, thin films are not promising for large-scale applications due to their relatively low productivity and the high cost. Therefore, the synthesis of large amounts of materials by using solid-state methods is ultimately more practical and desirable for practical applications.

In this respect, various new approaches and concepts for the preparation of nanostructured thermoelectric materials at large scale have been proposed and developed. For example, there have been reports on nanoscale inhomogeneous compounds of $AgPb_mSbTe_{2+m}$ (LAST-*m*: lead antimony silver telluride) family and $NaPb_mSbTe_{2+m}$ (SALT-*m*: sodium antimony lead telluride) family, as potential thermoelectric materials with high *ZT* values that can be attributed to the substantial reduction in thermal conductivity as compared with PbTe. Important methods to prepare nanostructure PbTe-based thermoelectric materials include precipitation and growth, spinodal decomposition, matrix encapsulation and the synthesis of nano-sized polycrystalline precursors, which will be discussed in this subsection.

4.6.1.1 $AgPb_mSbTe_{2+m}$

PbTe has several derivatives that are among the most promising bulk thermoelectric materials. $AgPb_mSbTe_{2+m}$ (LAST-*m*) is one of the most important derivatives. It is composed of PbTe and $AgSbTe_2$, with a figure of merit of $ZT > 1$.

According to initial studies, it has been considered to be a solid solution between PbTe and $AgSbTe_2$, both of which are of rock salt NaCl-type structure, have *p*-type properties and possess unusually low lattice thermal conductivity. Another reason is that the dependence of the lattice constant with composition of LAST-*m* series follows the Vegard's law. The latest development has shown that this system can be made to have *n*-type properties by introducing Ag deficiency, which can be used to control the carrier concentration of the materials.

The LAST materials are thermally stable up to their melting points of > 1200 K. The doping effect is generally controlled by the nonstoichiometry of Ag, Pb, or Sb fractions, so that it can be expressed as $Ag_{1-x}Pb_{m+y}Sb_{1+z}Te_{2+m}$. LAST-*m* with m = 18–22 has high power factor and pretty low lattice thermal conductivity, which lead to ZT = 1.7 at ~ 700 K [253]. During the growth of bulk LAST-*m* system, nanostructures are spontaneously formed when it is cooled from the melts, which is an additional reason that explains the high ZT values of the materials. High ZT values and the presence of nanostructured inclusions in *n*-type $AgPb_mSbTe_{2+m}$ system have been confirmed experimentally [306, 307]. The phase diagram of the material system is very complex, so that its thermoelectric properties are highly sensitive to the synthesis processing and conditions [308, 309]. Similar to its ender member $AgSbTe_2$, LAST-*m* system is also prone to decompose. Therefore, special carefulness should be given when synthesizing these materials so as to achieve desirable thermoelectric performances.

It has been recently that LAST-*m* can be prepared by using mechanical alloying combined with SPS. The materials prepared in this way are of nano-sized structures, thus having shown promising thermoelectric performance [307]. For example, polycrystalline $Ag_{0.8}Pb_{18+x}SbTe_{20}$ that is very similar to LAST can be synthesized from elemental powder by using mechanical alloying. The mechanically alloyed powders are consolidated by using SPS, which possesses a ZT value of 1.5 at 673 K [306]. This thermoelectric performance is very close to that of those materials grown by using the conventional melting method [253, 310]. The average grain size of the $Ag_{0.8}Pb_{18+x}SbTe_{20}$ nanocomposite is about 1 μm, with a relative density of ~95 %. All grains consist of nano-sized precipitates of ~20 nm, which is believed to play an important role in reducing lattice thermal conductivity of the materials. The presence of Sb is found to be crucial to obtaining high ZT values. For example, if the Sb is replaced by Bi, the power factor will be lowered and the lattice thermal conductivity will be increased, although there is similar nanostructure [311]. In other words, nanostructuring alone is not sufficient to reduce the thermal conductivity, there have to have other factors together to enhance the phonon scattering, such as proper mass contrast and the presence of strain at the interfaces.

The power factor of these materials is still slightly lower than that of the optimized PbTe. This is mainly because electron scattering of the former is stronger than that of the latter. However, due to the lowered lattice thermal conductivity, which is about 30 % of that of PbTe, the LAST-*m* can be optimized to have higher ZT values than PbTe, for example, LAST-18. Therefore, the presence of the nano-sized inclusions in the LAST-18 is the reason for the enhancement in *ZT* value, when compared with PbTe. In addition, its lattice thermal conductivity is also lower than that of its counterparts prepared by using the conventional solid solution approach.

The nano-sized inclusions are present as minor phases in $AgPb_mSbTe_{2+m}$, which have coherent or semicoherent interfaces with the matrix, as revealed by the cross-sectional TEM images. The nanostructures are often the regions of one composition endotaxially embedded inside a matrix of another composition. The nano-sized inclusions in the LAST-*m* materials are believed to be form through the solid-state phenomena, such as thermodynamic spinodal decomposition and nucleation and growth events during cooling from high temperatures. These solid-state phenomena are ideal mechanisms for the formation of the sites of phonon-scattering, because the precipitate phases experience a series of coarsening steps with coherent or semicoherent interfaces during the early stages, which can be well retained after the fabrication process [256]. Because their composition, structure, size, and size distribution can be readily controlled through controlling the cooling rate and postsynthesis heat treatment, further improvement in thermoelectric performance is expectable.

It has interestingly been found that thermal annealing has a positive effect on performance of thermoelectric materials, by reducing both electrical conductivity and thermal conductivity [306]. Figure 4.31 shows XRD patterns of the $Ag_{0.8}Pb_{22.5}SbTe_{20}$ samples treated at different conditions. The as-milled powder has a β-PbTe crystal structure (NaCl structure), with grain size of hundreds of nanometers. Thermal annealing has no observable effect on phase composition of the samples.

Figure 4.32 shows TEM images of the sample before and after annealing for 30 days. A large number of nano-seized particles are embedded in the PbTe-rich matrix for all the samples before and after annealing. Both the number and size of these particles are increased with increasing annealing time. These nanoscopic regions can only be probed through a detailed examination of the structure at the atomic scale. Figure 4.32c indicates that there are localized wavy patterns with sizes of 5–10 nm in almost all the crystal grains. These Ag/Sb-rich nanostructures are coherent with their surrounding crystal matrix, which do not affect the electronic conductivity. The annealed samples contain larger and more particles with sizes of 10–50 nm, as shown in Fig. 4.32d. Therefore, the thermal annealing leads to a reduction in thermal conductivity. As a result, the sample annealed for 30 days possesses a large *ZT* value of 1.5 at 700 K [306].

4.6.1.2 $AgPb_mSn_nSbTe_{2+m+n}$ (LASTT)

$AgPb_mSn_nSbTe_{2\ m+n}$ (LASTT) is a *p*-type analog of the LAST-*m* family based on $Pb_{1-x}Sn_xTe$ solid solution, which has been reported to show high ZT values of 1.4 near 700 K for certain *m* and *n* values [312]. The matrix of these nanocomposites has been considered to be comprised of $Pb_{1-x}Sn_xTe$ solid solution, with the

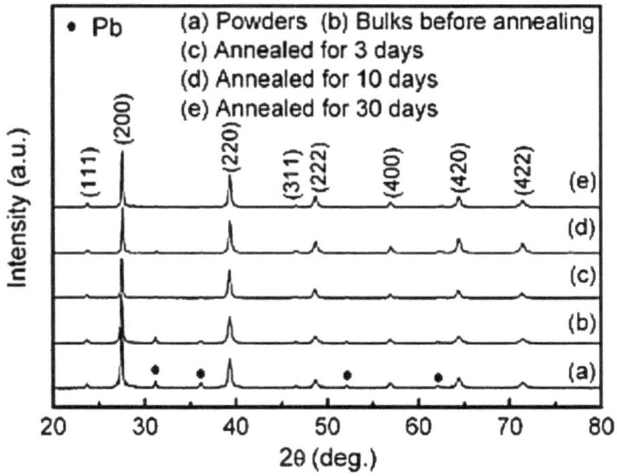

Fig. 4.31 XRD patterns of $Ag_{0.8}Pb_{22.5}SbTe_{20}$ powders (**a**) and bulks before (**b**) and after annealing for different times (**c–e**). Reproduced with permission from [306], Copyright @ 2008, American Chemical Society

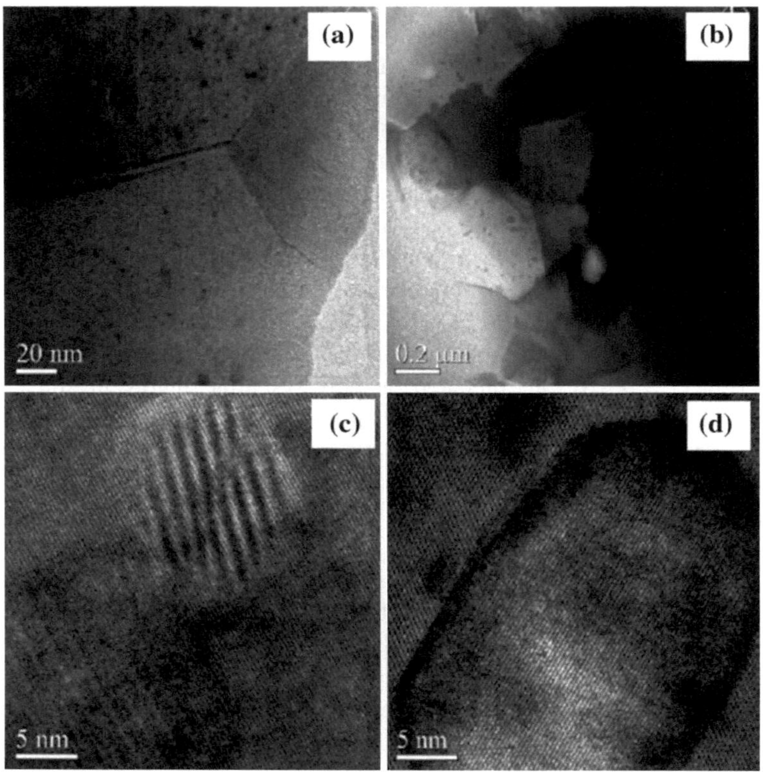

Fig. 4.32 Low magnification TEM images of (**a**) unannealed sample and (**b**) the sample annealed for 30 days. HRTEM images of (**c**) unannealed sample and (**d**) the sample annealed for 30 days. Reproduced with permission from [306], Copyright @ 2008, American Chemical Society

nano-sized precipitates being rich in Ag and Sb, similar to that observed in LAST. However, it is relatively difficult to tune the transport properties of these materials by varying the concentrations of Ag or Sb. This aspect is different from the case of the Sn-free *n*-type material. Fortunately, the transport behavior of LASTT can be easily tuned through the Pb/Sn ratio. For example, Pb/Sn ratio of 9/9 is an optimum composition in terms of high values of thermopower and electrical conductivity. Preliminary results have demonstrated that high *ZT* value *p*-type materials for device applications can be synthesized though further tuning and optimization [313]. In this family, several new compositions have been developed, for example, $AgSnSbTe_3$, which has a large Seebeck coefficient of 160 mV K^{-1} observed with nearly metallic carrier concentrations ($\sim 5 \times 10^{21}$ cm^{-3}).

4.6.1.3 $NaPb_mSbTe_{2+m}$

The sodium-substituted system $NaPb_mSbTe_{2+m}$ (SALT-*m*) is also a high-performance system. For example, SALT-*m* with $m = 20$ has a *ZT* of ~ 1.6 at 675 K and is a *p*-type semiconductor [257, 314]. This group of materials has very low thermal conductivity of 0.85 W m^{-1} K^{-1}, with 0.5 W m^{-1} K^{-1} contributed by lattice. According to HRTEM images, the materials are characterized by nano-sized particles distributed in the matrix of main phase, similar to those observed in the LAST materials. An example is shown in Fig. 4.33. Na-Sb-rich clusters are usually formed in the $NaPb_mSbTe_{m+2}$ system. The *ZT* value of $Na_{0.95}Pb_{20}SbTe_{22}$ is found to increase quickly with temperature, reaching 1 at about 475 K and 1.7 at 650 K. This is one of the widest temperature ranges in which a single-phase material exhibits a *ZT* value above 1. Thermoelectric performance of $NaPb_mSn_nSbTe_{m+n+2}$ is still lower than that of the SALT-*m* systems [315].

The effect of Sn doping on microstructure, thermal, electrical, and thermoelectric properties of $NaPb_{18}MTe_{20}$ (M = Sb and Bi) have been systematically studied. The materials, with compositions of $NaPb_{18-x}Sn_xSbTe_{20}$ and $NaPb_{18-x}Sn_xSbTe_{20}$ (x = 0–16), are prepared as polycrystalline ingots in silica tubes. Figure 4.34 shows TEM and HRTEM images of representative samples of $NaPb_{18-x}Sn_xSbTe_{20}$ and $NaPb_{18-x}Sn_xSbTe_{20}$. They have similar features in microstructure, with nano-sized particles of 10–20 nm in size being dispersed in the main matrix.

A lamellar feature is formed during the sample preparation of $NaPb_{13}Sn_5BiTe_{20}$, as shown in Fig. 4.35. Sawtooth-shaped features are formed at the edge of the sample during the thinning with ion beam. This feature is also closely related to the local ordering between PbTe and SnTe. Because the Pb–Te bond is stronger than the Sn–Te one, it is easier to mill away the SnTe layer than the PbTe layer, during sample milling. Such nanostructure and interface-induced matrix elastic perturbations could have played a decisive role in enhancing the phonon scattering, thus leading to significant reduction in thermal conductivity in these materials.

Figure 4.36 shows figure of merit of the $NaPb_{18-x}Sn_xMTe_{20}$ materials as a function of temperature. The content of Sn has a visible effect on the values of *ZT*

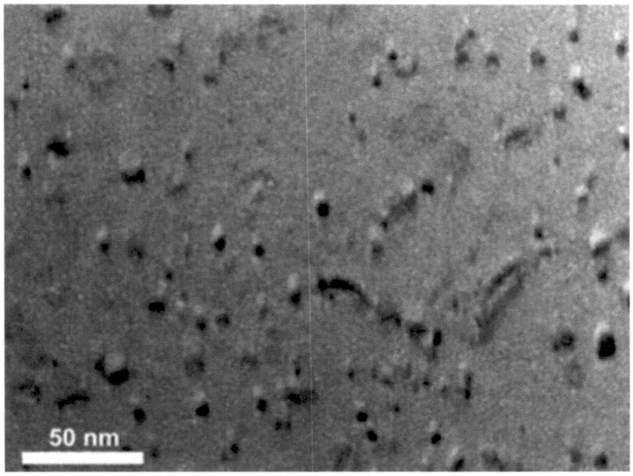

Fig. 4.33 A low magnification TEM image of a $Na_{0.95}Pb_{20}SbTe_{22}$ sample, with evenly distributed nano-sized particles dispersed in the matrix. Reproduced with permission from [257], Copyright @ 2006, John Wiley & Sons

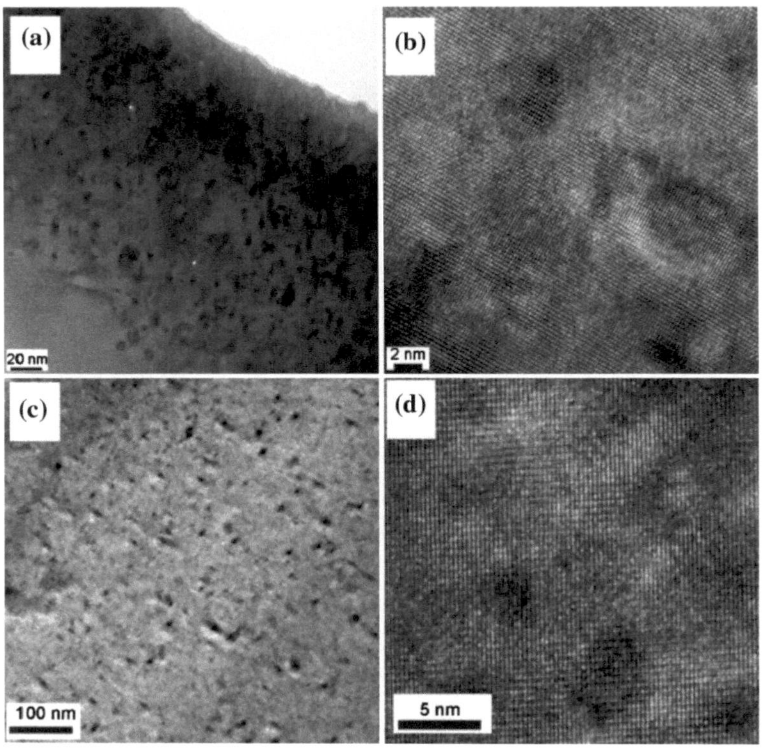

Fig. 4.34 Representative TEM and HRTEM images of (**a**), (**b**) $NaPb_{13}Sn_5BiTe_{20}$ and (**c**, **d**) $NaPb_{13}Sn_5SbTe_{20}$. Reproduced with permission from [315], Copyright @ 2009, American Chemical Society

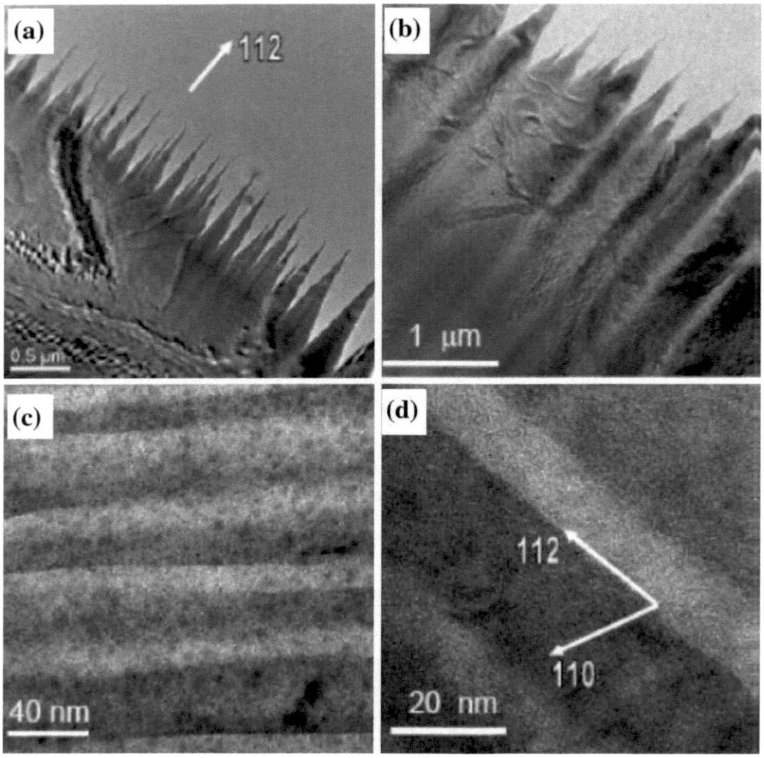

Fig. 4.35 TEM images of $NaPb_{13}Sn_5BiTe_{20}$ with lamellar structures. **a** Saw shape at edge of the sample, **b** further ion milling for the sample, with the lamella profile appearing in the region close to edge, **c** Strong contrast in the area far from the edge, and **d** high-resolution image, with the lamella being indexed as 112 direction. Reproduced with permission from [315], Copyright @ 2009, American Chemical Society

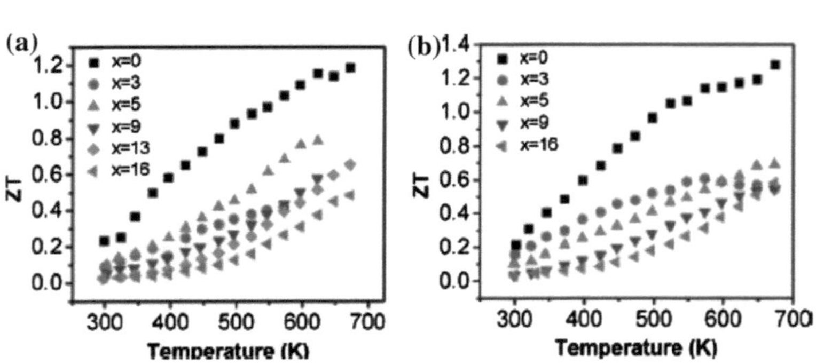

Fig. 4.36 Figure of merit *ZT* as a function of temperature of the samples. **a** $NaPb_{18\text{-}x}Sn_xSbTe_{20}$ (x = 0, 3, 5, 9, 13, and 16) and **b** $NaPb_{18-x}Sn_xBiTe_{20}$ (x = 0, 3, 5, 9, and 16). Reproduced with permission from [315], Copyright @ 2009, American Chemical Society

and their dependences on temperature. For the samples with Sn-rich compositions, the slope of *ZT* at near room temperature is not very steep, but it increases quite fast at higher temperatures. The Sn-free sample $NaPb_{18}MTe_{20}$ exhibits the highest *ZT* over the temperature range of 300–675 K with a *ZT* of as high as ~1.3 at 675 K. Both groups of samples doped with Sn have *ZT* values of <1. The introduction has positive effect in reducing thermal conductivity of the $NaPb_{18-x}Sn_xMTe_{20}$ materials, but decreases the power factor at the sample time. This is the reason they lower *ZT* values than $Na_{1-x}Pb_mSbyTe_{m+2}$. The introduction of Sn results in higher carrier concentrations and thus lower thermopower values, which leads to lower the power factor.

4.6.1.4 PbTe–PbS

It has been found that in the PbTe–PbS system there are three levels of inhomogeneity. The system $(PbTe)_{1-x}(PbS)_x$ is not a solid solution. It consists of PbTe-rich and PbS-rich regions, forming coherent nanoscaled heterogeneities [258, 316]. With $x > 0.03$, the materials have orders at three submicron length scales. Figure 4.37 shows TEM and HRTEM images of the $(PbTe)_{1-x}(PbS)_x$ materials with different concentrations of PbS [139]. The TEM images reveal that the systems contain well-dispersed nanocrystals of Sb, Bi, and InSb. The content and size of these nanocrystals have a close relation to the composition of the materials. High content of PbS leads to larger and more nanocrystals. Most of the nanoparticles have sizes of 2–6 nm, which are evenly dispersed throughout the matrix. High-resolution TEM images of PbTe–Sb(4 %) and PbTe–Bi(4 %), in Fig. 4.37e and f, indicate that the nanoparticles are coherently imbedded in the matrix.

Further studies have illustrated that the 16 % sample contains coexisting regions with spinodal decomposition, together with nucleation and growth phenomena, as shown in Fig. 4.38 [258]. There are regions of arbitrary shape, with lengthes of 300–600 nm and widths of 90–300 nm, which are made of parallel stripes that have the width of the patch of about 2 nm wide. These are the characteristics of spinodal decomposition. In addition, there are also regions that contain particles with sizes 3–10 nm. The presence of these nanoparticles is attributed to the nucleation and growth mechanism. The well-defined patch of ~100 nm wide and ~600 nm long, which partly consists of stripes (Fig. 4.38a), stripes mixed with nanoparticles (Fig. 4.38b), and nanoparticles (Fig. 4.38c and d). EDS the elongated narrow patch is S-rich and Te-poor. Therefore, it is concluded that the narrow patch consists of PbS-rich phase which is separated from the PbTe-rich matrix.

The phase segregation could occur through either the spinodal decomposition mechanism at nanoscale or the nucleation and growth mechanism. Both mechanisms of nanostructuring have the same characteristics, providing a high degree of coherence between the precipitated phase and the matrix phase. Due to this coherence, the presence of phase separation does not affect the charge scattering at the interfaces. A schematic structure of the nanostructuring process is shown in Fig. 4.39 [258]. The PbS-rich stripes are formed by spinodal decomposition and

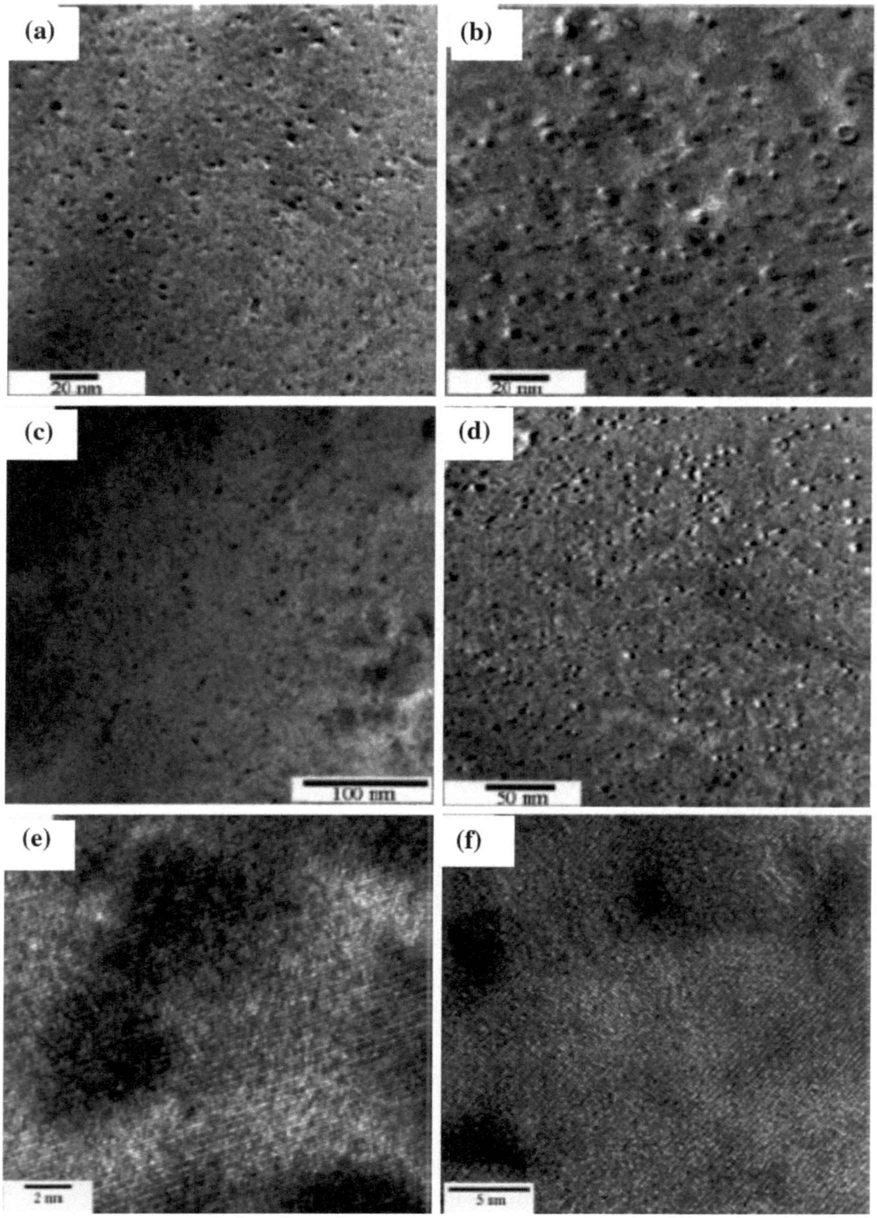

Fig. 4.37 TEM images showing dispersed nanoparticles in the samples. **a** PbTe–Sb(2 %), **b** PbTe–Sb(4 %), **c** PbTe–Sb(8 %) and **d** PbTe–Sb(16 %). Because the 8 and 16 % samples contain distinct Sb regions the images shown in **c** and **d** are from the PbTe-rich regions. **e** High-resolution TEM image showing several nanoparticles of Sb coherently embedded within the matrix of PbTe. **f** High-resolution TEM image of the PbTe–Bi(4 %). Reproduced with permission from [316], Copyright @ 2006, American Chemical Society

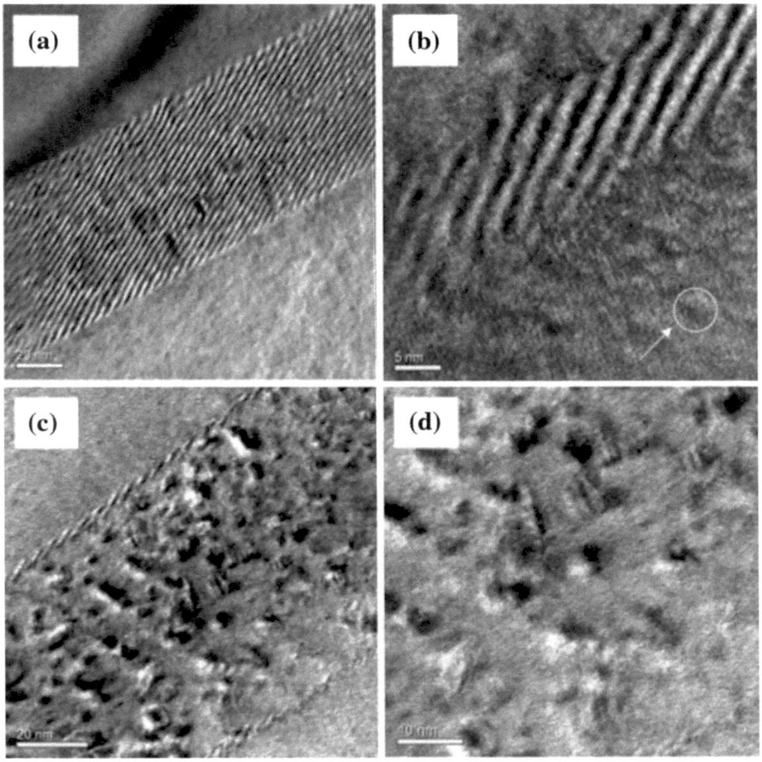

Fig. 4.38 Representative HRTEM images of the sample with 16 % PbS. **a** Spinodal decomposition of PbTe into PbS. The parallel lines correspond to periodical compositional fluctuations of PbS (*bright*) and PbTe (*dark*) phases. **b** Depending on the local S concentration, parts of the patch exit the spinodal region of the phase diagram and the characteristic compositional fluctuation fades away. **c** and **d** Outside the spinodal region where the PbTe–PbS mixture is still unstable, but the two phases separate by nucleation and growth, thus leading embedded nanoparticles with *arbitrary* shape. Reproduced with permission from [258], Copyright @ 2007, American Chemical Society

the PbS nanocrystals are precipitated from the PbTe matrix. The degree of spinodal decomposition increases with increasing content of PbS in the materials.

Because the presence of coherent nano-sized inclusions in $(Pb_{0.95}Sn_{0.05}Te)_{1-x}(PbS)_x$ does not lead to excessive electron scattering and thus the electrical conductivity is not affected, high electron mobilities of over 100 $cm^2\ V^{-1}$ s have been observed at 700 K in these materials. With $x = 0.08$, the materials have a very low room temperature lattice thermal conductivity (~30 % that of PbTe) and thus possess a high $ZT = 1.5$ at 650 K [258].

It is necessary to have a quantitative estimation on the contribution of the various factors, including the presence of the nano-sized structures, defects, and solid solution behavior, on the reduction in thermal conductivity. This is because the other factors also inhibit heat flow. The lattice thermal conductivities of

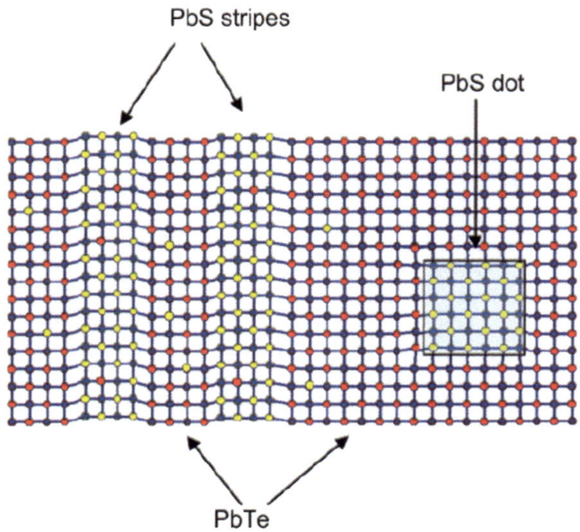

Fig. 4.39 Schematic of the $(PbTe)_{1-x}(PbS)_x$ system showing the nanophase separation occurring due to spinodal decomposition (*stripes*) and nucleation and growth (*dot*). Reproduced with permission from [258], Copyright @ 2007, American Chemical Society

PbTe-based materials have been evaluated, as comparison with pure PbTe. It has been demonstrated that of the reduction in thermal conductivity is 30–40 %, due to the solid solution point-defect scattering and the corresponding contribution of the nanostructures. Therefore, nanostructuring is the most effective approach to achieve high performance of thermoelectric materials.

4.6.1.5 Nanostructured PbTe

Several strategies, including precipitation, direct growth, and matrix encapsulation, have been used to prepare nanocomposite materials based on PbTe as matrix. Similar to that in lattice thermal conductivity of the PbTe–PbS system, significant reduction in thermal conductivity has been observed in the PbTe samples containing <3 % Sb nanoparticles [317]. Doping of Bi at the same level of content has no such effect, because Bi has almost the same atomic mass as Pb ions in the rock salt lattice [317, 318]. However, it has been found that the presence of Pb nanosized inclusions can lead to an increase in Seebeck coefficient in PbTe, although it is accompanied by a reduction in the mobility and a decreased *ZT* value.

PbTe with conanostructure precipitates of both Pb and Sb has a *ZT* value to 1.4 at 673 K, due to the unique temperature-dependent behavior of the electron mobility and its enhancements at high temperature [3]. It has been noticed that in this nanostructure material there are an increase in power factor and a decrease in thermal conductivity at the same time. At a given carrier concentration, an increased electrical conductivity at high temperatures will lead to an increase in power factor. This increase in electrical conductivity is responsible for the fact that the mobility of the material at high temperatures is higher than the expected value. The temperature dependence of mobility in the PbTe follows a power law, as $\mu \approx T^{-2.5}$.

The larger mobilities at high temperatures are directly related to this variation trend, such a decrease in mobility more slowly than in PbTe. However, this behavior is not observed in PbTe with only single type of precipitates of Pb or Sb. Therefore, there must be a synergy effect from the two precipitates. In addition, it has been found that the sample with a composition of PbTe–Pb(0.5 %)–Sb(2 %) has a substantially low lattice thermal conductivity, which cannot be explained by using the currently available mechanisms. Nevertheless, further understanding on the mechanism of this composite could be used as guidance for the development of high-performance thermoelectric materials.

4.6.1.6 InGaAs–ErAs

InGaAs–ErAs system has been widely studied mainly in the form of thin films, with behavior of lattice thermal conductivity similar to that of the PbTe system. However, it is believed that the conclusions can be applied to three-dimensional systems. The size distribution of ErAs nanoparticles in the matrix of InGaAs is not strongly influenced by the growth parameters. The average particle size is 2–4 nm. The volume fraction of the embedded nanoparticles can be readily controlled in the concentration range of 0.01–6 % without introducing defects or dislocations. The nanocomposite has a reduction in thermal conductivity by a factor of as much as 3, as compared with that of the bulk alloy. Calculation has shown that the significant reduction in thermal conductivity is attributed to the reduction in lattice thermal conductivity, which is due to the scattering of mid-and long-wavelength phonons caused by the nanoparticles. Comparatively, these phonons are not effectively scattered by the point defects in bulk alloy, because they are more effective to scatter short wavelength phonons. Theoretical studies indicate that the reduction in thermal conductivity can be further enhanced if a wide size distribution of nanoparticles is available.

4.6.2 Polycrystalline Nanocomposites

Polycrystalline nanocomposites have emerged as a new group of nanostructure thermoelectric materials. Novel sintering techniques, including hot-pressing and SPS, are usually used to fabricate the polycrystalline nanocomposites, with fine powders that are formed by grinding and milling or wet-chemistry processing as precursors. As compared with the nanoparticles in a matrix system discussed above, the nanocomposites prepared in this have more extensive interfaces between adjacent nanoparticles that are compacted together. The presence of such interfaces has been responsible for the lower thermal conductivity of these materials. These materials have several advantages, such as include reduced thermal conductivity due to the phonon scattering at grain boundaries, increased power factor due to the electron filtering at grain boundaries, better mechanical properties and improved isotropy.

These techniques can be used to make samples with a relatively low-cost at large scale [319], which have been applied to many other thermoelectric materials including PbTe [320, 321] and skutterudites [322–324]. However, a major challenge of this approach is the removal of the binders or organic agents used in the grinding, milling or wet-chemistry processes. Another challenge is to achieve full densification of the samples, because the carrier mobility could be substantially reduced at a small level of porosity.

An *n*-type nano/polycrystalline Bi_2Te_3 material fabricated by using this method has a high *ZT* value of 1.25 at 420 K, which is much higher than that of normal bulk Bi_2Te_3. This material is prepared by hot-pressing a mixture of nanometer-sized and micron-sized *n*-type Bi_2Te_3 powders. The improvement in *ZT* value has been attributed to the fact that although the electrical conductivity is slightly increased, the thermal conductivity is reduced by ~ 25 %. Another example is *p*-type BiSbTe alloy that is fabricated by using ball milling and hot-pressing. It has a *ZT* value of 1.2 at room temperature and 1.4 at 373 K [319]. These values are higher, by 20 and 40 %, respectively, than those of the ingot BiSbTe alloy counterpart, prepared by the conventional processing method.

It is found that electrical conductivity of the nanostructure materials is higher than that of the ingot counterpart. The Seebeck coefficient is dependent on temperature, yielding a slightly enhanced power factor for the nanostructure material. It is not confirmed whether the nanostructure materials have same carrier concentration as the ingot materials or not. The significant enhancement in *ZT* of the nanostructure material is due to the greatly reduced thermal conductivity and the slightly increased power factor, which is especially pronounced at elevated temperatures.

Nanocomposite boron-doped Si/Ge materials fabricated in this way have been reported to possess increased Seebeck coefficients and only slightly reduced electrical conductivities, as compared to those of the bulk SiGe alloys (JIMO). As a result, they have a high power factor over 300–1000 K. Furthermore, thermal conductivity of the Si/Ge nanocomposite is significantly reduced, thus leading to a peak *ZT* value of ~ 0.72 at 1000 K, which higher than that (~ 0.6) of the bulk SiGe alloy counterpart. Therefore, nanostructuring has been a promising approach to achieving high-performance thermoelectric materials, which will be still a focus of research in this important area [325, 326].

4.6.3 Superlattice Nanostructures

4.6.3.1 Fabrication and Characterization

Two-dimensional layered thermoelectric materials—quantum well superlattice nanostructures, including PbTe, SiGe, and $SrTiO_3$, have been fabricated by using sophisticated technologies, such as molecular beam epitaxy (MBE), which offer

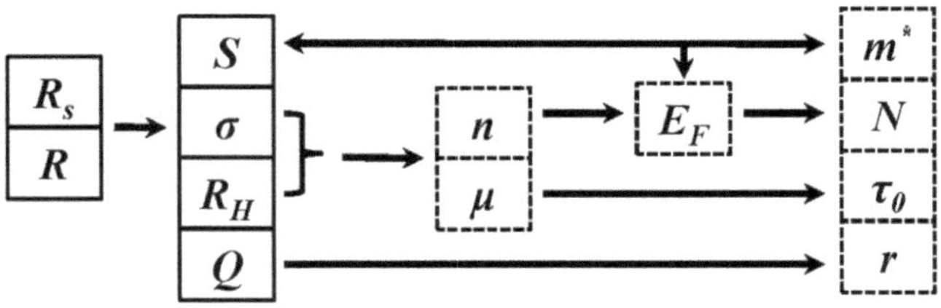

Fig. 4.40 Determination of carrier concentration (n) and mobility (m) from the measured resistance (R), or sheet resistance (R_s) or electrical conductivity (s), along with the Hall (R_H) coefficient. Determination of the effective masses (m^*) and number of conduction valleys (N), the scatterings constants (r and τ_0) and the Fermi level (E_F) can be determines with the measurement of the Seebeck (S) and Nernst (Q) coefficients

high-quality atomic scale control. Superlattice structures also allow for larger sample size along with a large temperature gradient, due to the relatively easy characterization in the in-plane direction, for example, the measurement of S and σ. Moreover, due to the thin film geometry, n and μ can be readily measured by using the van der Paw from σ and the Hall coefficient (R_H) [327].

Figure 4.40 show flowchart of interrelationship among all the relevant parameters and their characterization [4]. With the values of the number of conduction valleys (N) and the density of states effective mass (m_d), the Fermi energy (E_F) can be derived from n. After that, S can be calculated from E_F, which can be used to compare with experimental results. Once the variation of either μ or S, as a function of temperature (T) is available, the two relaxation time parameters, τ_0 and r, can be readily estimated. Additionally, the values of r can be directly obtained if the Nernst coefficient (Q) is determined.

Superlattice are heterostructure structures, their properties are the sum of those of active layers and the barrier layers, which can be characterized by using Eqs. (4.50–4.52). Therefore, it is necessary to further determine the characteristics of the individual active layers and the barrier layers. If the barrier layers are insulating, the electrical properties of the active layers can be more easily derived from those of the superlattices.

4.6.3.2 PbTe Superlattice

Bulk PbTe is a direct band gap semiconductor, with E_g of 0.31 eV. It has both the CB minimum and the VB maximum at the L-point of the Brillouin zone. With a face-centered cubic (*fcc*) crystal structure, there are eight equivalent L-point in the 1st Brillouin zone for bulk PbTe. PbTe can be doped to form either p-type with Bi [224]) or n-type with Na [109], both of which have four full conduction valleys or eight half-full valleys that across the edge of the first Brillouin zone at the L-point, thus with N = 4. Due to the coupling of CB and VB, nonparabolicity is induced in the band, which leads to effective masses that are small, anisotropic, and temperature-dependent, with $m_L = 0.35m_0$ and $m_T = 0.034m_0$ at 300 K [214]. Due to the large difference between m_L and m_T, the transport coefficients of the materials are strongly anisotropic. With a relatively large μ and a low k_L, PbTe has a high *ZT* of about 0.6 at temperatures of 700–800 K. It has been reported that the k_L of PbTe can be further reduced by doping with Sn to form *n*-type semiconductor to obtain higher values of *ZT* [4, 8].

It is demonstrated that *n*-type [111] PbTe/$Pb_{0.927}Eu_{0.073}Te$ superlattice nanostructures, deposit by using MBE, have shown significant enhancement in $S^2\sigma$ [208, 224]. In these superlattice, electrons are confined in the PbTe layers, because the PbEuTe barrier layers have increased band gap and effective mass due the doping of Eu. Both the band gap and the effective mass are found to increase with increasing concentration of Eu. Infrared transmission analysis results confirm the presence of the effects of quantum confinement. Because the oblique valleys have smaller confinement effective mass in the [111] PbTe quantum wells, they possess larger confinement potential, thus the degeneracy of the single longitudinal valley is lifted from the three oblique valleys.

The value of S^2n can be increased beyond the highest value of bulk materials, with the multilayered quantum wells, up to a thickness of 5 nm in one study and 1.7 nm in the other [208, 224]. If the relaxation time is assumed to be constant ($r = 0$) and there are $m_L = 0.62m_0$ and $m_T = 0.053m_0$, the discrepancy between experimental results and theoretical prediction is very small. The effective mass values of superlattice structures are much higher than the values of their bulk counterparts, which has been attributed to penetration of electron wave function into the PbEuTe barrier layers, whose effective masses are larger as compared with that of bulk PbTe [328]. In addition, it is shown that the enhancement in $|S|$ is also due to the confinement effect not the variation in n and E_F. For example, although a 2 nm sample and a 4 nm one have same value of n, the former has a much larger $|S|$ than the latter, while the latter has a close value to that of the bulk materials.

The value of μ of the PbTe superlattice is 640–1,420 $cm^2\ V^{-1}\ s^{-1}$, which is close to those of doped PbTe thin film (1200 $cm^2\ V^{-1}\ s^{-1}$) and bulk PbTe (1,000–1,700 $cm^2\ V^{-1}\ s^{-1}$), but much greater than that of PbEuTe (45 $cm^2\ V^{-1}\ s^{-1}$). The large variation in mobility values of the superlattice has been attributed to the wider spread of processing parameters. The presence of possible interface scattering can be used to explain the decreased μ of the superlattice and the thin films, as compared with that of the bulk materials. However, the presence of a potential barrier of 171 meV has not provided an interface scattering that is strong enough to significantly affect μ [208]. The highest value of μ of 1,420 $cm^2\ V^{-1}\ s^{-1}$ for the superlattice, which is higher than that of doped PbTe thin film, implies a possibility to increase μ by the modulation doping, through carrier transfer from the barrier layers to the quantum well layers. With the value of k of bulk PbTe, the PbTe quantum wells are estimated to have *ZT* values of 1.2–2 at 300 K, together with the enhancement of the power factor, whereas the effective figure of merit of the whole structure, ZT_{SL}, is much lower, due to the parasitic heat conduction through the PbEuTe layers.

Another example is quantum well superlattice made of n-type [100] PbTe quantum well (5 nm)/EuTe barrier layer (0.66 nm), with similar enhancement in S^2n [202]. Because EuTe has larger band gap than PbTe, which leads to a large band offset of about 1 eV. Therefore, the barrier layer can be much thinner to meet the requirement of preventing tunneling. The growth of the thin films along the [100] is much better than that along the [111] direction, in terms of maintaining a density of states, because the former type of growth ensure to preserve the valley degeneracy (N), which has been confirmed by the value of DOS. As a result, the [100] superlattice structures have lower E_F and higher $|S|$ than the [111] oriented ones, n and a are same.

Without available value of μ, it is not able to show the effect of the superlattice structure on the power factor. However, due to the presence of the large potential barrier offset, the use of the very thin EuTe barrier layers may also reduce μ, parasitic heat conduction can be significantly decreased. For example, the value of μ of a 5 nm PbTe quantum well can be reduced by 40 % from the value of bulk materials, which has been attributed to the increased interface scattering [41]. If the values of μ and k of bulk materials are used, the value of ZT of the [100] PbTe/EuTe superlattice is estimated to be 0.66 at 300 K.

4.6.3.3 Si/SiGe Superlattice

Bulk $Si_{0.7}Ge_{0.3}$ has once been found to have a ZT close to ~1 at 1,100 K, for potential applications in radioisotope thermoelectric generators (RTG) in deep-space satellites. SiGe is an indirect band gap semiconductor, with large m_d but relatively small μ. The band structure and electronic properties of $Si_{1-x}Ge_x$ alloys, with $x < 0.8$, are similar to those of Si, with the lowest conduction minima near the X-point ($N = 6$) in the Brillouin zone. At $x > 0.8$, the $Si_{1-x}Ge_x$ alloys have a band structure that is very close that of Ge, with the conduction band minima at the L-point ($N = 4$). The $Si_{1-x}Ge_x$ alloys exhibit the advantages of the high $S^2\sigma$, a characteristic of Si, and reduced k_L, for example, from 133 W m^{-1} K^{-1} for Si to about 10 W m^{-1} K^{-1}, for $Si_{0.7}Ge_{0.3}$ alloy at 300 K. The value of k_L of $Si_{0.7}Ge_{0.3}$ is further decreased to about 4 at 1,100 K.

There have also been studies reported for $Si_{1-x}Ge_x$/Si superlattice nanostructures [329]. One of the examples is $Si_{0.8}Ge_{0.2}$/Si superlattice on [100] Si substrates, deposited by using MBE, where the uniformly doped p-type $Si_{0.8}Ge_{0.2}$ is the quantum well layer and Si is the barrier layer. The other example is $Si_{0.7}Ge_{0.3}$/Si superlattice, with n-type Si as the quantum well layer and the $Si_{0.7}Ge_{0.3}$ the barrier layer. In the two studies, similar values of carrier concentrations ($n \approx 10^{19}$ cm^{-3}), quantum well thicknesses (a = 1–5 nm), barrier thickness (b = 20–30 nm) and number of periods (10–15), are used. The Si/SiGe superlattice heterostructures exhibit a Type II band alignment, where holes and electrons can be confined in the SiGe and Si layers, respectively.

Significant barrier conduction has been observed in both studies. The potential barrier is relatively low due to the small band offset of 0.1–0.2 eV and thin due to the significant band bending as a consequence of heavy doping in the barrier layer, which can lead to both thermionic emissions over the barrier and tunneling effect. Therefore, the quantum confinement effects are reduced due to the presence of the thermionic and tunneling currents. The former is a classical effect in which thermally excited electrons gain sufficient amount of energy to overcome the potential barrier, whereas the latter is a quantum mechanical effect, during which the tunneling electrons penetrate the barrier although they have lower energy than the barrier.

The reason for the increase in S^2n observed in the superlattice structures has not been fully clarified. It has in some cases been simply attributed to the increased proportion of the Si layer, because the thickness of the Si layer is increased while that of the barrier layer is kept unchanged and Si has higher $|S|$ than $Si_{0.8}Ge_{0.2}$. Transport coefficients of the quantum wells cannot be determined because it is not possible to measure the individual layers. Theoretical studies are based on the assumption of optimal dopant concentration and constant relaxation time, thus leading to overestimation, due to the variation in thicknesses of the actual layers and the strain caused by the lattice mismatch.

The highest value of S^2n is observed in the superlattice having the thinese layers, which is up to 2–3 times larger than the highest value reported for the bulk counterparts. Therefore, the enhancement can be attributed mainly to the increase in DOS, instead of the increase in n, because n increases with decreasing a, as demonstrated in Eq. (4.39). The degree of the increase in S^2n of superlattice nanostructures is also affected by the conduction of the barrier layers.

Although the value of μ of the n-type $Si_{0.7}Ge_{0.3}$/Si superlattice decreases with increasing a, due to increased interface scattering, it is much higher than that of the bulk Si or $Si_{0.7}Ge_{0.3}$, as long as the values of n and a are not too high. For instance, the superlattice with $n = 1.8 \times 10^{19}$ cm^{-3} have a value of μ of about 535 cm^{-2} V^{-1} s^{-1}, where as the value of μ of the bulk n-Si with a similar n is about 100 cm^{-2} V^{-1} s^{-1}. However, if n is increased to 6.1×10^{19} cm^{-3}, the value of μ of the Si/$Si_{0.7}Ge_{0.3}$ quantum well superlattices is decreased to about 105 cm^{-2} V^{-1} s^{-1}, which is close to the value of the bulk Si.

The effect of carrier pocket engineering has been observed in n-type Si/Ge superlattice, as evidenced by the variation of $S^2\sigma$ as a function of period of the superlattice [68]. For the superlattice with d of >10 nm, the slightly higher $S^2\sigma$ as compared with the values of the bulk components can be attributed to the increase in the DOS due to confinement effect. As the value of d is reduced from 10 to 4 nm, there is a sharp increase in $S^2\sigma$, which has been explained by the transfer or redistribution of electrons from the L-conduction valleys with lower m_d to the X-conduction valleys with larger m_d, induced by the confinement potential. A maximum $S^2\sigma$ of 163 mW cm^{-1} K^{-2} is observed at $d = 4$ nm. Further decrease in d to <4 nm, the value of $S^2\sigma$ starts to decrease, which can be ascribed to the reduction in μ, due to the significant increment in interface scattering. In addition, the superlattices have a lower in-plane thermal diffusivity of about 0.02 cm^2 s^{-1}, as compared with the value (0.055 cm^2 s^{-1}) of the bulk materials. Highest ZT of 0.75 at 300 K has been reported in the superlattice structures with d of 7.5 nm.

However, further theoretical and experimental evidence is necessary to attribute the increase in thermoelectric performances of the superlattice nanostructures to the confinement effect.

Carrier pocket engineering has also been reported for [100] oriented *n*-type Si/Ge (2/2 nm) superlattice [215, 330]. The measured values of $S^2\sigma$ are lower than the theoretical predictions, which is explained by including an additional contribution from acoustic phonon and defect scattering. The superlattice have a maximum ZT of about 0.1 at 300 K, with a k_L of 5 W m^{-1} K^{-1}. This value is nearly seven fold increase, as compared with the value of 0.014 for bulk Si, which has been used as an evidence to verify the concept of conduction valley/carrier pocket engineering. In fact, the value of k_L of bulk Si is about 133 W m^{-1} K^{-1}, which is much higher than the value for the superlattice, which in turn implies that the $S^2\sigma$ of the superlattice is much smaller. Experimental results indicate that the $S^2\sigma$ value of the superlattice nanostructures is smaller by about one order of magnitude. Theoretical studies also predict that the lowest *L*-valley sub-bands of Ge are higher than the lowest *X*-valley sub-bands of Si [215]. In this case, carrier pocket engineering cannot be solely used to explain the experimental results.

In summary, while some experimental work seems to show an enhancement of S^2n over bulk value, the lack of detailed theoretical calculations that take into account the influence of the barrier conduction makes clear interpretation difficult. Consequently the claims of $S^2\sigma$ enhancement due to quantum confinement, and evidence of carrier pocket engineering is still unclear. It should also be noted that the optimal temperature of SiGe based alloys is over 1,000 K, and the ZT at room temperature is much smaller than that of Bi_2Te_3. Hence, SiGe should be operated or characterized at much higher temperature. However, the possible degradation of the nanostructures due to atomic diffusion may render operation at such high temperatures difficult or impractical.

4.6.3.4 Oxide Superlattice

Materials, with smaller band gaps of $\leq$1 eV, are limited to low-temperature applications because of their earlier onset of bipolar conduction. In comparison, metal oxide thermoelectric materials have relatively large band gaps, so that they are suitable for applications at high temperatures, such as waste heat recovery from industrial processes, due to their chemical and thermal stability. Among various thermoelectric oxides, $SrTiO_3$ is of particular interest, especially as superlattice structures.

$SrTiO_3$ has a large band gap, with E_g of 3.5 eV. According to the $10k_BT$ rule, the bipolar conduction onset should be above 4000 K. The conduction band of $SrTiO_3$ has a minimum at the *G*-point of the Brillouin zone (N = 1), which are formed by the triply degenerate Ti *d*-orbitals, with very large and anisotropic effective masses, $m_T = 1.1m_0$ and $m_L = (20 \sim 50)m_0$, leading to $m_d = (6 \sim 13)m_0$ [331]. Because it has large m_d and DOS, the carrier concentration and the value $|S|$ can be very high. For example, the optimized carrier concentrations of *n*-type Nb-doped [176] and

La-doped [176, 210] bulk $SrTiO_3$ are 3.3×10^{20} and 1.0×10^{21} cm^{-3}, respectively, with the value of $|S|$ to be 167 mV K^{-1}, which is very close to the value predicted theoretical modeling.

Nb-doped thin films, grown by using pulsed-laser deposition (PLD), have n of 4×10^{21} cm^{-3}, which increases with temperature. One of the advantages of using PLD to deposit $SrTiO_3$ thin films is that the thermodynamic dopant solubility limit of 4×10^{20} cm^3 can be overcome [38]. The maximum ZT $SrTiO_3$ thin film is 0.4 at 1,000 K. Therefore, it is a potential candidate of thermoelectric material for high temperature applications, if higher ZT.

The value of $S^2\sigma$ has been found to be increased by using the quantum confinement effects, for example, in $Sr(Ti_{0.8}Nb_{0.2})O_3/SrTiO_3$ multiquantum well (MQW) superlattice structures [332, 333]. The $Sr(Ti_{0.8}Nb_{0.2})O_3$ layers are quantum wells, which is due to the donor doping of Nb, with higher valence than Ti. The confinement effect has been confirmed by using electron energy loss spectroscopy (EELS) and optical absorption measurements [332]. As compared with that of the doped bulk $SrTiO_3$, the value of S^2n of the superlattice nanostructures is significantly increased. The value of ZT is increased from 0.09 of bulk material to 2.4 of the quantum well, according to the values of S^2n. When the parasitic heat conduction is considered, the effective ZT of the nanostructures will be about 0.24.

Although the S^2n value of $SrTiO_3$ is much higher than those of PbTe and SiGe, their $S^2\sigma$ values are comparable, becuase $SrTiO_3$ has a pretty low μ. The low μ $SrTiO_3$ is due to the localized nature of the d-orbital electrons and the very large effective masses. In addition, $SrTiO_3$ has relatively comparable $k_L \approx 3$ W m^{-1} K^{-1} at 300 K, which increases to 10 W m^{-1} K^{-1} at 1,000 K [176]. The slight increase in k_L with temperature has been attributed to the increased number of active phonon modes associated with the oxygen atoms. The increased k_L will decrease the value of ZT at high temperatures, which is a problem for $SrTiO_3$ as a thermoelectric material for high temperature applications.

4.6.4 Bi Nanowires

Bi has been acknowledged to be a potential candidate to have high Z because of its intrinsic high μ and low k, if it is converted from SM to SC. This can be accomplished by using quantum confinement effects, in the form of quantum wells and nanowires. Bi has a relatively low effective mass and large carrier mean free path, quantum confinement effects can be observed at much larger length scales. Theoretical studies have predicted that, a 10 nm Bi quantum well can have a ZT of as high as 3 [207]. In contrast, to reach this value of ZT with PbTe quantum well, the thickness cannot be larger than 1 nm [334].

Bi has a very low melting point of 271 °C, which makes it easy to fabricate Bi nanowires, by using melt-injection and vapor deposition methods. For example, Bi nanowires with diameters of about 7 nm have been fabricated by using electrolytic processing method, with porous alumina as templates. The diameters of the nanowires can be simply controlled by selecting templates with desired sizes [335]. There is a critical diameter (d_c) for Bi nanowires to show the SM–SC transition. Theoretical calculations indicate that the values of d_c for circular Bi wires orientated along the trigonal, binary, bisectrix, $[10\bar{1}\bar{1}]$ and $[01\bar{1}\bar{2}]$ directions are 55.1, 39.8, 48.5, 48.7, and 49 nm at 77 K, respectively. It is also found Bi nanowires with large diameters ($d > 60$ nm) have preferred growth orientation along $[10\bar{1}\bar{1}]$ direction, while thinner nanowires ($d < 50$ nm) are grown along $[01\bar{1}\bar{2}]$ direction [336]. The d_c is also a function of temperature. As the temperature is increased, d_c decreases.

With large diameters, *p*-type Bi nanowires are expected to show inferior performances as compared with *n*-type ones, due to unfavorable orientation of the dominant *T*-point valence band valleys. However, if the diameter is <5 nm, the *T*-valley is pushed to below the *L*-point valence band valley due to the confinement effects. Because the *L*-point valence band valley is a mirror image of the *L*-point conduction band valley, there is no difference in electronic properties between *p*-type and *n*-type Bi nanowires, both of which are expected to have a *ZT* of as high as 6 at 77 K [337].

The behavior of resistance versus temperature, $R(T)$, is usually used to show the SM–SC transition [335, 338, 339]. Because it is difficult to determine the number of the nanowires in an array, normalized resistance, $R_n(T) = R(T)/R(300\text{ K})$, is thus used. R_n increases with T for bulk Bi and Bi nanowires with $d > 70$ nm, which is typical metallic behavior, whereas nanowires with $d < 50$ nm demonstrate semiconductor-like behavior, with R_n decreasing with temperature. Therefore, this behavior can be used as the evidence of the SM–SC transition. However, to further confirm the SM–SC transition, it is necessary to have the absolute values of σ and μ.

However, reports on size dependence of SM–SC transition of Bi nanowries are not consistent, especially bwteeen theoretical prediction and experimental results in some cases. Theoretical calculations have indicated that the critical diameter of d_c for the SM–SC transition of Bi nanowires is about 15 nm at room temperature. There have been experiment results that are consistent with this prediction. For example, magnetoresistance characteristics of Bi nanowire arrays, with diameters of 80 and 30 nm, at 5 K are in metallic state [340]. In constrast, 40 nm × 15 nm rectangular polycrystalline Bi nanowires have been found to show semimetallic behavior, according to the measurement of $R(T)/R$, S and σ as a function of gate voltage at 20 K, as shown in Fig. 4.41 [341].

Due to the dependence on various parameters, including variability in wire diameter, crystallinity, crystal growth orientation, concentration of dopants, and doping type (electorn or hole) and temperature of measurement, the experimental data of S of Bi nanowires are also scattered in the open literature. Therefore, systematic studies on the effects of various parameters on thermoelectric properties of Bi nanowires are highly demanded, in order to clarify the underlying reasons for the enhancement in S. It has been shown that an array of 40 nm Bi wires with orientation along the $[10\bar{1}\bar{1}]$ direction has a value of S to be about -60 mV K^{-1} at

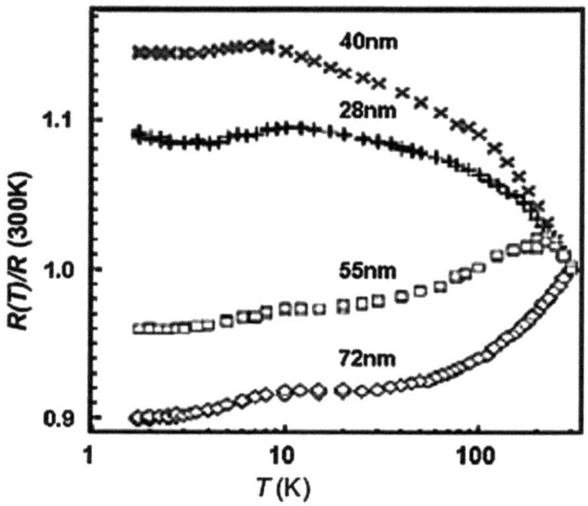

Fig. 4.41 Temperature-dependent resistances of the Bi NWs, normalized to the resistance at 300 K. Reproduced with permission from [341], Copyright @ 2006, John Wiley & Sons

300 K [335], which inbetween values of S of bulk Bi at 300 K, -100 and -50 mV K^{-1}, in the directions of parallel and perpendicular to the trigonal direction, respectively [208]. Comparatively, a value of only 15 mV K^{-1} has been reported for a Bi nanowire array with wire diameter of 200 nm over 77–300 K. The Bi nanowires are undoped with unknown crystal orientation.

There also have been reports on thermoelectric properties of single Bi nanowires. For instance, a single undoped Bi nanowire has been characterized by using gate bias to control the carrier concentration. For the nanowires with diameters of 28 nm < d < 72 nm, the value S is as low as 10 mV K^{-1} [340]. Because these nanowires are not doped, neither electrons nor holes dominate the conduction process; as a result, the value of $|S|$ cannot be increased by inducing band gaps. An extremely high $|S|$ of 300 mV K^{-1} has been observed in a Bi nanowire array of 9 nm nanowires, which is attributed to the SM–SC transition [342]. Such a value is far beyond the expectation for an electrical carrier based diffusion process. According to Eq. (4.43), the value corresponds to an energy separation between $R\overline{3}m$ and E_F to be as high as 30 eV, which is obviously unreasonable, according to the conventional electron diffusion theory. It can only be explained in terms of a peculiar band structure with spikes in the DOS [343].

Currently, no effective measurement technologies have been developed to accurately determine the individual values of n and μ of Bi nanowires, which is the main reason why definitive conclusions are still lacking on thermoelectric properties of Bi nanowires. Relevant parameters (including S, n, μ, and σ), along with the characterization of the energy levels/sub-bands, have to be determined, in order to characterize thermoelectric performances of Bi nanowires, for practical applications [344, 345].

4.6.5 Si Nanowires

Si is the most abundant element on earth and has been widely used in semiconductor technologies. Therefore, the advantage of large industrial infrastructure for low-cost and high-yield processing makes it attractive, if Si has promising thermoelectric performance. However, due to its high thermal conductivity, $k \approx 150\ \mathrm{W\ m^{-1}\ K^{-1}}$ at room temperature, bulk Si however, has a high a ZT of < 0.01 at 300 K, which is too low for real applications [346]. Recent progresses have shown that thermal conductivity of Si can be reduced by nearly two orders of magnitude while maintain electrical conductivity almost unchanged by making it into nanowires. As a result, a high $ZT \approx 1$ has been achieved in Si nanowires [299, 301]. With this breakthrough, Si nanowires have attracted increasing interest as thermoelectric materials [347].

Si nanowires have been synthesized by using various methods. One example is aqueous EE [299, 348]. The technique is based on the galvanic displacement of Si by $Ag^+ \rightarrow Ag$ reduction on B-doped Si wafer surface. The reaction is usually conducted in an aqueous solution of $AgNO_3$ and HF. During the reaction, Ag^+ is reduced on the surface of the Si wafer by injecting holes into the Si valence band to oxidize surrounding lattices, which is etched due to the reaction with HF. Ag nanoparticles are then formed on the wafer surface due to the reduction of Ag^+ ions, which delimit the spatial extent of the oxidation and etching process. The Ag nanoparticles act as active cathode by electron transfer from the underlying wafer. Consequently, Si nanowire arrays are formed. The Ag nanoparticles are washed off with nitric acid and deionized water after the synthesis. Representative SEM image of the Si nanowire arrays and TEM image of a single Si nanowire are shown in Fig. 4.42 [203].

Due to their rough suface, the Si nanowires prepared in this way have much lower k than other nanowires. The value of k of the Si nanowires with a diameter of 52 nm is only about $1.6\ \mathrm{W\ m^{-1}\ K^{-1}}$ at room temperature, leading to a ZT of about 0.6, which is expected to be further increased through optimized doping, diameter reduction and roughness control, as discussed above [299].

Another method that has been used to synthesize Si nanowires with improved thermoelectric properties is called superlattice nanowire pattern transfer (SNAP) [301, 349]. The method is able to control over the layer thickness of a superlattice, so as to control over the width and spacing of Si nanowires at atomic level. Si nanowires made via SNAP inherit their impurity dopant concentrations directly from the single-crystal Si epilayers of the silicon-on-insulator substrates from which they are fabricated. The epilayers are Si (100) films with thickness of 20 or 35 nm on 150 nm thick SiO_2. The Si wafers are p-type impurity doped with B by using diffusion-based doping technology. Si nanowires can be several micrometers long, with which arrays with lateral width × thickness dimensions of 10×20 nm, 20×20 nm and 520×35 nm have been fabricated. The last approximates the bulk and, in fact, measurements on the sample obtained bulk values for S, s, and k. Measurements of k for our nanowires were consistent with

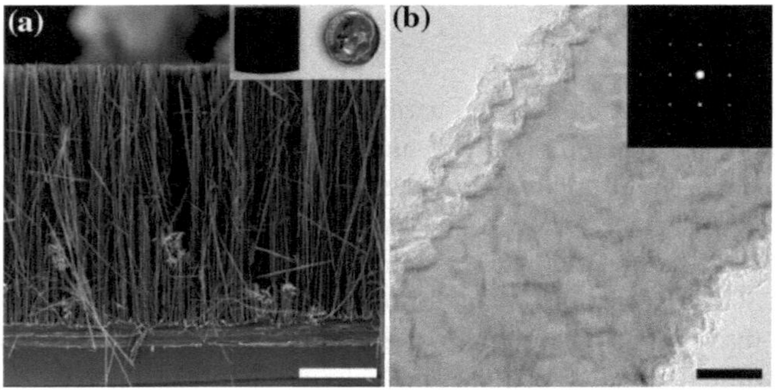

Fig. 4.42 Structural characterization of the Si nanowires. **a** Cross-sectional SEM of a Si nanowire array, with visible dendritic Ag nanoparticles within the array and Si wafer sample in the inset and **b** bright-field TEM image of a segment of a Si nanowire, with the roughness being clearly seen on surface of the nanowire and a selected area electron diffraction pattern (inset) showing that the nanowire is single crystalline all along its length. Scale bars for **a** and **b** are 10 and 20 nm. Reproduced with permission from [299], Copyright @ 2008, Nature Publishing Group

literature values for materials grown (round) Si nanowires16. Four-point probe conductivity measurements of the silicon-on-insulator films were used to extract dopant concentrations.

The Si nanowires have electrical conductivity of between 10 and 90 % of that of the bulk material, depending on dimension of the nanowires. The reduced conductivity has been attributed to the surface scattering of charge carriers [350]. However, most of the nanowires possess metallic-like conductivity, with σ increasing with decreasing temperature. Thermal conductivity k greatly decreases with decreasing cross-sectional area of the nanowires. For example, the 10 nm wide nanowires have k value of as low as 0.76 W m^{-1} K^{-1}, which is even lower than the theoretical limit of Si bulk. A very high ZT value of about 1 has been observed at 200 K for such Si nanowires, which is the result of the significant reduction in thermal conductivity due to the size confinement effect, as analyzed in the previous section.

4.6.6 Nanocarbon Thermoelectric Materials

Recently, nanocarbon based thermoelectric materials, including single-wall carbon nanotubes (SWCNTs), multiwalled carbon nanotubes (MWCNTs), graphene, graphene oxide (GO) and carbon nanofibers (CNFs), as well as their nanocomposites with conductive polymers, have attracted much attention by the research community in this area [292–294, 351–354], in line with the hot topics of these materials for varous other applications.

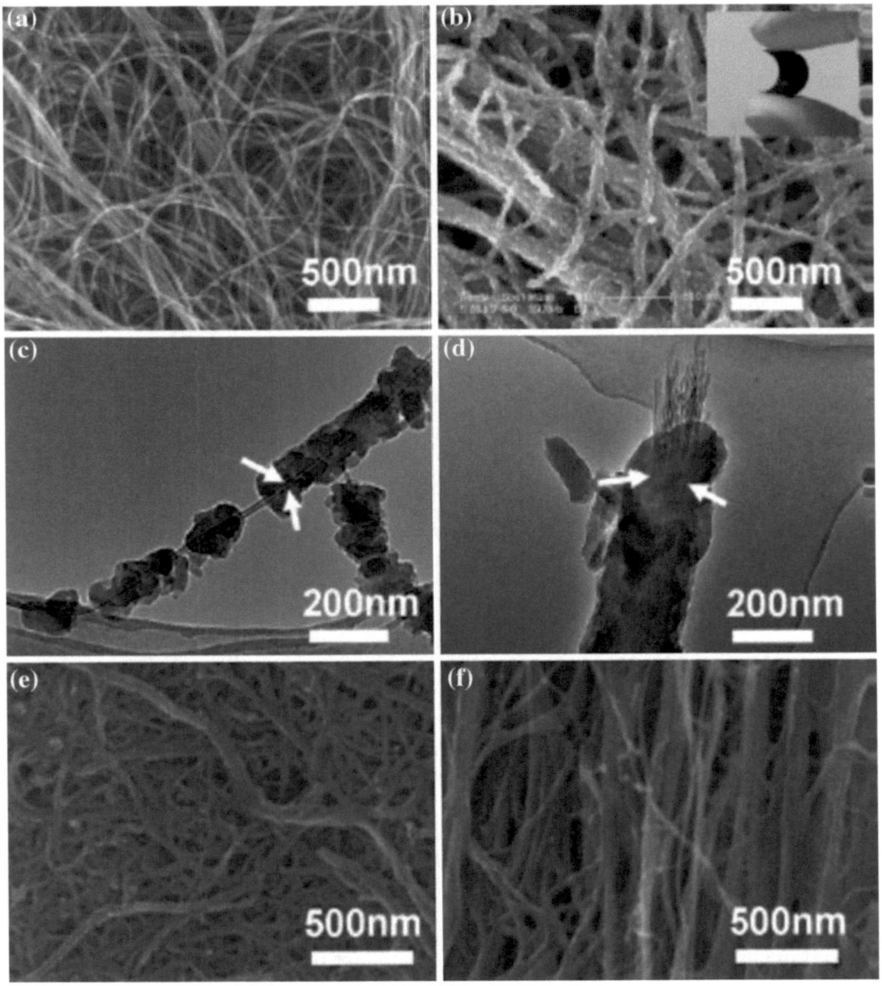

Fig. 4.43 **a**, **b** SEM images of the thick CNT sheet before and after coating with PANI. The inset in (**b**) shows the flexibility of the nanocomposite. **c**, **d** TEM images of individual CNTs and CNT bundles, respectively, coated with a layer of PANI. **e**, **f** SEM images of the PANI-coated CNT pellet (*top view*) and the PANI-coated CNT array (*side view*). Reproduced with permission from [351], Copyright @ 2010, John Wiley & Sons

One example is MWCNT-PANI nanocomposite fabricate with a two-step method, preparation of freestanding CNT network of randomly entangled individual CNTs and CNT bundles by using filtration, followed by coating PANI through in situ polymerization [351]. Figure 4.43 shows microstructural characterizations of the nanocomposites.

The individual CNTs and their bundles randomly are intertwined together to form a CNT network structure, as shown in Fig. 4.43a. The original diameters of CNTs are 15–30, which are increased to 60–200 nm after coating a layer of PANI

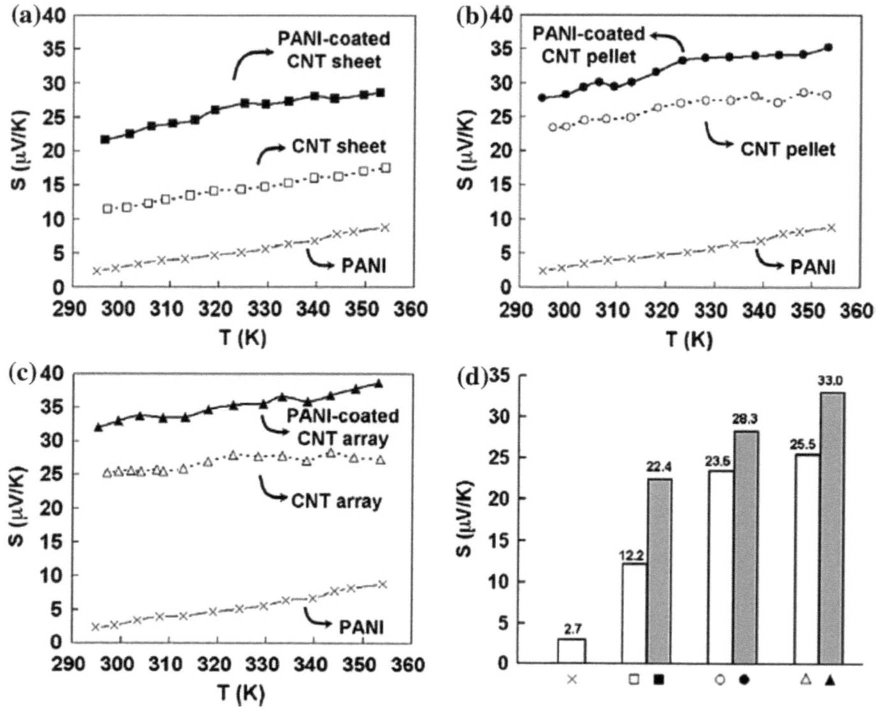

Fig. 4.44 **a–c** Seebeck coefficient as a function of temperature for the CNT/PANI nanocomposites made from different CNT sources with different forms, as well as those of the pristine CNTs and PANI. **d** Seebeck coefficient of the nanocomposites, compared with those of the pristine CNTs and PANI at 300 K. Reproduced with permission from [351], Copyright @ 2010, John Wiley & Sons

through in situ polymerization. The PANI layers are uniform on the CNTs, without altering the flexibility of the nanocomposites, as shown in Fig. 4.43b. TEM images in Fig. 4.43c and d further demonstrate the core-shell nanostructure, with clear interfaces. The thickness of the PANI layer is 50–90 nm.

Seebeck coefficients of nanocomposites, in forms of sheet, array, and pellet, together with CNT sheet and PANI, as a function of temperatures, are shown in Fig. 4.44 [351]. Seebeck coefficients of PANI and the CNTs are 2.74 mV K^{-1} and 12.2 mV K^{-1} at 300 K, increasing with increasing temperature. The Seebeck coefficients of all the CNT-PANI nanocomposites are higher than those of pure PANI and the CNTs. The enhancement in Seebeck coefficient of the CNT-PANI nanocomposites has been attributed to the quantum confinement effect.

Effect of thickness of the PANI layer on Seebeck coefficient enhancement of the nanocomposites has been studied, where the thicknee is controlled by adjusting the amount of the aniline monomer precursor [351]. Figure 4.45d shows that there

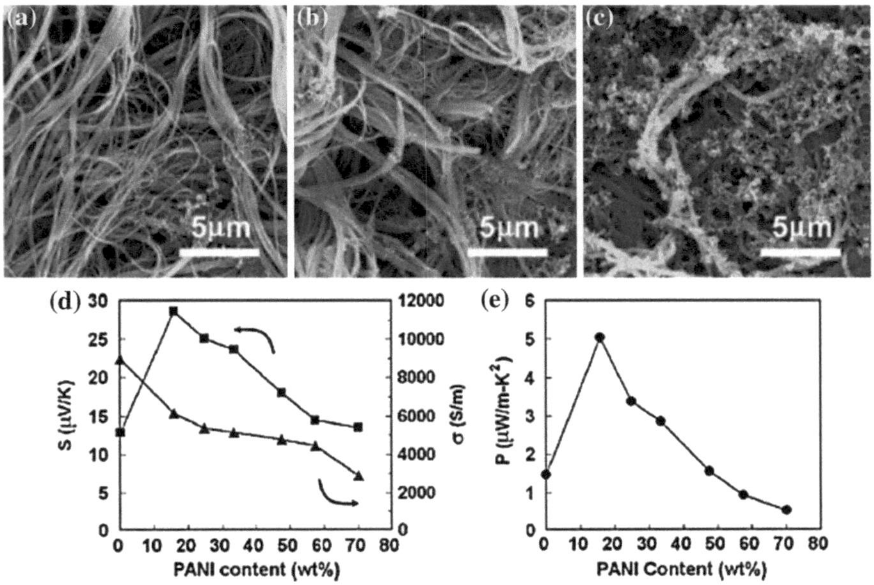

Fig. 4.45 a–c SEM images of the PANI-coated CNT sheet prepared with different aniline concentrations (0.02, 0.2, and 1.0 M). **d, e** Thermoelectric properties of the PANI-coated CNT sheets as a function of PANI composition at 300 K: Seebeck coefficient ■, electrical conductivity▲, and power factor ●. Reproduced with permission from [351], Copyright @ 2010, John Wiley & Sons

is an optimized Seebeck coefficient at about 20 wt% of PANI, which implies the presence of quantum confinement effect in the nanocomposites. Such optimized Seebeck coefficient leads to an optimized power factor.

Although it is believed that nanocarbon based thermoelectric materials will be of continuously increasing research interest and significant progress has been made in increasing some of the thermoelectric paramerters, such as Seebeck coefficient, the overall performance of such materials or nanocomposites are far below the requirement for device applications. For example, the *ZT* value of a MWCNT-PANI nanocomposite, with CNT content of 30 wt%, prepared by using cryogenic grinding, is only $7.6 \times 10{-}5$ at 300 K [353]. Therefore, enhancement of performance is an urgent issue of nanocarbon based thermoelectric materials.

4.7 Processing of Bulk Nanocomposites

With the above discussion, it is readily expected that bulk nanocomposites will be the most promising candidates of thermoelectric materials for large-scale applications, because large quantity nanocomposites can be prepared in a cost-effective way, as compared with thin film superlattice nanostructures and template derived nanowires. In this respect, it is important to emphasize the processing of bulk nanocomposite thermoelectric materials. In order to produce bulk nanocomposites, the first step is to synthesize nanoparticles, which are then assembled or consolidated into dense bulk solids in the second step. Synthetic methods of the nanoparticles and consolidation techniques will be briefly described as follows.

4.7.1 Synthesis of Nanoparticles

Various methods, including hydrothermal synthesis [355–358], wet chemical precipitation [359–361], and mechanical ball milling [98, 362–365], have been employed to synthesize thermoelectric nanoparticles to fabricate bulk nanocomposites with improved thermoelectric performances.

Mechanical ball milling is top-down method to prepare ultrafine or nano-sized particles, which has potential for large-scale industrial applications. Powders treated by using the conventional ball milling, with plastic, agate or ceramic milling demia, have particle sizes of not <1 μm. Comparatively, high-energy ball milling, with heavy milling media, such as hardened stainless steel and tungsten carbide (WC), has been widely used to prepare ultrafine or nano-sized particles [366, 367]. More recently, it has been acknowledged to be an effective and a powerful technique to produce large quantities of thermoelectric nanoparticles in a short time [364, 365]. Thermoelectric nanoparticles can be obtained either from ingots with given compositions or directly from component elements [368, 369].

4.7.2 Consolidation Technologies

To make bulk thermoelectric nanocomposites materials, the nanoparticles have to be consolidated. Various consolidating techniques have been used to fabricate bulk thermoelectric nanocomposites, such as spark plasma sintering [370, 371], cold-pressing [372–374], conventional sintering [375], hot-pressing sintering [365, 376], and extrusion methods [373]. Cold-pressing is to mechanically compact the nanoparticles, thus leading to nanocomposites with relatively low density and poor mechanical strength. Therefore, cold-pressed nanocomposites need to be further densified by using high temperature sintering. To ensure high electrical conduction, bulk thermoelectric samples should have relative densties of 95–100 %. In this regard, hot-pressing and spark plasma sintering methods are more effective [98, 377].

It has been found that thermal conductivity of the thermoelectric materials, prepred by using the combination of ball milling and hot-pressing sintering, can be effectively reduced, as compared with their single-crystal counterparts [362]. However, the reduction in thermal conductivity of the materials with micrometer grain size is not sufficiently high to achieve high values of *ZT*. For example, the peak *ZT* of SiGe composites consisting of micrometer-sized grains can be increased by only <20 %. In contrast, high-energy ball milling, when combined with hot-pressing, is able to produce thermoelectric composites with nano-sized grains, due to the nano-sized precursors. Bulk thermoelectric nanocomposites prepared by using this combined method include Bi_2Te_3 [98], SiGe alloys [365], and skutterudite $CoSb_3$ [378, 379].

SPS has emerged recently as a rapid method to prepare bulk thermoelectric nanocomposites. Hot-press sintering uses mechanical pressure to assist sintering process, while during SPS process, not only pressure but also an electric field is applied, so that the sintering can be accomplished in a very short-time duration.

4.8 Thermoelectric Effect for Waste Energy Harvesting

4.8.1 Principle

Two main appilcations can be expected by using thermoelectric effect: Seebeck effect thermoelectric generator (TEG) and Peltier effect thermoelectric cooling (TEC). Besides, thermoelectric effect can be used for applications as thermal energy sensors, such as cryogenic heat flux sensor, ultrasonic intensity sensor, detection of water condensation, fluid flow sensor, infrared sensor, and thin film thermoelectrid sensor [380].

Thermoelectric generators or thermogenerators are devices which convert heat (temperature differences) directly into electrical energy, through the phenomenon called the Seebeck effect or thermoelectric effect (Fig. 4.2 right). Thermoelectric generators can be used for a wide range of applications. For example, the harvesting of the waste heat in combustion engines could be a potential application field. Space aircrafts have solar system to make use of the effect in radioisotope thermoelectric generators for electrical power. Solar cells use only the high frequency part of the radiation, while the low frequency heat energy is wasted. The idea is to increase the efficiency of the combined solar-thermoelectric system to convert the solar radiation into useful electricity. Various thermoelectric modules have been assembled for energy recovery of otherwise wasted heat.

Seebeck effect is actually an inverse effect of Peltier effect. When heat is supplied at one side of the junction, an electric current is produced to flow in the circuit and electrical power is thus delivered. In practice, a large number of such thermocouples are connected electrically in series to form a module. Within the module each semiconductors is called a thermoelement and a pair of thermoelements is called a thermocouple.

A typical thermoelectric device is composed of two ceramic substrates that serve as a foundation and electrical insulation for the *p*-type and *n*-type thermoelements that are connected electrically in series and thermally in parallel between the ceramics. Conventional thermoelectric devices have various specifications for

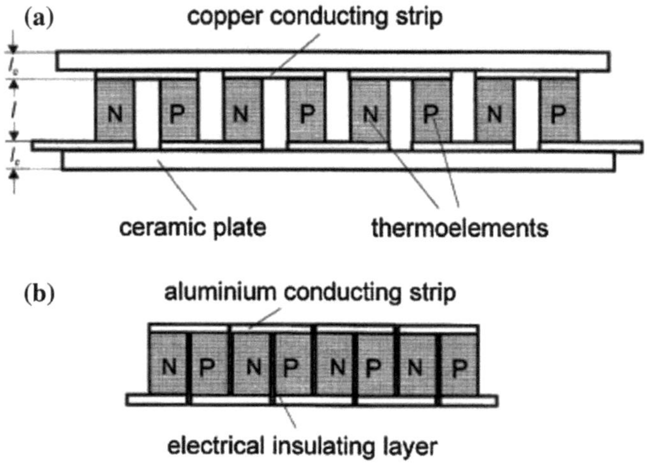

Fig. 4.46 Schematic diagrams of multicouple thermoelectric modules: **a** Type A configuration with ceramic insulating plates and large interthermoelement separation and **b** Type B configuration with very small interthermoelement separation but without ceramic insulating plate. Reproduced with permission from [381], Copyright @ 1998, Elsevier

different applications. Typical dimensions range from $3 \times 3 \times 4$ mm^3 to $60 \times 60 \times 5$ mm^3, with maximum heat-pumping rates of from 1 to 125 W. The maximum temperature difference between the hot and cold side can be up to 70 °C. The number of thermocouples is from 3 to 127. The maximum temperature difference has been up to 130 °C, which is realized by using multistage (cascade) series thermoelectric devices. Because the cold side of the device contracts while the hot side expands, devices with large dimensions of >50 mm^2 suffer from thermally induced stresses, which can cause short at the electrical connection points inside the module. Therefore, this is a challenge for pracatical applications of thermoelectric energy harvestors.

Two types of commercially available multicouple thermoelectric devices are shown in Fig. 4.46 [381]. Type A is designed for cooling applications, with large interthermoelement separation. In this type of devices, *n*-type and *p*-type semiconductor thermoelements are connected electrically in series by highly conductive metal strips and sandwiched between thermally conductive but electrically insulating plates. Type B is for power generation, with much dense structure and very small interthermoelement separation to increase the power density of the devices.

There are two general problems in such thermoelectric devices: high output resistance and adverse thermal characteristics. High output resistance menas that, in order to achieve a high output voltage, a very high Seebeck coefficient is requried (high V °C^{-1}). A common approach is to assemble a number of thermo elements in series, thus leading a generator to have very high effective output resistance of >10 kΩ. Becuase power is only efficiently transferred to loads with high resistance, most of the power is lost across the output resistance. Therefore, a device with too high output impedance can only be used as a temperature sensor, instead of a generator.

Adverse thermal characteristics are attributed to the fact that low thermal conductivity is required for a good thermoelectric generaor, which in turn severely dampens the heat dissipation of the device. In other words, the thermoelectric generator is a poor heat sink. In this respect, a thermoelectric generator is suitable to be integrated with an essential IC chip that requires effective cooling. The low thermal conductivity of a thermoelectric generator lowers the cooling rate.

Thermoelectric cooling uses the Peltier effect to create a heat flux across the junction of two types of materials (Fig. 4.2 left). A Peltier cooler or thermoelectric heat pump is a solid-state active heat-pumping device, which transfers heat from one side of the device to the other, through consumption of electrical energy, depending on the direction of the current. This device has several otherr names, such as Peltier heat pump, solid-state refrigerator, and thermoelectric cooler. The Peltier device is a heat pump, which means that, when direct current runs through it, heat is moved from one side to the other. Therefore it can be used either for heating or for cooling (refrigeration), although in practice the main application is cooling. It can also be used as a temperature controller that either heats or cools.

Theoretically, the amount of heat absorbed at the cold end and the heat dissipated at the hot end are dependent on the product of the Peltier coefficient and the current flowing through the semiconductor thermoelectric materials. Practically, the net amount of heat absorbed at the cold end due to the Peltier effect is decreased by two sources, conducted heat, and Joule heat. Due to the temperature differential between the cold and hot ends of the semiconductor materials, heat will be conducted through the materials from the hot to cold end. As the current is increased, the temperature difference, and thus the conducted heat, increases because the Peltier cooling effect increases. However, the other loss, Joule heat, is proportional to the square of the current and thus eventually becomes the dominant factor. At any given current, thermal equilibrium that is at the hot end is the sum of the net heat absorbed at the cold end plus the applied electric power. The coefficient of performance (COP) used to define the cooling ''efficiency'' is defined as the net heat absorbed at the cold end divided by the applied electric power [380].

The application of thermoelectric effect to refrigeration is still under development, as compared with the vapor-compression refrigeration. Two main reasons are responsible for this situation, high cost and low power efficiency. However, TEC devices have various advantages, such as absence of circulating liquid, small size and flexible shape. Therefore, development of Peltier coolers that are both cheap and efficient is still of practical significances.

A Peltier cooler can also be used as a thermoelectric generator. When operated as a cooler, a voltage is applied across the device, and as a consequence, a difference in temperature is built up between the two sides. When operated as a generator, one side of the device is heated to a temperature that is higher than the other side, and thus a difference in voltage is built up between the two sides (the Seebeck effect). However, a well-designed Peltier cooler should not be a good thermoelectric generator and vice versa, due to their different design and packaging requirements.

Thermoelectric junctions are generally only 5–10 % as efficient as the ideal refrigerator (Carnot cycle), compared with 40–60 % achieved by conventional compression cycle systems (reverse Rankine systems using compression and expansion). Due to the relatively low efficiency, thermoelectric cooling is generally only used in environments where the solid-state nature (no moving parts, maintenance-free, and compact size) is more importanat than the requirement of high efficiency. Peltier (thermoelectric) cooler performance is a function of ambient temperature, hot and cold side heat exchanger (heat sink) performance, thermal load, Peltier module (thermopile) geometry and Peltier electrical parameters.

Thermoelectric devices can not be used independently. They should be connected with heat exchangers to dissipate heat, which consist of thermoelectric systems. The basic theory and operation of thermoelectric systems have been developed. Thermoelectric systems are usually small heat pumps or power generators, which follow the laws of thermodynamics in the same manner as mechanical heat pumps, vapor compressors associated with conventional refrigerators and any other apparatus used to transfer energy.

The thermoelectric devices offer several distinct advantages over other technologies. For example, they have no moving parts, so that they need less maintenance. Thermoelectric devices have long-term stability. In these devices, no chlorofluorocarbons or similar materials are required and thus there is no requirement of periodic replenishment. The direction of heat-pumping in a thermoelectric system is fully reversible. Therefore, changing the polarity of the DC power supply causes heat to be pumped in the opposite direction, i.e., a cooler can then become a heater. Thermoelectric devices can be used in harsh environments where the conventional facilities do not function. Because of these advantages, thermoelectric devices have very bright future and thus deserve to be further studied and developed [380].

4.8.2 Examples of Applications

As discussed above, thermoelectric generators are unique heat engines, in which charge carriers function as working fluids. They have no moving parts and are silent in operation and highly reliable. However, due to their relatively low efficiency of typically about 5 %, they are currently used only in some special applications, such as radioisotope power for deep-space probes, remote power states, deep-sea oil pipelines, and sea buoys, where cost is not a critical requirement. Recent increase in the concern of environmental issue and particular global warming has made the research of thermoelectric effect to be renascent. Thermoelectrics has attracted increasing attention as a green and versatile technology to generator electricity from waste thermal energies [381].

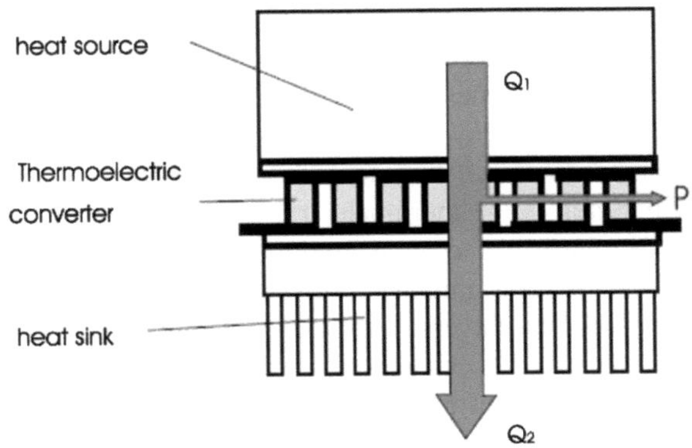

Fig. 4.47 Conventional arrangement for thermoelectric power generation. Q1 is the heat supplied by the heat source. P is the electrical power generated. Q2 is the heat dissipated to the heat sink, which is the thermal energy wasted. Reproduced with permission from [382], Copyright @ 2002, Elsevier

The working principle of thermoelectric energy harvester is shown in Fig. 4.47 [382]. The thermoelectric module is sandwiched between a heat source and a heat sink. Heat from the source flows through the module and then is rejected through the heat sink into the ambient. When a temperature difference is maintained across the module, electrical power can be generated constantly.

4.8.2.1 Low Power Generator

Energy to power small, independent and wireless system for remote sensing, control, safety surveillance, and metering is mostly provided by using batteries. There are a number of disadvantages when using batteries as energy supplier. Firstly, batteries have lifetime, so that the systems have to be maintained or replaced after certain duration of time. Moreover, batteries usually contain chemical substances that are harmful to the environment after they are used and abandoned. Therefore, this kind of systems cannot be disposed directly without any treatment, whereas such treatment precudures introduce additional cost. Although solar cell has been increasingly used to power small electronic devices, such as calculators and watches, if there is no light, it will be a problem. In this respect, thermoelectric generators can be a useful alternative to replace batteries [383]. With a temperature difference of 10 °C, a normal thermoelectric harverstor can generate an electrical power of 1.5 μW, which is sufficiently high to power small devices, like preamplifier and sensor control system.

4.8.2.2 High Power Generators

As mentioned earlier, thermoelectric energy harversters with the currently avaible thermoelectric materials have the problem of low efficiency. This is mainly because the values of dimensionless figure of merit of the state-of-the-art thermoelectric materials are still too low ($ZT \leq 1$). This problem is not very serious for low power generation. In contrast, for high power generation, the low efficiency becomes a severe disadvantage, which has limited their applications in various areas. However, if the supply of heat is cheap or free, for example, waste heat, efficiency of thermoelectric harvesters becomes a less critical requirement, which makes the thermoelectric technology commercially competitive [383].

Similarly, the cost of thermoelectrically harvested electricity also consists of two main part, running cost, and module cost [380]. The running cost is determined by the conversion efficiency, while the module cost is dependent on the total cost of material consumption and manpower expenditure. However, when thermoelectric harvesters are used to convert waste heat to electricity, the running cost can be neglected, as compared with the module cost because there is no cost on fuels. As a result, reducing the module cost becomes the most important objective in thermoelectric power generation from waste heat. The module cost can be reduced by optimizing the device design, improving the manufacture quality and simply by operating the devices under large temperature differences. For instance, the power density can be greatly increased by decreasing the interthermoelement separation.

The potential of waste heat thermoelectric power generation for diesel cycle and gas turbine cogeneration in the manufacturing industrial sector in some region has been assessed [384]. It spreads over different sectors, including chemical product, food processing, oil refining, palm oil mills, petrochemical, pulp and paper rice mills, sugar mills, and textiles. It is shown that gas turbine and diesel cycle cogeneration systems can recover electricity to be 33 and 40 % of fuel input, respectively.

Due to the low efficiency, the thermoelectric harvesters still dissipate a large amount of unconverted heat from their cold side. In order to further improve the effectiveness, a new concept called "symbiotic" has been proposed. In this concept, the thermoelectric generators are uses as a dual function device, heat exchanger, and generator. When heat flows through a thermoelectric generator, part of the heat is converted into electricity, while the rest is collected and reused for preheating, instead of being discharged to the ambient. Figure 4.48 shows a schematic diagram of a thermoelectric symbiotic cogeneration system [382]. In this arrangement, the thermoelectric device is connected to a fluid heater by attaching the cold side of the thermoelectric device to the cold fluid inlet and the hot side to the hot fluid outlet. The main purpose of such a cogeneration system is to produce hot fluid. However, when a thermoelectric device is incorporated, a small portion of heat (Q1–Q3) from the outlet of the fluid heater will flow through a bypass, consisting of the thermoelectric module, which is thus converted to electricity. The heat dissipated from the cold side of the thermoelectric generator (Q2) returns to the fluid heater inlet and preheats the cold fluid [382].

It is understandable that the overall efficiency, consisting of the heat production and theelectricity generation, of a symbiotic generation system is the same as that of heat production by a conventional heating system. In principle, the input energy

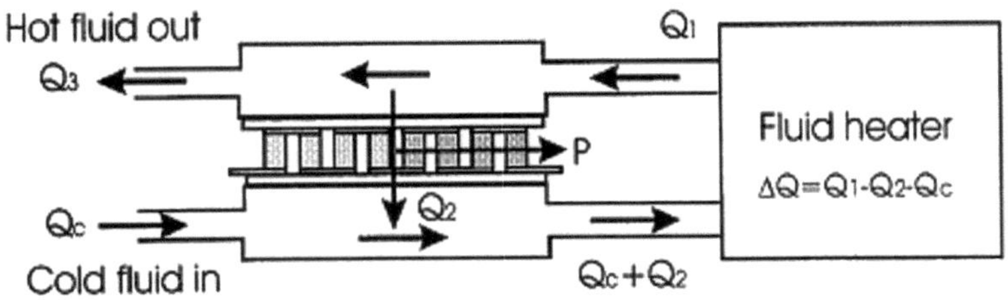

Fig. 4.48 Schematic of thermoelectric generators functioning as fluid preheating/parasite generation devices, with the heat dissipated from the thermoelectric generator being reused to preheat the fluid, i.e., Q2 is not wasted. Reproduced with permission from [382], Copyright @ 2002, Elsevier

can be completely utilized through the conversion into heat and electricity, i.e., 100 % efficiency. In this case, the low conversion efficiency problem of thethermoelectric harvesters does not matter, because such a system can generate heat and electricity simultaneously, with negligible or no loss in terms of overall efficiency of the system. This makes such systems especially potential for applications with the requirement of both electricity and heat. Typical applications are domestic central heating boilers and conventional fuel or biomass combustion units. These systems require fuel supply for heat production and electricity supply for powering pumps, fans or control panels. With the symbiotic thermoelectric systems, only the supply of fuel is necessary for energy input, whereas electricity can be generated in situ. This is a significant advantage for applications where an electricity supply is not available or unreliable, due to severe weather and environmental conditions [382].

Solar energy can be considered to be a special type of waste energies, which is the most promising energy resources on earth and in space, because it is clean and inexhaustible. Solar thermoelectric energy haversters have been proposed and developed. Solar thermoelectric generators are usually a combination of a solar thermal collector and a thermoelectric generator. To improve the energy conversion efficiency, various strategies have been applied. One example is solar-powered high-efficiency thermionic/thermoelectric conversion systems [385]. The system combines a thermionic converter (TIC) with a thermoelectric generator, to increase the thermal energy collection and high-efficiency conversion to electricity. The TIC emitter must uniformly heated up to 1,800 K, by using thermal radiation from a solar receiver. A cylindrical cavity-type solar receiver is constructed with graphite and is heated in vacuum by using the solar concentrator as shown in Fig. 4.49 [385]. The maximum temperature of the solar receiver that is enclosed by a molybdenum cup can be 1965 K, which is thus sufficiently high to heat the TIC emitter using thermal radiation from the receiver. The system has a conversion efficiency of about 40 %.

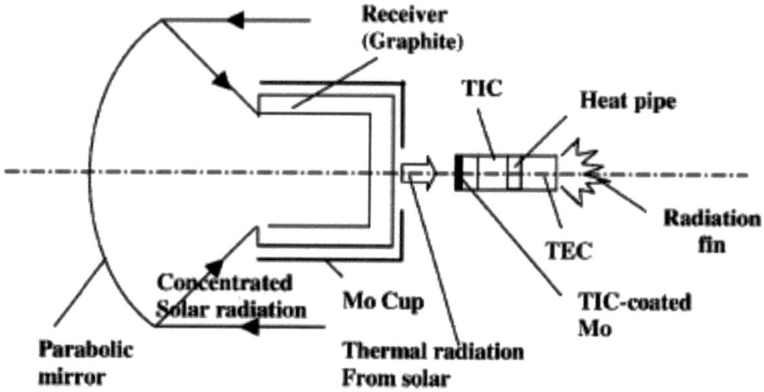

Fig. 4.49 Schematic view of a solar-powered conversion system. Reproduced with permission from [385], Copyright @ 2003, Elsevier

4.9 Challenge in Thermoelectric Nanocomposites

There have been several types of structures that can be described by using the term of nanocomposite, which can be either nanoparticles embedded in a host matrix or nanoparticles of different materials forming heterostructure geometries [297, 386, 387]. The nanocomposites with heterostructure geometries ususlly consist of grains with nanometer sizes, such as $Si_{80}Ge_{20}$ nanocomposite [388]. Its grain size is 10–20 nm, which is much smaller than that of normal $Si_{80}Ge_{20}$ bulk alloys, >1 μm [362]. The size of the nanograins is smaller than the phonon mean free path while larger than the electron (hole) mean free path, so that phonons are strongly scattered by the interfaces while the transports of electrons or holes are not affected. As a result, their thermal conductivity is significantly reduced while electrical conductivity is maintained a high value, thus leading to an increase in *ZT*.

Although significant progress has made in developing nanocomposites with enhanced *ZT* by reducing the thermal conductivity of the bulk material, the values of *ZT* are not sufficiently high for large-scale applications. However, the state-of-the-art processing is base on trial and error, which becomes a bottleneck of the development of thermoelectric materials. Therefore, it is imperative to have more controlled design processes, with which desired thermoelectric properties can be achieved by designing nanocomposites with necessary nanostructures. The prerequisit is fundamental understanding of the relationship between transport characteristics and the nanoscale structures of the materials. For example, as the grains are continuously decreased, the volume fraction of grain boundaries increases. Once the volume fraction of grain boundaries is comparable or even higher than that of the grains, the grain boundaries should be treated as another phase. In this case, the function of the grain boundaries is not just scattering but also contributing to transport.

4.9.1 Phonon Transport

It has been recognized that the high values of *ZT* of most nanocomposites are attributed to the reduction in lattice thermal conductivity. However, the phonon transport characteristics in these nanocomposites are still not clearly understood. It is also a question as to how the thermal conductivity in nanocomposites is reduced due to the phonon scattering by the interfaces, such as grain boundaries. This situation has made some theoretical predictions of thermal conductivity in nanostructured to be incorrect, because ultra lower thermal conductivity than theoretical limit in 2D nanostructures has been been observed [389].

Bulk materials have a limit in thermal conductivity that can be achieved by using alloys, which is called alloy limit. Such a limit has been observed in varous alloys, such as $Si_{1-x}Ge_x$, which is broken through by forming nanocomposites [364, 388]. This achievement can be readily explained in terms of phonon scattering [390]. As mentioned earlier, phonons in materials have a spectrum of wavelengths, as shown in Fig. 4.50, which have different contributions to total thermal conductivity. By forming alloys, impurity atoms are introduced, which scatter those phonons with short wavelength. However, the mid-to long-wavelength phonons are almost not affected, which are still able to transport heat. Because the short wavelength phonon has relatively low contribution to thermal conductivity, reduction in thermal conductivity by using alloys is quite limited [387, 391]. To further decrease the thermal conductivity, it is important to creta structures with longer characteristic lengthes than that of atomic impurities, which leads to interface scattering mechanism to scatter mid- to long-wavelength phonons.

This is the underlying reason why nanocomposites are able to breakthrough the alloy limit. Phonon models by taking into account of the interface scattering effect have been established, based on the Callaway theory of lattice thermal conductivity [392, 393]. These models can be used to explain the thermal conductivity of $In_{0.53}Ga_{0.47}As$/ErAs superlattice nanostructure [318] and Si nanowires [299] very well. However, it is found that although the simple Callaway model can be adjusted to fit the total thermal conductivity, it cannot predict quantitative parameters, such as phonon mean free path and wavelength, even for bulk materials. In other words, phonon transport characteristics of bulk thermoelectric materials are not well understood. Recent studies indicate that the unusually low thermal conductivity in bulk cubic $AgSbTe_2$ is due to the extremely anharmonic bonds [394].

Compared with normal microsized bulk alloys, nanocomposites contain much more interfaces with thickness that is smaller than the phonon mean free path. These interfaces provide a thermal boundary resistance between every two adjacent regions within the nanocomposite. Accurate prediction of such thermal boundary resistance becomes an important subject of research on nanostructured thermoelectric materials [395–397]. Phonon transmissivity at a single interface is a fundamental parameter in predicting thermal boundary resistance, which is still not

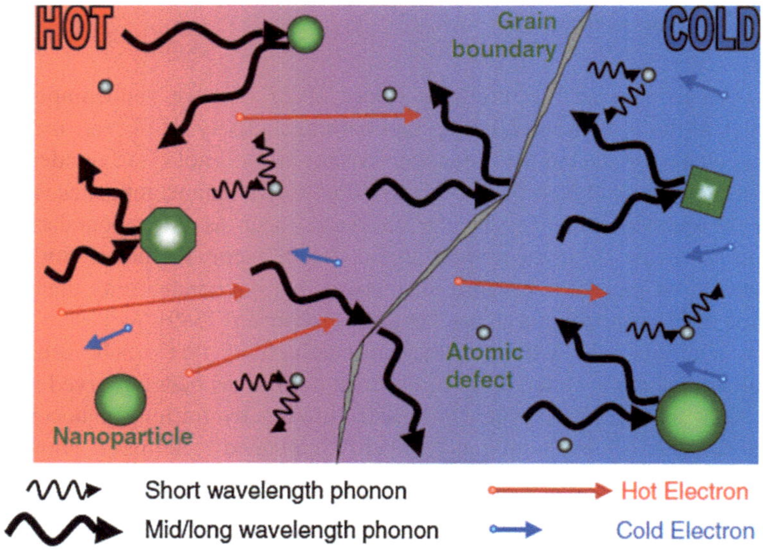

Fig. 4.50 Schematic diagram illustrating various phonon scattering mechanisms within a thermoelectric material, along with electronic transport of hot and cold electrons. Atomic defects are effective to scatter short wavelength phonons, whereas larger embedded nanoparticles are required to scatter mid-to long-wavelength phonons. Grain boundaries are also found to play an effective role in scattering the longer wavelength phonons. Reproduced with permission from [390], Copyright @ 2009, John Wiley & Sons

predictable by any available model. Nanoparticles within nanocomposites have multiple interfaces, which should be taken into account when discussing their contributions to phonon scattering. The current models do not include effects of interfacial roughness [318, 398], while the computation power is not sufficient to support the molecular dynamics to deal with nanoparticles [399].

There are several questions that currently cannot be answered for thermoelectric materials, which has hindered the progress in further increasing the value of *ZT* by reducing the thermal conductivity. For example, it is not clear which phonon modes dominate the phonon transport in nanocomposites, what are the optimal size and size distribution of the nanostructures, what is the relationship between interface properties and phonon scattering, and so on.

The thermal conductivity limit of bulk materials has been determined by using the criterion that the phonon mean free path must be at least half the phonon wavelength [235]. However, this criterion cannot be simply applied to nanocomposites. Nevertheless, the minimum thermal conductivity of nanostructures can be lower than that of bulk material counterparts, because the theory of minimum thermal conductivity of bulk materials considers only isotropic scatterings, whereas the interfacial scatterings in nanostructured materials are highly anisotropic [387]. For instance, the thermal conductivity of layered WSe_2 is lower than the theoretical value by six times, which means that the factors to influence the phonon transport of nanostructures are not entirely accounted for by the conventional theories [389]. Therefore, it is urgent to develop new theoretical models.

More recently, a model based on nonequilibrium Green's function for phonons has been developed to predict the phonon transport through nanotubes with defects, which allows the thermal conductivity to be calculated without using any adjustable parameters [400, 401]. Another example is the molecular dynamics simulation, which can be used to calculate the thermal conductivity of various structured thermoelectric nanocomposites [402, 403]. Experimental progress includes the capability to probe the phonon distribution by using a variable frequency time-domain thermoreflectance technique [404]. With these significant achievements, is confident to expect that the lattice thermal conductivity of the current thermoelectric materials can be further reduced.

4.9.2 Electronic Transport

The qualitative or even semi-quantitative understanding in certain aspests on phonon transport has been used as guidance to enhance thermoelectric performance by reducing thermal conductivity through nanostructuring. Comparatively, electronic transport is much less well understood even in bulk materials. In most cases, the reduction in thermal conductivity is usually accompanied by a slight decrease in electrical conductivity. This could be one of the barriers to further increase the value of *ZT* of thermoelectric nanocomposites.

The electron transport is relatively complicated and thus there is no theory that is able to describe it. One of the reasons is that most of the characteristic lengths in thermoelectric materials are smaller than the de Broglie wavelength of electron, which means that the standard analysis with the Boltzmann equation would not be valid entirely [387]. For example, in heavily doped $Si_{80}Ge_{20}$ consolidated nanocomposite with an average grain size of 10–20 nm, the electron wavelength is 5–10 nm, electron mean free path is 2–5 nm and interimpurity distance is 1–2 nm [388]. Therefore, the electron wavelength is quite comparable to the grain size and larger than its mean free path and the interimpurity distance. According to the Boltzmann equation, the mean free path should be much larger than the wavelength. As a consequence, predictions with the Boltzmann equation would lead to large errors. Also, thermoelectric materials usually require very high concentration of dopant and thus their electrical conductivities could be much higher than that of semiconductors while close to those of metals. However, most theoretical models are derived with the assumption of low level doping.

The presence of various interfaces or interfacial microstructures makes it even difficult to describe the electron transport of thermoelectric nanocomposites. It has once been predicted that nanocomposites would not possess high value of *ZT* because their electrical conductivity should be greatly decreased due to the interface scattering of electron [405, 406]. However, if nanocomposites consist of components with similar band structures or even same materials, interfacial scattering of electrons can be minimizes, due to the absence of bandedge offset [362]. With this understanding, it is expected to develop more accurate theoretical models to predict or describe the electronic transport of thermoelectric materials with great improvement in performances.

4.10 Concluding Remarks

Thermoelectric electrical generators are solid-state devices that convert thermal energy (heat) directly into electricity by using a temperature gradient based on "Seebeck effect". The thermoelectric power cycle, in which charge carriers (electrons) act as the working fluid, abide by the fundamental laws of thermodynamics, so that it is analogous to the power cycle of the conventional heat engines. Several distinct advantages can be listed for thermoelectric power generators. Firstly, they have extremely high reliability, with exceeding 100,000 hours of steady–state operation and thus require considerably less maintenance. Because they have no mechanical moving parts, they are silent during operation. Secondly, they are suitable for applications at small–scales in remote rural areas, where electricity is difficult to supply. Also, they are simple and compact in structure and safe to human being and the environment.

For thermoelectric powder generation, the key prerequisite is the availability of thermoelectric materials with as high as possible figure of merit, *ZT*. During the early study of thermoelectric effect, efforts have been mainly on the study crystalline single-phase materials with narrow band gaps, containing heavy elements, and presence of point defects that are formed through the preparation of solid solutions (e.g., $Bi_{2-x}Sb_xTe_3$). The creation of defects in solid solutions has been considered to be the key to achieving low thermal conductivity, by using the phonon scattering behavior of the defects. The more recent discovery of stable nanostructures in LAST-m and its derivatives, together with the achievement in the reduction of thermal conductivity, has changed our views on the development of thermoelectric materials. While finding and synthesizing new thermoelectric materials with higher values of *ZT* are still the direction of research in this area, it is worth mentioning that nanostructured multiphase systems could be a promising strategy toward high *ZT* by using the materials that are currently available.

In the concept of nanostructuring to improve the thermoelectric performance, the main contribution is the significant decrease in lattice thermal conductivity, rather than the increase in electrical power factor. This is the reason why the presence of nanodots has played has become a very effective way to reduce the lattice thermal conductivity, because they can scatter the phonons that have relatively long mean free paths. It has been clearly demonstrated that the contribution to the increment of *ZT* dur to reduction in thermal conductivity far exceeds the concomitant reduction in power factor caused by the scattering of electronic carriers.

Further strategies beyond nanostructuring is the coupling of other mechanisms, for example, that of resonance states in the PbTe:Tl system. This is because resonance states can be used to increase the Seebeck coefficient. The use of complex nanostructures could be another way, because it allows for the changes in carrier mobility, as demonstrated in the conanostructured PbTe–Pb–Sb system. More importantly, theoretical understanding, together with new synthesis techniques and state-of-the-art measurements, will be indispensable for this field to be further advanced.

Currently, it is not a difficult task to achieve $ZT = 1.6–1.7$ by using bulk materials at temperatures of 700–800 K. These values are still far below the desired limit value of $ZT = 3$, for widespread practical applications of this technology. It is expected that further reduction in the thermal conductivity alone may be able to raise to ZT value to 2. However, to reach $ZT = 3$ or even high, the power factor must be enhanced at the same time, rather than decreased, which remains a severe challenge.

References

1. Website: http://www.energyefficiencyasia.org/docs/ee_modules/Chapter - Waste%20Heat%20Recovery.pdf
2. C. Wood, Materials for thermoelectric energy conversion. Rep. Prog. Phys. **51**, 459–539 (1988)
3. J.R. Sootsman, D.Y. Chung, M.G. Kanatzidis, New and old concepts in thermoelectric materials. Angew. Chem. Int. Ed. **48**, 8616–9639 (2009)
4. P. Pichanusakorn, P. Bandaru, Nanostructured thermoelectric. Mater. Sci. Eng. R **67**, 19–63 (2010)
5. R. Amatya, R.J. Ram, Trend for thermoelectric materials and their earth abundance. J. Electron. Mater. **41**, 1011–1019 (2012)
6. G. Chen, M.S. Dresselhaus, G. Dresselhaus, J.P. Fleurial, T. Caillat, Recent developments in thermoelectric materials. Int. Mater. Rev. **48**(1), 54–66 (2003)
7. C.J. Vineis, A. Shakouri, A. Jajumdar, M.G. Kanatzidis, Nanostructured thermoelectric: big efficiency gains from small features. Adv. Mater. **22**, 3970–3980 (2010)
8. Y. Lan, A.J. Ninnich, G. Chen, Z. Ren, Enhancement of thermoelectric figure-of-merit by a bulk nanostructuring approach. Adv. Funct. Mater. **20**, 357–376 (2010)
9. S.K. Bux, J.P. Fleurial, R.B. Kaner, Nanostructured materials for thermoelectric applications. Chem. Commun. **46**, 8311–8324 (2010)
10. B.C. Sales, D. Mandrus, B.C. Chakoumakos, V. Keppens, J.R. Thompson, Filled skutterudite antimonides: electron crystals and phonon glasses. Phys. Rev. B **56**, 15081–15089 (1997)
11. G.S. Nolas, D.T. Morelli, T.M. Tritt, Skutterudites: a phonon-glass-electron crystal approach to advanced thermoelectric energy conversion applications. Annu. Rev. Mater. Sci. **29**, 89–116 (1999)
12. B.C. Sales, D. Mandrus, R.K. Williams, Filled skutterudite antimonides: a new class of thermoelectric materials. Science **272**, 1325–1328 (1999)
13. G.S. Nolas, M. Kaeser, R.T. Littleton, T.M. Tritt, High figure of merit in partially filled ytterbium skutterudite materials. Appl. Phys. Lett. **77**, 1855–1857 (2000)
14. X. Tang, L. Chen, T. Goto, T. Hirai, Effects of Ce filling fraction and Fe content on the thermoelectric properties of Co-rich $Ce_yFe_xCo_{4-x}Sb_{12}$. J. Mater. Res. **16**, 837–843 (2001)

15. D.T. Morelli, G.P. Meisner, B. Chen, S. Hu, C. Uher, Cerium filling and doping of cobalt triantimonide. Phys. Rev. B **56**, 7376–7383 (1997)
16. L.D. Chen, T. Kawahara, X.F. Tang, T. Goto, T. Hirai, J.S. Dyck, W. Chen, C. Uher, Anomalous barium filling fraction and *n*-type thermoelectric performance of $Ba_yCo_4Sb_{12}$. J. Appl. Phys. **90**, 1864–1868 (2001)
17. X. Tang, Q. Zhang, L. Chen, T. Goto, T. Hirai, Synthesis and thermoelectric properties of *p*-type- and *n*-type-filled skutterudite $R_yM_xCo_{4-x}Sb_{12}$ (R: Ce, Ba, Y; M: Fe, Ni). J. Appl. Phys. **97**, 093712 (2005)
18. M. Puyet, A. Dauscher, B. Lenoir, M. Dehmas, C. Stiewe, E. Muller, J. Hejtmanek, Beneficial effect of Ni substitution on the thermoelectric properties in partially filled $Ca_yCo_{4-x}Ni_xSb_{12}$ skutterudites. J. Appl. Phys. **97**, 083712 (2005)
19. J.S. Dyck, W. Chen, C. Uher, L. Chen, X. Tang, T. Hirai, Thermoelectric properties of the *n*-type filled skutterudites $Ba_{0.3}Co_4Sb_{12}$ doped with Ni. J. Appl. Phys. **91**, 3698–3705 (2002)
20. M. Puyet, B. Lenoir, A. Dauscher, M. Dehmas, C. Stiewe, E. Muller, High temperature transport properties of partially filled $Ca_xCo_4Sb_{12}$ Skutterudites. J. Appl. Phys. **95**, 4852–4855 (2004)
21. G.A. Slack, V.G. Tsoukala, Some properties of semiconducting $IrSb_3$. J. Appl. Phys. **76**, 1665–1671 (1994)
22. G.S. Nolas, M. Beekman, J. Gryko, G.A. Lamberton Jr, T.M. Tritt, P.F. McMillan, Thermal conductivity of elemental crystalline silicon clathrate Si_{136}. Appl. Phys. Lett. **82**, 910–912 (2003)
23. M. Beekman, G.S. Nolas, Inorganic clathrate-II materials of group 14: synthetic routes and physical properties. J. Mater. Chem. **18**, 842–851 (2008)
24. N.P. Blake, S. Latturner, J.D. Bryan, G.D. Stucky, H. Metiu, Band structures and thermoelectric properties of the clathrates $Ba_8Ga_{16}Ge_{30}$, $Sr_8Ga_{16}Ge_{30}$, $Ba_8Ga_{16}Si_{30}$, and $Ba_8In_{16}Sn_{30}$. J. Chem. Phys. **115**, 8060 (2001)
25. G.S. Nolas, J.L. Cohn, G.A. Slack, S.B. Schujman, Semiconducting Ge clathrates: promising candidates for thermoelectric applications. Appl. Phys. Lett. **73**, 178–180 (1998)
26. V.L. Kuznetsov, L.A. Kuznetsova, A.E. Kaliazin, D.M. Rowe, Preparation and thermoelectric properties of $A_8^{II}B_{16}^{III}B_{30}^{IV}$ clathrate compounds. J. Appl. Phys. **87**, 7785–7871 (2000)
27. A. Saramat, G. Svensson, A.E.C. Palmqvist, C. Stiewe, E. Mueller, D. Platzek, S.G.K. Williams, D.M. Rowe, J.D. Bryan, G.D. Stucky, Large thermoelectric figure of merit at high temperature in Czochralski-grown clathrate $Ba_8Ga_{16}Ge_{30}$. J. Appl. Phys. **99**, 023708 (2006)
28. J.-H. Kim, N.L. Okamoto, K. Kishida, K. Tanaka, H. Inui, High thermoelectric performance of type-III clathrate compounds of the Ba-Ge-Ga system. Acta Mater. **54**, 2057–2062 (2006)
29. A. Bentien, V. Pacheco, S. Paschen, Y. Grin, F. Steglich, Transport properties of composition tuned alpha- and beta-$Eu_8Ga_{16-x}Ge_{30+x}$. Phys. Rev. B **71**, 165206 (2005)
30. A. Bentien, M. Christensen, J.D. Bryan, A. Sanchez, S. Paschen, F. Steglich, G.D. Stucky, B.B. Iversen, Thermal conductivity of thermoelectric clathrates. Phys. Rev. B **69**, 045107 (2004)
31. G.K.H. Madsen, K. Schwarz, P. Blaha, D.J. Singh, Electronic structure and transport in type-I and type-VIII clathrates containing strontium, barium, and europium. Phys. Rev. B **68**, 125212 (2003)
32. S.J. Poon, Electronic and thermoelectric properties of half-Heusler alloys. Semicond. Semimetals **70**, 37–75 (2001)
33. S. Ogut, K.M. Rabe, Band gap and stability in the ternary intermetallic compounds NiSnM (M = Ti, Zr, Hf): a first-principles study. Phys. Rev. B **51**, 10443–10453 (1995)
34. F.G. Aliev, N.B. Brandt, V.V. Moschalkov, V.V. Kozyrkov, R.V. Scolozdra, A.I. Belogorokhov, Gap at the Fermi level in the intermetallic vacancy system RBiSn (R = Ti, Zr, Hf). Z. Phys. B **75**, 167–171 (1989)
35. J. Tobola, J. Pierre, S. Kaprzyk, R.V. Skolozdra, M.A. Kouacou, Crossover from semiconductor to magnetic metal in semi-Heusler phases as a function of valence electron concentration. J. Phys.: Condens. Matter **10**, 1013–1032 (1998)

36. C. Uher, J. Yang, S. Hu, D.T. Morelli, G.P. Meisner, Transport properties of pure and doped MNiSn (M=Zr, Hf). Phys. Rev. B **59**, 8615–8621 (1999)
37. Y. Xia, S. Bhattacharya, V. Ponnambalam, A.L. Pope, S.J. Poon, T.M. Tritt, Thermoelectric properties of semimetallic (Zr, Hf)CoSb half-Heusler phases). J. Appl. Phys. **88**, 1952–1955 (2000)
38. S. Bhattacharya, A.L. Pope, R.T. Littleton IV, T.M. Tritt, V. Ponnambalam, Y. Xia, S.J. Poon, Effect of Sb doping on the thermoelectric properties of Ti-based half-Heusler compounds, $TiNiSn_{1-x}Sb_x$. Appl. Phys. Lett. **77**, 2476–2478 (2000)
39. H. Hohl, A.P. Ramirez, C. Goldmann, G. Ernst, B. Wolfing, E. Bucher, Efficient dopants for ZrNiSn-based thermoelectric materials. J. Phys. Condens. Matter **11**, 1697–1709 (1999)
40. S. Sakurada, N. Shutoh, Effect of Ti substitution on the thermoelectric properties of (Zr, Hf)NiSn half-Heusler compounds. Appl. Phys. Lett. **86**, 082105 (2005)
41. Q. Shen, L. Chen, T. Goto, T. Hirai, J. Yang, G.P. Meisner, C. Uher, Effects of partial substitution of Ni by Pd on the thermoelectric properties of ZrNiSn-based half-Heusler compounds. Appl. Phys. Lett. **79**, 4165–4167 (2001)
42. J. Yang, H. Li, T. Wu, W. Zhang, L. Chen, J. Yang, Evaluation of half-Heusler compounds as thermoelectric materials based on the calculated electrical transport properties. Adv. Funct. Mater. **18**, 2880–2888 (2008)
43. J. Yang, G.P. Meisner, L. Chen, Strain field fluctuation effects on lattice thermal conductivity of ZrNiSn-based thermoelectric compounds. Appl. Phys. Lett. **85**, 1140–1142 (2004)
44. S.R. Culp, S.J. Poon, N. Hickman, T.M. Tritt, J. Blumm, Effect of substitutions on the thermoelectric figure of merit of half-Heusler phases at 800 °C. Appl. Phys. Lett. **88**, 042106 (2006)
45. P. Larson, S.D. Mahanti, S. Sportouch, M.G. Kanatzidis, Electronic structure of rare-earth nickel pnictides: narrow-gap thermoelectric materials. Phys. Rev. B **59**, 15660–15668 (1999)
46. H.W. Mayer, I. Mikhail, K. Schubert, Über einige phasen der Mischungen $ZnSb_N$ und $CdSb_N$. J. Less-Common Met. **59**, 43–52 (1978)
47. G.J. Snyder, M. Christensen, E. Nishibori, T. Caillat, B.B. Iversen, Disordered zinc in Zn_4Sb_3 with phonon-glass and electron-crystal thermoelectric properties. Nat. Mater. **3**, 458–463 (2004)
48. F. Cargnoni, E. Nishibori, P. Rabiller, L. Bertini, G.J. Snyder, M. Christensen, C. Gatti, B.B. Iversen, Interstitial Zn atoms do the trick in thermoelectric zinc antimonide, Zn_4Sb_3: a combined maximum entropy method *X*-ray electron density and *ab initio* electronic structure study. Chem. Eur. J. **10**, 3861–3870 (2004)
49. T. Caillat, J.P. Fleurial, A. Borshchevsky, Preparation and thermoelectric properties of semiconducting Zn_4Sb_3. J. Phys. Chem. Solids **58**, 1119–1125 (1997)
50. M. Tsutsui, L.T. Zhang, K. Ito, M. Yamaguchi, Effects of in-doping on the thermoelectric properties of beta-Zn_4Sb_3. Intermetallics **12**, 809–813 (2004)
51. V.L. Kuznetsov, D.M. Rowe, Solid solution formation in the Zn_4Sb_3-Cd_4Sb_3 system. J. Alloys Compd. **372**, 103–106 (2004)
52. S.R. Brown, S.M. Kauzlarich, F. Gascoin, G.J. Snyder, $Yb_{14}MnSb_{11}$: new high efficiency thermoelectric material for power generation. Chem. Mater. **18**, 1837–1873 (2006)
53. O. Yamashita, N. Sadatomi, Thermoelectric properties of $Si_{1-x}Ge_x$ ($x \leq 0.10$) with alloy and dopant segregations. J. Appl. Phys. **88**, 245 (2000)
54. I.R. Fisher, S.L. Bud'ko, C. Song, P.C. Canfield, T.C. Ozawa, S.M. Kauzlarich, $Yb_{14}ZnSb_{11}$: charge balance in Zintl compounds as a route to intermediate Yb valence. *Phys. Rev. Lett.* **85**, 1120–1123 (2000)
55. S.M. Kauzlarich, S.R. Brown, G.J. Snyder, Zintl phases for thermoelectric devices. Dalton Trans. **2007**, 2099–2107 (2007)
56. A. Akrap, N. Barisic, L. Forro, D. Mandrus, B.C. Sales, High-pressure resistivity and thermoelectric power in $Yb_{14}MnSb_{11}$. Phys. Rev. B **76**, 085203 (2007)

57. A. Bentien, S. Johnsen, G.K.H. Madsen, B.B. Iversen, F. Steglich, Colossal Seebeck coefficient in strongly correlated semiconductor $FeSb_2$. EPL **80**, 17008 (2007)
58. A. Bentien, G.K.H. Madsen, S. Johnsen, B.B. Iversen, Experimental and theoretical investigations of strongly correlated $FeSb_{2-x}Sn_x$. Phys. Rev. B **74**, 205105 (2006)
59. D.A. Wright, Thermoelectric properties of bismuth telluride and its alloys. Nature **181**, 834 (1958)
60. M.W. Oh, D.M. Wee, S.D. Park, B.S. Kim, H.W. Lee, Electronic structure and thermoelectric transport properties of AgTlTe: first-principles calculations. Phys. Rev. B **77**, 165119 (2008)
61. B. Wolfing, C. Kloc, J. Teubner, E. Bucher, High performance thermoelectric Tl_9BiTe_6 with an extremely low thermal conductivity. Phys. Rev. Lett. **86**, 4350–4353 (2001)
62. J.W. Sharp, B.C. Sales, D.G. Mandrus, Thermoelectric properties of Tl_2SnTe_5 and Tl_2GeTe_5. Appl. Phys. Lett. **74**, 3794–3796 (1999)
63. D. Paccard, L. Paccard, G. Brun, J.C. Tedenac, A new phase in the Ti-Ag-Te system: crystal structure of $Tl_2Ag_{16}Te_{11}$. J. Alloys Compd. **184**, 337–342 (1992)
64. K. Kurosaki, A. Kosuga, H. Muta, M. Uno, S. Yamanaka, Ag_9TlTe_5: a high-performance thermoelectric bulk material with extremely low thermal conductivity. Appl. Phys. Lett. **87**, 061919 (2005)
65. E.M. Godzhaev, R.A. Kerimova, Thermoelectric properties of $TlIn_{1-x}Yb_xTe_2$ ($0.01 \leq x \leq 0.09$) crystals. Inorg. Mater. **40**, 1153–1155 (2004)
66. T.J. McCarthy, T.A. Tanzer, M.G. Kanatzidis, A new metastable three-dimensional bismuth sulfide with large tunnels: synthesis, structural characterization, ion-exchange properties, and reactivity of KBi_3S_5. J. Am. Chem. Soc. **117**, 1294–1301 (1995)
67. M.G. Kanatzidis, T.J. McCarthy, T.A. Tanzer, L.H. Chen, L. Iordanidis, T. Hogan, C.R. Kannewurf, C. Uher, B. Chen, Synthesis and thermoelectric properties of the new ternary bismuth sulfides $KBi_{6.33}S_{10}$ and $K_2Bi_8S_{13}$. Chem. Mater. **8**, 1465–1474 (1996)
68. B.X. Chen, C. Uher, L. Iordanidis, M.G. Kanatzidis, Transport properties of Bi_2S_3 and the ternary bismuth sulfides $KBi_{6.33}S_{10}$ and $K_2Bi_8S_{13}$. Chem. Mater. **9**, 1655–1658 (1997)
69. T.J. McCarthy, S.P. Ngeyi, J.H. Liao, D.C. Degroot, T. Hogan, C.R. Kannewurf, M.G. Kanatzidis, Molten salt synthesis and properties of three new solid-state ternary bismuth chalcogenide, β-$CsBiS_2$, γ-$CsBiS_2$, and $K_2Bi_8Se_{13}$. Chem. Mater. **5**, 331–340 (1993)
70. D.Y. Chung, K.S. Choi, L. Iordanidis, J.L. Schindler, P.W. Brazis, C.R. Kannewurf, B.X. Chen, S.Q. Hu, C. Uher, M.G. Kanatzidis, High thermopower and low thermal conductivity in semiconducting ternary K-Bi-Se compounds: synthesis and properties of β-$K_2Bi_8Se_{13}$ and $K_{2.5}Bi_{8.5}Se_{14}$ and their Sb analogues. Chem. Mater. **9**, 3060–3071 (1997)
71. L. Iordanidis, M.G. Kanatzidis, Redox-induced "zipper" action in $Rb_2Bi_4Se_7$ and $Cs_2Bi_4Se_7$: coupling of slabs to a three-dimensional framework through single-crystal to single-crystal conversion. Angew. Chem. Int. Ed. **39**, 1927–1930 (2000)
72. D.Y. Chung, S. Jobic, T. Hogan, C.R. Kannewurf, R. Brec, J. Rouxel, M.G. Kanatzidis, Oligomerization versus polymerization of Te_x^{n-} in the polytelluride compound $BaBiTe_3$: structural characterization, electronic structure, and thermoelectric properties. J. Am. Chem. Soc. **119**, 2505–2515 (1997)
73. D.Y. Chung, T. Hogan, P. Brazis, M. Rocci-Lane, C. Kannewurf, M. Bastea, C. Uher, M.G. Kanatzidis, $CsBi_4Te_6$: a high-performance thermoelectric material for low-temperature applications. Science **287**, 1024–1027 (2000)
74. L. Iordanidis, J.L. Schindler, C.R. Kannewurf, M.G. Kanatzidis, $ALn_{1\pm}xBi_{4\pm}xS_8$ (A = K, Rb; Ln = La, Ce, Pr, Nd): new semiconducting quaternary bismuth sulfides. J. Solid State Chem. **143**, 151–162 (1999)
75. K.S. Choi, L. Iordanidis, K. Chondroudis, M.G. Kanatzidis, $KThSb_2Se_6$ and $BaLaBi_2Q_6$ (Q = S, Se) adopt a new structure type stabilized with dichalcogenide groups. Inorg. Chem. **36**, 3804–3805 (1997)
76. D.Y. Chung, L. Iordanidis, K.K. Rangan, P.W. Brazis, C.R. Kannewurf, M.G. Kanatzidis, First quaternary A-Pb-Bi-Q (A = K, Rb, Cs; Q = S, Se) compounds: synthesis, structure,

and properties of α- and β-CsPbBi3Se6, APbBi3Se6, (A = K, Rb), and APbBi3S6 (A = Rb, Cs). Chem. Mater. **11**, 1352–1362 (1999)

77. A. Mrotzek, D.Y. Chung, T. Hogan, M.G. Kanatzidis, Structure and thermoelectric properties of the new quaternary tin selenide $K_{1-x}Sn_{5-x}Bi_{11+x}Se_{22}$. J. Mater. Chem. **10**, 1667–1672 (2000)
78. K.S. Choi, D.Y. Chung, A. Mrotzek, P. Brazis, C.R. Kannewurf, C. Uher, W. Chen, T. Hogan, M.G. Kanatzidis, Modular construction of A1+xM4-2xM'7+xSe15 (A = K, Rb; M = Pb, Sn; M' = Bi, Sb): a new class of solid state quaternary thermoelectric compounds. Chem. Mater. **13**, 756–764 (2001)
79. K. Adouby, C. Perez Vicente, J.C. Jumas, R. Fourcade, A. Abba Toure, Structure and temperature transformation of SnSe: stabilization of a new cubic phase $Sn_4Bi_2Se_7$. *Z. Kristallogr.* **213**, 343–349 (1998)
80. C. Perez Vicente, J.L. Tirado, K. Adouby, J.C. Jumas, A. Abba Toure, G. Kra, X-ray Diffraction and ^{119}Sn mössbauer spectroscopy study of a new phase in the Bi_2Se_3-SnSe system: $SnBi_4Se_7$. *Inorg. Chem.* **38**, 2131–2135 (1999)
81. W. Choe, S. Lee, P. O'Connell, A. Covey, Synthesis and structure of new Cd-Bi-S homologous series: a study in intergrowth and the control of twinning patterns. Chem. Mater. **9**, 2025–2030 (1997)
82. Y.C. Wang, F.J. DiSalvo, Synthesis and characterization of $Ba_3Bi_{6.67}Se_{13}$ and its filled variants $Ba_3Bi_6PbSe_{13}$ and $Ba_3Bi_6SnSe_{13}$. Chem. Mater. **12**, 1011–1017 (2000)
83. F.J. DiSalvo, Thermoelectric cooling and power generation. Science **285**, 703–706 (1999)
84. T. Kyratsi, J.S. Dyck, W. Chen, D.Y. Chung, C. Uher, K.M. Paraskevopoulos, M.G. Kanatzidis, Highly anisotropic crystal growth and thermoelectric properties of $K_2Bi_{8-x}Sb_xSe_{13}$ solid solutions: band gap anomaly at low x. J. Appl. Phys. **92**, 965–975 (2002)
85. T. Kyratsi, E. Hatzikraniotis, K.M. Paraskevopoulos, C.D. Malliakas, J.S. Dyck, C. Uher, M.G. Kanatzidis, Structure inhomogeneities, shallow defects, and charge transport in the series of thermoelectric materials $K_2Bi_{8-x}Sb_xSe_{13}$. J. Appl. Phys. **100**, 123704 (2006)
86. T. Kyratsi, E. Hatzikraniotis, M. Paraskevopoulous, J.S. Dyck, H.K. Shin, C. Uher, M.G. Kanatzidis, Lattice thermal conductivity of $K_2(Bi_{1-z}Sb_z)_8Se_{13}$ solid solutions. J. Appl. Phys. **95**, 4140–4146 (2004)
87. T. Kyratsi, D.Y. Chung, M.G. Kanatzidis, Bi/Sb distribution and its consequences in solid solution members of the thermoelectric materials $K_2Bi_{8-x}Sb_xSe_{13}$. J. Alloys Compd. **338**, 36–42 (2002)
88. T. Kyratsi, D.Y. Chung, J.R. Ireland, C.R. Kannewurf, M.G. Kanatzidis, Chem. Mater. **15**, 2040–3035 (2003)
89. T. Kyratsi, M.G. Kanatzidis, Synthesis, crystallographic studies, and characterization of $K_2Bi_8Se_{13-x}S_x$ solid solutions. Z. Anorg. Allg. Chem. **629**, 2222–2228 (2003)
90. J.F. Meng, N.V.C. Shekar, D.Y. Chung, M. Kanatzidis, J.V. Badding, Improvement in the thermoelectric properties of pressure-tuned β-$K_2Bi_8Se_{13}$. J. Appl. Phys. **94**, 4485–4488 (2003)
91. D.I. Bilc, S.D. Mahanti, T. Kyratsi, D.Y. Chung, M.G. Kanatzidis, P. Larson, Electronic structure of $K_2Bi_8Se_{13}$. Phys. Rev. B **71**, 085116 (2005)
92. D.Y. Chung, T.P. Hogan, M. Rocci-Lane, P. Brazis, J.R. Ireland, C.R. Kannewurf, M. Bastea, C. Uher, M.G. Kanatzidis, A new thermoelectric material: $CsBi_4Te_6$. J. Am. Chem. Soc. **126**, 6414–6428 (2004)
93. V.A. Greanya, W.C. Tonjes, R. Liu, C.G. Olson, D.Y. Chung, M.G. Kanatzidis, Angle-resolved photoemission study of the high-performance low-temperature thermoelectric material $CsBi_4Te_6$. Phys. Rev. B **65**, 205123 (2002)
94. L. Lykke, B.B. Iversen, G.K.H. Madsen, Electronic structure and transport in the low-temperature thermoelectric $CsBi_4Te_6$: semiclassical transport equations. Phys. Rev. B **73**, 195121 (2006)
95. P. Larson, S.D. Mahanti, D.Y. Chung, M.G. Kanatzidis, Electronic structure of $CsBi_4Te_6$: a high-performance thermoelectric at low temperatures. Phys. Rev. B **65**, 045205 (2002)

96. X.B. Zhao, X.H. Ji, Y.H. Zhang, T.J. Zhu, J.P. Tu, X.B. Zhang, Bismuth telluride nanotubes and the effects on the thermoelectric properties of nanotube-containing nanocomposites. Appl. Phys. Lett. **86**, 062111 (2005)
97. X. Tang, W. Xie, H. Li, W. Zhao, Q. Zhang, Preparation and thermoelectric transport properties of high-performance *p*-type Bi_2Te_3 with layered nanostructure. Appl. Phys. Lett. **90**, 012102 (2007)
98. B. Poudel, Q. Hao, Y. Ma, Y. Lan, A. Minnich, B. Yu, X. Yan, D. Wang, A. Muto, D. Vashaee, X. Chen, J. Liu, M.S. Dresselhaus, G. Chen, Z.F. Ren, High-thermoelectric performance of nanostructured bismuth antimony telluride bulk alloys. Science **320**, 634–638 (2008)
99. R. Venkatasubramanian, E. Siivola, T. Colpitts, B. O'Quinn, Thin-film thermoelectric devices with high room-temperature figures of merit. Nature **413**, 597–602 (2001)
100. Z.H. Dughaish, Lead telluride as a thermoelectric material for thermoelectric power generation. Phys. B **322**, 205–223 (2002)
101. F. Ren, E.D. Case, J.R. Sootsman, M.G. Kanatzidis, H. Kong, C. Uher, E. Lara-Curzio, R.M. Trejo, The high-temperature elastic moduli of polycrystalline PbTe measured by resonant ultrasound spectroscopy. Acta Mater. **56**, 5954–5963 (2008)
102. Y. Gelbstein, Z. Dashevsky, M.P. Dariel, The search for mechanically stable PbTe based thermoelectric materials. J. Appl. Phys. **104**, 033702 (2008)
103. Y. Gelbstein, G. Gotesman, Y. Lishzinker, Z. Dashevsky, M.P. Dariel, Mechanical properties of PbTe-based thermoelectric semiconductors. Scr. Mater. **58**, 251–254 (2007)
104. K. Hoang, S.D. Mahanti, Electronic structure of Ga-, In-, and Tl-doped PbTe: a supercell study of the impurity bands. Phys. Rev. B **78**, 085111 (2008)
105. S.D. Mahanti, K. Hoang, S. Ahmad, Deep defect states in narrow band-gap semiconductors. Phys. B **401–402**, 291 (2007)
106. K. Hoang, S.D. Mahanti, P. Jena, Theoretical study of deep-defect states in bulk PbTe and in thin films. Phys. Rev. B **76**, 115432 (2007)
107. S. Ahmad, S.D. Mahanti, K. Hoang, M.G. Kanatzidis, Theoretical study of deep-defect states in bulk PbTe and in thin films. Phys. Rev. B **74**, 155205 (2006)
108. V. Jovovic, S.J. Thiagarajan, J.P. Heremans, T. Komissarova, D. Khokhlov, A. Nicorici, Low temperature thermal, thermoelectric, and thermomagnetic transport in indium rich $Pb_{(1-x)}Sn_{(x)}Te$ alloys. J. Appl. Phys. **103**, 053710 (2008)
109. J.P. Heremans, V. Jovovic, E.S. Toberer, A. Saramat, K. Kurosaki, A. Charoenphakdee, S. Yamanaka, G.J. Snyder, Enhancement of thermoelectric efficiency in PbTe by distortion of the electronic density of states. Science **321**, 554–557 (2008)
110. P.F.P. Poudeu, J. D'Angelo, H. Kong, A. Downey, J.L. Short, R. Pcionek, T.P. Hogan, C. Uher, M.G. Kanatzidis, Nanostructures versus solid solutions: low lattice thermal conductivity and enhanced thermoelectric figure of merit in $Pb_{9.6}Sb_{0.2}Te_{10-x}Se_x$ bulk materials. J. Am. Chem. Soc. **128**, 14347–14355 (2006)
111. S. Geller, J.H. Wernick, Ternary semiconducting compounds with sodium chloride-like structure: $AgSbSe_2$, $AgSbTe_2$, $AgBiS_2$, $AgBiSe_2$. Acta Crystallogr. **12**, 46–54 (1959)
112. J.H. Wernick, S. Geller, K.E. Benson, Constitution of the $AgSbSe_2$-$AgSbTe_2$-$AgBiSe_2$-$AgBiTe_2$ system. J. Phys. Chem. Solid **7**, 240–248 (1958)
113. S.S. Ragimov, S.A. Aliev, $\alpha \rightarrow \beta$ phase transition of Ag_2Te in the $AgSbTe_2$ alloy of the Ag-Sb-Te system. Inorg. Mater. **43**, 1184–1186 (2007)
114. T. Ishihara, Lattice thermal conductivity in $AgSbTe_2$-SnTe as deduced from the calculation of electronic thermal conductivity. J. Phys. Soc. Jpn. **17**, 719–720 (1962)
115. T. Irie, Thermoelectric properties of the single-phase of $AgSbTe_2$. J. Phys. Soc. Jpn. **17**, 1810–1811 (1962)
116. R. Wolfe, J.H. Wernick, S.E. Haszko, Anomalous hall effect in $AgSbTe_2$. J. Appl. Phys. **31**, 1959–1964 (1960)
117. K. Hoang, S.D. Mahanti, J.R. Salvador, M.G. Kanatzidis, Atomic ordering and gap formation in Ag-Sb-based ternary chalcogenide. Phys. Rev. Lett. **99**, 156403 (2007)

118. L.H. Ye, K. Hoang, A.J. Freeman, S.D. Mahanti, J. He, T.M. Tritt, M.G. Kanatzidis, First-principles study of the electronic, optical, and lattice vibrational properties of $AgSbTe_2$. Phys. Rev. B **77**, 245203 (2008)
119. D.T. Morelli, V. Jovovic, J.P. Heremans, Intrinsically minimal thermal conductivity in cubic I-V-VI(2) semiconductors. Phys. Rev. Lett. **101**, 035901 (2008)
120. V. Jovovic, J.P. Heremans, Measurements of the energy band gap and valence band structure of $AgSbTe_2$. Phys. Rev. B **77**, 245204 (2008)
121. S.K. Plachkova, Thermoelectric figure of merit of the system $(GeTe)_{1-x}(AgSbTe_2)_x$. Phys. Status Solidi A **83**, 349–355 (1984)
122. B.A. Cook, M.J. Kramer, X. Wei, J.L. Harringa, E.M. Levin, Nature of the cubic to rhombohedral structural transformation in $(AgSbTe_2)_{15}(GeTe)_{85}$ thermoelectric material. J. Appl. Phys. **101**, 053715 (2007)
123. B.A. Cook, X. Wei, J.L. Harringa, M.J. Kramer, In situ elevated-temperature TEM study of $(AgSbTe_2)_{15}(GeTe)_{85}$. J. Mater. Sci. **42**, 7643–7646 (2007)
124. S.H. Yang, T.J. Zhu, T. Sun, S.N. Zhang, X.B. Zhao, J. He, Nanostructures in high-performance $(GeTe)_x(AgSbTe_2)_{100-x}$ thermoelectric materials. Nanotechnol. **19**, 245707 (2008)
125. G.C. Christakudis, S.K. Plachkova, L.E. Shelimova, E.S. Avilov, Thermoelectric figure of merit of some compositions in the system $(GeTe)_{1-x}[(Ag_2Te)_{1-y}(Sb_2Te_3)_y]_x$. Phys. Status Solidi A **128**, 465–471 (1991)
126. J.W. Fergus, Oxide materials for high temperature thermoelectric energy conversion. J. Eur. Ceram. Soc. **32**, 525–540 (2012)
127. K. Koumoto, R. Funahashi, E. Guilmeau, Y. Miyazaki, A. Weidenkaff, Y. Wang, C. Wan, Thermoelectric ceramics for energy harvesting. J. Am. Ceram. Soc. **96**, 1–23 (2013)
128. M. Karppinen, H. Fjellvåg, T. Konno, Y. Morita, T. Motohashi, H. Yamauchi, Evidence for oxygen vacancies in misfit-layered calcium cobalt oxide $[CoCa_2O_3]_xCoO_2$. Chem. Mater. **16**, 2790–2793 (2004)
129. H. Ohta, K. Sugiura, K. Koumoto, Recent progress in oxide thermoelectric materials: *p*-type $Ca_3Co_4O_9$ and *n*-Type $SrTiO_3$. Inorg. Chem. **47**, 8429–8436 (2008)
130. C.D. Ling, K. Aivazian, S. Schmid, P. Jensen, Structural investigation of oxygen non-stoichiometry and cation doping in misfit-layered thermoelectric $(Ca_2CoO_{3-x})(CoO_2)_\delta$ $\delta \sim 1.61$. J. Solid State Chem. **180**, 1446–1455 (2007)
131. H. Fukutomi, Y. Konno, K, Okayasu, M. Hasegawa, H. Nakatsugawa, Texture development of $Ca_3Co_4O_9$ thermoelectric oxide by high temperature plastic deformation and its contribution to the improvement in electric conductivity. *Mater. Sci. Eng. A* **527**, 61–64 (2009)
132. Y. Morita, J. Poulsen, K. Sakai, T. Motohashi, T. Fujii, I. Terasaki et al., Oxygen nonstoichiometry and cobalt valence in misfit-layered cobalt oxides. J. Solid State Chem. **177**, 3149–3155 (2004)
133. Y. Takahashi, J. Akimoto, N. Kijima, Y. Gotoh, Structure and electron density analysis of $Na_{0.74}CoO_2$ by single-crystal X-ray diffraction. Solid State Ionics **172**, 505–508 (2004)
134. N. Li, Y. Jiang, G. Li, C. Wang, J. Shi, D. Yu, Self-ignition route to Ag doped $Na_{1.7}Co_2O_4$ and its thermoelectric properties. J. Alloys Compd. **467**, 444–449 (2009)
135. D. Dutta, J. Battogtokh, D. McKewon, I. Vidensky, N. Dutta, I.L. Pegg, Thermoelectric properties of $NaCo_{2-x}Fe_xO_y$. J. Electron. Mater. **36**, 746–752 (2007)
136. T. Sun, H.H. Hng, Q. Yan, J. Ma, Effects of pulsed laser deposition conditions on the microstructure of $Ca_3Co_4O_9$ thin films. J. Electron. Mater. **39**, 1611–1615 (2010)
137. M. Tahashi, T. Tanimoto, H. Goto, M. Takahashi, T. Ido, Sintering temperature dependence of thermoelectric performance and crystal phase of calcium cobalt oxides. J. Am. Ceram. Soc. **93**, 3046–3048 (2010)
138. O.J. Kwon, W. Jo, K.E. Ko, J.Y. Kim, S.H. Bae, H. Koo et al., Thermoelectric properties and texture evaluation of $Ca_3Co_4O_9$ prepared by a cost-effective multisheetcofiring technique. J. Mater. Sci. **46**, 2887–2894 (2011)

139. T. Yin, D. Liu, Y. Ou, F. Ma, S. Xie, J.F. Li et al., Nanocrystalline thermoelectric $Ca_3Co_4O_9$ ceramics by sol-gel based electrospinning and spark plasma sintering. J. Phys. Chem. C **114**, 10061–10065 (2010)
140. N.V. Nong, S. Yanagiya, S. Monica, N. Pryds, M. Ohtaki, High-temperature thermoelectric and microstructural characteristics of cobalt-based oxides with Ga substituted on the Co-site. J. Electron. Mater. **40**, 716–722 (2011)
141. N.V. Nong, N. Pryds, S. Linderoth, M. Ohtaki, Enhancement of the thermoelectric performance of *p*-type layered oxide $Ca_3Co_4O_{9+\delta}$ through heavy doping and metallic nanoinclusions. Adv. Mater. **23**, 2484–2490 (2011)
142. H. Su, Y. Jiang, X. Lan, X. Liu, H. Zhong, D. Yu, $Ca_{3-x}Bi_xCo_4O_9$ and $Ca_{1-y}Sm_yMnO_3$ thermoelectric materials and their power-generation devices. Phys. Status Solidi A **208**, 147–155 (2011)
143. S. Li, R. Funahashi, I. Matsubara, K. Ueno, S. Sodeoka, H. Hamada, Synthesis and thermoelectric properties of the new oxide materials $Ca_{3-x}Bi_xCo_4O_{9+\delta}$ ($0.0 < x < 0.75$). Chem. Mater. **12**, 2424–2427 (2000)
144. G. Xu, R. Funahashi, M. Shikano, I. Matsubara, Y. Zhou, Thermoelectric properties of the Bi-and Na-substituted $Ca_3Co_4O_9$ system. Appl. Phys. Lett. **80**, 3660–3662 (2002)
145. Y. Wang, Y. Sui, J. Cheng, X. Wang, W. Su, Comparison of the high temperature thermoelectric properties for Ag-doped and Ag-added $Ca_3Co_4O_9$. J. Alloys Compd. **477**, 817–821 (2009)
146. Y. Wang, Y. Sui, J. Cheng, X. Wang, W. Su, Efficient room temperature thermoelectric characteristics of $Ca_{3-x}Ag_xCo_4O_{9+\delta}/Ag_y$ composites. J. Phys. D Appl. Phys. **41**, 045406 (2008)
147. K. Park, K.K. Kim, S.J. Kim, N. Lee, B. Yang, High-temperature thermoelectric properties of Cu-doped $Ca_{3-x}Cu_xCo_4O_9$ ($0 \leq x \leq 0.4$). J. Korean Phys. Soc. **49**, 1553–1557 (2006)
148. M. Mikami, R. Funahashi, The effect of element substitution on hightemperature thermoelectric properties of $Ca_3Co_2O_6$ compounds. J. Solid State Chem. **178**, 1670–1674 (2005)
149. D. Wang, L. Chen, Q. Yao, J. Li, High-temperature thermoelectric properties of $Ca_3Co_4O_{9+\delta}$ with Eu substitution. Solid State Commun. **129**, 615–618 (2004)
150. D. Wang, L. Chen, Q. Wang, J. Li, Fabrication and thermoelectric properties of $Ca_{3-x}Dy_xCo_4O_{9+\delta}$ system. J. Alloys Compd. **376**, 58–61 (2004)
151. J. Pei, G. Chen, N. Zhou, D.Q. Lu, F. Xiao, High temperature transport and thermoelectric properties of $Ca_{3-x}Er_xCo_4O_{9+\delta}$. Phys. B **406**, 571–574 (2011)
152. M. Prevel, E.S. Reddy, O. Perez, W. Kobayashi, I. Teraski, C. Goupil, J.G. Noudem, Thermoelectric properties of sintered and textured Nd-substituted $Ca_3Co_4O_9$ ceramics. Jpn. J. Appl. Phys. **46**, 6533–6538 (2007)
153. N.V. Nong, C.J. Liu, M. Ohtaki, High-temperature thermoelectric properties of late rare earth-doped $Ca_3Co_4O_9$. J. Alloys Compd. **509**, 977–981 (2011)
154. Y.H. Lin, C.W. Nan, Y. Liu, J. Li, T. Mizokawa, Z. Shen, High-temperature electrical transport and thermoelectric power of partially substituted $Ca_3Co_4O_9$-based ceramics. J. Am. Ceram. Soc. **90**, 132–136 (2007)
155. P.S. Liu, G. Chen, J. Pei, Y. Cui, D.Q. Lu, N. Zhou, H.Z. Xian, Preparation and characterization of the new oxides $Ca_{2-x}Na_xCo_2O_5$. Phys. B **403**, 1808–1812 (2008)
156. J. Pei, G. Chen, D.Q. Lu, P.S. Liu, N. Zhou, Synthesis and high temperature thermoelectric properties of $Ca_{3.0-x-y}Nd_xNa_yCo_4O_{9+\delta}$. Solid State Commun. **146**, 283–286 (2008)
157. M. Senthilkumar, R. Vijayaraghavan, High-temperature resistivity and thermoelectric properties of coupled substituted $Ca_3Co_2O_6$. Sci. Technol. Adv. Mater. **10**, 015007 (2009)
158. Y. Fujishiro, M. Miyata, M. Awano, K. Maeda, Characterization of thermoelectric metal oxide elements prepared by the pulse electric-current sintering method. J. Am. Ceram. Soc. **87**, 1890–1894 (2004)
159. T. Nagira, M. Ito, S. Katsuyama, K. Majima, H. Nagai, Thermoelectric properties of (Na1-yMy)xCo2O4 (M = K, Sr, Y, Nd, Smand Yb; y = 0.01–0.35). J. Alloys Compd. **348**, 263–269 (2003)

160. L. Wang, M. Wang, D. Zhao, Thermoelectric properties of *c*-axis oriented Ni-substituted $NaCoO_2$ thermoelectric oxide by the citric acid complex method. J. Alloys Compd. **471**, 519–523 (2009)
161. P.H. Tsai, T. Norby, T.T. Tan, R. Donelson, Z.D. Chen, S. Li, Correlation of oxygen vacancy concentration and thermoelectric properties in $Na_{0.73}CoO_{2-\delta}$. Appl. Phys. Lett. **96**, 141905 (2010)
162. M. Ito, D. Furumoto, Microstructure and thermoelectric properties of $Na_xCo_2O_4$/Ag composite synthesized by the polymerized complex method. J. Alloys Compd. **450**, 517–520 (2008)
163. P. Liu, G. Chen, Y. Cui, H. Zhang, F. Xiao, L. Wang, H. Nakano, High temperature electrical conductivity and thermoelectric power of Na_xCoO_2. Solid State Ionics **179**, 2308–2312 (2008)
164. D.J. Singh, D. Kasinathan, Thermoelectric properties of Na_xCoO_2 and prospects for other oxide thermoelectrics. J. Electron. Mater. **36**, 736–739 (2007)
165. T. Seetawan, V. Amornkitbamrung, T. Burinprakhon, S. Maensiri, K. Kurosaki, H. Muta, M. Uno, S. Yamanaka, Thermoelectric power and electrical resistivity of Ag doped $Na_{1.5}Co_2O_4$. J. Alloys Compd. **407**, 314–317 (2006)
166. T. Seetawan, V. Amornkitbamrung, T. Burinprakhon, S. Maensiri, K. Kurosaki, H. Muta, M. Uno, S. Yamanaka, Thermoelectric properties of $Na_xCo_2O_4$/Ag composites. J. Alloys Compd. **414**, 293–297 (2006)
167. M. Ito, D. Furumoto, Effects of noble metal addition on microstructure and thermoelectric properties of $Na_xCo_2O_4$. J. Alloys Compd. **450**, 394–398 (2008)
168. K. Park, K.U. Jang, H.C. Kwon, J.G. Kim, W.S. Cho, Influence of partial substitution of Cu for Co on the thermoelectric properties of $NaCo_2O_4$. J. Alloys Compd. **419**, 213–219 (2006)
169. K. Park, K.Y. Ko, J.G. Kim, W.S. Cho, Microstructure and high-temperature thermoelectric properties of CuO and NiO co-substituted $NaCo_2O_4$. Mater. Sci. Eng. B **129**, 200–206 (2006)
170. K. Park, J.H. Lee, Enhanced thermoelectric properties of $NaCo_2O_4$ by adding ZnO. Mater. Lett. **62**, 2366–2368 (2008)
171. P.H. Tsai, T.S. Zhang, R. Donelson, T.T. Tan, S. Li, Power factor enhancement in Zn-doped $Na_{0.8}CoO_2$. J. Alloys Compd. **509**, 5183–5186 (2011)
172. K. Park, K.U. Jang, Improvement in high-temperature thermoelectric properties of $NaCo_2O_4$ through partial substitution of Ni for Co. Mater. Lett. **60**, 1106–1110 (2006)
173. M. Ito, T. Nagira, S. Hara, Thermoelectric properties of $Na_xCo_2O_4$ with rare-earth metals doping prepared by polymerized complex method. J. Alloys Compd. **408–412**, 1217–1221 (2006)
174. D. Kenfaui, D. Chateigner, M. Gomina, J.G. Noudem, Anisotropy of the mechanical and thermoelectric properties of hot-pressed single-layer and multilayer thick $Ca_3Co_4O_9$ ceramics. Int. J. Appl. Ceram. Technol. **8**, 214–226 (2011)
175. D. Kenfaui, G. Bonnefont, D. Chateigner, G. Fantozzi, M. Gomina, J.G. Noudem, $Ca_3Co_4O_9$ ceramics consolidated by SPS process: optimization of mechanical and thermoelectric properties. Mater. Res. Bull. **45**, 1240–1249 (2010)
176. S. Ohta, T. Nomura, H. Ohta, K. Koumoto, High-temperature carrier transport and thermoelectric properties of heavily La-or Nb-doped $SrTiO_3$ single crystals. J. Appl. Phys. **97**, 034106 (2005)
177. J. Ravichandran, W. Siemons, D.W. Oh, J.T. Kardel, A. Chari, H. Heijmerikx, M.L. Scullin, A. Majumdar, R. Ramesh, D.G. Cahill, High-temperature thermoelectric response of double-doped $SrTiO_3$ epitaxial films. Phys. Rev. B **82**, 165126 (2010)
178. H. Muta, K. Kurosaki, S. Yamanaka, Thermoelectric properties of rare earth doped $SrTiO_3$. J. Alloys Compd. **350**, 292–295 (2003)
179. A. Kinaci, C. Sevik, T. Cagin, Electronic transport properties of $SrTiO_3$ and its alloys: $Sr_{1-x}La_xTiO_3$ and $SrTi_{1-x}M_xO_3$ (M = Nb, Ta). Phys. Rev. B **82**, 155114 (2010)

180. Y.F. Wang, K.H. Lee, H. Ohta, K. Koumoto, Fabrication and thermoelectric properties of heavily rare-earth metal-doped $SrO(SrTiO_3)_n$ (n = 1, 2) ceramics. Ceram. Int. **34**, 849–852 (2008)
181. K. Koumoto, Y. Wang, R. Zhang, A. Kosuga, R. Funahashi, Oxide thermoelectric materials: a nanostructuring approach. Ann. Rev. Mater. Res. **40**, 363–394 (2010)
182. H. Muta, K. Kurosaki, S. Yamanaka, Thermoelectric properties of doped $BaTiO_3$-$SrTiO_3$ solid solution. J. Alloys Compd. **368**, 22–24 (2004)
183. L. Bocher, M.H. Aguirre, D. Logvinovich, A. Shkabko, R. Robert, M. Trottmann, A. Weidenkaff, $CaMn_{1-x}Nb_xO_3$ ($x \leq 0.08$) perovskite-type phases as promising new high-temperature n-type thermoelectric materials. Inorg. Chem. **47**, 8077–8085 (2008)
184. J. Lan, Y.H. Lin, H. Fang, A. Mei, C.W. Nan, Y. Liu, S.L. Xu, M. Peter, High-temperature thermoelectric behaviors of fine-grained Gd-doped $CaMnO_3$ ceramics. J. Am. Ceram. Soc. **93**, 2121–2124 (2010)
185. J.W. Park, D.H. Kwak, S.H. Yoon, S.C. Choi, Thermoelectric properties of Bi, Nb co-substituted $CaMnO_3$ at high temperature. J. Alloys Compd. **487**, 550–555 (2009)
186. G. Xu, R. Funahashi, Q. Pu, B. Liu, R. Tao, G.S. Wang, Z.J. Ding, High-temperature transport properties of Nb and Ta substituted $CaMnO_3$ system. Solid State Ionics **172**, 147–151 (2004)
187. Y. Wang, Y. Sui, W. Su, High temperature thermoelectric characteristics of $Ca_{0.9}R_{0.1}MnO_3$ (R = La, Pr, …, Yb). J. Appl. Phys. **104**, 093703 (2008)
188. H. Cheng, X.J. Xu, H.H. Hng, J. Ma, Characterization of Al-doped ZnO thermoelectric materials prepared by RF plasma powder processing and hot press sintering. Ceram. Int. **35**, 3067–3072 (2009)
189. M. Ohtaki, T. Tsubota, K. Eguchi, H Arai, High-temperature thermoelectric properties of $(Zn_{1-x}Al_x)O$. *J. Appl. Phys.* **79**, 1816–1818 (1996)
190. J.P. Wiff, Y. Kinemuchi, H. Kaga, C. Ito, K. Watari, Correlations between thermoelectric properties and effective mass caused by lattice distortion in Al-doped ZnO ceramics. J. Eur. Ceram. Soc. **29**, 1413–1418 (2009)
191. K. Park, J.K. Seong, G.H. Kim, NiO added $Zn_{1-x}Ni_xO$ ($0 \leq x \leq 0.05$) for thermoelectric power generation. J. Alloys Compd. **473**, 423–427 (2009)
192. K. Park, J.K. Seong, Influence of simultaneous addition of Sb_2O_3 and SnO_2 on thermoelectric properties of $Zn_{1-x-y}Sb_xSn_yO$ prepared by tape casting. J. Alloys Compd. **464**, 1–5 (2008)
193. K. Park, J.K. Seong, S. Nahm, Improvement of thermoelectric properties with the addition of Sb to ZnO. J. Alloys Compd. **455**, 331–335 (2008)
194. N.W. Ashcroft, N.D. Mermin, *Solid State Physics* (Saunders College, Orlando, 1976)
195. H.J. Goldsmid, *Applications of Thermoelectricity* (Butler and Tanner Ltd, London, 1960)
196. A. Bulusu, D.G. Walker, Quantum modeling of thermoelectric properties of Si/Ge/Si superlattices. IEEE Trans. Electron Devices **55**, 423–429 (2008)
197. A. Bulusu, D.G. Walker, Modeling of thermoelectric properties of semi-conductor thin films with quantum and scattering effects. J. Heat Transf. **129**, 492–499 (2007)
198. M.J. Kelly, *Low-Dimensional Semiconductors: Materials, Physics, Technology, Devices (Series on Semiconductor Science and Technology)* (Oxford University Press, Oxford, 1996)
199. D.L. Young, T.J. Coutts, V.I. Kaydanov, A.S. Gilmore, W.P. Mulligan, Direct measurement of density-of-states effective mass and scattering parameter in transparent conducting oxides using second-order transport phenomena. J. Vac. Sci. Technol. A **18**, 2978–2985 (2000)
200. M. Lundstrom, *Fundamentals of Carrier Transport* (Cambridge University Press, Cambridge, 2000)
201. D. Vashaee, A. Shakouri, Improved thermoelectric power factor in metal-based superlattices. Phys. Rev. Lett. **92**, 106103 (2004)
202. A. Ishida, D. Cao, S. Morioka, M. Veis, Y. Inoue, T. Kita, Enhanced Seebeck coefficient in EuTe/PbTe [100] short-period superlattices. Appl. Phys. Lett. **92**, 182105 (2008)
203. Y.I. Ravich, *CRC Handbook of Thermoelectrics*, ed. by D.M. Rowe. Selective carrier scattering in thermoelectric materials (CRC Press, Boca Raton, 1995)

204. N.F. Mott, H. Jones, *The Theory of the Properties of Metals and Alloys* (Dover Publications, New York, 1936)
205. G.D. Mahan. J.O. Sofo, *The Best Thermoelectric*, vol. 93 (National Academy of Science, Washington, 1996), pp. 7436
206. P. Pichanusakorn, P. Bandaru, The optimal Seebeck coefficient for obtaining the maximum power factor in thermoelectrics. Appl. Phys. Lett. **94**, 223108 (2009)
207. R. Kim, S. Datta, M.S. Lundstrom, Influence of dimensionality on thermoelectric device performance. J. Appl. Phys. **105**, 034506 (2009)
208. T.C. Harman, D.L. Spears, M.J. Manfra, High thermoelectric figures of merit in PbTe quantum wells. J. Electron. Mater. **25**, 1121–1127 (1996)
209. J.P. Dismukes, L. Ekstrom, E.F. Steigmeier, I. Kudman, D.S. Beers, Thermal and electrical properties of heavily doped GeSi alloys up to 1300 K. J. Appl. Phys. **35**, 2899–2907 (1964)
210. T. Okuda, K. Nakanishi, S. Miyasaka, Y. Tokura, Large thermoelectric response of metallic perovskites: $Sr_{1-x}La_xTiO_3$ ($0 \leq x \leq 0.1$). Phys. Rev. B **63**, 113104 (2001)
211. S. Ohta, T. Nomura, H. Ohta, M. Hirano, H. Hosono, K. Koumoto, Large thermoelectric performance of heavily Nb-doped $SrTiO_3$ epitaxial film at high temperature. Appl. Phys. Lett. **87**, 092108 (2005)
212. G.D. Mahan, *Good Thermoelectrics. Solid State Physics*, vol. 51 (Academic Press, New York, 1997), pp. 82–157
213. L.D. Hicks, M.S. Dresselhaus, Effect of quantum-well structures on the thermoelectric figure of merit. Phys. Rev. B **47**, 16631–16634 (1993)
214. D.A. Broido, T.L. Reinecke, Theory of thermoelectric power factor in quantum well and quantum wire superlattices. Phys. Rev. B **64**, 045324 (2001)
215. T. Koga, X. Sun, S.B. Cronin, M.S. Dresselhaus, Carrier pocket engineering applied to "strained" Si/Ge superlattices to design useful thermoelectric materials. Appl. Phys. Lett. **75**, 2438–2440 (1999)
216. T. Koga, X. Sun, S.B. Cronin, M.S. Dresselhaus, Carrier pocket engineering to design superior thermoelectric materials using GaAs/AlAs superlattices. Appl. Phys. Lett. **73**, 2950–2952 (1998)
217. L.D. Hicks, T.C. Harman, M.S. Dresselhaus, Use of quantum-well superlattices to obtain a high figure of merit from nonconventional thermoelectric materials. Appl. Phys. Lett. **63**, 3230–3232 (1993)
218. C.F. Gallo, B.S. Chandrasekhar, P.H. Sutter, Transport properties of bismuth single crystals. J. Appl. Phys. **34**, 144–152 (1963)
219. Y.M. Lin, O. Rabin, S.B. Cronin, Y.Y. Jackie, M.S. Dresselhaus, Semimetal-semiconductor transition in $Bi_{1-x}Sb_x$ alloy nanowires and their thermoelectric properties. Appl. Phys. Lett. **81**, 2403–2405 (2002)
220. J.O. Sofo, G.D. Mahan, Thermoelectric figure of merit of superlattices. Appl. Phys. Lett. **65**, 26902692 (1994)
221. D.A. Broido, T.L. Reinecke, Effect of superlattice structure on the thermoelectric figure of merit. Phys. Rev. B **51**, 13797–13800 (1995)
222. D.A. Broido, T.L. Reinecke, Thermoelectric figure of merit of quantum wire superlattices. Appl. Phys. Lett. **67**, 100–102 (1995)
223. D.A. Broido, T.L. Reinecke, Thermoelectric transport in quantum well superlattices. Appl. Phys. Lett. **70**, 2384–2386 (1997)
224. L.D. Hicks, T.C. Harman, X. Sun, M.S. Dresselhaus, Experimental study of the effect of quantum-well structures on the thermoelectric figure of merit. Phys. Rev. B **53**, R10493–R10496 (1996)
225. C. Kittel, *Introduction to Solid State Physics* (Wiley, New York, 1996)
226. H. Kroemer, C. Kittel, *Thermal Physics* (W.H. Freeman and Co, New York, 1980)
227. D.W. Oliver, G.A. Slack, Ultrasonic attenuation in insulators at room temperature. J. Appl. Phys. **37**, 1542–1548 (1966)
228. M. Roufosse, P.G. Klemens, Thermal conductivity of complex dielectric crystals. Phys. Rev. B **7**, 5379–5386 (1973)

229. Y.P. Joshi, G.S. Verma, Analysis of phonon conductivity: application to Si. Phys. Rev. B **1**, 750–755 (1970)
230. R.A.H. Hamilton, J.E. Parrott, Variational calculation of the thermal conductivity of germanium. Phys. Rev. **178**, 1284–1292 (1969)
231. I. Pomeranchuk, On the thermal conductivity of dielectrics. Phys. Rev. **60**, 820–821 (1941)
232. C. Herring, Role of low-energy phonons in thermal conduction. Phys. Rev. **95**, 954–965 (1954)
233. G.S. Nolas, J. Sharp, H.J. Goldsmid, *Thermoelectrics-Basic Principles and New Materials Development* (Springer, New York, 2001)
234. H.B. Callen, *Thermodynamics* (Wiley, New York, 1960)
235. D.G. Cahill, S.K. Watson, R.O. Pohl, Lower limit to the thermal conductivity of disordered crystals. Phys. Rev. B **46**, 6131–6140 (1992)
236. C. Kittel, Interpretation of the thermal conductivity of glasses. Phys. Rev. **75**, 972–974 (1949)
237. G.D. Mahan, F. Claro, Nonlocal theory of thermal conductivity. Phys. Rev. B **38**, 1963–1969 (1988)
238. G. Chen, Nonlocal and nonequilibrium heat conduction in the vicinity of nanoparticles. J. Heat Transf. **118**, 539–545 (1996)
239. A.A. Joshi, A. Majumdar, Transient ballistic and diffusive phonon heat transport in thin films. J. Appl. Phys. **74**, 31–39 (1993)
240. A. Majumdar, Microscale heat conduction in dielectric thin films. J. Heat Transf. **115**, 7–16 (1993)
241. S. John, Localization of light. Phys. Today **44**, 32–40 (1991)
242. G. Chen, Size and interface effects on thermal conductivity of superlattices and periodic thin-film structures. J. Heat Transf. **119**, 220–229 (1997)
243. L.G.C. Rego, G. Kirczenow, Quantized thermal conductance of dielectric quantum wires. Phys. Rev. Lett. **81**, 232–235 (1998)
244. K. Schwab, E.A. Hendriksen, J.M. Worlock, M.L. Roukes, Measurement of the quantum of thermal conductance. Nature **404**, 974–977 (2000)
245. J.S. Dugdale, D.K.C. MacDonald, Lattice thermal conductivity. Phys. Rev. **98**, 1751–1752 (1955)
246. M.M. Koza, M.R. Johnson, R. Viennois, H. Mutka, L. Girard, D. Ravot, Breakdown of phonon glass paradigm in La- and Ce-filled Fe_4Sb_{12} skutterudites. Nat. Mater. **7**, 805–810 (2008)
247. B.C. Sales, B.C. Chakoumakos, D. Mandrus, Thermoelectric properties of thallium-filled skutterudites. Phys. Rev. B **61**, 2475–2481 (2000)
248. V. Keppens, D. Mandrus, B.C. Sales, B.C. Chakoumakos, P. Dai, R. Coldea, M.B. Maple, D.A. Gajewski, E.J. Freeman, S. Bennington, Localized vibrational modes in metallic solids. Nature **395**, 876–878 (1998)
249. M. Christensen, A.B. Abrahamsen, N.B. Christensen, F. Juranyi, N.H. Andersen, K. Lefmann, J. Andreasson, C.R.H. Bahl, B.B. Iversen, Avoided crossing of rattler modes in thermoelectric materials. Nat. Mater. **7**, 811–815 (2008)
250. M.S. Jeng, R. Yang, R. Song, G. Chen, Modeling the thermal conductivity and phonon transport in nanoparticle composites using Monte Carlo simulation. J. Heat Transf. **130**, 042410 (2008)
251. S. Mazumder, A. Majumdar, Monte Carlo study of phonon transport in solid thin films including dispersion and polarization. J. Heat Transf. **123**, 749–759 (2001)
252. R.S. Prasher, Phonon transport in anisotropic scattering particulate media. J. Heat Transf. **125**, 1156–1162 (2003)
253. K.F. Hsu, S. Loo, F. Guo, W. Chen, J.S. Dyck, C. Uher, T. Hogan, E.K. Polychroniadis, M.G. Kanatzidis, Cubic $AgPb_mSbTe_{2-m}$: bulk thermoelectric materials with high figure of merit. Science **303**, 818–821 (2004)
254. D.A. Porter, K.E. Easterling, *Phase Transformations in Metals and Alloys* (Nelson Thornes, Cheltenham, 1992)

255. I. Bonev, On the terminology of the phenomena of mutual crystal orientation. Acta Crystallogr. A **28**, 508–512 (1972)
256. J. Androulakis, K.F. Hsu, R. Pcionek, H. Kong, C. Uher, J.J. D'Angelo, A. Downey, T. Hogan, M.G. Kanatzidis, Nanostructuring and high thermoelectric efficiency in *p*-type $Ag(Pb_{1-y}Sn_y)_mSbTe_{2+m}$. Adv. Mater. **18**, 1170–1173 (2006)
257. P.F.P. Poudeu, J. D'Angelo, A.D. Downey, J.L. Short, T.P. Hogan, M.G. Kanatzidis, High thermoelectric figure of merit and nanostructuring in bulk *p*-type $Na_{1-x}Pb_mSb_yTe_{m+2}$. Angew. Chem. Int. Ed. **45**, 3835–3839 (2006)
258. J. Androulakis, C.H. Lin, H.J. Kong, C. Uher, C.I. Wu, T. Hogan, B.A. Cook, T. Caillat, K.M. Paraskevopoulos, M.G. Kanatzidis, Spinodal decomposition and nucleation and growth as a means to bulk nanostructured thermoelectrics: enhanced performance in $Pb_{1-x}Sn_xTe$-PbS. J. Am. Chem. Soc. **129**, 9780–9788 (2007)
259. C.R.M. Grovenor, Grain boundaries in semiconductors. J. Phys. C: Solid State Phys. **18**, 4079–4119 (1985)
260. G. Chen, Thermal conductivity and ballistic phonon transport in cross-plane direction of superlattices. Phys. Rev. B **57**, 14958–14973 (1998)
261. G.L. Pollack, Kapitza resistance. Rev. Mod. Phys. **41**, 48–81 (1969)
262. E.T. Swartz, R.O. Pohl, Thermal boundary resistance. Rev. Mod. Phys. **61**, 605–668 (1989)
263. W.A. Little, The transport of heat between dissimilar solids at low temperatures. Can. J. Phys. **37**, 334–349 (1959)
264. S.M. Lee, D.G. Cahill, Heat transport in thin dielectric films. J. Appl. Phys. **81**, 2590–2595 (1997)
265. S.M. Lee, D.G. Cahill, R. Venkatasubramanian, Thermal conductivity of Si–Ge superlattices. Appl. Phys. Lett. **70**, 2957–2959 (1997)
266. Y.K. Koh, Y. Cao, D.G. Cahill, D. Jena, Heat-transport mechanisms in superlattices. Adv. Funct. Mater. **19**, 610–615 (2009)
267. M.J. Kelly, Acoustic phonon transmission in superlattices. J. Phys. C: Solid State Phys. **18**, 5963–5973 (1985)
268. Y. Chen, D. Li, J. Yang, Y. Wu, J.R. Lukes, A. Majumdar, Molecular dynamics study of the lattice thermal conductivity of Kr/Ar superlattice nanowires. Phys. B: Conden. Matt. **349**, 270–280 (2004)
269. S. Volz, G. Chen, Molecular dynamics simulation of thermal conductivity of silicon nanowires. Appl. Phys. Lett. **75**, 2056–2058 (1999)
270. S. Volz, J.B. Saulnier, G. Chen, P. Beauchamp, Computation of thermal conductivity of Si/Ge superlattices by molecular dynamics techniques. Microelectron. J. **31**, 815–819 (2000)
271. L. Braginsky, N. Lukzen, V. Shklover, H. Hofmann, High-temperature phonon thermal conductivity of nanostructures. Phys. Rev. B **66**, 134203 (2002)
272. G. Chen, C.L. Tien, X. Wu, J.S. Smith, Thermal diffusivity measurement of GaAs/AlGaAs thin-film structures. J. Heat Transf. **116**, 325–331 (1994)
273. T. Yao, Thermal properties of AlAs/GaAs superlattices. Appl. Phys. Lett. **51**, 1798–1800 (1987)
274. W.S. Capinski, H.J. Maris, Thermal conductivity of GaAs/AlAs superlattices. Phys. B: Conden. Matt. **219–220**, 699–701 (1996)
275. D.G. Cahill, Thermal conductivity measurement from 30 to 750 K: the 3ω method. Rev. Sci. Instrum. **61**, 802–808 (1990)
276. S.T. Huxtable, A.R. Abramson, C.L. Tien, A. Majumdar, C. LaBounty, X. Fan, G. Zeng, J.E. Bowers, A. Shakouri, E.T. Croke, Thermal conductivity of Si/SiGe and SiGe/SiGe superlattices. Appl. Phys. Lett. **80**, 1737–1739 (2002)
277. T. Borca-Tasciuc, W. Liu, J. Liu, T. Zeng, D.W. Song, C.D. Moore, G. Chen, K.L. Wang, M. Goorsky, T. Radetic, R. Gronsky, T. Koga, M.S. Dresselhaus, Thermal conductivity of symmetrically strained Si/Ge superlattices. Superlatt. Microstruct. **28**, 199–206 (2000)
278. R. Venkatasubramanian, Lattice thermal conductivity reduction and phonon localization like behavior in superlattice structures. Phys. Rev. B **61**, 3091–3097 (2000)

279. A.L. Moore, S.K. Saha, R.S. Prasher, L. Shi, Phonon backscattering and thermal conductivity suppression in sawtooth nanowires. Appl. Phys. Lett. **93**, 083112 (2008)
280. J.M. Ziman, *Electrons and Phonons: The Theory of Transport Phenomena in Solids* (Clarendon Press, Oxford, 1960)
281. T.C. Harman, P.J. Taylor, D.L. Spears, M.P. Walsh, Thermoelectric quantum-dot superlattices with high ZT. J. Electron. Mater. **29**, L1–L4 (2000)
282. K. Imamura, Y. Tanaka, S. Tamura, Group velocities of phonons in superlattices. Phys. Rev. B **65**, 174301 (2002)
283. V. Narayanamurti, H.L. Stormer, M.A. Chin, A.C. Gossard, W. Wiegmann, Selective transmission of high-frequency phonons by a superlattice: the "dielectric" phonon filter. Phys. Rev. Lett. **43**, 2012–2016 (1979)
284. D.C. Hurley, S. Tamura, J.P. Wolfe, K. Ploog, J. Nagle, Angular dependence of phonon transmission through a Fibonacci superlattice. Phys. Rev. B **37**, 8829–8836 (1988)
285. D.C. Hurley, S. Tamura, J.P. Wolfe, H. Morkoc, Imaging of acoustic phonon stop bands in superlattices. Phys. Rev. Lett. **58**, 2446–2449 (1987)
286. R. Orbach, Phonon localization and transport in disordered-system. J. Non-Cryst. Solids **166**, 917–922 (1993)
287. A. Majumdar, Lower limit of thermal conductivity: diffusion versus localization. Microscale Thermophys. Eng. **2**, 5–9 (1998)
288. A.A. Balandin, K.L. Wang, Significant decrease of the lattice thermal conductivity due to phonon confinement in a free-standing semiconductor quantum well. Phys. Rev. B **58**, 1544–1549 (1998)
289. B. Taylor, H.J. Maris, C. Elbaum, Phonon focusing in solids. Phys. Rev. Lett. **23**, 416–419 (1969)
290. A.A. Balandin, Nanophononics: phonon engineering in nanostructures and nanodevices. J. Nanosci. Nanotechnol. **5**, 1015–1022 (2005)
291. X. Lu, Lattice thermal conductivity of Si nanowires: effect of modified phonon density of states. J. Appl. Phys. **104**, 054314 (2008)
292. P. Kim, L. Shi, A. Majumdar, P.L. McEuen, Thermal transport measurements of individual multiwalled nanotubes. Phys. Rev. Lett. **87**, 215502 (2001)
293. H. Maune, H.Y. Chiu, M. Bockrath, Thermal resistance of the nanoscale constrictions between carbon nanotubes and solid substrates. Appl. Phys. Lett. **89**, 013109 (2006)
294. R.S. Prasher, Thermal boundary resistance and thermal conductivity of multiwalled carbon nanotubes. Phys. Rev. B **77**, 075424 (2008)
295. F.M. Ross, J. Tersoff, M.C. Reuter, Sawtooth faceting in silicon nanowires. Phys. Rev. Lett. **95**, 146104 (2005)
296. R.S. Prasher, Thermal conductivity of tubular and core/shell nanowires. Appl. Phys. Lett. **89**, 063121 (2006)
297. R. Yang, G. Chen, M.S. Dresselhaus, Thermal conductivity of simple and tubular nanowire composites in the longitudinal direction. Phys. Rev. B **72**, 125418 (2005)
298. E.P. Pokatilov, D.L. Nika, A.A. Balandin, A phonon depletion effect in ultrathin heterostructures with acoustically mismatched layers. Appl. Phys. Lett. **85**, 825–827 (2004)
299. A.I. Hochbaum, R. Chen, R.D. Delgado, W. Liang, E.C. Garnett, M. Najarian, A. Majumdar, P. Yang, Enhanced thermoelectric performance of rough silicon nanowires. Nature **451**, 163–167 (2008)
300. P. Carruthers, Theory of thermal conductivity of solids at low temperatures. Rev. Mod. Phys. **33**, 92–138 (1961)
301. A.I. Boukai, Y. Bunimovich, J. Tahir-Kheli, J.K. Yu, W.A. Goddard III, J.R. Heath, Silicon nanowires as efficient thermoelectric materials. Nature **451**, 168–171 (2008)
302. C. Herring, Theory of the thermoelectric power of semiconductors. Phys. Rev. **96**, 1163–1187 (1954)
303. H.P.R. Frederikse, Thermoelectric power of germanium below room temperature. Phys. Rev. **92**, 248–252 (1953)

304. E. Behnen, Quantitative examination of the thermoelectric power of n-type Si in the phonon drag regime. J. Appl. Phys. **87**, 287–292 (1990)
305. R. Lifshitz, M.L. Roukes, Thermoelastic damping in micro-and nanomechanical systems. Phys. Rev. B **61**, 5600–5609 (2000)
306. M. Zhou, J.F. Li, T. Kita, Nanostructured $AgPb_mSbTe_{m+2}$ system bulk materials with enhanced thermoelectric performance. J. Am. Chem. Soc. **130**, 4527–4532 (2008)
307. H. Wang, J.F. Li, C.W. Nan, M. Zhou, W.S. Liu, B.P. Zhang, T. Kita, High-performance $Ag_{0.8}Pb_{18+x}SbTe_{20}$ thermoelectric bulk materials fabricated by mechanical alloying and spark plasma sintering. Appl. Phys. Lett. **88**, 092104 (2006)
308. T.J. Zhu, F. Yan, S.N. Zhang, X.B. Zhao, Microstructure and electrical properties of quenched $AgPb_{18}Sb_{1-x}Te_{20}$ thermoelectric materials. J. Phys. D: Appl. Phys. **40**, 3537–3540 (2007)
309. H.S. Dow, M.W. Oh, S.D. Park, B.S. Kim, B.K. Min, H.W. Lee, D.M. Wee, Thermoelectric properties of $AgPb_mSbTe_{m+2}$ ($12 \leq m \leq 26$) at elevated temperature. J. Appl. Phys. **105**, 113703 (2009)
310. M.K. Han, K. Hoang, H. Kong, R. Pcionek, C. Uher, K.M. Paraskevopoulos, S.D. Mahanti, M.G. Kanatzidis, Substitution of Bi for Sb and its role in the thermoelectric properties and nanostructuring in $Ag_{1-x}Pb_{18}MTe_{20}$ (M = Bi, Sb) (x = 0, 0.14, 0.3). Chem. Mater. **20**, 3512 (2008)
311. E. Quarez, K.F. Hsu, R. Pcionek, N. Frangis, E.K. Polychroniadis, M.G. Kanatzidis, Nanostructuring, compositional fluctuations, and atomic ordering in the thermoelectric materials $AgPb_mSbTe_{2+m}$: the myth of solid solutions. J. Am. Chem. Soc. **127**, 9177–9190 (2005)
312. T.P. Hogan, A. Downey, J. Short, J. D'Angelo, C.I. Wu, E. Quarez, J. Androulakis, P.F.P. Poudeu, J.R. Sootsman, D.Y. Chung, M.G. Kanatzidis, S.D. Mahanti, E.J. Timm, H. Schock, F. Ren, J. Johnson, E.D. Case, Nanostructured thermoelectric materials and high-efficiency power-generation modules. J. Electron. Mater. **36**, 704–710 (2007)
313. J. Androulakis, R. Pcionek, E. Quarez, J.H. Do, H. Kong, O. Palchik, C. Uher, J.J. D'Angelo, J. Short, T. Hogan, M.G. Kanatzidis, Coexistence of large thermopower and degenerate doping in the nanostructured material $Ag_{0.85}SnSb_{1.15}Te_3$. Chem. Mater. **18**, 4719–4721 (2006)
314. K. Hoang, K. Desai, S.D. Mahanti, Charge ordering and self-assembled nanostructures in a fcc Coulomb lattice gas. Phys. Rev. B **72**, 064102 (2005)
315. A. Gueguen, P.F.P. Poudeu, C.P. Li, S. Moses, C. Uher, J. He, V. Dravid, K.M. Paraskevopoulos, M.G. Kanatzidis, Thermoelectric properties and nanostructuring in the *p*-type materials $NaPb_{18-x}Sn_xMTe_{20}$ (M = Sb, Bi). Chem. Mater. **21**, 1683–1694 (2009)
316. J.R. Sootsman, R.J. Pcionek, H. Kong, C. Uher, M.G. Kanatzidis, Strong reduction of thermal conductivity in nanostructured PbTe prepared by matrix encapsulation. Chem. Mater. **18**, 4993–4995 (2006)
317. J.P. Heremans, C.M. Thrush, D.T. Morelli, Thermopower enhancement in lead telluride nanostructures. Phys. Rev. B **70**, 115334 (2004)
318. W. Kim, J. Zide, A. Gossard, D. Klenov, S. Stemmer, A. Shakouri, A. Majumdar, Thermal conductivity reduction and thermoelectric figure of merit increase by embedding nanoparticles in crystalline semiconductors. Phys. Rev. Lett. **96**, 045901 (2006)
319. M.S. Dresselhaus, G. Chen, M.Y. Tang, R.G. Yang, H. Lee, D.Z. Wang, Z.F. Ren, J.P. Fleurial, P. Gogna, New directions for low-dimensional thermoelectric materials. Adv. Mater. **19**, 1043–1053 (2007)
320. J. Martin, G.S. Nolas, W. Zhang, L. Chen, PbTe nanocomposites synthesized from PbTe nanocrystals. Appl. Phys. Lett. **90**, 222112 (2007)
321. X. Ji, B. Zhang, T.M. Tritt, J.W. Kolis, A. Kumbhar, Solution-chemical syntheses of nano-structured Bi_2Te_3 and PbTe thermoelectric materials. J. Electron. Mater. **36**, 721–726 (2007)
322. P.N. Alboni, X. Ji, J. He, N. Gothard, T.M. Tritt, Thermoelectric properties of $La_{0.9}CoFe_3Sb_{12}$-$CoSb_3$ skutterudite nanocomposites. J. Appl. Phys. **103**, 113707 (2008)

323. X. Ji, J. He, P. Alboni, Z. Su, N. Gothard, B. Zhang, T.M. Tritt, J.W. Kolis, Thermal conductivity of $CoSb_3$ nano-composites grown via a novel solvothermal nano-plating technique. Phys. Status Solidi RRL **1**, 229–231 (2007)
324. W.X. Tian, R.G. Yang, Effect of interface scattering on phonon thermal conductivity percolation in random nanowire composites. Appl. Phys. Lett. **90**, 263105 (2007)
325. J.L. Mi, X.B. Zhao, T.J. Zhu, J.P. Tu, Improved thermoelectric figure of merit in *n*-type $CoSb_3$ based nanocomposites. Appl. Phys. Lett. **91**, 172116 (2007)
326. H. Li, X.F. Tang, X.L. Su, Q.J. Zhang, C. Uher, Electronic structure and transport coefficients of binary skutterudite antimonide. J. Phys. D: Appl. Phys. **42**, 145409 (2009)
327. Annual Book of ASTM Standards, *American Society of Testing and Materials* (1996), pp. 70–82
328. A. Casian, I. Sur, H. Scherrer, Z. Dashevsky, Thermoelectric properties of *n*-type $PbTe/Pb_{1-x}Eu_xTe$ quantum wells. Phys. Rev. B **61**, 15965–15974 (2000)
329. X. Sun, J. Liu, S.B. Cronin, K.L. Wang, G. Chen, T. Koga, M.S. Dresselhaus, Experimental study of the effect of the quantum well structures on the thermoelectric figure of merit in $Si/Si_{1-x}Ge_x$ system. Mater. Res. Soc. Symp. **545**, 369–374 (1999)
330. T. Koga, S.B. Cronin, M.S. Dresselhaus, J.L. Liu, K.L. Wang, Experimental proof-of-principle investigation of enhanced $Z_{3D}T$ in (001) oriented Si/Ge superlattices. Appl. Phys. Lett. **77**, 1490–1492 (2000)
331. H.P.R. Frederikse, W.R. Thurber, W.R. Hosler, Electronic transport in strontium titanate. Phys. Rev. **134**, A442–A445 (1964)
332. H. Ohta, S. Kim, Y. Mune, T. Mizoguchi, K. Nomura, S. Ohta, T. Nomura, Y. Nakanishi, Y. Ikuhara, M. Hirano, H. Hosono, K. Koumoto, Giant thermoelectric Seebeck coefficient of two-dimensional electron gas in $SrTiO_3$. Nat. Mater. **6**, 129–134 (2007)
333. Y. Mune, H. Ohta, K. Koumoto, T. Mizoguchi, Y. Ikuhara, Enhanced Seebeck coefficient of quantum-confined electrons in $SrTiO_3/SrTi_{0.8}Nb_{0.2}O_3$ superlattices. Appl. Phys. Lett. **91**, 192105 (2007)
334. L.D. Hicks, M.S. Dresselhaus, Effect of quantum-well structures on the thermoelectric figure of merit. Phys. Rev. B **47**, 12727–12731 (1993)
335. Z. Zhang, R. Zou, L. Yu, J. Hu, Recent research on one-dimensional silicon-based semiconductor nanomaterials: synthesis, structures, properties and applications. Crit. Rev. Solid State Mater. Sci. **36**, 148–173 (2011)
336. Y.M. Lin, S.B. Cronin, J.Y. Ying, M.S. Dresselhaus, J.P. Heremans, Transport properties of Bi nanowire arrays. Appl. Phys. Lett. **76**, 3944–3946 (2000)
337. Y.M. Lin, X. Sun, M.S. Dresselhaus, Theoretical investigation of thermoelectric transport properties of cylindrical Bi nanowires. Phys. Rev. B **62**, 4610–4623 (2000)
338. Z. Zhang, X. Sun, M.S. Dresselhaus, J.Y. Ying, J. Heremans, Electronic transport properties of single-crystal bismuth nanowire arrays. Phys. Rev. B **61**, 4850–4861 (2000)
339. J. Heremans, C.M. Thrush, Y.M. Lin, S.B. Cronin, Z. Zhang, M.S. Dresselhaus, J.F. Mansfield, Bismuth nanowire arrays: Synthesis and galvanomagnetic properties. Phys. Rev. B **61**, 2921–2930 (2000)
340. T.E. Huber, A. Nikolaeva, D. Gitsu, L. Konopko, C.A. Foss Jr, M.J. Graf, Confinement effects and surface-induced charge carriers in Bi quantum wires. Appl. Phys. Lett. **84**, 1326–1328 (2004)
341. A. Boukai, K. Xu, J.R. Heath, Size-dependent transport and thermoelectric properties of individual polycrystalline bismuth nanowires. Adv. Mater. **18**, 864–869 (2006)
342. J.P. Heremans, C.M. Thrush, D.T. Morelli, M.C. Wu, Thermoelectric power of bismuth nanocomposites. Phys. Rev. Lett. **88**, 216801 (2002)
343. D. Dragoman, M. Dragoman, Giant thermoelectric effect in graphene. Appl. Phys. Lett. **91**, 203116 (2007)
344. M.R. Black, P.L. Hagelstein, S.B. Cronin, Y.M. Lin, M.S. Dresselhaus, Optical absorption from an indirect transition in bismuth nanowires. Phys. Rev. B **68**, 235417 (2003)

345. M.R. Black, M. Padi, S.B. Cronin, Y.M. Lin, O. Rabin, T. McClure, G. Dresselhaus, P.L. Hagelstein, M.S. Dresselhaus, Intersubband transitions in bismuth nanowires. Appl. Phys. Lett. **77**, 4142–4144 (2000)
346. L. Weber, E. Gmelin, Transport properties of silicon. Appl. Phys. A **53**, 136–140 (1991)
347. D. Davila, A. Tarancon, C. Calaza, M. Salleras, M. Fernandez-Regulez, A. Sanpaulo, L. Fonseca, Improved thermal behavior of multiple linked arrays of silicon nanowires integrated into planar thermoelectric microgenerators. J. Electron. Mater. **42**, 1918–1925 (2013)
348. K.Q. Peng, Y.J. Yan, S.P. Gao, J. Zhu, Synthesis of large-area silicon nanowire arrays via self-assembling nanochemistry. Adv. Mater. **14**, 1164–1167 (2002)
349. N.A. Melosh, A. Boukai, F. Diana, B. Gerardot, A. Badolato, P.M. Petroff, J.R. Heath, Ultra-high density nanowire lattices and circuits. Science **300**, 112–115 (2003)
350. D. Wang, B.A. Sheriff, J.R. Heath, Complementary symmetry silicon nanowire logic: power-efficient inverters with gain. Small **2**, 1153–1158 (2006)
351. C.Z. Meng, C.H. Liu, S.S. Fan, A promising approach to enhanced thermoelectric properties using carbon nanotube networks. Adv. Mater. **22**, 535–539 (2010)
352. Q. Wang, Q. Yao, J. Chang, L.D. Chen, Enhanced thermoelectric properties of CNT/PANI composite nanofibers by highly orienting the arrangement of polymer chains. J. Mater. Chem. **22**, 17612–17618 (2012)
353. Q.L. Zhang, W.J. Wang, J.L. Li, J.J. Zhu, L.J. Wang, M.F. Zhu, W. Jang, Preparation and thermoelectric properties of multi-walled carbon nanotube/polyaniline hybrid nanocomposites. J. Mater. Chem. A **1**, 12109–12114 (2013)
354. C.H. Yu, A. Murali, K.W. Choi, Y.T. Ryu, Air-stable fabric thermoelectric modules made of N- and P-type carbon nanotubes. Energy Environ. Sci. **5**, 9481–9486 (2012)
355. H.L. Ni, T.J. Zhu, X.B. Zhao, Hydrothermally synthesized and hot-pressed Bi2(Te, Se)3 thermoelectric alloys. Phys. B **364**, 50–54 (2005)
356. T.J. Zhu, Y.Q. Liu, X.B. Zhao, Synthesis of PbTe thermoelectric materials by alkaline reducing chemical routes. Mater. Res. Bull. **43**, 2850–2854 (2008)
357. W.Z. Wang, B. Poudel, J. Yang, D.Z. Wang, Z.F. Ren, High-yield synthesis of single-crystalline antimony telluride hexagonal nanoplates using a solvothermal approach. J. Am. Chem. Soc. **127**, 13792–13793 (2005)
358. J.L. Mi, X.B. Zhao, T.J. Zhu, J.P. Tu, Nanosized La filled $CoSb_3$ prepared by a solvothermal-annealing method. Mater. Lett. **62**, 2363–2365 (2008)
359. L. Bertini, C. Stiewe, M. Toprak, S. Williams, D. Platzek, A. Mrotzek, Y. Zhang, C. Gatti, E. Muller, M. Muhammed, M. Rowe, Nanostructured $Co_{1-x}Ni_xSb_3$ skutterudites: synthesis, thermoelectric properties, and theoretical modeling. J. Appl. Phys. **93**, 438–447 (2003)
360. T. Sun, X.B. Zhao, T.J. Zhu, J.P. Tu, Aqueous chemical reduction synthesis of Bi_2Te_3 nanowires with surfactant assistance. Mater. Lett. **60**, 2534–2537 (2006)
361. J. Martin, G.S. Nolas, W. Zhang, L. Chen, PbTe nanocomposites synthesized from PbTe nanocrystals. Appl. Phys. Lett. **90**, 222112 (2007)
362. D.M. Rowe, V.S. Shukla, N. Savvides, Phonon scattering at grain boundaries in heavily doped fine-grained silicon-germanium alloys. Nature **290**, 765–766 (1981)
363. A.A. Joraide, Thermoelectric properties of fine-grained sintered $(Bi_2Te_3)_{25}$-$(Sb_2Te_3)_{75}$ p-type solid solution. J. Mater. Sci. **30**, 744–748 (1995)
364. G. Joshi, H. Lee, Y.C. Lan, X.W. Wang, G.H. Zhu, D.Z. Wang, R.W. Gould, D.C. Cuff, M.Y. Tang, M.S. Dresselhaus, G. Chen, Z.F. Ren, Enhanced thermoelectric figure-of-merit in nanostructured *p*-type silicon germanium bulk alloys. Nano Lett. **8**, 4670–4674 (2008)
365. Y. Ma, Q. Hao, B. Poudel, Y.C. Lan, B. Yu, D.Z. Wang, G. Chen, Z.F. Ren, Enhanced thermoelectric figure-of-merit in *p*-type nanostructured bismuth antimony tellurium alloys made from elemental chunks. Nano Lett. **8**, 2580–2584 (2008)
366. C. Suryanarayana, Mechanical alloying and milling. Prog. Mater. Sci. **46**, 1–184 (2001)
367. L.B. Kong, T.S. Zhang, J. Ma, Y.C.F. Boey, Progress in synthesis of ferroelectric ceramic materials via high-energy mechanochemical techniques. Prog. Mater. Sci. **53**, 207–322 (2008)

368. P.S. Gilman, J.S. Benjamin, Mechanical alloying. Ann. Rev. Mater. Res. **13**, 279–300 (1983)
369. R. Davis, C. Koch, Mechanical alloying of brittle components: silicon and germanium. Scr. Metall. **21**, 305–310 (1987)
370. W.S. Liu, B.P. Zhang, L.D. Zhao, J.F. Li, Improvement of thermoelectric performance of $CoSb_{3-x}Te_x$ Skutterudite compounds by additional substitution of IVB-group elements for Sb. Chem. Mater. **20**, 7526–7531 (2008)
371. H. Li, X.F. Tang, X.L. Su, Q.J. Zhang, Preparation and thermoelectric properties of high-performance Sb additional $Yb_{0.2}Co_4Sb_{12+y}$ bulk materials with nanostructure. Appl. Phys. Lett. **92**, 202114 (2008)
372. K. Sridhar, K. Chattopadhyay, Synthesis by mechanical alloying and thermoelectric properties of Cu_2Te. J. Alloy. Compd. **264**, 293–298 (1998)
373. R. Martin-Lopez, A. Dauscher, H. Scherrer, J. Hejtmanek, H. Kenzari, B. Lenoir, Thermoelectric properties of mechanically alloyed Bi–Sb alloys. Appl. Phys. A **68**, 597–602 (1999)
374. S.S. Kim, S. Yamamoto, T. Aizawa, Thermoelectric properties of anisotropy-controlled *p*-type Bi-Te-Sb system via bulk mechanical alloying and shear extrusion. J. Alloy. Compd. **375**, 107–113 (2004)
375. S.C. Ur, I.H. Kim, P. Nash, Thermoelectric properties of Zn_4Sb_3 processed by sintering of cold pressed compacts and hot pressing. J. Mater. Sci. **42**, 2143–2149 (2007)
376. C.J. Liu, H. Yamauchi, Thermoelectric power and resistivity of $La_{1.8}Sr_{0.2}CaCu_2O_{6-\delta}$ and the effects of O_2 hot-isostatic-press annealing. Phys. Rev. B **51**, 11289–11826 (1995)
377. K. Kishimoto, T. Koyanagi, Preparation of sintered degenerate n-type PbTe with a small grain size and its thermoelectric properties. J. Appl. Phys. **92**, 2544–2549 (2002)
378. J. Yang, Q. Hao, H. Wang, Y.C. Lan, Q.Y. He, D.Z. Wang, J.A. Harriman, V.M. Varki, M.S. Dresselhaus, G. Chen, Z.F. Ren, Solubility study of Yb in n-type skutterudites $Yb_xCo_4Sb_{12}$ and their enhanced thermoelectric properties. Phys. Rev. B **80**, 115329 (2009)
379. Q.Y. He, Q. Hao, X.W. Wang, J. Yang, Y.C. Lan, X. Yan, B. Yu, Y. Ma, B. Poudel, G. Joshi, D.Z. Wang, G. Chen, Z.F. Ren, Nanostructured thermoelectric Skutterudite $Co_{1-x}Ni_xSb_3$ alloys. J. Nanosci. Nanotechnol. **8**, 4003–4006 (2008)
380. S.B. Riffat, X.L. Ma, Thermoelectrics: a review of present and potential applications. Appl. Therm. Eng. **23**, 913–935 (2003)
381. D.M. Rowe, G. Min, Evaluation of thermoelectric modules for power generation. J. Power Sources **73**, 193–198 (1998)
382. G. Min, D.M. Rowe, "Symbiotic" application of thermoelectric conversion for fluid preheatiing/power generation. Energ. Convers. Manag. **43**, 221–228 (2002)
383. D.M. Rowe, Thermoelectrics, an environmentally-friendly source of electrical power. Renew. Energ. **16**, 1251–1256 (1999)
384. P. Yodovard, J. Khedari, J. Hirunlabh, The potential of waste heat thermoelectric power generation from diesel cycle and gas turbine cogeneration plants. Energy Sources **23**, 213–224 (2001)
385. H. Naito, Y. Kohsaka, D. Cooke, H. Arashi, Development of a solar receiver for a high-efficiency thermionic/thermoelectric conversion system. Solar Energy **58**, 191–195 (1996)
386. R. Yang, G. Chen, M.S. Dresselhaus, Thermal conductivity modeling of core-shell and tubular nanowires. Nano Lett. **5**, 1111–1115 (2005)
387. A.J. Minnich, M.S. Dresselhaus, Z.F. Ren, G. Chen, Bulk nanostructured thermoelectric materials: current research and future prospects. Energy Environ. Sci. **2**, 466–479 (2009)
388. X.W. Wang, H. Lee, Y.C. Lan, G.H. Zhu, G. Joshi, D.Z. Wang, J. Yang, A.J. Muto, M.Y. Tang, J. Klatsky, S. Song, M.S. Dresselhaus, G. Chen, Z. Ren, Enhanced thermoelectric figure of merit in nanostructured *n*-type silicon germanium bulk alloy. Appl. Phys. Lett. **93**, 193121 (2008)
389. C. Chiritescu, D.G. Cahill, N. Nguyen, D. Johnson, A. Bodapati, P. Keblinski, P. Zschack, Ultralow thermal conductivity in disordered, layered WSe_2 crystals. Science **315**, 351–353 (2007)

390. C.J. Vineis, A. Shakouri, A. Majumdar, M.G. Kanatzidis, Nanostructured thermoelectrics: big efficiency gains from small features. Adv. Mater. **22**, 3970–3980 (2010)
391. A.S. Henry, G. Chen, Spectral phonon transport properties of silicon based on molecular dynamics Simulations and lattice dynamics. J. Comput. Theor. Nanosci. **5**, 141–152 (2008)
392. J. Callaway, Model for lattice thermal conductivity at low temperatures. Phys. Rev. **113**, 1046–1051 (1959)
393. E.F. Steigmeier, B. Abeles, Scattering of phonons by electrons in germanium-silicon alloys. Phys. Rev. **136**, A1149–A1155 (1964)
394. D.T. Morelli, V. Jovovic, J.P. Heremans, Intrinsically minimal thermal conductivity in cubic I-V-VI_2 semiconductors. Phys. Rev. Lett. **101**, 035901 (2008)
395. R. Prasher, Thermal boundary resistance of nanocomposites. Int. J. Heat Mass Transf. **48**, 4942–5952 (2005)
396. H. Zhong, J.R. Lukes, Interfacial thermal resistance between carbon nanotubes: molecular dynamics simulations and analytical thermal modeling. Phys. Rev. B **74**, 125403 (2006)
397. A. Minnich, G. Chen, Modified effective medium formulation for the thermal conductivity of nanocomposites. Appl. Phys. Lett. **91**, 073105 (2007)
398. A. Khitun, K.L. Wang, G. Chen, Thermoelectric figure of merit enhancement in a quantum dot superlattice. Nanotechnology **11**, 327–331 (2000)
399. N. Zuckerman, J.R. Lukes, Acoustic phonon scattering from particles embedded in an anisotropic medium: A molecular dynamics study. Phys. Rev. B **77**, 094302 (2008)
400. W. Zhang, T.S. Fisher, N. Mingo, The atomistic Green's function method: an efficient simulation approach for nanoscale phonon transport. Numer. Heat Transfer, Part B: Fundam. **51**, 333–349 (2007)
401. N. Mingo, D.A. Stewart, D.A. Broido, D. Srivastava, Phonon transmission through defects in carbon nanotubes from first principles. Phys. Rev. B **77**, 033418 (2008)
402. Y. Chen, D. Li, J.R. Lukes, Z. Ni, M. Chen, Minimum superlattice thermal conductivity from molecular dynamics. Phys. Rev. B **72**, 174302 (2005)
403. Y. Chen, D. Li, J. Yang, Y. Wu, J.R. Lukes, A. Majumdar, Molecular dynamics study of the lattice thermal conductivity of Kr/Ar superlattice nanowires. Phys. B: Condens. Matter **349**, 270–280 (2004)
404. K.Y. Koh, D.G. Cahill, Frequency dependence of the thermal conductivity of semiconductor alloys. Phys. Rev. B **76**, 075207 (2007)
405. N. Scoville, C. Bajgar, J. Rolfe, J. Fleurial, J. Vandersande, Thermal conductivity reduction in SiGe alloys by the addition of nanophase particles. Nanostruct. Mater. **5**, 207–223 (1995)
406. C.B. Vining, A model for the high-temperature transport properties of heavily doped n-type silicon-germanium alloys. J. Appl. Phys. **69**, 331–341 (1991)

Chapter 5
Waste Thermal Energy Harvesting (II): Pyroelectric Effect and Others

5.1 Introduction

As discussed in Chap. 2, pyroelectric materials are a subgroup of piezoelectric materials and ferroelectric materials are a subgroup of pyroelectric materials. Therefore, ferroelectric materials have much better pyroelectric performance among all pyroelectric materials, according to the grouping with crystal structures [1–3]. This is the reason why the study of pyroelectric effects only focuses on ferroelectric materials. The prefix "pyro-" is a word originating from the Greek phrase "pyr," meaning "fire," that implies a relation to fire or heat. Pyroelectricity or pyroelectric effect describes a relation between thermal and electrical entity.

Pyroelectricity, known as a change in net dipole moment, has been a scientific achievement in modern times, while as a phenomenon, the pyroelectric effect has been known for 24 centuries, since it was mentioned by the Greek philosopher Theophrastus [4, 5]. He described a stone, called *lyngourion* in Greek or *lyncurium* in Latin, that had the property of attracting straws and bits of wood. Those attractions were actually the effects of electrostatic charges created due to the changes in temperature in the mineral tourmaline.

The initial interest was in the origin of the stone and its possible therapeutic functions, rather than physical effect itself. Theophrastus proposed that *lyngourion* was formed from the urine of a wild animal, which was identified later by Pliny the Elder as a lynx [6]. Two thousand years after Theophrastus, the unique physical properties of tourmaline were reintroduced to Europe, due to the publication of a book entitled *Curiöse Speculationes bey Schlaflosen Nächten* (*Curious Speculations During Sleepless Nights*) in 1707, authored by Johann Georg Schmidt, using the pen name Immer Gern Speculirt (Always Gladly Speculating). Among a series of 48 dialogs, there was one that contained a section describing this phenomenon.

In 1717, the physician and chemist Louis Lemery wrote the first scientific paper on pyroelectricity in a journal. The naturalist Carl von Linné (Linnaeus) was the first who related the pyroelectric property of tourmaline to electricity. The mineral *lapis electricus* was called electric stone. During the eighteenth century, various interesting experiments were carried out on tourmaline by a number of pioneers,

L. B. Kong et al., *Waste Energy Harvesting*, Lecture Notes in Energy 24,
DOI: 10.1007/978-3-642-54634-1_5,

including Franz Ulrich Theodor Aepinus, Johann Karl Wilcke, Benjamin Wilson, Joseph Priestley, John Canton, and Torben Bergman, which made major contributions to the field of electrostatics.

Due to the availability of more sophisticated techniques, a more quantitative understanding of pyroelectricity emerged during the nineteenth century. The term "pyroelectricity" was first used in 1824 by David Brewster, a famous scientist for his achievements in optics. The most famous compound, Rochelle salt, was one of the materials he studied.

A short while later, electrometer was developed by Antoine Becquerel and others, which made John Mothée Gaugain to be able to precisely measure pyroelectric charges for the first time in 1859, thus leading to several important conclusions. For example, it was accepted that the total quantity of electricity produced by a crystal of tourmaline was dependent upon the variation of temperature and the amount of electricity produced during heating was the same as that produced during cooling, but with the signs of the charges being reversed. It was also confirmed that the amount of charge produced was proportional to the crosssectional area of the crystal, but independent of its length.

The first theoretical work on pyroelectricity was reported by William Thomson (Lord Kelvin) in 1878, which predicted the presence of electrocaloric effect. Shortly after that, in 1880, Jacques and Pierre Curie speculated that the electrical effects of quartz crystals due to the nonuniform heating was caused by pressure, which led to the discovery of piezoelectricity. During the latter part of the nineteenth century and the early decades of the twentieth century, a large number of papers on pyroelectricity were published by authors including seven Nobel laureates whose expertise were actually in other fields.

In the beginning of 1920s, Rochelle salt was found to shown strong ferroelectricity by Joseph Valasek. After that, there was short period that the interest in pyroelectricity seemly vanished until 1938. It was the paper published by Yeou Ta that initiated the great growth of research on pyroelectricity which continues until today. Ta proposed that tourmaline crystals could be used as IR sensors in spectroscopy, which triggered some research on pyroelectric IR detectors during and immediately after World War II in the UK, the US, and Germany. In 1962, J. Cooper made the first detailed analysis of the behavior of fast IR detectors and conducted experiments using barium titanate, $BaTiO_3$, the first ferroelectric ceramics. In that year, S. B. Lang proposed the use of pyroelectric devices to measure the change of temperature with a magnitude as small as 0.2 mK. An explosive growth in theoretical studies, basic measurements and applications was thus begun.

The earlier works on pyroelectricity have been mainly on the applications as detectors [7, 8]. Recent development in the researches focusing on renewable energy and environment has shifted from the applications of pyroelectric materials in detector-like devices to emphasizing their potential for energy harvesting [9].

Contrary to thermoelectric generators, pyroelectric materials do not need a temperature gradient (spatial gradient), but temporal temperature changes [9]. This opens different applications fields, where temperature gradients are not possible and where temperature is not static. Small-scale microgenerators with dimensions smaller than the temperature spatial fluctuation length may with difficulty be subjected to temperature gradients. Natural temperature time variations occur due to convection process, and this thermal energy is difficult to transform in a stable temperature gradient. On the other hand, it is possible to transform a temperature gradient into a temperature variable in time using a caloric fluid pumping between hot and cold reservoirs. The pumping unit may require much less energy than the total produced energy (depending on the scale of the device) and may produce temperature variations of 1–20 °C at 2 Hz for example. To optimize energy harvesting from temperature, the first step should be the optimization of energy conversion. Then, the problems of electric loading (modifying the cycles shape) should be addressed.

The objective of this chapter is to review the progress in the research of pyroelectric materials for waste energy applications. Principles of pyroelectric effect, including definition, pyroelectric coefficients, coefficient measurement, phase transition, and so on, will be presented first, which will be followed by a brief description on type and characteristics of the state-of-the-art pyroelectric materials. After that, the latest progress in pyroelectric waste heat energy harvesters based on the Olsen cycle and other types of principles, with pyroelectric polymers, perovskite ceramics, and single crystals, will be summarized. The chapter will be ended with some concluding remarks.

5.2 Principle of Pyroelectric Effect

5.2.1 Definition

As stated earlier, pyroelectric is defined as the phenomenon of change in polarization of a material induced by a change in temperature. For a single domain ferroelectric material, there are localized charges that are present on the two end surfaces, due to the alignment of the polarization. At thermal equilibrium state, these localized charges are shielded by the free charges with same quantity but opposite signs, so that there is no net electricity produced in the ferroelectric material. However, when there is a change in temperature, the polarization of the material is changed. As a result, the free charges cannot completely shield the localized charges and thus there will be free charges at the surfaces, which leads to the formation of an electric field nearby. The presence of the electric field is indicated by the factor that the material can attract or repel charged particles. If the surfaces are connected with external circuit, there will be electric current flowing through it. The directions of the electrical currents corresponding to heating and cooling are opposite.

Similar to the piezoelectric of ferroelectric materials, the charge of current of pyroelectric effects are closely related to the change in polarization, which thus breaks the equilibrium with free charges, so that the amount of surface charges are increased and decreased, corresponding to heating and cooling, respectively. The difference is that the change in polarization of pyroelectric effect is caused by the change in temperature, while that of piezoelectric effect is caused by mechanical stress. Physics of pyroelectric effect has been well described in the open literature, for instance, Ref. [8] and references therein. A brief description of various physical quantities related to pyroelectric effect and pyroelectric materials is present in this subsection.

Pyroelectric effect of a pyroelectric material is characterized by pyroelectric coefficient, a parameter to measure pyroelectric efficiency of the material. With a small change in temperature, ΔT, the variation induced can be given by:

$$\Delta P = p\Delta T, \tag{5.1}$$

where p is pyroelectric coefficient, which is vector and has three nonzero components, with unit of C m^{-2} K^{-1}:

$$p_m = \frac{\partial P_m}{\partial T}, \quad m = 1, 2, 3. \tag{5.2}$$

The sign of pyroelectric coefficient is determined with respect to the piezoelectric axis of piezoelectric crystal. According to IRE standards, the end a crystal axis with positive charge under tension is defined as the positive end. When a pyroelectric crystal is heated, if there is positive produced at the positive side, pyroelectric coefficient is defined to be positive, otherwise, it is negative. Generally, because the spontaneous polarization of ferroelectric materials decreases with increasing temperature, pyroelectric coefficient is usually negative. However, there are also exceptions, for example, the spontaneous polarization of Rochell salt increases with increasing temperature slightly below its Curie point.

Because best pyroelectric materials are piezoelectric materials, the deformation of a pyroelectric material due to the change in temperature can induce variation in polarization, which also has contribution to pyroelectric effect. Therefore, it is important to pay an attention to mechanical boundary conditions and the ways of change variation.

Under uniform heating or cooling, depending on mechanical boundary conditions, there are two types of pyroelectric effects. Under constant strain (sample is clapped), only the variation in polarization induced by temperature change contributes to pyroelectric effect. It is called primary pyroelectric effect or constant strain pyroelectric effect. Usually, it is difficult to maintain a constant strain. In other words, samples are in free-strain state or constant stress state. In this case, the variation in polarization induced by the mechanical deformation of the sample due the thermal expansion will be added on top of the primary pyroelectric effect. This additional pyroelectric effect at uniform temperature is called secondary pyroelectric effect. Therefore, the pyroelectric coefficient of a material under constant stress is the summation of primary and the secondary coefficients, which is known as total pyroelectric coefficient. However, in practice, the pyroelectric element is a pyroelectric device is neither under constant strain or constant stress [10]. For example, it is only strained in one specific direction and free in other directions. In this case, the pyroelectric coefficient is called partially strained coefficient.

If a pyroelectric sample is heated or cooled in a nonuniform manner, it will experience stress gradient, which also contributes to pyroelectric effect, through piezoelectric effect. This nonuniform heating or cooling induced pyroelectric effect is called tertiary effect or false effect. It is named false effect, because piezoelectric materials, including those outside the ten polar point group, have such an effect under nonuniform heating or cooling. Those piezoelectric materials actually have no pyroelectric effect. As a result, the heating or cooling should be uniform to exclude the false pyroelectric effect.

5.2.2 Pyroelectric Coefficient and Electrocaloric Coefficient

Thermodynamic states of elastic dielectrics can be described with temperature T, entropy S, electric field E, electric displacement D, strain X, and stress x. With T, E and X as independent variables, the differential form of electric displacement can be expressed as:

$$\begin{aligned} \mathrm{d}D_m &= \left(\frac{\partial D_m}{\partial X_i}\right)_{E,T}\mathrm{d}X_i + \left(\frac{\partial D_m}{\partial E_n}\right)_{X,T}\mathrm{d}E_n + \left(\frac{\partial D_m}{\partial T}\right)_{E,X}\mathrm{d}T \\ &= \mathrm{d}_{mi}^{E,T}\mathrm{d}X_i + \varepsilon_{mn}^{X,T}\mathrm{d}E_n + p_m^{E,X}\mathrm{d}T, \end{aligned} \tag{5.3}$$

where, the subscript $m = 1 - 3$ and $i = 1 - 6$, while the superscript means that the physical quantities are kept constant. On the right hand side, the first and the second terms are piezoelectric and dielectric characteristics, whereas the third term is pyroelectric property of a material.

When $\mathrm{d}X_i = 0$ and $\mathrm{d}E_n = 0$, there is:

$$\mathrm{d}D_m = p_m^{E,X}\mathrm{d}T. \tag{5.4}$$

With independent variables, T, E and X, the characteristic functions of Gibbs free energy become:

$$G = U - TS - X_i x_i - E_m D_m. \tag{5.5}$$

According to thermodynamic first and second laws, there is:

$$\mathrm{d}G = -x_i\mathrm{d}X_i - D_m\mathrm{d}E_m - S\mathrm{d}T. \tag{5.6}$$

The differential form of the Gibbs free energy, with respect to T, E and X can be expressed as:

$$\mathrm{d}G = \left(\frac{\partial G}{\partial X_i}\right)_{E,T}\mathrm{d}X_i + \left(\frac{\partial G}{\partial E_m}\right)_{X,T}\mathrm{d}E_m + \left(\frac{\partial G}{\partial T}\right)_{E,X}\mathrm{d}T. \tag{5.7}$$

By comparing Eqs. (5.6) and (5.7), there are:

$$\left(\frac{\partial G}{\partial E_m}\right)_{X,T} = -D_m, \tag{5.8}$$

$$\left(\frac{\partial G}{\partial T}\right)_{E,X} = -S, \tag{5.9}$$

$$\left(\frac{\partial^2 G}{\partial E_m \partial T}\right)_X = -\left(\frac{\partial D_m}{\partial T}\right)_{E,X} = -p_m^{E,X}, \tag{5.10}$$

$$\left(\frac{\partial^2 G}{\partial E_m \partial T}\right)_X = -\left(\frac{\partial S}{\partial E_m}\right)_{X,T}. \tag{5.11}$$

Pyroelectric coefficient is thus defined in Eq. (5.10), whereas Eq. (5.11) describes the change in entropy induced by external electric field, which is known as electrocaloric coefficient. Therefore, electrocaloric effect is inverse pyroelectric effect, and thus there is:

$$p_m^{E,X} = -\left(\frac{\partial S}{\partial E_m}\right)_{X,T}. \tag{5.12}$$

It means that the pyroelectric coefficient under constant electric field and stress is equal to the electrocaloric effect under constant stress and temperature.

With independent variables, T, E, and x, the differential form of electric displacement can be expressed as:

$$\begin{aligned} \mathrm{d}D_m &= \left(\frac{\partial D_m}{\partial x_i}\right)_{E,T} \mathrm{d}x_i + \left(\frac{\partial D_m}{\partial E_n}\right)_{x,T} \mathrm{d}E_n + \left(\frac{\partial D_m}{\partial T}\right)_{E,x} \mathrm{d}T \\ &= e_{mi}^{E,T}\mathrm{d}x_i + \varepsilon_{mn}^{x,T}\mathrm{d}E_n + p_m^{E,x}\mathrm{d}T. \end{aligned} \tag{5.13}$$

Under constant strain and electric field, there is:

$$\mathrm{d}D_m = p_m^{E,x}\mathrm{d}T. \tag{5.14}$$

Similarly, with independent variables, T, E, and x, the characteristic functions of Gibbs free energy are written as:

$$\mathrm{d}G_2 = X_i\mathrm{d}x_i - D_m\mathrm{d}E_m - S\mathrm{d}T, \tag{5.15}$$

Similar to Eq. (5.7), there is:

$$\mathrm{d}G_2 = \left(\frac{\partial G_2}{\partial x_i}\right)_{E,T} \mathrm{d}x_i + \left(\frac{\partial G_2}{\partial E_m}\right)_{x,T} \mathrm{d}E_m + \left(\frac{\partial G_2}{\partial T}\right)_{E,x} \mathrm{d}T, \tag{5.16}$$

As a result, there are:

$$\left(\frac{\partial G_2}{\partial E_m}\right)_{x,T} = -D_m, \tag{5.17}$$

$$\left(\frac{\partial G_2}{\partial T}\right)_{E,x} = -S. \tag{5.18}$$

According to partial derivatives, there is:

$$p_m^{E,x} = -\left(\frac{\partial S}{\partial E_m}\right)_{x,T}. \tag{5.19}$$

Therefore, the pyroelectric coefficient under constant electric field and strain is equal to the electrocaloric effect under constant strain and temperature.

The change in entropy of a ferroelectric material is due to the change in ordering of polarization. The higher the degree of ordering, the lower the value of entropy of the system will be. As a consequence, depolarization of a ferroelectric material leads to an increase in value of entropy, because the degree of ordering of polarization is decreased. Under adiabatic conditions, depolarization results in a decrease in temperature, which is the reason why electrocaloric effect can be used for refrigeration through depolarization.

5.2.3 Primary and Secondary Pyroelectric Coefficient

At zero electric field (constant electric field), electric displacement is only the function of strain and temperature, while strain is only the function of stress and temperature, i.e.:

$$\mathrm{d}D_m = \left(\frac{\partial D_m}{\partial x_i}\right)_T \mathrm{d}x_i + \left(\frac{\partial D_m}{\partial T}\right)_x \mathrm{d}T, \tag{5.20}$$

$$\mathrm{d}x_i = \left(\frac{\partial x_i}{\partial X_j}\right)_T \mathrm{d}X_j + \left(\frac{\partial x_i}{\partial T}\right)_X \mathrm{d}T. \tag{5.21}$$

With $\mathrm{d}X_j = 0$, combination of Eqs. (5.20) and (5.21) leads to:

$$\mathrm{d}D_m = \left[\left(\frac{\partial D_m}{\partial x_i}\right)_T \left(\frac{\partial x_i}{\partial T}\right)_X + \left(\frac{\partial D_m}{\partial T}\right)_x\right] \mathrm{d}T. \tag{5.22}$$

Rearranging Eq. (5.22), there is:

$$\left(\frac{\partial D_m}{\partial T}\right)_X = \left(\frac{\partial D_m}{\partial T}\right)_x + \left(\frac{\partial D_m}{\partial x_i}\right)_T \left(\frac{\partial x_i}{\partial T}\right)_X, \tag{5.23}$$

where the left hand side term is total pyroelectric coefficient, while the first term on the right hand side is primary pyroelectric coefficient and the second term is secondary coefficient.

In Eq. (5.23), there are:

$$\left(\frac{\partial D_m}{\partial x_i}\right)_T = e_{mi}, \tag{5.24}$$

$$\left(\frac{\partial x_i}{\partial T}\right)_X = \alpha_i, \tag{5.25}$$

where e_{mi} and α_i are piezoelectric constant and thermal expansion coefficient, respectively. Therefore, Eq. (5.23) becomes:

$$p_m^X = p_m^x + e_{\mathrm{mi}}^T \alpha_i^X, \tag{5.26}$$

where the second term on the right hand side means that the secondary pyroelectric coefficient is the product of piezoelectric constant and thermal expansion coefficient.

If Eq. (5.23) is rewritten as:

$$\left(\frac{\partial D_m}{\partial T}\right)_X = \left(\frac{\partial D_m}{\partial T}\right)_x + \left(\frac{\partial D_m}{\partial X_i}\right)_T \left(\frac{\partial X_i}{\partial x_j}\right)_T \left(\frac{\partial x_j}{\partial T}\right)_X, \tag{5.27}$$

there is following expression for total pyroelectric coefficient:

$$p_m^X = p_m^x + d_{\mathrm{mi}}^T c_{ij}^T \alpha_j^X, \tag{5.28}$$

where the second term on the right hand side means that the secondary pyroelectric coefficient is the product of piezoelectric constant, elastic stiffness, and thermal expansion coefficient. In most cases, the primary coefficient is the main contributor of pyroelectric materials.

5.2.4 Tertiary Pyroelectric Coefficient and Other Aspects

As stated before, under the condition of nonuniform heating or cooling without mechanical stress, the total pyroelectric coefficient consists of primary, secondary and tertiary coefficients. The tertiary pyroelectric coefficient is attributed to thermal stress, whose contribution to the change in polarization is $d_{mnp}X_{np}(\boldsymbol{r}, t)$, where d_{mnp} and X_{np} are piezoelectric constant and thermal stress component, while $\boldsymbol{r}$ and t are displacement and time, respectively [11]. Due to the difficulty in characterization of thermal stress which is a function of location and time, it is a challenge to accurately measure the tertiary pyroelectric coefficient.

When elastic Gibbs free energy G_1 is used as characteristic function, with variables, such as stress and electric field, to be one dimensional, there is:

$$dG_1 = -SdT + xdX + EdD. \quad (5.29)$$

Taking derivative of Eq. (5.29), there are:

$$\frac{\partial G_1}{\partial D} = E, \quad (5.30)$$

$$\frac{\partial^2 G_1}{\partial D^2} = \frac{1}{\varepsilon}. \quad (5.31)$$

At temperatures near Curie point, there is:

$$\frac{\partial G_1}{\partial D} = \alpha D + \beta D^3 + \gamma D^5, \quad (5.32)$$

which can be rewritten as:

$$E = \alpha_0 (T - T_0) D + \varphi(D), \quad (5.33)$$

where $\varphi(D)$ represents all higher order terms and $\alpha_0 = 1/(\varepsilon_0 C)$, with C to be the Curie constant, which leads to:

$$\frac{\partial E}{\partial T} = \alpha_0 D. \quad (5.34)$$

Therefore, if $E = 0$ and $D = P_s$, at temperatures near the Curie point, there is;

$$\alpha_0 (T - T_0) P_s + \varphi(P_s) = 0, \quad (5.35)$$

i.e.:

$$\left.\frac{\partial G_1}{\partial D}\right|_{D=P_s} = \alpha_0 (T - T_0) P_s + \varphi(P_s) = 0, \quad (5.36)$$

$$\frac{d}{dT}\left(\frac{\partial G_1}{\partial D}\right)_{D=P_s} = \left.\frac{\partial^2 G_1}{\partial D^2}\right|_{D=P_s} \frac{dD}{dT} + \left.\frac{\partial^2 G_1}{\partial T \partial D}\right|_{D=P_s} = 0. \quad (5.37)$$

By comparing Eqs. (5.31) and (5.34), it leads to:

$$\left.\frac{\partial^2 G_1}{\partial T \partial D}\right|_{D=P_s} = \left.\frac{\partial E}{\partial T}\right|_{D=P_s} = \alpha_0 P_s. \quad (5.38)$$

By putting Eqs. (5.38) to (5.37), there is:

$$\frac{p}{\varepsilon_r} + \alpha_0 P_s = 0, \quad (5.39)$$

where $p = dD/dT$, which is pyroelectric coefficient, thus leading to:

$$\frac{p}{\varepsilon_r} = \frac{P_s}{C}, \quad (5.40)$$

which correlates the pyroelectric coefficient with Curie constant, spontaneous polarization and permittivity of a pyroelectric material. With this relationship, pyroelectric coefficient can be estimated by using the Curie constant, and vice versa. It is necessary to mention the temperature range of validity of the formula should not be too wide.

It has been found that although there is a significant difference in pyroelectric coefficient and permittivity among various pyroelectric/ferroelectric materials, the value of the ratio, $p\varepsilon_r^{-1/2}$, is nearly constant for most ferroelectric materials, with a temperature range across room temperature. This phenomenon has been well explained by using ferroelectric phenomenological theories. With Devonshire's assumption, the elastic Gibbs free energy G_1 can be expressed as:

$$\frac{\partial G_1}{\partial D} = \alpha_0(T - T_0)D + \beta D^3 + \gamma D^5 = E. \tag{5.41}$$

Near the Curie point, with $E = 0$ and $D = P_s$, there is:

$$\alpha_0(T - T_0)P_s + \beta P_s^3 + \gamma P_s^5 = 0, \tag{5.42}$$

The solutions of this equation are:

$$P_s^2 = -\frac{\beta}{2\gamma}\left\{1 + \left[1 - 4\alpha_0\gamma\beta^{-2}(T - T_0)\right]^{1/2}\right\}, \tag{5.43}$$

$$P_s^2 = -\frac{\beta}{2\gamma}\left\{1 - \left[1 - 4\alpha_0\gamma\beta^{-2}(T - T_0)\right]^{1/2}\right\}, \tag{5.44}$$

where Eq. (5.43) is for $\beta < 0$ and Eq. (5.44) is for $\beta > 0$. For second-order phase transition with $\beta > 0$, Eq. (5.43) becomes:

$$P_s^2 = -\frac{\beta}{2\gamma} - \frac{\beta}{2\gamma}\left[1 + \frac{4\gamma(T_0 - T)}{\varepsilon_0 C\beta^2}\right]^{1/2}. \tag{5.45}$$

For first-order phase transition, it is:

$$P_s^2 = -\frac{|\beta|}{2\gamma} - \frac{|\beta|}{2\gamma}\left[1 - \frac{4\gamma(T_0 - T)}{\varepsilon_0 C\beta^2}\right]^{1/2}. \tag{5.46}$$

Pyroelectric coefficient can be derived from their derivatives with respect to temperature, given by:

$$p = (2P_s\varepsilon_0 C|\delta|)^{-1}\left[1 + \frac{4\gamma(T_0 - T)}{C\beta^2}\right]^{1/2}. \tag{5.47}$$

With Eq. (5.40), timing the two sides of Eq. (5.47) by p/ε_r and P_s/C, respectively, there is:

$$\frac{p^2}{\varepsilon_r} = P_0^2(2\varepsilon_0 C T_0)^{-1}\left[1 + \frac{4P_0^2\gamma}{|\beta|}\left(1 - \frac{T}{T_0}\right)\right]^{1/2}, \tag{5.48}$$

where

$$P_s^2 = T_0 \frac{\partial P_s^2}{\partial T}\bigg|_{T \le T_0} = \frac{2T_0}{\varepsilon_0 C|\beta|}. \tag{5.49}$$

When $T \approx T_0$, Eq. (5.48) can be simplified as:

$$p\varepsilon_r^{-1/2} \approx P_0(2CT_0)^{-1/2}. \tag{5.50}$$

The equation can be used to understand the above mentioned phenomenon, which has been taken as a reference to estimate the performance of pyroelectric materials.

5.2.5 Pyroelectric Effect Versus Phase Transition

For ferroelectric pyroelectric materials, there are two types of phase transition, where a peak is observed in pyroelectric coefficient curve with temperature [12]. Ferroelectric-paraelectric transition is the most important phase occurring in ferroelectric pyroelectric materials. During this phase transition, there is a sharp change in spontaneous polarization.

For first-order ferroelectrics, at temperatures slightly higher than Curie point T_C, external electric field may induce the appearance of ferroelectric phase, where the temperature of the maximized variation in polarization, corresponding to the peak of pyroelectric coefficient, increases with increasing magnitude of external applied electric field [13]. This phenomenon can be readily explained with Eq. (5.41). According to reduced electric displacement d, electric field e and temperature t, there are:

$$d = -(2\gamma/|\beta|)^{1/2} D, \tag{5.51}$$

$$e = -8\left(2\gamma^3/|\beta|^5\right)^{1/2} E, \tag{5.52}$$

$$t = 4\alpha_0 \beta^{-2}(T - T_0). \tag{5.53}$$

Then, Eq. (5.41) can be simplified as:

$$e = 2d^5 - 4d^3 + 2td. \tag{5.54}$$

According to the definition of pyroelectric coefficient, we have reduced pyroelectric coefficient, given by:

$$\frac{\partial d}{\partial t} = \left(4d - 4d^3 - \frac{e}{2d^2}\right)^{-1} = \frac{d}{6d^2 - 5d^4 - t}. \tag{5.55}$$

For second-order phase transition ferroelectrics, there is no electric-field-induced phase transition [12]. The application of an external electric field only suppresses the curve of dielectric constant versus temperature and thus the temperature corresponding to the pyroelectric coefficient peak is kept unchanged. However, the magnitude of pyroelectric coefficient decreases with increasing magnitude of electric field. By omitting the higher order term, Eq. (5.41) is simplified as:

$$E = \alpha_0(T - T_0)D + \beta D^3, \tag{5.56}$$

where $T_0 = T_C$. Because electric field is constant, derivative of Eq. (5.56) is:

$$0 = \left[\alpha_0(T - T_0) + 3\beta D^2\right]\mathrm{d}D + \alpha_0 D\mathrm{d}T. \tag{5.57}$$

Therefore, pyroelectric coefficient is given by:

$$p = \left(\frac{\partial D}{\partial T}\right)_{E,X} = \frac{-\alpha_0 D_0}{\alpha_0(T - T_0) + 3\beta D^2}, \tag{5.58}$$

where D_0 is the electric displacement induced by the electric field E at temperature T, which can be calculated by using Eq. (5.56). There are three special cases, where we have:

$$D_0 \approx E[\alpha_0(T - T_0)]^{-1}, \quad T \gg T_0, \tag{5.59}$$

$$D_0 \approx \left[\alpha_0\beta^{-1}(T_0 - T)\right]^{-1/2}, \quad T \ll T_0, \tag{5.60}$$

$$D_0 \approx (\alpha_0/\beta)^{1/3}, \quad T = T_0. \tag{5.61}$$

Putting Eqs. (5.59)–(5.61) to (5.58), with p_0 to be the pyroelectric coefficient at $T = T_0$, we have:

$$p = E\alpha_0^{-1}(T - T_0)^{-2}, \quad T \gg T_0, \tag{5.62}$$

$$p = \frac{1}{2}\alpha_0^{1/2}\beta^{-1/2}(T_0 - T)^{-1/2}, \quad T \ll T_0, \tag{5.63}$$

$$p_0 = \frac{1}{3}\alpha_0/\beta^{-2/3}E^{-1/3}, \quad T = T_0. \tag{5.64}$$

Therefore, below Curie point, pyroelectric coefficient increases with temperature with $(T_0 - T)^{-1/2}$, while above Curie point, pyroelectric coefficient decreases with $(T - T_0)^{-2}$. At Curie point, pyroelectric coefficient is maximized, which is affected by electric field in terms of magnitude, but without shift in temperature.

Another phase transition is ferroelectric–ferroelectric phase transition, where both magnitude and direction of the spontaneous polarization in most cases. There are also seldom cases, where the direction of the polarization does change. In this case, other phase transition related parameters will experience only negligible variation, but pyroelectric effect still has a peak value. Therefore, pyroelectric coefficient has been a sensitive indicator of ferroelectric–ferroelectric phase transition of ferroelectric materials.

When a pyroelectric material is heated or cooled, from one temperature to another, it takes time for the polarization to reach a new equilibrium state. Therefore, the response of pyroelectric charges to the variation in temperature is not instantaneous, but with a certain degree of delay, depending on a number of factors, including thermal conductivity and heat capacity of the material, as well as sample dimension and shape. Generally, thermal relaxation time can be expressed as:

$$\tau_T = \frac{L^2 c'}{\sigma_T}, \tag{5.65}$$

where c' is the heat capacitance per unit volume in J K^{-1} m^{-3}, L is the length in the direction of heat conduction in meter (m) and σ_T is thermal conductivity in J K^{-1} m^{-1} s^{-1}. For example, a piece of TGS ($(NH_2CH_2COOH)_3{\cdot}H_2SO_4$) crystal with a thickness of 1 mm has a relaxation time τ_T of about 1 s [14], at near room temperature, while it is about 0.4 s for a 0.75 mm thick x-cut $LiNbO_3$ crystal. Therefore, it is important to have an estimate on τ_T before discussing pyroelectric response. It has been acknowledged that, during both the ferroelectric–ferroelectric and ferroelectric–paraelectric phase transitions, ferroelectric materials usually experience a peak in the value of their specific heat, which will lead to a significant increase in τ_T also [15].

5.2.6 Measurement of Pyroelectric Coefficient

There are four methods that have been widely used to measure pyroelectric coefficient of pyroelectric materials, namely, equivalent circuit method, voltage method, charge integration method, and current method.

5.2.6.1 Equivalent Circuit Method

Figure 5.1 shows schematic of equivalent circuit method. The sample, with thickness of a and surface area A, is poled in the direction perpendicular to the surfaces with electrode. C_L and R_L are loading capacitance and resistance, respectively, which include the contribution from both circuit and measurement equipment.

A voltage is produced across the load, forming an inner electric field $E = V/a$, which has a current density:

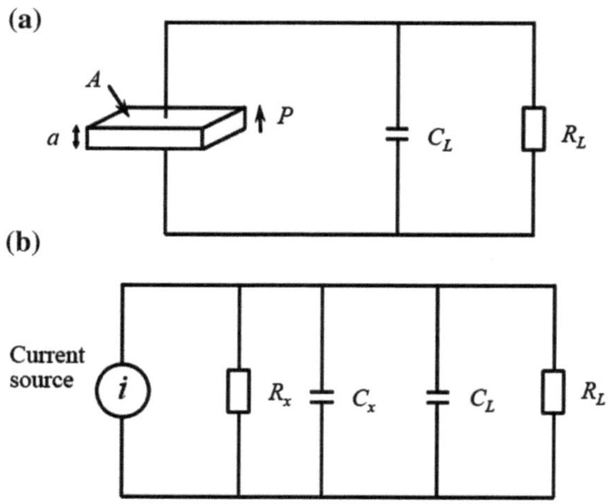

Fig. 5.1 Schematic diagram of equivalent circuit method (**a**) and equivalent circuit (**b**)

$$J = \sigma E + \frac{\partial D}{\partial t}, \tag{5.66}$$

where σ is conductivity. The first term on the right-hand side is conductive current density, while the second one is displacement current density. If the sample has resistance of R_x, the conductive current density is given by:

$$\sigma E = \frac{V}{R_x A}. \tag{5.67}$$

If the sample is not subject to external stress, the change in electric displacement can be expressed as:

$$\Delta D = \varepsilon^{X,T} \Delta E + p^{E,X} \Delta T, \tag{5.68}$$

Therefore, displacement current density is given by:

$$\frac{\partial D}{\partial t} = \frac{C_x}{A} \frac{\partial V}{\partial t} + p \frac{\partial T}{\partial t}, \tag{5.69}$$

where C_x is capacitance of the sample.

As a result, the total current flowing through the sample can be calculated with Eqs. (5.67) and (5.69) as:

$$I_{\text{tot}} = AJ = \frac{V}{R_x} + C_x \frac{\partial V}{\partial t} + pA \frac{\partial T}{\partial t}. \tag{5.70}$$

This equation means that the sample can be described by using a current source $p\partial T/\partial t$, a parallel-connected resistor R_x and a capacitor C_x, as shown by the left hand side half part in Fig. 5.1. Because the loading resistor R_L and capacitor C_L are also in parallel connection with the sample, the voltage across them is V and thus the currents flowing through them are V/R_L and $C_L \partial V/\partial t$, respectively.

Noting that all the currents flowing through C_x, R_x, C_L, and R_L are contributed by the pyroelectric current of the sample, we have:

$$(C_L + C_x)\frac{\partial V}{\partial t} + \left(\frac{1}{R_x} + \frac{1}{R_L}\right)V = -pA\frac{\partial T}{\partial t}. \tag{5.71}$$

Let:

$$C = C_L + C_x, \tag{5.72}$$

$$R = \frac{R_x R_L}{R_x + R_L}, \tag{5.73}$$

Equation (5.71) can be simplified as:

$$\frac{\partial V(t)}{\partial t} + \frac{V(t)}{RC} = -\frac{pA}{C}\frac{\partial T}{\partial t}. \tag{5.74}$$

Solving this differential equation, with initial conditions, $t = 0$, $V(0) = 0$, we have:

$$V(t) = \frac{pA}{C}\exp\left(-\frac{t}{RC}\right)\int_0^t \exp\left(\frac{t'}{RC}\right)\left(\frac{\mathrm{d}T}{\mathrm{d}t'}\right)\mathrm{d}t'. \tag{5.75}$$

This is the principle of the equivalent circuit method, with which pyroelectric coefficient p can be derived, since the time-dependent pyroelectric voltage $V(t)$ can be measured experimentally.

5.2.6.2 Voltage Method

Equation (5.75) describes a general time-dependent pyroelectric voltage, which should be specified to be compared with experimental data. For experimental convenience, only several special cases are considered.

If the temperature of the sample is changed in a very rapid fashion, the time-dependent pyroelectric voltage can be described by using time constant RC:

$$V(t) = \frac{pA\Delta T}{C}\exp\left(-\frac{t}{RC}\right). \tag{5.76}$$

For sufficient time, Eq. (5.76) can be further simplified as:

$$V = \frac{pA\Delta T}{C}. \tag{5.77}$$

This equation can be used to measure pyroelectric coefficient of pyroelectric materials with high resistivity. An effective strategy to maintain a high value of time constant is to use electrostatic voltmeter that has high input impedance, without connection of any other external loading.

If the temperature change is very small, Eq. (5.75) can be simplified as:

$$V(t) = pAR\left(\frac{\mathrm{d}T}{\mathrm{d}t}\right)_0\left[1 - \exp\left(-\frac{t}{RC}\right)\right], \tag{5.78}$$

where $\left(\frac{\mathrm{d}T}{\mathrm{d}t}\right)_0$ is the rate of temperature change at $t = 0$.

This equation has been used to measure pyroelectric coefficient, as well as direct current (DC) permittivity and resistivity [16–18], by continuously changing temperature, at two extreme conditions.

When the loading resistance is much smaller than the resistance of the sample, $R_L \ll R_x$, Eq. (5.78) becomes:

$$V(t) = pAR_L\left(\frac{\mathrm{d}T}{\mathrm{d}t}\right)_0\left[1 - \exp\left(-\frac{t}{R_LC}\right)\right]. \tag{5.79}$$

If the time is sufficiently long, pyroelectric voltage is given by:

$$V = pAR_L\left(\frac{\mathrm{d}T}{\mathrm{d}t}\right). \tag{5.80}$$

If the measurement is within a very short time span, so that $t \ll RC$, Eq. (5.78) can be simplified as:

$$V(t) = pA\left(\frac{\mathrm{d}T}{\mathrm{d}t}\right)_0\left(\frac{t}{C} - \frac{t^2}{2RC^2}\right). \tag{5.81}$$

During experiment, loading resistors with various values of RL are used to obtain curves of pyroelectric voltage versus time, $V - t$, with Eqs. (5.78), (5.80), and (5.81).

For example, if Eq. (5.78) is valid, the $V - t$ data can be fitted with the following equation:

$$V = A_1[1 - \exp(B_1t)]. \tag{5.82}$$

If Eq. (5.81) is valid, the $V - t$ data can be fitted with the following equation:

$$V = A_2 + B_2t + C_2t^2 \tag{5.83}$$

With the values of A_1, B_1, A_2, B_2, and C_2, pyroelectric coefficient p can be obtained. At the same time, R_x and C_x can be derived, from which resistivity and DC permittivity can be calculated. If pyroelectric coefficient is the only parameter to be measured, valid Eq. (5.80) is sufficient, with measured values of voltage.

There has been an alternative voltage method, where the temperature of the pyroelectric sample is varied in a cosine waveform at frequencies of 0.01–1 Hz [14]. From the changes of temperature and pyroelectric voltage, pyroelectric coefficient can be derived. At temperatures near T_0, T is changed with a magnitude of ΔT, i.e.,

$$T = T_0 + \Delta T \cos \omega t. \tag{5.84}$$

The change in temperature is converted into electrical signal. Between $T + \Delta T$ and $T - \Delta T$, if pyroelectric coefficient has not significant variation, the pyroelectric current can be written as:

$$I = pA \frac{\partial T}{\partial t} = pA\Delta T \omega \sin \omega t. \tag{5.85}$$

As shown in Fig. 5.1, the output pyroelectric voltage is given by:

$$V = pA\Delta T \omega R \left(1 + \omega^2 C^2 R^2\right)^{-1/2} \sin(\omega t + \varphi), \tag{5.86}$$

where $\varphi = \tan^{-1}(\omega CR)$. If $\omega CR \ll 1$ in experiment, pyroelectric voltage will be:

$$V = V_0 \sin \omega t, \tag{5.87}$$

where

$$V_0 = pA\Delta T \omega R. \tag{5.88}$$

By comparing Eqs. (5.84) and (5.87), it is found that the functions of pyroelectric voltage and temperature versus time are orthogonal, with which pyroelectric coefficient can be derived.

The three types of voltage method have their own characteristics. However, voltage methods have an obvious disadvantage, with two problems. First, it is difficulty to meet the requirement of zero electric field. This is because when there is a voltage across the sample, there must be an electric field, $E = V/a$. Since the sample has a nonzero resistance, there should be conductive current flowing through it, $I = \sigma E$. Because this current flows in the opposite direction of the pyroelectric current, it causes errors of the measured pyroelectric coefficient. The higher the voltage and lower the resistance of the sample, the more pronounced the error will be. Second, measurement results of pyroelectric coefficient are affected by the impedance of the sample, as demonstrated in Eqs. (5.76) and (5.78), which contain not only p but also resistance R_x and capacitance C_x. Pyroelectric coefficient is always measured with a certain range of temperature, for example, from $T - \Delta T$ to $T + \Delta T$, which has effects on resistance and capacitance and thus the pyroelectric coefficient of the sample.

5.2.6.3 Charge Integration Method

With a small change in temperature, ΔT, pyroelectric charge is cumulated with increasing time, which can be measured through:

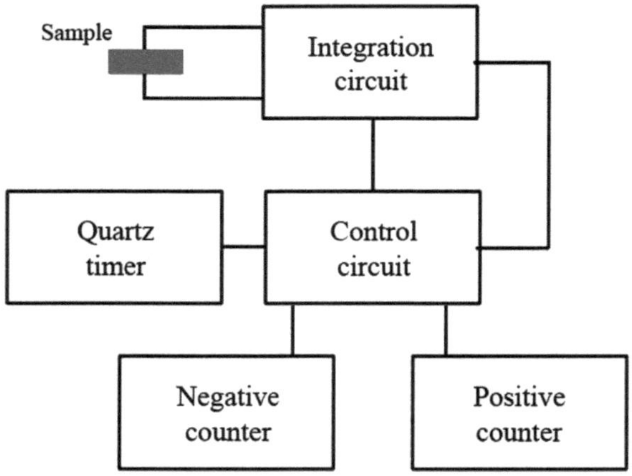

Fig. 5.2 Schematic diagram of charge integration method with operational amplifier

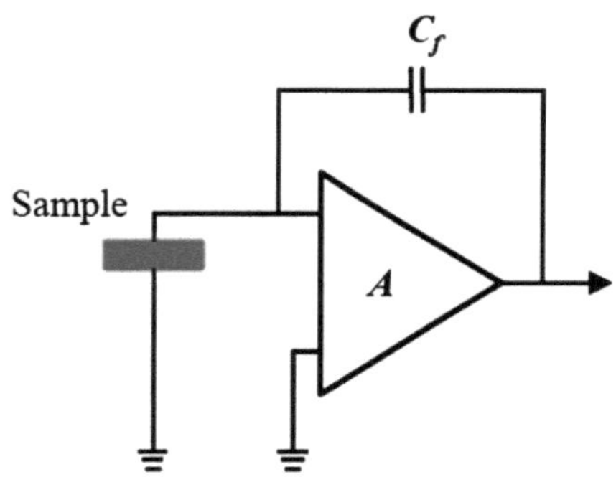

Fig. 5.3 Circuit of charge integration method with discharger and counter

$$Q = \int_0^t AJ\mathrm{d}t. \tag{5.89}$$

With the value of charge, pyroelectric coefficient can be derived as:

$$p = \frac{Q}{A\Delta T} = \frac{1}{A\Delta T}\int_0^t AJ\mathrm{d}t. \tag{5.90}$$

If the pyroelectric charge is accumulated at the two sides of the sample, it will destroy the zero electric field condition. Therefore, a charge integration method is proposed, in which the pyroelectric charge is transferred immediately when it is produced. The transferred charge is integrated and measured as total pyroelectric charge. By doing so, the zero electric field condition can be maintained at the same time.

Figure 5.2 shows schematic flow chart of charge integration method with the use of an operational amplifier [19]. A circuit of charge integration method with discharger is shown in Fig. 5.3 [8]. A is operational amplifier and C_f is feedback capacitor with known capacitance, $C_f = 1$ μF. Once produced, the pyroelectric charges will be transferred into the feedback capacitor, so that there is accumulated charge at the two sides of the sample, thus ensuring zero electric field condition. Pyroelectric coefficient can be calculated with the voltage across the capacitor C_f, by

$$V = \frac{Q}{C_f} = \frac{A}{C_f} \int_0^t J\mathrm{d}t. \tag{5.91}$$

Since $E = 0$, with Eqs. (5.66) and 5.68), we have:

$$\int_0^t J\mathrm{d}t = p \int_0^t \frac{\partial T}{\partial t} \mathrm{d}t = p\Delta T, \tag{5.92}$$

from which pyroelectric coefficient is derived as:

$$p = \frac{VC_f}{A\Delta T}. \tag{5.93}$$

With the method shown in Fig. 5.3, the sample is heated from T_1 to T_2, at which the sample is kept for a sufficiently long time. During this process, a positive or negative integration voltage is creased due to the integration of the pyroelectric charges. Once the voltage reaches a certain value, an opposite pulse discharge is triggered by the control circuit. The time of the discharging is controlled by the Quartz timer (0.5 ms). The discharging voltage is provided by using standard positive or negative power supplier, ensuring that every charging has almost the same value, $\delta Q \approx 10^{-11}$ C. The number of charging is counted with counter. Due to the small value of δQ, the voltage across the sample can be kept to be below μV level, offering a near zero electric field condition.

When stable readings of the positive and negative counters are N_+ and N_-, respectively, total pyroelectric charge is:

$$\Delta Q = (N_+ - N_-)\delta Q = \Delta N \delta Q. \tag{5.94}$$

Therefore, pyroelectric coefficient can be given by:

$$p = \frac{\Delta Q}{A\Delta T} = \frac{\Delta N \delta Q}{A(T_2 - T_1)}. \tag{5.95}$$

5.2.6.4 Current Method

From Eqs. (5.70) and (5.85), with $E = 0$, pyroelectric current and pyroelectric coefficient can be related by the following equation:

$$AJ = Ap\frac{\partial T}{\partial t}. \tag{5.96}$$

Equation (5.96) means that pyroelectric coefficient can be obtained from the current AJ, due to the change in temperature. One way for this method is to use heating at constant rate, by using differential thermal analyzer [20, 21]. The heating rate can be kept almost constant over a wider range of temperature, so that the variation of pyroelectric coefficient can be observed by using the response of pyroelectric current.

Another method is to heat the sample to make its temperature change periodically, by using shining the sample with a modulated square waved IR which is absorbed by the sample [45]. To ensure that all the energy is absorbed by the sample within the period of shining, there should be heat conduction between sample and sample holder. With this requirement, the pulse width of the incident wave should be shorter than the thermal time constant of the sample, which is:

$$\tau'_T = \frac{M}{G_T}, \tag{5.97}$$

where M is heat capacitance (J K^{-1}) of the sample and G_T is heat conductivity (J K^{-1} s^{-1}) between the sample and sample holder. This requirement can be met by using thermal insulating sample holder or inserting an insulating layer between the sample and the holder.

If the incident energy is W within unit time, the absorption coefficient of the sample is e, the rate of temperature change of the sample is given by:

$$\frac{dT}{dt} = \frac{eW}{M}. \tag{5.98}$$

Putting Eqs. (5.98) into (5.95), we have:

$$AJ = \frac{ApeW}{M}. \tag{5.99}$$

Although the heat capacitance and absorption coefficient are accurately measured, if they have no significant variation over the measurement temperature range, pyroelectric current is almost linearly proportional to pyroelectric coefficient.

Generally, pyroelectric samples are kept to be free of mechanical stress, so the pyroelectric coefficient measured is the summation of primary and secondary coefficients. To measure primary coefficient, it is necessary to keep the sample to be strain free (clapped). Otherwise, it can be calculated by using piezoelectric constant and thermal expansion coefficient through Eqs. (5.26) and (5.28). Both primary and secondary coefficients are functions of temperature. Furthermore, in some pyroelectric materials, they have opposite sign. In this case, the total pyroelectric coefficient could be zero if they are equal in quantity. Similarly, it is possible to observe a change in sign of pyroelectric coefficient [22, 23]. In addition, to avoid the presence of the tertiary coefficient, the sample should be heated or cooled uniformly, so as to prevent the presence of gradient strain.

In summary, voltage method is relatively simple, but the presence of voltage across the sample destroys the zero field condition. Therefore, it is necessary to use parallel capacitors with large capacitances, to reduce the inner electric field. The higher the conductivity of the sample, the larger the negative effect of the nonzero electric field will be. As a result, this method is more suitable for the pyroelectric materials with higher resistivity. The advantage of charge integration method is that zero electric field can be maintained. However, to have sufficiently amount of charge to be accumulated, the temperature change should be sufficiently large ($\Delta T > 2$ K). As a consequence, this method has relatively poor temperature resolution. Current method has high sensitivity and can be used for continuous temperature change. However, it is a challenge to measure the rate of temperature change experimentally. Also, pyroelectric current response is related to thermal properties of the materials, so that the measured current is a relative quantity and the measurement should be calibrated by using other methods.

5.2.7 Pyroelectric Figure of Merit

It is general to pyroelectric detector to show the response characteristics of pyroelectric materials. In Eq. (5.97), τ'_T is a measure of how fast a pyroelectric element reaches an equilibrium state between its surrounding, which is different from the relaxation time, τ_T, given in Eq. (5.65). τ_T is the time needed for the pyroelectric element to reach an equilibrium within the material itself.

Generally, a pyroelectric element is connected with a high input impedance amplifier. If the pyroelectric element has a capacitance of C_x, while the amplifier has an input resistance of R_g and capacitance of C_g, electrical time constant can be written as:

$$\tau_E^{'} = R_g(C_x + C_g). \tag{5.100}$$

The selection of thermal time constant and electrical time constant is determined by the requirements of detector, such as response speed and frequency dependence. Thermal time constant is usually between 0.01 and 10 s, while electrical time constant is over a wide range of 10^{-12}–100 s [24].

Let the incident power be W, the temperature change ΔT of the pyroelectric element is given by:

$$\eta W = M\frac{\mathrm{d}T}{\mathrm{d}t} + G_T\Delta T, \tag{5.101}$$

where η is the absorption percentage of W. If the incident radiation is expressed by:

$$W = W_0 \exp(j\omega t), \tag{5.102}$$

there is:

$$\Delta T = \eta W_0 (G_T + j\omega M)^{-1} \exp(j\omega t). \tag{5.103}$$

Pyroelectric charge produced by the element is:

$$q = pA\Delta T, \tag{5.104}$$

where p is pyroelectric coefficient and A is effective area of the element.

The current response is the current induced by the incident power, given by:

$$R_i = \frac{i}{W_0}, \tag{5.105}$$

where i is pyroelectric current, $i = \mathrm{d}q/\mathrm{d}t$, then we have:

$$R_i = \frac{\eta p A \omega}{G_T\left(1 + \omega^2 \tau_T'^2\right)^{1/2}}. \tag{5.106}$$

At frequencies that are much higher than $1/\tau'_T$, Eq. (5.106) can be simplified as:

$$R_i = \frac{\eta p}{c' d}, \tag{5.107}$$

where c' is heat capacitance per unit volume and d is thickness of the pyroelectric element. In Eq. (5.107), there is a parameter, which is called current figure of merit:

$$F_i = \frac{p}{c'}. \tag{5.108}$$

The voltage response can be derived from the pyroelectric current and admittance. With the ignorance of conduction of the pyroelectric element, the admittance can be written as:

$$Y = R_g^{-1} + j\omega C = R_g^{-1} + j\omega\left(C_x + C_g\right). \tag{5.109}$$

Then, the voltage response is given by:

$$R_v = \frac{i}{YW_0} = \frac{R_g \eta p A \omega}{G_T\left(1 + \omega^2 \tau_T'^2\right)^{1/2}\left(1 + \omega^2 \tau_E'^2\right)^{1/2}}. \tag{5.110}$$

When $\omega = (\tau'_E \tau'_T)^{-1/2}$, R_v has a maximum:

$$R_v(\max) = \eta p A \frac{R_g}{G_T(\tau_T' + \tau_E')}. \tag{5.111}$$

Therefore, a low G_T, i.e., a good insulation between the pyroelectric element and the element holder, means a high voltage response.

When $\omega \gg (\tau'_E)^{-1}$ and $\omega \gg (\tau'_T)^{-1}$, there is:

$$R_v = \frac{\eta p}{c'd(C'_x + C'_g)\omega}. \tag{5.112}$$

If $C_x \gg C_g$, there is:

$$R_v = \frac{\eta p}{c'\varepsilon A\omega}. \tag{5.113}$$

In Eq. (5.113), another material parameter, called voltage figure of merit, is given by:

$$F_v = \frac{p}{c'\varepsilon}. \tag{5.114}$$

It is difficult to measure specific heat, with $c' \approx 2.5 \times 10^6$ J m^{-3} K^{-1} for most widely used pyroelectric materials. As a simple measure, p/ε or p/ε_r, are used to characterize pyroelectric performance, which is called pyroelectric figure of merit.

According to Eq. (5.40), $p/\varepsilon_r \approx P_s/C$. With available ferroelectric database, for displacive ferroelectrics, the maximum value of p/ε_r is 4×10^{-6} C m^{-2} K^{-1}, whereas for order–disorder ferroelectrics, it is 20×10^{-6} C m^{-2} K^{-1} [25].

5.3 Pyroelectric Materials

Since ferroelectric materials have been found to show the highest pyroelectric figures of merit, the following discussion is only limited to ferroelectric pyroelectric materials. A thorough review on pyroelectric materials can be found in Refs. [7, 8, 10].

5.3.1 Triglycine Sulfate (TGS)

Triglycine sulfate (TGS), with a formula of $(NH_2CH_2COOH)_3H_2SO_4$, together with the large family of isomorphous compounds based on it, has been demonstrated to possess the highest pyroelectric figures of merit, among various ferroelectric pyroelectric materials. TGS crystals can be readily grown from aqueous solution [26]. TGS has a Curie temperature, $T_C = 49$ °C, above which it is centrosymmetric, class 2/m, and below which it is polar, point group 2 with the polar axis along the monoclinic ***b*** axis. The crystal structure has been extensively studied [27]. The glycine (HN_2CH_2COOH) groups are polar, which are located in the crystal structure as shown in the projection in Fig. 5.4. Reversal of the polarization in the material is largely associated with the rotation of the Glycine I group about the crystallographic ***a*** axis to change into its mirror image.

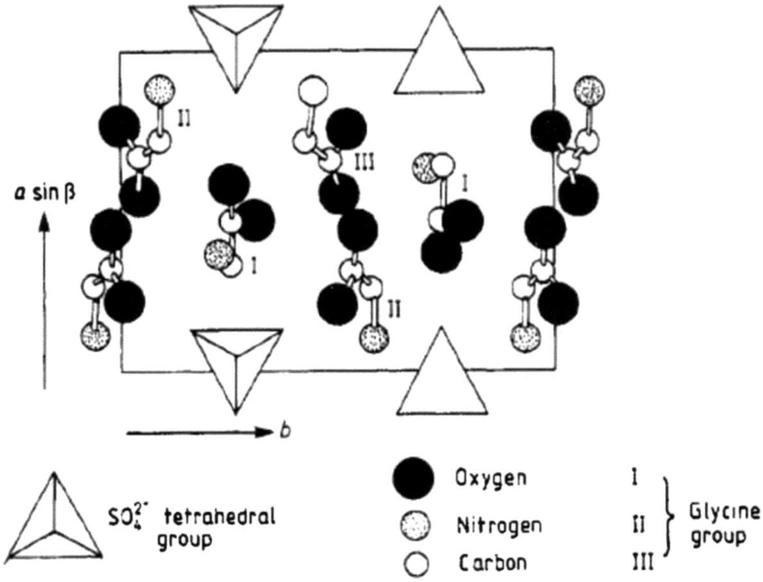

Fig. 5.4 Projection of the TGS crystal structure down the crystallographic c axis. Reproduced with permission from [8]. Copyright @ 1986, IOP Publishing

TGS has been modified to find out new compounds with higher pyroelectric performances. The modified compounds still maintain the same basic crystal structure of TGS. Examples include deuteration (DTGS), substitution of sulfuric with fluoroberyllic acid to form triglycine fluoroberyllate (TGFB) without and with deuteration (DTGFB) [28], and substitution of glycine either by L-alanine or D-alanine (ATGS). In the last example, there is an effect of "internal" electric bias which stabilizes the spontaneous polarization of the materials, because the alanine molecules have an extra methyl group than can prevent them from rotating within the lattice [29]. As a result, the dipole within it is fixed with respect to the crystal structure even at T_C. This is the reason why ATGS crystals can be thermally cycled through their T_C, while retaining spontaneous polarization. There have also been solution grown crystals doped with alanine and phosphoric or arsenic acids, which are ATGSP or ATGSAS [30, 31].

Performance figures of these materials at their respective optimum temperatures have been reported [32]. It can be shown that deuteration of TGS leads to materials with significant improvement in pyroelectric properties, by increasing T_C, which is desirable because as such the temperature range of operation can be increased [33]. For example, T_C of TGFB is higher than that of TGS by 12 °C. The As-doped materials have performance improvements by about 30 % in F_v, as compared with DTGS. It is interesting to mention that the TGS family offers much higher values of F_v than any of the oxide or polymer ferroelectric materials. However, at the same time, most of them also exhibit dielectric losses that are an order of magnitude higher. As a consequence, their values of F_D are comparable with those of oxide- and polymer-based materials.

The TGS family pyroelectric materials have large anisotropy in the principal dielectric constant tensor coefficients, which can be used to manipulate the p'/ε ratio, so as to optimize the pyroelectric performance [34, 35]. One example is to take oblique crystal cuts in which the normal to the electrode faces of the detector is not parallel to the polar axis, thus offering higher p'/ε ratio. This is because the effective pyroelectric coefficient has a cosine dependence on the angle of rotation, while the dielectric constant is dependent on a cosine squared relationship. Although TGS family pyroelectrics have exhibited the highest pyroelectric voltage responses, they have several serious problems in practical applications, such as water solubility, hygroscopic nature, and fragility.

5.3.2 Polyvinylidene Fluoride

As discussed in Chap. 2, polyvinylidene fluoride (PVDF) is a ferroelectric polymer, which possesses strong piezoelectric effects. The PVDF molecules have a repeat unit of ($-CH_2-CF_2-$), in which the carbon–hydrogen and the electrically polar carbon–fluorine bonds can take up a number of stable configurations dependent on the processing of the polymers. The nonpiezoelectric α-phase is formed from slow-cooled melts or acetone solution, in which the bonds are in a trans-gauche–trans-gauche configuration and the molecules are stacked to form a nonpolar unit cell. As the α-phase is stretched and electrically poled, it becomes β-phase, in which the molecular groups are in an all-trans configuration, so that are assembled to form a polar unit cell.

As a result, this form of PVDF is also a pyroelectric material, but with pyroelectric coefficients lower than those of the TGS family materials. Although this is offset to some extent by its low dielectric constant, the response figure of merit is lower than that of DTGS by a factor of six, whereas its high dielectric loss also leads to a low F_D. In addition, PVDF is ferroelectric, without showing a Curie point up to its melting point of about 180 °C. However, its polar properties start to degrade at temperatures of above 80 °C for extended periods of time, which can be considered as the upper limit to its operational temperature. As pyroelectric materials, PVDF-based ferroelectrics have major advantages, including the flexibility of large-area production and low cost. They are robust and do not require the lapping and polishing, which is different from their single crystals and ceramics counterparts.

Copolymers of vinylidene fluoride with trifluoroethylene (VDF-TRFE) are important ferroelectric and pyroelectric materials. They can crystallize into the β-phase directly from melts or solutions, without the requirement of stretching [36]. This offers the potential for depositing thin film elements directly onto a substrate by using spin-casting method. PVDF-TRFE has higher pyroelectric coefficients than PVDF [37, 38].

5.3.3 Lithium Tantalate

Lithium tantalite, $LiTaO_3$, is one of the most important pyroelectric materials. Its crystal structure consists of layers of oxygen ions in approximately hexagonal close packing, with Li^+ and Ta^{5+} ions occupying two thirds of the octahedral interstices between the layers. Its point group is $R\bar{3}m$ at room temperature, which transforms to R3m (nonpolar paraelectric) at $T_C = 665$ °C. It also has a high melting point of 1650 °C and is insoluble in water. These properties make the material one of the most stable pyroelectrics, with a very wide temperature range of operation. It is generally used in the form of single crystals, grown by using the Czochralski method. An example of $LiTaO_3$ single crystal is shown in Fig. 5.5 [39].

It exhibits a moderate pyroelectric coefficient and dielectric constant, which combine to offer a response figure of merit of about one quarter of TGS. However, the very low dielectric loss ($\sim 10^{-4}$) of TGS leads to a potentially high F_D, which can be four times greater than that of DTGS. $LiTaO_3$ has a serious disadvantage, i.e., it has a high thermal diffusivity, which reduces the minimum resolvable temperature difference at high spatial frequencies when it is used for array applications. One possible way to address this problem is the use of ion beam reticulation to physically separate the individual detector elements by cutting slots [40]. This material has a very high potential for commercial applications, due to its robustness, good performance, and wide availability.

5.3.4 Strontium Barium Niobate

Strontium barium niobate, $Sr_xBa_{1-x}Nb_2O_6$, with $0.25 \leq x \leq 0.75$, is a tungsten bronze ferroelectric, which has been extensively studied for pyroelectric applications [19]. Its crystal structure consists of corner-sharing NbO_6 octahedra, forming a framework within which there are three types of interstitial sites, with two of them being occupied by Ba/Sr ions [41]. The structure is tetragonal both above and below T_C, which varies from 195 °C for $x = 0.72$ to 53 °C for $x = 0.25$, with the point groups of 4 mm below T_C and 4/mmm above T_C. The composition with $x = 0.5$ (SBN-50) has been one of the most studied pyroelectric materials, which is characterized by a high pyroelectric coefficient, high dielectric constant, and low dielectric loss. These properties lead to a low response figure of merit, but a high D^* figure of merit, comparable with that of DTGS.

It has been found that the material can be doped with low levels of lanthanum (~ 1 atm%) or other rare earth elements, which have a significant effect on the electrical properties, by reducing T_C and increasing the pyroelectric coefficient and dielectric constant at room temperature [42]. Additionally, dopant-free and La-doped crystals can be cycled repeatedly up to T_C within a few degrees of temperature, without losing performance [43]. Therefore, this material with relatively low

Fig. 5.5 An example of LiTaO3 single crystal. Reproduced with permission from [39], Copyright @ 2005, Elsevier

T_C can be used in harsh environments (~70 °C). SBN-50 is robust and can be grown as single crystals by using the Czochralski method. Growth of high quality SBN single crystal is still a challenge.

5.3.5 Perovskite Structure

Ferroelectrics with perovskite structure have a general formula of ABO_3, which has been discussed in detail in Chap. 2. The structure can be described as a network of corner-sharing BO_6 octahedra, within which there is a 12-fold coordinated site occupied by the A cation. The A and B sites can be occupied by a wide range of ions, creating distortions to form rhombohedral, tetragonal, or orthorhombic structures. Two groups of perovskites have been extensively studied for their pyroelectric properties, rhombohedral structures based on modified lead zirconate ($PbZrO_3$) and tetragonal structures based on modified lead titanate ($PbTiO_3$). Although small pieces of single crystals of the two compounds can be grown from a variety of fluxes, the studies have been conducted mainly on ceramics [44]. This is because the distortion from the cubic aristotype provides with a large number of choices for the polar axis in any crystallite and at the same time a large percentage of the theoretical single-crystal polarization can be readily obtained in the ceramic forms. Moreover, a wide range of dopants can be introduced in a ceramic without altering the structure, so that single phase materials can be retained, but with high flexibility to optimize the material properties. These characteristics cannot be achieved by using single-crystal materials.

Modified lead zirconate ($PbZrO_3$ or PZ) is first described. PZ is an antiferroelectric. However, when PZ is modified with lead titanate ($PbTiO_3$ or PT), a number of ferroelectric phases can be obtained, as discussed in Chap. 2. According to the PZ-PT phase diagram, the $F_{R(LT)}$ phase in the region of PZ90:PT10 has a lower dielectric constant than the compositions in the center of the phase diagram, which thus has been of more interest for pyroelectric applications. A wide range of dopants, such as La^{3+} and $PbX_{1/2}Nb_{1/2}O_3$ (with X = Fe, Ni or Cr), have been attempted to optimize the pyroelectric properties [45]. It has been reported that the phase transition from $F_{R(LT)}$ to $F_{R(HT)}$ is associated with a step-like change in spontaneous polarization, giving rise to high values of dP_s/dT, with a very small change in dielectric constant, which has been exploited doping with bismuth and $Pb(Fe_{1/2}Nb_{1/2})O_3$ or $Pb(Fe_{1/2}Ta_{1/2})O_3$ [46, 47].

However, this transition is of first-order characteristic, so that the thermal hysteresis reduces the effective reversible pyroelectric coefficient, which is only a fraction of what has been expected from the dP_s/dT [48]. This problem has been addressed by using multicomposition materials, for example, $PbZrO_3$-$PbTiO_3$-$PbFe_{1/2}Nb_{1/2}O_3$ ceramics, which have pretty high pyroelectric coefficients [49, 50]. Pyroelectric response of these materials can be optimized by controlling their compositions, while UO_3 can be used to reduce their dielectric constant and loss tangent. In addition, the dopant allows the DC resistivity, ρ(DC), to be adjusted over a quite wide range, offering high flexibility in device design.

$PbTiO_3$ (PT) is a ferroelectric, with tetragonal structure at room temperature and a high Curie point of ~490 °C. It has a very high spontaneous polarization of up to 75 μC cm^{-2}. PT single crystals can be grown by using flux and top-seeded solution method, but they can be only small pieces [44, 51]. Therefore, it has been most widely studied in the form of ceramics. The high spontaneous lattice strain associated with the phase transition prohibits the fabrication of pure $PbTiO_3$ ceramics, as they are mechanically unstable when the samples are cooled down to room temperature after sintering. As a result, a wide range of dopants has been employed to modify the properties of PT-based ceramics. Dopants include manganese (Mn), rare earth ions, calcium (Ca) and $Pb(Co_{1/2}W_{1/2})O_3$ [52, 53].

There have been reports on pyroelectric properties of such materials, for instance $(Pb_{1-x}Ca_x)[(Co_{1/2}W_{1/2})_yTi_{1-y}]O_3$, with $x = 0.24$ and $y = 0.04$ (PCWT-4/24) [54]. Their pyroelectric coefficient is similar to those the modified PZ materials, but with lower dielectric constant, thus leading to a slightly higher value of F_v. However, their dielectric loss is higher, thus resulting in a lower value of F_D. Furthermore, the loss increases rapidly with decreasing frequency. This problem can be solved by using hot pressing to fabricate the samples [55, 56].

5.3.6 Lead Germanate

Lead germinate, $Pb_5Ge_3O_{11}$, is a hexagonal ferroelectric material, with a Curie temperature of 178 °C. It melts congruently, so that single crystals can be grown from the melt by using the Czochralski method [57, 58]. It can be doped with barium to replace lead) or silicon to substitute germanium [59], both of which reduce its T_C. Pyroelectric properties of single crystals of both $Pb_5Ge_3O_{11}$ (PGO) and $Pb_{4.7}Ba_{0.3}Ge_3O_{11}$ (PGO:Ba_3) have been studied. The values of F_v and F_D of PGO are close to that of $LiTaO_3$. The doping of barium can increase all p', ε, and tanδ, thus leading a slight enhancement in F_v and degradation in F_D. Both pure PGO and doped one can be made into thin films, with similar pyroelectric performances to bulk materials.

5.3.7 Improper Ferroelectrics

The ferroelectrics that have been discussed above exhibit the phase transition from paraelectric to ferroelectric that is brought about due to the instability in a polar zone-center optic phonon mode. This polar distortion leads to the divergence in the dielectric susceptibility as T_C is approached either from above or below. These are called "proper" ferroelectrics. There is a group of ferroelectrics, in which the phase transition is caused by the freezing out of a nonpolar, usually zone-boundary, phonon mode. In such materials, the dielectric susceptibility does not diverge at T_C and there is the possibility of the p'/ε ratio increasing significantly as the temperature approaches T_C from below. These ferroelectrics are called improper ferroelectrics. Examples of such materials include the boracites ($M_3B_7O_{13}X$, with $M = Fe^{2+}$, Ni^{2+}, Cr^{2+} and $X = Cl^-$, Br^-, I^-, NO^-), the rare-earth molybdates, such as $Tb_2(MoO_4)_3$ (TMO) and dicalcium lead propionate ($Ca_2Pb(CH,CH_2COO)_6$—DLP.

The dielectric, pyroelectric, and thermal properties of FeI boracite (FIB), NiBr boracite, TMO, and DLP, with $T_C = 72.4$, 123, 161 and 54 °C, respectively, have been assessed as functions of temperature [60, 61]. FeI boracite shows the greatest promise, with the F_v figure of merit approaching that of DTGS near T_C. However, the dielectric losses are rather high (tan$\delta \approx 0.06$), limiting F_D. Also, the growth of large single crystals of boracites by using vapor transport method is difficult, so that they have not been available for commercial applications [62]. Nevertheless, the principle of using improper ferroelectrics for pyroelectric applications has been proved. For example, a broadband thermal detector with CuCl boracite in a "mixed" pyroelectric/resistance bolometer mode has been demonstrated [63]. It is believed that if new improper ferroelectrics with higher performance can be found in the future, they will be important pyroelectric materials for practical applications.

5.3.8 Materials Optimization

Besides doping with various elements, new strategies have been developed to further enhance the performances of pyroelectric materials, including employment of composites, making use of improper ferroelectric phase transition related properties, use of ferroelectric–ferroelectric phase transition related properties and controlling the cutting angle of pyroelectric crystals.

Because dielectric permittivity and pyroelectric coefficient of composites are closely related to the phase compositions of the individual components, it is readily expected that the voltage figure of merit and detection figure of merit of composite could be higher than those of its individual components. For example, a composite is assumed to consist of ferroelectric component with ellipsoidal shape and permittivity of ε_1 that are embedded in a continuous nonferroelectric matrix of ε_2. The volume of the individual ferroelectric particle is v and the dipole moment M is in the X-axis direction. The potential of M at point A (x, y, z) is:

$$\varphi_1 = \frac{M}{4\pi\varepsilon_0[\varepsilon_c + (\varepsilon_2 - \varepsilon_c)n_x]} \cdot \frac{x}{r^3}, \tag{5.115}$$

where ε_c is permittivity of the composite, n_x is depolarization factor of the ellipsoidal ferroelectric particle in the X-axis direction, r is distance of the dipole away from point A, with $r \gg v^{1/3}$. It is assumed that there are N ellipsoidal ferroelectric particles homogeneously distributed in the matrix with a volume of V, $V \ll r^3$, the electric potential of point A is given by:

$$\varphi = N\varphi_1 = \frac{NM}{4\pi\varepsilon_0[\varepsilon_c + (\varepsilon_2 - \varepsilon_c)n_x]} \cdot \frac{x}{r^3}. \tag{5.116}$$

If the polarization of the composite is P, the electric potential of the N ellipsoidal ferroelectric particles at point A is:

$$\varphi = \frac{PV}{4\pi\varepsilon_0\varepsilon_c} \cdot \frac{x}{r^3}. \tag{5.117}$$

By comparing Eqs. (5.116) and (5.117), we have:

$$P = \frac{q\varepsilon_c P_2}{\varepsilon_c + (\varepsilon_2 - \varepsilon_c)n_x}, \tag{5.118}$$

where $P_2 = M/v$ is polarization of the ellipsoidal ferroelectric particle, $q = Nv/V$ is volume fraction of the ferroelectric phase. When the polarization ratio of the ferroelectric particles to the composite is represented by α, pyroelectric coefficient of the composite, $p_c = \mathrm{d}P/\mathrm{d}T$, can be given by:

$$p_c = \frac{\alpha q\varepsilon_c p_2}{\varepsilon_c + (\varepsilon_2 - \varepsilon_c)n_x}, \tag{5.119}$$

where $\alpha p_2 = \mathrm{d}P_2/\mathrm{d}T$ and p_2 is pyroelectric coefficient of the pure ferroelectric phase. Voltage figure of merit of the composite is proportional to p_c/ε_c, which is given by:

$$\frac{p_c}{\varepsilon_c} = \frac{\alpha q}{(1 - n_x)\varepsilon_c/\varepsilon_2 + n_x} \cdot \frac{p_2}{\varepsilon_2}. \tag{5.120}$$

At the same time, permittivity of the composite can be expressed as:

$$\varepsilon_c = \varepsilon_1 \left[1 + \frac{q(\varepsilon_2 - \varepsilon_1)}{\varepsilon_1 + (\varepsilon_2 - \varepsilon_1)(1 - q)n_x} \right]. \tag{5.121}$$

The values of α are related to the volume fraction of the ferroelectric phase and external electric field. When the external electric field is saturated, α is not dependent on the volume fraction. The saturation value of α is nearly equal to the saturation value of pure ferroelectric phase, i.e., $\alpha = 1$.

In special situations, Eqs. (5.119) and (5.120) can be simplified. In parallel connection, $n_x = 0$, there is:

$$p_c = \alpha q \cdot p_2, \tag{5.122}$$

$$\frac{p_c}{\varepsilon_c} = \frac{\alpha q \varepsilon_2}{q\varepsilon_2 + (1 - q)\varepsilon_2} \cdot \frac{p_2}{\varepsilon_2}. \tag{5.123}$$

In series connection, $n_x = 1$, we have:

$$p_c = \frac{\alpha q \varepsilon_c}{\varepsilon_2} \cdot p_2, \tag{5.124}$$

$$\frac{p_c}{\varepsilon_c} = \alpha q \cdot \frac{p_2}{\varepsilon_2}. \tag{5.125}$$

For spherical particles, $n_x = 1/3$, pyroelectric coefficient of the composite is:

$$p_c = \frac{3\alpha q \varepsilon_c}{2\varepsilon_c + \varepsilon_2} \cdot p_2, \tag{5.126}$$

With Eqs. (5.119)–(5.121), pyroelectric coefficient and p_c/ε_c can be readily calculated. Although the pyroelectric coefficients of the composites are always lower than that of phase pure ferroelectric component, the values of p_c/ε_c of the composites can be higher than that of pure ferroelectric phase, p_2/ε_2, when an appropriate volume fraction is selected and there is a sufficiently high value of $\varepsilon_2/\varepsilon_1$ [64].

A good example is the composite made of TGS and PVDF, the former has higher pyroelectric properties but poor mechanical strength, whereas the latter has good mechanical property but weak electrical performance. The composite has exhibited combined advantage of the two components. When the volume fraction of TGS particles is about 0.8, the value of p_c/ε_c is very close to that of pure TGS, but the composite has provided more flexibility in the fabrication of devices.

Improper ferroelectric phase transition is usually accompanied by a step-like change in permittivity, instead of a peak of typical ferroelectric phase transition. However, because there is sudden change in polarization, pyroelectric coefficient can have peak value during the phase transition. Therefore, it is still possible to obtain high voltage figure of merit, $F_v = p/(c'\varepsilon)$. Three representative improper ferroelectrics, terbium (rare-earth) molybdates (TMO), double calcium lead propionates (DLP) and iron iodine boracites (FIB), have been studied for such a purpose [60]. As discussed earlier, Curie temperatures of the three crystals are 161,

54, and 72.4 °C, respectively. As they are heated to their T_C, their pyroelectric coefficients increase with increasing temperature. For DLP, if the sample is not annealed, it has inner bias electric field, so its pyroelectric coefficient is still not zero, even above T_C. There is a broad pyroelectric coefficient peak near T_C. The three crystals have very small variation in permittivity across T_C, with permittivity values of 7, 13, and 19 for DLP, TMO, and FIB, respectively. Their pyroelectric figures of merit are all higher than that of TGS near T_C.

Ferroelectric–ferroelectric phase transition is also accompanied by a pyroelectric coefficient peak, with permittivity showing a slight variation. If the direction of polarization does not change, there is only a small jump in permittivity. Therefore, ferroelectric–ferroelectric phase transition provides another opportunity to achieve high pyroelectric responses. For instance, $PbZr_xTi_{1-x}O_3$, with $0.65 < x < 0.95$, has a ferroelectric–ferroelectric phase transition from R3 m (high temperature phase) to R3c (low temperature phase) [65]. For the high temperature phase (R3 m), the triad axis of the O_6 octahedral is slightly deviated, while there no such a deviation observed for the low temperature phase (R3c). However, because the displacement direction of the Zr^{4+}/Ti^{4+} ions are the same, the direction of polarization is not changed during the phase transition. However, as stated before, there is a thermal hysteresis of 15 °C, due to its first-order characteristic. To use this phase transition as a method to achieve high pyroelectric response, this problem should be addressed, by using dopant to modify this hysteresis or applying bias electric fields [65].

There have been reports on ferroelectric–ferroelectric phase transition of multicomponent phases. One example is $xPbTiO_3$-$yPbZn_{1/3}Nb_{2/3}O_3$-$zPbMn_{1/3}Nb_{2/3}O_3$-$(1\text{-}x\text{-}y\text{-}z)PbZrO_3$, which has a ferroelectric–ferroelectric phase transition from R3 m to R3c at about 67 °C, when x = 0.07, y = 0.05, and z = 0.07. At room temperature, it has pyroelectric coefficient $p = 5.0 \times 10^{-4}$ C m^{-2} K^{-1}, $\varepsilon_r = 316$, $\tan\delta = 0.01$, and voltage figure of merit $F_v = 0.08$ m^2 C^{-1}. Over 0–100 °C, the variation of F_v with respect to the room temperature value is only 4.7 %, which is a very good performance parameter for practical applications.

The permittivity and pyroelectric coefficient in the formula of various figures of merit are actually the components that are perpendicular to the electrode surfaces of the pyroelectric elements. Since permittivity and pyroelectric coefficient are second-order tensor and first-order tensor (vector), respectively, they are isotropic. As a result, samples with different cutting angles from a crystal should have different figures of merit. By selecting appropriate cutting angles, it is possible to achieve high figure of merit, through the values of permittivity and pyroelectric coefficient. Relationships between pyroelectric response and cutting angles of different ferroelectrics have been well investigated [34, 35, 66].

The ferroelectric phase of the TGS family crystals has point group $2(C_2)$, with spontaneous polarization along the diad axis (b-axis). According to crystal physics, y-axis is parallel to b-axis, while z-axis is parallel to c-axis, so that pyroelectric coefficient and permittivity can be expressed as:

$$p = (0, p_2, 0), \tag{5.127}$$

$$\varepsilon = \begin{pmatrix} \varepsilon_1 & 0 & 0 \\ 0 & \varepsilon_2 & 0 \\ 0 & 0 & \varepsilon_3 \end{pmatrix}. \tag{5.128}$$

The component of pyroelectric coefficient in an arbitrary direction is given by:

$$p = p_2 \cos\theta, \tag{5.129}$$

where θ is the angle of the direction and y-axis. The corresponding component of permittivity is:

$$\varepsilon = \varepsilon_1 l_1^2 + \varepsilon_2 l_2^2 + \varepsilon_3 l_3^2, \tag{5.130}$$

where l_1, l_2, and l_3 are cosine values of its angles with x-, y-, and z-axis, respectively.

The voltage figure of merit of the sample cut along the direction perpendicular to the ferroelectric polarization direction, F_0, is given by:

$$F_0 = \frac{p_2}{\varepsilon_2}. \tag{5.131}$$

If the cutting direction is arbitrary, figure of merit is:

$$\varepsilon = \frac{p}{\varepsilon} = \frac{p_2 \cos\theta}{\varepsilon_1 l_1^2 + \varepsilon_2 l_2^2 + \varepsilon_3 l_3^2}. \tag{5.132}$$

Therefore, the ratio of the two figures of merit is:

$$G = \frac{F}{F_0} = \frac{\varepsilon_2 \cos\theta}{\varepsilon_1 l_1^2 + \varepsilon_2 l_2^2 + \varepsilon_3 l_3^2}. \tag{5.133}$$

If $G > 1$, $F > F_0$. For TGS family ferroelectrics, among the three values of permittivity, ε_2 is the maximum, while ε_3 is the minimum, so the direction with maximum G and F should be within the y-z plane. As a consequence, there are $l_1 = 0$, $l_2 = \cos\theta$, and $l_3 = -\sin\theta$. When all these parameters are put into Eqs. (5.132) and (5.133), there are:

$$F = \frac{p_2 \cos\theta}{\varepsilon_2 \cos^2\theta + \varepsilon_3 \sin^2\theta} = \frac{p_2 \cos\theta}{(\varepsilon_2 - \varepsilon_3)\cos^2\theta + \varepsilon_3}, \tag{5.134}$$

$$G = \frac{\varepsilon_2 \cos\theta}{\varepsilon_2 \cos^2\theta + \varepsilon_3 \sin^2\theta} = \frac{n \cos\theta}{(n-1)\cos^2\theta + 1}, \tag{5.135}$$

where $n = \varepsilon_2/\varepsilon_3$ is called anisotropic ratio of permittivity. According to Eqs. (5.134) and (5.135), the optimum cutting direction can be obtained as follows:

$$\theta_{op} = \cos^{-1}\left[(n-1)^{-1/2}\right]. \tag{5.136}$$

When $n < 2$, Eq. (5.136) has no solution, while $n = 2$, $\theta_{op} = 0$, which means that if a pyroelectric crystal has $n \leq 2$, cutting along any angle deviated from the polarization direction cannot achieve high value of F [66].

Obviously, the higher the anisotropic ratio of permittivity, the high the value of G that can be achieved, and thus the higher the optimum cutting angle will be. The cutting angle with $G = 1$ is called critical angle. With the range of between 0° and the critical angle, there is $G > 1$. The TGS family ferroelectrics usually have $n = 6$ at room temperature, it is possible to get enhanced pyroelectric figure of merit by using optimum cutting angle.

5.4 Olsen Cycle and Pyroelectric Energy Harvesting

Before evaluating energy harvesting from heat, it is natural to consider thermodynamic cycles to understand the maximum potential efficiency [9]. Among various cycles, the so-called Olsen cycle is the most effective and practical one in realizing pyroelectric energy conversion with pyroelectric materials, which is the electric analog of the Ericsson heat engine in the charge–voltage diagram (q-V), as opposed to the pressure–volume (p-v) diagram [67–77]. Figure 5.6 shows q–V diagram of a typical pyroelectric element (PE), where the Olsen cycle is represented by the 1-2-3-4 loop [78]. A pyroelectric element consists of a pyroelectric thin film and its metallic electrodes deposited on both faces. The Olsen cycle begins by charging the PE at low temperature T_L by increasing the voltage from V_L to V_H (process 1–2). Next, the PE is discharged by heating the element from T_L to T_H at constant voltage V_H where ideally $T_H > T_{\text{Curie},\uparrow}$ (process 2–3). The material is further discharged by reducing the voltage from V_H to V_L at constant temperature T_H (process 3–4). Finally, the PE is recharged by cooling it from T_H to $T_L < T_{\text{Curie},\downarrow}$ at constant voltage V_L (process 4–1). The shaded area in Fig. 5.6 corresponds to the electrical work output W_E in J l^{-1}, which is given by,

$$W_E = \oint V_{\text{PE}} \mathrm{d}q_{\text{PE}}. \tag{5.137}$$

It is not necessary to heat and cool the pyroelectric element above and below the Curie temperatures during the cycles. However, to achieve a larger electrical energy production, that is required.

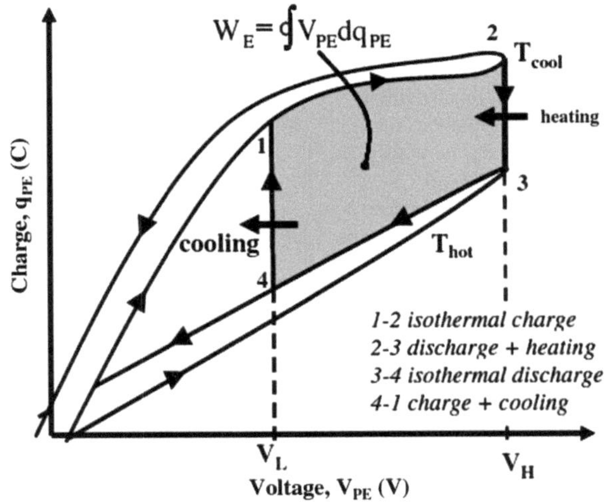

Fig. 5.6 Charge versus voltage (C–V) behavior of a typical pyroelectric material at two different temperatures. The Olsen power cycle is represented by the area enclosed by *1-2-3-4*. Reproduced with permission from [78], Copyright @ 2010, Elsevier

5.5 Pyroelectric Thermal Energy Harvesters with Olsen Cycle

5.5.1 Thermal Subsystem

To analyze the efficiency of pyroelectric conversion, prototype direct pyroelectric converter has been designed and fabricated [76, 79, 80]. It consists of thermal and electrical subsystems. The thermal subsystem is used to create a time-dependent temperature oscillation to heat the PEs to a maximum temperature T_H and cool them to a minimum temperature T_C (low temperature). The electrical subsystem serves to control the electric field applied to the PEs to perform the Olsen cycle. It also collected the charges generating by cooling and heating the PEs.

Figure 5.7 shows crosssectional view of a prototypical two stage pyroelectric converter, along with the dimensions and the associated coordinate system for simulation [80]. The working fluid oscillates vertically between N stationary equidistant walls forming $(N + 1)$ channels. All walls are identical and comprised of four plates placed vertically along the direction of the fluid flow. The top and bottom plates are made of Al_2O_3 ceramics, while the center two plates are made of lead zirconate stannate titanate (PZST) films, forming the active region of the device. A heat exchanger located at the bottom of the test section cools the working fluid at constant temperature T_C. In addition, a thin heating band is located in the middle of the channel at the top of the test section, which can heat the working fluid and operates at a constant heat flux q_{in}''. The half-widths of the flow channel and the wall are denoted by w_f and w_w, respectively. The total length of the channel is denoted by L, while L_{cr} is the length of each of the lower and upper Al_2O_3 plates. The bottom of the heating band is located at a height of z_h,

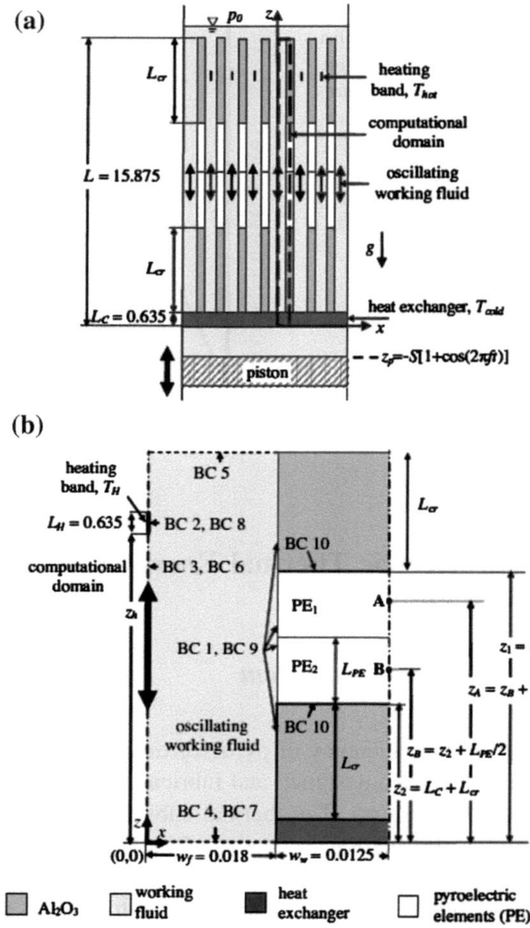

Fig. 5.7 **a** Schematic of the overall pyroelectric converter of length L. **b** Schematic of the computational domain along with the coordinate system and the boundary conditions, with all dimensions in centimeter. Reproduced with permission from [80], Copyright @ 2010, Elsevier

whereas the lengths of the heat exchanger and heating band are denoted by L_C and L_H, respectively. The heating band is simulated as a line source located in the middle of the channel.

Additionally, points A and B are located at the center of the pyroelectric elements PE1 and PE2, at heights z_A and z_B, respectively. The heights from the bottom of the test section to the top of PE1 and bottom of PE2 are $z_1 = L - L_{cr}$ and $z_2 = L_C + L_{cr}$, respectively. The entire pyroelectric converter has a depth d = 3.8 cm. The oscillation of the fluid flow through the minichannel is achieved by using a piston-in-a-cylinder pump with adjustable piston amplitude S and frequency f. At time t, the piston top surface is located at $z_p = -S[1 + \cos(2\pi ft)]$, where, initially, i.e., t = 0, the piston is at its lowest position at $z_p(0) = -2S$ and the liquid free surface is located at $z = L$.

5.5.2 Electrical Subsystem

The converter also has an electric subsystem, which is essential for achieving the desired thermodynamic cycles shown in Fig. 5.6 [77]. Both faces of each pyroelectric thin film are plated with electrodes and connected to the electrical subsystem. The charge from the pyroelectric elements and the applied voltage are measured from the voltage drop across loaded capacitor and resistor. Voltages V_L and V_H are supplied in order to maintain the desired low or high voltages required for the Olsen cycle. The temperature and voltage of each pyroelectric element are synchronized by using a switch to trigger the cycle. Each switch is actuated whenever the temperature of the pyroelectric element reaches a maximum or a minimum with a control system. In Fig. 5.6, the isothermal charge 1-2 of the pyroelectric element at constant temperature T_C is provided by the power supply switching from V_L to V_H. The pyroelectric element is then heated (2-3) at constant voltage V_H up to temperature constant T_H. It is then isothermally discharged in the load (3-4) by switching the voltage from V_H to V_L. Finally, the cycle is closed by cooling the pyroelectric element at constant voltage V_L.

5.5.3 Assumptions for Numerical Study

To make the problem mathematically trackable, a list of assumptions have been made [79, 80]. Briefly, the entire pyroelectric converter is well insulated, so that heat losses to the surrounding and between channels are negligible. All material properties for the wall and the working fluid are assumed to be constant and isotropic. The working fluid is silicone oil, which is treated as a Newtonian and incompressible fluid. The spacing between the walls is much smaller than the depth of the plate, so that two-dimensional flow is dominant. The Reynolds number of the periodically oscillating fluid flow is less than a critical value for all simulations, so that fluid flow is laminar [81, 82]. The relationship between the charge q and the open-circuit voltage V at constant temperature forms a hysteresis curve as shown in Fig. 5.6. Edge effects can be neglected, i.e., all channels are equivalent, so that it is necessary to consider only one of them. As a first-order approximation, the electrodes deposited on the PE plates to collect the generated current have assumedly no effect on the heat transfer and fluid flow. Due to the lack of thermophysical properties for various doping levels of PZST, the PEs at the two stages of the converter are assumed to be identical.

5.5.4 Governing Equations

The components of the velocity vector and the pressure field of the working fluid are calculated by solving the two-dimensional mass and momentum conservation equations expressed for an incompressible and Newtonian fluid with constant properties in Cartesian coordinates as [79, 80],

$$\frac{\partial u_f}{\partial x} + \frac{\partial v_f}{\partial z} = 0, \tag{5.138}$$

$$\frac{\partial u_f}{\partial t} + u_f \frac{\partial u_f}{\partial x} + v_f \frac{\partial u_f}{\partial z} = -\frac{1}{\rho_f}\frac{\partial p_f}{\partial x} + \nu_f \left(\frac{\partial^2 u_f}{\partial x^2} + \frac{\partial^2 u_f}{\partial z^2} \right), \tag{5.139}$$

$$\frac{\partial v_f}{\partial t} + u_f \frac{\partial v_f}{\partial x} + v_f \frac{\partial v_f}{\partial z} = -\frac{1}{\rho_f}\frac{\partial p_f}{\partial z} + \nu_f \left(\frac{\partial^2 v_f}{\partial x^2} + \frac{\partial^2 v_f}{\partial z^2} \right) - g, \tag{5.140}$$

where ρ_f and ν_f are the density and kinematic viscosity of the working fluid, while uf and v_f are the components of the fluid velocity vector in the x- and z-directions, respectively. The pressure of the working fluid is denoted by p_f, while g is the gravitational acceleration that is equal to 9.81 m s^{-2}, acting along the z-axis oriented upward. The two-dimensional energy equation solved to determine the temperature distribution within the working fluid is given by [79, 80],

$$\rho_f c_{p,f} \left(\frac{\partial T_f}{\partial t} + u_f \frac{\partial T_f}{\partial x} + v_f \frac{\partial T_f}{\partial z} \right) = k_f \left(\frac{\partial^2 T_f}{\partial x^2} + \frac{\partial^2 T_f}{\partial z^2} \right), \tag{5.141}$$

where $c_{p,f}$ is the heat capacity of the fluid, T_f is the local fluid temperature, and k_f is the fluid thermal conductivity, which is assumed to be a constant.

In a similar way, the two-dimensional heat diffusion equation for the Al_2O_3 plates and the pyroelectric PZST plates constituting the channel walls is given by,

$$\rho_w c_{p,w} \frac{\partial T_w}{\partial t} = k_{w,x} \frac{\partial^2 T_w}{\partial x^2} + k_{w,z} \frac{\partial^2 T_w}{\partial z^2}, \tag{5.142}$$

where the subscript w refers to either the Al_2O_3 plates or the pyroelectric PZST elements. The thermal conductivities are assumed to be independent of temperature, but are possibly anisotropic, as represented by the subscripts in $k_{w,x}$ and $k_{w,z}$.

5.5.5 Initial and Boundary Conditions

At time $t = 0$, the working fluid is assumed to be at rest, i.e., for $0 \leq x \leq w_f$ and $0 \leq z \leq L$, there is:

$$u_f(x, z, 0) = v_f(x, z, 0) = 0. \tag{5.143}$$

Initially, there is no variation in temperature throughout both the walls and the working fluid in the x-direction, i.e., for $0 \leq x \leq w_f + w_w$, and the temperature is linear between the heating band at location $z = z_h$ and temperature T_H and the heat exchanger at location $z = L_C$ and temperature T_C. In addition, the temperature profile of the working fluid remains constant along and above the heat exchanger and along the heating band, so that for $0 \leq x \leq w_f + w_w$, there is:

$$T_f(x, z, 0) = T_w = \begin{cases} T_C & 0 \leq z \leq L_C \\ T_C + \frac{(T_H + T_C)(z - L_C)}{z_h - L_C} & L_C \leq z \leq z_h \\ T_H & z_h \leq z \leq L \end{cases}. \quad (5.144)$$

At time t, the no-slip condition is dominant at the wall ($x = w_f$) and heating band ($x = 0$). Therefore, the velocity components $u_f(x, z, t)$ and $v_f(x, z, t)$ become:

$$u_f(w_f, z, t) = v_f(w_f, z, t) = 0, \quad 0 \leq z \leq L, \quad (5.145)$$

$$u_f(0, z, t) = v_f(0, z, t) = 0, \quad z_h \leq z \leq z_h + L_H. \quad (5.146)$$

On the other hand, the symmetry boundary condition is imposed at the channel centerline ($x = 0$), i.e.,

$$u_f(0, z, t) = \frac{\partial v_f}{\partial x}(0, z, t) = 0, \quad 0 \leq z \leq z_h, \quad z_h + L_H \leq z \leq L. \quad (5.147)$$

In addition, the velocity of the working fluid at the piston is equal to the velocity of the piston, given by $v_p = 2\pi f S \sin(2\pi f t)$. From mass conservation considerations and assuming that the fluid is incompressible, the velocity at the bottom of the microchannels is uniform and a sinusoidal function of time with amplitude, $S' = SA_P/(N + 1)A_c$, i.e.,

$$u_f(x, 0, t) = 0, \quad v_f(x, 0, t) = 2\pi f S' \sin(2\pi f t), \quad 0 \leq x \leq w_f, \quad (5.148)$$

where A_p and A_c are the crosssectional area of the piston and a single channel, respectively.

The incompressible working fluid above the channel oscillates with the same amplitude and frequency as the piston, in such a way that the free surface oscillates between $z = L$ and $z = L + 2S$, as the piston moves from $z_p = -2S$ to $z_p = 0$, respectively. In addition, since the crosssectional area of the working fluid at the top of the reservoir above $z = L$ is large and equal to that of the piston A_p, the pressure at $z = L$ can be assumed to be dominated by hydrostatic pressure, which is given by,

$$p_f(x, L, t) - p_0 = \rho_f g S[1 - \cos(2\pi f t)], \quad 0 \leq x \leq w_f, \quad (5.149)$$

where p_0 is the atmospheric pressure at the free surface.

The symmetry boundary condition is used for the temperature at the centerline of the wall and of the flow channel, resulting in,

$$\frac{\partial T_f}{\partial x}(0, z, t) = \frac{\partial T_f}{\partial x}\left(w_f + w_w, z, t\right) = 0, \quad 0 \le z \le L \tag{5.150}$$

In addition, the temperature of the working fluid at z = 0 is constant, which is equal to the temperature of the heat exchanger T_C across the channel, so that there is:

$$T_C = T_f(x, 0, t), \quad 0 \le x \le w_f\,, \tag{5.151}$$

Likewise, the heat flux along the heating band surface is constant, which is equal to q_{in}'', i.e.,

$$q''(0, z, t) = q_{\text{in}}'', \quad z_h \le z \le z_h + L_H\,, \tag{5.152}$$

where q″ in is adjusted to achieve the desired temperature T_H at the heating band. Moreover, the heat fluxes normal to the flow direction at the interfaces between the working fluid and either the PE or the Al_2O_3 plates can be expressed as:

$$-k_f \frac{\partial T_f}{\partial x}\left(w_f, z, t\right) = -k_w \frac{\partial T_w}{\partial x}\left(w_f, z, t\right) = 0, \quad 0 \le z \le L. \tag{5.153}$$

Therefore, the axial heat transferred by conduction between the PEs and the Al_2O_3 plates is accounted for, by considering that the axial heat fluxes on both sides of the interfaces located at z_1 and z_2 are equal, i.e.,

$$-k_{cr} \frac{\partial T_{cr}}{\partial x}\left(x, z_1/z_2, t\right) = -k_{\text{PE}} \frac{\partial T_{\text{PE}}}{\partial x}\left(x, z_1/z_2, t\right) = 0, \quad w_f \le x \le w_f + w_w. \tag{5.154}$$

5.5.6 Material Properties

The thermophysical properties of Al_2O_3, PZST, and the working fluid are assumed to be constant over the temperature range simulated. This assumption is justified by the fact that the relative differences in the materials properties are less than 10 %, between T_C = 145 °C and T_H = 185 °C, except for the kinematic viscosity of the working fluid ν_f, which varies by up to 37 %. As a first-order approximation, ν_f has been assumed to be a constant of 8.29 mm^2 s^{-1}, which is taken at the arithmetic mean temperature of 165 °C. The density, specific heat, and thermal conductivity of Al_2O_3 and silicone oil are also estimated at the arithmetic mean temperature of 165 °C [79, 80]. The density, specific heat, and thermal conductivity of PZST are all the values at room temperature, even though its charge-voltage curve is estimated at the temperature computed in the pyroelectric elements PE1 and PE2.

5.5.7 Solution Method

The local velocity, pressure, and temperature throughout the computational domain are determined by solving the mass, momentum, and energy equations (Eqs. (5.138)–(5.141)), along with their associated initial boundary conditions (Eqs. (5.139)–(5.153)). At every time step, the mass and momentum conservation equations are solved simultaneously [79, 80]. Then, the energy equation is solved for the fluid, as well as Al_2O_3 and pyroelectric elements.

The internal time step is chosen arbitrarily to ensure numerical stability in solving the mass, momentum, and energy equations. On the other hand, the result is recorded at time intervals Δt, in such a way that $2\pi f\Delta t = \pi/2$ for all frequencies. Numerical convergence is established by solving the governing equations with a coarse grid and a grid with size to be 1.3 times smaller. The results are assumed to be numerically converged when the maximum relative difference in the local velocity, pressure, and temperature throughout the computational domain is less than 0.5, 0.1, and 1.8 %, respectively, between two consecutive grid refinements for all frequencies.

For validation purposes, the numerical results for velocity and pressure for vertically oscillating adiabatic flow have been compared with the analytical solution for fully developed laminar flow, expressed as [81],

$$u_f(x,z,t) = 0, \quad v_f(x,z,t) = 3\pi f S' \left[1 - \left(\frac{x}{w_f}\right)^2\right] \sin(2\pi f t). \tag{5.155}$$

The numerical results fall within 1.05 % of the analytical solutions at all points for $0 \le x \le w_f$ and $0 \le z \le L$. The maximum relative differences for pressure and temperature are 0.007 and 0.43 %, respectively, validating the numerical simulation and proper implementation of the governing equations and boundary conditions.

5.5.8 Performance Analysis

The total electrical power, generated by the two PEs in all channels, can be determined as,

$$\dot{W}_E = Nf\left(\oint V_1 A_1 \mathrm{d}q_1 + \oint V_2 A_2 \mathrm{d}q_2\right), \tag{5.156}$$

where N is the total number of internal walls and f is the frequency in Hz. The surface areas A_1 and A_2 of each pyroelectric plate are 9.652 cm^2. To determine the charge–voltage curve at the temperatures T_C and T_H computed at points A and B (Fig. 5.7), linear interpolation of the charge–voltage curves at different temperatures for PZST are used [82]. PE1 and PE2 are assumed to have same charge–voltage curves. The device operated between voltages V_L and V_H are equal to 100 and 700 V, respectively [76].

The pumping power required to pump the working fluid through the channels is assumed to be only due to frictional losses within the channel. As a result, the total pumping power time averaged over the period, $\tau = 1/f$, is defined as,

$$\dot{W}_E \approx \frac{N+1}{\tau} \int_0^\tau \left(\int_{A_c} \rho_f v_f dA \right) \mathrm{d}t, \tag{5.157}$$

where the local instantaneous fluid velocity v_f and pressure ρ_f are determined by solving the mass and momentum conservation equations. The integrations in Eqs. (5.156) and (5.157) can be performed numerically by using the trapezoidal rule.

The total heat transfer rate offered by the heating band is given by:

$$\dot{Q}_{\text{in}} = 2(N+1) q''_{\text{in}} A_{hb}, \tag{5.158}$$

where A_{hb} is the surface area of the heating band in one channel, which is equal to 2.413 cm^2.

The average thermodynamic energy efficiency of the system over one cycle was defined as [78],

$$\eta = \frac{\dot{W}_{\text{cycle}}}{\dot{Q}_{\text{in}}} = \frac{\dot{W}_E - \dot{W}_p}{\dot{Q}_{\text{in}}}. \tag{5.159}$$

In this case, all other losses, such as heat losses to the surrounding and current leakages in the electrical circuit, have been neglected for simplicity.

The power density of the device, representing the total amount of electrical power generated per unit volume of PEs, can be expressed as:

$$P_D = \frac{\dot{W}_E}{2N\forall_{\text{PE}}}. \tag{5.160}$$

Each internal wall supported two PEs of volume $\forall_{\text{PE}}$, so that the total volume of PEs in the device was 2 $N\forall_{\text{PE}}$. The number of internal walls can be experimentally determined [76, 78].

5.6 Olsen Harvesters with Pyroelectric Polymers

5.6.1 Device Assembly and Characterization

A pyroelectric converter has been designed and fabricated to directly convert thermal energy into electrical energy based on the Olsen cycle [78]. Figure 5.8 shows a crosssection of the device, together with a photograph. It consists of a thermal and an electrical subsystems.

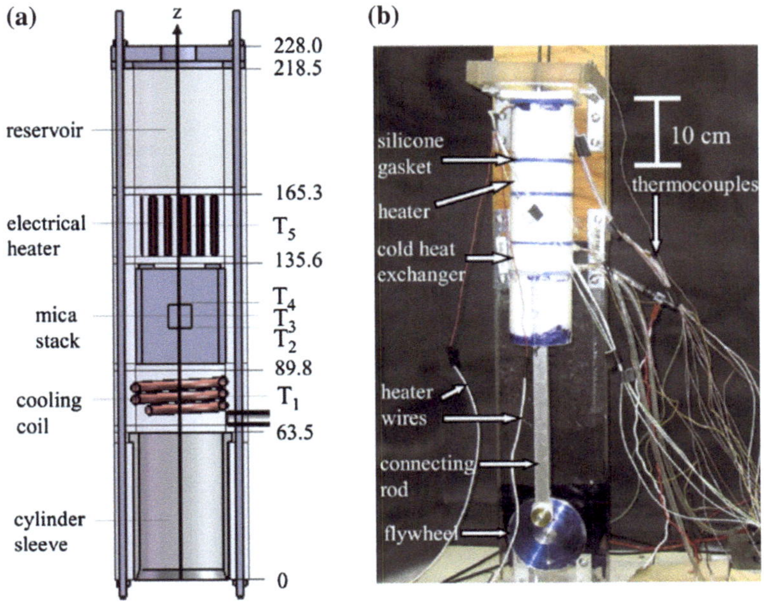

Fig. 5.8 **a** Schematic of crosssectional view of the P(VDF-FrFE) pyroelectric energy harvester, with corresponding components. **b** Photograph of the harvester, with all units in millimeter. Reproduced with permission from [78], Copyright @ 2010, Elsevier

The thermal subsystem consists of five major parts, including (i) the piston-cylinder assembly, (ii) the pyroelectric elements, (iii) the heating source, (iv) the cold heat exchanger, and (v) the working fluid [78]. The piston oscillates the working fluid vertically between the heater located at the top of the device and the cold heat exchanger located at the bottom. Teflon enclosure cylinders are used to guide the motion of the working fluid and serve also as both a thermal insulation and a support. Inner and outer diameters of the cylinder are 38.10 and 57.15 mm, respectively. There are two holes of 3.20 mm diameter on the part where they are offset from the center by 25 mm. Two M3 threaded rods of 25 cm in length are inserted into the holes to hold the device together. The outer cylinders are made with Teflon because of its low density and thermal conductivity. Teflon is stable up to 260 °C and be easily machined into various shapes, as compare with Al_2O_3 ceramic plates. Silicone gaskets are used to enclose the cylinders, leakage of working fluid can be prevented.

For piston-cylinder assembly, a piston and a cylinder sleeve assembly are commercial products. The sleeve has an inner diameter of 3.5 cm and a height of 6.4 cm. Both the sleeve and the piston are made of stainless steel. The fluid located within the cylinder sleeve is sealed from via the piston ring mounted onto the piston head. The piston stroke length is varied from 2.5 to 4.7 cm by changing the mounting radius of the connection between rod and the flywheel.

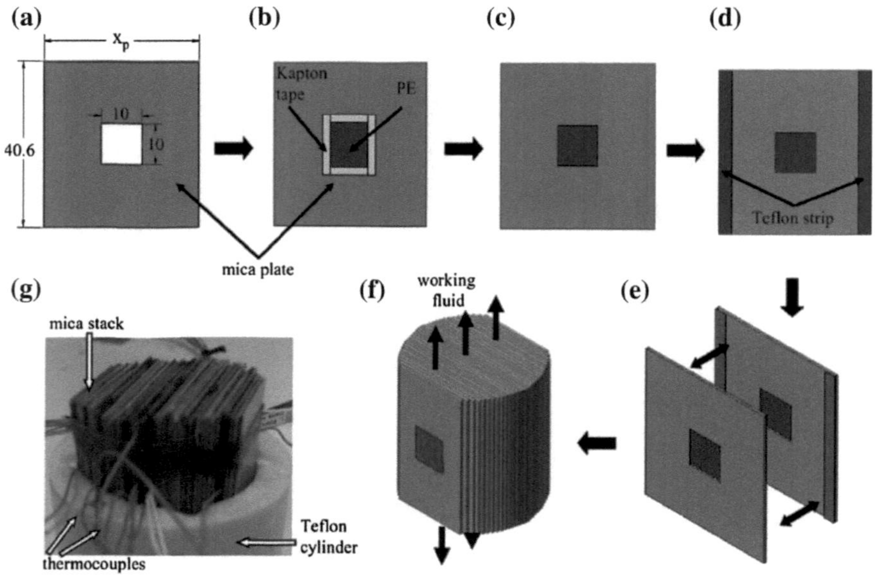

Fig. 5.9 Schematic of the pyroelectric stack assembling process: **a** a mica plate with a 10 mm × 10 mm window, **b** a mica plate with PE, **c** PE sandwiched between two mica plates, **d** pyroelectric assembly with Teflon strips, **e** mating of two mica plates, **f** entire PE stack, and **g** photograph of the real stack. All dimensions are in millimeter. Reproduced with permission from [78], Copyright @ 2010, Elsevier

Each pyroelectric element in the pyroelectric assembly is sandwiched between two mica plates, so that they will not move horizontally and vertically. Figure 5.9 shows schematic of the pyroelectric assembly and stack integration process. First, a 310 μm thick mica plate is laid flat on a smooth surface, with a 10 mm × 10 mm window at the center. The length of the mica plates is 40.6 mm while their widths are from 20.8 to 38.1 mm, so that they can be stacked and fit into the Teflon cylinder.

The PE is then placed onto the plate and four pieces of 89 μm thick Kapton tape are used to tape the edges of the PE to the mica plate. After that, an electrical wire was connected to each electrode of the PE by using 89 μm thick copper tape. The wires are then guided outside of the device and connected to the electrical circuit. Next, an identical mica plate is pressed down on the edges of the PE. The two plates are bonded together via a thin layer of thermal epoxy at four corners. Two pieces of 3.18 mm wide and 330 μm thick double-sided Teflon tape are then adhered to the vertical edges of the mica plate where they are bonded to another pyroelectric assembly to create a 330 μm wide microchannel. Finally, a stack of 38 pyroelectric elements is formed by repeating the process.

The heat source is a flexible 50 W AC Kapton electrical heater. It is taped to a thin copper sheet, with dimensions of 12.7 cm × 2.54 cm × 0.20 mm. It is bent to fit inside the Teflon cylinder above the PE stack. The power input can be continuously increased up to a maximum of 50 W by increasing the voltage across the heater. The cold heat exchanger is made of a copper tube with inner and outer diameters of 1.59 mm and 3.18 mm, respectively. The pipe is 34.3 cm long, which can be bent into a helical shape before being mounted onto the outer cylinder of cold heat exchanger. Cold water is fed into the coil by using a DC electrical pump. The flow rate can be adjusted to achieve desired operating conditions.

The working fluid is a silicone oil, which is a dielectric fluid compatible with electronic components. Its relatively large viscosity ensures that the fluid flow is laminar in the microchannels. The silicone oil has a high dielectric strength of 15.7×10^6 V m^{-1} at 25 °C and a boiling point of 200 °C at atmospheric pressure, so that it is very attractive for pyroelectric power harvesting applications.

The device has an electrical subsystem to control the voltage applied to the pyroelectric element and impose V_L or V_H at during the Olsen cycle. It is also used to measure the power generated by the pyroelectric element through the measurement of the voltage V_{PE} and the charge q_{PE}.

A triple DC output power supply is used to provide input voltages V_L* and V_H* to the high voltage power supply. Two rocker switches are connected to the high voltage power supply terminals to provide either voltage V_L* or V_H*. The input voltage $V_{L/H}$* is converted to output voltage, with $V_{L/H} = 250V_{L/H}$*. Throughout the operation, whenever one switch is opened, the other should be closed. A resistor $R_L = 7.80$ MΩ is used as a voltage divider to scale down the voltage across another series-connected resistor $R_2 = 21.8$ kΩ, in order to match the maximum voltage input of 10 V of the data acquisition system (DAQ). A capacitor $C_1 = 1.0$ mF is series-connected with the PE. Voltage V_1 across the capacitor and voltage V_2 across the resistor R_2 can be measured through the DAQ, which are used to calculate magnitude of the electric displacement of the PE D_{PE} and electric field E_{PE}. The magnitude of the electric displacement D_{PE} is defined as,

$$D_{PE} = \frac{q_{PE}}{A} = \frac{C_1 V_1}{A}, \tag{5.161}$$

where A is surface area of the pyroelectric element. The magnitude of the electric field across the pyroelectric element can be derived from Ohm's and Kirchhoff's laws, given by,

$$E_{PE} = \frac{V_{PE}}{b} = \frac{V_2(1 + R_L/R_2) - V_1}{b}, \tag{5.162}$$

where b is the thickness the pyroelectric film. Therefore, the D-E plots (equivalent to q-V plots) for the pyroelectric element undergoing the Olsen cycle can be generated.

Resistivity of the prepoled pyroelectric element can be measured as a function of time before the Olsen cycle. A constant voltage is applied across the pyroelectric element at 85 °C. The leakage current I_L can be expressed as:

$$I_L = C_1 \frac{dV_1}{dt_1}, \tag{5.163}$$

where V_1 is the registered voltage across capacitor C_1. Therefore, electrical resistivity of the pyroelectric film, in Ω m, is given by,

$$r_{PE} = \frac{V_{PE}A}{I_L b} = \frac{E_{PE}A}{I_L}. \tag{5.164}$$

The oscillating frequency of the piston is measured by using a transmission sensor, consisting of an infrared-emitting diode and a bipolar negative-positive-negative (NPN) silicon phototransistor. In order to determine the local temperatures within the device, type J thermocouples were securely mounted at various locations within the working fluid. Thermocouples used to measure the temperatures of different locations are shown in Fig. 5.8. T_1 is the temperature at the center of the cold heat exchanger, while T_2, T_3, and T_4 are temperatures at bottom, middle, and top of the pyroelectric element. *T*5 is the temperature at the center of the heater cylinder. In addition, the inlet and outlet temperatures of the cold heat exchanger, $T_{c,i}$ and $T_{c,o}$, are also measured.

5.6.2 Performance

The evolution of temperature (T_1) of the fluid near the cold heat exchanger for piston has a frequency of 0.025 Hz, with minimum and maximum values of 21.4 and 49.7 °C. The temperature swing ΔT_i at location "i" is defined as,

$$\Delta T_i = T_{i,\max} - T_{i,\min}, \tag{5.165}$$

where $T_{i,\max}$ and $T_{i,\min}$ are the respective maximum and minimum temperatures during one cycle. The arithmetic mean temperature $\bar{T}_i$ at location "i" is defined as,

$$\bar{T}_i = \frac{T_{i,\max} + T_{i,\min}}{2}. \tag{5.166}$$

For example, T_3, measured in the middle of the PE, shows an oscillation between 90.7 and 66.7 °C, leading to a temperature swing $\Delta T_3 = 24$ °C. The corresponding values for other temperatures are also available. Therefore, the maximum temperature corresponds to the piston at the bottom, whereas the minimum temperature means the piston at the top of its stroke. The temperature oscillations have the same frequency as that of the piston.

After the pyroelectric element, a piece of P(VDF-TrFE) of 1 cm × 1 cm × 45.7 μm, is poled at $E_{PE} = 201$ kV cm^{-1} and at $T_3 = 85$ °C for 180 min, the Olsen cycle is applied. The device has a stroke length of 4.7 cm and 330 μm wide channels operating at 0.061 Hz. The values of the low and high electric fields imposed in the Olsen cycle are $E_L = 202$ kV cm^{-1} and $E_H = 379$ kV cm^{-1}, respectively. The low electric field E_L is set to keep the film properly poled during process 4–1 of the Olsen cycle. The heat input provided by the heater is 30.9 W. The minimum and maximum temperatures measured at the center of the PE, $T_{3,\min}$ and $T_{3,\max}$ are 66.4 and 83.0 °C, respectively, corresponding to $\Delta T_3 = 16.6$ °C and $\bar{T}_3 = 74.7$ °C. The corresponding energy and power densities are 130 J l^{-1} and 8.12 W l^{-1}, with respect to the pyroelectric material used.

A maximum energy density of 83 J l^{-1} with respect to the pyroelectric material can be achieved at $T_3 = 74 \pm 1$ °C and $E_H = 329$ kV cm^{-1}. The presence of the maximum energy density can be explained in two aspects. At higher temperatures, the energy density of the pyroelectric element is decreased due to the possible increase in leakage current. However, if the temperature is low, there will be smaller energy density due to the smaller temperature swing.

The temperature swings, ΔT_3, at the center of the pyroelectric element, with channel widths of 0.33 mm and 1 mm, have been studied, over a frequency range of 0.025–0.16 Hz [78]. The stroke length is equal to 3.4 cm. Generally, the temperature swing decreases with increasing frequency, which can be attributed to thermal inertia of both the working fluid and the pyroelectric element. As the frequency is increased, there is less time for the thermal energy to be exchanged between the heating band and the oscillating working fluid and between the working fluid and the pyroelectric elements. This problem can be addressed by reducing the thermal time constant of the pyroelectric element, which is defined as [83],

$$\tau_t = \frac{\rho_{\mathrm{PE}} c_{p,\mathrm{PE}} \forall}{h_\omega A}, \tag{5.167}$$

where h_ω is the oscillatory flow convective heat transfer coefficient. The thermal time constant can be reduced by increasing h_ω and surface area A of the pyroelectric element and decreasing thickness of the pyroelectric element $v = \forall / A$.

It is also found that a larger temperature swing is achieved with thinner wide channels. According to Ozawa et al. [84], there is a correlation between the Strouhal, Reynolds, and Nusselt numbers for laminar oscillating flow in a circular channel, which is given by,

$$Nu_\omega = \frac{h_\omega l_0}{k_f} = 0.92 St^{-0.2} St_L^{-0.26} \mathrm{Pr}_f^{0.4} \mathrm{Re}_\omega^{0.44}, \tag{5.168}$$

where l_0 and k_f are the amplitude of fluid displacement and the thermal conductivity of the working fluid, respectively. The Strouhal numbers are represented by St and St_L, which are defined as [84],

$$St = \frac{D_h}{S}, \tag{5.169}$$

$$St_L = \frac{L}{S}, \tag{5.170}$$

where D_h is the channel hydraulic diameter, S is the piston stroke length, and L is the channel length. In addition, the Reynolds and Prandtl numbers are represented by Re_ω and Pr_f, which are given by [84],

$$\mathrm{Re}_\omega = \frac{u_0 l_0}{v_f} = \frac{u_0^2}{v_f}, \tag{5.171}$$

$$\Pr_f = \frac{v_f}{\alpha_f}, \tag{5.172}$$

where v_f, $\omega = 2\pi \mathrm{f}$, $\mathrm{u}_0 = \mathrm{l}_0\omega$, and α_f are the fluid kinematic viscosity, piston angular frequency, amplitude of velocity oscillation, and fluid thermal diffusivity, respectively. As the channel width is decreased, the hydraulic diameter decreases, while the velocity of the working fluid increases. As a result, there is an increase in the Strouhal number St and the Reynolds number Re_ω, which, in turns, leads to larger Nusselt number and heat transfer coefficient between the working fluid and the pyroelectric material and thus larger temperature swing in the pyroelectric element.

The energy density harvested per cycle (N_D) can be estimated by using the area enclosed by the Olsen cycle in the D-E diagram (Fig. 5.6), which is calculated by integrating the elementary electrical work, EdD, over the cycle, given by [77]:

$$N_D = \oint E\mathrm{d}D. \tag{5.173}$$

Powder density, P_D, in mW l^{-1}, is the amount of energy generated by the pyroelectric element per unit volume per unit time, which is expressed as $P_D = N_D/\tau$, with τ to the cycle period.

When the pyroelectric material is not clamped and the polar axis is in the direction perpendicular to the planar electrodes, the elementary change in the electric displacement can be derived from Eq. (5.3) in a simplified form, given by:

$$\mathrm{d}D = \varepsilon \mathrm{d}E + p\mathrm{d}T. \tag{5.174}$$

The dielectric permittivity is expressed as $\varepsilon = \varepsilon_r\varepsilon_0$, where ε_r is the relative permittivity of the material and ε_0 is the vacuum permittivity, which is 8.854×10^{-12} F m^{-1}. p is pyroelectric coefficient in μC cm^{-2}, which is defined as $p = \partial D/\partial T$. As discussed earlier, for unclamped samples, it consists of two parts, i.e., the primary pyroelectric coefficient that accounts for the changes in crystal dipole moment and the secondary pyroelectric coefficient, due to the dimensional change in the crystal, both of which are caused by the variation in temperature within the devices. The unclamped pyroelectric coefficient under constant stress (X) and electric field (E) is given by:

$$p = \left(\frac{\partial D}{\partial T}\right)_{X,E} = p^{x,E} - \frac{d_{33}\alpha_3}{s_{33}}, \tag{5.175}$$

where $p^{x,E}$ is the primary pyroelectric coefficient under constant strain (x) and electric field (E). The second term is the secondary pyroelectric coefficient due to thermal expansion. It is the contribution of piezoelectric effect to the pyroelectric effect, as mentioned before. $\alpha_3 = x_3/\Delta T$ is the coefficient of thermal expansion (CTE). d_{33} and s_{33} are the piezoelectric coefficient in C N^{-1} and elastic compliance in $\mathrm{m}^2\ \mathrm{N}^{-1}$. The secondary pyroelectric coefficient is zero when the pyroelectric element is clamped and thus cannot expand. For isoelectric field processes 2–3 and 4–1, the first term is zero, whereas for isothermal processes 1–2 and 3–4 the second term on the right hand side of Eq. (5.174) is zero. Therefore, Eq. (5.173) can be integrated over the four processes independently to get the energy density per cycle, expressed as:

$$
\begin{aligned}
N_D = & \int_{E_L}^{E_H} \varepsilon_0 \varepsilon_r(E, T_{\text{cold}}) E dE + E_H \int_{T_{\text{cold}}}^{T_{\text{hot}}} p(E_H, T) \mathrm{d}T \\
& + \int_{E_H}^{E_L} \varepsilon_0 \varepsilon_r(E, T_{\text{hot}}) E dE + E_L \int_{T_{\text{hot}}}^{T_{\text{cold}}} p(E_L, T) \mathrm{d}T.
\end{aligned}
\tag{5.176}
$$

For the Olsen cycle, there is a difficulty to model the energy generation because the properties are strongly dependent on electric field and temperature. Also, the pyroelectric materials exhibit electrical and thermal hysteresis characteristics, the properties measured are usually not matching very well with operating conditions. Therefore, further assumption is necessary to derive a workable formula of energy density.

Near phase transition, the properties of pyroelectric materials are dependent on temperature, so that temperature effects on material properties must be taken into account. The change in electric displacement with respect to temperature is the primary pyroelectric coefficient, given by:

$$
p^{x,E} = \left(\frac{\partial D}{\partial T}\right)_{x,E} = \varepsilon_0 \frac{\partial \varepsilon_r}{\partial T} E + \frac{\partial P_s}{\partial T}.
\tag{5.177}
$$

The first and second terms on the right hand side of Eq. (5.176) are the dielectric and dipole contributions to the primary pyroelectric coefficient, respectively. By including the secondary pyroelectric coefficient, the unclamped pyroelectric coefficient can be expressed as:

$$
p = p^{x,E} - \frac{d_{33}\alpha_3}{s_{33}} = \varepsilon_0 \frac{\partial \varepsilon_r}{\partial T} E + \frac{\partial P_s}{\partial T} - \frac{d_{33}\alpha_3}{s_{33}}.
\tag{5.178}
$$

Combining Eqs. (5.176) and (5.178), the energy density for Olsen cycle can be derived as:

$$
\begin{aligned}
N_D = & (E_H - E_L)\left\{\frac{\varepsilon_0}{2}[\varepsilon_r(T_{\text{cold}}) - \varepsilon_r(T_{\text{hot}})](E_H + E_L)\right. \\
& \left. + \left[P_s(T_{\text{cold}}) - P_s(T_{\text{hot}}) + \frac{d_{33}x_3}{s_{33}}\right]\right\},
\end{aligned}
\tag{5.179}
$$

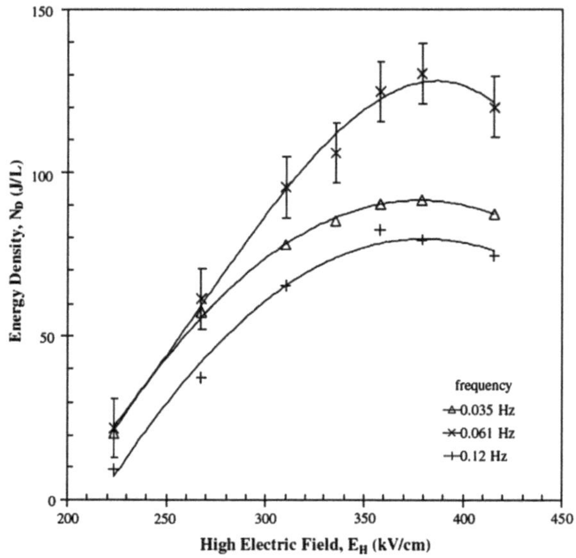

Fig. 5.10 Energy densities as a function of high electric field E_H at different frequencies. The channel width, stroke length, low field, and mean temperature $\bar{T}_3$ were constant and equal to 330 μm, 4.7 cm, 202 kV cm^{-1} and 74 ± 1 °C, respectively. Reproduced with permission from [78], Copyright @ 2010, Elsevier

where $x_3 = \alpha_3\ (T_{\text{hot}} - T_{\text{cold}})$. If the pyroelectric coefficient is not dependent on temperature and electric field, Eq. (5.179) is reduced to:

$$N_D = (T_{\text{hot}} - T_{\text{cold}}) \int_{E_L}^{E_H} p dE = p(T_{\text{hot}} - T_{\text{cold}})(E_H - E_L). \tag{5.180}$$

Further study indicates that the temperature swing decreases as the frequency is increased regardless of the stroke length [78]. At a given frequency, the temperature swing increases as the stroke length is increased. A longer stroke length allows the fluid particle to be located at the heating band when the piston is at the highest position, when the thermal energy is carried down to the bottom of the pyroelectric element as the piston reaches its lowest position. This results in a larger temperature swing in the pyroelectric element. Also, the heat input to maintain the desired temperature of the pyroelectric element increases with increasing frequency and increasing stoke length.

Figure 5.10 shows energy densities (N_D) as a function of high electric field (E_H), between 220 and 415 kV cm^{-1} at frequencies of 0.035, 0.061, and 0.12 Hz [78]. The piston stroke length, channel width, and low electric field are 4.7 cm, 330 mm, and 202 kV cm^{-1}, respectively. It is observed that a maximum energy density of 130 J l^{-1} is obtained at 0.061 Hz at a high electric field E_H of 379 kV cm^{-1}. At E_H of >379 kV/cm, the energy density decreases due to the increased leakage current. However, at lower E_H (<379 kV cm^{-1}), a smaller difference in electric field (E_H − E_L) leads to a lowered energy density. Therefore, the optimum value of E_H is 379 kV cm^{-1}.

With constant piston stroke length, stack channel width, and arithmetic mean temperature at the center of the pyroelectric element, at given low and high electric fields, the maximum energy and power densities are affected by frequency. It is found that although the temperature swing at 0.035 Hz is higher than that at 0.061 Hz, the energy density is lower. The longer cycle time at 0.035 Hz offers an opportunity for the charges at surface of the pyroelectric element to conduct through its body (leakage current), thus dissipating the energy as Joule heating. However, at higher frequencies of >0.061 Hz, the energy density is decreased because the temperature swing is reduced. As a result, there is a trade-off between high temperature swing and short cycle time and leakage current.

It is important to reduce the heat losses of the device to the environment. This is the reason why Teflon is used for the device outer cylinders and fiberglass wool is used to entirely insulate the device. Experimental results indicate that heat losses of the device are 15–20 % of the total heat input, which is increased as the operating temperature and heat input are increased. In addition, the efficiency of the device is still quite low, 0.045 and 0.053 % at 0.061 and 0.12 Hz, respectively.

5.6.3 Heat Conduction

Another example is to use heat conduction as a way to heat and cool the pyroelectric element to facilitate the Olsen cycle [85]. Figure 5.11 shows a schematic of the experimental setup. It consists of a thermal and an electrical subsystem. The thermal system consisted of two 3 cm × 2.5 cm × 1.27 cm cold and hot aluminum blocks.

They are used as the cold and hot sources, at the temperatures T_C and T_H, respectively. A 50 W Omega CS-10150 cartridge heater is embedded in the hot block as power supplier. The temperature T_H is controlled with a temperature controller. The cold block is at room temperature by free convection to the surrounding at atmospheric pressure. A thermal conductive epoxy layer with high electrical resistance and high thermal conductivity is placed on top surfaces of the two blocks. It is used to electrically isolate the electrodes of the pyroelectric element from the metallic blocks, while minimizing the thermal contact resistance. It has a thermal conductivity of 1.384 W m^{-1} K^{-1} and an electrical resistivity of on the order of 10^{15} Ω cm. The pyroelectric element sample is taped to the flat base of a wooden stamp. The wooden stamp is designed for convenient and safe handling of the film under various applied pressures. J-type thermocouples are embedded at center of the hot and cold blocks. In addition, a J-type thermocouple is attached to top of the film in such a way that it is in thermal contact but not electrical contact with the electrodes. Because the film has a thickness of just tens of micrometers, the temperatures measured by the thermocouple are assumed to be the same across the film. The electrical subsystem is similar to that discussed above.

The P(VDFeTrFE) film is poled at 200 kV cm^{-1} to increase its electrical resistivity. The temperature of the hot block is set to be $T_H = 90$ °C. Then, the stamp assembly supporting the pyroelectric element is made in thermal contact

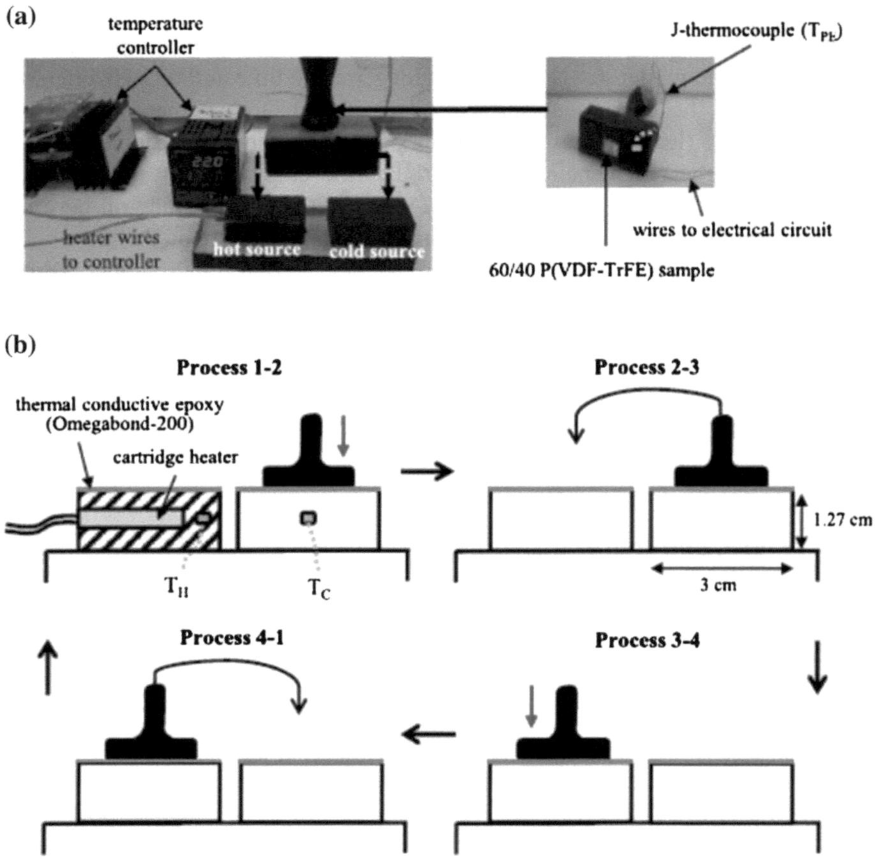

Fig. 5.11 **a** Thermal subsystem used to create periodic temperature oscillations during the Olsen cycle as well as the PE-stamp assembly **b** Schematic of each process in the Olsen cycle during stamping experiments. Reproduced with permission from [85], Copyright @ 2012, Elsevier

with the hot block for 70 min. The resistivity of the pyroelectric element is monitored until it reaches a steady-state value, 5.22×10^{10} Ω m. Once the poling is finished, the Olsen cycle can be carried out.

To create time-dependent temperature oscillations required in the Olsen cycle, the pyroelectric element is successively heated and cooled by manually stamping it against the hot and cold blocks. The time required for the film to reach the temperatures, T_H and T_C, of the hot and cold reservoirs (Processes 2–3 and 4–1), is longer than that to completely charge or discharge the pyroelectric element (Processes 1–2 and 3–4). The Olsen cycle is conducted with high electric fields E_H of 290–475 kV cm^{-1}. The low electric field E_L is 200 kV cm^{-1} to prevent depoling of the film during the cycle. The cold and hot source temperatures are maintained at 25 and 110 °C, respectively.

Fig. 5.12 Temperature of the PE over seven consecutive Olsen cycles. The operating conditions were $T_C = 25$ °C, $T_H = 110$ °C, $E_L = 200$ kV cm^{-1}, and $E_H = 350$ kV cm^{-1} while the cycle frequency ranged from 0.066 to 0.077 Hz. Reproduced with permission from [85], Copyright @ 2012, Elsevier

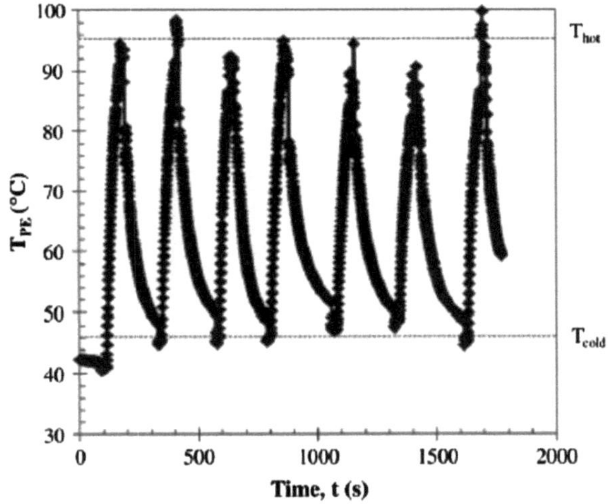

Figure 5.12 shows temperatures of the pyroelectric element during seven consecutive Olsen cycles at frequencies ranging from 0.066 to 0.078 Hz [85]. The cold source temperature T_C is 25 °C and the hot source temperature T_H was 110 °C. The average minimum and maximum temperatures of the pyroelectric element during one cycle are represented by T_{cold} and T_{hot}, respectively. The temperature of the pyroelectric element oscillates averagely between $T_{cold} = 45.2$ °C and $T_{hot} = 94.8$ °C. Because there is thermal contact resistance between the pyroelectric element and the hot or cold blocks and all the contacts remain within a relatively short time span, it is understood that $T_H > T_{hot}$ and $T_C < T_{cold}$.

Figure 5.13 shows D-E plot obtained for an Olsen cycle performed under electric fields between $E_L = 200$ and $E_H = 350$ kV cm^{-1} and cold and hot sources at $T_C = 25$ °C and $T_H = 110$ °C. It is found that the Olsen cycle starts and ends at different points. The offset has been attributed to the leakage current through the pyroelectric film [86]. The current-carrying impurities become mobile and drift towards the electrodes under high temperatures and high electric potentials. It is also noted that the Process 3–4 is not a smooth path, because the phase transition from ferroelectric to paraelectric is incomplete during the Process 2–3. The energy density is estimated to be 155 J l^{-1} $cycle^{-1}$ by applying the trapezoidal rule on the 1–4 region of the D-E curve.

Curie temperatures of 60/40P(VDFeTrFE) are about 66 °C at zero applied electric field [87], 92.5 °C at 300 kV cm^{-1}, and 120 °C at 527 kV cm^{-1} [88]. When the hot source temperature T_H is below 110 °C, the temperature of the film T_{hot} reached during the Olsen cycles is below the Curie temperature as the applied high electric field is above 350 kV cm^{-1}. As a result, the ferroelectric to paraelectric transition is not complete at $E_H > 350$ kV cm^{-1}. However, when the hot source temperature is increased from 110 to 130 °C, T_{hot} is further increased, short-circuit occurs due to the reduction in dielectric strength of air near the hot block with increasing temperature. In addition, increasing the hot source

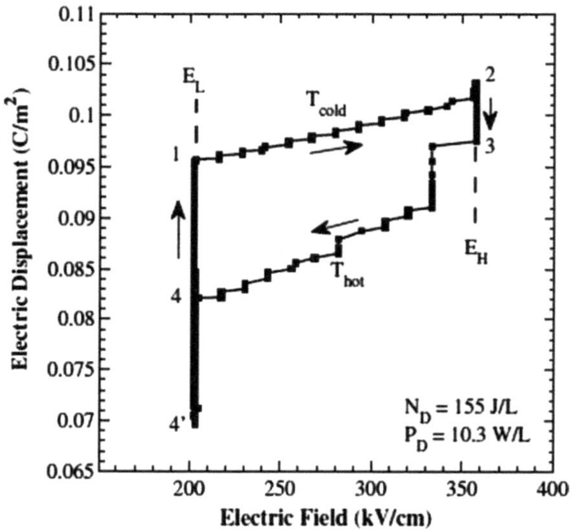

Fig. 5.13 Experimental Olsen cycle in the electric displacement versus electric field (D-E) diagram obtained with a 1 cm × 1 cm area and 60.45 μm thick 60/40P(VDFeTrFE) film, between $T_C = 25$ °C and $T_H = 110$ °C with $E_L = 200$ kV cm^{-1} and $E_H = 350$ kV cm^{-1}. Reproduced with permission from [85], Copyright @ 2012, Elsevier

temperature beyond 110 °C resulted in excessive leakage current. Under these conditions, the rate of surface charges conducting through the film exceeded the rate of electrical discharge during Process 2–3 over time. This resulted in crossovers in the D-E curves between Processes 1–2 and 3–4. Therefore, the hot source temperature TH of 110 °C was found to be optimum to maximize energy density for 60/40P(VDFeTrFE).

Figure 5.14 shows energy density as a function of applied high electric field E_H over 290–475 kV cm^{-1}. The low electric field E_L is 200 kV cm^{-1} and the cycles are operated at frequencies between 0.060 and 0.077 Hz. It is found that the energy density increases with increasing E_H up to 350 kV cm^{-1}, above which it starts to decrease with further increase in electric field. The highest energy density is 155 J l^{-1} cycle^{-1} at $E_H = 350$ kV cm^{-1} at 0.066 Hz, corresponding to a power density of 10.3 W l^{-1}. As the high electric field E_H is increased, the electric field span $(E_H - E_L)$ is enlarged, thus leading to large energy and power densities. However, leakage current also increases with increasing electric field.

The thermal contact resistance between the pyroelectric element and aluminum blocks can be reduced by increasing the pressure applied to the film. This in turn can reduce the time for the materials to reach the cold and hot source temperatures, thus increasing the power density. However, excessive pressure may cause electrical short-circuits in the pyroelectric materials especially when they are in contact with the hot block. A possible reason is the formation of microcracks between the electrodes when pressed against the uneven aluminum blocks. Nevertheless, no significant effect has been observed when the applied pressures are lower than 200 kPa.

Figure 5.14 also indicates that the heat transfer mechanism has an important role in determining the efficiency of the pyroelectric heat energy harvesters. For example, the largest energy density generated in the "stamping experiments"

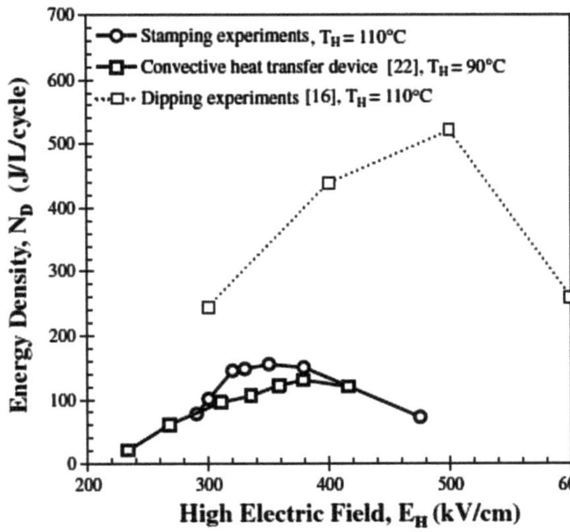

Fig. 5.14 Energy density as a function of high electric field E_H for different types of heat transfer. Operating conditions are (1) $E_L = 200$ kV cm^{-1}, $E_H = 290$–475 kV cm^{-1}, $T_C = 25$ °C (present study), (2) $E_L = 200$ kV cm^{-1}, $E_H = 300$–600 kV cm^{-1}, $T_C = 25$ °C (dipping experiments) [86], and (3) $E_L = 202$ kV cm^{-1}, $E_H = 233$–475 kV cm^{-1}, $T_C = 25$ °C (convective heat transfer device) [78]. Reproduced with permission from [85], Copyright @ 2012, Elsevier

(155 J l^{-1} cycle^{-1}) is significantly lower than that obtained in "dipping experiments" [86]. It has been attributed to the higher temperatures ($T_{\text{hot}} = 100$ °C) and higher applied electric fields ($E_H = 600$ kV cm^{-1}) imposed in the "dipping experiments" [86], as compared with $T_{\text{hot}} = 87$–96 °C. Also, the electric breakdown field strength of silicone oil is higher than that of air [88].

The Olsen cycle requires high electric fields across the pyroelectric materials. Therefore, for bulk materials or thick films, such high electric fields correspond to extremely high voltages, which might be prohibitively high for implementation in devices. In this regard, the use of very thin films would substantially reduce the voltages. However, on the other hand, leakage current may increase with decreasing film thickness [89, 90].

5.7 Olsen Harvesters with Perovskite Ceramics and Single Crystals

5.7.1 Relaxor–Ferroelectric PLZT Ceramics

As mentioned earlier, ferroelectric materials with perovskite crystal structure are promising candidates for pyroelectric applications [91]. Among them, relaxor–ferroelectric systems, consisting of x mol% lanthanum-doped 65 mol% lead zirconate and 35 mol% lead titanate solid solution, $Pb_{1-x}La_x(Zr_{0.65}Ti_{0.35})_{1-x/4}O_3$ or x/65/35PLZT, have been systematically studied. According to the phase diagram of x/65/35 PLZT ceramics at room temperature and zero electric field, the compositions with 5 mol% $\leq x \leq$ 9 mol% have rhombohedral crystal symmetry and are near the rhombohedral ferroelectric-mixed ferroelectric/cubic phase boundary [92].

They are paraelectric beyond the Burns temperature $T_B \approx 350$ °C, corresponding to the Curie temperature of the La-free compositions [93–96]. Upon cooling below T_B, the material transforms from paraelectric to ergodic relaxor, resulting in domain alignment and enlargement of the domain correlation radius. In the relaxor state, interactions between randomly oriented polar nanodomains cause distortion in the crystal structures, which gives rise to large dielectric and piezoelectric constants [97].

As domains grow and coalesce into polar clusters, random fields induced in the relaxor state become less active. This marks a phase transition between ergodic relaxor (nanodomain) and ferroelectric (macrodomain) that is field dependent and occurs at the Curie temperature T_{Curie}. T_{Curie} can be derived from the maximum of the pyroelectric coefficient $(\partial D/\partial T)_{\sigma,E}$ versus temperature T curve measured at a specified electric field T [99]. The materials will revert phase from ferroelectric to ergodic relaxor when they are heated above their T_{Curie} and/or de-poled below a critical electric field $E_{cr}(T)$. In x/65/35PLZT, the ferroelectric phase cannot be established spontaneously when cooling at zero electric field [93]. Application of an external electric field E greater than $E_{cr}(T)$ is required to stabilize the ferroelectric phase from the ergodic relaxor phase [99].

The isothermal D-E loops of x/65/35PLZT degenerate into narrow and linear loops as the temperature is increased above their T_{Curie}. As a result, the remnant polarization $P_r(T)$ and the coercive field $E_C(T)$ become negligible gradually in the relaxor state, as compared with their values in the ferroelectric state. The relaxor–ferroelectric phase transition is typically accompanied by a large change in electric displacement D. In order to produce the maximum energy and power output by using a relaxor–ferroelectric material, the relaxor–ferroelectric phase transition must occur during the Olsen cycle.

x/65/35 PLZT ceramics, with $x = 5, 6, 7, 8, 9$, and 10 mol%, are fabricated by using the conventional solid state reaction method, with starting powders of lead carbonate ($PbCO_3$), zirconium dioxide (ZrO_2), lanthanum oxide (La_2O_3), and titanium dioxide (TiO_2). The precursor powders are mixed and then calcined at 900 °C for 6 h. The calcined powders are compacted into disk samples, which are sintered at 1300 °C for 2 h in an Pb rich atmosphere to compensate the volatilization of Pb during the sintering process [100]. The sintered ceramics are subject to electrode deposition and then electrical and thermal characterization.

Figure 5.15a shows *D-E* diagrams of 9/65/35 PLZT for six different Olsen cycles, at low electric fields, E_L, 0, 0.1, 0.2, 0.3, 0.4, and 0.5 MV m^{-1}. The high electric field E_H is 4.0 MV m^{-1} while the cold and hot source temperatures are 3 and 150 °C. Figure 5.15b shows energy density and power density as a function of E_L.

The energy density reaches maximum at $E_L = 0.4$ MV m^{-1}. As E_L is decreased from 0.4 to 0 MV m^{-1}, the average energy density abruptly decreases from 368.2 to 143.0 J l^{-1} cycle^{-1}. This is because the sample is unable to re-polarize at zero electric field when the temperature is decreased from T_{hot} to

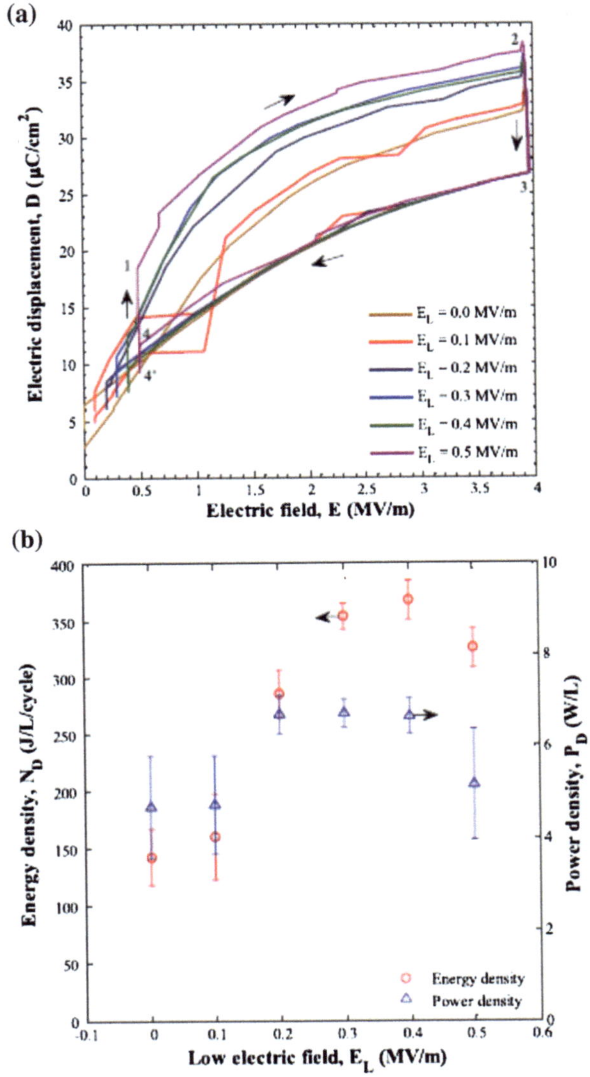

Fig. 5.15 **a** Experimental Olsen cycle in the D-E diagram performed on 9/65/35 PLZT Sample 10 between $T_{\text{cold}} = 3$ °C and $T_{\text{hot}} = 150$ °C for $E_H = 4.0$ MV m^{-1} and E_L varying from 0 to 0.5 MV m^{-1}. The Olsen cycles were vertically displaced to coincide at T_{hot} and E_H (state 3). **b** The corresponding energy and power densities for five different Olsen cycles performed under the above conditions. A peak in energy density is observed at $E_L{}^* = 0{:}4$ MV m^{-1}, while a peak in power density is obtained at $E_L^{\dagger} = 0.2$ MV m^{-1}. Reproduced with permission from [91], Copyright @ 2013, IoP Publishing

T_{cold} (Process 4–1). As the sample is de-poled below or near the coercive field $E_C = 0{:}09$ MV m^{-1} at $T_{\text{hot}} = 150$ °C, there are crossovers in the Olsen cycle D-E curve between processes 1–2 and 3–4, as shown in Fig. 5.15a ($E_L = 0$ MV m^{-1}) and a lower energy density. As the low electric field E_L is increased from 0.4 to 0.5 MV m^{-1}, the average energy density is reduced from 368.2 to 325.7 J l^{-1} cycle^{-1}, because the electric field span ($E_H - E_L$) imposed in the Olsen cycle becomes smaller. The largest power density is obtained for the low electric field $E_L = 0.2$ MV m^{-1}, as shown in Fig. 5.15b The power densities achieved at low electric fields $E_L = 0.2$, 0.3, and 0.4 MV m^{-1} are almost the same. Similar results are observed for the PLZT ceramics with other compositions.

Figure 5.16 shows energy densities generated of the x/65/35PLZT ceramics as a function of high electric field E_H. As the electric field span is increased, the energy densities are increased. It has been found that the maximum value of E_H is limited by dielectric strength of the materials. Cracks can be formed within the samples when the applied electric field is beyond the threshold. The formation of cracks may be attributed to mechanical stresses in the material along the grain boundaries, due to the spatially variant electric fields, thus causing preferential domain wall motion [101]. Those microcracks will propagate along the grain boundaries of the materials under the cyclic electric field loading/unloading, thus leading to sample failure ultimately.

The sample 7/65/35 PLZT exhibits a maximum energy density of 1013.5 ± 16.2 J l^{-1} per cycle, corresponding to a power density of 25.9 ± 0.8 W l^{-1}, as shown in Fig. 5.16c. The Olsen cycles are carried out under quasi-equilibrium conditions at 0.0256 Hz with electric field cycled between $E_L = E_L^* = 0{:}2$ MV m^{-1} and EH D 7.0 MV m^{-1}. The cold and hot source temperatures are $T_{\text{cold}} = 30$ °C and $T_{\text{hot}} = 200$ °C, respectively. This is the largest energy density achieved by using the Olsen cycle in the open literature.

Figure 5.17 shows the energy and power densities of the sample 9/65/35 PLZT as a function of cycle frequency, with cold and hot temperatures of $T_{\text{cold}} = 3$ °C and $T_{\text{hot}} = 150$ °C. The four different processes in the Olsen cycle performed at frequencies below 0.036 Hz are under quasi-equilibrium conditions, where the energy density has a maximum of 509.4 ± 29.6 J l^{-1} per cycle at 0.036 Hz. The increase in N_D has been attributed to the excessive leakage current in the material at frequencies below 0.036 Hz. At lower frequencies, there is sufficient time for the charges at the surface of the pyroelectric element to conduct through its body (leakage current), especially at high applied electric fields and/or high operating temperatures.

For 9/65/35 PLZT, the losses in energy density associated with the leakage current are 10 and 20 % when the Olsen cycle is carried out at 0.036 and 0.02 Hz, respectively. As the cycle frequency is above 0.036 Hz, the energy density decreases, due to the reduction in electric displacement span, because the processes are not performed under quasi-equilibrium conditions, i.e., the Olsen cycle do not span the isothermal D-E loops.

The power density P_D of 9/65/35PLZT has a maximum value of 32.4 ± 0.8 W l^{-1} at the peak power frequency of 0.096 Hz. The decrease in power output at higher frequencies is related to the relaxation and heat transfer mechanisms. The dipole realignment during the isoelectric field heating and cooling processes 2–3 and 4-1 is not complete at the high frequencies, due to the slow dipole relaxation inherent to ferroelectric–relaxor materials, especially at temperatures below T_{Curie}, because of the high energy barrier required to reorient the frozen nanodomains at low temperatures [102]. Also, the samples have not sufficient time to reach thermal equilibrium when the Olsen cycle is carried out at

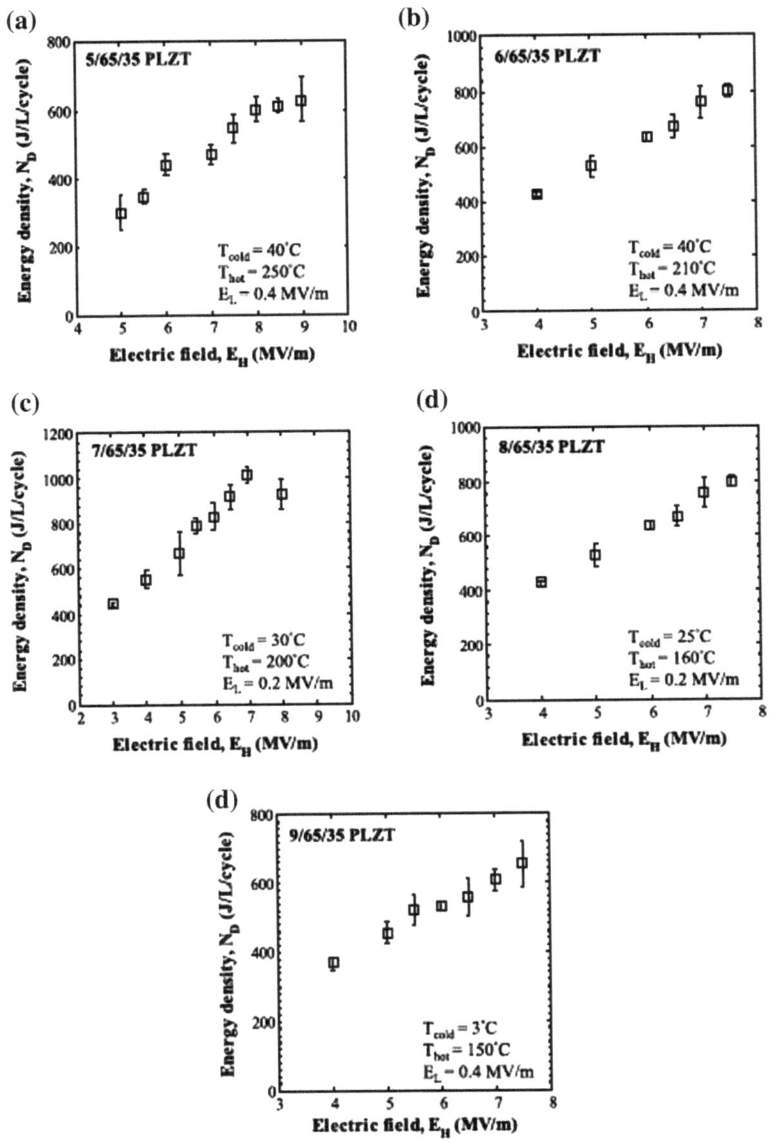

Fig. 5.16 Experimentally measured energy density produced by **a** 5/65/35 PLZT, **b** 6/65/35 PLZT, **c** 7/65/35 PLZT, **d** 8/65/35 PLZT, and **e** 9/65/35 PLZT, as a function of high electric field for Olsen cycles performed under quasi-equilibrium conditions, with $E_L = E_L*$ and their respective optimum temperatures T_{cold} and T_{hot}. Reproduced with permission from [91], Copyright @ 2013, IoP Publishing

the high cycle frequencies. As a result, phase transitions are incomplete, thus leading to smaller electric displacement span and lower electrical energy and power output.

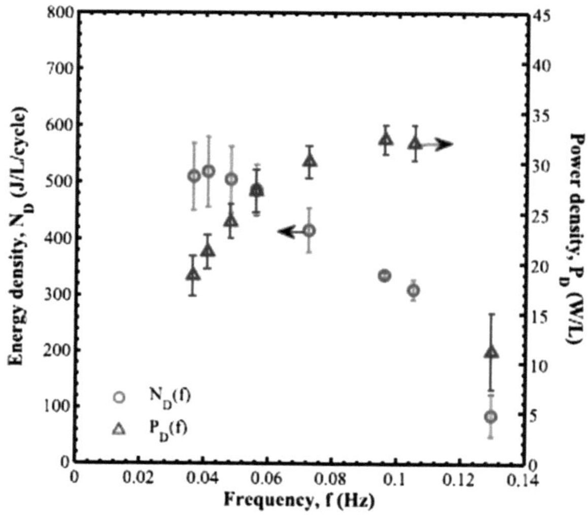

Fig. 5.17 Energy and power densities of the 9/65/35 PLZT ceramics as a function of cycle frequency, with $E_L = 0.2$ MV m^{-1} and $E_H = 6{:}0$ MV m^{-1}, $T_{\text{cold}} = 3$ °C and $T_{\text{hot}} = 150$ °C. Both the durations, τ_{12} and τ_{34}, are 1.5 s. The durations, τ_{23} and τ_{41}, of the isoelectric field processes 2–3 and 4–1 are in the range from 2 to 15 s. The energy density and the power density reached a maximum at 0.0364 and 0.0961 Hz, respectively. Reproduced with permission from [91], Copyright @ 2013, IoP Publishing

Figure 5.18 shows power density of 9/65/35 PLZT as a function of cycle frequency between 0.02 and 0.13 Hz. The peak power density increases with increasing high electric field E_H, which is attributed to the increase in energy density N_D due to the larger electric field span and the increase in the time rate of change in the electric field at a given cycle frequency. Peak power densities are observed at 0.0859, 0.0961, and 0.0709 Hz for high electric field E_H of 5.0, 6.0, and 7:0 MV m^{-1}, respectively.

Figure 5.19 shows energy densities of 8/65/35 PLZT as a function of high electric field with four samples at four different temperatures [103]. The energy harvested represents the averaged values over five cycles performed under quasi-equilibrium. It is found that variation is more pronounced at low temperatures (T_{hot}) and electric field (E_H). For instance, the largest sample variation is at $T_{\text{hot}} = 100$ °C and $E_H = 0.4$ MV m^{-1}, with a maximum relative difference among the samples to be 19.7 %. At $T_{\text{hot}} = 130$ °C and $E_H = 1.5$ MV m^{-1}, the variation is the lowest, with a maximum relative difference among the samples to be 9.1 %.

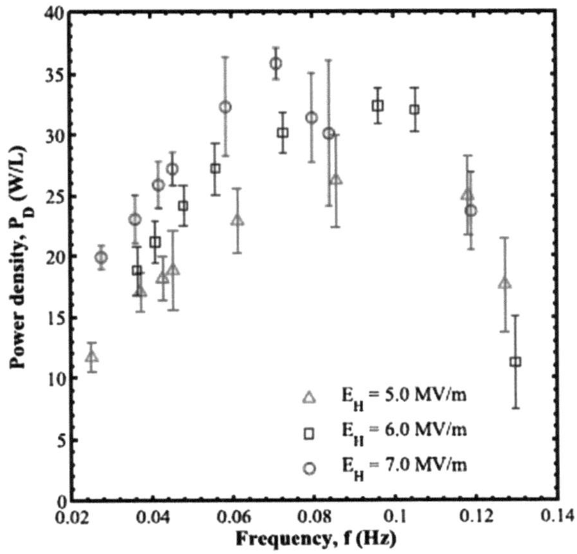

Fig. 5.18 Power density of the 9/65/35 PLZT ceramics as a function of cycle frequency, with low electric field $E_L = 0.2$ MV m^{-1} and high electric field $E_H = 5.0$, 6.0 or 7.0 MV m^{-1}, cold and hot source temperatures were equal to cold source temperature $T_{cold} = 3$ °C and hot source temperature $T_{hot} = 150$ °C. Reproduced with permission from [81], Copyright @ 2013, IoP Publishing

5.7.2 *Relaxor–Ferroelectric Single Crystals*

There have been several reports on relaxor–ferroelectric single crystals (1-x)$PbMg_{1/3}Nb_{2/3}$-xPT (PMN-PT) and (1-x)$PbZn_{1/3}Nb_{2/3}O_3$-x$PbTiO_3$ (PZN-PT), to be used for applications as pyroelectric waste heat energy harvesters [104–109]. For example, a 0.75PMN-0.5PT single crystal has been demonstrated to show a maximum energy density of 186 mJ l^{-1} cycle^{-1}, when it is subject to dipping experiments between 35 and 85 °C with electric field cycling between 0 and 30 kV cm^{-1} [104]. PZN-4.5PT single crystals can be used to harvest waste heat energy, due to the phase transition through rhombohedral, orthorhombic, and tetragonal phases during the Ericsson cycle. An energy density of 101.8 mJ l^{-1}cycle^{-1} has been achieved, when the single crystal device is operated at temperatures of between 100 and 130 °C and electric fields between 0 and 20 kV cm^{-1} [106].

Figure 5.20 shows energy densities of the device with 68$PbMg_{1/3}Nb_{2/3}O_3$-32$PbTiO_3$ (PMN-32PT) single crystal as a function of electric field E_H for different temperatures (T_{hot}) [109]. The electric field E_L is fixed at 2.0 kV cm^{-1} and the cold temperature T_{cold} is maintained at 80 °C. The highest energy density is 100 mJ l^{-1} cycle, corresponding to 4.92 mW l^{-1}, measured at temperatures between $T_{cold} = 80$ °C and $T_{hot} = 170$ °C and electric fields between $E_L = 2$ kV cm^{-1} and $E_H = 9$ kV cm^{-1}.

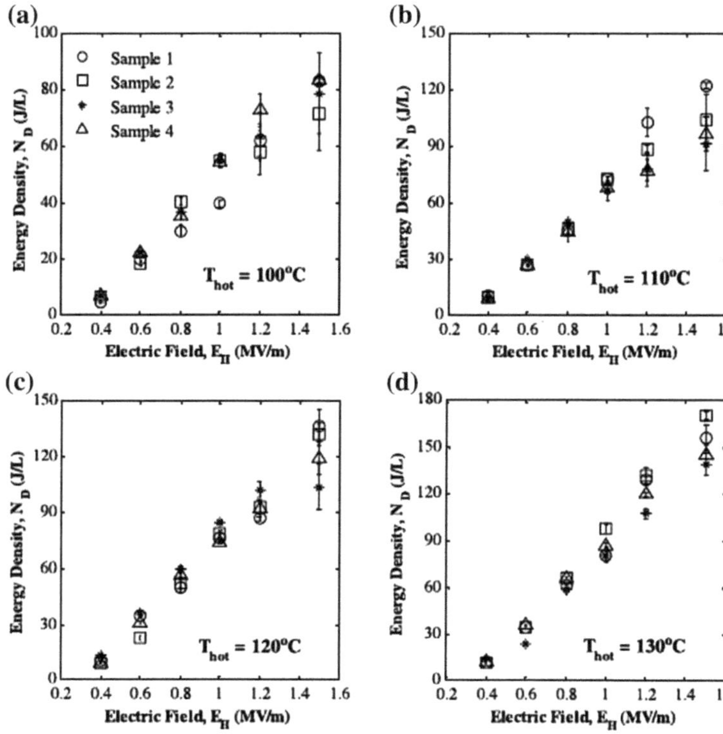

Fig. 5.19 Experimental energy densities of 8/65/35 PLZT as a function of high electric field E_H with four samples, at four hot source temperatures, T_{hot} = **a** 100 °C, **b** 110 °C, **c** 120 °C, and **d** 130 °C. The high electric field EH is ranged from 0.4 to 1.5 MV m^{-1}. The cold source temperature T_{cold} and low electric field E_L are 65 °C and 0.2 MV m^{-1}. Sample variability more pronounced at low temperatures and low electric fields. Reproduced with permission from [104], Copyright @ 2013, IoP Publishing

PMN-32PT single crystal with [001] orientation has two solid phase transitions: rhombohedral to tetragonal structures at about 80 °C and tetragonal to cubic structures at about 150 °C [110]. Thermal expansion is significant only when it undergoes a phase transition. The strain x_3 related to the thermal expansion between 80 and 120 °C is 0.11 %, while it is negligible when at temperatures away from the phase transition. Therefore, between T_{cold} = 80 °C and T_{hot} = 130 °C, the sample undergoes phase transition from rhombohedral to tetragonal.

Equilibrium D-E curves have been measured at different temperatures in order to compare with the Olsen cycle [108]. At given electric fields, e.g., E_L or E_H, the difference in the electric displacement ΔD between the D-E loop at T_{cold} and that at T_{hot} corresponds to the primary pyroelectric coefficient in Eq. (5.177). The iso-electric field processes in the Olsen cycle include both primary and secondary

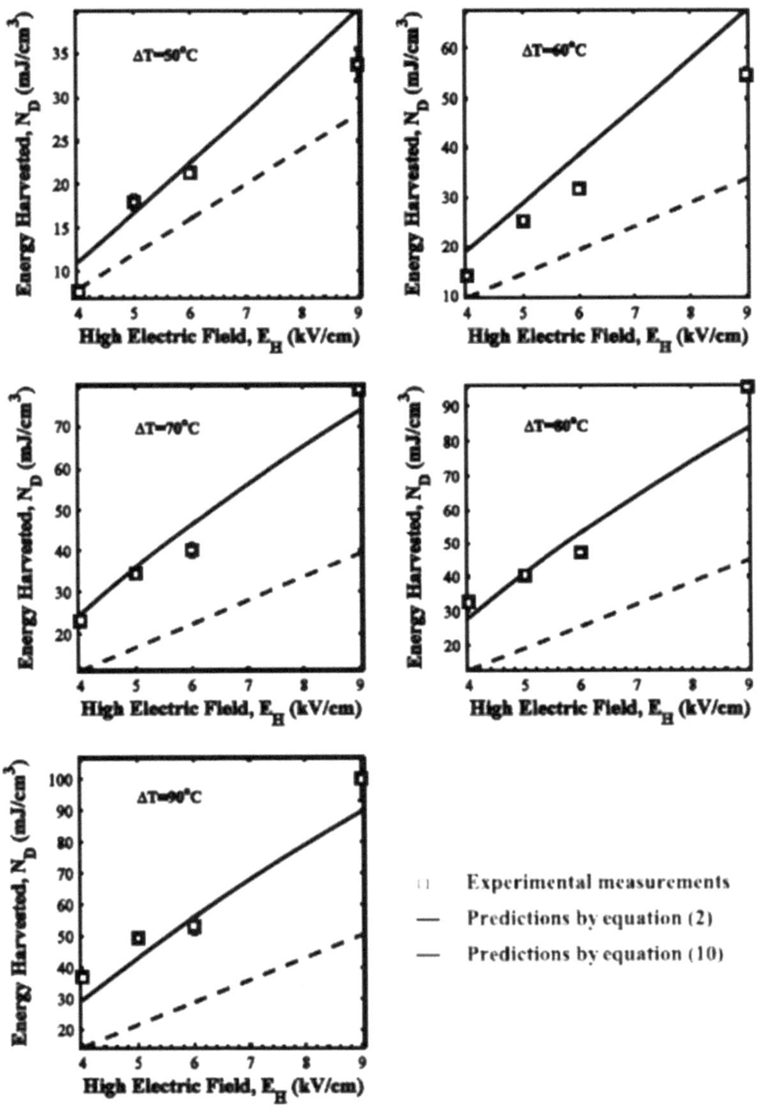

Fig. 5.20 Energy densities of the pyroelectric harvester with the PMN-32PT single crystals versus high electric field E_H for different values of $\Delta T = T_{\text{hot}} - T_{\text{cold}}$. The temperature T_{cold} is maintained at 80 °C and the low electric field E_L is 2.0 kV cm^{-1}. The dashed and solid lines correspond to predictions by Eqs. (5.180) and (5.179), respectively. Reproduced with permission from [108], Copyright @ 2011, IoP Publishing

pyroelectric effects, which are accounted for by Eq. (5.178). By comparing ΔD between the D-E loops and the Olsen cycle, it is possible to differentiate the primary and the secondary pyroelectric effects.

Displacements versus electric field for bipolar hysteresis curves have been measured at temperatures $T_{cold} = 80$ °C and $T_{hot} = 130$, 140, 150, and 170 °C for electric fields $E_L = 3$ kV cm^{-1} and $E_H = 6$ kV cm^{-1}, together with the Olsen cycle operating between the two temperatures.

For $T_{hot} = 130$ and 140 °C, the change in electric displacement observed during the Olsen cycle is larger than the change in displacement between the two D-E curves at temperatures of T_{hot} and T_{cold}. In this case, the secondary pyroelectric effect has a contribution to about 40 % of the total change in displacement. When T_{hot} is higher than $T_{Curie} = 150$ °C, the strain polarization due to thermal expansion disappears, due to the thermal hysteresis. As the sample is heated from 80 °C to temperatures below T_{Curie}, the strain due to tetragonal to rhombohedral phase transition is 0.11 %. When is cooled from above T_{Curie} down to 80 °C, the strain due to tetragonal to rhombohedral transition is 0.02 % [110]. The piezoelectric coefficient of unpoled material is significantly less than that of poled material, which can be neglected near the Curie point. In addition, there is no significant difference in relative permittivity in the linear region with respect to electrical or thermal hysteresis. As a result, change in relative permittivity should not be the cause for the change in the displacement. Therefore, above the Curie temperature, the strain polarization is significantly low and thus can be neglected. In other words, under constant strain conditions (unclamped), the performance of the device is not degraded, when T_{hot} is higher than T_{Curie}.

The largest change in polarization of PMN-32PT single crystal is observed between 80 and 170 °C, corresponding to largest energy density harvested. The energy density harvested N_D has a slight nonlinear relation with temperature T_{hot}, as shown in Fig. 5.20 [108]. The energy density increases by about 24 mJ l^{-1} cycle^{-1} when T_{hot} is increased from 140 to 150 °C, as compared with 9 mJ l^{-1} cycle^{-1} when T_{hot} is increased from 160 to 170 °C. This means energy density exhibits maximum value at T_{hot} is near T_{Curie}, confirming the importance of the ferroelectric to paraelectric phase transition on the efficiency of pyroelectric energy conversion.

The effects of electric fields, E_L and E_H, on energy density have also been studied. An optimal energy density of 24.8 mJ l^{-1} cycle^{-1} is observed at $E_L = 2.0$ kV cm^{-1}. The experimental results have been compared with theoretical predictions by using Eq. (5.180) with assumption of constant properties (p = 800 μC m^{-2} K^{-1} at 80 °C) [111] and Eq. (5.117) with temperature-dependent properties. The predictions of the constant–property model deteriorates as the temperature difference $\Delta T = T_{hot} - T_{cold}$ is increased. Because the pyroelectric coefficient of PMN-32PT increases by an order of magnitude as the temperature is increased from 80 to 150 °C [111]. As a result, the constant–property model has underestimated the energy density harvested during the Olsen cycle.

The second example of single crystal used for pyroelectric energy conversion is $94.5PbZn_{1/3}Nb_{2/3}O_{3}$-$5.5PbTiO_3$ (PZN-5.5PT) [109]. The properties of PZN-PT single crystals have been studied extensively studied due to their important applications in sensors and actuators. Poled PZN-4.5PT single crystal exhibits a phase transformation sequence from rhombohedral (R) to tetragonal (T) and to cubic (C), when it is heated from 25 to 160 °C at zero electric field [112]. A small fraction of rhombohedral domains are present in the tetragonal phase. According to the phase diagram of PZN-xPT, PZN-5.5PT has phase transitions from rhombohedral to tetragonal between 122 and 130 °C and tetragonal to cubic at $T_{Curie} = 165$ °C [113]. During field cooling from 177 to 27 °C, [001]-oriented PZN-4.5PT single crystal has an additional orthorhombic (O) phase and experiences a C-T-O-R phase transition sequence [114]. It is also found that the transition temperatures can be affected by the application of external electric field.

The crystal PZN-5.5PT samples used for study, with a surface area and thickness of 1 × 1 cm and 200 μm, are poled in the [001] direction [110]. Both surfaces of the samples are coated with a 10 nm NiCr bond layer and a 1 μm thick $Au_{0.68}Pd_{0.32}$ electrode layer, deposited by using an RF sputter-deposition facility. Electrical wires are attached to the electrodes with conductive silver paste. Isothermal bipolar displacement versus electric field hysteresis curves are collected at various temperatures by applying a triangular voltage with frequency of 0.1 Hz across the single crystal samples. The samples are measured in a silicone oil bath at temperatures of 100, 125, 150, 175, or 190 °C and electric fields from −1 to 1 MV m^{-1}.

The Olsen cycle is carried out on the PZN-5.5PT samples at different low and high electric fields, E_L and E_H, at various hot operating temperatures T_{hot} and various cycle frequency f. The cold operating temperature T_{cold} is fixed at 100 °C, while hot operating temperatures are from 125 to 190 °C. The electric fields E_L and E_H are ranged from 0 to 0.2 MV m^{-1} and from 0.5 to 1.5 MV m^{-1}, respectively. The overall cycle frequency is varied from 0.021 to 0.15 Hz by changing the duration of the isoelectric field heating and cooling processes 2–3 and 4–1. The time rate of change in the electric field during the isothermal processes 1–2 and 3–4 remains the same at 0.4 MV m^{-1} s^{-1}. The energy and power densities can be estimated and evaluated.

It has been observed that the Olsen cycles measured at frequencies of 0.021 and 0.034 Hz overlap each other. This is because over this frequency range the electric displacement is at steady state, i.e., $\partial D/\partial t = 0$, before the electric field is changed. Therefore, processes 1–2 and 3–4 possess a relatively smooth path, indicating that the four processes in the Olsen cycle are under quasi-equilibrium conditions. If the frequency is increased to be higher than 0.034 Hz, the isoelectric field processes 2–3 and 4–1 are not under quasi-equilibrium conditions. In this case, the electric displacement does not reach steady state before the electric field is changed to conduct the processes 1–2 and 3–4. In other words, the phase transition is incomplete. Additionally, the Olsen cycles are smooth between E_L and E_H during processes 1–2 and 3–4 in the D-E diagram [109].

Figure 5.21 shows energy density and power density as functions of frequency when the Olsen cycles is performed for a representative sample [109]. The energy density reaches a plateau of 140 J l^{-1} $cycle^{-1}$ at frequencies below 0.034 Hz and decreases with increasing cycle frequency. Reducing the cycle frequency below 0.034 Hz, by increasing the duration of the isoelectric field processes 2–3 and 4–1,

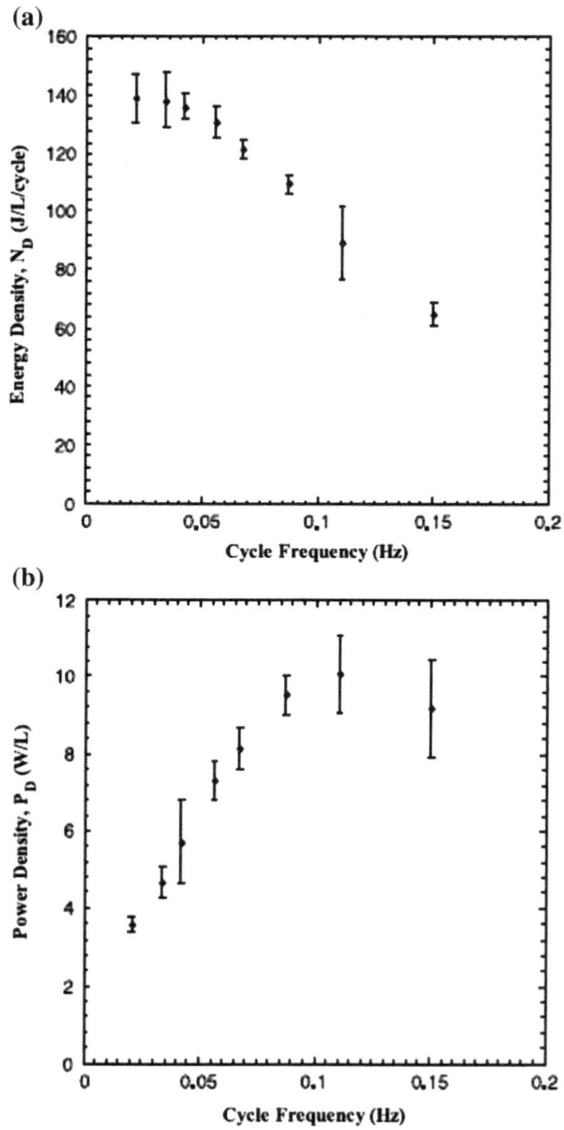

Fig. 5.21 Experimentally results of **a** energy density and **b** power density generated with the PZN-5.5PT single crystal as a function of cycle frequency. The temperatures T_{cold} and T_{hot} are 100 and 190 °C, respectively. The low and high electric fields E_L and E_H are 0 and 1.0 MV m^{-1}, respectively. Reproduced with permission from [109], Copyright @ 2012, IoP Publishing

does not lead to higher energy density, because each process of the Olsen cycle is carried out under quasi-equilibrium state.

In comparison, the power density increases with increasing frequency and reaches a maximum of 10.1 W l^{-1} at 0.1 Hz. The decrease in P_D higher frequencies can be explained with the expression $P_D = N_D(f)f$. At frequencies of less than 0.1 Hz, the decrease in $N_D(f)$ is compensated by the increase in frequency, so that P_D increases. Above 0.1 Hz, N_D decreases significantly with increasing

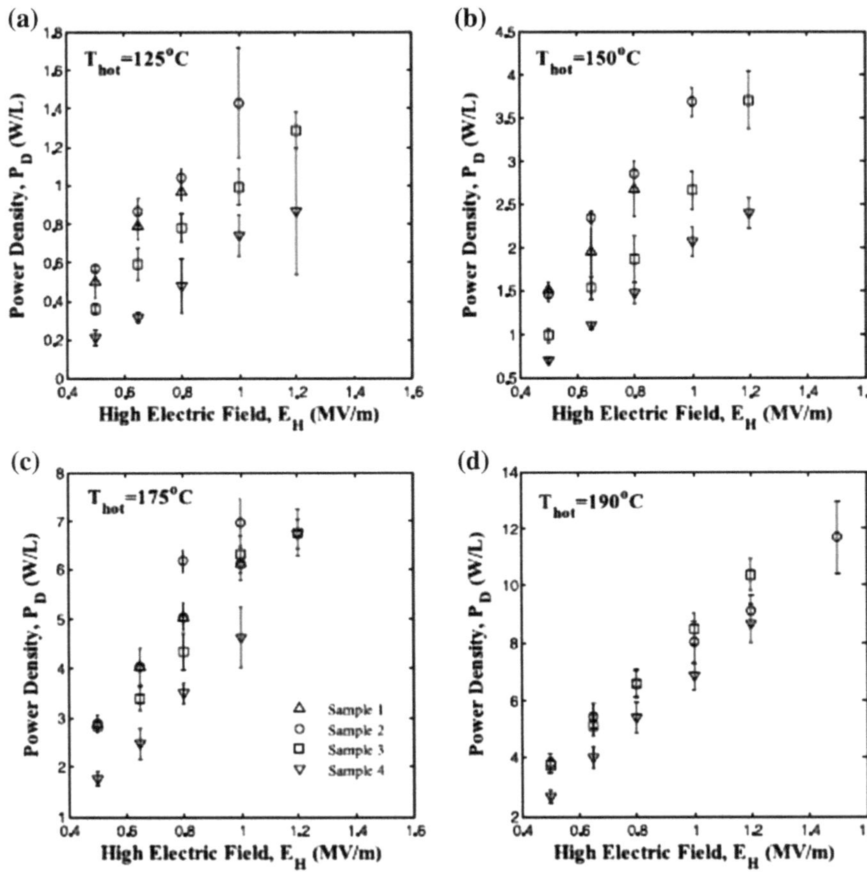

Fig. 5.22 Experimentally measured power density generated at 0.1 Hz from four different samples of PZN-5.5PT single crystal as a function of high electric field E_H for T_{hot} of **a** 125 °C, **b** 150 °C, **c** 175 °C, and **d** 190 °C, as well as T_{cold} of 100 °C and low electric field E_L of 0.2 MV m^{-1}. Reproduced with permission from [109], Copyright @ 2012, IoP Publishing

frequency, thus leading to lower values of P_D. This can be used as reference to determine the operating frequency in practical applications.

It is also necessary to mention that, for practical applications, the sample variability should be as small as possible. Figure 5.22 shows power density as a function of high electric field E_H at cycle frequency of 0.1 Hz for four different samples. For all samples, the power density increases with increasing electric field E_H and increasing hot source temperature T_{hot}. The maximum power is 11.7 W l^{-1} for T_{cold} = 100 °C, T_{hot} = 190 °C, E_L = 0:2 MV m^{-1}, and E_H = 1:5 MV m^{-1}. The samples are failure at electric fields of higher than 1:5 MV m^{-1}, due to the excessive thermoelectro mechanical stress. The maximum relative error for P_D among the samples for all values of E_H is 18.3 % for T_{hot} = 125 °C, which decreases gradually as the T_{hot} is increased from 150 to 190 °C. The larger variability observed at lower temperatures can be attributed to the large differences in the rhombohedral and tetragonal volume fractions among the samples.

5.8 Other Types of Pyroelectric Harvesters

Besides the pyroelectric thermal energy harvesters based on the Olsen cycle, there are other types of converters that have been reported in the open literature [115–117]. For example, it has been demonstrated in a study that Ericsson-based thermodynamic cycles are also suitable for pyroelectric thermal energy harvesting [115]. The harvested energy is 100 times higher than that with the simple linear pyroelectric effect. It is found that the principle of pyroelectric energy harvesting with the Ericsson cycle is analogous to the electrocaloric effect. Therefore, in order to identify the best material for pyroelectric energy harvesting, it is necessary to search for materials that have highest electrocaloric efficiency. The harvested energy per unit of volume is the product of the Carnot efficiency and the electrocaloric effect. Generally, there is a potential harvested energy capability ranging from 50 to 600 mJ l^{-1} for a temperature variation of 10 °C. The Ericsson cycle on a $0.90Pb(Mg_{1/3}Nb_{2/3})O_3$-$0.10PbTiO_3$ ceramics can yield a harvested energy of 186 mJ l^{-1} with a temperature variation of 50 °C at 3.5 kV mm^{-1}.

A simple method has been used fabricate pyroelectric thermal energy harvesting cells with commercial lead zirconate titanate (PZT) and PVDF, by using screen printing technique [116]. The PZT cells with a square pyroelectric layer (4 cm × 4 cm) sandwiched between two PdAg electrodes are printed on Al_2O_3 ceramic substrates, while the PVDF ones are deposited on flexible plastic substrates.

Currents in the order of 10^{-7} A and charges in the order of 10^{-5} C have been achieved for temperature fluctuations from 300 to 360 K in a time period of the order of 100 s. The harvesters have a similar behavior for heating and cooling processes. The charge generation is dependent on the difference of temperatures between two instant times but is not independent of the temporal evolution of the temperature. For PZT cells, the generated current increases with increasing thickness of the pyroelectric layer and poling electrical field during the fabrication process. Stored energy can be as high as 0.5 mJ, which is enough to power an autonomous sensor node during a measurement and transmission cycle [116].

A pyroelectric nanogenerator has been designed by using a PZT thin film [117]. An optical photograph of the fabricated pyroelectric nanogenerator is shown in Fig. 5.23a, which has a length of 21 mm and a width of 12 mm. Figure 5.23b shows a crosssectional scanning electron microscopy (SEM) image of the device, in which the thickness of the PZT layer is about 175 μm. High magnification cross-sectional SEM image indicates that the PZT film is polycrystalline, as shown in Fig. 5.23c.

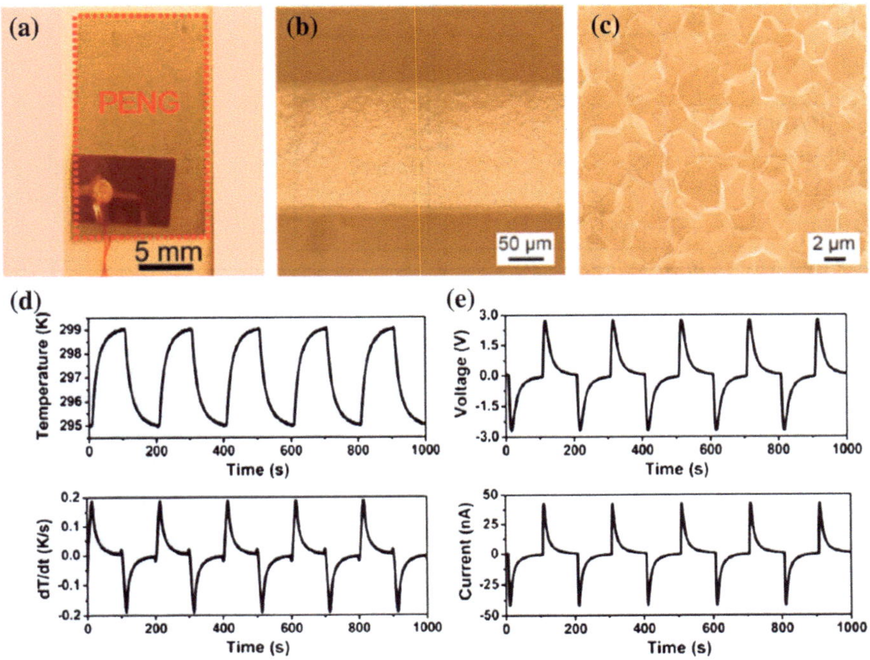

Fig. 5.23 **a** Photograph of a fabricated PENG. **b** Crosssectional SEM image of the PENG. **c** Enlarged Crosssectional SEM image of the PENG. **d** The cyclic change in temperature of the PENG and the corresponding differential curve. **e** Measured output voltage and current of the PENG under the change in temperature in (**d**). Reproduced with permission from [117], Copyright @ 2012, American Chemical Society

The output voltage and current of the nano-harvester are measured over a temperature range from 295 to 299 K, as shown in Fig. 5.23d. The peak value of the temperature changing rate is about 0.2 K s^{-1}. With forward connection, sharp negative voltage, and current pulses (2.8 V and 42 nA) are observed, as the temperature is quickly increased from 295 to 299 K, while positive pulses are present when the temperature is back to 295 K (Fig. 5.23e). With reverse connection (backward), the signs of the signals are reversed accordingly, proving that the signals are all generated by the nano-harvester [117].

The electrical energy generated by the nano-harvester can be stored for potential applications [117]. Figure 5.24a shows a schematic diagram, where a Li-ion battery is charged by the nanogenerator that is able to drive a LED. Figure 5.24b shows an optical image of the Li-ion coin battery. In this Li-ion battery, the TiO_2 nanotube array grown on Ti foil acts as the anode, with tube diameter and length of about 150 nm and 10 μm, as shown in Fig. 5.24c. Figure 5.24d shows the charging and the subsequent constant-current discharging curves of the battery, with nanogenerator working at a frequency of 0.005 Hz. The battery can be charged by the nanogenerator from 650 to 810 mV in about 3 h.

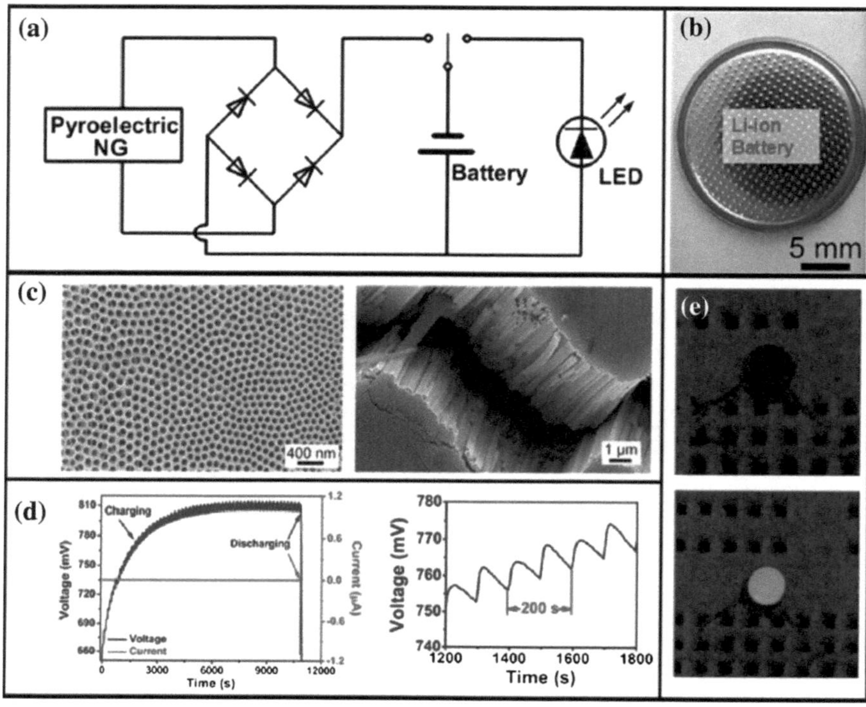

Fig. 5.24 **a** Schematic that illustrates the process for charging the Li-ion battery and driving a LED. **b** Photograph of the fabricated Li-ion battery. **c** SEM image of the obtained TiO2 nanotubes. **d** The charging and discharging curves of the Li-ion battery by using the PENG. e Photograph of a green LED driven by the Li-ion batteries. Reproduced with permission from [118], Copyright @ 2012, American Chemical Society

The obtained voltage peaks are associated with the universally existing self-discharge of the battery. The time interval of the two adjacent peaks of the charging is about 200 s. Under a constant current of 1 μA, the discharging of the battery lasts for 84 s before it gets back to its original state of 650 mV. Therefore, the stored electric capacity is about 0.023 μAh. Figure 5.24e shows that the charged Li-ion batteries can be used to power a green LED. It is also found that the harvested energy can be used as power sources for wireless sensors [117].

5.9 Concluding Remarks

Pyroelectric effect has been employed to harvest waste thermal energy by using pyroelectric materials in several ways. Pyroelectric energy harvesting requires temporal temperature variations, i.e., time gradients of temperatures, whereas thermoelectric energy harvesting requires spatial gradients of temperatures.

Although wasted heat more likely creates spatial gradients rather than time gradients, the conversion efficiency, which is defined as the ratio of net harvested energy to the heat taken from the hot reservoir, is much higher when using pyroelectric energy harvesting. In theory, it could reach the conversion ratio of the Carnot cycle, regardless of the materials properties. In contrast, the conversion efficiency of thermoelectric effect is highly limited by the materials properties, as discussed in previous chapter.

Although, among various cycles the Carnot cycle is the most efficient, it is the least practical according to the laws of thermodynamics. Olsen cycle has been proved to be applicable for practical thermal energy harvesting, with the largest energy density as compared other cycles. According to theoretical prediction, it is possible to increase the energy density by increasing the electric field span, which evidently requires pyroelectric materials to have sufficiently high dielectric breakdown strength. In this respect, bulk ceramics and single crystals have limits to certain degrees. To address this problem, it is suggested to use thin films and composites.

A theoretical study has indicated that pyroelectric response of ferroelectric nanowires can be tuned and maximized through the controlling of their dimension [118]. The pyroelectric coefficient of PZT nanowires strongly increases with decreasing wire radius before approaching a critical radius, which is corresponding to the size-driven phase transition from ferroelectric to paraelectric. Size-driven enhancement of pyroelectric coupling leads to giant pyroelectric current and voltage generation by the polarized ferroelectric nanoparticles in response to the temperature fluctuation. The maximum efficiency of the pyroelectric energy harvesting and bolometric detection is shown to approach the Carnot limit for low temperatures. This can be another direction of pyroelectric effect for thermal energy harvesting applications.

References

1. G.H. Haertling, Ferroelectric ceramics: history and technology. J. Am. Ceram. Soc. **82**, 797–818 (1999)
2. D. Damjanovic, Ferroelectric, dielectric and piezoelectric properties of ferroelectric thin films and ceramics. Rep. Prog. Phys. **61**, 1267–1324 (1998)
3. L.B. Kong, T.S. Zhang, J. Ma, Y.C.F. Boey, Progress in synthesis of ferroelectric ceramic materials via high-energy mechanochemical techniques. Prog. Mater Sci. **53**(2), 207–322 (2008)
4. S.B. Lang, The history of pyroelectricity : from ancient Greece to space missions. Ferroelectrics **230**, 99–108 (1999)
5. S.B. Lang, Pyroelectricity: from ancient curiosity to modern imaging tool. Phys. Today **2005**(8), 31–36 (2005)
6. S.B. Lang, A 2400 year history of pyroelectricity: from ancient Greece to exploration of the solar system. Br. Ceram. Trans. **103**, 65–70 (2004)
7. R.W. Whatmore, Pyroelectric devices and materials. Rep. Prog. Phys. **49**, 1335–1386 (1986)

8. W.L. Zhong, Physics of Ferroelectricity, Chapter 8 (Science Press, Beijing, 1996)
9. G. Sebald, E. Lefeuvre, D. Guyomar, Pyroelectric energy conversion: optimization principles. IEEE Trans. Ultrason. Ferroelect. Freq. Contr. **55**, 538–551 (2008)
10. S.T. Liu, D. Long, Pyroelectric detectors and materials. Proc. IEEE **66**, 14–26 (1978)
11. V.F. Kosorotov, L.S. Dremenchugski, L.V. Levash, L.V. Shchedrina, Tertiary pyroelectric effect in lithium-niobate and lithium tantalite crystals. Ferroelectrics **70**, 27–37 (1986)
12. R. Poprawski, Investigation of phase transitions in NH_4HSeO_4 crystals by pyroelectric method. Ferroelectrics **33**, 23–24 (1981)
13. A.F. Devonshire, Theory of ferroelectrics. Adv. Phys. **3**, 85–130 (1954)
14. N.P. Hartley, P.T. Squire, E.H. Putley, A new method of measuring pyroelectric coefficients. J. Phys. E: Sci. Instrum. **5**, 787–790 (1972)
15. R.R. Zeyfang, W.H. Sehr, K.V. Kiehl, Enhanced pyroelectric properties at a F.E.-F.E. phase transition. Ferroelectrics **11**, 355–358 (1976)
16. S.B. Lang, F. Stechel, Study of the ultrasensitive pyroelectric thermometer. Rev. Sci. Instru. **36**, 1817–1822 (1965)
17. S.B. Lang, S.A. Shaw, L.H. Rice, K.D. Timmerhaus, Pyroelectric thermometer for use at low temperatures. Rev. Sci. Instru. **40**, 274–284 (1969)
18. S.B. Lang, L.H. Rice, S.A. Shaw, Pyroelectric effect in barium titanate ceramic. J. Appl. Phys. **40**, 4335–4340 (1969)
19. A.M. Glass, Investigation of the electrical properties of $Sr_{1-x}Ba_xNb_2O_6$ with special reference to pyroelectric detection. J. Appl. Phys. **40**, 4699–4713 (1969)
20. R.L. Byer, C.B. Roundy, Pyroelectric coefficient direct measurement technique and application to a nsec response time detector. Ferroelectrics **3**, 333–338 (1972)
21. A.G. Chynoweth, Dynamic method for measuring the pyroelectric effect with special reference to barium titanate. J. Appl. Phys. **27**, 78–84 (1956)
22. S.B. Lang, Pyroelectric coefficient of lithium sulfate monohydrate (4.2–320 °K). Phys. Rev. B **4**, 3603–3609 (1971)
23. A. Halliyal, A.S. Bhalla, L.E. Cross, X-ray structure refinement and pyroelectric investigation of fresnoite, $Ba_2TiSi_2O_8$. Ferroelectrics **62**, 3–9 (1985)
24. S.G. Porter, A brief guide to pyroelectric detectors. Ferroelectrics **33**, 193–206 (1985)
25. M. Daglish, A.J. Moulson, Dependence of the dielectric response of $Pb(Sc_{0.5}Ta_{0.5})O_3$ on applied DC field: implications for the choice of target materials in thermal imaging devices. Ferroelectrics **126**, 215–220 (1992)
26. E.A.D. White, J.D.C. Wood, V.M. Wood, The growth of large area, uniformly doped TGS crystals. J. Cryst. Growth **32**, 149–156 (1976)
27. K. Itoh, T. Mutsui, Studies of the crystal structure of triglycine sulfate in connection with its ferroelectric phase transition. Ferroelectrics **5**, 235–251 (1973)
28. G.M. Loiacono, W.N. Osborne, M. Delfino, G. Kostecky, Single crystal growth and properties of deuterated triglycine fluoroberyllate. J. Cryst. Growth **46**, 105–111 (1979)
29. E.T. Keve, K.L. Bye, P.W. Whipps, A.D. Annis, Structural inhibition of ferroelectric switching in triglycine sulphate-I. additives. Ferroelectrics **3**, 39–48 (1971)
30. A.S. Bhalla, C.S. Fang, L.E. Cross, Pyroelectric properties of alanine and deuterium substituted TGSP and TGSAs single crystals. Mater. Lett. **3**, 75–77 (1985)
31. A.S. Bhalla, C.S. Fang, X. Yao, L.E. Cross, Pyroelectric properties of the alanine and arsenic-doped triglycine sulfate single crystals. Appl. Phys. Lett. **43**, 932–934 (1983)
32. P. Felix, P. Gamot, P. Lacheau, Y. Raverdy, Pyroelectric, dielectric and thermal properties of TGS, DTGS and TGFB. Ferroelectrics **17**, 543–551 (1977)
33. A.J. Goss, R.D. Nixon, R. Watton, W.M. Wreathall, Inrearred television using the pyroelectric vidicon. GEC J. Res. **2**, 198–203 (1984)
34. A. Shaulov, Improved figure of merit in obliquely cut pyroelectric crystals. Appl. Phys. Lett. **39**, 180–181 (1981)
35. A. Shaulov, W.A. Smith, Optimum cuts of monoclinic m crystals for pyroelectric detectors. Ferroelectrics **49**, 223–228 (1983)

36. T. Yamada, T. Kitayama, Ferroelectric properties of vinylidene fluoride-trifluoroethylene copolymers. J. Appl. Phys. **52**, 6859–6865 (1981)
37. H. Yamazaki, J. Ohwaki, T. Yamada, T. Kitayama, Temperature dependence of the pyroelectric response of vinylidene fluoride trifluoroethylene copolymer and the effect of its poling conditions. Appl. Phys. Lett. **39**, 772–773 (1981)
38. E. Fukada, T. Furukawa, Piezoelectricity and ferroelectricity in polyvinylidene fluoride. Ultrasonics **19**, 31–39 (1981)
39. S. Kumaragurubaran, S. Takekawa, M. Nakamura, K. Kitamura, Growth of 4-in diameter near-stoichiometric lithium tantalate single crystals. J. Cryst. Growth **285**, 88–95 (2005)
40. N.E. Byer, A. Van der Jagt, Monolithic pyroelectric arrays. Ferroelectrics **27**, 11 (1980)
41. P.B. Jamieson, S.C. Abrahams, J.L. Bernstein, Ferroelectric tungsten bronze type crystal structures. I. barium strontium niobate $Ba_{0.27}Sr_{0.75}Nb_2O_{5.78}$. J. Chem. Phys. **48**, 5048–5057 (1968)
42. S.T. Liu, R.B. Maciolek, Rare-earth-modified $Sr_{0.5}Ba_{0.5}Nb_2O_6$, ferroelectric crystals and their applications as infrared detectors. J. Electron. Mater. **4**, 91–100 (1975)
43. S.T. Liu, A.S. Bhalla, Some interesting properties of dislocation-free and La-modified $Sr_{0.5}Ba_{0.5}Nb_2O_6$. Ferroelectrics **15**, 47–51 (1983)
44. R.C. Clarke, R.W. Whatmore, The growth and characterization of $PbZr_xTi_{1-x}O_3$ single crystals. J. Cryst. Growth **33**, 29–39 (1976)
45. D. Luff, R. Lane, K.R. Brown, H.J. Marshallsy, Ferroelectric ceramics with high pyroelectric properties. Trnas. J. Br. Ceram. Soc. **73**, 251–264 (1974)
46. D. Appleby, S.G. Porter, F.W. Ainger, The pyroelectric properties of the $F.E._{RL}/F.E._{RH}$ phase transitions lead zircon ate ferric niobate and related compounds. Ferroelectrics **14**, 715–717 (1976)
47. R.M. Henson, R.R. Zeyfang, E. Linhart, Pyroelectric properties of $Pb(Ti, Zr, [Fe_{1/2}Ta_{1/2}])O_3$ polycrystalline solid solutions. Phys. Status Solidi A **46**, 511–515 (1978)
48. R.C. Clarke, A.M. Glazer, F.W. Ainger, D. Appleby, N.J. Poole, S.G. Porter, Phase transitions in lead zirconate-titanate and their applications in thermal detectors. Ferroelectrics **11**, 359–364 (1976)
49. R.W. Whatmore, A.J. Bell, Pyroelectric ceramics in the lead zirconate-lead titanate-lead iron niobate system. Ferroelectrics **35**, 155–160 (1981)
50. R.W. Whatmore, High performance, conducting pyroelectric ceramics. Ferroelectrics **49**, 201–210 (1985)
51. B.C. Grabmaier, $PbTiO_3$ grown from melt. Ferroelectrics **13**, 501–503 (1976)
52. J.G. Bergman, G.R. Crane, E.H. Turner, The tetragonal deformation of the TiO_6 octahedron in ferroelectric $PbTiO_3$. J. Solid State Chem. **21**, 127–133 (1977)
53. H. Takeuchi, S. Jyomura, Y. Ito, K. Nagatsuma, Rare-earth substituted piezoelectric $PbTiO_3$ ceramics for acoustic wave applications. Ferroelectrics **51**, 71–80 (1983)
54. N. Ichinose, Electronic ceramics for sensors. Am. Ceram. Soc. Bull. **64**, 1581–1585 (1985)
55. R.W. Whatmore, P.C. Osbond, N.M. Shorrocks, Ferroelectric materials for thermal IR detectors. Ferroelectrics **76**, 351–367 (1987)
56. P.C. Osbond, R.W. Whatmore, Improvements to pyroelectric ceramics via strontium doping of the lead zirconate-lead iron niobate-lead titanate system. Ferroelectrics **118**, 93–101 (1991)
57. M.R. Houlton, G.R. Jones, D.S. Robertson, A study of growth defects in lead germanate crystals. J. Phys. D Appl. Phys. **8**, 219–225 (1975)
58. W.K. Zwicker, J.P. Dougherty, M. Delfino, J. Ladell, Growth of high quality lead germanate crystals for pyroelectric applications. Ferroelectrics **11**, 347–350 (1991)
59. R. Watton, C. Smith, G.R. Jones, Pyroelectric materials: operation and performance in the pyroelectric camera tube. Ferroelectrics **14**, 719–721 (1976)
60. A. Shaulov, W.A. Smith, G.M. Loiacono, M.I. Bell, Y.H. Tsuo, Improper ferroelectrics for pyroelectric detection of infrared radiation. Ferroelectrics **27**, 117–121 (1980)
61. R.W. Whatmore, C.J. Brierley, F.W. Ainger, The dielectric properties of NiBr and FeI boracite single crystals. Ferroelectrics **27**, 67–70 (1980)

62. R.W. Whatmore, C.J. Brierley, F.W. Ainger, Nucleation control during the growth of boracite single crystals. Ferroelectrics **27**, 329–332 (1980)
63. A. Shaulov, Broad band infrared thermal detector. Sens. Actuat. **5**, 207–215 (1984)
64. Y.G. Wang, W.L. Zhong, P.L. Zhang, Pyroelectric properties of ferroelectric-polymer composite. J. Appl. Phys. **74**, 521–524 (1993)
65. R. Clarke, A.M. Glazer, The observation of critical behaviour in the thermal expansion of $PbZr_{0.9}Ti_{0.1}O_3$. J. Phys. C: Solid State Phys. **7**, 2147–2157 (1974)
66. M. Wang, C.S. Fang, The optimum cut direction of ADTGSP crystal for the pyroelectric detector. Ferroelectrics **118**, 341–348 (1989)
67. R.B. Olsen, D.A. Bruno, J.M. Briscoe, Pyroelectric conversion cycle of vinylidene fluoride-trifluoroethylene copolymer. J. Appl. Phys. **57**, 5036–5042 (1985)
68. M. Ikura, Conversion of low-grade heat to electricity using pyroelectric copolymer. Ferroelectrics **267**, 403–408 (2002)
69. L. Kouchachvili, M. Ikura, Improving the efficiency of pyroelectric conversion. Int. J. Energy Res. **32**, 328–335 (2008)
70. L. Kouchachvili, M. Ikura, Pyroelectric conversion-effects of P(VDF-TrFE) preconditioning on power conversion. J. Electrostatics **65**, 182–188 (2006)
71. V. Sencadas, S. Lanceros-Mendez, J.F. Mano, Characterization of poled and non-poled b-PVDF films using thermal analysis techniques. Thermochim. Acta **424**, 201–207 (2004)
72. B. Chu, X. Zhou, K. Ren, B. Neese, M. Lin, Q. Wang, F. Bauer, Q.M. Zhang, A dielectric polymer with high electric energy density and fast discharge speed. Science **313**, 334–336 (2006)
73. R.B. Olsen, D.A. Bruno, J.M. Briscoe, W.F. Butler, A pyroelectric energy converter which employs regeneration. Ferroelectrics **38**, 975–978 (1981)
74. R.B. Olsen, Ferroelectric conversion of heat to electrical energy—a demonstration. J. Energy **6**, 91–95 (1982)
75. R.B. Olsen, D.D. Brown, High-efficiency direct conversion of heat to electrical energy related pyroelectric measurements. Ferroelectrics **40**, 17–27 (1982)
76. R.B. Olsen, D.A. Bruno, J.M. Briscoe, Cascaded pyroelectric energy converter. Ferroelectrics **59**, 205–219 (1984)
77. R.B. Olsen, D.A. Bruno, J.M. Briscoe, Pyroelectric conversion cycles. J. Appl. Phys. **58**, 4709–4716 (1985)
78. H. Nguyen, A. Navid, L. Pilon, Pyroelectric energy converter using co-polymer P(VDF-TrFE) and Olsen cycle for waste heat energy harvesting. Appl. Therm. Eng. **30**, 2127–2137 (2010)
79. D. Vanderpool, J.H. Yoon, L. Pilon, Simulations of a prototypical device using pyroelectric materials for harvesting waste heat. Int. J. Heat Mass Transf. **51**, 5052–5062 (2008)
80. A. Navid, D. Vanderpool, A. Bah, L. Pilon, Towards optimization of a pyroelectric energy converter for harvesting waste heat. Int. J. Heat Mass Transf. **53**, 4060–4070 (2010)
81. H. Atabek, C. Chang, Oscillatory flow near the entry of a circular tube. Z. Angew. Math. Phys. **12**, 185–201 (1961)
82. M.O. Çarpinlioglu, M.Y. Gündogdu, A critical review on pulsite pipe flow studies directing towards future research topics. Flow Meas. Instrum. **12**, 163–174 (2001)
83. F.P. Incropera, D.P. DeWitt, *Heat and Mass Transfer*, 5th edn. (John Wiley and Sons, New York, 2002)
84. M. Ozawa, M. Shinoki, K. Nagoshi, E. Serizawa, Scaling of heat transfer characteristics in an oscillating flow. J. Enh. Heat Transf. **10**, 275–285 (2003)
85. F.Y. Lee, A. Navid, L. Pilon, Pyroelectric waste heat energy harvesting using heat conduction. Appl. Therm. Eng. **37**, 30–37 (2012)
86. A. Navid, L. Pilon, Pyroelectric energy harvesting using Olsen cycles in purified and porous poly(vinylidene fluorideetrifuoroethylene) thin films. Smart Mater. Struct. **20**, 025012 (2011)

87. A. Navid, C.S. Lynch, L. Pilon, Purified and porous poly(vinylidene fluoride trifluoroethylene) [P(VDFeTrFE)] thin films for pyroelectric infrared sensing and energy harvesting. Smart Mater. Struct. **19**, 055006 (2010)
88. E.O. Forster, H. Yamashita, C. Mazzetti, C. Pompili, L. Caroli, S. Patrissi, The effect of electrode gap on breakdown in liquid dielectrics. IEEE Trans. Dielect. Elect. Insulat. **1**, 440–446 (1994)
89. F.Y. Chen, Y.K. Fang, C.Y. Hsu, J.R. Chen, Time response analysis of a pyroelectric detector. Ferroelectrics **200**, 257–268 (1997)
90. M.C. Kao, C.M. Wang, H.Z. Chen, M.S. Lee, Y.C. Chen, Thickness-dependent leakage current of (polyvinylidene fluoride/lead titanate) pyroelectric detectors. IEEE Trans. Ultrason. Ferroelect. Freq. Control. **50**, 958–964 (2003)
91. F.Y. Lee, H.R. Jo, C.S. Lynch, L. Pilon, Pyroelectric energy conversion using PLZT ceramics and the ferroelectric-ergodic relaxor phase transition. Smart Mater. Struct. **22**, 025038 (2013)
92. D.E. Dausch, G.H. Haertling, The domain switching and structural characteristics of PLZT bulk ceramics and thin films chemically prepared from the same acetate precursor solutions. J. Mater. Sci. **31**, 3409–3417 (1996)
93. B. Vodopivec, C. Filipic, A. Levstik, J. Holc, Z. Kutnjak, E-T phase diagram of the 6.5/65/35 PLZT incipient ferroelectric. J. Eur. Ceram. Soc. **24**, 1561–1564 (2004)
94. V. Bobnar, Z. Kutnjak, R. Pirc, A. Levstik, A 1999 Electric-field-temperature phase diagram of the relaxor ferroelectric lanthanum-modified lead zirconate titanate. Phys. Rev. B **60**, 6420–6427 (1999)
95. S. Kamba, V. Bovton, J. Petzelt, I. Rychetsky, R. Mizaras, A. Brilingas, J. Banys, J. Gringas, M. Kosec, Dielectric dispersion of the relaxor PLZT ceramics in the frequency range 20 Hz–100 THz. J. Phys.: Condens. Matter **12**, 497–519 (2000)
96. D. Viehland, X.H. Dai, J.F. Li, Z. Xu, La-modified lead zirconate titanate: long and short-range ordered structurally incommensurate phases, and glassy polar clusters. J. Appl. Phys. **84**, 458–471 (1998)
97. A.A. Bokov, Z.G. Ye, Recent progress in relaxor ferroelectrics with perovskite structure. J. Mater. Sci. **41**, 31–52 (2006)
98. B. Vodopivec, C. Filipic, A. Levstik, J. Holc, Z. Kutnjak, Dielectric properties of partially disordered lanthanum-modified lead zirconate titanate relaxor ferroelectrics. Phys. Rev. B **69**, 224208 (2004)
99. A.S. Divya, V. Kumar, A novel mechanism for relaxor-ferroelectric transition in PLZT (8/65/35). J. Am. Ceram. Soc. **92**, 2029–2032 (2009)
100. S. Shah, M.S. Ramachandra Rao, Preparation and dielectric study of high-quality PLZT x/65/35 (x = 6, 7, 8) ferroelectric ceramics. Appl. Phys. A **71**, 65–69 (2000)
101. D. Wang, Y. Fotinich, G.P. Carman, Influence of temperature on the electromechanical and fatigue behavior of piezoelectric ceramics. J. Appl. Phys. **83**, 5342–5350 (1998)
102. M. Pavel, I. Rychetsky, J. Petzelt, Polar clusters in relaxor (Pb, La)(Zr, Ti)O3 revealed by second harmonic generation. J. Appl. Phys. **89**, 5036–5039 (2001)
103. F.Y Lee, S. Goljahi, I.M. McKinley, C.S. Lynch, L. Pilon, Pyroelectric waste heat energy harvesting using relaxor ferroelectric 8/65/35 PLZT and the Olsen cycle. Smart Mater. Struct. 21, 025021 (2012)
104. G. Sebald, L. Seveyrat, D. Guyomar, L. Lebrun, B. Guiffard, S. Pruvost, Electrocaloric and pyroelectric properties of 0.75Pb(Mg1/3Nb2/3)O3–0.25PbTiO3 single crystals. J. Appl. Phys. **100**, 124112 (2006)
105. D. Guyomar, S. Pruvost, G. Sebald, Energy harvesting based on FE-FE transition in ferroelectric single crystals. IEEE Trans. Ultrason. Ferroelectr. Freq. Control **55**, 279–285 (2008)
106. A. Khodayari, S. Pruvost, G. Sebald, D. Guyomar, S. Mohammadi, Nonlinear pyroelectric energy harvesting from relaxor single crystals. IEEE Trans. Ultrason. Ferroelectr. Freq. Control **56**, 693–699 (2009)

107. H. Zhu, S. Pruvost, D. Guyomar, A. Khodayari, Thermal energy harvesting from $Pb(Zn_{1/3}Nb_{2/3})_{0.955}Ti_{0.045}O_3$ single crystals phase transitions. J. Appl. Phys. **106**, 124102 (2009)
108. R. Kandilian, A. Navid, L. Pilon, The pyroelectric energy harvesting capabilities of PMN-PT near the morphotropic phase boundary. Smart Mater. Struct. **20**, 055020 (2011)
109. I.M McKinley, R. Kandilian, L. Pilon, Waste heat energy harvesting using the Olsen cycle on $0.945Pb(Zn_{1/3}Nb_{2/3})O_3$-$0.055PbTiO_3$ single crystals. Smart Mater. Struct. 21, 035015 (2012)
110. Z. Li, Z. Xu, Z. Xi, F. Xiang, X. Yao, Thermal expansion characteristics in [001]-oriented PMN-0.32PT single crystals. Ferroelectrics **355**, 245–251 (2007)
111. P. Kumar, S. Sharma, O.P. Thakur, C. Prakash, T.C. Goel, Dielectric, piezoelectric and pyroelectric properties of PMN-PT (68:32) system. Ceram. Int. **30**, 585–589 (2004)
112. W.S. Chang, L.C. Lim, P. Yang, H.O. Moser, F.T. Wang, C.T. Tseng, C.S. Tu, Phase transformations in poled PZN-4.5 %PT single crystals revealed by combined property measurements and high-resolution diffraction technique. J. Appl. Phys. **104**, 054102 (2008)
113. J.J. Lima-Silva, I. Guedes, J. Mendes Filho, A.P. Ayala, M.H. Lente, J.A. Eiras, D. Garcia, Phase diagram of the relaxor $(1-x)Pb(Zn_{1/3}Nb_{2/3})O_3$-$xPbTiO_3$ investigated by dielectric and Raman spectroscopies. Solid State Commun. 131, 111–114 (2004)
114. A.E. Renault, H. Dammak, G. Calvarin, P. Gaucher, Electric-field-induced orthorhombic phase in $Pb[(Zn_{1/3}Nb_{2/3})_{0.955}Ti_{0.045}]O_3$ single crystals. J. Appl. Phys. **97**, 044105 (2005)
115. G. Sebald, S. Pruvost, D. Guyomar, Energy harvesting based on Ericsson pyroelectric cycles in a relaxor ferroelectric ceramic. Smart Mater. Struct. **17**, 015012 (2008)
116. A. Cuadrasa, M. Gasulla, V. Ferrari, Thermal energy harvesting through pyroelectricity. Sens. Actuat. A **158**, 132–139 (2010)
117. Y. Yang, S. Wang, Y. Zhang, Z.L. Wang, Pyroelectric nanogenerators for driving wireless sensors. Nano Lett. **12**, 6408–6413 (2012)
118. A.N. Morozovska, E.A. Eliseev, G.S. Svechnikov, S.V. Kalinin, Pyroelectric response of ferroelectric nanowires: size effect and electric energy harvesting. J. Appl. Phys. **108**, 042009 (2010)

Chapter 6
Waste Thermal Energy Harvesting (III): Storage with Phase Change Materials

6.1 Introduction

In last two chapters, both methods to harvest waste thermal energy through the conversion to electricity. In this chapter, energy storage (ES) as an alternative method to harvest waste thermal energy, especially by using phase change materials (PCMs), will be presented.

ES, as suggested by the name, is to store a certain form of energy, which thus be used later when necessary. A device that can be used to store any form of energy is generally called an accumulator. There are various forms of energy, commonly including kinetic energy, potential energy (e.g., gravitational or chemical), electrical energy, thermal energy, and so on. All these forms of energy could be stored by using an appropriate method [1, 2]. For instance, mechanical energy can be stored by using hydrostorage, compressed air storage, and flywheels, while electrical energy is stored by using various batteries and supercapacitors. Thermal energy storage (TES) is the main content of this chapter.

This chapter is arranged as follows. A brief description on various TES conceptions, together with a comparison, will be presented first. Among them, PCMs for TES will be emphasized. After that, PCMs with their classification according to their composition will be discussed. Design criteria, heat transfer phenomena, configuration of various PCM TES systems, which have been reported in open literature, will be summarized. Specific attention will be paid to the exergy analysis. The main applications of the PCM storage systems will be listed, analyzed, and compared. Strategies that have been used to improve the performance of such storage systems will be thoroughly evaluated. It will be ended with concluding remarks and perspectives of development of this technology.

L. B. Kong et al., *Waste Energy Harvesting*, Lecture Notes in Energy 24,
DOI: 10.1007/978-3-642-54634-1_6, © Springer-Verlag Berlin Heidelberg 2014

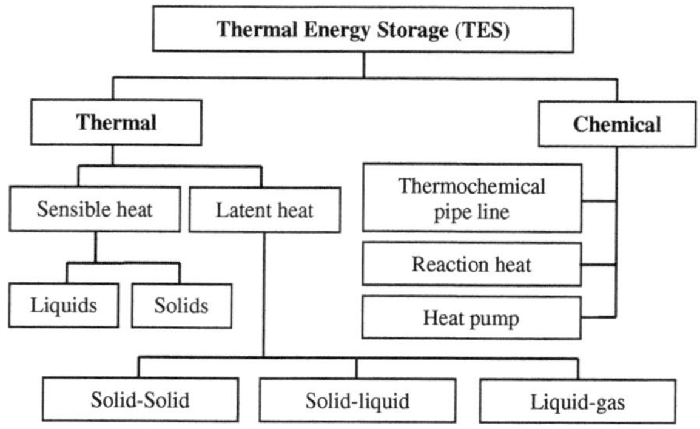

Fig. 6.1 Schematic showing the types of thermal storage [3]

6.2 Thermal Energy Storage

Thermal energy can be stored directly as internal energy of a material through sensible heat, latent heat, and thermochemical heat or their combinations, as shown in Fig. 6.1 [3].

Sensible heat storage (SHS) is to store thermal energy by raising the temperature of a solid or liquid without the occurrence of a phase change. The amount of heat that can be stored is dependent on factors, like the amount of the materials used, its specific heat and the temperature difference experienced by the systems, through the following equation:

$$Q = \int_{T_L}^{T_H} mC_p \mathrm{d}T \tag{6.1}$$

where m is mass and C_p is specific heat at constant pressure of the SHS materials, T_L and T_H are the low and high temperatures between which the TES is conducted. If C_p is independent of temperature, Eq. (6.1) can be simplified as:

$$Q = mC_p(T_H - T_L) \tag{6.2}$$

In applications below 100 °C, water is the best SHS liquid, because of its wide availability, low cost, and relatively high specific heat. Above 100 °C, oils, molten salts, liquid metals, and so on are used. For air heating applications, rock bed type storage materials are more suitable. Besides physical properties, such as density and specific heat of the storage materials, there are other properties that are important for the efficiency of SHS, including operational temperatures, thermal conductivity and diffusivity, vapor pressure, compatibility among materials stability, heat loss coefficient as a function of the surface areas to volume ratio, and cost of the materials and systems.

A SHS system usually consists of storage medium, container tank, and inlet/outlet connection devices. Tanks are used to not only retain the storage material but also prevent any potential loss of thermal energy. The presence of a

Fig. 6.2 Comparison of ES behaviors with PCMs and non PCMs

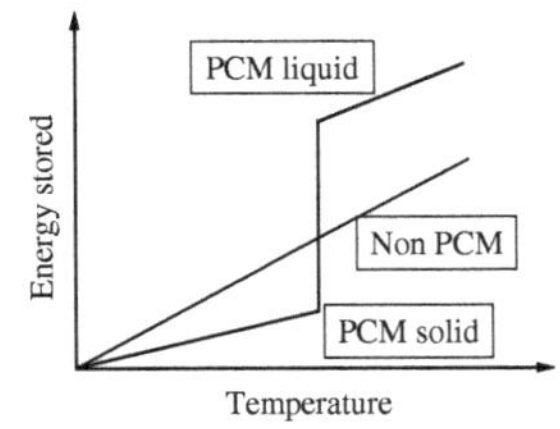

temperature gradient across the storage system is desirable. The storage medium can be solid or liquid. Solid media, such as concrete and castable ceramics, are usually used in the form of packed beds, which require a fluid for effective heat exchanging. If the fluid is a liquid, the system is called dual storage system, because the heat capacity of the solid in the packed bed cannot be neglected. One of the most significant advantages of the dual system is the use of inexpensive solids, such as rock, sand, or concrete as storage materials.

Liquid media mainly include molten salts, mineral oils, and synthetic oils. They maintain natural thermal stratification due to the difference in density between hot and cold layers. The hot fluid is supplied to the upper part of storage during charging, whereas the cold fluid is extracted from the bottom part during discharging.

Latent heat storage (LHS) is to use the heat absorbed or released by the storage material when it experiences a phase change between solid and liquid or liquid and gas. Storage capacity of a LHS system with PCMs is given by:

$$Q = \int_{T_L}^{T_m} mC_{\mathrm{sp}}\mathrm{d}T + m\Delta H_m + \int_{T_m}^{T_H} mC_{\mathrm{lp}}\mathrm{d}T \tag{6.3}$$

where T_m is the melting point of the PCM, C_{sp} and C_{lp} are the specific heat of the PCM in solid and liquid state respectively. If C_{sp} and C_{lp} are not a function of temperature, the heat storage capability can be calculated by using the following equation:

$$Q = m[C_{\mathrm{sp}}(T_m - T_L) + \Delta H_m + C_{\mathrm{lp}}(T_H - T_m)] \tag{6.4}$$

As shown in Fig. 6.2, PCMs have a much higher ES capability than a non PCM. Due to the latent heat of fusion, there is a steep jump in energy stored at the temperature of melting point of the PCM.

Thermochemical storage is to use the heat energy absorbed and released by a completely reversible chemical reaction. In this case, the amount of heat stored is proportional to the amount of storage material (m), endothermic heat of the reaction (ΔH_r) and the reaction extent (ξ_r), given by:

$$Q = m\xi_r\Delta H_r \tag{6.5}$$

Among the three thermal heat storages, LHS with PCMs is the most promising one, because it exhibits high ES density and can store heat at constant temperatures, i.e., phase transition temperatures of the materials.

Phase transition may occur in forms of solid–solid, solid–liquid, solid–gas, and liquid–gas. Among these phase changes, only solid–liquid phase change has been used for LHS. Liquid–gas phase changes are not practical for use as thermal storage due to the large volumes involved or high pressures required to store the energy when they are in their gaseous phase, although liquid–gas transitions have a higher latent heat of transformation than solid–liquid transitions. Solid–solid phase changes are usually too slow and have a very low value of latent heat of transformation.

6.3 PCMs for Thermal Storage

Below phase transition temperature, the solid–liquid PCMs behave like the conventional SHS materials, i.e., their temperature rises as they absorb heat. However, when PCMs reach their phase transition points (their melting temperatures), they can absorb a large amount of heat without increase in temperature. At this temperature, they continue to absorb heat until the materials are completely transformed from solid to liquid phase. Conversely, when the temperatures of the materials are decreased, phase transition from liquid to solid occurs, i.e., the PCMs solidify, releasing the latent heat stored in them.

PCMs should meet various requirements for practical applications, which are listed as follows [3]:

- Appropriate melting temperature to match the specific working requirement of TES applications,
- High specific heat and density to ensure high TES capacity,
- Stability in stoichiometric composition to withstand the repeated phase transformation cycles,
- High heat conductivity to eliminate the presence of large temperature gradients,
- Small change in volume at transition from one phase to another allowing for simple design of containers and heat exchangers,
- Low overcooling,
- High chemical stability to ensure long life-span,
- Good compatibility with constructional materials,
- Nontoxic for safety consideration,
- Flame and fire retardant capability, and
- Wide availability and cost-effectiveness.

There have been various PCMs available currently with phase change temperatures ranging from 10 to 200 °C. According to their compositions, PCMs can be classified into three groups: organic, inorganic, and eutectic, which are shown in Fig. 6.3 [3].

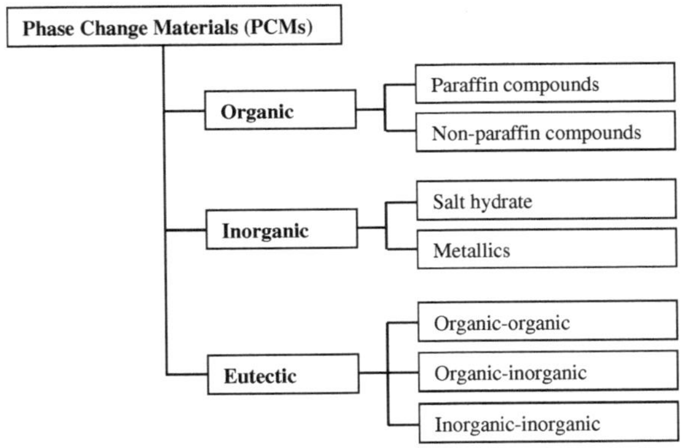

Fig. 6.3 Classification of PCMs according to their compositions [3]

6.3.1 Organic PCMs

Organic PCMs mainly include paraffin and non-paraffin. Paraffin, also known as paraffin wax, is a mixture of mostly straight chain *n*-alkanes, with a general formula of CH_3–(CH_2)–CH_3 or C_nH_{2n+2}. The crystallization of the (CH_3)-chain releases a large amount of latent heat. Both the melting point and latent heat of fusion increase generally with increasing length of the chains. Paraffin has various advantages for waste ES applications, such as nontoxic, widely available, and noncorrosive. They are chemically inert and stable up to a temperature of as high as 500 °C. They experience very small change in volume on melting or cooling and have very low vapor pressure in melting state. Other important characteristics as PCMs include freezing without significant supercooling, melting congruently, self-nucleating, being compatible with conventional materials, nonsegregating, being chemically stable, being recyclable, and so on. They also have shortcomings, including relatively low thermal conductivity in their solid state, not sufficiently high volumetric latent heat and flammable. Physical properties, melting point and latent heat of fusion, of paraffin with different numbers of carbon atom are listed in Table 6.1 [3].

Non-paraffin organic compounds that can be used for PCM TES applications include esters, fatty acids, alcohols, and glycols. These organic PCMs can be further subcategorized into fatty acids and other non-paraffins. Fatty acids have a general formula of $CH_3(CH_2)_{2n}COOH$. They have a heat of fusion comparable to that of paraffin. Fatty acids also demonstrate reliable melting and freezing behaviors and freeze without the occurring of supercooling. The main disadvantages of fatty acids are their relatively high cost and mild corrosive problem. Representative fatty acids suitable for low temperature LHS applications are listed in Table 6.2 [3].

Table 6.1 Melting point and latent heat of fusion of paraffins

Number of carbon atom	Melting point (°C)	Latent heat of fusion (kJ/kg)
14	5.5	228
15	10.0	205
16	16.7	237.1
17	21.7	213
18	28.0	244
19	32.0	222
20	36.7	246
21	40.2	200
22	44.0	249
23	47.5	232
24	50.6	255
25	49.4	238
26	56.3	256
27	58.8	236
28	61.6	253
29	63.4	240
30	65.4	251
31	68.0	242
32	69.5	170
33	73.9	268
34	75.9	269

Table 6.2 Melting point and latent heat of fusion of representative fatty acids

Compound	Formula	Melting point (°C)	Latent heat (kJ/kg)
Acetic acid	CH3COOH	16.7	184
Polyethylene glycol 600	$H(OC_2H_2)_n$–OH	20–25	146
Capric acid	$CH_3(CH_2)_8$–COOH	36	152
Eladic acid	$C_8H_7C_9H_{16}$–COOH	47	218
Lauric acid	$CH_3(CH_2)_{10}$–COOH	49	178
Pentadecanoic acid	$CH_3(CH_2)_{13}$–COOH	52.5	178
Tristearin	$(C_{17}H_{35}COO)C_3H_5$	56	191
Myristic acid	$CH_3(CH_2)_{12}$–COOH	58	199
Palmatic acid	$CH_3(CH_2)_{14}$–COOH	55	163
Stearic acid	$CH_3(CH_2)_{16}$–COOH	69.4	199
Acetamide	CH_3CONH_2	81	241
Methyl fumarate	$(CHCO_2NH_3)_2$	102	242

6.3.2 Inorganic PCMs

Inorganic PCMs are mainly salt hydrates (M_nH_2O). They usually have high volumetric LHS capacity, wide availability, cost-effectiveness, high thermal conductivity, and nonflammability. Their problems include large volumetric change, supercooling, and relatively poor recyclability.

Salt hydrates can be regarded as alloys of inorganic salts and water, forming a typical crystalline solid of general formula $AB{\cdot}nH_2O$. The solid–liquid transformation of salt hydrates is actually a dehydration or hydration of the salts. It is thermodynamically very similar to melting or freezing. A salt hydrates usually melts to either to a salt hydrate with fewer moles of water, i.e. $AB{\cdot}nH_2O \rightarrow AB \cdot mH_2O + (n - m)H_2O$ $(n > m)$, or to its anhydrous form, i.e., $AB \cdot nH_2O \rightarrow AB + nH_2O$.

Because of the two possible reactions, hydrate crystals decompose into either an anhydrous salt and water or a lower hydrate and water, at their melting points. In most cases, salt hydrates melt incongruently and the amount of the released water is not sufficient to dissolve all the solid phase. Therefore, due to density difference, the anhydrous salt or lower hydrate precipitate at bottom of the container tanks of the storage systems.

Generally, salt hydrates have poor nucleating capability, thus resulting in supercooling of the liquid before crystallization beings. To address this problem, it is necessary to use external nucleating agents, which provide effective nucleion to trigger the formation of crystals, or to use a small cold region with a certain number of crystals to serve as nuclei.

Salt hydrates possess high latent heat of fusion per unit volume, relatively high thermal conductivity which is almost double that of the paraffins and small volume changes on melting or solidifying. They are compatible with plastics, not very corrosive and only slightly toxic. As a result, salt hydrates have formed the most important group of PCMs, which have been extensively studied for applications in various latent heat TES systems.

Molten salts can be classified into three groups: congruent, incongruent, and semi-congruent melting. Congruent melting means that the anhydrous salt is completely soluble in the water of hydration at the melting temperature, while incongruent melting occurs when the salt is not entirely soluble in the water of hydration at the melting point. Semi-congruent melting takes place in the equilibrium of the liquid and solid phases during a phase transition, when there are different melting compositions due to the conversion of the hydrate to a lower hydrate state through an incomplete loss of water.

Incongruent melting is a major problem salt hydrates as PCMs. Since the amount of water of a hydrate is usually not sufficient to completely dissolve the salt, the resulting solution is supersaturated at the melting temperature. Therefore, due to its higher density, the solid salt precipitates at bottom of the container and is unavailable to recombine with water to become hydrate during the reverse process of freezing, thus leading to an irreversible melting-freezing of the salt hydrate, which decreases the ES capability with further charge–discharge cycles.

Because the rate of nucleation of salt hydrates is generally very low at the fusion temperature, supercoiling occurs in most cases. To achieve a reasonable rate of nucleation, the solution has to be in a supercooled state. In this case, the energy is discharged at much lower temperatures, as compared with the fusion temperature, thus decreasing the storage capability.

Various strategies have been employed to address the problem of incongruent melting of salt hydrate PCMs, including mechanical stirring for uniform distribution, encapsulating the PCMs to reduce separation, using thickening agents to prevent the precipitation of solid salts by trapping in the suspension, using of excessive water to avoid the formation of supersaturated solution and modifying the chemical composition of the system to transfer the incongruent state to congruent [4–6].

An example is the design of rolling cylinder heat storage system with salt hydrate PCMs [7]. The system consists of a cylindrical vessel mounted horizontally with two sets of rollers. A rotation rate of 3 rpm produced sufficient motion of the solid content, so that it is maintained at effective chemical equilibrium and the nucleation of solid crystals on the walls can be prevented. This rolling cylinder system has several advantages, such as complete phase change, latent heat release of 90–100 % of the theoretical value, repeatable performance over 200 cycles, high internal heat transfer rates and uniform occurrence of freezing.

Metallic PCMs form the second most important group of inorganic PCMs, including low melting metals and metal eutectics. Metallic PCMs have not been widely appliedin PCM technologies due to heavy weight problem. However, when volume becomes more important consideration, they are promising candidates owing to their high heat of fusion per unit volume. They have high thermal conductivities, which can be taken as compensation, because the addition of high thermal conductivity fillers is exempted. The characteristics of metallic PCMs include low heat of fusion per unit weight, yet high heat of fusion per unit volume, high thermal conductivity, low specific heat, and significantly low vapor pressure.

6.3.3 Eutectic PCMs

A eutectic system is defined as the composition of two or more components with minimum- melting point. Each of the components melts and freezes congruently, thus forming a mixture of the component crystals during crystallization. On melting, all components liquefy simultaneously, whereas on cooling, the components nearly always melt, both without phase segregation. Eutectic PCMs may be of organic–organic, organic–inorganic, inorganic–inorganic compound combinations. The most distinctive advantages of eutectic PCMs are their sharp melting points which are very similar to pure compounds and relatively high volumetric ES density as compared with their organic counterparts. Thermophysical properties of eutectic PCMs still remain to be further studied.

6.4 Design of TES Systems with PCMs

6.4.1 Design Criteria

PCMs themselves cannot be used as heat transfer medium. A separate heat transfer medium must be employed with heat exchanger in between to transfer energy from the source to the PCMs and from PCMs to the load. The heat exchanger to be used has to be designed specially, in view of the low thermal diffusivity of PCMs in general. The volume changes of the PCMs on melting should also be specifically taken into account when designing the whole system. These volume changes must be effectively absorbed during the phase change of PCMs used. Therefore, a latent heat ES system usually possesses at least following three components: (i) a suitable PCM with its melting point in the temperature range of interest of applications, (ii) a suitable heat exchange surface/interface, and (iii) a suitable container compatible with the PCM used.

There are several factors that have to be considered when deciding on the type and the design of a thermal storage system. One of the key parameters of a TES system is its thermal capacity. However, the criteria of the design of an appropriate system include cost-effectiveness considerations, technical evaluations, and environmental issues. The cost of an ES system depends mainly on the cost of the storage material itself, the heat exchanger for charging and discharging the system and the cost of the space and/or enclosure for the whole system.

Technically, the PCMs selected should have high energy density or storage capacity, there should be a good heat transfer between heat transfer fluid (HTF) and storage medium (high efficiency), the PCMs must be mechanically and chemically stable so as to ensure multiple charging/discharging cycles, there must be compatibility between HTF, heat exchanger, and/or storage medium for safety considerations, there should be complete reversibility of a number of charging/discharging cycles and thermal losses must be minimized. Further technical considerations include convenient operation, maximum loading, suitable temperature, specific enthalpy drop in load, and integration capability into power plant.

Storage systems with PCMs have smaller physical dimensions as compared with the single-phase sensible storage systems. This is because PCMs allow large amounts of energy to be stored in relatively small volumes, due to their high latent heat during the phase transitions. However, heat transfer design and media selection are more complicated and thus more difficult. For example, it has been shown that the performance of the materials can degrade after moderate number of freeze-melt cycles, when some low temperature salts are used as PCMs.

6.4.2 Storage Systems

TES systems can be classified into two types: active and passive systems. In an active storage system, the convection heat transfer is forced into the storage materials. The storage medium is circulated through a heat exchanger, which can also be used as a solar receiver or a steam generator. This system uses one or two tanks as storage media. Active systems have two sub-groups: direct and indirect systems. In a direct system, the HTF acts also as the storage medium, while in an indirect system, a second medium is used to store the heat.

Passive storage systems are generally dual medium storage systems, where the HTF passes through the storage only for charging and discharging a solid material. The HTF carries energy from the energy source to the storage medium during charging, and takes energy from the storage when discharging. They are also called regenerators. In these systems, the temperature of the HTF decreases during discharging when the storage material cools down. Also, the heat transfer efficiency is relatively low, because there is no direct contact between the HTF and the storage material as the heat is transferred via a heat exchanger.

6.4.2.1 Active Systems

In an active direct system, the HTF is also used as storage material, which means that PCMs selected must meet the requirements of good HTF and a good storage material simultaneously. In this case, the use of expensive heat exchangers can be eliminated, which allows for operations at high temperatures. For the same reason, this combination results in a substantial reduction in the cost of the systems, thus improving the performance of the plant and suppressing the risen electrical cost. However, when molten salts are used as PCMs, they would freeze at relatively high temperatures (120–220 °C) when they are cooled down from a higher temperature. For instance, a solar thermal storage system will experience low temperatures at night. In this case, special care must be taken to ensure that the salts do not freeze in the solar field piping during the night. Therefore, routine freeze protection operation must be conducted for the thermal storage system, for example, an auxiliary heater must be used to maintain the minimum temperature, thus increasing maintenance and operational costs. This problem can be readily tackled by combining a solar plant and a thermal storage system [1, 8].

One solution to this problem is the use of two-tank direct system, in which the HTF is directly stored in a hot tank, which can be used during cloudy periods or nights. The cooled HTF is pumped to the other tank, which is cold tank, where needs to be heated one more time [9]. The two-tank system has several advantages. First, the cold and heat storage materials are stored separately, which enables to increase the solar field output temperature to 450–500 °C, thus increasing the Rankine cycle efficiency of the power block steam turbine to the 40 % level [8]. Second, the HTF temperature rise in the collector field can be increased, leading directly reduction in the physical size of the thermal storage system [8, 9]. However, on the other hand, this kind of system increases the cost of the material used as HTF and storage medium and the cost of the heat exchangers. Therefore, in practical design, these factors must be well considered.

Active indirect storage systems can be either two-tank systems or single-tank systems. A single-tank system is also called thermocline system. In both cases, the HTF which circulates in the solar field is different from the one for the TES medium.

A two-tank indirect system is similar to its direct counterpart, in terms of configuration. One of the tanks is used to store hot storage material, while the other one is to receive the cold storage material. During a thermal storage charge cycle, a portion of the fluid from the collector field is directed to the fluid-to-salt heat exchanger, where the fluid is cooled for a high temperature to a low temperature. If nitrate salt is used, it flows from the cold storage tank in a counter current arrangement through the heat exchanger. The salt is heated from an inlet temperature (low) to an outlet temperature (high), which is then stored in the hot storage tank. During the discharge cycle, the fluid and the salt flow paths are reversed in the fluid-to-salt heat exchanger. Heat is then transferred from the salt to the fluid to provide the thermal energy for the steam generator.

For a single-tank active indirect storage system, hot and cold fluids are stored in the same tank. The single-tank system is more cost-effective than two-tank storage system. The hot and cold fluids are separated because of the stratification and the zone between the hot and cold fluids is called the thermocline. In this thermocline system, the hot fluid is located on top and the cold is at the bottom. When the HTF flows through the heat exchanger, it heats the thermal storage fluid media. Generally, a filler material is used to promote the thermocline effect [10].

6.4.2.2 Passive Systems

In passive storage systems, the heat transfer medium passes through the storage only for charging or discharging of the systems. The storage medium itself is not circulated. Passive systems are generally dual medium storage systems, which are also called regenerators. Passive storage systems are mainly solid storage systems. Typical PCMs used in these systems are binary $NaNO_3/KNO_3$. Graphite fin tube-“sandwich” design concept has been adopted for scale up production.

6.4.3 Heat Transfer in PCMs and Storage Systems

6.4.3.1 General Consideration

Heat transfer characteristics of melting and solidification process are underlying mechanisms that govern the performances of TES systems, which however has not been well studied and completely understood. The freezing of the ground, of the heat conducting material occupying the space $-\infty < x < \infty$, has been posed and solved [11]. At the initial time, the liquid phase fills the domain $0 < x < \infty$ at temperature $T_1 > 0$ °C, while the solid phase occupies the domain $-\infty < x < 0$ at temperature $T_S < 0$ °C. At $x = 0$, it is maintained at constant temperature, $T < 0$ °C, with the assumption that the solidification occurs isothermally at temperature, $T_m = 0$ °C, without supercooling and the conduction is the mode of heat transfer in the material. The energy equation at the solid–liquid interface is thus given by:

$$\lambda\rho\left(\frac{\mathrm{d}s(t)}{\mathrm{d}t}\right) = k_s\left(\frac{\delta T_s}{\delta t}\right) - k_l\left(\frac{\delta T_l}{\delta t}\right). \tag{6.6}$$

This equation is valid when there is phase change and the energy conservation is across the interface of the two phases. In literature, this solid–liquid interface boundary is known as Stefan or moving boundary problem.

The prediction of the behavior of a system with phase change is very difficult, due to the nonlinear characteristic at the moving interface between the two phases. In addition, the two phases have different thermophysical properties. The Stefan problem has been approached as a pure conduction in semi-infinite medium initially and natural convection has been considered during the melting and solidification of PCMs later [12, 13].

Both analytical and numerical solutions have been available for the Stefan problem as new progresses. Various approximate analytical techniques, such as heat balance integral, variation technique, isothermal migration, source and sink method, and periodic solution, have been developed to solve the problem [14–18]. However, these approximate techniques are limited to one-dimensional analysis and they become very complicated when applied to multidimensional cases. Numerical methods, including both finite difference and finite element, have been found to be more powerful in solving the problems with moving boundary [19–21]. Other approaches include the time variant mesh approach which offers promising accuracy but is limited only to simple problems and geometries [22, 23] and the fixed mesh method, in which the latent heat of fusion is combined into the specific heat or enthalpy the material, making it much simpler in practical applications [24, 25].

The low conductivity of the PCMs that are widely available has become a common problem in latent heat thermal storage technologies. The solid–liquid interface constantly moves away from the heat transfer surface, during the phase change process. One way to increase the area of effective heat transfer is to space the PCMs from the side walls of containers that are made of high thermal conducting materials. Due to the moving boundary and the unbalanced heat and mass conditions in the region, it is a challenge to solve the phase change problem. There have been two numerical approaches available: the finite difference and finite element techniques. One of the methods to solve the moving boundary problem is called enthalpy formulation [26–29]. In this method, the only unknown variable is temperature of the PCMs and the solidification occurs at a uniform temperature.

Enthalpy is treated as a temperature dependent variable, with which the latent heat flow through the volume is integrated into the whole system [30, 31].

Heat transfer with moving boundary during the phase change is also an important subject of studies in various applications, such as ice formation, freezing of food, castings, metallurgy, crystal growth, and various others involving solidification. To design TES systems with high performances, it is necessary to accurately predict the temperature distribution and the rate of melting or solidification [32–36]. Currently, all theoretical works are limited to study the moving boundary problem for specific conditions and designs. No information has been available about the melting process with respect to time.

Also, the analysis of LHS systems has been performed with a number of assumptions. One assumption is that the thermophysical properties of the PCMs and fin materials are independent of temperature. However, they are in fact different for solid and liquid phases of the PCMs. Other assumptions include that PCMs are in solid phase initially, PCMs are homogeneous and isotropic and the mode of heat transfer is conduction only.

For a phase change process of either melting or freezing, energy conservation can be expressed in terms of total volumetric enthalpy and temperature when constant thermophysical properties are assumed, which is given by [37]:

$$\frac{\partial H}{\partial t} = \nabla[k_k(\nabla T)], \tag{6.7}$$

where H is the total volumetric enthalpy, which is the sum of sensible and latent heat of the PCM, i.e.,

$$H(T) = h(T) + \rho_l f(T)\lambda, \tag{6.8}$$

where h is given by:

$$h = \int_{T_1}^{T_2} \rho_k c_k \mathrm{d}T. \tag{6.9}$$

For isothermal phase change, the fraction of the liquid phase in the melt is given by:

$$f = \begin{cases} 0, & T < T_m \quad \text{(solid)}, \\ 0-1, & T = T_m \quad \text{(melting)}, \\ 1, & T > T_m \quad \text{(liquid)}. \end{cases} \tag{6.10}$$

According to Eqs. (6.8) and (6.9), the enthalpies of the PCM at different states are expressed as:

$$H = \int_{T_1}^{T_m} \rho_s c_s \mathrm{d}T, \quad T < T_m \quad \text{(solid)}, \tag{6.11}$$

$$H = \rho_l f \lambda, \quad T = T_m \quad \text{(melting)}, \tag{6.12}$$

$$H = \int_{T_m}^{T_2} \rho_l c_l \,\mathrm{d}T + \rho_l \lambda, \quad T > T_m \quad \text{(liquid)}. \tag{6.13}$$

Equation (6.11) accounts the sensible enthalpy from a low temperature (T_1) to the melting temperature (T_m), while Eq. (6.13) represents the sensible enthalpy from the melting temperature to a high temperature (T_2). The temperatures of the PCM can be derived by solving the above three equations, which are:

$$T = \tfrac{T_m + H}{\rho_s c_s}, \quad H < 0 \quad \text{(solid)}, \tag{6.14}$$

$$T = T_m, \quad 0 \le H \le \rho_l \lambda \quad \text{(interface)}, \tag{6.15}$$

$$T = \tfrac{T_m + (H - \rho_l \lambda)}{\rho_l c_l}, \quad H > \rho_l \lambda \quad \text{(liquid)}. \tag{6.16}$$

From Eqs. (6.8) and (6.9), Eq. (6.7) can be expressed with an alternative form for two-dimensional heat transfer in the PCM as:

$$\frac{\partial h}{\partial t} = \frac{\partial}{\partial x}\left(\alpha \frac{\partial h}{\partial x}\right) + \frac{\partial}{\partial y}\left(\alpha \frac{\partial h}{\partial y}\right) - \rho_l \lambda \frac{\partial f}{\partial x}. \tag{6.17}$$

While for the material of the heat exchanger container, there is:

$$\frac{\partial h_f}{\partial t} = \frac{\partial}{\partial x}\left(\alpha_f \frac{\partial h_f}{\partial x}\right) + \frac{\partial}{\partial y}\left(\alpha_f \frac{\partial h_f}{\partial y}\right). \tag{6.18}$$

By dividing the domain into elementary control volumes and then integrating the equation in these control volumes, the algebraic equations can be derived [37]. Equation (6.17) can be solved by using a fully implicit finite difference solution method, with a finite difference equation for the PCM obtained by integrating it over each control volume. The discretization of Eq. (6.17) for $\Delta x = \Delta y$ leads to the following expressions:

$$h_P = h_P^{\circ} + \alpha R(h_E - 4h_P + h_W + h_N + h_S) + \rho_l \lambda (f_P^{\circ} - f_P^k), \tag{6.19}$$

$$a_E h_E + a_W h_W + a_P h_P + a_N h_N + a_S h_S) = Q, \tag{6.20}$$

where there are:

$$a_E = a_W = a_N = a_S = -\alpha R, \tag{6.21}$$

$$a_P = 1 - a_E - a_W - a_P - a_N - a_S, \tag{6.22}$$

$$Q = h_P^{\circ} + \rho_l \lambda (f_P^{\circ} - f_P^k), \tag{6.23}$$

$$R = \frac{\mathrm{d}t}{(\mathrm{d}x)^2}. \tag{6.24}$$

It has been shown that Eq. (6.19) can be solved by using a tri-diagonal matrix algorithm (TDMA). Q is the source term of the fixed grid enthalpy method. The terms, h_P° and f_P°, are the enthalpy and the fraction of melting state, respectively. The source term Q can be used to track the latent heat evolution, driven by the melting fraction. The value of Q is determined iteratively from the solution of the enthalpy equation. Therefore, after $(k + 1)$th numerical solution of the enthalpy equation at the Pth node, Eq. (6.19) can be rearranged as:

$$a_P h_P = -a_E h_E - a_W h_W - a_N h_N - a_S h_S + h_P^{\circ} + \rho_l \lambda (f_P^{\circ} - f_P^k). \tag{6.25}$$

If the phase change takes place at about the Pth node (i.e. $0 \leq f \leq l$), the $(k + 1)$th estimate of the melting fraction needs to be updated such that the left hand side of Eq. (6.25) is zero, i.e.,

$$0 = -a_E h_E - a_W h_W - a_N h_N - a_S h_S + h_P^{\circ} + \rho_l \lambda (f_P^{\circ} - f_P^{k+1}), \tag{6.26}$$

which can be further rewritten as:

$$f_P^{k+1} = \frac{-a_E h_E - a_W h_W - a_N h_N - a_S h_S + h_P^{\circ}}{\rho_l \lambda + f_P^{\circ}}. \tag{6.27}$$

The melting fraction Eq. (6.27) is applied at every node after the $(k + 1)$th solution of Eq. (6.20) for sensible volumetric enthalpy (h), together with under or over correction, i.e.,

$$f = \begin{cases} 0, & f^{k+1} \leq 0, \\ 1, & f^{k+1} \geq 1. \end{cases} \tag{6.28}$$

Convergence at a given time step is achieved when the difference in the total enthalpy fields falls below a given tolerance, which is usually set to be 10^{-4} [27].

A two-dimensional theoretical model has been developed to predict the melting interface characteristics of PCMs, basing on the enthalpy formulation [38–40]. There has been report on the effects of thermophysical properties of PCMs and heat exchanger materials on the thermal performance of the storage systems [41]. Numerical results have demonstrated that the time taken for the melting of the PCMs is decreased, if the thermal conductivity, specific heat, and density of the materials used to construct the heat exchanger container are increased.

Through the enthalpy modeling, several conclusions have been arrived for the melting fraction. First, the melt fraction is closely related to thermal conductivity of the heat exchanger container material and effective thermal conductivity of the PCMs. Second, the higher the thermal conductivity of container material, the shorter the time is required for complete melting of the PCMs. Third, the thickness of the heat exchanger container material has no obvious effect on melting fraction of PCMs. Lastly, the melting fraction is not affected by the initial temperature of

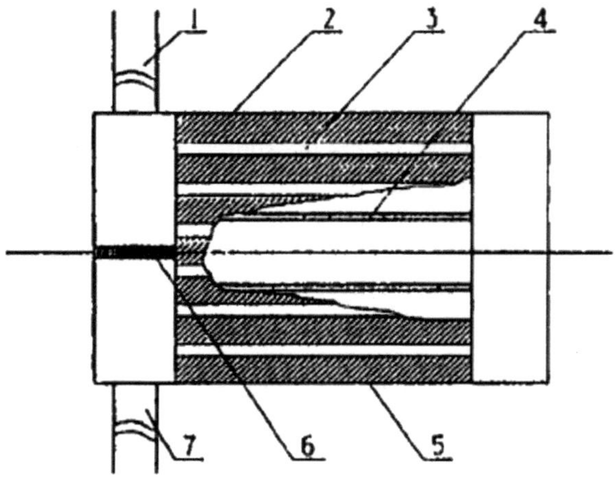

Fig. 6.4 Schematic draw of main outlets of the exchanger used to collect data for the verification of the modeling. *1* Tube for gas entrance, *2* PCM, *3* tube for heat transfer, *4* electrical heater, *5* shell, *6* baffle-board, and *7* tube for gas exit. Reproduced with permission from [42], Copyright @ 2001, John Wiley & Sons

the PCMs, while the temperature of the boundary wall plays an important role during the melting process and thus has a strong effect on the melting fraction [41]. Therefore, thermophysical properties of the materials for the heat exchanger container materials should be selected carefully to maximize the performance of the thermal storage systems with PCMs. Further and extensive theoretical studies are still the future challenges in the field of TES with PCMs.

6.4.3.2 Specific Configurations

Two specific configurations for TES systems with PCMs, cylinder-tube geometry, and packed bed, have been adopted in TES systems. Cylinder-tube geometry is more widely suited than packed bed arrangement in the open literature. Theoretical modeling or simulation of both types of configurations will be summarized with selected examples in this subsection.

Under the category of cylinder-tube geometry, there are actually several sub-types, which will not be intentionally classified. Instead, examples will be presented in a random way. A mathematical model has been developed to describe the unsteady freezing problem of a TES system with Al–Si alloy as PCM, which is coupled with forced convection [42]. Because the Al–Si alloy is a high temperature PCM and has a large variation in temperature, the sensible heat transfer of the PCM cannot be neglected. Thermal conductivity of the Al–Si alloy is very high, the radial temperature variation of the PCM is thus small, so that the natural convection in liquid phase of the PCM is negligible. Therefore, the mechanism of heat transfer in the Al–Si PCM is conduction, while a forced convection exists between the HTF and tube walls. Heat transfer equations of the conduction and convection should be solved simultaneously. The analysis has been simplified with several assumptions, such as isotropy of the PCM, neglected axial direction thermal conduction of the HTF and neglected heat losses. The numerical results have been found to be in a good agreement with values in experiments, as shown in Fig. 6.4.

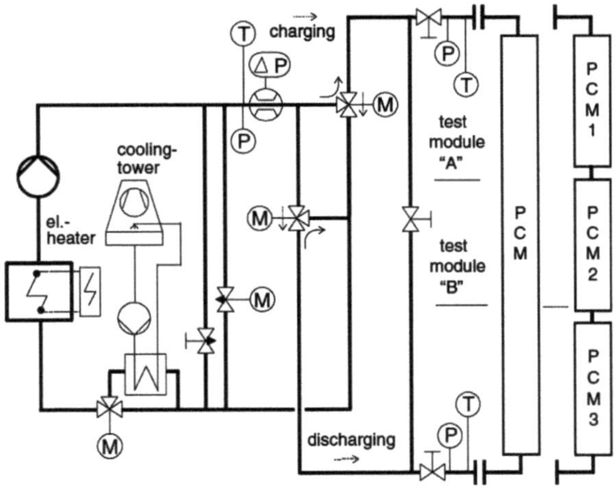

Fig. 6.5 Simplified flow diagram of the thermal oil test facility to validate the theoretical simulated results. Reproduced with permission from [43], Copyright @ 2007, Elsevier

Cascaded LHS systems for parabolic trough solar power plants have been modeled by using the standard library Tech-Thermo [43]. To simplify the simulation, various assumptions have been applied. The PCM is considered as a lumped mass with a uniform temperature throughout. The overall heat transfer coefficient is thought to be independent of time and axial position along the storage system. The model does not reflect the cyclic thermal process of continual charging/discharging. All the PCM sections are assumed to run out of energy (freeze solid) at exactly the same time. Phase change temperature and heat of fusion of the PCM are available as desired, neglecting the property limitations of real materials. The simulated results are well compared with the experimental data derived from the setup as shown in Fig. 6.5 [43].

Another example is a storage system with multicomponent PCMs, which has been simulated with a finite element model [44]. The cyclic thermal process of the alternating melting and freezing processes of the multiphase PCMs is specifically modeled. The physical module consists of a tube which is surrounded by an external coaxial cylinder, which contains the PCMs with different melting points in the segments, as shown in Fig. 6.6. Semidiscrete equations using standard Galerkin finite element method are used. The numerical results indicate that the heat transfer rates can be greatly enhanced due to the use of the multiple PCMs, as compared with those made of single-phase PCMs. This model has been successfully applied to design storage system for space-based applications. Numerical results indicate that variation in temperature of the fluid outlet can be significantly reduced by appropriate selection of multiple component PCMs [45].

The energy transfer of the heat receiver cavity of a TES system, as shown in Fig. 6.7, has been simulated by using finite volume method [46]. A heat balance model, a cavity radiation mathematical model, and a working fluid tube heat transfer model are developed to describe the solar heat receiver, the cavity, and the HTF, respectively. The whole system is studied as a three-dimensional cylindrical coordinate with a simple explicit scheme. In Fig. 6.7, the incident aperture is

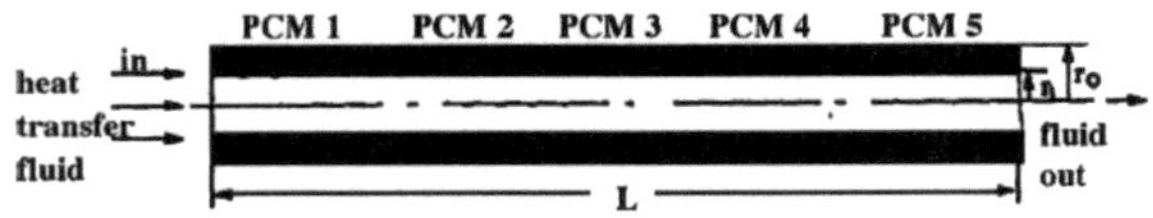

Fig. 6.6 Physical model of the tube-segmented PCM cylinders structure. Reproduced with permission from [44], Copyright @ 1996, Elsevier

Fig. 6.7 The radiation model of the heat receiver cavity of the storage system. Reproduced with permission from [46], Copyright @ 2003, Elsevier

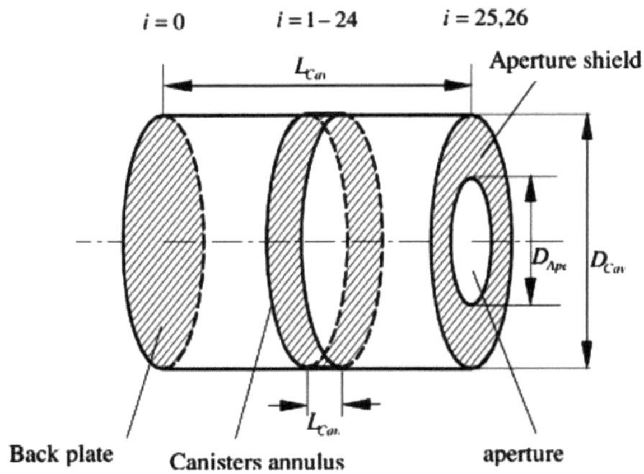

Fig. 6.8 Schematic of encapsulated PCM tube configuration with annular gas flow. Reproduced with permission from [46], Copyright @ 2003, Elsevier

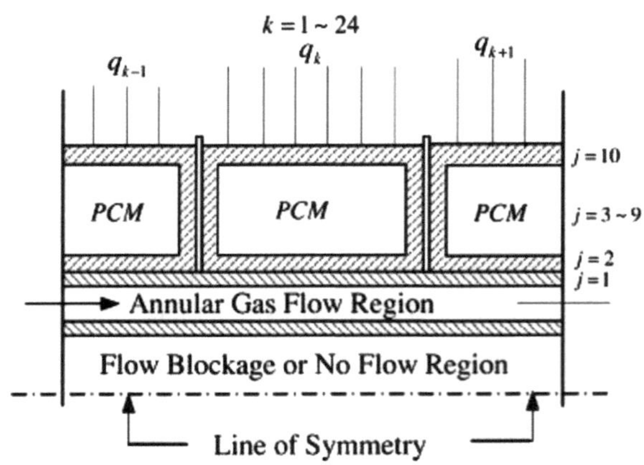

assumed to be a black body in the radiation model of the heat receiver, while all other surfaces are treated as gray bodies. The physical domain is discretized into a scheme of $12\theta \times 10r \times 24z$ grids, as shown in Fig. 6.8. Theoretical results indicate that the design of the working fluid tube can be further improved to increase the liquid PCM fraction and reduce both the gas outlet temperature variation and the total mass of the system [46].

A numerical model has been developed for the unit heat exchanger tube of a solar energy receiver by Cui et al. [47]. The simple physical model of high temperature TES system is shown in Fig. 6.9. Phase change is thought to take place at a distinct

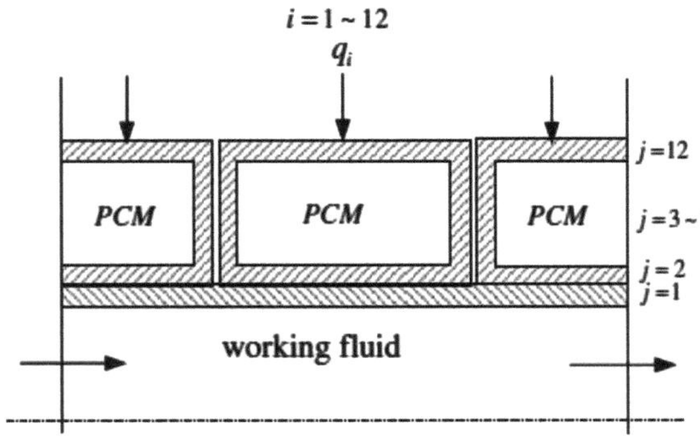

Fig. 6.9 Schematic of PCM tube configuration. Reproduced with permission from [47], Copyright @ 2003, Elsevier

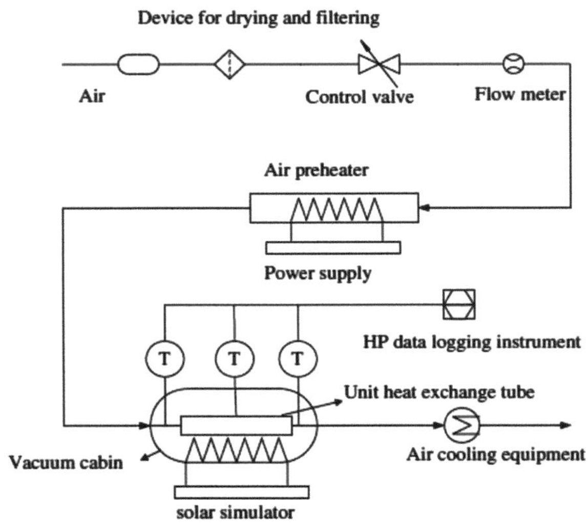

Fig. 6.10 Scheme of experiment system to collect data for comparison with the theoretical results. Reproduced with permission from [47], Copyright @ 2003, Elsevier

temperature. The PCM is optically opaque. There are no PCM convective motions due to buoyancy, while the marangoni effect or solidification shrinkage/melt expansion can be negligible. The working fluid flow is hydrodynamically and thermally fully developed. The incident and absorbed solar flux distributions are circumferentially uniform. It has been shown that the simulated results are slightly higher than those of experiments. The experimental facility is schematically shown in Fig. 6.10, in which 12 canisters are stacked to the working fluid tube.

A new type of storage system with KNO_3–$NaNO_3$ as PCMs has been reported by Guo and Zhang [48]. Configuration and layout for simulation of the storage system are shown in Figs. 6.11 and 6.12, respectively. It is a typical foil-tube arrangement. Al foils are arranged orthogonal to the axis of the steam tubes and the PCMs are filled in between the spaces of the foils and the tubes, while water/steam flows inside the tubes. The tubes are staggered, so that there is a symmetry circle (dotted lines) around each of them, through which heat conduction fluxes remain null. Therefore, the system to be analyzed can be represented by one tube as shown

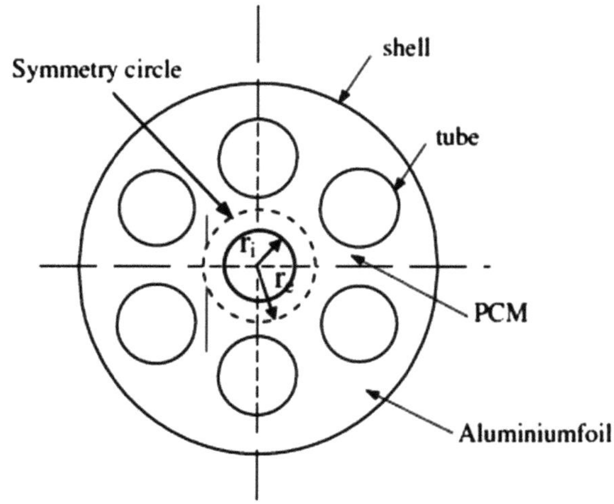

Fig. 6.11 Configuration of the storage system. Reproduced with permission from [48], Copyright @ 2008, Elsevier

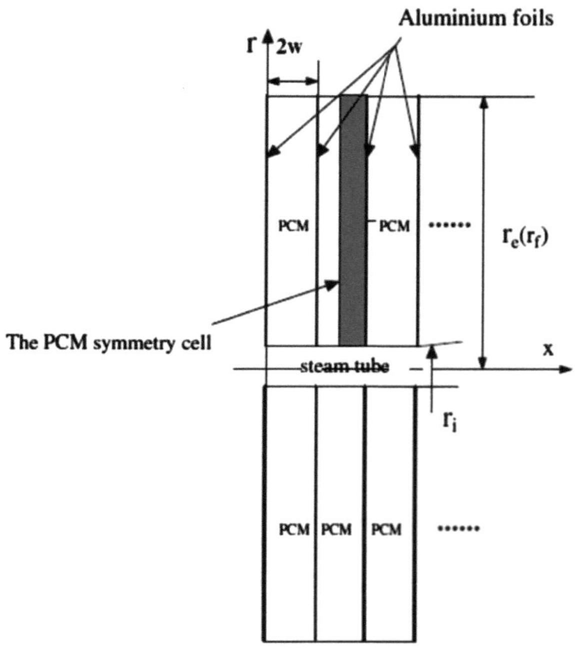

Fig. 6.12 Layout of the problem for simulation. Reproduced with permission from [48], Copyright @ 2008, Elsevier

in Fig. 6.13. Because the storage system has a symmetrical structure, the computational domain can be further simplified with one symmetry unit cell where the planes of symmetry are in the middle of the foil and midway between every two adjacent foils. The computational results have well predicted the ways in which the heat transfer and discharge time are affected by the changes in geometry of the Al foil, tube radius, boundary conditions, and thermal conductivity of the PCMs. These results can be used to design high performance TES systems.

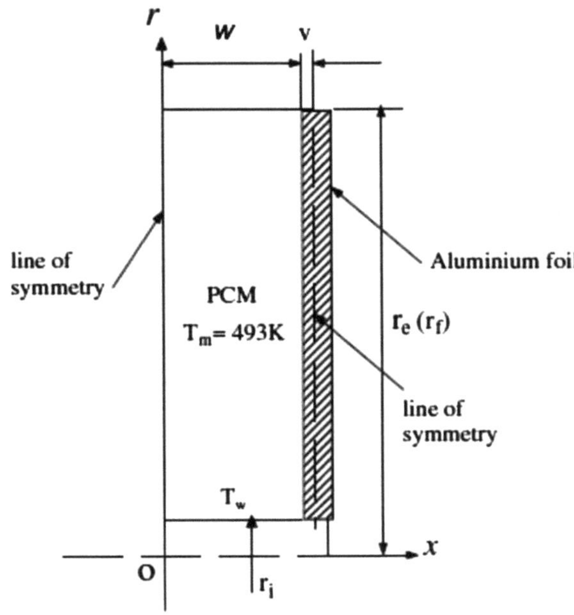

Fig. 6.13 Computational domain of the storage unit. Reproduced with permission from [48], Copyright @ 2008, Elsevier

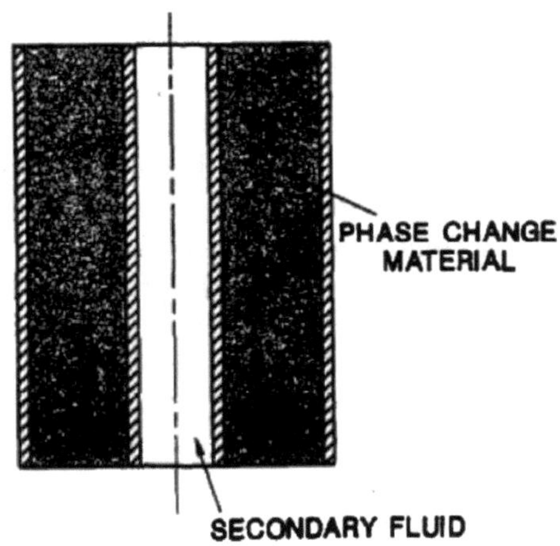

Fig. 6.14 Configuration of the 2D system for simulation. Reproduced with permission from [49], Copyright @ 1997, Elsevier

A two-dimensional transient analytical model based on the enthalpy method has been reported by Yimer and Adami [49]. The system consists of a tube and shell arrangement, in which LiH is used as the PCM, is contained in an annulus bounded by an inner tube and an outer shell, as shown in Fig. 6.14. They presented a dimensionless analysis. Two-dimensional finite difference representations are used: radial and angular variations, and radial and axial variations. The nonlinear simultaneous difference equations are solved by using the Gauss–Seidel iterative method with successive over-relaxation (SOR).

Simulation results indicate that the amount of energy stored and the melting front depth increase with increasing tube and outer shell diameters. Charging rates

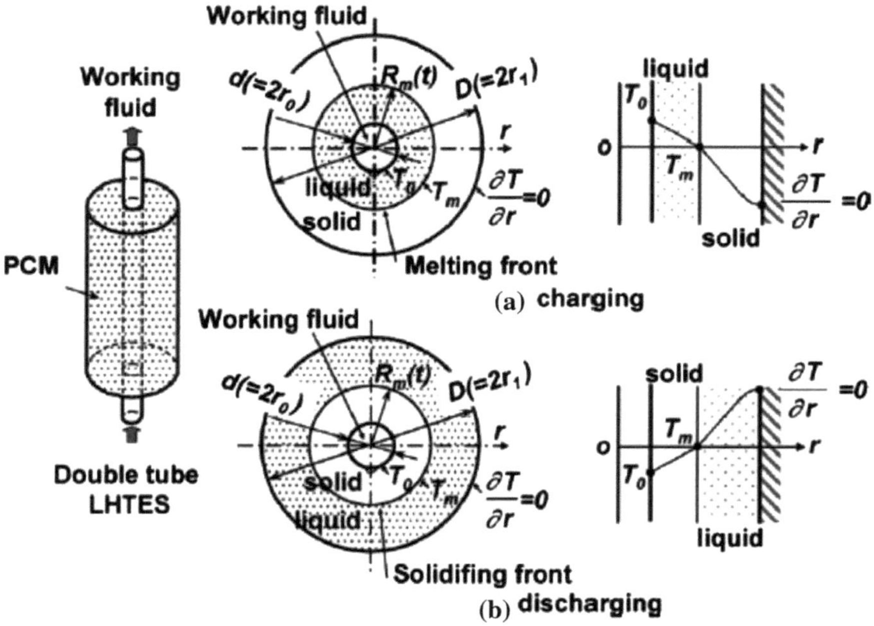

Fig. 6.15 Schematic of PCM tube configuration. **a** Charging, **b** discharging. Reproduced with permission from [50], Copyright @ 2005, Elsevier

and heat storage increase as the initial system temperature is decreased, but penetration of the melting front is slower. The use of fins can increase the amount of the heat stored and the melting front penetration. Heat storage, charging rate, and melting front depth all increase with increasing effective thermal conductivity of the melted medium due to the presence of acceleration forces. The effect of liquid effective thermal conductivity is insignificant during the discharging mode. These conclusions should be useful references for the practical design of TES systems.

A double tube latent heat TES model system has been developed by Hoshi et al. [50]. The model system is shown in Fig. 6.15, with the assumptions of constant physical properties of the materials, negligible volumetric expansion during melting and negligible natural convection in the liquid region. The numerical results have provided several useful suggestions. For example, in storage systems with high melting point substances as PCMs, the thermal conductivity of the PCMs is a significant factor. Materials with high melting points usually possess high thermal conductivity.

The design approach is shown schematically in Fig. 6.16. The storage unit and turbine are situated on the ground and air is used as the primary circulating fluid in the tower receivers. The use of a single storage and turbine unit for multiple towers is to increase the efficiency, due to the reduction in part-load turbine operation. Also, the efficiency of Brayton cycle turbines is increased rapidly with turbine size.

A detailed model of heat transfer and fluid flow for numerical simulation of LHS unit has been available, which takes into account the solid/liquid and water/

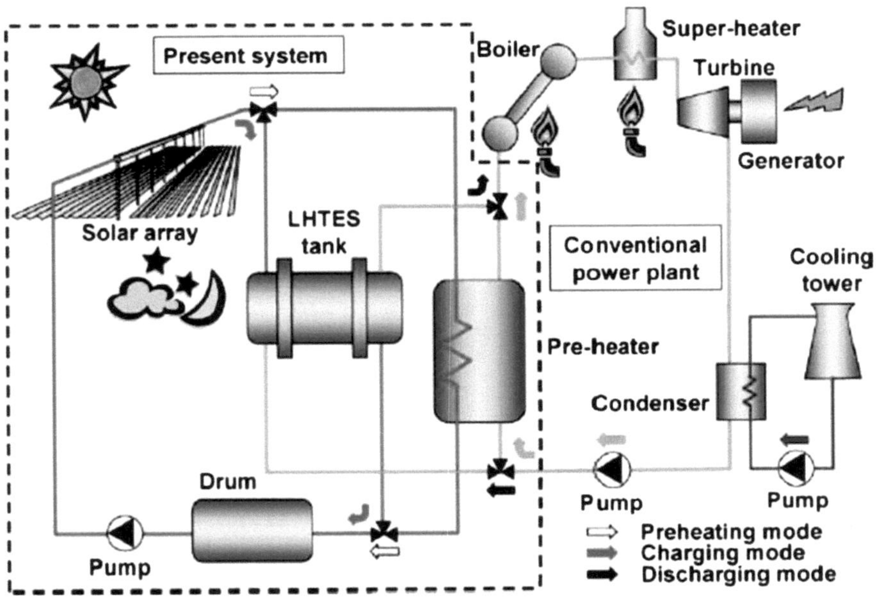

Fig. 6.16 Power plant design incorporating high temperature storage system with PCMs. Reproduced with permission from [50], Copyright @ 2005, Elsevier

vapor phase change processes occurring simultaneously in the PCM and HTF, with appropriate coupling between them [51]. Simplified schematic diagrams for the control schemes implemented during charging and discharging modes of operation are shown in Fig. 6.17. In this configuration, the overall storage capacity of the system is represented by a number of parallel blocks to be used at a given time. The mass flow rate per block can be optimized independently of the fluctuations of the load and of solar radiation. During charging mode of operation (Fig. 6.17a), the vapor produced by the solar collectors is sent to the storage system where it is condensed, while the energy of condensation is transferred to the PCM which undergoes a solid–liquid phase transition. In the discharging mode, Fig. 6.17b, the water coming from the power cycle is vaporized when passing through the PCM storage system which undergoes a liquid–solid transformation.

A typical storage block is composed of parallel tubes, where the water/steam can pass through the unit, as shown in Fig. 6.18. The tubes are embedded in the PCMs. Theoretical modeling is carried out with various assumptions. Heat transfer within the PCM is controlled by conduction, with convection to be negligible due to the high viscosity of the PCM salts in the liquid state. The model can be further simplified with initially obtained parameters together with further assumptions.

A model has been developed to simulate a TES system composed of three PCMs with different phase change temperatures [52]. Theoretical prediction indicates that, the use multiple PCMs instead of a single PCM can increase the ES rate and decrease the fluctuation of the gas exit temperature. Therefore, it is possible to increase the performance of the receiver and thus reduce the weight of heat receiver.

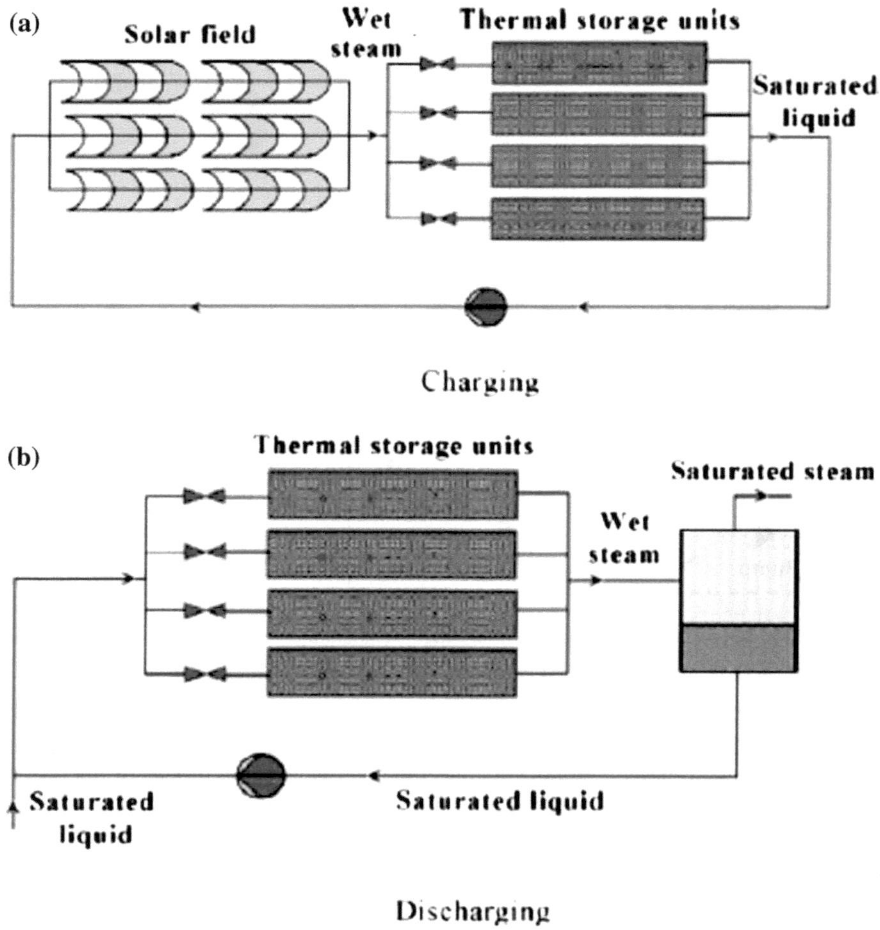

Fig. 6.17 Simplified control schemes during charging and discharging modes of operation for the storage system. **a** Charging, **b** discharging. Reproduced with permission from [51], Copyright @ 2008, Elsevier

Fig. 6.18 2D configuration of the system for simulation. Reproduced with permission from [51], Copyright @ 2008, Elsevier

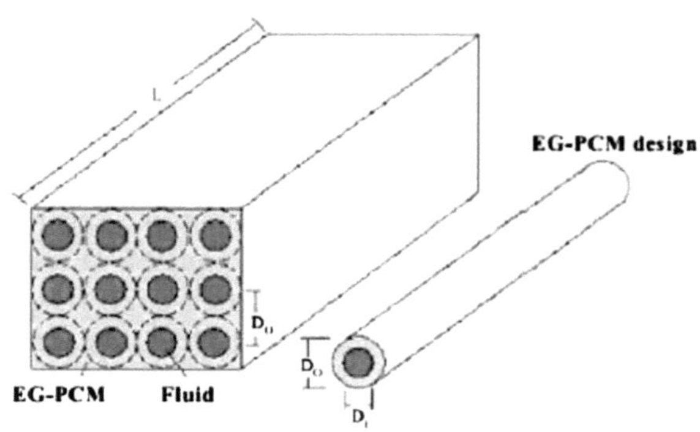

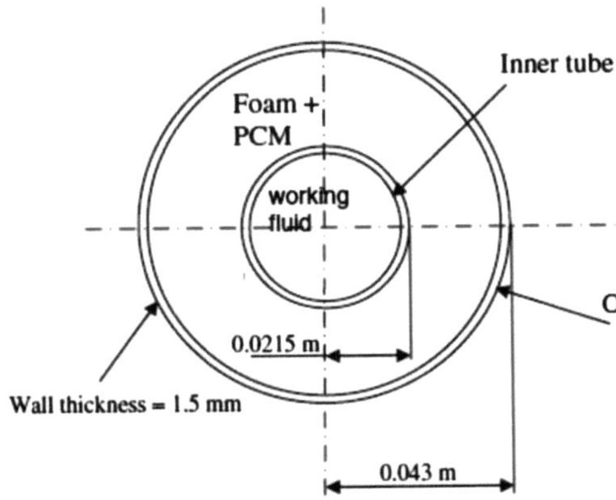

Fig. 6.19 Computational model of a PCM TES system. Reproduced with permission from [53], Copyright @ 2008, Elsevier

A model has been developed to study the system with carbon foams saturated with PCMs as thermal storage materials [53]. The numerical model is based on a volume averaging technique while a finite volume method is used to discretize the heat diffusion equation. A line-by-line solver based on tri-diagonal matrix algorithm is used to iteratively solve the algebraic discretization equations. Two-dimensional general coordinates are used, with the configuration the dimensions of one canister being shown in Fig. 6.19. Convection motion of the molten PCM inside the foam is not considered, because the gravity in space can be neglected. A mathematical formulation of the governing equations on nonorthogonal curvilinear system is employed. The standard SIMPLE (Semi-Implicit Method for Pressure Linked Equations) algorithm is used to solve the coupled continuity and momentum equations. The model has validated with experimental data obtained separately [54].

As mentioned above, packed bed configuration is less widely used until now. One example is reported by Yagi and Akiyama [55]. A model for a single spherical capsule has been developed and the heat transfer of the packed bed system with the spherical capsules is simulated, by using the enthalpy method. For assumptions, convection and radiation in the surface are considered and fully implicit scheme based on control volume (with SOR method) is used. The model results have been compared with experimental data showing a certain degree of difference. Another computational model for packed bed system, developed by Jalalzadeh-Azar [56], has been validated experimentally. In this simulation, second law thermodynamic analyses along with material stability tests are employed as criteria for the assessment of the PCMs, consisting of PCM components and sensible heat pellets.

In summary, the prerequisite to use a TES system for practical applications is that its performance is known. The performance assessment of any storage system requires a thorough understanding on the thermal behavior of the PCMs. Although a great amount of efforts have been made both numerically and experimentally on the study of thermal behavior of different configurations of PCM TES systems, there is no a standard available for different types of systems. It is still a long way to go for the research community in this increasingly important area.

6.4.4 Exergy

6.4.4.1 Energy Analysis

The performance of a thermal system can be evaluated in terms of efficiency or effectiveness. The efficiency is a measure of how effectively the heat or cold energy is stored or recovered. In other words, it indicates how much heat or cold that could not be stored or recovered. As a result, there different types of efficiency, including charging efficiency, discharge efficiency, and overall (complete cycle) efficiency. The terms charging and discharging are used, respectively, for melting and solidification processes of a PCM with solid–liquid phase transition [57]. A system is always expected to produce the maximum possible efficiency, but this possibility is achievable only when certain conditions are met. Therefore, the influence of various design and operating parameters on the efficiency has to be addressed.

There has been the use of ratios of heat storage and heat release as performance parameters in the open literature [58], which are defined as:

$$r_{\text{char}} = \frac{\text{Total energy stored}}{\text{Maximum storage capacity}}, \tag{6.29}$$

$$r_{\text{dis}} = \frac{\text{Total energy released}}{\text{Maximum storage capacity}}. \tag{6.30}$$

The heat stored and released can be calculated from the heat absorbed and released, respectively, by the HTF. The maximum heat storage capacity is composed of the sensible and latent heat capacity of the PCM. Therefore, both the ratios increase always with increasing mass flow rate of the HTF.

Charging and discharge efficiencies have been proposed for solar cookers by Mawire and McPherson [59] as follows:

$$\eta_{\text{char}} = \frac{\text{Total energy stored}}{\text{Total energy provided by the HTF}}, \tag{6.31}$$

$$\eta_{\text{dis}} = \frac{\text{Total energy provided by the HTF}}{\text{Total energy stored in the system}}. \tag{6.32}$$

With these definitions, the performances of the PCM ES systems are all assessed by using only energy balance, which are based on the first law of thermodynamics and generally employed to identify the ways to increase the quantity of heat stored or recovered. However, as stated by Bejan [60], "the primary purpose of a TES system is not, as the name implies, to store the energy, rather, to store useful work." In other words, analysis based on the first law of thermodynamics is inadequate, because the usefulness of the energy is not evaluated. To address this problem, it is necessary to apply the second law of thermodynamics.

Furthermore, energy analysis becomes complicated and confusing, when it comes to cold thermal storage systems, such as those for building cooling [61]. This is because the energy analysis with first law only caters to heat flows but not cold flows. In contrast, second law analysis can treat both heat and cold flows. In this respect, second law analysis is better, especially when cold thermal storage should be considered. Moreover, second law analysis offers thermoecomomics analysis of a thermal system [62]. This is because thermoeconomics analysis combines second law analysis and engineering economics. Because the second law analysis takes into account the true potential of energy, it is possible to estimate the cost of a system through thermoeconomics.

6.4.4.2 Definition of Exergy

As compared with first law of thermodynamics, second law quantifies the amount of energy. The quality of energy is measured according to the state of system in relation to the surrounding conditions. This quality or usefulness of the energy is termed as exergy. There are many other equivalent names, like availability, available energy, essergy, work capability, utilizable energy, etc., that are used in open literature. However, exergy is more commonly used than others [63]. Therefore, the analysis based on second law is called exergy analysis.

Exergy is defined as the maximum quantity of work that can be done by a system, when it reaches an equilibrium state from a nonequilibrium state. In other words, exergy is a measure of potential of the system to cause change, due to its nonequilibrium condition with surrounding. A system can carry exergy only if it is away from the equilibrium state. Once the equilibrium is approached, exergy becomes zero. This means that exergy cannot be conserved, but can be consumed or destroyed, which is the reason why energy always tends to be degraded. A thermal storage system is to store useful work, which is also recovered to useful work. Therefore, the quantity of exergy transfer is more appropriate than that of energy transfer during charging and discharging processes.

Exergy content of a system or a matter can be derived by using the thermodynamic relations, entropy relations. Unlike energy, exergy cannot be conserved, but can be destroyed, because all real processes are irreversible The exergy analysis is not only concerned with exergy, but also with its destruction or degradation. This means that it is also necessary to know that to what extent exergy can be degraded, such that the net potential of the system or the process can be quantified.

Entropy thus is a measure of irreversibility. An irreversible process is always accompanied by energy degradation and when an equilibrium state is approached, the exergy content becomes zero. Meanwhile, the irreversible process is accompanied by an increase in entropy which reaches a maximum when the system becomes equilibrium. Therefore, irreversible processes always proceed in the direction of increase in entropy and decrease in exergy. In a real system, entropy is always generated and exergy is degraded and the degraded exergy is proportional to the amount of the generated entropy.

6.4.4.3 Second Law Efficiency

First law-based efficiency can be stated as the ratio of energy output to energy input. In practical thermal systems, the energy output is less than energy input, due to the presence of energy loss. The first law efficiency indicates the amount of loss that occurs during the process of the system. Therefore, improving efficiency is through the reduction of losses. However, the losses do not reflect the degradation of energy.

Specifically, the efficiency of a well-insulated PCM ES system, during the cycle (melting—solidification process), can be defined as the ratio of total energy recovered during the solidification to total energy stored during the melting. Because the latent heats of both melting and solidification are the same, the efficiency of the system should be 100 %. Due to the adiabatic condition, no energy loss can occur. However, energy is possibly degraded due to internal irreversibility. Even considering the realistic condition of nonadiabatic, the energy efficiency cannot reflect the degraded quantity, since it accounts only for the energy loss. In contrast, the second law or exergy efficiency involves the exergy, so it can identify and quantify both the degraded quantity and lost quantity. In this regard, the exergy efficiency is always lower than the energy efficiency [61, 64–70].

The second law or exergy efficiency of a thermal system is thus defined as the ratio of the output exergy to the input exergy. For a given quantity of exergy, the output exergy is less due to exergy destruction. Because exergy degradation is due to the irreversibility, exergy efficiency is a measure of the irreversibility, given by:

$$\psi = 1 - \frac{\text{Exergy destroyed}}{\text{Exergy input}}. \tag{6.33}$$

As stated earlier, the more the exergy is degraded, the more the entropy is generated. Therefore, the ratio of the destroyed exergy to the input exergy can be represented by a factor called entropy generation number (N_s), i.e.:

$$\psi = 1 - N_s. \tag{6.34}$$

Equation (6.34) indicates that the performance of a system can be improved by reducing the generation number of entropy. The entropy minimization can thus be used as a criterion to optimize the performance of thermal systems. It is thus very clear that exergy and entropy are closely related to each other [71].

6.4.4.4 Exergy Efficiency

It has been accepted that although energy based efficiencies for TES systems are reasonable and widely applied, exergy efficiencies are more meaningful to practical applications [72]. Similar to energy efficiencies, there are also exergy efficiencies charging period, discharging period, and complete cycle.

During charging process, the HTF transfers exergy to PCM and part of the exergy is stored in the PCM. Therefore, the exergy stored is considered as desired output in the definition of exergy efficiency. Accordingly, exergy efficiency is defined as [73]:

$$\psi_{\text{char}} = \frac{\text{Exergy stored in PCM}}{\text{Exergy provided by HTF}}. \tag{6.35}$$

In PCM ES systems, it is also important to evaluate the exergy efficiency at different times during melting, because the heat transfer is time dependent. Therefore, exergy efficiency can also be defined as exergy rate, given by:

$$\psi_{\text{char}} = \frac{\text{Rate of exergy stored in PCM}}{\text{Rate of exergy provided by HTF}}. \tag{6.36}$$

However, the power input consumed when processing the HTF, such as pump work, should be included in the exergy efficiency evaluation [74]. It has been found that the difference between the exergy efficiency which neglects the pump power and that with considering the pump work is very significant. Moreover, this difference is more pronounced for exergy efficiency than for energy efficiency. In this respect, exergy efficiency can be further defined with considering the pump work (working power), given by [66]:

$$\psi_{\text{char}} = \frac{\text{Rate of exergy stored in PCM}}{\text{Rate of exergy providedby HTF} + \text{working power}}. \tag{6.37}$$

In the above definitions, the exergy supplied by HTF is the change in flow exergy of HTF during heat transfer with PCM. The change in flow exergy can be derived from the difference between the inlet and outlet temperatures of HTF along with the environment temperature. As a result, the exergy efficiency reflects the amount of exergy stored from the supplied exergy during the heat transfer. Because the maximum exergy that can be supplied by the HTF is not considered, the maximum possible exergy storage capacity of the storage system cannot be well described. This problem can be readily addressed by defining the exergy efficiency with the following equation [75, 76]:

$$\psi_{\text{char}} = \frac{\text{Rate of exergy stored in PCM}}{\text{Rate of exergy possessed by HTF before reaching PCM}}. \tag{6.38}$$

In this expression, the exergy efficiency involves the temperature at which the exergy is available to be stored, since the difference between the inlet temperature of HTF and environment temperature, rather than the difference between the inlet and outlet temperatures of HTF, is used. In this case, it quantifies the maximum exergy that can be stored within the maximum available potential work.

For discharging (solidification) process, similar definitions can be applied. In the discharging process, the exergy output is the exergy obtained by the HTF which is the change of flow exergy of HTF. The exergy input is also the exergy available with the PCM. Therefore, the exergy efficiency of the system during the discharging process can be defined by using the following expression [73]:

$$\psi_{\text{dis}} = \frac{\text{Exergy gained by HTF}}{\text{Initial exergy available with PCM}}. \tag{6.39}$$

This expression means that the total exergy extracted by the HTF from the maximum available exergy with the PCM has been considered. With considering time dependent operation, the exergy efficiency can be further defined as [64, 75]:

$$\psi_{\text{dis}} = \frac{\text{Rate of exergy gained by HTF}}{\text{Rate of exergy released by PCM}}. \tag{6.40}$$

The operation of PCM TES systems can be treated as a cycle, comprising of ES process followed by energy releasing process. If the cyclic is not correctly characterized, there could be errors in the analysis, with regardless charging or discharging processes [75, 77, 78]. In other words, it is necessary to calculate overall cycle exergy efficiency of a system. The evaluation of exergy efficiency can readily be done from the exergy efficiencies of charging and discharging processes. The total exergy efficiency can be calculated by using the following expression [64]:

$$\psi_{\text{overall}} = \psi_{\text{char}} \cdot \psi_{\text{dis}}. \tag{6.41}$$

The overall efficiency can also be evaluated as [65]:

$$\psi_{\text{overall}} = \frac{\text{Exergy extracted from PCM by HTF during discharging}}{\text{Exergy input to PCM during charging}}. \tag{6.42}$$

Several forms of exergy efficiencies have been developed which are valid at different conditions. The overall exergy efficiency is more useful approach to evaluate the performance of a PCM TES system.

6.4.4.5 Calculation of Exergy

To calculate the exergies of storage system, the rate of exergy supplied by HTF during charging process is expressed as [76]:

$$\dot{\text{E}}\text{x}_{\text{input}} = \dot{m}_{\text{HTF}} c_{\text{HTF}} \left[\left(T_{\text{HTF, in}} - T_{\text{HTF, out}} \right) - T_0 \ln \left(\frac{T_{\text{HTF, in}}}{T_{\text{HTF, out}}} \right) \right]. \tag{6.43}$$

In this expression, the exergy input is the change in flow exergy of HTF as a result of the heat transfer. However, as mentioned before, the exergy content of HTF can be used to the exergy input. Because the exergy content of HTF is the minimum useful energy available when the system is approaching an equilibrium state, the rate of exergy input can be expressed as following:

$$\dot{E}x_{input} = \dot{m}_{HTF} c_{HTF} \left[\left(T_{HTF,in} - T_0 \right) - T_0 \ln \left(\frac{T_{HTF,in}}{T_0} \right) \right]. \tag{6.44}$$

The total exergy input to PCM within any time interval can be readily derived from the time integration of the relevant expressions.

During the charging process, the exergy stored at any time can be calculated with the instantaneous heat transfer rate. The instantaneous heat transfer rate (Q) can be obtained by establishing the energy balance, i.e., the heat gained by PCM = the heat transferred by HTF, and thus, there is:

$$Q = \dot{m}_{HTF} c_{HTF} \left(T_{HTF,in} - T_{HTF,out} \right). \tag{6.45}$$

Therefore, the rate of exergy stored in PCM is given by:

$$\dot{E}x_{stored} = Q \left(1 - \frac{T_0}{T_{PCM}} \right). \tag{6.46}$$

The melting temperature of PCM has been used as the temperature of the PCM (T_{PCM}) by some researchers [75], because the PCM melts completely at the melting temperature throughout the melting process. This means that the sensible heat of the PCM has been neglected. However, in practical applications, the sensible heat contributes considerably to the total heat, because the PCM is subjected to sensible heating prior to melting and there is sub cooling after solidification [79]. In other words, the PCM temperature is most likely different from its melting point before and after the charging or discharging processes.

Figure 6.20 shows the temperature profile of the PCM throughout the phase transition with considering sensible heating and cooling [57]. In the beginning of charging process, it is at its subcooled temperature in solid state and the charging is terminated when the PCM reaches its superheated temperature. In this case, the average of the initial and final temperatures is used as the temperature of PCM [76]. This simple approximation obviously overlooks the instantaneous temperatures of PCM at every point. Moreover, at a given time, the temperature of the PCM is also different at different locations, because of the continuous movement of solid/liquid interface. Therefore, it is necessary to consider the temperature variation of the PCM, when evaluating the value of T_{PCM}.

The energy balance used in Eq. (6.45) implies that the system perfectly insulated and thus has no heat loss during the heat transfer, so the exergy balance can be expressed as: *exergy input = exergy output + exergy degraded*. However, this exergy balance reflects only the degraded exergy due to internal irreversibility. In fact, the heat loss or heat gain is the primary parameter of thermal behavior of a storage system and both the energy and exergy efficiencies are dependent on the heat loss or heat gain [68]. Therefore, the heat gain should also be included when evaluating the exergy efficiency of a thermal storage system, i.e., *energy transferred by HTF = energy gained by PCM + energy lost to environment and exergy associated with HTF = exergy stored + exergy degraded + exergy lost to environment*. In this case, the rate of exergy stored is thus given by:

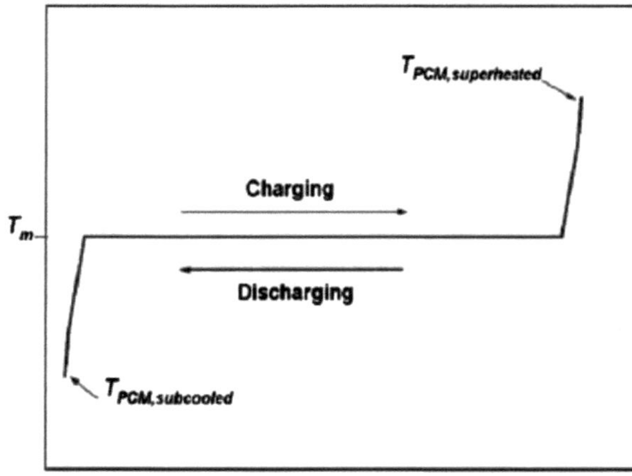

Fig. 6.20 Temperature profile of the PCM during phase transition with considering sensible heating and cooling. Reproduced with permission from [57], Copyright @ 2010, Elsevier

$$\dot{\mathrm{E}}\mathrm{x}_{\mathrm{stored}} = Q\left(1 - \frac{T_0}{T_{\mathrm{PCM}}}\right) - Q_{\mathrm{loss}}\left(1 - \frac{T_0}{T_{\mathrm{PCM}}}\right), \tag{6.47}$$

where Q_{loss} is the rate of heat loss of the whole storage system due to the difference in temperature between the system and environment.

The total exergy stored in the PCM during charging can be obtained through time integration of Eq. (6.47), with considering the heat loss of the system to the environment. In practical applications, a simpler form is usually used for such purpose [69], which is:

$$\begin{aligned}\dot{\mathrm{E}}\mathrm{x}_{\mathrm{stored}} = & ML\left(1 - \frac{T_0}{T_m}\right) + Mc_{\mathrm{PCM,s}}\left[\left(T_m - T_{\mathrm{PCM,\,init}}\right) - T_0 ln\left(\frac{T_m}{T_{\mathrm{PCM,\,init}}}\right)\right] \\ & + Mc_{\mathrm{PCM,l}}\left[\left(T_{\mathrm{PCM,\,final}} - T_m\right) - T_0 ln\left(\frac{T_{\mathrm{PCM,\,final}}}{T_m}\right)\right].\end{aligned} \tag{6.48}$$

The first term on the right hand side of Eq. (6.48) is the exergy stored during melting, while the second and third terms are the exergy stored due to the sensible heating before and after the melting, respectively. With including the exergy stored due to sensible heat transfer before and after phase transition process, this formula is more accurate and useful. For discharging process, the exergy balance can be evaluated similarly, i.e., *exergy of PCM + exergy gain from environment = exergy gained by HTF + exergy degraded.* The exergy balance of a storage system can be schematically shown in Fig. 6.21 [57].

6.4.4.6 Generation of Entropy

During the calculation of the exergy efficiency with the exergy output and exergy input, between the exergy stored in the PCM and the exergy supplied by the HTF,

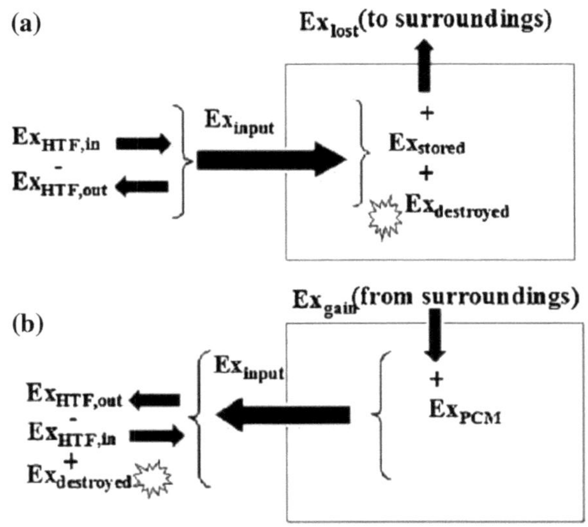

Fig. 6.21 Exergy balance of a TES system. **a** Charging process and **b** discharging process. Reproduced with permission from [57], Copyright @ 2010, Elsevier

there is a difference, which is the exergy that has been degraded. However, all the expressions till now have not included the calculation of exergy degradation. It has been mentioned earlier that the exergy degradation is due to the internal irreversibility of the systems. If the irreversibility of a system can be clearly identified, it is more targetable to optimize or enhance the performance of the system, which is actually to minimize the irreversibility.

The difference between a reversible and an irreversible process can be described quantitatively by using the amount of entropy generation. Thus, entropy generation can be used as a direct measure of internal irreversibility or exergy degradation of a thermal storage system. The strategy of optimization of a system is to minimize the entropy generation. Compared with exergy analysis, the optimization through the minimization of entropy generation involves not only thermodynamics but also heat transfer and fluid mechanics principles [70].

With a cycle of charging or discharging process, the net entropy generation (S_{gen}) can be expressed as [68, 80–82]:

$$S_{gen} = S_{flow} + \Delta S_{PCM} + \Delta S_{HTF}, \tag{6.49}$$

where S_{flow} is the entropy transfer due to the heat transfer between PCM and HTF and the second and third terms are the entropy variations in the PCM and HTF, respectively.

The lasted studies indicate that the entropy generation due to the heat gain by the system from the environment should also be taken into account [3, 83, 84]. Therefore, the general form of the expression for charging and discharging processes can be written as:

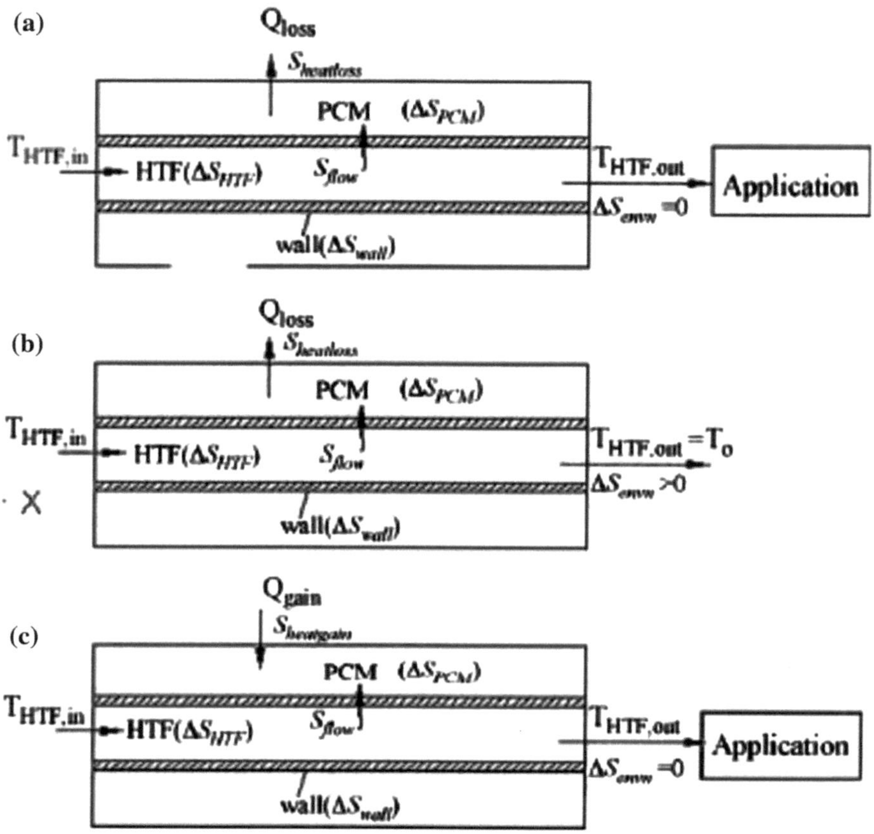

Fig. 6.22 Entropy generation of a TES system. **a** Charging-HTF output is used in the application after storage, **b** charging-HTF output is dumped into atmosphere and **c** discharging for application. Reproduced with permission from [57], Copyright @ 2010, Elsevier

$$S_{\text{gen}} = S_{\text{flow}} + \Delta S_{\text{PCM}} + \Delta S_{\text{HTF}} + A\Delta S_{\text{wall}} \pm \left(\frac{Q_{\text{loss}} \text{ or } Q_{\text{gain}}}{T_0}\right). \tag{6.50}$$

In Eq. (6.50), the positive and negative signs are for Q_{loss} and Q_{gain}, respectively.

TES is useful in two cases. In case that the available heat source is excessive than the demand, the excessive energy can be stored for later use. If the heat source is available but there is no demand, the entire energy should be stored. Figure 6.22 shows a schematic of entropy generation in different cases [57].

For the first case, the HTF output is used in the application after the storage. Therefore, there will be no entropy generation due to the difference between the HTF outlet temperature and environment temperature (Fig. 6.22a). For the second case, the HTF output is discharged into the atmosphere (Fig. 6.22b). In this case, the additional entropy generation should be included in the net entropy generation, given by:

$$S_{\text{gen}} = S_{\text{flow}} + \Delta S_{\text{PCM}} + \Delta S_{\text{HTF}} + A\Delta S_{\text{wall}} \pm \left(\frac{Q_{\text{loss}} \text{ or } Q_{\text{gain}}}{T_0} \right) + \Delta S_{\text{envn}}. \quad (6.51)$$

The last term on the right hand side of Eq. (6.51) represents the entropy generation due to the discharge of HTF output into the atmosphere. For discharging process, HTF output is always used in application after the energy recovery from the system (Fig. 6.22c). Therefore, the term ΔS_{envn} can be always excluded for discharging process.

As discussed above, the term S_{flow} is the entropy transfer due to the heat transfer between PCM and HTF, which can be derived by using the following equations. The first one is:

$$S_{\text{flow}} = \dot{m}_{\text{HTF}} c_{\text{HTF}} \int\limits_0^t \ln\left(\frac{T_{\text{HTF, out}}}{T_{\text{HTF, in}}} \right) \mathrm{d}t. \quad (6.52)$$

In Eq. (6.52), the irreversibility related to pressure drop due to the flow of HTF is not considered. When this irreversibility is taken into account, there is [68]:

$$S_{\text{flow}} = \dot{m}_{\text{HTF}} c_{\text{HTF}} \int\limits_0^t \ln\left(\frac{T_{\text{HTF, out}}}{T_{\text{HTF, in}}} \right) \mathrm{d}t + \dot{m}_{\text{HTF}} R \int\limits_0^t \ln\left(\frac{T_{\text{HTF, in}}}{T_{\text{HTF, out}}} \right) \mathrm{d}t. \quad (6.53)$$

The second term on the right hand side of Eq. (6.53) stands for the pressure drop irreversibility. However, the HTF is in most cases a liquid, the assumption of an ideal gas is invalid for TES systems. Therefore, the entropy generation due to the pressure drop irreversibility has been replaced by the entropy generation due to viscous dissipation [34], thus there is:

$$S_{\text{flow}} = \dot{m}_{\text{HTF}} c_{\text{HTF}} \int\limits_0^t \ln\left(\frac{T_{\text{HTF, out}}}{T_{\text{HTF, in}}} \right) \mathrm{d}t + \frac{\dot{m}_{\text{HTF}} \Delta T \left(P_{\text{HTF, in}} - P_{\text{HTF, out}} \right)}{\rho_{\text{HTF}} T_{\text{HTF}}}, \quad (6.54)$$

where T_{HTF} is the average of the initial and final temperatures of HTF and the second term is the entropy generation due to the pressure drop irreversibility.

According to thermodynamics, it can be assumed that during the phase change process, the specific Gibbs free energies for the solid and the liquid phases are same at the melting points [85]. As a result, the entropy variation in the PCM can be given by:

$$h_{\text{PCM}, s} - T_m S_{\text{PCM}, s} = h_{\text{PCM}, l} - T_m S_{\text{PCM}, l}. \quad (6.55)$$

Rearranging Eq. (6.55) leads to:

$$\Delta S_{\mathrm{PCM}} = \frac{h_{\mathrm{PCM},s} - h_{\mathrm{PCM},l}}{T_m} = \frac{ML}{T_m}. \tag{6.56}$$

Obviously, Eq. (6.56) is valid only when the temperature of the PCM is always equal to its melting point throughout the phase transition period. If there is sensible heat transfer involved, the entropy variation should be [82]:

$$\Delta S_{\mathrm{PCM}} = \rho_{\mathrm{PCM},l} L V_{\mathrm{PCM}} \int_0^t f(t)\mathrm{d}t + \rho_{\mathrm{PCM}} c_{\mathrm{PCM}} V_{\mathrm{PCM}} \int_0^t \ln\left[\frac{T_{\mathrm{PCM}}(t)}{T_{\mathrm{PCM}}(t=0)}\right] \mathrm{d}t. \tag{6.57}$$

In Eq. (6.57), the first term on the right hand side is the entropy variation of the PCM due to the phase transition while the second term is the entropy variation of the PCM due to the sensible heating or cooling.

The entropy variations of HTF and wall can be obtained similarly, i.e.:

$$\Delta S_{\mathrm{HTF}} = \rho_{\mathrm{HTF}} c_{\mathrm{HTF}} V_{\mathrm{HTF}} \int_0^t \ln\left[\frac{T_{\mathrm{HTF}}(t)}{T_{\mathrm{HTF}}(t=0)}\right] \mathrm{d}t, \tag{6.58}$$

$$\Delta S_{\mathrm{wall}} = \rho_{\mathrm{wall}} c_{\mathrm{wall}} V_{\mathrm{wall}} \int_0^t \ln\left[\frac{T_{\mathrm{wall}}(t)}{T_{\mathrm{wall}}(t=0)}\right] \mathrm{d}t. \tag{6.59}$$

With the availability of entropy generation, the degraded exergy ($\mathrm{Ex}_{\mathrm{degraded}}$) can be written as [86]:

$$\mathrm{Ex}_{\mathrm{degraded}} = T_0 S_{\mathrm{gen}}. \tag{6.60}$$

In this case, the exergy efficiency is given by:

$$\psi_{\mathrm{Ex}} = 1 - \frac{T_0 S_{\mathrm{gen}}}{\mathrm{Ex}_{\mathrm{input}}}. \tag{6.61}$$

It has been mentioned that the ratio of the degraded exergy ($T_0 S_{\mathrm{gen}}$) to the exergy input can be replaced by the entropy generation number, so the expression of exergy efficiency is similar to the form of Eq. (6.34). According to Eq. (6.34), the system achieves the maximum efficiency if entropy generation number is zero. That is to say that the system is completely reversible, i.e., no exergy is degraded. This is obviously an ideal condition, which the target of system optimization [85].

6.4.4.7 Exergy for System Optimization

Significant progress has been achieved in performance optimization of PCM TES systems with the consideration of exergy. Optimization is the selection of materials and structures for design of a system with optimized performances. Influent parameters include thermophysical properties of PCMs and HTFs and structural dimension of the systems [57].

Thermophysical properties of PCMs include melting temperature, latent heat, thermal conductivity, chemical stability, and so on. Among these, the melting point plays a crucial role in determining proper operation of the systems. Therefore, the melting temperature of PCM is the main criterion. In practice, the melting point should be within the operating temperature range of interest.

The operating cycle of a TES consists of heat storage and removal processes, which can be assumed to follow the Carnot cycle [86]. Accordingly, the heat is supplied at a constant temperature (melting point) and is rejected into a low temperature reservoir (environment). In this case, the rate of exergy extracted from the PCM is given as:

$$\dot{\mathrm{E}}\mathrm{x} = m_{\mathrm{HTF}} c_{\mathrm{HTF}} \left(T_{\mathrm{HTF,in}} - T_m\right)\left(1 - e^{-\mathrm{NTU}}\right)\left(1 - \frac{T_0}{T_m}\right). \tag{6.62}$$

Maximizing Eq. (6.62) with respect to melting temperature results in:

$$T(\text{optimum}) = \sqrt{T_{\mathrm{HTF,in}}\, T_0}. \tag{6.63}$$

It means that the optimum melting temperature of PCM is the geometric mean of the HTF inlet temperature and the temperature of the environment. The optimum melting temperature can be obtained either by minimizing the entropy generation or by maximizing exergy efficiency [87, 88]. Although Eq. (6.63) is derived with the assumption that PCM is at melting temperature during the cycling, it is valid even sensible heat transfer is involved [57].

The validity of Eq. (6.63) has been proved by some experimental results, showing the minimum irreversibility at the optimum melting temperature of the PCM [89]. However, it is also found that the equation is only valid for charging process [90]. This is mainly because the entropy generation is decreased if the melting temperature approaches the HTF inlet temperature, during the discharging process. Therefore, it is more appropriate to use the optimization of melting temperature for the complete cycle (charging and discharging), i.e.:

$$T(\text{optimum}) = \frac{1}{2}\left(T_{\mathrm{HTF,in}}\, T_0\right). \tag{6.64}$$

It means that the melting temperature of PCM can be the arithmetic mean of the HTF inlet temperature and the temperature of environment for maximum exergy performance. When selecting PCM with optimum melting temperature, operating conditions, such as the initial temperature for PCM, should also be taken into account. The initial temperature of PCM should be compatible with the inlet temperature of HTF. The inlet temperature of HTF should be as low as possible for charging and as high as possible for discharging. Accordingly, the initial temperature for PCM should also be low for charging and high for discharging. In other words, the PCM should be at subcooled and superheated conditions, respectively, at the beginning of charging and discharging processes [57, 64].

Regarding to HTF in system performance optimization, the inlet temperature of HTF is very crucial in initiating the charging/discharging process. If the temperature of PCM (initial/melting temperature) is fixed, the inlet temperature of HTF depends on the heat source during heat storage and is dependent on the environment temperature during heat recovery. For maximizing the exergy performance, they should be properly controlled.

During charging, a high HTF inlet temperature results in a large difference in temperature between HTF and PCM, leading to a large entropy generation, because the entropy generation is directly proportional to the temperature difference. Therefore, to maintain a minimum entropy generation for a maximum exergy performance, the HTF inlet temperature should be as low as possible, i.e. close to the initial temperature of the PCM. However, the smaller temperature difference means a low heat transfer rate and thus a low charging rate. In some applications, like waste heat recovery, the charging rate should be sufficiently high, because the source availability is usually intermittent [86]. In this case, the selection of HTF inlet temperature should be a tradeoff between the required charging rate and exergy performance. However, for solar thermal applications, high charging rate is not very important, so that a low HTF inlet temperature is allowed and high exergy efficiency can be emphasized. For discharging process, in the beginning, the PCM is at a high temperature, due to stored energy. Similarly, the finite temperature difference between PCM and HTF should be reduced to decrease the entropy generation and exergy degradation. Also, if a high discharging rate is required, a low temperature is not appropriate.

It has also been found that the exergy performance of a PCM TES system is dependent on mass flow rate/velocity of HTF. It is general to use Reynolds (Re) number, the well-known dimensionless number, to evaluate the effect of mass flow rate/velocity of a fluid. High Re number means high mass flow rate/velocity and vice versa. An increase in Re number (mass flow rate/velocity) results in an increase in pressure drop and thus increased entropy generation number. However, extensive studies are required to elucidate the influence of Re number on exergy performance [57].

A TES module usually consists of two compartments separated by a wall which serves as the heat transfer interface. One compartment is loaded with PCM and the other is for HTF to flow through. The volume of the PCM unit is determined by the quantity of PCM to be loaded, which in turn determines the quantity of heat to be stored or recovered. For a given volume, the dimension of the unit can still be adjusted to achieve maximum exergy performance. The dimension of the PCM unit is expected to affect the heat transfer mechanism during the phase transition process. Because the solidification of a PCM is always governed by conduction heat transfer [91], it is not affected by the dimension of the unit. In contrast, the melting process is controlled by natural convection in the liquid PCM, although it is initially driven by conduction. As a consequence, an increase in vertical dimension of the PCM unit means an increase in the natural convection in the liquid PCM, which increases the heat transfer rate. However, a high rate of entropy generation is also expectable, due to the faster interface motion and liquid viscosity caused by the enhanced natural convection.

In case of tube unit for HTF, a decrease in length and an increase in diameter could decrease the pressure drop in the HTF flow. This can reduce the entropy generation due to the pressure drop irreversibility. However, the decrease in length of the tube reduces the heat transfer surface area. It has been reported that entropy generation increases as heat transfer surface area is decreased [85]. It is also found that the number of transfer unit (NTU) which is a measure of heat transfer surface area (proportional to length) affects the entropy generation in two ways [87]. The entropy generation due to heat transfer irreversibility decreases exponentially as the NTU increases. The entropy generation due to pressure drop irreversibility increases linearly as the NTU increases. It means that there is an optimum value for NTU, i.e. heat transfer area or length. Therefore, the length of the tube should selected according to the optimum NTU.

6.4.4.8 Thermoeconomics

The above discussion has shown that the second law techniques are very useful to design thermodynamically efficient TES systems [57]. When a component of a thermal storage system has degradation in performance due to irreversibility, not only its own performance but also those of the remaining components/units of the whole system are affected. It is difficult to identify the effect of a degraded component on the performance of the overall system. It is also necessary to understand how the overall performance of the system is enhanced, if a specific irreversibility is eliminated. These issues are related to thermoeconomics.

Thermoeconomic analysis combines economic and thermodynamic analysis by applying the concept of cost (originally an economic property) to exergy (an energetic property) [92]. It has been acknowledged that exergy is the thermodynamic property most closely related to cost since it accounts for energy quality. Thermoeconomic analysis is based on the representation of a thermal system by using a thermoeconomic model [93]. The productive structure, also known as functional diagram or structural scheme, is a graph with a set of relations defining the interaction among various components and the environment. Each subsystem has one or more entering flows that represent the resources or fuel and exiting flows (i.e., product). Due to irreversibility, in every component, product is less than fuel in cost. Therefore, the amount of resources per unit of exergy increases throughout the system.

The main feature of the thermoeconomics methods is that they propose a cost balance equation applying the exergetic unit cost to the exergy balance equation according to a specific principle [94]. The cost structure of the overall system is mostly influenced by the entropy production rate (i.e., irreversibility) at each component and is dependent on the level of aggregation of the system [95, 96]. The level of aggregation provides a list of the total irreversibility among the

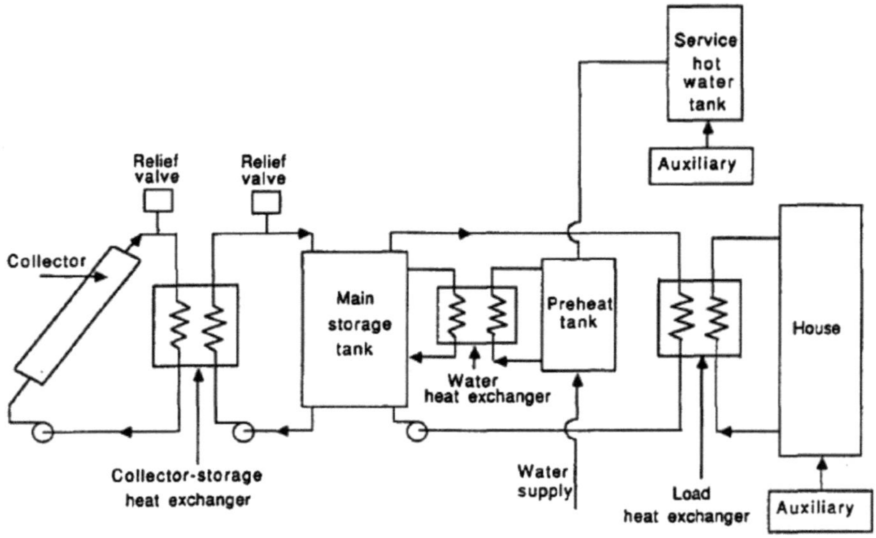

Fig. 6.23 Schematic representation of a solar water-heating system. Reproduced with permission from [100], Copyright @ 1989, Elsevier

components in the system. The selection of the level of aggregation affects the results of the analysis.

For PCM TES systems, the analysis with thermoeconomics is still in infancy [97–99]. In general, the objective of thermoeconomic optimization is to minimize the annualized total cost in owning and operating the systems. The total cost is composed of additive terms that measure the costs associated with the lost exergy and the annualized capital costs of the whole equipment. The aim of thermoeconomic optimization based on exergetic analysis must be directed toward minimizing the exergetic cost of the system. So, the problem of thermoeconomics analysis applied to ES systems can be formulated in certain ways. The results of thermoeconomic analysis of the ES systems are mainly dependent on the storage period required and the defined limits of the system.

6.5 Application Examples

6.5.1 Solar Water-Heating Systems

Solar water heaters are relatively inexpensive and simple to fabricate and maintain [100–102]. Figure 6.23 shows schematic diagram of a solar water-heating system [100]. This system consists of three independently operating subsystems, i.e., the collector loop, the load and the domestic hot water system. Whenever solar energy

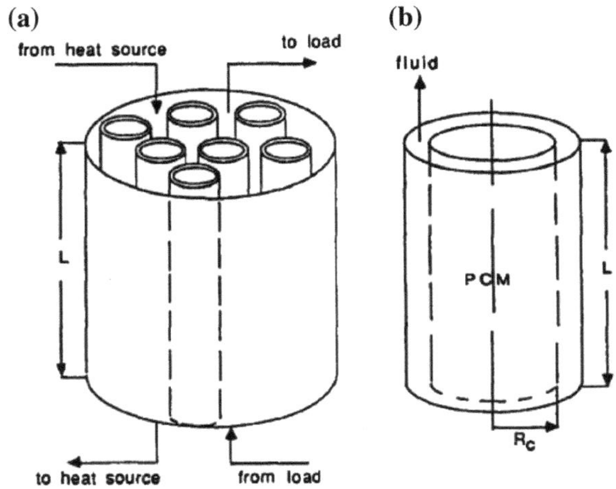

Fig. 6.24 Cylinder-packed storage vessel (**a**) and the model for one cylinder (**b**). Reproduced with permission from [100], Copyright @ 1989, Elsevier

is available, it is collected and transferred to the storage unit. Whenever a space heating load is present, it is met by the storage unit and the auxiliary energy source. The domestic hot water system takes energy from the storage unit when it is necessary. Relief valves are used for dumping excess energy.

Figure 6.24 shows configuration the storage vessel, which consists of a vessel packed in the vertical direction with cylindrical tubes. The storage material is inside the tubes and the HTF flows parallel to it. To have as much PCM as possible, the vessel has to be closely packed. The void fraction, i.e., the ratio between the fluid volume and the storage tank volume, is designed to be about 0.3. The inside volume and inside surface area of the tank are V_t and A_t. The number of cylinders inside the vessel is N_c. The radius of the cylinder tubes is R_e, the length of the cylinders as well as the inside height of the vessel is L. R_c is set to 0.02 m and L is 2.0 m, so that $R_c/L = 0.01$, which is small enough to minimize radial heat conduction in the storage material.

A simple solar water-heating system consists of a rectangular galvanized iron tank of 9 cm depth and 1 m^2 exposed surface area, filled with water [101]. The top surface of the tank, which works as the absorber plate, is coated with ordinary black paint. A glass cover 3 mm thick is placed over the black surface of the tank with an air gap of 2.5 cm. A layer of PCM is filled at the bottom. During the sunshine hours, the water is heated up, which in turn transfers the heat to the PCM below it. The PCM collects energy in the form of latent heat and melts. During off sunshine hours, the hot water is withdrawn and is substituted by cold water, which gains energy from the PCM. The energy is released by the PCM when it transfers from liquid to solid phase. This type of system is not applicable in large scale due to the poor heat transfer between the PCM and water.

There has reported a cylindrical storage unit in the closed loop with a flat plate collector, with charging and discharging modes, in which paraffin wax ($p - 116$) and stearic acid are used as PCMs [102]. Theoretical study of the LHS system

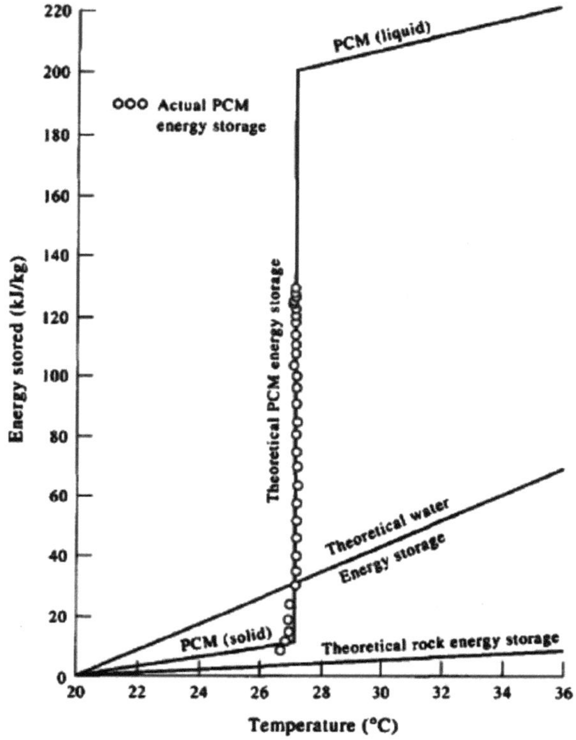

Fig. 6.25 Performances of the systems with storage media of PCM, water, and rock storage. Reproduced with permission from [103], Copyright @ 1995, Elsevier

coupled with the solar flat plate collector shows that the basic advantage of this type of storage is the ability to collect solar heat energy at constant temperature. This advantage is not available in SHS system, because the withdrawal of heat is necessarily accompanied by a decrease in temperature in this case. The added and envisaged advantage of the LHS system is the relatively small storage volume (only 25 % of the sensible case).

$CaCl_2{\cdot}6H_2O$ has also been used as PCM for solar water-heating system [103]. Both experimental and theoretical studies have been conducted to determine the performance of such phase change ES materials for solar water-heating systems. The performances of the PCM, water-, and rock-based storage systems have been compared with results shown in Fig. 6.25 [103]. Clearly, the performance of the system with PCM as storage media is much better than other systems.

An integrated collector module that uses water/organic liquid (oil) heat exchanger instead of water/PCM surface heat exchanger is shown schematically in Fig. 6.26 [104]. The system consists of a thermally insulated plastic container, containing a layer of stationary heat transfer liquid (SHTL), floating over a layer of an immiscible salt hydrate PCM. The density of the SHTL is lower than that of the liquid salt hydrate. A finned heat exchanger located in the SHTL is used to heat the cold water during the discharging process. The SHTL is covered with a metal plate coated with a selective absorbing film and with a transparent insulating lid. As the

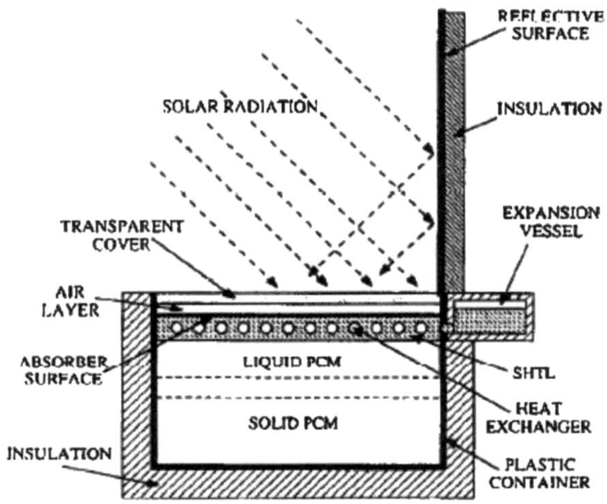

Fig. 6.26 Schematic presentation of the integrated solar collector storage system based on a phase change material (ICS-PCM). Reproduced with permission from [104], Copyright @ 1995, Elsevier

PCM melts, its density decreases, i.e., its volume increases. The SHTL displaced as a consequence of this expansion is stored in a small interconnecting chamber, designated the expansion vessel. A removable cover made of a material with a low thermal conductivity and coated on one side with a reflective material is used as an optional extra to increase solar energy absorption during the daytime. It can also be used as additional cover to reduce the heat loss during the night.

The concept and the mathematical model in the stage of the charging process have been compared with experimental results. A schematic cross-section view of the components of the experimental model is shown in Fig. 6.27. The system consists of a 10 mm thick Plexiglas container, 0.3 × 0.3 m in cross-section and 0.131 m in height. The sides of the container are thermally insulated with a 40 mm thick layer of polyurethane. The container is filled with 13.4 kg of a salt hydrate PCM, composed of a eutectic mixture (in wt%) of 48.0 CaCl, 4.5 KCl, 0.4 NaCl, and 47.1 H_2O, with the addition of 1 wt% $BaCl_2{\cdot}2H_2O$ as a nucleating agent. The mixed PCM has a phase transition temperature interval of 27–29 °C, a latent heat of 164.5 kJ kg^{-1} and solid and liquid densities of 1,610 and 1490 kg m^{-3}, respectively. Paztherm 22 oil (1460 cm^3), used as the SHTL, is poured onto the PCM in the solid state.

A compact PCM solar collector based on LHS has been reported [105]. As shown in Fig. 6.28, in this collector, the absorber plate-container unit has the function of both absorbing the solar energy and storing the PCM. The solar energy is stored in paraffin wax and is discharged to cold water flowing in the pipes located inside the wax. The collector has an effective area of 1 m^2 and its total volume is divided into five sectors. One of the collector sectors, with an apparatus-absorber effective area of 0.2 m^2, has been simulated.

Outdoor experiments confirm the applicability of the compact solar collector for water-heating. Various experimental data have been available, including time-wise temperatures of the PCM and the solar intensity during the processes of charging and discharging, effect of water flow rate(8.3–21.7 kg h^{-1}) on the useful

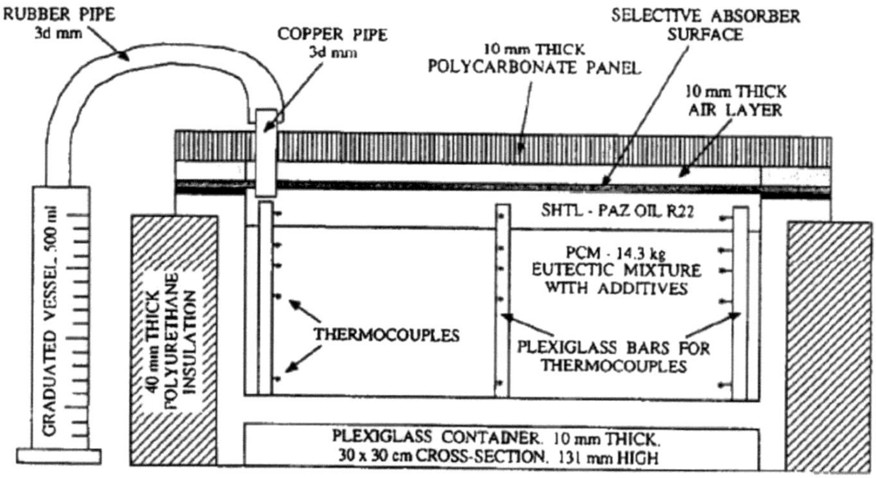

Fig. 6.27 Schematic cross-section of the experimental setup for the integrated solar collector storage system. Reproduced with permission from [104], Copyright @ 1995, Elsevier

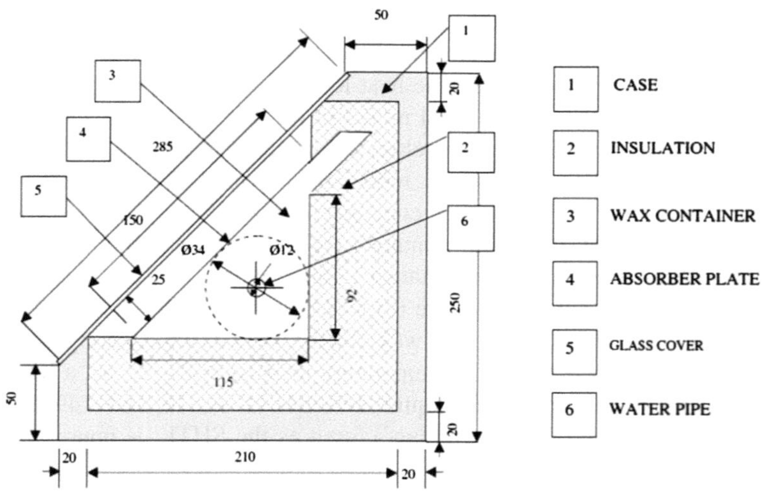

Fig. 6.28 Cross-sectional view of the experimental system. Reproduced with permission from [105], Copyright @ 2006, Elsevier

heat gain, heat transfer coefficients during the charging process, and propagation of the melting and freezing front during the charging and discharging processes. It is found that during the charging process, the average heat transfer coefficient increases sharply with increasing molten layer thickness, since the natural convection becomes strong. In the discharge process, the useful heat gain increases as the water mass flow rate is increased.

A solar pilot plant can be used to test the PCM behavior in real conditions. It can work continuously with a solar system and can also work with an electrical

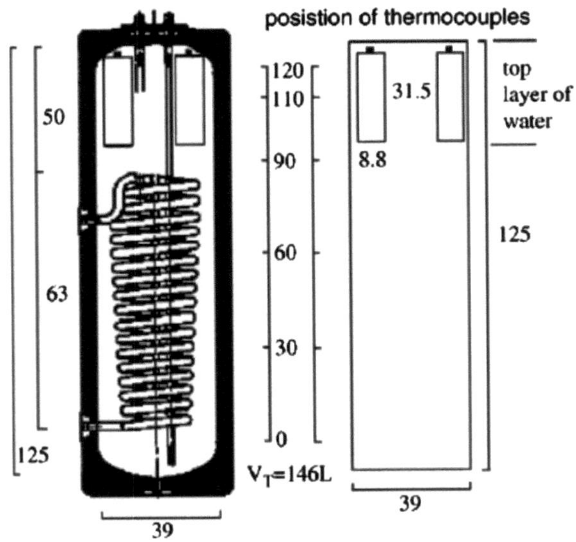

Fig. 6.29 Configuration of the water tank and PCM module. Reproduced with permission from [106], Copyright @ 2006, Elsevier

heater [106]. The PCM modules are cylinders at the top of water tanks, as shown in Fig. 6.29. Experiments with two, four, and six PCM modules have been carried out in the real installation. 90 vol.% of sodium acetate trihydrate and 10 vol.% graphite are mixed to form the PCM. Such a PCM module in water tanks for domestic hot water supply is a very promising technology. It can supply hot water for longer periods of time even without exterior energy supply.

Another example is an open-loop passive solar water-heating system combined with sodium thiosulfate pentahydrate as the PCM [107]. The experimental setup is an open-loop conventional passive solar energy system with natural circulation, generally used to provide domestic hot water requirements in buildings, consisting of solar collectors, hot water and cold water tanks, measurement, and data logger systems, as shown in Fig. 6.30. Flat parallel aluminum solar collectors with fixed-plate of $1.94 \times 0.94 \times 0.10\ m^3$ have single cover glass and black-painted absorber plate. Solar collectors faced to the south also have a net absorption area of $1.65\ m^2$ for each one and the tilt angle of 30° with horizontal plane. Insulated cold water feeding tank has a volume of 18 l. It has a buoyancy ball to prevent inlet of cold water by controlling maximum water level in the heat storage tank when tank is filled. All of the pipes and fittings used in the experimental system are insulated.

Figure 6.31 shows detailed cross-sectional view of a well-insulated cylindrical galvanized steel heat storage tank having a volume of 190 l, which is loaded with PCM. The insulation material is glass wool. The polyethylene bottles of 0.44 l each, containing PCM mass of 0.7347 kg, are filled into the tank as three rows. The volume of bottles used in the heat storage tank has a relatively small value of 0.44 l to minimize the problems detected during the melting and solidification of the PCM. The more the bottles, the larger the heat transfer surface for the PCM will be. The volumes of the PCM and water in the heat storage tank are 107.8 and

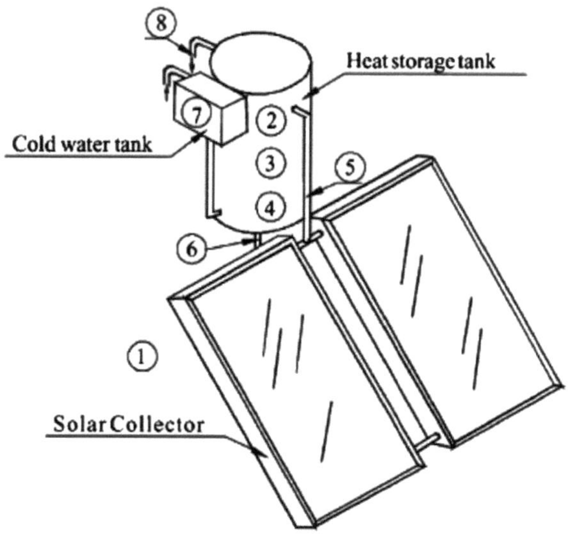

Fig. 6.30 Side view of the experimental open-loop passive solar water-heating system. Temperature measurement points: 1 – ambient temperature, 2 – water temperature at the upper point of heat storage tank, 3 – water temperature at the midpoint of heat storage tank, 4 – water temperature at the lower point of heat storage tank, 5 – collector outlet temperature, 6 – collector inlet temperature, 7 – water temperature at the midpoint of cold water tank and 8 – outlet temperature of the water from heat storage tank. Reproduced with permission from [107], Copyright @ 2005, Elsevier

Fig. 6.31 Detailed cross-sectional view of the cylindrical heat storage tank combined with PCM. Reproduced with permission from [107], Copyright @ 2005, Elsevier

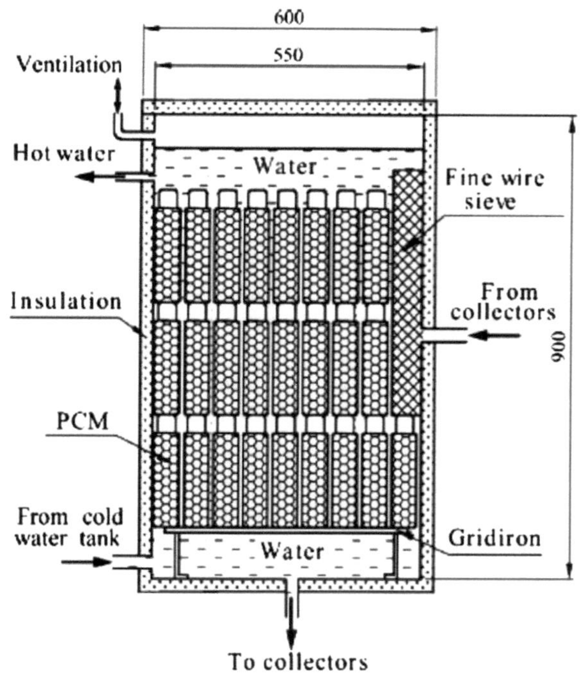

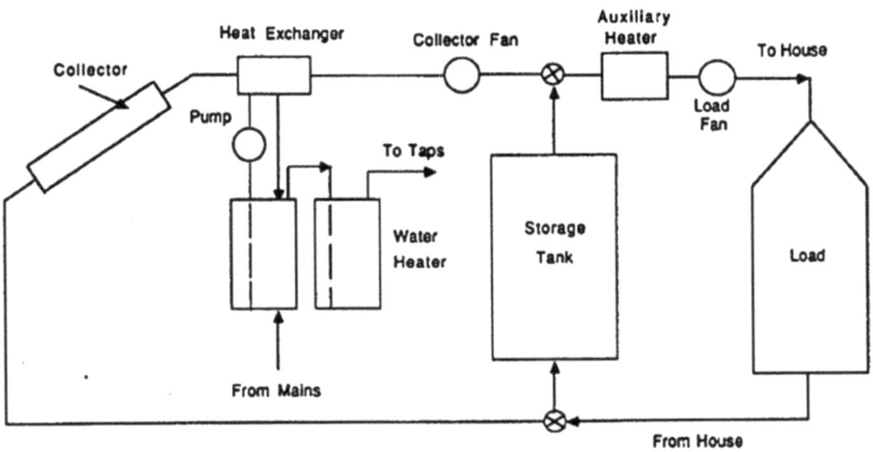

Fig. 6.32 Schematic representation of a solar air heating system. Reproduced with permission from [100], Copyright @ 1989, Elsevier

82.2 l, respectively. The total mass of PCM is about 180 kg. The bottles do not let the water flow in horizontal direction due to their close location. Therefore, the heat storage tank has cavities in the upper and lower parts to allow a more homogenous heat distribution horizontally in the heat storage tank due to the water flow through the vertical cavities between the bottles, although the upper part of the heat storage tank had a thermal stratification of the higher level.

Heat storage performances of the solar water-heating system, with various salt hydrates-PCMs, such as zinc nitrate hexahydrate, disodium hydrogen phosphate do decahydrate, calcium chloride hexahydrate, and sodium sulfate decahydrate, have been examined theoretically by using meteorological data and thermophysical properties of PCMs. The storage time of hot water, the produced hot water mass and total heat accumulated in the solar water-heating system, having the heat storage tank combined with PCM, are about 2.59–3.45 times of that of the conventional solar water-heating system. Among the hydrated salts studied, disodium hydrogen phosphate do decahydrate and sodium sulfate decahydrate demonstrate the highest solar TES performance as PCMs.

6.5.2 Solar Air Heating Systems

Figure 6.32 shows a standard solar air space and domestic water system, which has three modes of operation [100]. The first mode occurs when solar energy is available for collection and the space heating load is nonzero. During this mode, the fluid is circulated between the collectors and the load, while the storage unit is isolated. The second mode occurs when solar energy is available for collection and

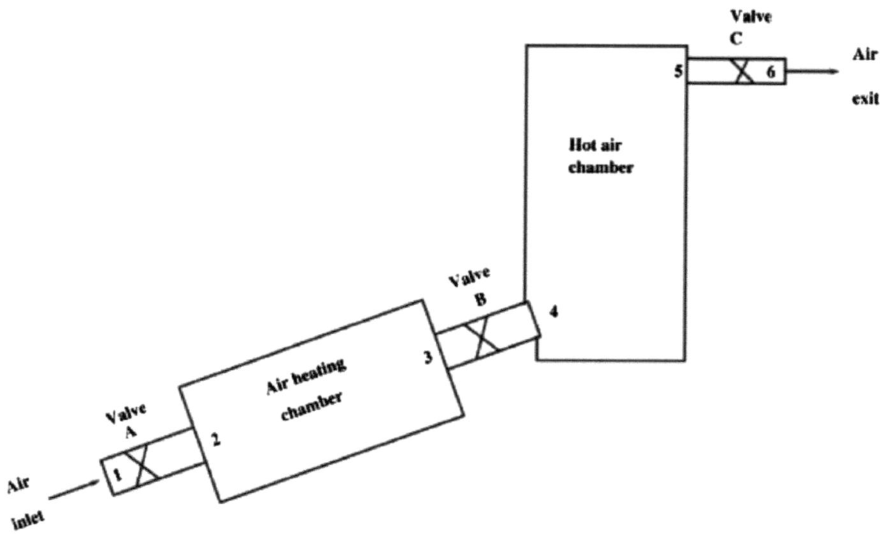

Fig. 6.33 Schematic view of a natural circulation air heater. The hot air (working chamber) holds the crops in a cabinet-type crop dryer and the eggs in an egg incubator. Reproduced with permission from [111], Copyright @ 2002, Elsevier

the space heating load is zero. In this case, air is circulated between the collectors and the storage unit. The final mode takes place when no solar energy is available for collection and there is a space heating load. In this case, air is circulated between the storage unit and the load. The system operation is controlled by a series of thermostats, fans, and flow diverters. Air flow to the space heating load and auxiliary heat is governed by a two-stage thermostat which monitors room temperature. It commands the first stage (solar source) heating when the room temperature drops below a certain value, while the second stage (auxiliary source) heating is triggered when the room temperature further drops below an even lower level.

The performances of similar air-based solar heating systems utilizing phase change ES unit have been well evaluated [108, 109]. The effects of the latent heat and melting temperature of PCMs on the thermal performances of air-based solar heating systems have been determined, by developing empirical model of significant phase change energy storage (PCES) units. It has been found that the PCM should be selected on the basis of melting point rather than its latent heat and the air-based system utilizing sodium sulfate decahydrate as a storage medium requires one-fourth the storage volume of a pebble bed and one half the storage volume of a water tank. Theoretical studies have been conducted to compare the performances of phase change and SHS for air and water-based solar heating systems [110]. With sodium sulfate decahydrate and paraffins as PCMs, similar results have been obtained as in Ref. [109].

Figure 6.33 shows the schematic diagram of a natural convection solar air heater [111]. It consists of a flat plate solar collector integrated with the heat

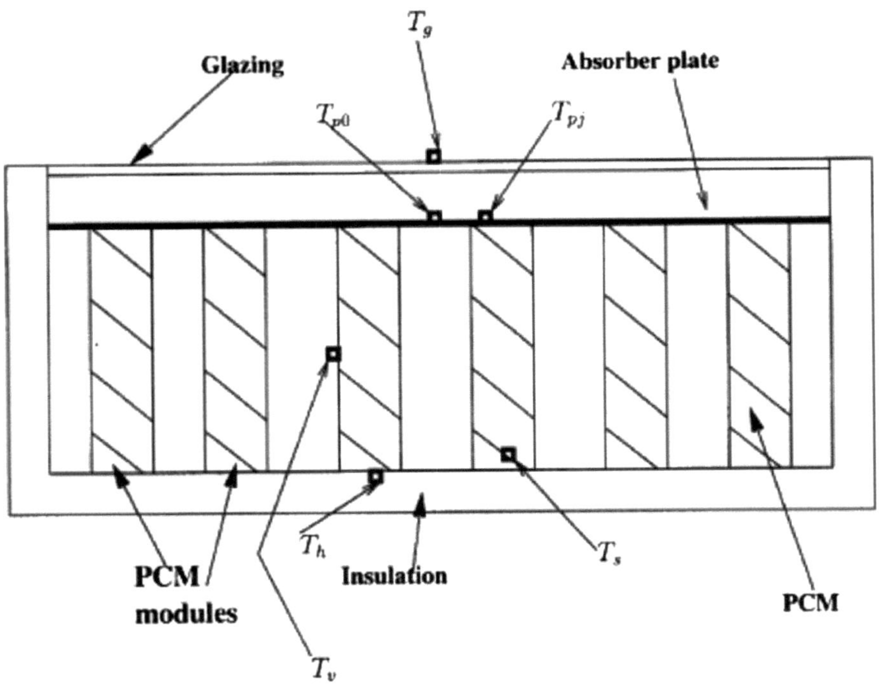

Fig. 6.34 Cross-sectional view of the collector assembly. The heated air flows perpendicular to the page and the little boxes indicate the approximate locations of the thermocouples. Reproduced with permission from [111], Copyright @ 2002, Elsevier

storage system, with a paraffin type PCM. The PCM is prepared in modules, with each module being made of thin rectangular blocks of the PCM encapsulated by the rectangular walls of a box-like structure which behaves like thin fins. The encapsulating box is made of a material with good thermal conductivity, which can be divided into a number of compartments with identical dimensions. A number of such modules are equispaced across the collector, as shown in Fig. 6.34 [111]. The space between each module pair serves as an air heater, with the heaters being connected to common air inlet and discharge header manifolds. The daytime performance of the system under no-load conditions has been tested under natural environmental conditions, with ambient temperature variations in the range of 19–41 °C and daily global irradiation in the range of 4.9–19.9 MJ m^{-2}. Peak temperature rise of the heated air is about 15 K, while peak cumulative useful efficiency is about 50 %. The system is suitable for use as a solar cabinet crop dryer for aromatic herbs, medicinal plants, and other crops, which do not require direct exposure to sunlight.

A hybrid heating system combined with thermal storage by shape-stabilized phase change material (SSPCM) plates has been studied numerically [112]. A photo and an SEM image of the SSPCM are shown in Fig. 6.35. A direct gain

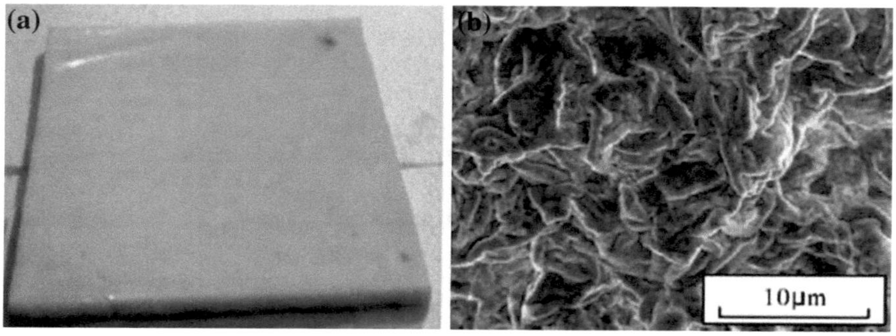

Fig. 6.35 A photograph of the shape-stabilized PCM plate (**a**) SEM image revealing its microstructure (**b**). Reproduced with permission from [112], Copyright @ 2007, Elsevier

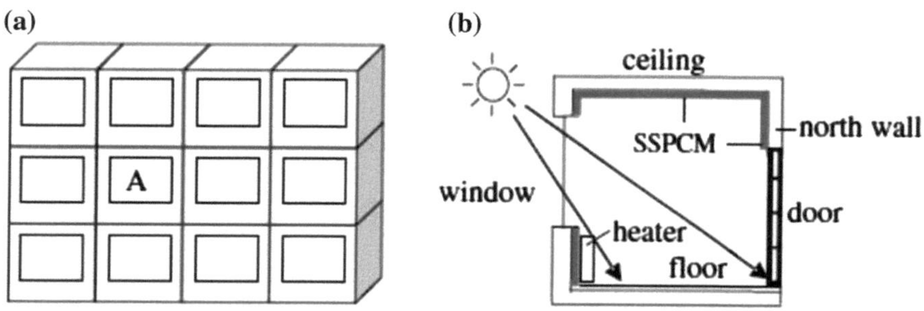

Fig. 6.36 Schematic of the simulated room. **a** Location of the simulated room A in the building and **b** profile of the room A with the SSPCM. Reproduced with permission from [112], Copyright @ 2007, Elsevier

passive solar house, with the SSPCM plates as inner linings of walls and the ceiling, has been considered. A typical south-facing middle room (room A shown in Fig. 6.36) in a multi-layer building in Beijing, China, is considered as the model room for analysis, which has only one exterior wall (the south wall) and others are all interior envelopes. Unsteady simulation is performed by using a verified enthalpy model, with a time period covering the winter heating-season.

Additional heat supply is employed during load hours at late night and early morning or during the whole day necessary to keep the minimum indoor air temperature above 18 °C. The results indicate that, due to the thermal storage effect of the SSPCM plates, the indoor thermal comfort level has been improved, corresponding to saves of about 47 % of normal- and-peak-hour energy use and 12 % of total energy consumption in winter. This hybrid heating system can level the electrical load for power plants and would provide significant economic benefits in areas where night and day electricity tariff policy is used.

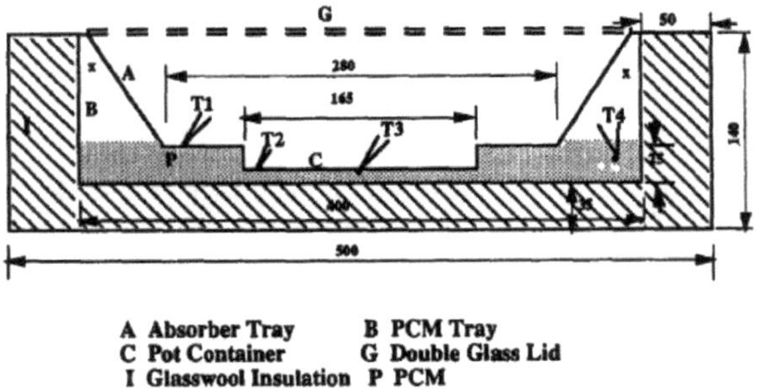

Fig. 6.37 Schematic diagram of the box type solar cooker with PCM TES. Reproduced with permission from [113], Copyright @ 1997, Elsevier

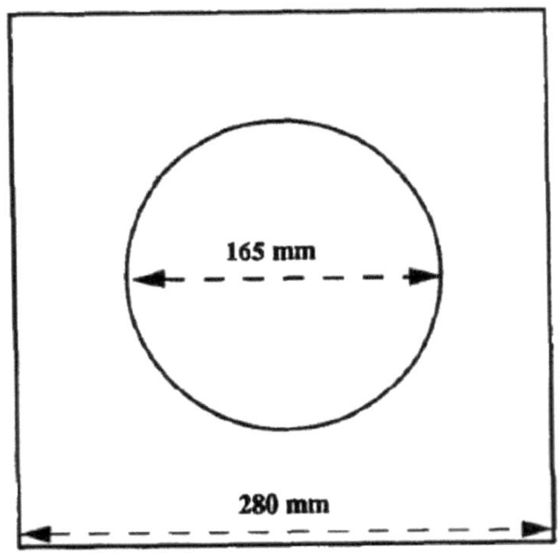

Fig. 6.38 Top view of the bottom of the absorbing plate. Reproduced with permission from [113], Copyright @ 1997, Elsevier

6.5.3 Solar Cookers

One of the major uses of solar energy is in cooking using different types of solar cookers [113–115]. Use of these solar cookers is limited, as cooking of food is not possible in the evening. If storage of solar energy is provided in a solar cooker, than the utility and reliability of these solar cookers would increase. Few studies have been conducted with the LHS materials in a box type solar cooker to cook the food in the late evening.

Figure 6.37 shows a box type of solar cooker for one vessel having a PCM to store the solar energy [113]. The cooker consists of an Al absorbing tray "A," with dimensions 0.28 × 0.28 m at the bottom, 0.40 × 0.40 m at the top, for the double glass lid. The vertical depth of the tray is 0.08 m. The thickness of the Al sheet is 0.006 m. At the center of the absorbing plate, a cylindrical container of 0.165 m in diameter and 0.02 m in depth has been welded, "C" in Fig. 6.37. The cooking pot is kept tightly in it. The top view of the bottom of the absorbing plate is shown in Fig. 6.38 [113]. This container provides a better heat transfer from the absorbing plate and PCM. Moreover, Al fins are also provided at the inner side of the tray and cylindrical container. The outer tray "B" with size of 0.40 × 0.40 × 0.108 m is also made of Al sheet. The distance between tray 'A' and tray 'B' is 0.025 m on the bottom side. Tray "B" is filled with 3.5 kg commercial grade stearic acid as PCM. The PCM is kept in good contact with the bottom side of tray "A." These two trays are encased in an Al box with dimensions of 0.5 × 0.5 × 0.14 m. The space between tray "B" and the casing is filled with glass wool to provide thermal insulation to the bottom and sides of the solar cooker.

The absorbing tray is provided with a double glass lid (0.4 × 0.4 m) hinged to one side of the casing at the top, which is similar to the case in a standard solar cooker. A flat glass mirror is fixed in an Al cover to serve as a booster and as a cover for the glass lid when the cooker is not exposed to solar radiation. The Al cooking pot, 0.16 m in diameter and 0.06 m in height, with a cover, is kept in the pot container "C." The Al tray "A," cooking pot and its cover are painted with a dull black paint on the outside.

Feasibility of the solar cooker has been experimentally demonstrated. It can be used to cook foods even in the evening. The plate temperature maintained by the cooker is nearly constant even in the late evening. The function of the solar cooker is comparable with the conventional solar cookers.

A schematic diagram of an evacuated tube solar collector (ETSC) for cooker, with a PCM storage unit, reported recently, is shown in Fig. 6.39 [116]. It consists of an ETSC, a closed loop pumping line containing water as the HTF, a PCM storage unit, cooking unit, pump, relief valve, flow meter, and a stainless steel tube heat exchanger. Two panels, each containing six collector tubes, are used. The absorbing plate of the ETSC is tilted at 20° facing due south.

The other part of the system is the PCM storage unit. This unit has two hollow concentric aluminum cylinders, with inner and outer diameters of 304 and 441 mm, respectively, and depth of 420 mm and thickness of 9 mm. The space between the cylinders is filled with 45 kg erythritol as the PCM. As the PCMs have positive volumetric expansion on melting, the storage unit is not filled completely and space should be left for volumetric expansion during melting of the PCM. The dimensions of the cooking vessel are 297 mm in diameter and 300 mm in height, which can be easily inserted inside the PCM storage unit. Sectional and vertical views are shown in Fig. 6.40a, b.

The system is able to cook twice, noon and evening, in a single day during summer months in Japan. Noon cooking is not affected the evening cooking. The evening cooking using the PCM heat storage is faster than noon cooking. The PCM did not melt in January (i.e., winter) in Japan. In summer, PCM temperatures can be higher than 110 °C during evening cooking. Therefore, erythritol is a promising PCM for solar cooking. The highest temperature of such solar cookers is 130 °C without tracking and thus allows cooking in the shade and also in a conventional kitchen during non-sunshine hours or in the evening [116].

Figure 6.41 shows a double reflector hot box solar cooker with a transparent insulation material (TIM) [117]. A 40 mm thick honeycomb made of polycarbonate capillaries is encapsulated between two glazing sheets of the cooker to minimize convective losses from the window, so that two meals can be prepared

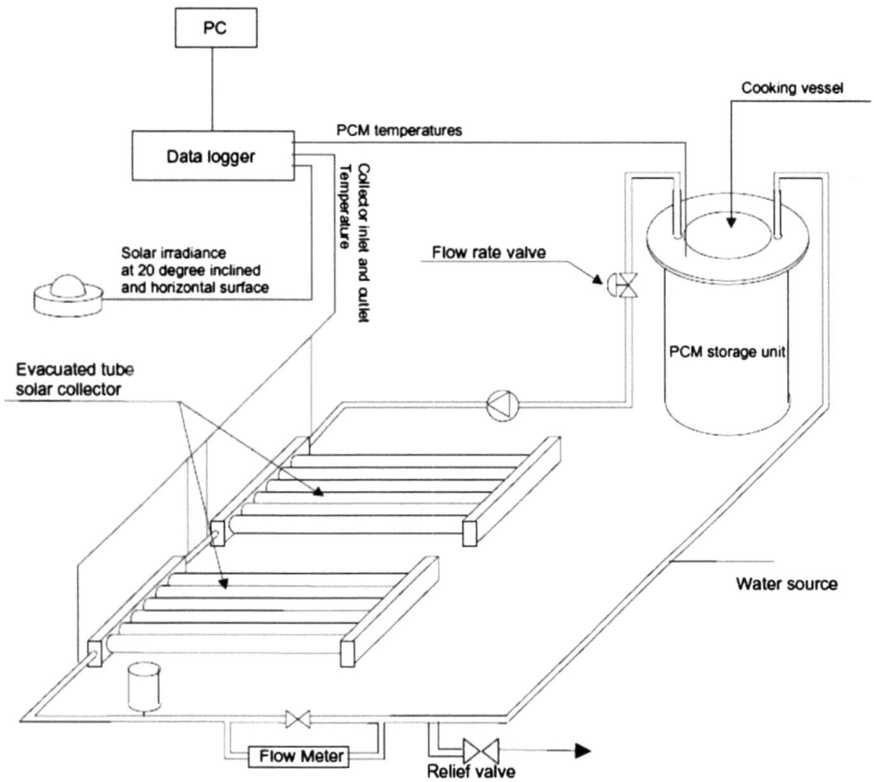

Fig. 6.39 Outline of the prototype solar cooker based on evacuated tube solar collector with PCM storage unit. Reproduced with permission from [116], Copyright @ 2005, Elsevier

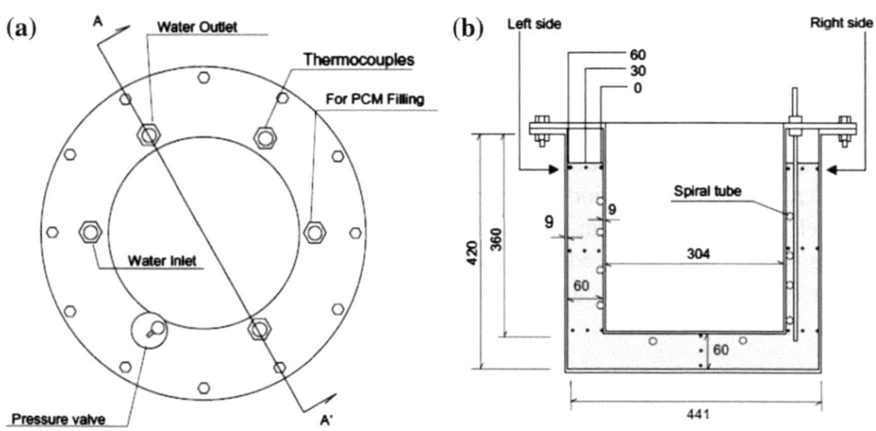

Fig. 6.40 Sectional view of the PCM storage unit (a) and A–A vertical section of the PCM storage unit (b). Reproduced with permission from [116], Copyright @ 2005, Elsevier

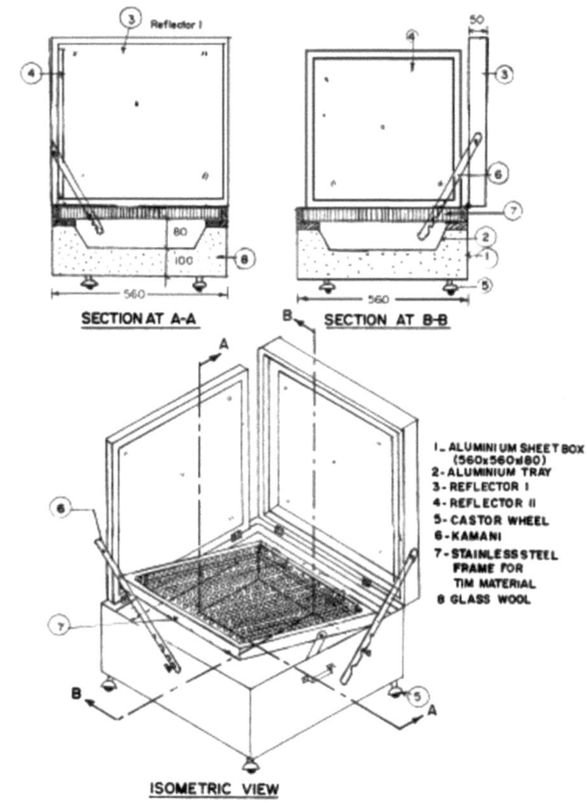

Fig. 6.41 Schematic diagrams of a double reflector TIM (transparent insulation material) solar cooker. Reproduced with permission from [117], Copyright @ 2001, Elsevier

even during an extremely cold but sunny day, which is not possible in a hot box solar cooker without TIM. The use of one more reflector can make it unnecessary toward sun for 3 h, so that cooking operations can be performed unattended, as compared to a hot box solar cooker where tracking ahead of the sun is required every hour. The efficiencies are 30.5 and 24.5 % for cookers with and without a TIM, respectively, during the winter season in Jodhpur. The energy saving by use of a solar cooker with TIM has been estimated to be 1485.0 MJ of fuel equivalent per year. The payback period varies between 1.66 and 4.23 years depending upon the fuel it replaces, with an increasing order of electricity, firewood, coal, LPG, and kerosene. The estimated life is about 15 years. Therefore, the kind of solar cooker is economical and can be used for energy savings.

6.5.4 Solar Greenhouses

PCMs have also been used in greenhouses by storing the solar energy for curing, drying process, and plant production [118]. Figure 6.42 shows a solar ES unit in

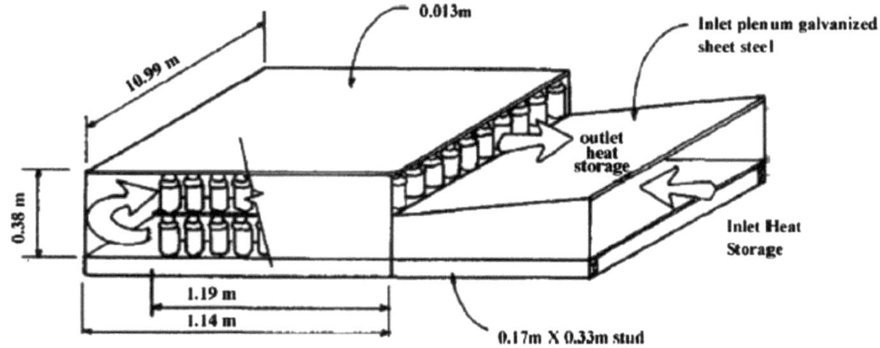

Fig. 6.42 ES unit inside the greenhouse. Reproduced with permission from [119], Copyright @ 2009, Elsevier

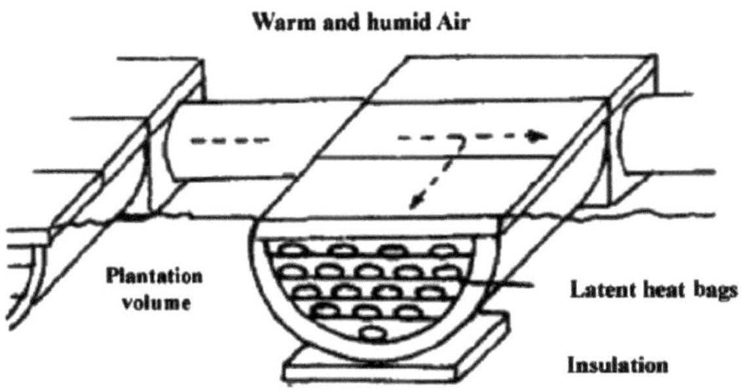

Fig. 6.43 Underground tunnel with PCM solar ES systems. Reproduced with permission from [120], Copyright @ 2009, Elsevier

greenhouse, in which 1,650 kg of $CaCl_2 \cdot 6H_2O$ in aerosol cans each weighing 0.74 kg is used to investigate ES possibilities both inside and outside a 36 m^2 ground area greenhouse covered with tedlar coated fiber-glass [119]. The PCM cans are placed in a store with 22.86 mm spacing and two stores containing different amounts of PCM are used, where one is inside and the other is outside the greenhouse. While the ES unit inside the greenhouse collect swarm air from the ridge of the greenhouse during the daytime, the direction of air flow is reversed for the energy releasing process at night. Therefore, the temperature inside the greenhouse can be maintained at a desired level.

There has appeared a LHS system with two different stacking configurations and air baffling as an integrated part of the greenhouse solar system, as shown in Fig. 6.43 [120]. Commercial cylindrical storage rods are used as the primary storage elements. This type of storage systems has significantly higher compact storage capacity than the rock or water storage systems. The solar heat available inside the greenhouse is transferred and stored by recycling the air through an underground of flat heat exchangers filled with PCMs. In comparison, the PCM greenhouse can save propane gas by 80 % (Fig. 6.44).

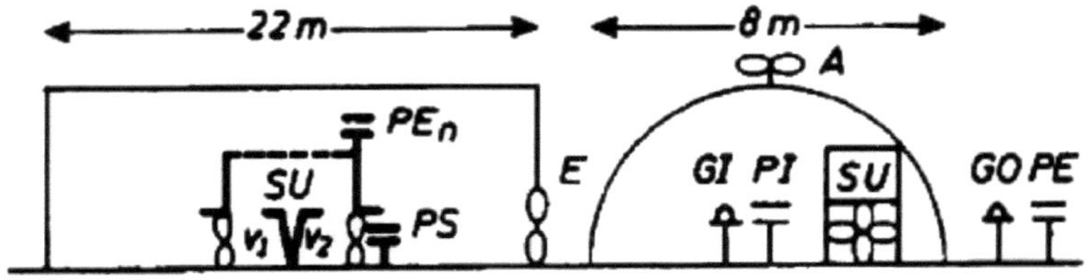

Fig. 6.44 Sketch of the greenhouse with its PCM storage system: *SU* storage unit, V_1 and V_2 storage unit fans, *E* greenhouse ventilator. Measurement devices: *PEn*, *PS*, *PI* and *PE* are psychrometers at the entrance and exit of the storage unit, inside and outside the greenhouse, *A* anemometer, *GO* and *GI* outside and inside pyrheliometer. Reproduced with permission from [121], Copyright @ 1990, Elsevier

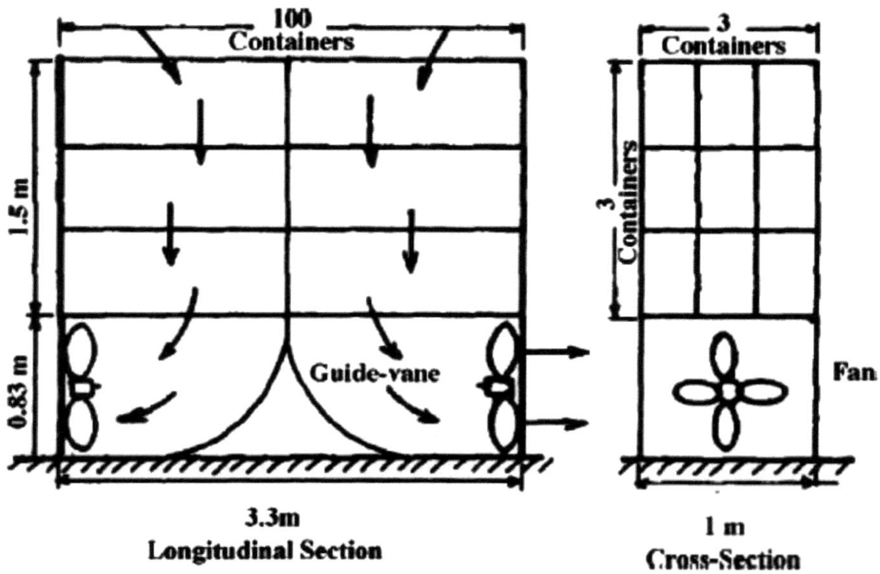

Fig. 6.45 Underground tunnel with PCM solar ES systems. Reproduced with permission from [121], Copyright @ 1990, Elsevier

Boulard et al. [121] $CaCl_2 \cdot 6H_2O$ melting at 21 °C was utilized in a greenhouse with 176 m^2 ground area, double polycarbonate-cover and forced ventilation (Fig. 6.45). Thermostat settings for night and day in February and March were, respectively, 12 and 14, and 22 and 26 °C. Air velocity in the greenhouse had an order of magnitude 1 m s^{-1}. It was calculated that while 0.260 kW h m^{-2} energy was stored during the daytime, 0.360 kW h m^{-2} was released at night in the greenhouse.

When the outside air temperatures were 3.8 and 6.6 °C, respectively, in February and March, inside air temperatures for the same periods of 10.9 and

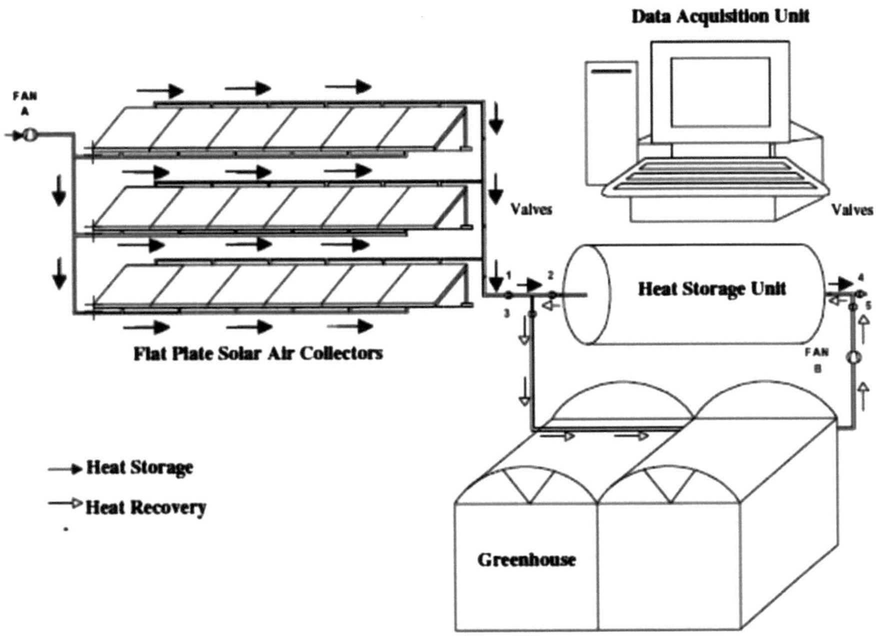

Fig. 6.46 The arrangement of the heat storage and greenhouse heating system. Reproduced with permission from [122], Copyright @ 2005, Elsevier

13.5 °C were obtained. With this method, instead of what would be 7.2 1 m^{-2} fuel requirement, 40 % of heating load was supplied and an overall 30 % in energy saving was achieved.

Figure 6.46 shows the schematic diagram of a PCM TES system for heating a greenhouse with a floor area of 180 m^2 [122, 123]. The system consists of five key units: (i) flat plate solar air collectors as heat collection unit, (ii) PCM storage unit, (iii) experimental greenhouse, (iv) heat transfer unit, and (v) data acquisition unit. The external heat collection unit consists of south-facing solar air heaters with an area of 27 m^2, which are all mounted at a tilt angle of 55°. The PCM storage unit is filled with 6,000 kg of paraffin, equivalent to 33.33 kg of PCM per square meter, according to the greenhouse ground surface area (Fig. 6.47).

The PCM unit is a cylindrical steel tank, with dimensions and the placement of the heat exchanger in the PCM unit shown in Fig. 6.48. The diameter and the total volume of the steel tank used as the LHS unit are 1.7 m and 11.6 m^3, respectively. The whole surface of the PCM unit is insulated with 0.05 m glass wool and 0.001 m steel plate, respectively. It is placed horizontally on the ground surface at a distance of 2 m from the experimental greenhouse. The PCM unit volume per square meter of the greenhouse ground surface area is 0.066 m^3 m^{-2}, while the storage volume per square meter of the solar air collector is about 0.44 m^3 m^{-2}. The unit is attached to the experimental greenhouse by using polyethylene (PE) pipes. Solar energy, collected by the solar air collectors, is transferred to the PCM

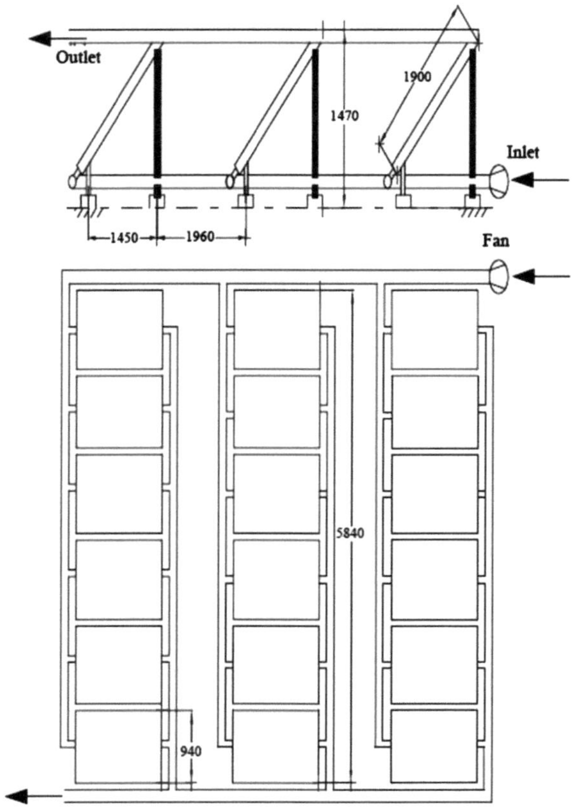

Fig. 6.47 External heat collection unit with all dimensions in mm. Reproduced with permission from [123], Copyright @ 2003, Elsevier

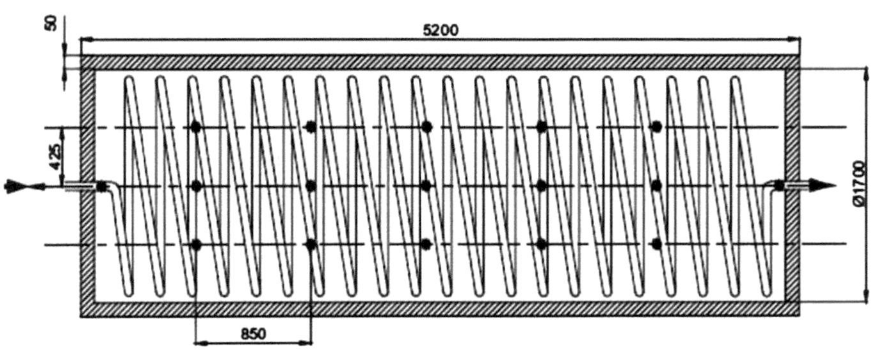

Fig. 6.48 The LHS unit and sensor placement with all dimensions in mm. Reproduced with permission from [122], Copyright @ 2005, Elsevier

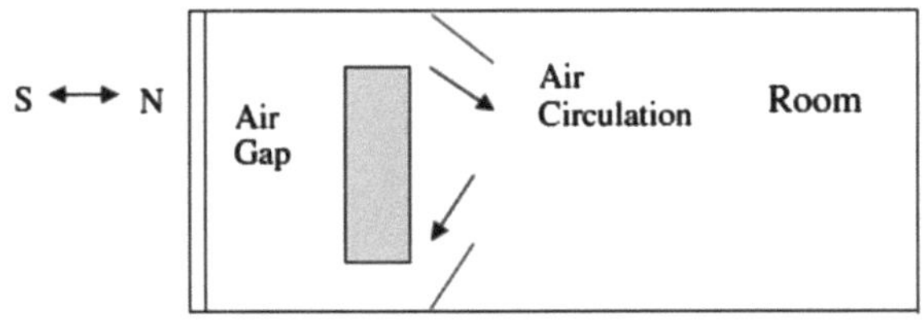

Fig. 6.49 Schematic diagram of a PCM Trombe wall. Reproduced with permission from [127], Copyright @ 2005, Elsevier

unit by circulating air through the PE pipes. Perforated PE pipes as the heat exchanger are installed in the PCM unit, to ensure direct contact between the HTF and the PCM in the unit. The total length and diameter of the perforated PE pipe are 97 and 0.1 m, respectively.

The efficiency of the system has been evaluated by using energy and exergy analysis. The rate of heat transferred in the PMC unit ranges from 1.22 to 2.63 kW, whereas the rate of heat stored in it is 0.65–2.1 kW. The average daily rates of thermal exergy transferred and stored in the LHS unit are 111.2 W and 79.9 W, respectively. During the experimental period, the average net energy and exergy efficiencies are 40.4 and 4.2 %, respectively. The effect of the difference in temperatures of the HTF at the inlet and outlet of the PCM unit on the computed values of the energy and exergy efficiency has also been evaluated during the charging period.

6.5.5 Solar Buildings

PCMs have been considered for thermal storage for applications in buildings for more than 30 years. The implementation of PCM thermal storage systems can be in Trombe wall, wallboards, shutters, under-floor heating systems, and ceiling boards, as a part of the buildings for purpose of heating or cooling. Development and testing of prototypes of PCM wallboard and concrete systems have been carried out, to enhance the TES capacity of standard gypsum wallboard and concrete blocks, with particular interest in peak load shifting and solar energy utilization.

PCM Trombe wall is one of such applications [124–126]. Figure 6.49 shows the schematic diagram of a PCM Trombe wall in a building [127]. PCMs have been employed masonry Trombe walls. Theoretical evaluation and experimental testing have been conducted to investigate the reliability and feasibility of PCMs as a Trombe wall. Obviously, for a given amount of heat storage, a PCM unit requires much less space and lighter weight than water walls or normal mass Trombe walls. This makes it much convenient for applications in retrofit

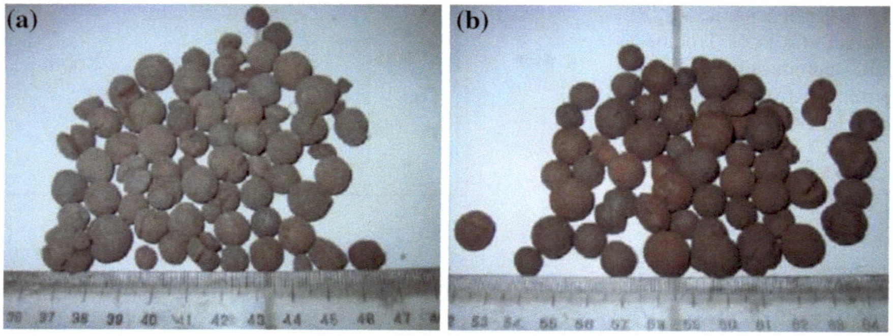

Fig. 6.50 Porous aggregates before (**a**) and after (**b**) absorbing PCM. Reproduced with permission from [137], Copyright @ 2004, Elsevier

Fig. 6.51 Schematic drawing of the vacuum impregnation setup to make PCM concrete. Reproduced with permission from [137], Copyright @ 2004, Elsevier

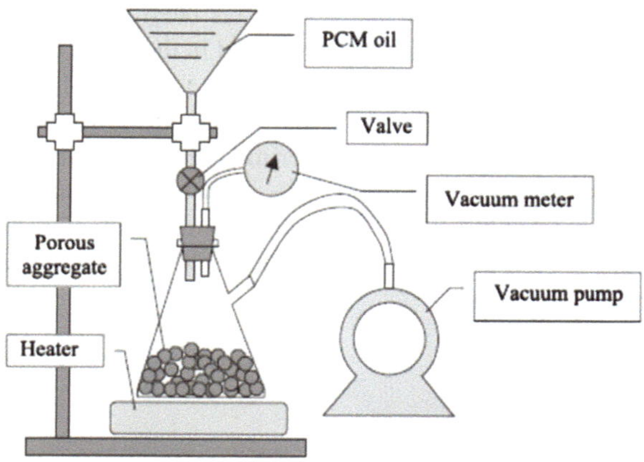

applications of buildings. Commonly used PCMs are salt hydrates and hydrocarbons. Metallic additives are used to increase the overall conductivity and efficiency.

Wallboards are cheap in cost and thus have been widely used in a variety of building applications. They are very suitable for encapsulation of PCMs [128–136]. PCMs can be incorporated into building materials in various ways. Take plasterboard as an example, PCMs can be either post leading from liquid phase into the pore space of the plasterboard and concrete or by directly mixing the plasterboard and PCMs at the wet stage of manufacture. Figure 6.50 shows an example of clay aggregates before and after absorbing a PCM. The absorption of PCM by concrete is conducted by using a setup as shown in Fig. 6.51. It is found that PCM can penetrate into the pore space with diameter of 1–2 μm and occupy up to 75 % of the total porosity of the porous materials tested. The method is a feasible and potential to develop PCM concrete for application in building energy conservation [137]. It is expected that this method should be applicable to other materials.

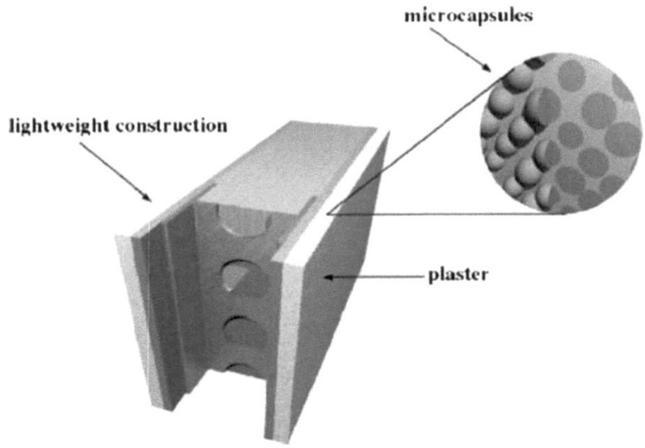

Fig. 6.52 Schematic view of a light weight wall, with PCM microcapsules being integrated into the interior plaster. Reproduced with permission from [138], Copyright @ 2005, Elsevier

The incorporation of PCMs with building materials or building has several problems, which hinder its practical applications. Non-encapsulated PCMs may interact with building structures and change the properties of the matrix materials. Leakage could be a problem over a relatively long time. Macrocapsules have to be protected against destruction while the building is used, such as drilling holes or nails in walls or ceilings. Also, the macrocapsules often need special carefulness, making the systems expensive. Another problem with macrocapsules is the drop in heat transfer rate during the solidification process for some PCMs, like paraffins, with poor heat transfer coefficients in the solid state. This may prevent the system from discharging completely overnight [138].

Instead, the emerging of microencapsulation technology makes the PCM building attractive again. For example, formaldehyde-free microencapsulation of paraffins allows PCMs to be readily integrated into conventional construction materials. These capsules, with diameters of a few of micrometers, can be integrated into specially optimized building materials, independent of the phase of the PCMs. Because of this, the problems mentioned above can be readily addressed. For instance, the capsule shell prevents any interaction between the PCMs and the matrix materials. There is no extra work at the building site to integrate the PCM products and the capsules are small enough that there is no need to protect them from destruction. The distribution of the small PCM capsules in the wall offers a much larger heat exchange surface, so that the heat transfer rate to charge and discharge the stored heat is increased significantly. Figure 6.52 shows a schematic drawing of the concept, with PCM microcapsules being integrated into plaster [138]. Products of this kind of building materials have been available in open market.

Floor is also an important part of a building, which has also been applied heating and cooling with PCMs. Figure 6.53 shows a schematic of a test room for discretization of floor heating system [139]. Both the cover layer and incident solar radiation have influence on temperature distribution of the floor and thus the energy consumption as a result. Complete and partial (area) carpets, as well as

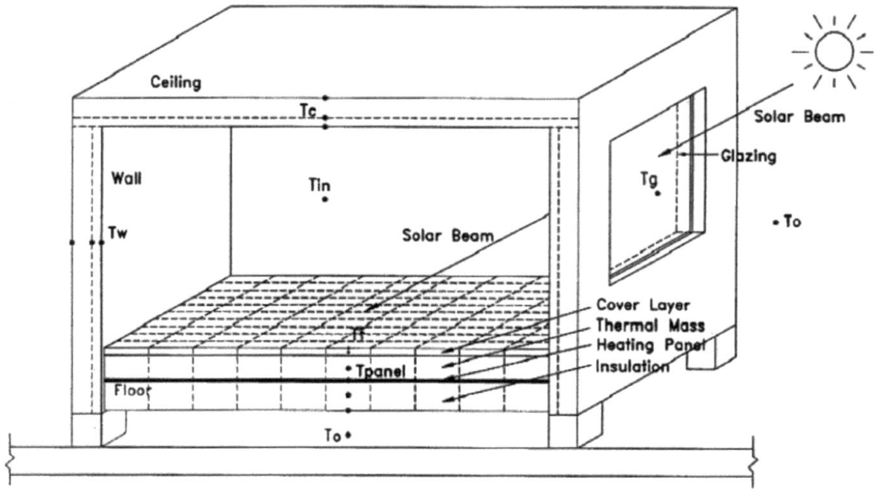

Fig. 6.53 Schematic of the test room with discretization scheme, with T_c ceiling surface temperature, T_o outside temperature, T_w wall surface temperature, T_f floor surface temperature, T_g window interior surface temperature. Reproduced with permission from [139], Copyright @ 2000, Elsevier

hardwood cover layers over concrete or gypcrete (gypsum-concrete mixture) thermal storage, are considered. According to experimental and simulation results for an outdoor test room, solar beam radiation can cause a local floor surface temperature in the illuminated area 8 °C higher than that in the shaded area. Partial carpet cover further increases surface temperature of the floor, with a difference of up to 15 °C when there is solar radiation absorbed. Solar radiation stored in the floor thermal mass can reduce the heating energy consumption significantly by >30 %.

Radiant heating has a number of advantages over convective air heating systems [140, 141]. It saves living and working space because it is integrated into the building envelope. Also, thermal mass integrated into a floor heating system can be used for off-peak storage of thermal energy. Thus, peak loads can be reduced and shifted to night-time when electricity costs are lower. From an application point of view, dense materials such as concrete may cause larger fluctuations of indoor temperature, while PCMs can provide large LHS over the narrow range of temperatures typically encountered in buildings, thus they can improve the thermal comfort level. Figure 6.54 shows the schematic of an under-floor electric heating system, with shape-stabilized PCM plates which include polystyrene insulation, electric heaters, PCM, air layer, and wood floor [141]. Electric heaters heat and melt the PCM layer by using cheaper night-time electricity and the system stores heat. During the daytime, the electric heaters are switched off and the PCM layer solidifies, discharging the heat stored. In countries or cities, the electricity tariff at night is much lower than that during the day. Therefore, the shift of electrical

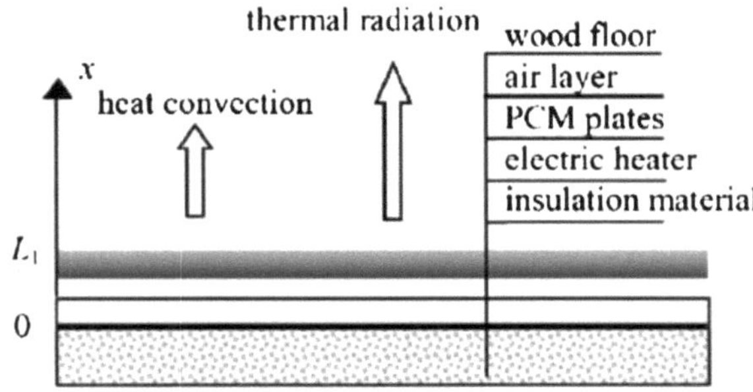

Fig. 6.54 Schematic of under-floor electric heating system with shape-stabilized PCM plates. Reproduced with permission from [141], Copyright @ 2004, Elsevier

consumption from peak periods to off-peak periods will provide significant economic benefit. Moreover, it is of importance to power plants by leveling the electrical load.

Ceiling boards are usually used in the roof of buildings, which can be readily employed for heating and cooling [142–145]. For example, PCMs can be used to store coolness during off-peak time and the energy is released in peak time, so that it possible to decrease the use of air conditioning systems. By using PCMs with melting point close to room temperature, this function can be realized [142]. It is also possible to develop space heating systems that incorporate PCMs located in ceiling voids, by using Sun reflectors to direct the solar energy via windows onto the PCMs [143]. This kind of systems allows for a large area for heat storage without the need for large volumes of storage medium.

Figure 6.55 shows the schematic of a night ventilation system, by using a PCM TES system with heat pipes embedded in a PCM [144]. Air from the room is drawn over the exposed ends of the heat pipes by using low powered fans. During the daytime, heat is transferred to the PCM, which is melted so that and the temperature in the room is reduced due to the absorption of energy for the melting. At night, shutters are opened, cool outside air is drawn over the heat pipes, so that the heat is extracted from the freezing PCM. As a result, the heat pipes in this design are reversible, as compared with the previous ones in terms of heat pipe-PCM exchangers. Because simple heat exchange geometries are used on the surface of the PCM exposed to the air, the systems can be modified flexibly. A one-dimensional mathematical model has been developed to simulate the heat transfer from air to PCM. This system stores coolness in night-time and releases it in daytime [144].

Figure 6.56 shows a sketch of the experimental setup, which consists of a duct, with dimensions of $1.5 \times 0.15 \times 0.60$ m^3 [144]. The width of the duct includes semicircular walls at either side. The ducting is lagged with 5 cm of foil-backed rockwool insulation. A fan is used to draw air through the duct. The exposed end of the heat pipe is sealed into the duct. Within the heat pipe-PCM unit, three stainless steel thermistors are used to measure spot temperatures in the PCM in the positions shown in Fig. 6.56, to assess the effectiveness and extent of the melting/freezing process. Figure 6.57 shows a side view of stand and units in the test room,

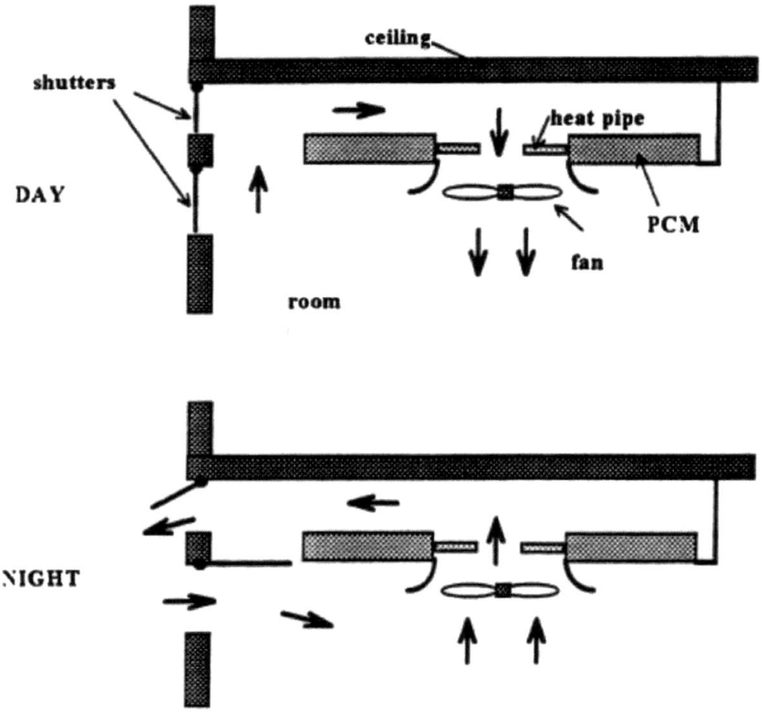

Fig. 6.55 Schematic of under-floor electric heating system with shape-stabilized PCM plates. Reproduced with permission from [144], Copyright @ 2000, Elsevier

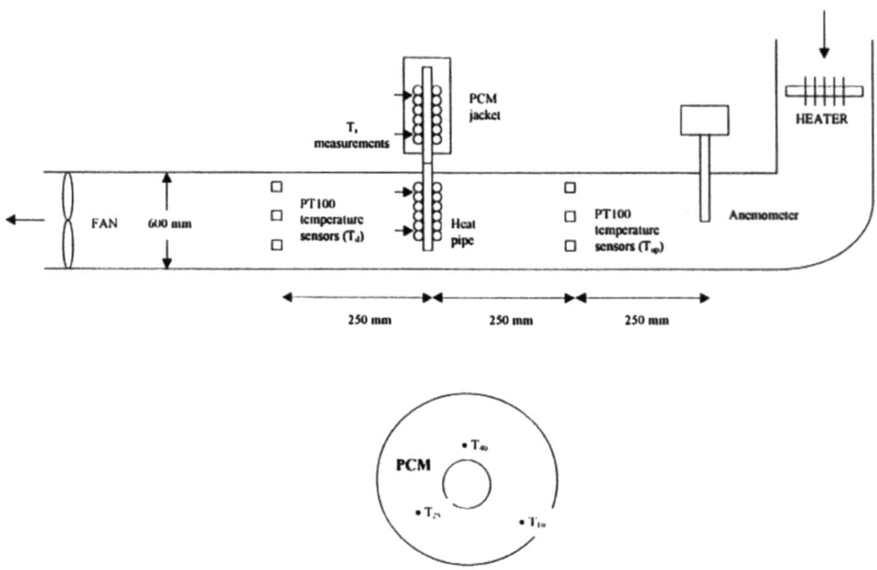

Fig. 6.56 Schematic plan view of experimental rig and cross-section of PCM container of the system. Reproduced with permission from [144], Copyright @ 2000, Elsevier

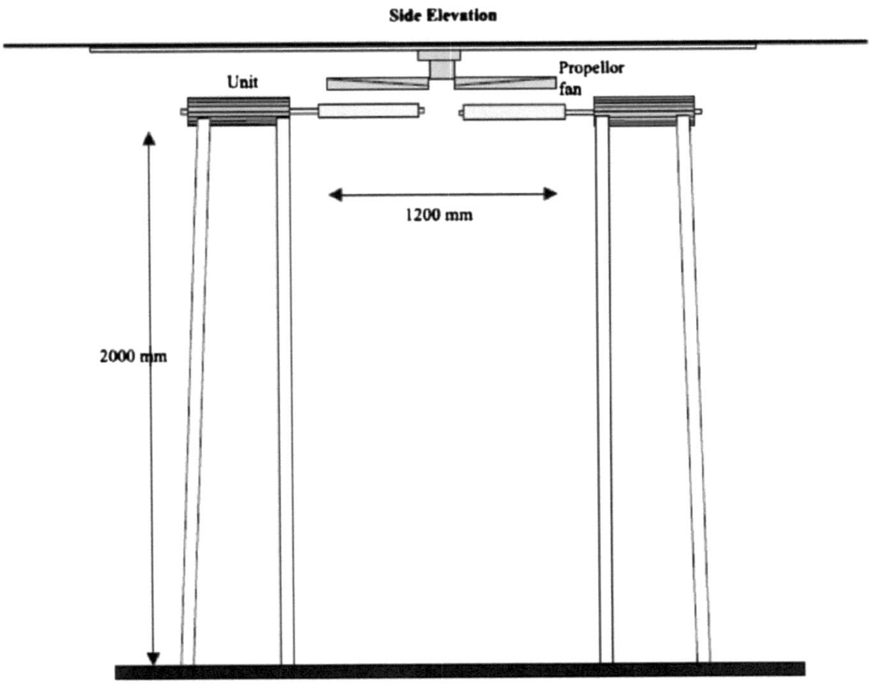

Fig. 6.57 Schematic side view of the stand and units in the text room. Reproduced with permission from [145], Copyright @ 2001, Elsevier

with a dimension of 6 × 4.8 m^2 and 2.5 m high. The test room has a large window extending the full length of one 4.8 m all, facing south–south-east and receiving large amount of solar energy in the summer. The venting arrangement is shown schematically in Fig. 6.58. At night, cool air is flowed in from the upper-level vents, guided above the units by a movable board and passed over the heat pipes by using the ceiling fan blowing downwards. The warmed air passes through the exit vent. During the day, the vents are blocked and the extract fan is switched off and the ceiling fan still blows downwards to cool the room through convention and the PCM storage unit. This system offers an alternative yet sustainable choice for building air conditioning system.

6.5.6 Off-Peak Electricity Storage

Electricity consumption varies during the day and night, according to the demand by industrial, commercial, and residential activities. The variation in electricity demand sometimes leads to a difference in price between peak and off-peak periods, usually after midnight until early morning. The shift of electricity usage

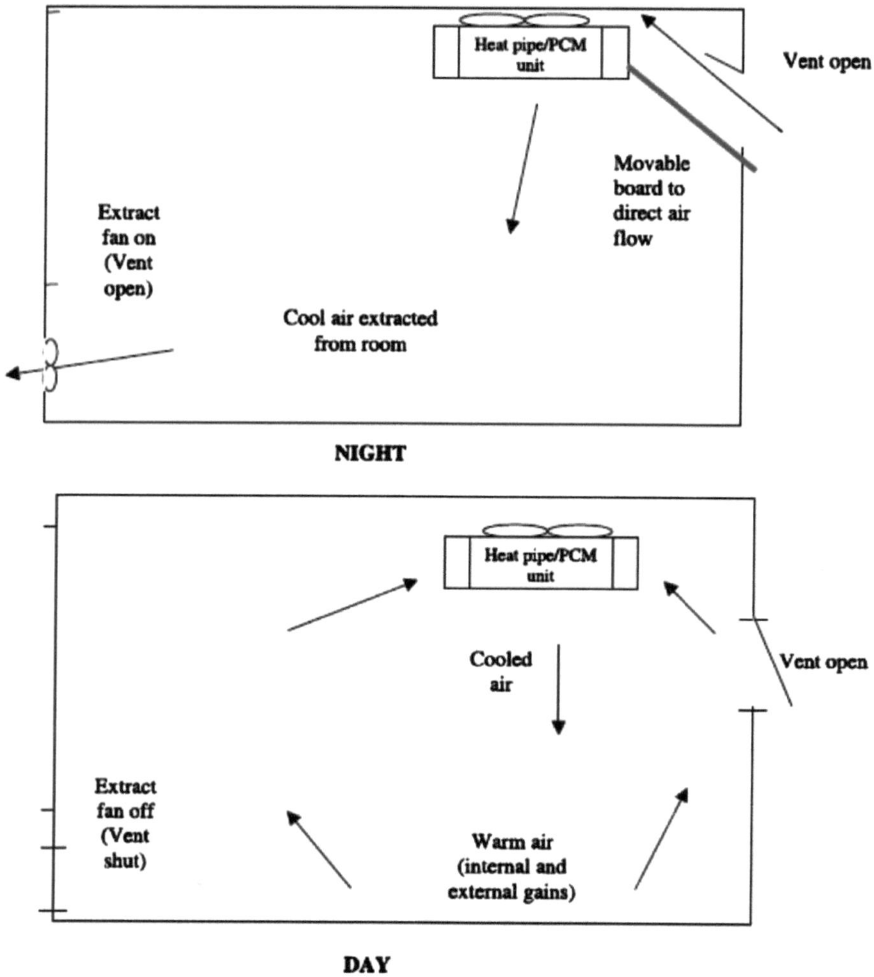

Fig. 6.58 Schematic view of venting system of the text room with modifications. Reproduced with permission from [145], Copyright @ 2001, Elsevier

from peak periods to off-peak periods will provide significant economic benefit. The development of an ES system may be one of the solutions to the problem, when electricity supply and demand are out of phase. ES systems will enable the surplus energy to be stored until such time as it is released when needed. The concept of off-peak electricity storage is to use the electricity during off-peak time to store coolness or heat, which is released during the peak time when large amount of electricity is needed. By doing this, the requirements of electricity during peak time can be balanced by the storage during the off-peak time, allowing for more flexible grid management [3, 146].

6.6 Strategies to Improve PCM Storage Efficiency

6.6.1 Principles

Although there are various advantages, such high energy density, nearly isothermal operation, wide availability of materials, and relatively low cost, PCM TES systems have a severe problem, i.e., the very low thermal conductivity of the PCMs currently available, which has limited the performance of such systems. The effect of the lower value of conductivity has been obviously reflected by the temperature drop during energy retrieval or withdrawal after storage. As a result, the rate of phase change process (melting/solidification of PCM) has not been up to the desired level for large-scale applications [147–152].

To address this problem, it is necessary to understand the mechanisms for enhancement of the PCM TES systems [153]. Both theoretical and experimental studies have been conducted during the last decades. The techniques employed to enhance the thermal performance of PCM TES systems mainly include increasing thermal exchanging surfaces, using multiphase PCMs, employing thermal conductivity enhancement additives and using microencapsulation of PCMs.

The studies on performance enhancement techniques have mainly focused on evaluation of the time required for melting/solidification, heat transfer rate, and amount of energy stored/retrieved with the application of enhancement techniques. According to the results obtained, it seems that all these studied techniques have are potential in enhancing the thermal performance of PCM TES systems. The enhancement mechanisms have been assessed by using energy and exergy analysis. Compared to energy analysis, exergy analysis is still much less presented in the open literature [57, 73, 75, 82, 154, 155].

Take multiple PCMs as an example, with exergy analysis, it is able to investigate the role of the multiple PCMs in improving the exergy performance, to identify the appropriate number of PCMs required and to evaluate the optimum distribution of melting points. Because the exergy performance increases with increasing number of PCMs, it is also important to identify the number of PCMs that can be used [75]. It is found that significant improvement in exergy efficiency can be obtained with larger number of NTU (number of thermal units). In addition, the difference between the highest and lowest melting temperatures has enormous effect on the charging and discharging rates and thus the exergy performance [73]. The authors have presented a simple expression for the optimum difference between the highest and lowest melting temperatures can be expressed, for charging, as:

$$T_{m(\text{first PCM})} - T_{m(\text{last PCM})} = \frac{\text{NTU}}{1 + (\text{NTU}/2)} \left(T_{\text{HTF, in}} - T_{m(\text{ave})}\right), \tag{6.65}$$

where $T_{m(\text{ave})}$ is the average melting temperature of the PCMs located at the middle. On the other hand, for discharging, there is:

$$T_{m(\text{first PCM})} - T_{m(\text{last PCM})} = \frac{\text{NTU}}{1 + (\text{NTU}/2)} \left(T_{m(\text{ave})} - T_{\text{HTF, in}} \right). \tag{6.66}$$

It is difficult to satisfy the two different conditions for charging and discharging. Furthermore, when multiple phase PCMs are used, it is also necessary to examine the optimum conditions for melting temperatures of intermediate PCMs, which can be achieved by maximizing the overall exergy efficiency of multiple PCMs module [75]. The optimum melting temperatures of PCMs are approximately a geometrical regression, i.e. $T_{m1}/T_{m2} \approx T_{m2}/T_{m3} \approx T_{m3}/T_{m4} \approx \ldots\ldots \approx T_{mn-1}/T_{mn}$. Moreover, the ratio between the melting temperatures of every two adjacent PCMs is found to be dependent on NTU of the system.

The use of additives, on the other hand, may lead to loss of storage capacity of pure PCM, due to the reduction in mass fraction of PCM. Therefore, it is important to evaluate the optimum levels. Such optimizations should be considered not only from storage capacity point of view but also from that of total entropy generation. Therefore, exergy analysis could play an important role in realizing these objectives.

6.6.2 Increasing Heat Transfer Surfaces

Fins have been widely used to increase heat transfer surfaces, so as to enhance the performances of PCM TES systems. Although the configurations and arrangements of fins can be different from one another, they are classified into two groups. One group is used in the systems packed with PCM which transfers heat with a hot/cold HTF, used in solar water heaters, space-based power plants, waste heat recovery systems, and so on. The other group is for the systems packed with PCM to transfer heat with hot/cold boundary wall, used in electronic cooling.

If no HTF is used a system, fins are obviously located at the PCM side, whereas if the PCM transfers the heat through HTF, the arrangement of the fins is according to the relative heat transfer coefficient. This is simply because the efficiency of fins increases with decreasing heat transfer coefficient [153]. For most systems, the heat transfer coefficients of PCMs are usually lower than those of the HTFs and thus fins are generally on the PCM side in practice. However, it is important to study on whether the presence of fins can enhance the performance of a given system, because the mechanisms governing the heat transfer can be different in different phase change processes, i.e., melting and solidification. Furthermore, the heat transfer mechanisms of phase change processes are also related to the configurations the system and orientations of the fins. Therefore, it is necessary to discuss the heat transfer mechanisms and the roles of fins in determining the modes of heat transfer during melting and solidification processes separately.

During melting, heat is transferred to the PCM first by conduction and later by natural convection. This is because as the heat transfer surface moves in the solid region is gradually disappearing, i.e., the thickness of the liquid region increases.

Since thermal conductivity of a liquid PCM is lower than that of its solid counterpart, the heat transfer by conduction becomes less and less as the melting process proceeds. Finally, natural convection due to the density gradient that exists within the liquid PCM becomes dominant.

The influence of natural convection on the location of melt front has been well investigated [36]. To study the melting characteristics in a semi-infinite PCM storage system with an internal fin, an analytical model is used, which consists of the well-known Newmann solution, with assumptions that the heat is transferred only by conduction and natural convection can be negligible. It is found that, it the location of the melt front has been underestimated although Newmann solution is the exact solution of the analysis.

In other similar studies, it is fount that if the natural convection effect is included, the theoretical predicted time for the PCM to reach the maximum temperature is double that of the experimental value [156, 157].

Experimental study on the melting behavior of n-octadecane has been reported by Zhang et al. [158]. The PCM is discretely heated at a constant rate from one side of an enclosure. During the melting process, there is a temperature gradient, decreasing from the upper region to the lower region of the liquid. This implies that natural convection in the liquid phase is present, so that the hot liquid with lower density goes up while the cold liquid with higher density precipitates. Homogeneous temperature distribution is only observable at the early stage of the melting process. This clearly indicates that the early stage of heat transfer is conduction controlled, whereas at a while convection heat transfer becomes to be dominant. Similar transition from conduction to convection in heat transfer is also observed when heating is applied at bottom of the enclosure of the PCM [159].

Like in rectangular enclosure, natural convection has also been found as dominant heat transfer mechanism in cylindrical and spherical enclosures. The melting of a PCM stored in a horizontal cylindrical annulus heated isothermally from inside shows natural convection flow in the liquid state [160]. For paraffin wax encapsulated in a cylindrical capsule surrounded by hot water, natural convection influences on the melting process in an axisymmetrical way, i.e., the melting in top region is much faster than that in bottom region [161]. It is also reported that better agreement between numerical results and experimental results could be achieved only after including the natural convection in the numerical modeling.

In an experimental and numerical study on melting of PCM in cylinder heated from side vertical wall, digital photographs taken during the melting process indicate that, only in earlier stages of melting, the melt layer thickness is uniform along the vertical direction, i.e., conduction is the dominant heat transfer mechanism [162]. Figure 6.59 shows schematic of the experimental setup. As the time elapsed, the melt layer thickness began to vary along the vertical direction. This is the indication of convection dominated melting. The shapes of the melt front at different times are similar to that of observed in rectangular enclosures. Representative experimental results are shown in Fig. 6.60.

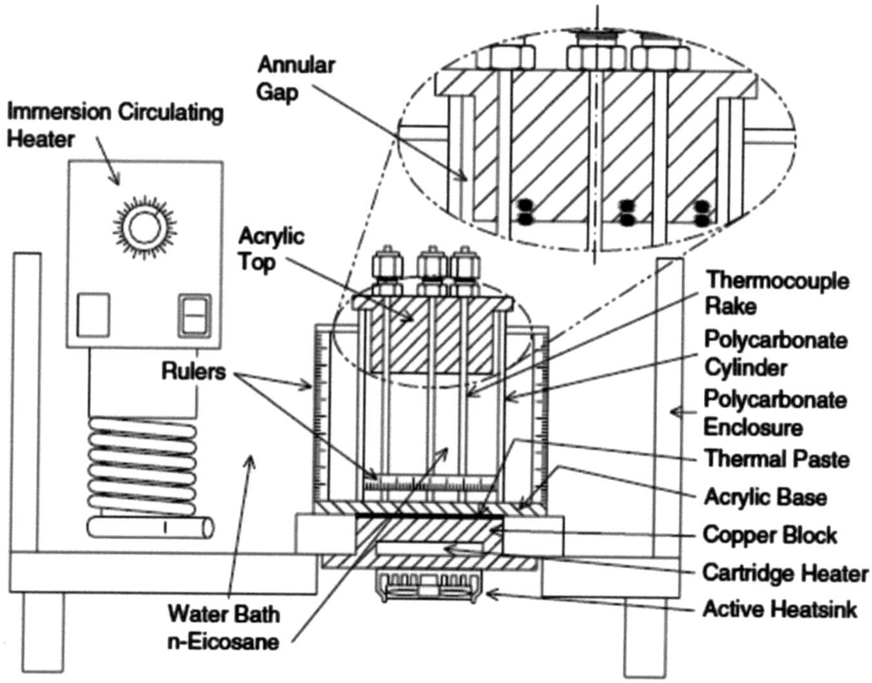

Fig. 6.59 Schematic diagram of the experimental facility. The inset shows an enlarged view of the top of the test cell and the locations of the thermocouples imbedded in the acrylic top. The thermocouples are approximately 0.5 and 3.8 mm from the lower surface of the acrylic top. The polycarbonate side wall is at a constant temperature TH on the outside, the acrylic bottom wall is at a constant temperature TB on the outside, and the top wall is adiabatic. The radial edge of the acrylic bottom wall is insulated as shown. Reproduced with permission from [163], Copyright @ 2006, Elsevier

Spherical systems offer largest volume per unit surface area. Melting of PCMs in sphere containers has two cases: unconstrained and constrained [163–165]. Experimental setup to measure and monitor the melting of n-octadecane in a sphere container is shown in Fig. 6.61 [164]. Experimental observation and theoretical simulation results, for unconstrained and constrained melting, are shown in Figs. 6.62 and 6.63, respectively. In case of unconstrained melting, at a very early stage, the solid–liquid interface is flat in the beginning of melting, but it becomes back to a dome-like shape for a short while just after melting. The dome-like shape of interface remains unchanged till the end of melting, indicating more melting at the upper portion of the spherical container. This is because the solid sinks to the bottom since the solid PCM has a higher density than the liquid one. Hence, solid PCM is in contact with the bottom wall of the sphere container throughout the melting process. In this case, the melting at the top portion is mainly due to natural convection. For observation of constrained melting, the solid PCM is held so that it never sinks to the bottom throughout the melting process. The melting becomes

Fig. 6.60 Photographs during the melting of the wax within the cylindrical enclosure at a wall temperature of 45 °C after four different times. **a** 1680 s, **b** 3120 s, **c** 7200 s, and **d** 10,800 s. The solid n-eicosane is opaque while the liquid n-eicosane is transparent. Photographs **c** and **d** are the vertically oriented thermocouple rakes. The *dashed vertical lines* indicate the location of the outside cylinder wall. Reproduced with permission from [163], Copyright @ 2006, Elsevier

more asymmetric. The melting front has an ovalshaped pattern during the melting, due to the existence of natural convection cells at the top and bottom halves of the sphere container. However, the melting of the top portion is faster, because a stronger natural convection prevails at the top portion.

In summary, during the melting process, the heat transfer is dominated by natural convection. Therefore, if fins are used on the side of PCMs, their function is to improve the natural convection process rather than conduction. However, when top wall of the container, for example, rectangular system, is used as heating wall, the conduction mechanism will dominate the melting for longer period of time, evidenced by the fact that the isotherms are parallel to the heated top wall [166].

Due to the absence of natural convection, in the case of top wall heating, the melting process takes longer time as compared to bottom wall heating, even though when HTF is allowed to flow downward [167]. Therefore, the role of natural convection in the melting process can be ruled out, especially in the case of a rectangular module with top wall heating. However, if the top wall heating is coupled with environment, i.e., the bottom wall is not at adiabatic condition,

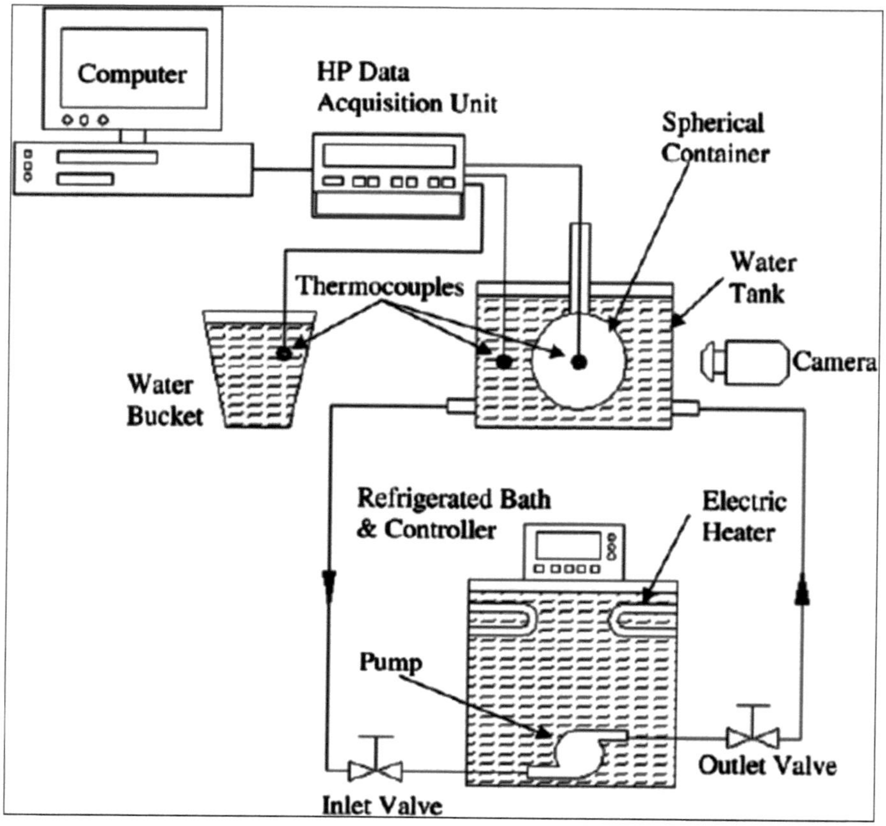

Fig. 6.61 Schematic diagram of the experimental setup to measure melting of PCM in sphere. Reproduced with permission from [164], Copyright @ 2008, Elsevier

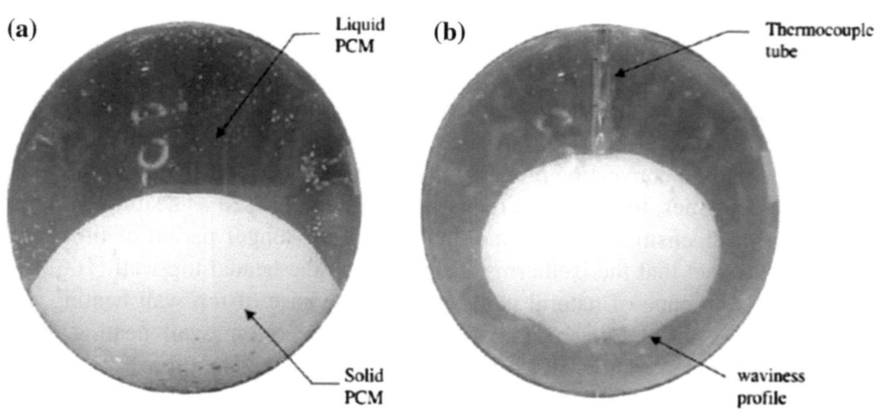

Fig. 6.62 Melting inside sphere. **a** Unconstrained melting and **b** constrained melting. Reproduced with permission from [164], Copyright @ 2008, Elsevier

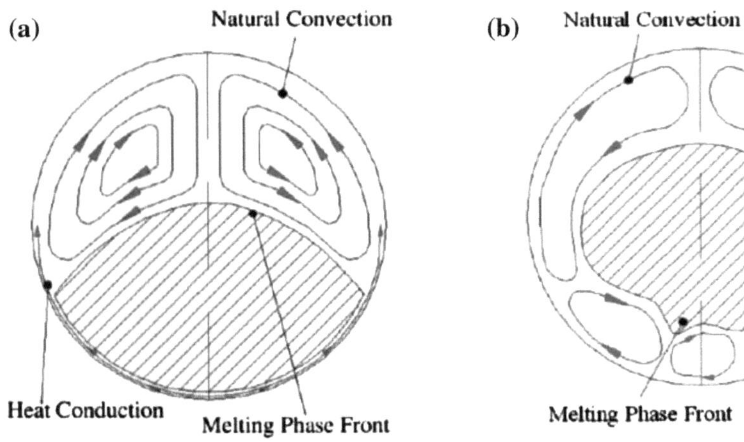

Fig. 6.63 Schematic sketches showing het conduction and natural convection in unconstrained (**a**) and constrained (**b**) melting. Reproduced with permission from [164], Copyright @ 2008, Elsevier

natural circulation in the liquid phase is observed [168]. In this case, there is disagreement between theoretical and experimental results, at advanced state of melting [169]. If natural convection is taken into account, numerical results will show significant quantitative improvement in the agreement with experimental data. This implies that natural convection is still present even in the case of top wall heating, when the system is coupled with environment.

The melting process can be enhanced by using fins, due to their ability to enhance the heat transfer. The efficiency of the enhancement is closely related to the configuration and number of the fins used [170–172]. Generally, more fins per unit length mean larger heat transfer area and thus energy stored. However, there is a critical value beyond which an increase in the number of fins does lead to an further enhancement in heat transfer efficiency [170].

It is also reported that vertical heated walls with horizontal fins are superior to horizontal heated walls with vertical fins [171]. This is because the horizontal fins offer double rate of heat transfer as compared to the vertical ones. The enhancement in heat transfer factor increases as the distance between the adjacent fins is decreased, because it is similar to the case that the system has more number of fins. The presence of a critical number of fins is due to the fact that as the number of fins is increased hampering effect becomes gradually pronounced, leading to the buoyancy driven flows. At the same time, the melting becomes conduction-dominated process [171].

The enhancement of melting behavior of a PCM is also affected by the length of fins [172]. If the fins are short, the melting rate is almost independent of the number of the fins. Also, once the melting front crosses the tip of the fins, the melting process becomes similar to that of the case without fins. If the heating temperatures of the walls are sufficiently high, an increase in the number of fins produces only

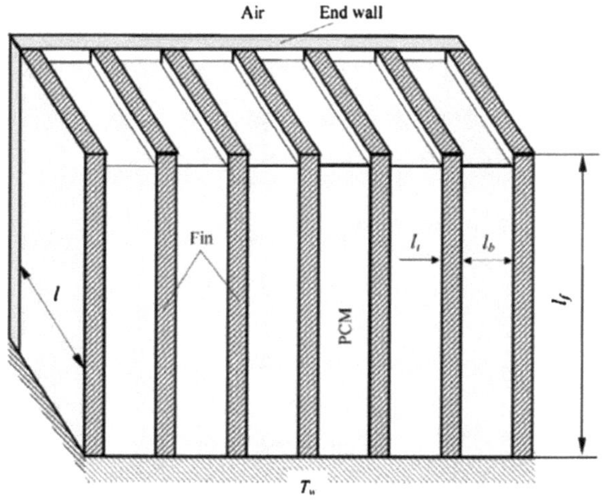

Fig. 6.64 Three-dimensional physical model of a rectangle storage unit with fins and the definition of the problem (front end wall is not shown). Reproduced with permission from [173], Copyright @ 2005, Elsevier

slight enhancement in melting rate, which can be explained also in terms of the hampering effect on the buoyancy driven flows. This also implies that the optimum number of fins is dependent on the heating temperatures of the walls. Therefore, longer fins are more efficient in enhancing the melting rate than shorter ones.

The effect of fins is also related to their distance. An example is shown in Fig. 6.64 and representative simulation results are shown in Fig. 6.65 [173]. The fluid motion becomes gradually stronger as the melting progresses for the thick fins with wider distance. However, in the case of thin fins with narrow distance, fluid motion is hardly observed even at advanced stage of melting. Thick fins have the same temperature of the heated surface uniformly along their length, while thin fins experience temperature gradient, for the same length. It is desired that the fins have steady performances in terms of heat transfer. However, if too thick fins are used, it means that the number of the fins should be reduced. As a result, the thickness of fins must also be optimized along to balance the number of fins and the space between adjacent fins should not be too narrow, in order to achieve the highest performance of the systems.

In terms of heat change between HTF and PCMs in a storage unit, shell and tube/concentric double pipe heat exchangers have the highest efficient for their minimum volume. An example is shown in Fig. 6.66, with which a three-dimensional numerical model for melting process of PCM has been developed [83]. In this case, annular fins are arranged around the tube. Natural convection is included through effective thermal conductivity as a function of Rayleigh number in the conduction equation. It has been found that large amount of heat is conducted through the fins along the radial direction. For all ranges of mass flow rate and inlet temperature of HTF, a significant increase in ES capacity is observed, due to the presence of the fins. An increasing in number of the fins leads to an increase in the rate of heat transfer and thus ES. However, the enhancement due to the presence of the fins is observed to be dependent on mass flow rate and inlet

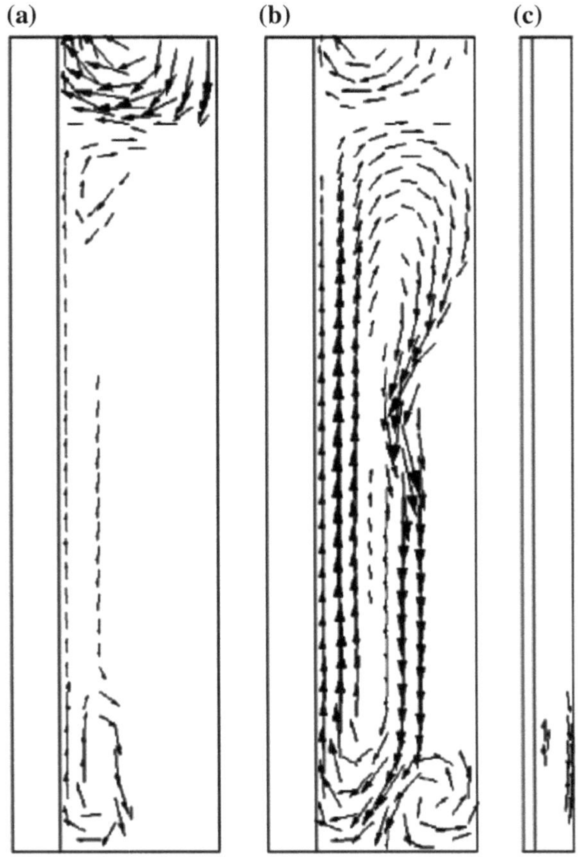

Fig. 6.65 Simulated velocity fields for the fins of different thickness at different stages of melting. **a** Case 1 ($l_f = 10$ mm, $l_t = 1.2$ mm and $l_b = 4$ mm) and melt fraction of 0.3, **b** case 1 and melt fraction of 0.6 and **c** case 3 ($l_f = 10$ mm, $l_t = 0.3$ mm and $l_b = 1$ mm) and melt fraction of 0.6. Reproduced with permission from [173], Copyright @ 2005, Elsevier

temperature of HTF. Maximum enhancement can only be achieved for moderate flow rates and relatively low inlet temperatures.

The use of fins is also an effect way to suppress the negative effect of subcooling on the performance of PCM TES systems. It has been shown that the liquid fraction of the PCM at any time during the melting can be increased by increasing the height of the fins [174]. This is because the melting fronts on both sides of the fins are greatly affected by the height of the fin, although the effect of fins on melt front is not very significant in between the fins. Generally, subcooling has a negative in the performances of thermal storage systems, which is a problem of many inorganic PCMs. It is interesting to find that the effect of subcooling on performance is nearly zero on both sides of the fins.

For solar space power systems, they in microgravity conditions, so that natural convection can be neglected [175]. It has been found that if a tube without fins is used, there is always some PCM nearer the exit of the tube in solid state. This is because the difference between the temperature of the HTF and the melting point of the PCMs is very small nearer the exit. If annular fins are used, a relatively high temperature difference between the HTF and the PCM melting point can be maintained. As a result, melting is throughout the tube in the axial direction. For a given size of the storage unit, significant increase in ES can be achieved, due to the presence of fins. However, this kind of increase is only significant in the initial stages of melting. The increase in ES becomes gradually less pronounced as the process continues, although the percentage increase in ES is directly proportional to the number of fins at any particular time. Obviously, the increase in number of fins leads to a decrease the amount of PCM that can be used.

In contrast to melting process, solidification process is dominated by conduction. During solidification, natural convection is present only in the beginning [176]. The analytical and experimental results of solidification process in a rectangular storage unit have shown that the natural convection is 10 times lower than that during the melting process [157]. According to Ettouney et al. [167], during the solidification of PCM in a shell and tube heat exchanger, the solidification isotherms are almost parallel to the heat transfer surface and solidified layers have the same shape of the heat transfer surface. Therefore, natural convection is negligible, no matter whether the direction of the HTF flow is upward or downward. Similar results are reported even in spherical storage device [165]. For the same module size, HTF temperature, and HTF velocity, the heat transfer coefficient during melting is much higher than that during solidification.

In all configurations of PCM TES systems, the solidified layers are formed from heat transfer surface and remain parallel to it during solidification. Although natural convection exists in the liquid PCM at earlier stages, it diminishes rapidly as the solidification progresses and the process becomes conduction dominant. This is because the volume fraction of the liquid phase decreases gradually and disappears finally. Therefore, if fins are used to enhance the rate of solidification, their function is to improve the heat transfer rate of conduction.

An example is about the effect of vertical fins attached to horizontal constant temperature wall on solidification of high temperature [48]. Without fins, the solidification starts on the wall and continuously moves away only in the vertical upward direction. However, if vertical fins are used, the movement of the solidified front is observed in both vertical and horizontal directions. Moreover, the rate of movement parallel to the fins is faster than that in the upward direction. It is also found that the discharge time with fins is decreased by almost 30 times when compared with that without fins. The time required for complete solidification decreases almost linearly with the number of fins. Similarly, an increase in the fin thickness also leads to decrease in time of solidification. There is also a critical thickness for the fins beyond which the effect on solidification becomes less pronounced.

Similar effect has been observed in systems with cylindrical geometrics, such as shell and tube or double pipe arrangements. Both radial and axial fins attached to the heat transfer surface have been studied. In a study to examine the heat transfer characteristics of PCM during solidification in cylindrical storage system, radial fins have been used [177]. The results of temperature distributions indicate that the temperature gradient at all radial positions are larger when fins are used, which

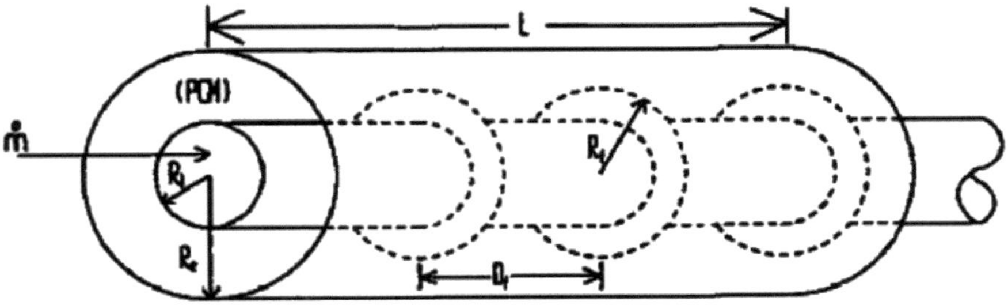

Fig. 6.66 A representative finned tube heat exchanger for modeling. Reproduced with permission from [83], Copyright @ 1993, Elsevier

means that that the heat transfer between the PCM and the tube is improved due to the use of the radial fins. Also, better axial conduction in the solid phase due to fins is observed, evidenced by the fact that the temperature distribution in the axial direction is more uniform of the finned system than that of the system without the fins.

However, the effectiveness of the fins is also dependent on mass flow rate of HTF. At low mass flow rate of HTF, the solidified layers are formed only on the small part of fins adjacent to the tube wall at earlier stages, although it is improved as the process progresses. At high rate of mass flow, the solidified layers are observed on larger parts of the fins throughout the process. According to Liu et al. [178], the use of fins increases not only the conduction heat transfer, but also natural convection, at earlier stages of solidification. When shell and tube or concentric double pipe heat exchanger are used, PCMs are stored in shell or annulus and the inner tube surface acts as heat transfer surface. During solidification, the solidified PCMs between the tube and liquid PCM increases the thermal resistance, thus leading to poor heat transfer to the liquid PCM. If fins are attached to the tube, the liquid PCMs will have a better thermal contact with the heat transfer surface, which in turn reduces the thermal resistance.

It has been reported that further enhancement is possible if the PCMs are stored in the inner tube when longitudinal fins are used [179]. This is because the thermal resistance experienced by the liquid PCM is uniformly reduced as the liquid PCM farthest away from the heat transfer surface can have maximum thermal contact with the fins in this configuration. An additional heat flow has been observed in the circumferential direction. Therefore, due to use of the longitudinal fins, higher surface heat flux can be achieved. Moreover, the surface heat flux increases with increasing number of the fins.

In addition, the use of vertical fins to the HTF side can also enhance the performance of the storage systems during solidification [180]. This is because vertical fins on the HTF side significantly promote the natural convection within the HTF. Simulation results of heat transfer coefficient indicate that the solidification time in the finned system is considerably less as compared with that in the system without fins. Therefore, it can be concluded that the system with fins can have the same heat transfer coefficient as that of the system without finsal though with smaller temperature difference. When longer fins are used, the heat transfer coefficient is decreased, due to the dampening effect on natural convection. However, the solidification time is still short because of the increase in heat transfer area.

In summary, the number or dimensions of the use of fins has been acknowledged to bea very important method to promote the natural convection during the phase change process of the PCMs used in TES systems. However, the number and dimensions of the fins should be optimized. Although different configurations combined with different fin geometries have been studied widely and extensively until now, standard criteria are lack for the selection of fins, which needs to bring further attention in the future study of this field.

6.6.3 Multiple PCMs

The use of multiple component PCMs in PCM TES systems has been shown to be an effective way for the enhancement in their performances. It means that in a system more than one PCM are used. These PCMs have different melting temperatures. The differences between the temperatures of HTF and the melting temperatures of the PCMs have significant effects on the heat transfer rate of the unit and thus enhance the performance of the system during charging (melting) and discharging (solidification) processes. In the case of single PCM, this temperature difference decreases in the flow direction of HTF, thus leading to a decrease in heat transfer rate and thus poor performance of the unit. If multiple PCMs with different melting temperatures are packed in the unit, in the decreasing order of their melting points, nearly a constant temperature difference can be maintained during melting process, even though the temperature of HTF decreases in the flow direction. This leads to an almost constant heat flux to the PCMs. During discharging, if the flow direction of the HTF is reversed, the PCMs remain in increasing order of their melting points. Also, nearly constant heat flux but from the PCM to HTF is maintained. An example of multiple PCMs in shell and tube is shown in Fig. 6.6 [44].

A new concept of homogeneous phase change process, in which the phase change occurs throughout the PCM, has been proposed [181]. It means that all the component PCMs have the same phase change rate, which remains constant during the process. Theoretical simulation results indicate that, if the PCMs of different phase change temperatures can be arranged in such a way that the phase change temperature distribution is parabolic, as shown in Fig. 6.67, a homogeneous phase change process can be realized. With such a nearly homogeneous phase change process, the phase change time can be reduced significantly.

An example of multiple PCM for experimental study is shown in Fig. 6.68 [43]. In real experiment, the shell side is loaded with three PCMs and synthetic oil is allowed to flow through the inner tube. Figure 6.69 shows the temperature

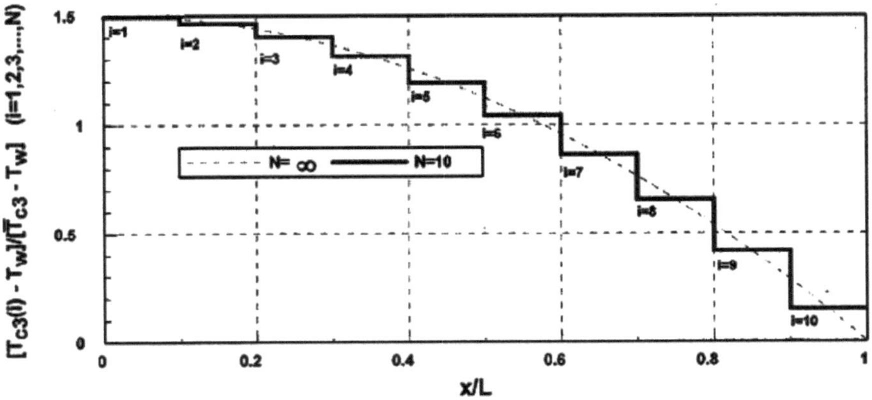

Fig. 6.67 Phase change temperature distribution of multiple PCMs. Reproduced with permission from [181], Copyright @ 1999, John Wiley & Sons

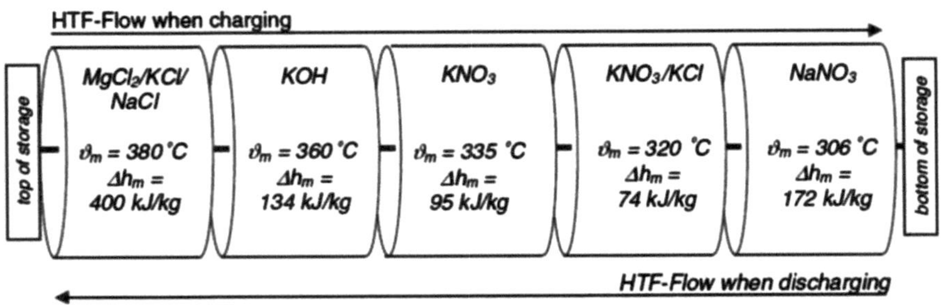

Fig. 6.68 Phase change temperature distribution of multiple PCMs. Reproduced with permission from [43], Copyright @ 2007, Elsevier

Fig. 6.69 Phase change temperature distribution of multiple PCMs. Reproduced with permission from [43], Copyright @ 2007, Elsevier

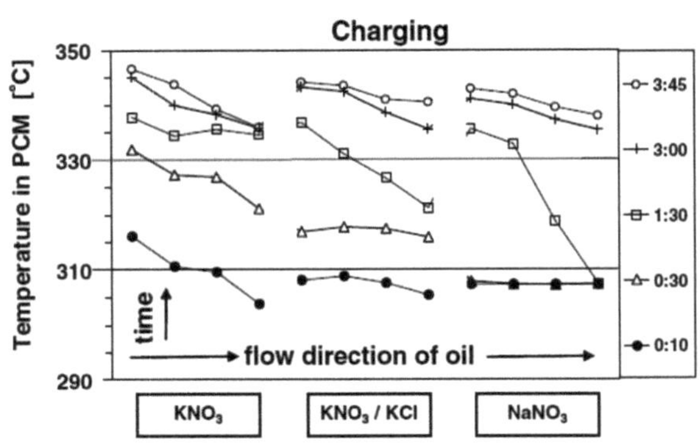

distribution during charging inside the PCMs from the top (left side of diagram) to the bottom (right side of diagram). The charging period lasts 4 h with heat transfer oil inlet temperature at the top of 350 °C and the discharging period lasts 2 h with a heat transfer oil inlet temperature at the bottom of 290 °C. Discharging periods are shorter due to additional heat losses to the ambient. Hence, over a period of time, the three PCMs unit possesses higher phase change fraction and thus higher storage or discharge capacity, as compared with the unit with single PCM of higher melting point. This observation has been predicted by using theoretical simulation [182].

In the configurations discussed above, HTF flows through the system and hence all the PCMs are arranged in the flow direction (axial direction) in contact with the HTF. In solar thermal applications, the systems are mostly kept in the storage tank, which contains the HTF [183]. The PCMs stored in the LHTS unit absorb excess heat energy from the HTF during charging process and release it to HTF during discharging. In this case, since the system unit is surrounded by the HTF, there is no axial temperature variation within the HTF. In such configurations, multiple PCMs should be arranged in radial direction instead of in axial direction. As the centrally located PCM material in the single PCM unit has minimum thermal contact with the HTF, melting is not expected immediately. With multiple PCM arrangement, it is possible to have lowest melting point PCM at the center and other PCMs arranged in the increasing order of their melting points from the center of the unit to the outer wall. This may lead to more or less simultaneous melting at all points, because a nearly constant temperature difference can be maintained in the radial direction.

A similar work, with three coaxial cylindrical tubes to store three PCMs and the entire unit being kept in water bath, has been reported by Wang et al. [184]. The PCMs are arranged in a similar way as discussed above. The experiments for charging process with different HTF temperatures indicate that the melting time in three PCMs system is decreased by 37–42 %, as compared with that of the single PCM unit, which has been attributed to the fact that the three PCMs unit terminates the melting process earlier by 15–25 %.

The concept of multiple PCMs has also been applied to the storage systems in which there is no HTF. It has been shown that, for the systems with single-phase PCM, the phase change temperature distribution is constant. With multiple phase PCMs, it is possible to produce variable distribution of phase change temperature [185, 186]. In the systems with constant temperature boundary conditions, the distributions of a linear phase change temperature can be realized by properly arranging those PCMs with different melting points. Simulation and experimental results indicate that a 25 % reduction in phase change time can be achieved by using multiple phase PCMs. It is also found that the efficiency in the enhancement is dependent on the number of the PCMs. For practical applications, the number of the PCMs should be in the range of 5–10 [45, 52].

When multiple PCMs are used to enhance the performance of PCM TES systems, it is important to understand what should be the optimized combination of the PCMs. In other words, the differences in melting temperature between every two adjacent component PCMs, which are required to achieve highest performances, should be identified before implementation. There has been an example of numerical study on the effect of combinations of PCMs on the performance of

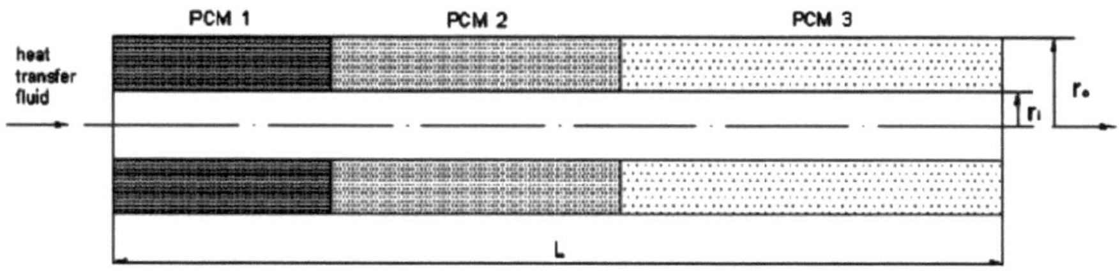

Fig. 6.70 Schematic of the LTES unit with multiple PCMs. Reproduced with permission from [187], Copyright @ 2007, Elsevier

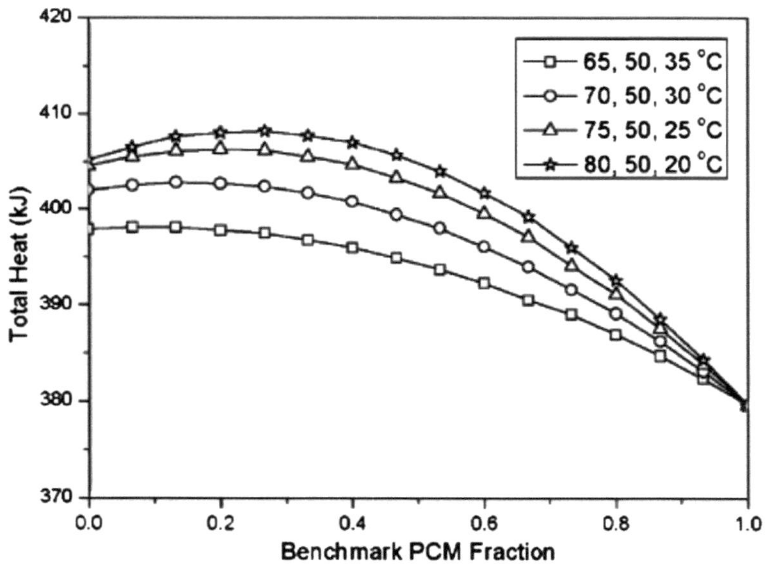

Fig. 6.71 Total heat stored in the storage units with different combinations of the PCMs after 7500 s. Reproduced with permission from [187], Copyright @ 2007, Elsevier

charging process [187]. Figure 6.70 shows schematic of the arrangement of the three PCMs for experiment. Total ESs of different PCM combinations are shown in Fig. 6.71, in which four groups of PCM combinations have been studied.

In each group, the difference in melting temperatures between two adjacent PCMs is kept unchanged, while the melting point of the middle PCM is fixed. It is found that, the larger the difference in melting point, the higher the melt fractions and thus higher heat storage that can be achieved with the same benchmark PCM proportion. The melt fraction and ES decrease when the benchmark PCM proportion is reduced. This is because if the proportion of the benchmark PCM is decreased, the proportion of the PCM with lowest melting point is increased, which means that less time is necessary for the PCMs to melt. There is an optimum value for the proportion of the benchmark PCM to achieve in terms of melt fraction and ES. Also, the optimum proportion of the benchmark PCM increases with increasing difference in melting point between the PCMs. Although this is a case of three PCMs, the conclusions should be applicable to those with more PCMs.

6.6.4 Increasing Thermal Conductivity

One of the serious problems of currently available PCMs is their relatively low thermal conductivity. This has limited their potential applications in TES systems in practice, although they can offer high energy density. The PCMs, no matter whether they are organic or inorganic, exhibit very low thermal conductivities, ranging from 0.1 to 0.6 W m^{-1} K^{-1} [147, 148, 188]. To increase the thermal conductivity of PCMs for real applications, various materials with high conductivity have been used to form composite structures, including impregnation of high conductivity porous material with PCMs, dispersion of high conductivity particles in PCMs, placing metal structures in the PCMs and incorporation of high conductivity low density materials.

Porous structure can be either metal matrices made of Al, Cu, etc., or naturally available porous material such as graphite. An example is numerical study the effect of impregnating high conductivity porous matrix of different thermal conductivities and different porosities with PCMs on the performance of TES systems, focusing on the melting process in a horizontal cylindrical annulus [189]. It is found that the enhancement in performance is dependent on both porosity and thermal conductivity of the matrix. For pure PCM, the rate of melting of the portion at the bottom of the container is slower than that at top portion, because the melting process is dominated by convection. With the incorporation of high conductivity porous matrix, the melting rate of the PCM at lower portion is enhanced. The enhancement due to porous structure is dependent on porosity of the matrix. Generally, low porosity means high effective thermal conductivity, so that in this respect, low porosity has a positive effect on the enhancement of performance. However, low porosity matrix dampens the motion of the liquid PCM and thus hinders the natural convection within the liquid PCM. Similar behaviors have been observed for PCMs impregnated with metal foams [190].

Graphite has high thermal conductivity, high electrical conductivity, and high absorbing capability, so that it can be used to increase the thermal conductivity of PCMs. The thermal conductivity of graphite is in the range of 24–470 W m^{-1} K^{-1} [153]. Therefore, if PCMs with thermal conductivity of 0.2–1 W m^{-1} K^{-1} are incorporated with graphite, their thermal conductivities can be increased by factors of 5–100, depending upon the content of graphite added into the PCMs [191]. The most available form of graphite is natural graphite flakes, which are composed of dense staked graphite layers [106]. Graphite flakes can be easily to form expanded graphite (EG), which is better than pristine graphite, due to the high porosity of EG. Generally, the expansion of graphite can be realized through chemical oxidation in

Fig. 6.72 Photographs of pure paraffin as PCM (**a**) and paraffin with 10 wt% EG as form-stable composite PCM (**b**). Reproduced with permission from [192], Copyright @ 2007, Elsevier

the presence of concentrated mixture of sulphuric and nitric acid, followed by drying and finally by rapid heating. When the EG is mixed with PCMs, the melted PCMs will be absorbed in the pores of the EG. The thermal conductivity and thermal capacity of the composite PCMs can be controlled through the adjustment of the composition.

The enhancements in thermal conductivity and thermal storage, due to the addition of EG in a paraffin based PCM storage system, have been observed, with focusing on the effect of the concentration of EG used [192]. As shown in Fig. 6.72, the composite PCM with 10 wt% EG is in a form-stable state, which means that the paraffin is well held by the capillary and surface tension forces of the porous EG. Also, the density of the 10 wt% EG composite is less than that of pure paraffin, thus leading to a lighter storage unit when the volume is the same. The thermal conductivity of the composites increases with increasing concentration of EG. However, the latent heat of the composites and the storage capacity of the unit decrease accordingly. The optimum content of EG is just 10 wt%, because this form-stable composite PCM exhibits higher thermal conductivity than that of pure paraffin by about four times, with only a slight decrease in latent heat. In addition, the melting temperature of the form-stable composite lower by 1.4 °C. The use of such form-stable composite leads to a significant reduction in melting time by about 32 %.

A separate study indicates that the optimum concentration of EG is 6.25 wt%, at which thermal conductivity is maximized [193]. The decrease in thermal conductivity at high concentrations of EG has been attributed to the presence of some lacunas and gases in the loose structure of composite PCM. The results of these two studies suggest that special carefulness should be taken during the preparation of the composite PCMs. The composite PCM with 6.25 wt% EG is also from-stable, resulting in a significant increase in the rate of heat transfer, during both charging and discharging processes, although the effect slightly more pronounced during the charging process. The melting and solidification times are decreased by

Fig. 6.73 SEM images of acid-intercalated graphite (**a**), expended graphite (EG) by microwave (**b**, **c**) and exfoliated graphite nanoplates (**d**). Reproduced with permission from [194], Copyright @ 2007, Elsevier

about 65.3 and 26.2 %, respectively, as compared with those in which pure paraffin is used.

Similar to EG, exfoliated graphite nanoplates have also been used to form graphite/paraffin composite PCM [194]. The nanoplatesusually have thickness of less than 10 nm, with dimension determined by the synthesis methods. In literature, if chemical methods are used to exfoliate EG, the nanoplates are called graphene oxide (GO). GO is reduced to form reduced graphene oxide (rGO) or graphene (G). Representative SEM images of the acid-intercalated graphite, expanded graphite, and exfoliated graphite nanoplates are shown in Fig. 6.73. Figure 6.74 shows SEM images of the paraffin based composite PCMs with low concentrations of graphite nanoplates. The optimum content of the nanoplates is 7 %, in terms of thermal conductivity. More importantly, the values of latent heat of the composite PCMs are very close to that of pure paraffin. This observation has been attributed to the fact that the graphite nanoplates have high surface area and thus homogeneously dispersed in the PCMs. It implies that graphite nanoplatesare more effective than EG to increase the thermal conductivity of PCMs.

The application of graphite to improve the thermal conductivity of low melting point PCMs has been extended to high melting point PCMs, such as salts and eutectic mixtures [195, 196]. Generally, such composite PCMs are made by

Fig. 6.74 SEM images of the paraffin based composite PCMs with 5 wt% (**a**) and 2 wt% (**b**) exfoliated graphite nanoplates. Reproduced with permission from [194], Copyright @ 2009, Elsevier

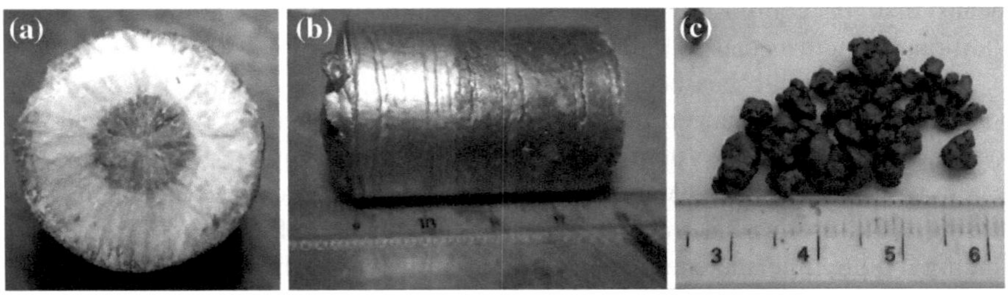

Fig. 6.75 Composites of $NaNO_3/KNO_3$ eutectic and dispersed graphite. **a** Bad graphite dispersion obtained under too low stirring, **b** SFG/75/10, **c** NG/400/22. Reproduced with permission from [196], Copyright @ 2007, Elsevier

mixing the graphites with molten PCMs. Due to the high viscosity of salt PCMs and the similarity in density between salts and graphite, graphite particles can be easily dispersed in molten salts. Therefore, PCMs made in this way are mostly isotropic composites. Figure 6.75 shows photographs of representative composite PCM samples made with different methods.

However, this method has several shortcomings. For instance, it not very cost-effective, because the melting of the PCMs with high melting points consumesis very energy consuming. Also, corrosion on equipment and safety issues have been considered to be quite serious. To address this problem, cold compressing has been proposed to prepare the PCM composites, in which graphite powders and the salt PCMs are first mixed physically and then compressed at room temperature. Because most of the pores can be removed during the cold compression, the PCM composites made this way are monolithic. Such monolithic composites can be easily shaped into any geometry for the storage systems. Comparatively, the PCM composites obtained by using cold compression method are anisotropic, due to the rearrangement of graphite layer orthogonally to the compression axis. For example, the radial thermal conductivity is much higher than the axial one. The cold compressed PCM composites usually have higher thermal conductivity than those prepared by mixing graphite and molten PCMs. In addition, due to the presence of impurities and mechanical stresses, cold compressed PCM composites are form-stable.

The effect of graphite particle size has been examined, in a study of PCM composites based on eutectic salts [196]. The thermal conductivity of the PCM composites is dependent not only on the amount of graphite but also on the particle size. At high concentrations of graphite, larger particle size is preferred in terms of thermal conductivity, because larger particles provide better conductive network. However, in terms of storage capacity, the smaller the particles the higher the performance of the systems. At low concentrations, the graphite particles should be relatively smaller when concerning thermal conductivity, whereas the TES capacity is not affected by the particle size. Also, the time of total melting is greatly reduced for the PCM composites, as compared with the pure PCMs, which means that the composites have almost a congruent melting point. This is an additional benefit because it allows for isothermal operation.

Although graphite has been shown to be a promising additive to form composite PCMs with enhanced thermal conductivity, thus leading to improved performance of PCM TES systems, the preparation of the graphite-PCM composites is relatively complicated. Various steps or procedures are involved in the processing, such as thermal treatment, drying, mixing, grinding, and compression, making it quite time and energy consuming. Furthermore, the characteristics of the porosity of graphite are crucial in determining the effectiveness of the PCM composites [197]. If the graphite has a small mean pore size, its presence could decrease the latent heat of the PCMs, because molecules can be trapped in small size pores, which hinders the molecular motion of the PCMs. However, large pore size means small capillary force, which results in leakage of liquid PCMs. To solve the problems, it has been suggested to use high conductivity metal particles, which are dispersed into the PCMs to have higher thermal conductivity.

One example is to use micro Al particles to increase thermal conductivity and thus the performance of PCM based solar collectors [198]. The heat transfer rate can be increased by increasing the mass fraction of Al particle, up to 0.5 mass fraction, above which the increase in heat transfer rate becomes negligible. Because the thermal conductivity is increased, the time for charging can be decreased by 60 %, as compared with that of the system with pure PCM. This effect is even more pronounced during the discharging process, which is because the conduction-dominated solidification is a homogeneous process.

A quantity, called mean daily efficiency, which is defined as the ratio of the heat gained by the water during the discharging to total incoming solar radiation during the charging, has been used to evaluate the overall benefit of the storage system with the embedding of Al particles. It is found that there is a significant increase in mean daily efficiency. The highest mean daily efficiency can be as high as 94 % for the system with the composite PCM, whereas the one with pure PCM has an efficiency of only 55 %. Unfortunately, there are no data on the effect of Al particle on storage capacity of the system [198].

There has been a report on the effect of Cu particle [199]. Theoretical study indicates that the latent heat of PCMs decreases with increasing mass fraction of embedded Cu particles. The solidification of nanofluid, water with nano Cuparticles, in a square storage model, has been simulated. The speed of the solid/liquid interface during solidification has been examined, for pure PCM and the PCMs embedded with different mass fractions of Cu particles. It is found that, at earlier stages, the increase in speed of the interface due to the addition of Cu particles is not obvious. However, a rapid increase can be obtained after a certain while, especially for the cases of higher mass fractions. As a result, the overall solidification time can be reduced considerably.

Ag has also been employed to fabricate composite PCMs [200, 201]. One example is the composite made of 1-tetradecanol and Ag nanoparticles, used to study the effect of Ag particles on thermal performance of the PCM [200]. It is found that the thermal conductivity of the PCM is increased with increasing content of Ag particles. Due to the inert property, Ag particles have no chemical reaction between the organic PCM. Therefore, thermal stability of the composite is almost same as that of pure PCM. However, the addition of Ag particles leads to a slight decrease in phase change temperature and latent heat. There is no report on TES efficiency of this kind of PCMs.

Obviously, an increase in fraction of the additive particle means a reduction in the volume occupied by the PCM, thus leading to a decrease in the amount of heat stored [201]. Therefore, it is important to determine the optimum particle fraction, so as to achieve the maximum heat transfer rate with minimum loss in LHS capacity. Theoretical analysis indicates that the optimum mass fraction for maximum ES is dependent on thermal conductivity of the particles and maximum allowable radius for interface location. The higher the thermal conductivity of the particle, the higher the allowable content of the particle will be. Similarly, the optimum mass fraction is increased if the maximum allowable radius for interface increase is increased. The maximum allowable radius for interface is determined by the dimensions of the thermal storage unit. Therefore, the particle thermal conductivity and particle mass fraction must be considered at the same time, in terms of the performance of a whole system.

Another strategy to increase the thermal conductivity of PCMs is to place metal structures into the PCMs. A thin walled hollow cylindrical steel structure, called lessing ring, has been used for such purpose [183]. Figure 6.76 shows photograph of the lessing rings. With the presence of such rings, solidification time can be considerably shortened. The time required for complete solidification with the lessing rings is about 1/9 of that with pure PCM. Comparatively, with longitudinal fins, the complete solidification time is just about 1/4 of the case without fins. Therefore, the use of the lesser rings is more effective than that of fins. However, the large time reduction of solidification is achieved as lesser rings occupy 20 % of the total volume, whereas the fins take just 7 % of the total volume. In other words, to get same reduction in solidification time, more volume fraction will be taken

Fig. 6.76 Photograph of the lessing rings. Reproduced with permission from [183], Copyright @ 1999, Elsevier

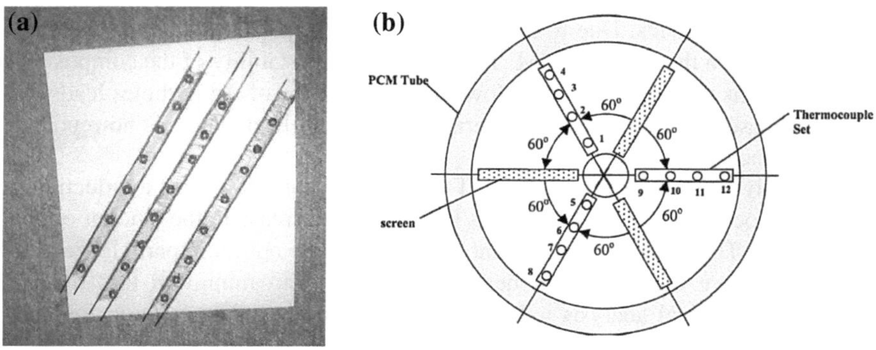

Fig. 6.77 Photograph of the screens with attached spheres (a) and cross-sectional (b) view showing location of the screens. Reproduced with permission from [202], Copyright @ 2004, Elsevier

when using lessing rings than the fins. Also, the enhancement factor due to lessing rings is closely related to the diameter of the cylindrical module. In contrast, in case of fins, it remains almost constant for a constant, irrespective of the diameter of the module, as long the number of fins is unchanged. Therefore, lessing rings are more effective than fins in large storage systems.

Other types of metal structures that have been applied to increase the thermal conductivity of PCMs include metal screens and metal beads [202, 203]. Figure 6.77 shows a photograph of such metal screes and schematic of the arrangement of the screens with other experimental setup. It is found that the enhancement in thermal conductivity due to screens with the metal balls is only dependent on diameter and number of the balls. With increasing diameter and number of balls, thermal conductivity is increased, with the enhancement factors of up to 3.

6.7 Use of High Conductivity and Low Density Materials

It can be expected that metal particles and structures tend to precipitate at the bottom of the container of PCMs, due to their relatively high density. Also, they bring in additional weight to the systems. Furthermore, not all metal particles are compatible with all PCMs. For instance, Al is compatible with paraffin, whereas Cu and Ni are not. Similarly, Al and Cu are not compatible with some salt hydrates. In this respect, additives with low density and high thermal conductivity, as well as compatibility with all PCMs, would be desirable. It is well-known that carbon has a density that is much lower than those of metals while it can offer thermal conductivities of almost equal to that of Al and Cu. As a result, carbon fibers have become most promising alternatives to enhance the thermal performance of PCM TES systems [204]. More importantly, carbon materials exhibit high corrosive resistance and thus compatible with most of the PCMs.

One type of carbon nanofibers, with an average diameter of 100 nm and an average length of 20 μm, has been used to improve the thermal conductivity of paraffin wax [197]. It is found that the thermal conductivity of the composite increases almost linearly with increasing content of the carbon nanofibers. As a result, high solidification rate has been observed. For instance, even with as low as 1 wt% of carbon fibers, the solidification time is decreased by 23 %. Moreover, the storage capacity is not reduced due to the addition of the carbon nanofibers, since the output power is increased linearly with increasing content of the carbon nanofibers. It is also observed that the effect of the carbon nanofibers on the solidification rate has close relation with their surface characteristics. Surface treated nanofibers provide higher heat transfer rate that those untreated ones. In addition, the distribution of the carbon nanofibers is also important.

To show the effect of the way for distribution, the carbon nanofibers are dispersed in the PCM, either randomly or orderly, to form cylindrical capsules [204], carbon fibers were added in the PCM in two ways. Figure 6.78 shows schematics of the ways in which the carbon nanofibers are distributed, random and ordered. Photographs of representative real samples are shown in Fig. 6.79. The effective thermal conductivity of theorderedone is three times higher than that of the random type. This is because all the fibers are arranged in radial direction in the case of ordered distribution, which is the direction of heat flow. At low concentrations of carbon nanofibers, the randomly distributed arrangement offers higher effective thermal conductivity, but does not lead to higher melting rate, as compared with that of pure paraffin. This has been attributed to the fact that the random distribution of carbon nanofibers has a negative effect on the natural convection. In contrast, this negative effect is not observed in the case of ordered distribution.

Further study reveals that 30 and 20 % higher solidification rate and melting rate have been achieved with 1 % mass fraction of orderly distributed carbon nanofibers, as compared with those in the case of pure PCM [205]. It is also found that the ordered distribution (fiber brush) has higher performance than fiber chips [206]. This is because, in the case of fiber chips, higher thermal resistance is

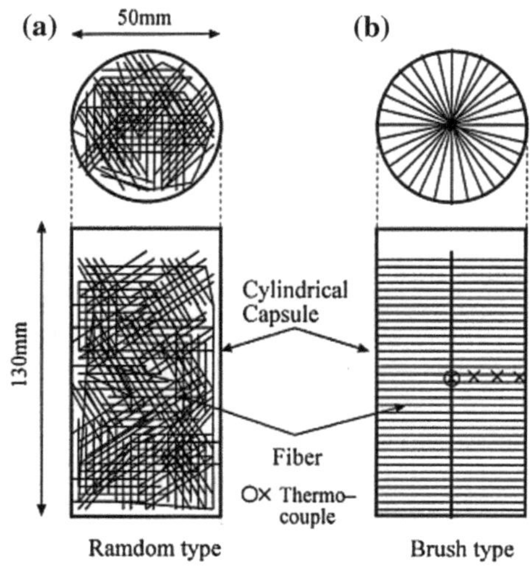

Fig. 6.78 Configurations of the carbon nanofibers within a cylindrical capsule. In **b**, the open circle indicates the location of the thermocouple for the measurement of transient temperature response, while the crosses indicate the locations for the measurement of thermal diffusivity. These thermocouples are also located at the corresponding locations in the random type. **a** Random type, **b** brush type. Reproduced with permission from [204], Copyright @ 2000, Elsevier

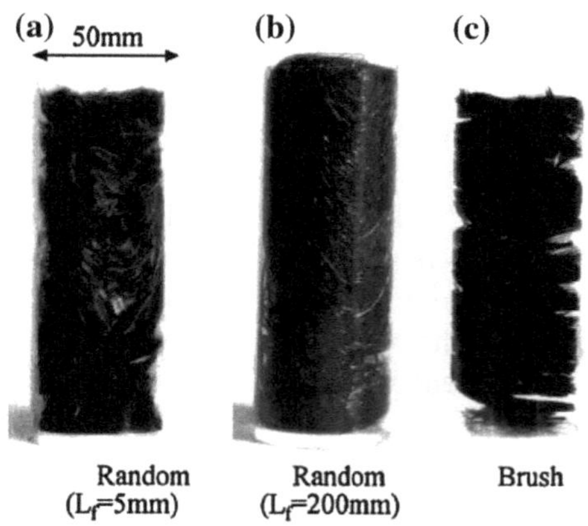

Fig. 6.79 Carbon nanofibers packed into a transparent cylindrical capsule in different ways with a volume fraction of the carbon nanofibers to 0.005. **a** Random (L_f = 5 mm), **b** random (L_f = 200 mm), **c** brush. Reproduced with permission from [204], Copyright @ 2000, Elsevier

observed near the heat transfer tube. As a result, the rate of overall heat transfer is relatively low. When carbon fiber brushes are used to surround the heat transfer tube, as shown in Fig. 6.80, there is a critical diameter for the brushes [207, 208]. Initially, the thermal response of the system increases with increasing brush diameter. As the diameter of the brushes reaches a critical value, beyond that the brushes cross the heat transfer tubes and there is no further improvement in thermal response. This is because beyond this critical diameter, the thermal resistance between the fibers and the tube surface has been increased. Actually, if

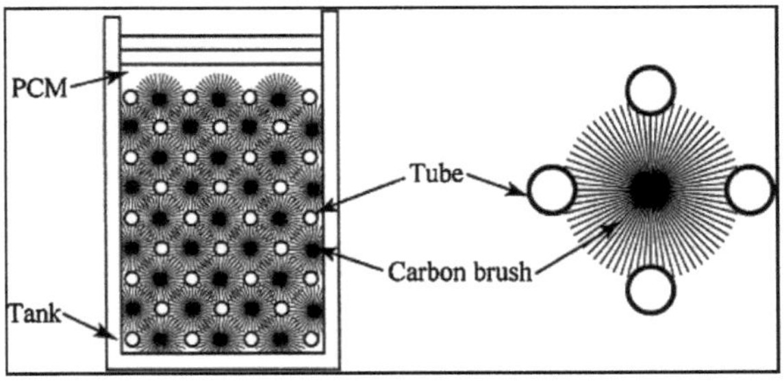

Fig. 6.80 TES units inserted with brushes of carbon nanofibers. Reproduced with permission from [207], Copyright @ 2002, Elsevier

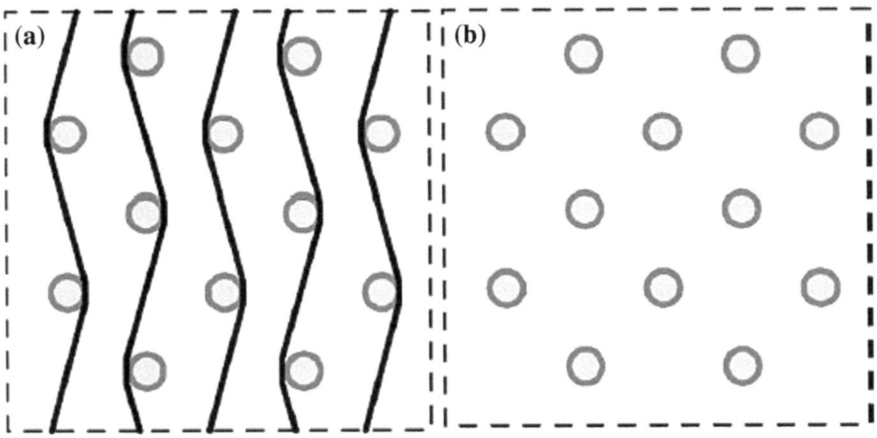

Fig. 6.81 Schematic illustrations of the additives installed in the tanks. **a** Carbon fiber cloths and **b** without additive. Reproduced with permission from [209], Copyright @ 2008, Elsevier

the diameter of the carbon fiber brush is too large, the fibers will leave the heat transfer surface to form regions with low density of the fibers [208]. The formation of such low density regions prevents further improvement in the rate of heat transfer.

A new strategy of using carbon fibers to improve the thermal performance of TES systems has been proposed, i.e., with carbon fiber cloths to replace carbon fiber brushes [209]. The carbon fibers cloths are highly stretched over the heat transfer surface by covering the whole tubes. The arrangement of the carbon fiber cloths is shown in Fig. 6.81. Photographs of representative carbon fiber cloths are shown in Fig. 6.82. It is found that these carbon fiber cloths with high surface density have the functions of fins. The fiber cloths with 0.42 % volume fraction has the same performance as the fiber brushes with 0.75 % volume fraction, in terms of increasing thermal conductivity. In is respect, carbon fiber cloths are superior to carbon fiber assemblies (bushes or chips).

Fig. 6.82 Photographs of carbon fiber cloths. **a** 142 g m^{-2} and **b** 304 g m^{-2}. Reproduced with permission from [209], Copyright @ 2008, Elsevier

Another group of nanosized carbon materials, carbon nanotubes (CNTs), has also been employed to improve the thermal performance of PCMs for TES applications [210–214]. Depending on the number of layers, CNTs are classified into three groups, single-walled (SW), double-walled (DW), and multiwalled (MW). SW and DW CNTs are much more expensive than MWCNTs, due to the difference in methods used to synthesize them. It is expected that more and more works will use CNTs as additives to fabricate high performance PCMs, with the advancement in their large-scale production.

A perceptive sense is that such nanosized carbon materials must be very expensive, so that the use of them in TES systems would definitely increase their cost. However, it is very fortunate to notice an opposite trend. For example, with the application of carbon fiber brushes, the storage unit will occupy less space and have lowered cost [215]. A study on carbon fiber brushes inserted PCM TES system used for air conditioning indicates that it takes smaller space, as compared with the one of sample storage capacity without using the fiber brushes. At the same time, for the same thermal output, the cost can be lowered by 25 %.

In summary, the rate of phase change of PCMs, melting and solidification, can be increased considerably by incorporating high conductivity components to form PCM composites. Although they are able to offer high thermal conductivity PCMs, these additives usually result in loss of storage capacity, as compared with the case of pure PCM. The loss of storage capacity is dependent on the usage level of the additives. Therefore, when using such additives, it is necessary to optimize their concentrations, so as not to lose TES capacity.

The last strategy is called microencapsulation, which means that PCMs are made into microsized particles in liquid/solid (core) form and then encapsulated within a solid structure (shell/wall). The shell can be made from a wide range of materials, mainly natural and synthetic polymers [216–218]. Microencapsulation of PCMs can be realized by two using methods, chemical (such as coacervation, complex coacervation and interfacial interaction) and physical or mechanical

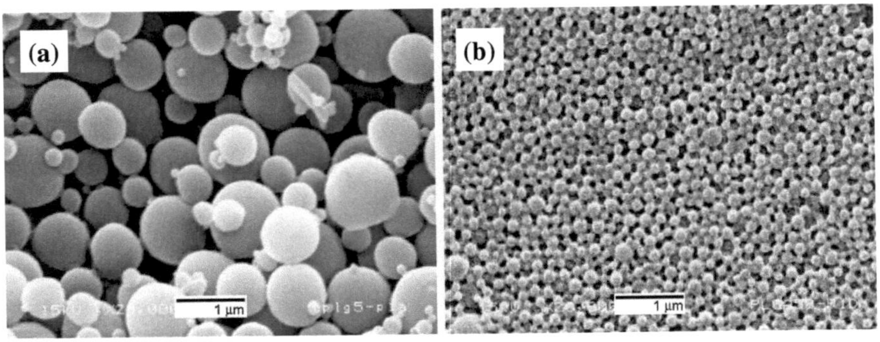

Fig. 6.83 SEM images of the microencapsulated PCM particles by using different methods. **a** Coacervated and **b** spray-dried. Reproduced with permission from [216], Copyright @ 2003, Elsevier

(such as spray-drying method and mechanical activation). The microscopic profiles of microencapsulated PCMs obtained from coacervation and spray-drying methods are shown in Fig. 6.83 [216]. Both are well-shaped spherical particles, with average particle sizes of 1 and 0.2 µm, respectively.

The microencapsulated PCMs are expected to have much better thermal performance than their original PCMs, because small PCM particles provide larger heat transfer area per unit volume and thus higher heat transfer rate. In addition, microencapsulated PCMs have exhibit various other advantages, such as prevention of reactivity of the PCMs with container materials, higher thermal stability, and higher mechanical strength to withstand the volume change during phase transition. An example is shown in Fig. 6.84 [219]. The encapsulated paraffin particles have almost no change in their geometrical profile after 1,000 thermal cycles. These advantages, together with their high capability of incorporation, make the microencapsulated PCMs very attractive in improving the thermal performance for various applications, including building materials, textile fabrics, and thermal fluids [220–223].

Due to the expected large heat transfer area per unit volume, microencapsulated PCMs should offer higher rate of heat transfer than their pure counterparts. Higher transfer rate means higher ES and release capacities [216, 224] However, it is observed in a recent study that a microencapsulated fatty acid only exhibits higher heat transfer rate during charging [225]. During discharging, pure PCM possesses higher heat transfer rate. Similar observation has never reported for other PCMs. Careful inspection reveals that the fatty acid mixture studied has a melting point range of 22–24 °C, which implies that un-capsulated fatty acid should be in liquid state at room temperature. In this case, during charging, the energy is stored in the form of sensible heat. In contrast, the microencapsulated fatty acid has slightly higher melting point range of 29–31 °C. Therefore, phase change takes place when it is heated from room temperature. It is this difference in melting point between the encapsulated and un-encapsulated cases. As a result, when evaluating the

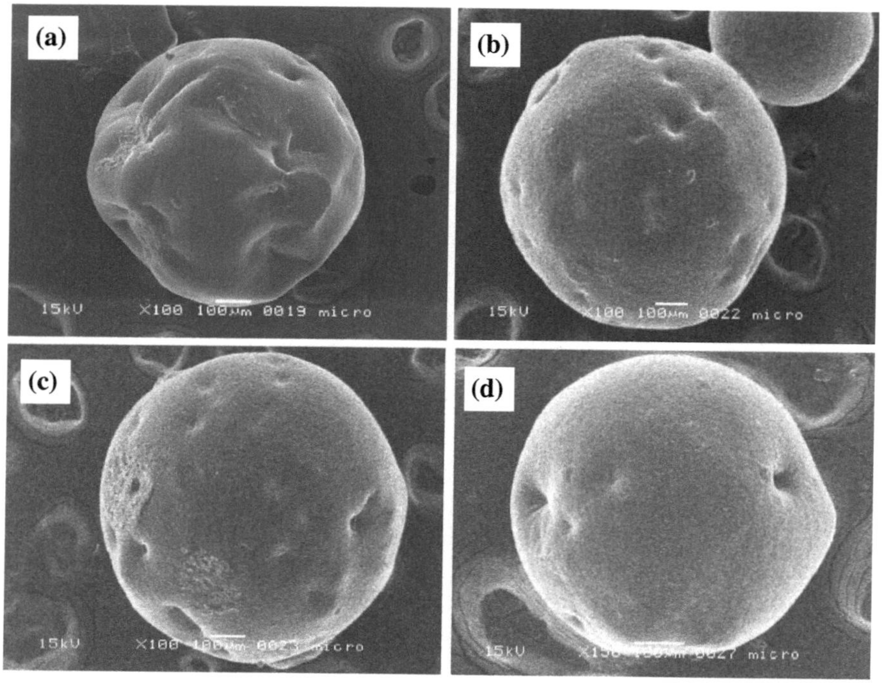

Fig. 6.84 SEM images of microencapsulated paraffin particles after different thermal cycles. **a** 0, **b** 200, **c** 500, and **d** 1000. Reproduced with permission from [217], Copyright @ 2002, John Wiley & Sons

thermal performance of any microencapsulated PCM, this difference must be taken into account. In a separate study, it is reported that the rigidity of the microencapsulated PCM matrix dampers the natural convection and thus reduces the heat transfer rate [226]. All these results, sometimes somehow contradictory, remind us that comprehensive and systematic studies are necessary to clarify the effect of encapsulation on the performance enhancement of PCM TES systems, which is of great significance for practical applications.

Besides the microencapsulation, nanoencapsulation has brought increasing attention of the research community in this area [227–229]. The principle of nanoencapulation is the same as that of microencapsulation, the only difference is their dimension. Because nanoencapsulation provided even larger surface area for heat transfer, it should be much more effective than microencapsulation, in terms of enhancing the thermal performances of PCMs.

For both micro and nanoencapsulation, the ratio of the PCM encapsulated and the encapsulating shell material plays an important role in determining the efficiency of the above mentioned enhancement [216, 224]. This is simply because the more the shell material used, the less the PCM that can be used. In this light, chemical method (complex coacervation) is more promising than physical method

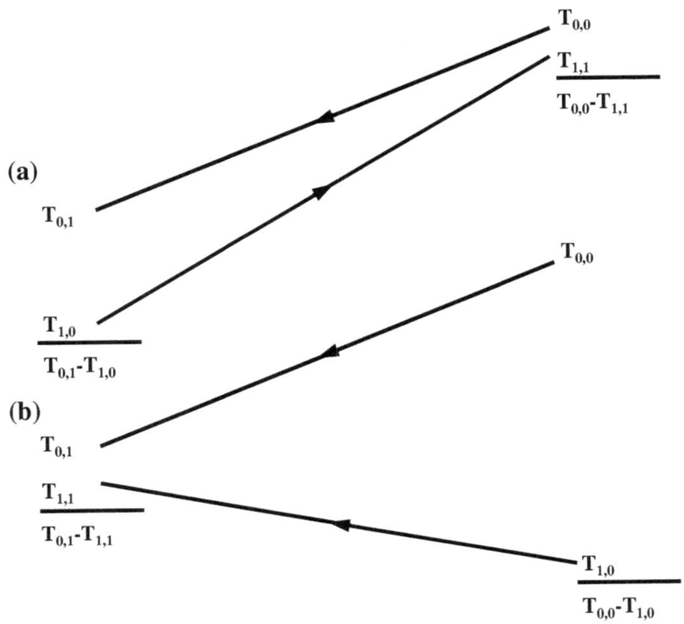

Fig. 6.85 Schematics of counter-flow (**a**) and parallel-flow (**b**) in heat exchangers

(spray-drying method), because the latter requires thinner shell. However, physical method is simpler and more cost-effective than chemical method. Therefore, it is also a process of optimization when selecting materials and methods to apply the encapsulation.

6.7.1 Effect of Heat Exchanger

PCMs not only have low thermal conductivity as discussed above, they but also expand on melting contract on consolidation. Therefore, efficiency of heat exchanger plays a significant role in determining the performance of a PCM TES system. General descriptions and design considerations of heat exchangers will be presented in this subsection.

According to flow arrangement, there are four primary categories of heat exchangers, counter-flow, and parallel-flow, as shown in Fig. 6.85. In counter-flow heat exchangers the fluids enter the exchanger from opposite ends. The counter current design is the most efficient, in that it can transfer the most heat from the heat (transfer) medium due to the fact that the average temperature difference along any unit length is greater. In parallel-flow heat exchangers, the two fluids enter the exchanger at the same end, and travel in parallel to one another to the other side. In a cross-flow heat exchanger, the fluids travel roughly perpendicular to one another through the exchanger.

To maximize the efficiency, heat exchangers are designed to have maximum surface area of the wall between the two fluids, while minimizing resistance to fluid flow through the exchangers. As discussed before, the performance a heat exchanger can also be affected by the addition of fins or corrugations in one or both directions, which increase surface area on the one hand and may channel fluid flow or induce turbulence on the other hand.

The driving temperature across the heat transfer surface varies with position, but an appropriate average temperature can be used to characterize the difference in temperature between the two fluids. In most simple systems, this is called log mean temperature difference (LMTD). Sometimes, direct knowledge of the LMTD is not available and the (number of transfer unite) NTU method is used.

According to configurations and arrangement, there are various types of heat exchangers, double pipe heat exchanger, shell and tube heat exchanger, plate heat exchanger, plate and shell heat exchanger, adiabatic wheel heat exchanger, plate fin heat exchanger, pillow plate heat exchanger, to name a few. The first four types of heat exchangers will be briefly described as follows.

Double pipe heat exchangers are the simplest exchangers that have been widely used in various industries. On one hand, these heat exchangers are cheap for both design and maintenance, making them a good choice for small industries. However, on the other hand, they have relatively low efficiency. Furthermore, these types of heat exchangers usually occupy large spaces, because of their large scales. To design a double pipe heat exchanger, the first step is to calculate its heat duty. Generally, for easier design, heat loss can be ignored for primary design. The heat duty is defined as the heat gained by cold fluid which is equal to the heat loss of the hot fluid.

Shell and tube heat exchangers consist of a series of tubes. One set of these tubes contains the fluid that are either heated or cooled. The second fluid runs over the tubes that are to be heated or cooled, so that it can either provide the heat or absorb the heat required. Schematics of a simple shell and tube heat exchanger with counter-flow and parallel-flow arrangements are shown in Fig. 8.86. A set of tubes is called the tube bundle and can be made up of several types of tubes: plain, longitudinally finned, etc. Shell and tube heat exchangers are typically used for high-pressure applications, with pressures greater than 30 bar and temperatures higher than 260 °C. This is because the shell and tube heat exchangers are robust due to their shape.

There are several features in thermal design that must be taken into account when designing the tubes in the shell and tube heat exchangers, such as tube diameter, tube thickness, length, pitch, layout, and buffer, which are described respectively as follows.

In terms of economical and compact consideration, the thinner the tube the better it will be. However, if a heat exchanger continuously works, it is very easy for the tubes to be dirty. Dirty tubes with small are very difficult to be cleaned. In this respect, the tubes cannot be too thin. Therefore, the determination of the

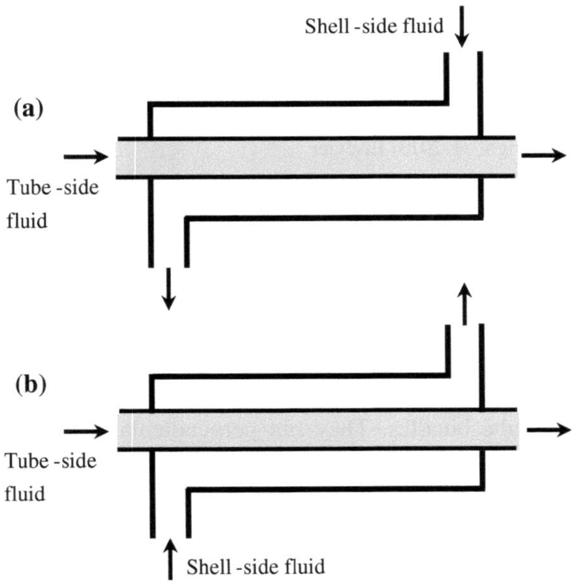

Fig. 6.86 Schematics of counter-flow (**a**) and parallel-flow (**b**) in a simple shell and tube heat exchanger

diameter of a tube must be comprehensively considered, with available space, budget, and the fouling nature of the fluids used. As for thickness of the wall of the tubes, it is usually determined to ensure that there is enough room for corrosion and the flow-induced vibration has resistance. Other factors related the tubes include axial strength, availability of spare parts, hoop strength to withstand internal tube pressure, and buckling strength to withstand overpressure in the shell.

To reduce the cost of a heat exchanger, an effective way is to use shell with smaller diameter and a tube with longer length. Therefore, one the design principles is to make the heat exchanger as long as physically possible as long as the production capabilities are not exceeded. However, in practice, there are various factors that limit this design, such as space availability at the installation site and the need to ensure that the tubes are available in length, which is twice the required length so that they can be withdrawn and replaced. Also, it is a difficult task to maintain or replace long and thin tubes.

Tube pitch, the center–center distance of adjoining tubes, is another important design parameter. It should not be less than 1.25 times the outside diameter of the tubes. A larger tube pitch leads to a larger overall shell diameter, which will increase the cost of the heat exchanger. For inner tubes, the use of corrugation can increase the turbulence of the fluids and thus enhance the efficiency of heat transfer.

Tube layout refers to how tubes are positioned within the shell. There are four types of tube layout, including triangular (30°), rotated triangular (60°), square (90°), and rotated square (45°). The triangular pattern has highest heat transfer efficiency because it forces the fluid to flow in a more turbulent fashion around the

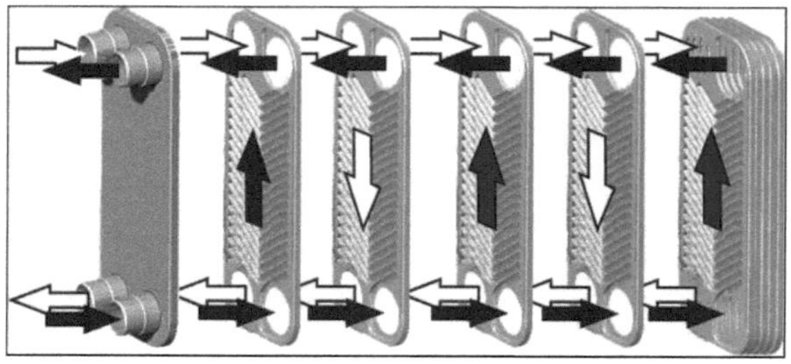

Fig. 6.87 Schematic of the design of a 1/1 pass countercurrent plate heat exchanger. Reproduced with permission from [230], Copyright @ 2013, Elsevier

piping. Square pattern is used where high fouling is experienced and cleaning is more regular.

Baffles are often used in shell and tube heat exchangers to direct fluids across the tube bundles. They run perpendicularly to the shells and hold the bundles, preventing the tubes from sagging when they are too long. They can also prevent vibration of the tubes. Segmental baffles are the most commonly used type of baffle. Semicircular segmental baffles are oriented by 180° to the adjacent baffles, in order to force the fluids to flow upwards and downwards between the tube bundles. Baffle spacing is a thermodynamic consideration in the design shell and tube heat exchangers, concerning the conversion of pressure drop and heat transfer. For thermoeconomic optimization, the baffles should be spaced more than 20 % of the inner diameter of the shells. If the baffles are spaced too closely, there will be a large pressure drop due to the flow redirection. However, if they are spaced too far apart, there will be cooler spots at the corners between baffles. Also, it is important to ensure that the baffles are spaced close enough so that the tubes will not sag. There are also disk and donut baffles, which consist of two concentric baffles. An outer and wider baffle has a shape of donut, while the inner baffle is disk-shaped. This type of baffle forces the fluid to pass around each side of the disk through the donut baffle, so as to generate a different type of fluid flow.

Another type of heat exchanger is the plate heat exchanger. A representative plate heat exchanger is shown in Fig. 6.87 [230]. It is composed of a set of thin and slightly separated plates that have very large surface areas and fluid flow passages for heat transfer. This stacked-plate arrangement makes them more effective than the shell and tube heat exchangers, in a given space. With the rapid advancement in gasket and brazing technology, the plate-type heat exchangers have become increasingly practical. In heating, ventilation and air conditioning (HVAC) applications, large heat exchangers of this type are called plate-and-frame. When used in open loops, these heat exchangers are normally of the gasket type to allow periodic disassembly, cleaning, and maintenance. There are many types of permanently bonded plate heat exchangers, such as dip-brazed, vacuum-brazed, and welded plate varieties, which are often specified for closed loop applications, such as refrigeration. Plate heat exchangers also can be different in the types of plates that are used and in the configurations of those plates. Some plates may be stamped with "chevron," dimpled, or other patterns, where others may have machined fins and/or grooves.

Plate and shell heat exchangers are a combination of plate heat exchanger and shell and tube heat exchanger. The key part of a heat exchanger contains a fully welded circular plate-pack made by pressing and cutting round plates and welding them together. Nozzles carry flow in and out of the plate-pack, i.e., the 'plate side' flowpath. The fully welded plate-pack is assembled into an outer shell that creates a second flow path, i.e., the 'shell side' flow path. The combined technology of plate and shell offers high heat transfer efficiency, high pressure, high operation temperature, compact size, low fouling, and close approach temperature. In particular, it can work completely without gaskets, which provides security against leakage at high pressures and temperatures.

A finned pipe heat exchanger, which consists of a container of a square cross-section, provided with heat pipe along its longitudinal axis, has been designed and demonstrated [231]. The outer surface of the tube, which comes in contact with PCM, was chromed due to considerations of material compatibility. The container and the heat pipe are divided into three regions, heat sources, storage, and heat sink. The heat pipe in the storage region is arranged with closely and equally spaced circular Al fins.

In an active storage system for solar space heating and solar space cooling, a small diameter close-spaced plastic tubing mat that is coiled into a spiral and inserted into a cylindrical tank containing PCM, is used as heat exchanger [232]. The PCM in the system is salt hydrate, consisting of 98 % sodium thiosulfate pentahydrate and 2 % sodium sulfate for space heating applications and magnesium chloride hexahydrate for space cooling units. The plastic heat exchanger has the advantage of low cost and anticorrosive property. The system can be operated for 1,000 cycles without any degradation in the performance of the whole system.

A heat exchanger consists of a rectangular container for the PCM with a rectangular flow channel underneath it has been studied by Smith et al. [233]. Gulf wax 33, with melting point of 50–55 °C, is used as the PCM. A limitation in the heat transfer due to resistance of the solid phase has been identified. However, the resistance can be significantly reduced by using an additional heat path provided by the conducting container.

A PCM TES, with a shell and tube type heat exchanger without fins, has been designed for recovery of low temperature industrial waste heat [234]. It has been found that the effective thermal conductivity of the system can be improved, when the radial distance among the tubes is kept 3–4 cm. The thermal performance of the heat exchanger, for charging and discharging process of PCM, at different mass flow rates and temperatures of the inlet water, has been systematically studied. Commercial grade stearic acid has been used as the PCM and filled up to about 90 % of the volume. Due to the poor thermal conductivity of the PCM, the overall heat transfer coefficient still needs to be further increased for practical applications.

In another experimental investigation of shell and tube heat exchanger, it has been found that the water inlet temperature and mass flow rate have great influences on the times of charging and discharging, as well as the temperature gradient for heat flow into or out of the two different storage substances used, i.e., paraffin wax and asphalt [235]. In this study, the sphalt is for the first time used as an ES material to test the storage capacity.

A rolling cylinder heat exchanger consists of a collection of smaller cylinder shaped sub-units, each of which is able to store energy [236]. In that system, a number of heat exchanger modules are operated in parallel. Each module consists of a number of cylinders mounted on a larger one, so that they can be rotated as a single unit. The objective of the rotation was to provide complete freezing. The problem in the coupling of the storage to the thermal system has been investigated, as a reference for practical system design. Another cylinder heat exchanger consists of cylindrical capsules that are filled with PCMs [237]. Three commercial waxes with melting temperatures of 44, 53, and 64 °C are used as PCMs. Longitudinal and transverse rectangular arrangements are used for the tubes.

In summary, there are no standardized criteria for the design of heat exchangers in PCM TES systems, probably because there are too many PCMs for different applications. A complete and systematic review on state-of-the-art of the heat exchangers for such purposes is also not available. Therefore, it is quite urgent to have such a review or similar for further advancement of this specific area.

6.7.2 Corrosion Aspects

The most important factor that have limited the widespread application of LHSs are the useful life of the PCMs-container systems and the number of cycles that they can withstand without any degradation in their properties. Insufficient long term stability of the storage materials has been related to two factors, poor stability of the materials properties and/or corrosion between the PCM and the container [238]. Most investigations on corrosion tests using PCMs were performed with salt hydrates [239, 240].

Stainless steels are usually used to make the hot tank in two-tank storage systems. In a test of resistance of stainless steel during thermal cycling in molten salts, three stainless steel samples are studied with four different mixtures of nitrate salts [241, 242]. In the experiment, three stainless steel coupons are immersed in crucibles that contained about 10 kg of the nitrate mixtures. The cycling period is in 7.5 h at the maximum temperature and 0.5 h cooling in ambient air, for a total cycle time of 8 h. This schedule allowed three cycles per day, for a total of 500 thermal cycles during 4,000 h of operation. For comparison, three more samples are tested under an isothermal immersion of molten nitrate salts. It is found that the three stainless steel samples have experienced thermal cycling exhibit higher corrosion rate, as compared with those isothermally immersed in the molten salt. Depending on the chloride content of the molten salt, the increase in corrosion rate is between 25 and 50 % [241, 242].

In applications where exposure to nitrate salts can be limited to 400 °C or less, carbon steels are promising candidate for cold tank. The effect of dissolved impurities, such as chloride and sulfate, on corrosion, has been studied, by comparing with those of pure nitrate melts. Experimental results indicate that the corrosion of mild steel at 400 °C increases nearly as a logarithm relation to the chloride concentration [243]. At 0.6 wt% NaCl, the corrosion rate increases by a factor of about three as compared with a chloride-free melt, after an 8 h test. Corrosion rates increases by a factor of about four after a 25 h test, when the chloride concentration is 0.7 wt%. The effect of dissolved sulfate in nitrate melts on corrosion rate of mild steel results has also been evaluated. An increase in corrosion rate is observed by about 20 % when 7.5 wt% Na_2SO_4 is added to the pure molten salt [244].

6.8 Concluding Remarks

Information that is available in the open literature regarding PCM TES systems for power generation by harvesting waste thermal energy has been summarized. The requirements for a high performance thermal storage system include high energy density in the storage material (storage capacity), good heat transfer between HTF and the storage medium, mechanical, and chemical stability of the storage materials and the materials used in the whole systems, compatibility between HTF, heat exchanger and/or storage medium (safety), complete reversibility of a number of charging/discharging cycles (lifetime), low thermal losses, ease of control and operation, maximum loading capability, nominal temperature, and specific enthalpy drop in load and integration into power plant when desired.

According to the storage concept, PCM storage systems can be classified into two groups, active and passive. In active systems, convection heat is transferred into the storage material, either directly with the storage media (direct systems) or indirectly, in which the HTF and the storage medium are different materials. In passive systems, the thermal storage medium is not circulated.

Although there are several investigations that have been published in the open literature, on the corrosion effect of molten salts on the container materials, more and extensive studies are urgent in this sub-area of PCM TES. Moreover, most commercially available PCMs have impurities and the impurities of the PCMs from different sources can have different types and concentrations. Therefore, one case study may not be applicable to other cases.

In terms of theoretical analysis, exergy based performance evaluation of PCM storage systems units and their performance optimization should be further studied, because it is a more perspective measure than energy based one, due to its reflection of true potential of a system and the economic value of the storage and recovery operation. It is necessary to mention that exergy analysis of any form should consider the complete cycle operation of a system rather than only consider either storage or recovery mode independently. For optimization of a storage module, exergy analysis should be conducted along with energy analysis. Entropy generation analysis has a significant importance especially in the optimization of design and operating parameters of the systems.

References

1. A. Gil, M. Medrano, I. Martorell, A. Lazaro, P. Dolado, B. Zalba, L.F. Cabeza, State of the art on high temperature thermal energy storage for power generation. Part 1—concepts, materials and modellization. Renew. Sustain. Energy Rev. **14**, 31–55 (2010)
2. S.M. Hasnain, Review on sustainable thermal energy storage technologies, part I: heat storage materials and techniques. Energy Convers. Manage. **39**, 1127–1138 (1998)
3. A. Sharma, V.V. Tyagi, C.R. Chen, D. Buddhi, Review on thermal energy storage with phase change materials and applications. Renew. Sustain. Energy Rev. **13**, 318–345 (2009)
4. G.A. Lane, H.E. Rossow, Encapsulation of heat of fusion storage materials, in *Proceedings of 2nd South Eastern Conference on Application of Solar Energy*, pp. 442–55 (1976)
5. R. Biswas, Thermal storage using sodium sulfate decahydrate and water. Sol. Energy **99**, 99–100 (1977)
6. B. Charlsson, H. Stymmeand, G. Wattermark, An incongruent heat of fusion system $CaCl_2 \cdot 6H_2O$ made congruent through modification of chemical composition of the system. Sol. Energy **23**, 333–350 (1979)
7. S. Herrick, A rolling cylinder latent heat storage device for solar heating/cooling. ASHRAE Tans. **85**, 512–515 (1979)
8. D. Kearney, B. Kelly, U. Herrmann, R. Cable, J. Pacheco, R. Mahoney, H. Price, D. Blake, P. Nava, N. Potrovitza, Engineering aspects of a molten salt heat transfer fluid in a trough solar field. Energy **29**, 861–870 (2004)
9. U. Herrmann, B. Kelly, H. Price, Two-tank molten salt storage for parabolic trough solar power plants. Energy **29**, 883–893 (2004)
10. D. Brosseau, J.W. Kelton, D. Ray, M. Edgar, K. Chisman, B. Emms, Testing of thermocline filler materials and molten-salt heat transfer fluids for thermal energy storage systems in parabolic trough power plants. J. Sol. Energy Eng. Trans. ASME **127**, 109–116 (2005)
11. J. Stefan, Uber einge problem der theoric der warmeleitung, S. B. Wein. Acad. Mat. Natur. **98**, 173–484 (1989)
12. H.S. Carslaw, J.C. Jager, *Conduction of Heat in Solids*, 2nd edn. (Oxford University Press, London, 1973)
13. V.J. Lauardini, *Heat Transfer in Cold Climates* (Van Nostrand, New York, 1981)
14. T.R. Goodman, The heat balance integral and its application in problems involving a change. J. Sol. Energy Eng. Trans. ASME **80**, 335–342 (1958)
15. L.T. Yeh, B.T. Chung, Solidification and melting of material subjected to convention and radiation. J. Space Cr. Rockets **12**, 329–334 (1975)
16. J. Crank, R. Gupta, Isothermal migration in two dimensions. Int. J. Heat Mass Transf. **18**, 1101–1117 (1975)
17. D. Buddhi, N.K. Bansal, R.L. Sawhney, M.S. Sodha, Solar thermal storage systems using phase change materials. Int. J. Energy Res. **12**, 547–555 (1988)
18. A. Lazaridas, A numerical solution of the multidimensional solidification (or melting) problem. Int. J. Heat Mass Transf. **13**, 1459–1477 (1970)
19. C. Bonacina, G. Comini, A. Fasano, M. Primicerio, Numerical solution phase change problems. Int. J. Heat Mass Transf. **16**, 1825–1832 (1973)
20. G. Comini, S. Del Guidice, R.W. Lewis, O.C. Zienkiewicz, Finite element solution of non-linear heat conduction with phase changes. Int. J. Numer. Methods Eng. **8**, 613–624 (1974)
21. III W. Rolph, K.J. Bathe, An efficient algorithm for analysis of nonlinear heat transfer with phase change. Int. J. Numer. Methods Eng. **18**, 119–134 (1982)
22. J. Yoo, B. Rubinsky, Numerical computation using finite elements for the moving interface in heat transfer problems with phase change transformation. Numer. Heat Transf. **6**, 209–222 (1983)
23. R. Bounerot, P. Janet, Numerical computation of free boundary for the two dimensional Stefan problems by space-time finite elements. J. Comput. Phys. **25**, 163–181 (1977)

24. N. Shamsunder, E. Sparrow, Analysis of multidimensional phase change via the enthalpy model. J. Heat Transf. Trans. ASME **19**, 333–340 (1975)
25. S.E. Hibbert, N.C. Markatos, V.R. Voller, Computer simulation of moving interface, convective, phase change process. Int. J. Heat Mass Transf. **31**, 1785–1795 (1988)
26. C. Bonacina, G. Cominl, A. Fasano, M. Primicerio, On the estimation of thermophysical properties in nonlinear heat-conduction problems. Int. J. Heat Mass Transf. **17**, 861–867 (1974)
27. M. Costa, D. Buddhi, A. Oliva, Numerical simulation of a latent heat thermal energy storage system with enhanced heat conduction. Energy Convers. Manage. **39**, 319–330 (1998)
28. A.J. Dalhuijsen, A. Segal, Comparison of finite element techniques for conduction problems. Int. J. Number. Meth. Eng. **23**, 1807–1829 (1986)
29. C. Wen, J.W. Sheffled, M.P. O'Dell, J.E. Leland, Analytical and experimental investigation of melting heat transfer. J. Thermophys. Heat Transf. **3**, 330–339 (1989)
30. C.R. Swaminathan, V.R. Vollar, On the enthalpy method. Int. J. Num. Meth. Heat Fluid Flow **3**, 233–244 (1993)
31. V.R. Voller, M. Cross, N.C. Markatos, An enthalpy method for convection/fusion phase change. Int. J. Numer. Methods. Eng. **24**, 271–284 (1987)
32. A. Laoud, M. Lacroix, Thermal performance of a latent heat energy storage ventilated panel for electric load management. Int. J. Heat Mass Transf. **42**, 275–286 (1999)
33. J. Bansszek, R. Domanski, M. Rebow, F. El-Sagier, Experimental study of solid-liquid phase change in a spiral thermal energy storage unit. Appl. Therm. Eng. **19**, 1253–1277 (1999)
34. M. Lacroix, Numerical simulation of a shell and tube latent heat thermal energy storage unit. Sol. Energy **50**, 357–367 (1993)
35. B. Zivkovic, I. Fujji, An analysis of isothermal phase change of phase change material within rectangular and cylindrical containers. Sol. Energy **70**, 51–61 (2001)
36. P. Lamberg, K. Siren, Analytical model for melting in semi-infinite PCM storage with an internal fin. Heat Mass Transf. **39**, 167–176 (2003)
37. V.R. Voller, Fast implicit finite-difference method for the analysis of phase change problems. Numer. Heat Transf. Part B **17**, 155–169 (1990)
38. A. Sharma, L.D. Won, D. Buddhi, J.U. Park, Numerical heat transfer studies of the fatty acids for different heat exchanger materials on the performance of a latent heat storage system. Renew. Energy **30**, 2179–2187 (2005)
39. C.R. Chen, A. Sharma, Numerical investigation of melt fraction of PCMs in a latent heat storage system. J. Eng. Appl. Sci. **1**, 437–444 (2006)
40. C.R. Chen, A. Sharma, S.K. Tyagi, D. Buddhi, Numerical heat transfer studies of PCMs used in a box type solar cooker. Renew. Energy **33**, 1121–1129 (2008)
41. A. Sharma, S.D. Sharma, D. Buddhi, L.D. Won, Effect of thermo physical properties of heat exchanger material on the performance of latent heat storage system using an enthalpy method. Int. J. Energy Res. **30**, 191–201 (2006)
42. Q. He, W.N. Zhang, A study on latent heat storage exchangers with the high temperature phase-change material. Int. J. Energy Res. **25**, 331–341 (2001)
43. H. Michels, R. Pitz-Paal, Cascaded latent heat storage for parabolic trough solar power plants. Sol. Energy **81**, 829–837 (2007)
44. Z.X. Gong, A.S. Mujumdar, Cyclic heat transfer in a novel storage unit of multiple phase change materials. Appl. Thermal Eng. **16**, 807–815 (1996)
45. Z.X. Gong, A.S. Mujumdar, A new solar receiver thermal store for space based activities using multiple composite phase-change materials. J. Sol. Energy Eng. Trans. ASME **117**, 215–220 (1995)
46. H. Cui, X. Hou, X. Yuan, Energy analysis of space solar dynamic heat receivers. Sol. Energy **74**, 303–308 (2003)
47. H. Cui, Y. Xing, Y. Guo, Z. Wang, H. Cui, X. Yuan, Numerical simulation and experimental investigation on unit heat exchange tube for solar heat receiver. Sol. Energy **82**, 1229–1234 (2008)

48. C. Guo, W. Zhang, Numerical simulation and parametric study on new type of high temperature latent heat thermal energy storage system. Energy Convers. Manage. **49**, 919–927 (2008)
49. B. Yimer, M. Adami, Parametric study of phase change thermal energy storage systems for space application. Energy Convers. Manage. **38**, 253–262 (1997)
50. A. Hoshi, D.R. Mills, A. Bittar, T.S. Saitoh, Screening of high melting point phase change materials (PCM) in solar thermal concentrating technology based on CLFR. Sol. Energy **79**, 332–339 (2005)
51. V. Morisson, M. Rady, E. Palomo, E. Arquis, Thermal energy storage systems for electricity production using solar energy direct steam generation technology. Chem. Eng. Process. **47**, 499–507 (2008)
52. H. Cui, X. Yuan, X. Hou, Thermal performance analysis for a heat receiver using multiple phase change materials. Appl. Therm. Eng. **23**, 2353–2361 (2003)
53. K. Lafdi, O. Mesalhy, A. Elyafy, Graphite foams infiltrated with phase change materials as alternative materials for space and terrestrial thermal energy storage applications. Carbon **46**, 159–168 (2008)
54. D.B. Khillarkar, Z.X Gong, A.S. Mujumdar, Melting of a phase change material in concentric horizontal annuli of arbitrary cross-section. Appl. Therm. Eng. **20**, 893–912 (2000)
55. J. Yagi, T. Akiyama, Storage of thermal energy for effective use of waste heat from industries. J. Mater. Process. Technol. **48**, 793–804 (1995)
56. A.A. Jalalzadeh-Azar, Performance comparison of high-temperature packed bed operation with PCM and sensible-heat pellets. Int. J. Energy Res. **21**, 1039–1052 (1997)
57. S. Jegadheeswaran, S.D. Pohekar, T. Kousksou, Exergy based performance evaluation of latent heat thermal storage system: a review. Renew. Sustain. Energy Rev. **14**, 2580–2595 (2010)
58. A. Kaizawa, H. Kamano, A. Kawai, T. Jozuka, T. Senda, N. Maruoka, T. Akiyama, Thermal and flow behaviors in heat transportation container using phase change material, Energy Convers. Manage. **49**, 698–706 (2008)
59. A. Mawire, M. McPherson, Experimental characterization of a thermal energy storage system using temperature and power controlled charging. Renew. Energy **33**, 682–693 (2008)
60. A. Bejan, Two thermodynamic optima in the design of sensible heat units for energy storage. J. Heat Transf. **100**, 708–712 (1978)
61. M.A. Rosen, N. Pedinelli, I. Dincer, Energy and exergy analyses of cold thermal storage systems. Int. J. Energy Res. **23**, 1029–1038 (1999)
62. M.A. Rosen, The exergy of stratified thermal energy storages. Sol. Energy **71**, 173–185 (2001)
63. K. Kaygusuz, T. Ayhan, Exergy analysis of solar-assisted heat-pump systems for domestic heating. Energy **18**, 1077–1085 (1993)
64. A. Venkataramayya, K.N. Ramesh, Exergy analysis of latent heat storage systems with sensible heating and subcooling of PCM. Int. J. Energy Res. **22**, 411–426 (1998)
65. A. Sari, K. Kaygusuz, Energy and exergy calculations of latent heat energy storage systems. Energy Sources **22**, 117–126 (2000)
66. H.H. Ozturk, Experimental evaluation of energy and exergy efficiency of a seasonal latent heat storage system for green house heating. Energy Convers. Manage. **46**, 1523–1542 (2005)
67. A. Koca, H.G. Oztop, T. Koyun, Y. Varol, Energy and exergy analysis of a latent heat storage system with phase change material for solar collector. Renew. Energy **33**, 567–574 (2008)
68. T. Kousksou, T. El Rhafiki, A. Arid, E. Schall, Y. Zeraouli, Power, efficiency, and irreversibility of latent energy systems. J. Thermophy. Heat Transf. **22**, 234–239 (2008)
69. A. Erek, I. Dincer, A new approach to energy and exergy analyses of latent heat storage unit. Heat Transf. Eng. **30**, 506–515 (2009)

70. D. MacPee, I. Dincer, Thermodynamic analysis of freezing and melting processes in a bed of spherical PCM capsules. J. Sol. Energy Eng. Trans. ASME **131**, 031017 (2009)
71. I. Dincer, Y.A. Cengel, Energy, entropy and exergy concepts and their roles in thermal engineering. Entropy **3**, 116–149 (2001)
72. M.A. Rosen, Appropriate thermodynamic performance measures for closed systems for thermal energy storage. J. Sol. Energy Eng. Trans. ASME **114**, 100–105 (1992)
73. T. Watanabe, A Kanzawa, Second law optimization of a latent heat storage system with PCMs having different melting points. Heat Recovery Syst. CHP **15**, 641–653 (1995)
74. M.A. Rosen, I. Dincer, Exergy methods for assessing and comparing thermal storage systems. Int. J. Energy Res. **27**, 415–530 (2003)
75. Z.X. Gong, A.S. Mujumdar, Thermodyanamic optimization of the thermal process in energy storage using multiple phase change materials. Appl. Therm. Eng. **17**, 1067–1083 (1996)
76. Y. Demirel, H.H. Ozturk, Thermoeconomics of seasonal latent heat storage systems. Int. J. Energy Res. **30**, 1001–1012 (2006)
77. R.J. Krane, A second law analysis of the optimum design and operation of thermal energy storage systems. Int. J. Heat Mass Transf. **30**, 43–57 (1987)
78. C. Bellecci, M. Conti, Phase change energy storage: entropy production, irreversibility, and second law efficiency. Sol. Energy **53**, 163–170 (1994)
79. M.M. Farid, A. Kanzawa, Thermal performance of a heat storage module using PCM's with different melting temperatures: mathematical modeling. J. Sol. Energy Eng. Trans. ASME **111**, 152–157 (1989)
80. F. Strub, J.P. Bedecarrats, Numerical second law analysis of a refrigeration phase-change storage. Int. J. Thermodyn. **2**, 133–138 (1999)
81. F. Strub F and J. P. Bedecarrats, Thermodynamics of phase-change energy storage: the effects of undercooling on entropy generation during solidification. Int. J. Thermodyn. **3**, 35–42 (2000)
82. T. Kousksou, F. Strub, J.S. Lasvignottes, A. Jamil, J.P. Bedecarrats, Second law analysis of latent thermal storage for solar system. Sol. Energy Mater. Solar Cells **91**, 1275–1281 (2007)
83. M. Lacroix, Study of the heat transfer behavior of a latent heat thermal energy storage unit with a finned tube. Int. J. Heat Mass Transf. **36**, 2083–2092 (1993)
84. A. Erek, I. Dincer, An approach to entropy analysis of a latent heat storage module. Int. J. Therm. Sci. **47**, 1077–1085 (2008)
85. H. El-Dessouky, F. Al-Juwayhel, Effectiveness of a thermal energy storage system using phase-change materials. Energy Convers. Manage. **38**, 601–617 (1997)
86. A. Bejan, *Entropy Generation Minimization* (CRC Press, London, 1996)
87. C. Charach, A. Zemel, Thermodynamic analysis of latent heat storage in a shell-and-tube heat exchanger. J. Sol. Energy Eng. Trans. ASME **114**, 93–99 (1992)
88. H. Bjustrom, B. Carlsson, An exergy analysis of sensible and latent heat storage. Heat Recovery Syst. CHP **5**, 233–250 (1985)
89. F. Aghbalou, F. Badia, J. Illa, Exergetic optimization of solar collector and thermal energy storage system. Int. J. Heat Mass Transf. **49**, 1255–1263 (2006)
90. C. Charach, Second-law efficiency of an energy storage-removal cycle in a phase-change material shell-and-tube heat exchanger. J. Sol. Energy Eng. Trans. ASME **115**, 240–243 (1993)
91. H. Ettouney, H. El-Dessouky, E. Al-Kandari, Heat transfer characteristics during melting and solidification of phase change energy storage process. Ind. Eng. Chem. Res. **43**, 5350–5357 (2004)
92. A. Valero, M.A. Lozano, An introduction of thermoeconomics. in *Developments in the Design of Thermal Systems*, ed. by R.F. Boehm (Cambridge, Cambridge University Press, 2005), pp. 203–223
93. C.A. Frangopoulos, Thermoeconomic functional and optimization. Energy **19**, 563–571 (1987)

94. D.J. Kim, A new thermoeconomic methodology for energy systems. Energy **35**, 410–422 (2010)
95. M.A. Lozano, A. Valero, Theory of the exergetic cost. Energy **18**, 939–960 (1993)
96. B. Erlach, L. Serra, A. Valero, Structural theory as standard for thermoeconomics. Energy Convers. Manage. **40**, 1627–1649 (1999)
97. M.A. Badar, S.M. Zubair, A.A. Al-Farayedhi, Second-law-based thermoeconomic optimization of a sensible heat thermal energy storage system. Energy **18**, 641–649 (1993)
98. M.A. Badar, S.M. Zubair, On thermoeconomic of a sensible heat thermal energy storage system. J. Sol. Energy Eng. Trans. ASME **117**, 225–259 (1995)
99. S.M. Zubair, M.A. Al-Naglah, Thermoeconomic optimization of a sensible heat thermal storage system: a complete cycle. ASME J. Energy Res. Techn. **121**, 286–294 (1999)
100. A.A. Ghoneim, Comparison of theoretical models of phase change and sensible heat storage for air and water solar heating systems. Sol. Energy **42**, 209–220 (1989)
101. J. Prakash, H.P. Garg, G. Datta, A solar water heater with a built-in latent heat storage. Energy Convers. Manage. **25**, 51–56 (1985)
102. N.K. Bansal, D. Buddhi, An analytical study of a latent heat storage system in a cylinder. Sol. Energy **33**, 235–242 (1992)
103. K. Kaygusuz, Experimental and theoretical investigation of latent heat storage for water based solar heating systems. Energy Convers. Manage. **36**, 315–323 (1995)
104. Y. Rabin, I. Bar-Niv, E. Korin, B. Mikic, Integrated solar collector storage system based on a salt hydrate phase change material. Sol. Energy **55**, 435–444 (1995)
105. E-B.S. Mettawee, G.M.R. Assassa, Experimental study of a compact PCM solar collector. Energy **31**, 2958–2968 (2006)
106. L.F. Cabeza, M. Ibanez, C. Sole, J. Roca, M. Nogues, Experimentation with a water tank including a PCM module. Sol. Energy Mater. Solar Cells **90**, 1273–1782 (2006)
107. S. Canbazoglu, A. Sahinaslan, A. Ekmekyapar, Y. Gokhan Aksoy, F. Akarsu, Enhancement of solar thermal energy storage performance using sodium thiosulfate pentahydrate of a conventional solar water-heating system. Energy Build. **37**, 235–242 (2005)
108. D.J. Morrison, S.I. Abdel Khalik, Effects of phase change energy storage on the performance of air-based and liquid-based solar heating systems. Sol. Energy **20**, 57–67 (1978)
109. J.J. Jurinak, S.I. Adbel Khalik, On the performance of air-based solar heating systems utilizing phase change energy storage. Energy **4**, 503–522 (1979)
110. A.A. Ghoneim, S.A. Klein, The effect of phase change material properties on the performance of solar air-based heating systems. Sol. Energy **42**, 441–447 (1989)
111. S.O. Enibe, Performance of a natural circulation solar air heating system with phase change material energy storage. Renew Energy **27**, 69–86 (2002)
112. G. Zhou, Y. Zhang, Q. Zhang, K. Lin, H. Di, Performance of a hybrid heating system with thermal storage using shape-stabilized phase-change material plates. Appl. Energy **84**, 1068–1077 (2007)
113. D. Buddhi, L.K. Sahoo, Solar cooker with latent heat storage design and experimental testing. Energy Convers.Manage. **38**, 493–498 (1997)
114. S.D. Sharma, D. Buddhi, R.L. Sawhney, A. Sharma, Design, development and performance evaluation of a latent heat unit for evening cooking in a solar cooker. Energy Convers. Manage. **41**, 1497–1508 (2000)
115. D. Buddhi, S.D. Sharma, A. Sharma, Thermal performance evaluation of a latent heat storage unit for late evening cooking in a solar cooker having three reflectors. Energy Convers. Manage. **44**, 809–817 (2003)
116. S.D. Sharma, T. Iwata, H. Kitano, K. Sagara, Thermal performance of a solar cooker based on an evacuated tube solar collector with a PCM storage unit. Sol. Energy **78**, 416–426 (2005)
117. N.M. Nahar, Design, development and testing of a double reflector hot box solar cooker with a transparent insulation material. Renew. Energy **23**, 167–179 (2001)

118. K. Hung C.F. Abrams Jr., L.L. Coasts, C.G. Bowers Jr., Development of greenhouse bulk drying systems for solar energy utilization and planted mechanization. AHARE paper no. 75–1018, Am. Soc. Agric. Eng. St. Joseph, MI, (1975)
119. M. Kern and R. A. Aldrich, Phase change energy storage in a greenhouse solar heating system. ASME paper no. 79–4028. Am. Soc. Agric. Eng. St. Joseph, MI, (1979)
120. K. Hung, M. Toksoy, Design and analysis of green house solar system in agricultural production. Energy Agric. **2**, 115–136 (1983)
121. T. Boulard, E. Razafinjohany, A. Baille, A. Jaffrin, B. Fabre, Performance of a greenhouse heating system with a phase change material. Agric. Forest Meteorol. **52**, 303–318 (1990)
122. H.H. Ozturk, Experimental evaluation of energy and exergy efficiency of a seasonal latent heat storage system for greenhouse heating. Energy Convers. Manage. **46**, 1523–1542 (2005)
123. H.H. Ozturk, A. Bascetincelik, Energy and exergy efficiency of a packed-bed heat storage unit for greenhouse heating. Biosyst. Eng. **86**, 231–245 (2003)
124. A.A. Ghoneim, S.A. Klein, J.A. Duffie, Analysis of collector-storage building walls using phase change materials. Sol. Energy **47**, 237–242 (1991)
125. S. Chandra, R. Kumar, S. Kaushik, S. Kaul, Thermal performance of a non A/C building with PCCM thermal storage wall. Energy Convers. Manage. **25**, 15–20 (1985)
126. T.R. Knowles, Proportioning composites for efficient-TSWs. Sol. Energy **31**, 319–326 (1983)
127. V.V. Tyagi, D. Buddhi, PCM thermal storage in buildings: a state of art. Renew. Sustain. Energy Rev. **11**, 1146–1166 (2007)
128. D. Feldman, M.A. Khan, D. Banu, Energy storage composite with an organic phase change material. Sol. Energy Mater. **18**, 333–341 (1989)
129. D. Feldman, M. Shapiro, D. Banu, C.J. Fuks, Fatty acids and their mixtures as phase change materials for thermal energy storage. Sol. Energy Mater. **18**, 201–216 (1989)
130. D. Feldman, D. Banu, D. Hawes, E. Ghanbari E, Obtaining an energy storing building material by direct incorporation of an organic phase change material in gypsum wallboard. Solar Energy Mater. **22**, 231–242 (1991)
131. D.W. Hawes, D. Feldman, D. Banu, Latent heat storage in building materials. Energy Build. **20**, 77–86 (1993)
132. A.K. Athienitis, C. Liu, D. Hawes, D. Banu, D. Feldman, Investigation of the thermal performance of a passive solar test-room with wall latent heat storage. Build. Environ. **32**, 405–410 (1997)
133. D.A. Neeper, Solar buildings research: what are the best directions? Passive Sol. J. **3**, 213–219 (1986)
134. D.A. Neeper, Thermal dynamics of wallboard with latent heat storage. Sol. Energy **68**, 393–403 (2000)
135. T.K. Stovall, J.J. Tomlinson, What are the potential benefits of including latent heat storage in common wall board. J. Sol. Energy Eng. Trans. ASME **117**, 318–325 (1995)
136. K. Peippo, P. Kauranen, P.D. Lund, A multi-component PCM wall optimized for passive solar heating. Energy Build. **17**, 259–270 (1991)
137. D. Zhang, Z. Li, J. Zhou, K. Wu, Development of thermal energy storage concrete. Cement Concrete. Res. **34**, 927–934 (2004)
138. P. Schossig, H.M. Henning, S. Gschwander, T. Haussmann, Micro-encapsulated phase-change materials integrated into construction materials. Sol. Energy Mater. Solar Cells **89**, 297–306 (2005)
139. A. Athienitis, Y. Chen, The effect of solar radiation on dynamic thermal performance of floor heating systems. Sol. Energy **69**, 229–237 (2000)
140. G. Bakos, Energy management method for auxiliary energy saving in a passive-solar-heated residence using low-cost off-peak electricity. Energy Build. **31**, 237–241 (2003)
141. K.P. Lin, Y.P. Zhang, X. Xu, H.F. Di, R. Yang, P.H. Qin, Modeling and simulation of under-floor electric heating system with shape stabilized PCM plates. Build. Environ. **39**, 1427–1434 (2004)

142. C. Benard, D. Gobin, M. Gutierrez, Experimental results of a latent heat solar roof used for breeding chickens. Sol. Energy **6**, 347–354 (1981)
143. J.M. Gutherz, M.E. Schiler, A passive solar heating system for the perimeter zone of office buildings. Energy Sources **13**, 39–54 (1991)
144. J.R. Turnpenny, D.W. Etheridge, D.A. Reay, Novel ventilation cooling system for reducing air conditioning in buildings, Part I. Testing and theoretical modeling. Appl. Therm. Eng. **20**, 1019–1037 (2000)
145. J.R. Turnpenny, D.W. Etheridge, D.A. Reay, Novel ventilation cooling system for reducing air conditioning in buildings, Part II. Testing of prototype. Appl. Therm. Eng. **21**, 1203–1217 (2001)
146. M.M. Farid, M.H. Rafah, An electrical storage heater using the phase change method of heat storage. Energy Convers. Manage. **30**, 219–230 (1990)
147. M.M. Farid, A.M. Khudhair, S.A.K. Razack, S. Al-Hallaj, A review on phase change energy storage: materials and applications. Energy Convers. Manage. **45**, 1597–1615 (2004)
148. B. Zalba, J.M. Marin, L.F. Cabeza, H. Mehling, Review on thermal energy storage with phase change: materials, heat transfer analysis and applications. Appl. Therm. Eng. **23**, 251–283 (2003)
149. A. Shukla, D. Buddhi, R.L. Sawhney, Solar water heaters with phase change material thermal energy storage medium: a review. Renew. Sustain. Energy Rev. **13**, 2119–2125 (2009)
150. M. Medrano, A. Gil, I. Martorell, X. Potau, L.F. Cabeza, State of the art on high-temperature thermal energy storage for power generation. Part 2—case studies. Renew. Sustain. Energy Rev. **14**, 56–72 (2010)
151. N. Gokon, D. Nakano, S. Inuta, T. Kodama, High-temperature carbonate/MgO composite materials as thermal storage media for double-walled solar reformer tubes. Sol. Energy **82**, 1145–1153 (2008)
152. M.F. Demirbas, Thermal energy storage and phase change materials: an overview. Energy Sources Part B **1**, 85–95 (2006)
153. S. Jegadheeswaran, S.D. Pohekar, Performance enhancement in latent heat thermal storage system: a review. Renew. Sust. Energy Rev. **13**, 2225–2244 (2009)
154. R. Domanski, G. Fellah, Exergy analysis for the evaluation of a thermal storage system employing PCMs with different melting temperatures. Appl. Therm. Eng. **16**, 907–919 (1996)
155. Z.X. Gong, A.S. Mujumdar, Finite element analysis of a multistage latent heat thermal storage system. Numer. Heat Transf. Part A **30**, 669–684 (1996)
156. P. Lamberg, R. Lehtiniemi, A.M. Henell, Numerical and experimental investigation of melting and freezing processes in phase change material storage. Int. J. Therm. Sci. **43**, 277–287 (2004)
157. U. Stritih, An experimental study of enhanced heat transfer in rectangular PCM storage. Int. J. Heat Mass Transf. **47**, 2841–2847 (2004)
158. Y. Zhang, Z. Chen, Q. Wang, Q. Wu, Melting in an enclosure with discrete heating at a constant rate. Exp. Therm. Fluid Sci. **6**, 196–201 (1993)
159. Y. Jellouli, R. Chouikh, A. Guizani, A. Belghith, Numerical study of the moving boundary problem during melting process in a rectangular cavity heated from below. Am. J. Appl. Sci. **4**, 251–256 (2007)
160. K.W. Ng, Z.X. Gong, A.S. Mujumdar, Heat transfer in free convection-dominated melting of a phase change material in a horizontal annulus. Int. Commun. Heat Mass Transf. **25**, 631–640 (1998)
161. A.F. Regin, S.C. Solanki, J.S. Saini, Latent heat thermal storage using cylindrical capsule: numerical and experimental investigations. Renew. Energy **31**, 2025–2041 (2006)
162. B.J. Jones, D. Sun, S. Krishnan, S.V. Garimella, Experimental and numerical investigation of melting in a cylinder. Int. J. Heat. Mass. Transf. **49**, 2724–2738 (2006)
163. P.A. Bahrami, Natural melting within a spherical shell. *NASA Technical Memorandum* (Ames Research Center, California, 1990) Report No. 102822

164. F.L. Tan, Constrained and unconstrained melting inside a sphere. Int. Commun. Heat Mass Transf. **35**, 466–475 (2008)
165. H. Ettouney, H. El-Dessouky H, A. Al-Ali, Heat transfer during phase change of paraffin wax stored in spherical shells. J. Solar Energy Eng. Trans. ASME **127**, 357–365 (2005)
166. M. Lacroix, M. Benmadda, Analysis of natural convection melting from a heated wall with vertically oriented fins. Int. J. Numer. Methods Heat Fluid Flow **8**, 465–478 (1998)
167. H. Ettouney, H. El-Dessouky, E. Al-Kandari, Heat transfer characteristics during melting and solidification of phase change energy storage process. Ind. Eng. Chem. Res. **43**, 5350–5357 (2004)
168. M. Pinelli, S. Piva, Solid/liquid phase change in presence of natural convection: a thermal energy storage case study. ASME J. Energy Res. Techn. **125**, 190–197 (2003)
169. M. Pinelli, G. Casano, S. Piva, Solid-liquid phase change heat transfer in a vertical cylinder heated from above. Int. J. Heat Techn. **18**, 61–67 (2000)
170. R. Akhilesh, C. Balaji, A. Narasimhan, Method to improve geometry for heat transfer enhancement in PCM composite heat sinks. Int. J. Heat Mass Transf. **48**, 2759–2770 (2005)
171. M. Akhilesh, I. Sezai, Enhancement of heat transfer in latent heat storage modules with internal fins. Numer. Heat Transf. Part A **53**, 749–765 (2008)
172. M. Lacroix, M. Benmadda, Numerical simulation of natural convection-dominated melting and solidification from a finned vertical wall. Numer. Heat Transf. Part A **31**, 71–86 (1997)
173. V. Shatikian, G. Ziskind, R. Letan, Numerical investigation of a PCM-based heat sink with internal fins. Int. J. Heat Mass Transf. **48**, 3689–3706 (2005)
174. Y. Zhang, A. Faghri, Heat transfer enhancement in latent heat thermal energy storage system by using an external radial finned tube. J. Enhanc. Heat Transf. **3**, 119–127 (1996)
175. R.V. Seeniraj, R. Velraj, N.L. Narasimhan, Thermal analysis of a finned-tube LHTS module for a solar dynamic power system. Heat Mass Transf. **38**, 409–417 (200)
176. P. Lamberg, Approximate analytical model for two-phase solidification problem in a finned phase-change material storage. Appl. Energy **77**, 131–152 (2004)
177. J.C. Choi, S.D. Kim, Heat-transfer characteristics of a latent heat storage system using $MgCl_2 \cdot 6H_2O$. Energy **17**, 1153–1164 (1992)
178. Z. Liu, X. Sun, C. Ma, Experimental study of the characteristics of solidification of stearic acid in an annulus and its thermal conductivity enhancement. Energy Convers. Manage. **46**, 971–984 (2005)
179. R. Velraj, R.V. Seeniraj, H. Hafner, C. Faber, K. Schwarzer, Experimental analysis and numerical modeling of inward solidification on a finned vertical tube for a latent heat storage unit. Sol. Energy **60**, 281–290 (1997)
180. A. Castell, C. Sole, M. Medrano, J. Roca, L.F. Cabeza, D. Garcia, Natural convection heat transfer coefficients in phase change material (PCM) modules with external vertical fins. Appl. Therm. Eng. **28**, 1676–1686 (2008)
181. J. Wang, G. Chen, H. Jiang, Theoretical study on a novel phase change process. Int. J. Energy Res. **23**, 287–294 (1999)
182. M.M. Farid, A. Kanzawa, Thermal performance of a heat storage module using PCMs with different melting temperatures: mathematical modeling. J. Sol. Energy Eng. Trnas. ASME **111**, 152–157 (1989)
183. R. Velraj, R.V. Seeniraj, B. Hafner, C. Faber, K. Schwarzer, Heat transfer enhancement in a latent heat storage system. Sol. Energy **65**, 171–180 (1999)
184. J. Wang, Y. Ouyang, Chen G, Experimental study on charging processes of a cylindrical heat storage capsule employingmultiple-phase-changematerials. Int. J. Energy Res. **25**, 439–447 (2001)
185. J. Wang, G. Chen, F. Zheng, Study on phase change temperature distributions of composite PCMs in thermal energy storage systems. Int. J. Energy Res. **23**, 277–285 (1999)
186. R.V. Seeniraj, N.L. Narasimhan, Performance enhancement of a solar dynamic LHTS module having both fins and multiple PCMs. Sol. Energy **82**, 535–542 (2008)
187. M. Fang, G. Chen, Effects of different multiple PCMs on the performance of a latent thermal energy storage system. Appl. Therm. Eng. **27**, 994–1000 (2007)

188. S.D. Sharma, K. Sagara, Latent heat storage materials and systems: a review. Int. J. Green Energy **2**, 1–56 (2005)
189. O. Mesalhy, K. Lafdi, A. Elgafi, K. Bowman, Numerical study for enhancing the thermal conductivity of phase change material (PCM) storage using high thermal conductivity porousmatrix. Energy Convers. Manage. **46**, 847–867 (2005)
190. S. Krishnan, J.Y. Murthy, S.V. Garimella, A two-temperature model for solidliquid phase change in metal foams. ASME J. Heat Transf. **127**, 995–1004 (2005)
191. D. Haillot, X. Py, V. Goetz, M. Benabdelkarim, Storage composites for the optimization of solar water heating systems. Chem. Eng. Res. Design **86**, 612–617 (2008)
192. A. Sari, A. Karaipekli, Thermal conductivity and latent heat thermal energy storage characteristics of paraffin/expanded graphite composite as phase change material. Appl. Therm. Eng. **27**, 1271–1277 (2007)
193. H. Yin, X. Gao, J. Ding, Z. Zhang, Experimental research on heat transfer mechanism of heat sink with composite phase change materials. Energy Convers. Manage. **49**, 1740–1746 (2008)
194. S. Kim, L.T. Drzal, High latent heat storage and high thermal conductive phase change materials using exfoliated graphite nanoplatelets. Sol. Energy Mater. Sol. Cells **93**, 136–142 (2009)
195. S. Pincemin, X. Py, R. Olives, M. Christ, O. Oettinger, Elaboration of conductive thermal storage composites made of phase change materials and graphite for solar power plant. J. Sol. Energy Eng. Trans. ASME **130**, 11005–11009 (2008)
196. S. Pincemin, R. Olives, X. Py, M. Christ, Highly conductive composites made of phase change materials and graphite for thermal storage. Sol. Energy Mater. Sol. Cells **92**, 603–613 (2008)
197. A. Elgafy, K. Lafdi, Effect of carbon nanofiber additives on thermal behavior of phase change materials. Carbon **43**, 3067–3074 (2005)
198. E.S. Mettawee, G.M.R. Assassa, Thermal conductivity enhancement in a latent heat storage system. Sol. Energy **81**, 839–845 (2007)
199. J.M. Khodadadi, S.F. Hosseinizadeh, Nanoparticle-enhanced phase change materials (NEPCM) with great potential for improved thermal energy storage. Int. Commun. Heat Mass Transf. **34**, 534–543 (2007)
200. J.L. Zeng, L.X. Sun, F. Xu, Z.C. Tan, Z.H. Zhang, J. Zhang, T. Zhang, Study of a PCM based energy storage system containing Ag nanoparticles. J. Therm. Anal. Calorim. **87**, 369–373 (2007)
201. R.V. Seeniraj, R. Velraj, N.L. Narasimhan, Heat transfer enhancement study of a LHTS unit containing dispersed high conductivity particles. J. Sol. Energy Eng. Trans. ASME **124**, 243–249 (2002)
202. H. Ettouney, I. Alatiqi, M. Al-Sahali, S.A. Al-Ali, Heat transfer enhancement by metal screens and metal spheres in phase change energy storage systems. Renew. Energy **29**, 841–860 (2004)
203. H. Ettouney, I. Alatiqi, M. Al-Sahali, K. Al-Hajirie, Heat transfer enhancement in energy storage in spherical capsules filled with paraffin wax and beads. Energy Convers. Manage. **47**, 211–228 (2006)
204. J. Fukai, M. Kanou, Y. Kodama, O. Miyatake, Thermal conductivity enhancement of energy storage media using carbon fibers. Energy Convers. Manage. **41**, 1543–1556 (2000)
205. J. Fukai, Y. Hamada, Y. Morozumi, O. Miyatake, Improvement of thermal characteristics of latent heat thermal energy storage units using carbonfiber brushes: experiments and modeling. Int. J. Mass Heat Transf. **46**, 4513–4525 (2003)
206. Y. Hamada, W. Ohtsu, J. Fukai, Thermal response in thermal energy storage material around heat transfer tubes: effect of additives on heat transfer rates. Sol. Energy **75**, 317–328 (2003)
207. J. Fukai, Y. Hamada, Y. Morozumi, O. Miyatake, Effect of carbon-fiber brushes on conductive heat transfer in phase change materials. Int. J. Mass Heat Transf. **45**, 4781–4792 (2002)

208. Y. Hamada, W. Otsu, J. Fukai J, Y. Morozumi, O. Miyatake, Anisotropic heat transfer in composites based on high-thermal conductive carbon fibers. Energy **30**, 221–233 (2005)
209. K. Nakaso, H. Teshima, A. Yoshimura, S. Nogami, Y. Hamada, J. Fukai, Extension of heat transfer area using carbon fiber cloths in latent heat thermal energy storage tanks. Chem Eng Process. **47**, 879–885 (2008)
210. Y. Cui, C. Liu, S. Hu, X. Yu, The experimental exploration of carbon nanofiber and carbon nanotube additives on thermal behavior of phase change materials. Sol. Energy Mater. Sol. Cells **95**, 1208–1212 (2011)
211. J. Wang, H. Xie, Z. Xin, Y. Li, L. Chen, Enhancing thermal conductivity of palmitic acid based phase change materials with carbon nanotubes as fillers. Sol. Energy **84**, 339–344 (2010)
212. J. Wang, H. Xie, Z. Xin, Y. Li, L. Chen, Thermal properties of paraffin based composites containing multi-walled carbon nanotubes. Thermochim. Acta **488**, 39–42 (2009)
213. J. Wang, H. Xie, Z. Xin, Y. Li, Increasing the thermal conductivity of palmitic acid by the addition of carbon nanotubes. Carbon **48**, 3979–3986 (2010)
214. L.W. Fan, X. Fang, X. Wang, Y. Zeng, Y. Q. Xiao, Z.T. Yu, X. Xu, Y.C. Hu, K.F. Cen, Effects of various carbon nanofillers on the thermal conductivity and energy storage properties of paraffin-based nanocomposite phase change materials. Appl. Energy, **110**, 163–172 (2013)
215. Y. Hamada, J. Fukai, Latent heat thermal energy storage tanks for space heating of buildings: comparison between calculations and experiments. Energy Convers. Manage. **46**, 3221–3235 (2005)
216. M.N.A. Hawlader, M.S. Uddin, M.M. Khin, Microencapsulated PCM thermal energy storage system. Appl. Energy **74**, 195–202 (2003)
217. B. Chen, X. Wang, R. Zeng, Y. Zhang, X. Wang, J. Niu, Y. Li, H. Di, An experimental study of convective heat transfer with microencapsulated phase change material suspension: laminar flow in a circular tube under constant heat flux. Exper. Therm. Fluid Sci. **32**, 1638–1646 (2008)
218. C. Alkan, A. Sari, A. Karaipekli, O. Uzun, Preparation, characterization, and thermal properties of microencapsulated phase change material for thermal energy storage. Sol. Energy Mater. Sol. Cells **93**, 143–147 (2009)
219. M.N.A. Hawlader, M.S. Uddin, H.J. Zhu, Encapsulated phase change materials for thermal energy storage: Experiments and simulation. Int. J. Energy Res. **26**, 159–171 (2002)
220. L.F. Cabeza, C. Castellon, M. Nogues, M. Medrano, R. Leppers, O. Zubillaga, Use of microencapsulated PCM in concrete walls for energy savings. Energy Build. **39**, 113–119 (2007)
221. L. Sanchez, P. Sanchez, A. Lucas, M. Carmona, J.F. Rodriguez, Microencapsulation of PCMs with a polystyrene shell. Colloid. Polymer Sci. **285**, 1377–1385 (2007)
222. Y. Rao, F. Dammel, P. Stephen, G. Lin, Convective heat transfer characteristics of microencapsulated phase change material suspensions in minichannels. Heat Mass Transf. **44**, 175–186 (2007)
223. Y. Zhang, A. Faghri, Analysis of forced convection heat transfer in microencapsulated phase change material suspensions. J. Thermophys. Heat Transf. **9**, 727–732 (1995)
224. M.N.A. Hawlader, M.S. Uddin, H.J. Zhu, Preparation and evaluation of a novel solar storage material: microencapsulated paraffin. Int. J. Sol. Energy **20**, 227–238 (2000)
225. Y. Ozonur, M. Mazman, H.O. Paksoy, H. Evliya, Microencapsulation of coco fatty acid mixture for thermal energy storage with phase change material. Int. J. Energy Res. **30**, 741–749 (2006)
226. A.F. Regin, S.C. Solanki, J.S. Saini, Heat transfer characteristics of thermal energy storage system using PCM capsules: a review. Renew. Sustain. Energy Rev. **12**, 2438–2458 (2008)
227. Y. Fang, S. Kuang, X. Gao, Z. Zhang, Preparation and characterization of novel nanoencapsulated phase change materials. Energy Convers. Manage. **49**, 3704–3707 (2008)

228. Z.H. Chen, F. Yu, X.R. Zeng, Z.G. Zhang, Preparation, characterization and thermal properties of nanocapsules containing phase change material n-dodecanol by miniemulsion polymerization with polymerizable emulsifier. Appl. Energy **91**, 7–12 (2012)
229. G. Fang, H. Li, F. Yang, X. Liu, S. Wu, Preparation and characterization of nano-encapsulated *n*-tetradecane as phase change material for thermal energy storage. Chem. Eng. J. **153**, 217–221 (2009)
230. A. Michel, A. Kugi, Accurate low-order dynamic model of a compact plate heat exchanger. Int. J. Heat and Mass Transf. **61**, 323–331 (2013)
231. A. Abhat, Performance studies of a finned heat pipe latent heat thermal energy storage system, in *Sun* (Pergamon Press, NY, 1981) pp. 541–546
232. C.D. Maccracken, PCM bulk storage, in *Proceedings of the international conference on energy storage*, pp. 159–165 (1981)
233. R.N. Smith, T.E. Ebersole, F.P. Griffin, Heat-exchanger performance in latent-heat thermal-heat thermal-energy storage. J. Sol. Energy Eng. Trans. ASME **102**, 112–118 (1980)
234. D. Buddhi, Thermal performance of a shell and tube PCM storage heat exchanger for industrial waste heat recovery, in *Solar World Congress*, Taejon, Korea, Aug 24–30 (1977)
235. V.M. Morcos, Investigation of a latent heat thermal energy storage system. Solar Wind Techn. **7**, 197–202 (1990)
236. M.J. Santamouris, C.C. Lefas, On the coupling of PCM stores to active solar systems. Int. J. Energy Res. **12**, 603–610 (1988)
237. M.M. Farid, A. Kanzawa, Thermal performance of heat storage module using PCMs with different melting temperatures: mathematical modeling. J. Sol. Energy Eng. Trans. ASME **111**, 152–157 (1989)
238. D.R. Biswas, Thermal energy storage using sodium sulphate decahydrate and water. Sol. Energy **19**, 99–100 (1977)
239. F.C. Porisini, Salt hydrates used for latent heat storage: corrosion of metals and reliability of thermal performance. Sol. Energy **41**, 193–197 (1988)
240. L. Cabeza, J. Illa, J. Roca, F. Badia, H. Mehling, S. Hiebler, F. Ziegler, Immersion corrosion tests on melt-salt hydrate pairs used for latent heat storage in the 32–36 °C temperature range. *Mater. Corrosion* **52**, 140–146 (2201)
241. R.W. Bradshaw, S.H. Goods, Corrosion resistance of stainless steels during thermal cycling in alkali nitrate molten salts, *SANDIA Report*, SAND2001-8518 (2001)
242. R.W. Bradshaw, S.H. Goods, Corrosion of alloys and metals molten nitrates, *SADIA Report*, SAND2000-8727 (2000)
243. A. Baraka, A.I. Abdel-Rohman, A.A. El Hosary, Corrosion of mild steel in molten sodium nitrate-potassium nitrate eutectic. Brit. Corrosion J. **11**, 44–46 (1976)
244. I.B. Singh, U. Sen, Influence of temperature and sulphate ion on corrosion of mild steel in molten $NaNO_3$. Brit. Corrosion J. **27**, 299–304 (1992)

MIX
Papier aus verantwortungsvollen Quellen
Paper from responsible sources
FSC® C105338

If you have any concerns about our products,
you can contact us on
ProductSafety@springernature.com

In case Publisher is established outside the EU,
the EU authorized representative is:
Springer Nature Customer Service Center GmbH
Europaplatz 3, 69115 Heidelberg, Germany

Printed by Libri Plureos GmbH
in Hamburg, Germany